# RADIOACTIVE WASTE

Edited by **Rehab Abdel Rahman**

**Radioactive Waste**

Edited by Rehab Abdel Rahman

**Published by InTech**

Janeza Trdine 9, 51000 Rijeka, Croatia



**Publishing Process Manager** Sasa Leporic

**Technical Editor** Teodora Smiljanic

**Cover Design** InTech Design Team

Additional hard copies can be obtained from orders@intechopen.com

Radioactive Waste, Edited by Rehab Abdel Rahman
p. cm.
ISBN 978-953-51-0551-0

# Meet the editor

Dr Rehab O. Abdel Rahman, PhD, is a nuclear engineering lecturer at Radioactive Waste Management Department, Hot Laboratory & Waste Management Center, Atomic Energy Authority of Egypt. Dr Abdel Rahman received her Ph.D. degree in Assessment of Unsaturated Zone as a Part of Waste Disposal Site from Nuclear Engineering Department, Faculty of Engineering, Alexandria University, Egypt. She has experience in developing safety, performance and risk assessment studies for waste management systems. She is noted for her work on the assessment and evaluation of different natural and synthetic materials for their potential application in waste management practice. Dr Abdel Rahman has lectured extensively and published more than 25 research and review papers in international journals and conferences. Her research interests include different aspects of radioactive waste management, including radioactive waste treatment, conditioning and disposal.

# Contents

# Preface

The safe management of nuclear and radioactive wastes is a subject that has recently received considerable recognition from public and different governmental, regional and international bodies. This recognition has not only stem from the huge volume of cumulative wastes from previous practice in both peaceful and military fields, but also because the public relate their acceptance for new nuclear power programs to their confidence in the waste management practice. These points impose new burdens on the workers in the waste management field, since they have to understand and deal with technical difficulties beside their reactions to non-technical issues. This book aims to cover the practice and research efforts that are currently conducted to deal with the technical difficulties in different radioactive waste management activities and to introduce to the non-technical factors that can affect the management practice. International experts have cooperated to summarize their practical experience and present advances in managing different types of radioactive wastes and their long-term behavior. The book is targeting professional people in the radioactive waste management industry and reader with technical background such as graduate and postgraduate students undertaking courses in Environmental Science and Environmental, Civil, Chemical and Nuclear Engineering.

The book consists of 20 chapters, organized into three sections that cover important topics in the radioactive waste management field. The first four chapters are introductory, that introduces to the waste management system, explain the interference between technical and non-technical factors, and illustrate how old management practices and radioactive accident can affect the environment. The opening chapter By Dr. R. O. Abdel Rahman, introduces the radioactive waste management system and refer to the technical and non-technical aspects for planning and implementing this system. Dr. P. Brunet presents the historical interference between the technical, social and political factors and summarizes the controversial management process for uranium mine in France. Prof. Dejenbeav et al. present methods and results of the recent identification of uranium tailings in Kyrgyzstan. And finally, the long-term environmental migration from accidentally contaminated site in southern urals in Russian federation was summarized by Prof. Kostyuchenko et al.

The second section is concerned with pre-disposal activities, it summarizes the knowledge gained from current radioactive waste management practice and results of research efforts for using some innovative technologies. The presented activities include reclassification, reduction of generated wastes, decontamination practice and advances, assessment of the performance of different solidified wastes and waste management facilities. The section is beginning with the presentation of a new analytical technique that is used in U.K. for the reclassification of the wastes at the Atomic Weapon Establishment (AWE) by Dr. Miller.

The reduction of the amount of generated radioactive wastes is one of the most important topics in radioactive waste management. Three chapters are presenting this topic, the first deals with the reduction of radioactive wastes in the design phase of Accelerator Driven Subcritical System. This chapter, by Prof. Nandy et al., was directed to estimate the induced activity in the two windows types, which separate between the target and beam pipe in Accelerator Driven Subcritical System. A method for reducing the radioactive wastes associated by medical isotope production is presented by Dr. Yu et al. This new low waste technology for the production of medical isotopes (Mo99 and Sr89) using a homogenous liquid nuclear fuel that were conducted in Kurchatov Institute Russian Federation. Where the third chapter is related to the transmutation of minor actinides in fast reactors as a technique to reduce the half life of these actinides and consequently reduce the requirements for managing these wastes. Prof. Yamawaki Michio et al. are presenting in this chapter the results of actinide-hydride target transmutation research.

The decontamination of contaminated facilities is one of the phases that led to the generation of radioactive wastes, as secondary wastes, good planning and implementation of this phase can led to the minimization of the generated secondary radioactive wastes. Two chapters in this section deals with this topic, Dr. Pavelescu Alexandru Octavian et al. addressed the practice of the clean up and decontamination of hot cells in IFIN-HH VVR-S research reactor, Romania. The second chapter by Prof. Park Kwangheon et al. introduces innovative technique to reduce the amount of generated secondary wastes.

Producing a stable radioactive waste form is an important activity in the management practice, three chapters were devoted to present the recent research effort in this activity. The utilization of fluidized bed steam reforming process to immobilize radioactive wastes is discussed by Dr. Neeway et al. The aim of this chapter is to assess the performance of this technique in immobilizing the radioactive wastes. The second chapter was directed to study the solidification stress induced in ceramic waste form produced from the immobilization of actinides and fission products by Dr. Solbrig Charles et al. The third chapter, by Dr. Clinton Wilson et al, aimed to assess the impact of accidental release of cadmium from immobilization facility. That assessment was conducting by estimating the airborne cadmium concentration caused by facility design base earthquake which damages the electrorefiner vessel. The last chapter in this section is summarizing the waste management practice in fusion technology. By Dr. Luigi Di Pace et al.

The third section provides the reader with an overview of the performance of the waste package, site characterization studies, and modelling ground water efforts. The first chapter by Dr. Mattigod et al., assess the performance of waste package under unsaturated conditions. The diffusivity measurements of Re, Tc and I in concrete containment and the surrounding vadose zone was presented. The effects of carbonation, presence of metallic iron, and fracturing of concrete and the varying moisture contents in soil on the diffusivities of Tc and I were summarized.

Within the site selection activity, the relative importance of the processes that occur within the site and can affect the disposal performance is evaluated and important parameters for assessing the system performance are collected and utilized to ensure that the host environment is having a good containment performance. One of the criteria that determine the suitability of a specific site to host high – level radioactive waste disposal is the radionuclides transport through the fracture of the host rock. The identification of the fracture network using a new approach was addressed by Dr. Reeves et al. The proposed approach is relaying on screening the candidate host rock according to relatively simple criteria obtained from fracture characterization. The characterization of elastic specific storativity and elastic effective porosity in claystone sites is an important topic to model the performance of the claystone as a host rock. Dr. Ababou et al summarizes the statistical methods utilized in quantifying the performance of the clay rock under barometric and earth tides fluctuations. Dr. Constantin summarized the effect of the presence of particulate phases on the radionuclides transport form disposal facilities. His review work is dedicated to provide an adequate definition of the particulate phases, and their effect on radiocolloids development and distribution.

Modeling groundwater contamination is an essential activity to ensure the long-term safety of the disposal facility, Dr. Michael is summarizing the modelling effort to predict the groundwater contamination above high-level waste disposal in three different host rocks. And finally the impact of using transmutation on the radioactive waste management the geological disposal was discussed by Dr. Marivot et al. The chapter is summarizing the results of important international projects that studied these topics.

**R. O. Abdel Rahman**
Hot Lab. Center, Atomic Energy Authority of Egypt
Egypt

# Section 1

## Introduction

# 1

# Planning and Implementation of Radioactive Waste Management System

R. O. Abdel Rahman
*Hot Lab. Center, Atomic Energy Authority of Egypt, Egypt*

## 1. Introduction

The application of radioactive and nuclear materials in power generation, industries, and research can lead to radioactive pollution. The sources of this pollution might include the discharge of radionuclides to the environment by nuclear power facilities, military establishments, research organizations, hospitals and general industry. Also, historical tests of nuclear weapons, nuclear and radioactive accidents and the deliberate discharge of radioactive wastes are representing major sources for this pollution (R.O. Abdel Rahman et. al 2012). Several international agreements and declarations were developed to control the radioactive pollution especially those related to the discharge of radionuclides to the environment. These agreements and declarations impose obligations on national policies to prevent the occurrence of radioactive pollution (IAEA 200a, 2010). On national scale, governments are responsible for protecting the public and environments; the manner at which this responsibility is implemented varies from country to country by using different legislative measures.

The protection of the environment and human health from the detrimental effects of radioactive wastes could be achieved through the effective development and implementation of radioactive waste management system. Recently, some trends that influence the practice of radioactive waste management have emerged worldwide. These trends include planning and application of radioactive waste policy and strategy, issue of new legislation and regulations, new waste minimization strategies, strengthen the quality assurance procedures, increased use of safety and risk assessment, strengthened application of physical protection and safeguards measures in designing and operation of waste management facilities, and new technological options (R.O. Abdel Rahman et. al 2011 a). In this chapter, the recent development in radioactive waste management planning and implementation will be overviewed, the prerequisites and elements for developing and implementing radioactive waste policy and strategy will be highlighted. The advances in the development and application of legal framework and different technical options for radioactive waste management activities will be briefly introduced.

## 2. Waste management policy and strategy development

Policy is defined as a plan or course of action, as of a government, political party, or business, intended to influence and determine decisions and actions (the three dictionary http://www.thefreedictionary.com/policy). In the beginning of the nuclear era, the countries that first started to utilize nuclear and radioactive materials did not have any radioactive waste policy or strategy. To address the radioactive waste issue, some countries had developed and implemented permanent disposal repositories for radioactive wastes and other countries placed radioactive wastes into on-site or off-site storage facilities without the development of national policy for dealing with these wastes.

Preventing risks, to human and the environment, associated with exposure to radioactive wastes was the primary reason to motivate the International Atomic Energy Agency (IAEA) to formulate and publish the policy principals statement in 1995 that deals with the environmental and ethical issue related to managing and disposing these wastes. This statement indicated that "*Radioactive waste should be managed in such a way as to secure an acceptable level of protection for human health, provide an acceptable level of protection for the environment, assure that possible effects on human health and the environment beyond national borders will be taken into account, ensure that the predicted impacts on the health of future generations will not be greater than relevant levels of impact that are acceptable today, and that the management practice will not impose undue burdens on future generations. Also, radioactive waste should be managed within an appropriate national legal framework including clear allocation of responsibilities and provision for independent regulatory functions, the generation of radioactive waste shall be kept to the minimum practicable, interdependencies among all steps in radioactive waste generation and management should be taken into account and the safety of facilities for radioactive waste management shall be appropriately assured during their lifetime*" (IAEA 1995).

These policy principles can be applied to all types of radioactive wastes, regardless their physical and chemical characteristics or origin. In addition to these principles, each country have its own policy principles that define the aims and requirements for the regulatory and legislative framework and might includes administrative and operational measure (R.O. Abdel Rahman et. al). These principals are reflecting the national priorities, circumstances, structures, and human and financial resources. In 2009, IAEA has identified the prerequisites and elements for the development of national radioactive waste management policy. These prerequisites and elements are summarized in Table 1 (IAEA 2009).

As indicated above, some countries started to build and operate radioactive waste disposal without the existence of national waste management policy. Nowadays, these countries started to develop national radioactive waste management policy principals. On the other hand, some existing national radioactive waste management policy principals may need to be updated to improve parts of the policy based on experience of its application and to reflect the changing circumstances in the country and in the world (IAEA 2009). Within this context, the South African policy and strategy document recently developed and was issued in 2005. It included beside the international principals proposed by the IAEA some national principals, that identify the financial and human resources, management transparency and public perception, nature of waste decision making process, international cooperation and national involvement (Department of minerals and energy 2005). In 2007, policy for the long-term management of the United Kingdom's solid low-level radioactive waste was developed following public consultation. That policy statement covers all management aspects for these wastes; it defines this waste category and the key requirements for the management plans. It identifies the importance of using risk informed decision making process, minimization of waste generation, transparency and public involvement, and the consideration of potential effect of climatic changes. Finally it outlines waste import and export and the national organization involvement (Defra 2007).

| Prerequisites | Elements |
|---|---|
| Existence of institutional structure (regulatory body, operational organization) | Allocation of responsibilities between the government, regulatory body and operational organizations |
| Existence of national legal structure and regulatory framework | Identification of safety measure in addition to physical protection and security of facilities |
| Availability of resources to implement the policy | Mechanisms for providing and maintaining the financial, technical and human resources |
| Applicable international conventions | Address the need to minimize the generation of radioactive waste at the design.<br>Identify the export/import of option for radioactive wastes. |
| Indicative national inventories (amounts and types) of existing and anticipated wastes should be identified | Decide whether the spent fuel is considered as resource or as waste, or returned to supplier<br>Identify the main sources of radioactive waste and the intended technical management arrangements.<br>Identify whether the nuclear regulations are applied to naturally occurring radioactive material (NORM) or not based on its radioactive properties. |
| The main parties concerned and involved with spent fuel and radioactive waste management in the country | Indicate the extent of public and stakeholder involvement |
| The existing relevant national policies and its applicable strategies, if any, should be available in response to any policy development | |

Table 1. Prerequisites and element for the development of national radioactive waste policy

After developing the waste management policy principals there is a need to have practical mechanisms to implement these principals, those practical mechanisms are forming the strategy. The first step in developing the waste management strategy is to assign the strategy development responsibility, then assess the availability of information that will be used to develop the strategy. The IAEA has developed a list of important information that should be taken into account during the development of waste management strategy. Those include the estimation of existing and anticipated waste inventory and waste management facilities, the existence of acceptable waste classification system and regulation, the evaluation of waste characteristics and available resources, the knowledge of waste management strategies in other countries and the identification of concerned parties (IAEA 2009). The second step in the development of waste management strategy is the identification of possible end point and technical options. Finally the optimal strategy is determined and the implementation responsibility is assigned. It is worthy to mention that in strategy development, there are two alternatives. The first is a one level method called national plan, which is formulated from a national perspective and often specify one waste operator who is responsible for coordinating the development of such plans. While in the second method, there are two levels for formulating the strategy. At the first level the principal strategy elements are prescribed in general terms as a national strategy by government. At the second level, the detailed implementation of the principal strategy elements is delegated to particular waste owners (company strategies).

To assist the member countries in the nuclear energy agency (NEA), in developing safe sustainable and broadly acceptable strategies for the long-term management of all types of radioactive wastes. NEA has published recently the strategic plan that identity the role of the radioactive waste management committee (RWMC) with respect to the challenges that face the member countries and describe the area of interest for the future work. The identified strategic areas of interest included the following (NEA 2011):

1. Organization of a comprehensive waste management system, including its financing
2. Development of robust and optimized roadmaps for spent fuel and radioactive waste management towards disposal, including transportation
3. Licensing the first geological repositories for high level wastes and /or spent fuel and for other long-lived wastes
4. Industrial implementation of deep geological disposal
5. Effective decommissioning
6. Management of low level wastes and special types of radioactive waste
7. Knowledge management and long-term preservation of records, knowledge and memory

## 3. Developments and implementation of legal framework

To ensure a safe practice for radioactive waste management, there is a need to develop and implement legal framework successfully (IAEA 2000 b). This framework is a part of the national legal system and usually has a hierarchy structure. IAEA has identified a four-level legal framework. The first level in this hierarchy is at the constitutional level, where the basic institutional and legal structure governing all relationships in the country is established. Below this level, there is the statutory level, at which specific laws are enacted by a parliament in order to establish necessary bodies and to adopt measures relating to the broad range of activities affecting national interests. At this level the independency of the regulatory body should be established and maintained. The third level comprises regulations for authorization, regulatory review and assessment, inspection and enforcement. And the final level consists of non-mandatory guidance instruments, which contain recommendations designed to assist persons and organizations in meeting the legal requirements (Stoiber et. al. 2003). In 2005, NEA identified the responsibility of each level development as follow: the first and second level is the responsibility of the main national legislative body. The third level is the responsibility of the government departments or ministries whose portfolios cover one or more aspects affected or influenced by the management of radioactive waste. Exceptionally, the third level in the form of binding rules or codes as distinct from standards may be the responsibility of other bodies such as EPA and NRC in the United States or SSI and SKI in Sweden. There are two philosophies that could be adopted to develop the third and fourth levels, at the first there is a need to develop specifications standards and guides to direct the implementer on how to implement the first and second legislations. At this philosophy, the regulator has some responsibilities and the operator elaborate the detailed specifications then the reviewer and decision is made by the regulator. In the second philosophy, the regulation system is based only on the primary and secondary legislations (NEA 2005, Norrby & Wingefors 1995).

After the establishment of the policy principles set, legal framework is created. To ensure the compliance with the legal framework, there is a need to acquire a formal legal instrument often described as license, permit or authorization. Depending on national legal framework, the licensing process may begin with some kind of decision on the site selection or site authorization or with the construction permit. Successful experiences in facility sitting have shown that active regulatory involvement is needed and is also possible without endangering the independence and integrity of the regulatory authorities (NEA 2003).

## 4. Technical option for radioactive waste management

Radioactive waste management schemes differ from country to country, but the philosophical approach adopted generally is to dispose these wastes in environmentally acceptable ways (R.O. Abdel Rahman et. al 2005 a). During the planning for such scheme, the collection and segregation of wastes, their volume reduction and appropriate conditioning into a form suitable for future handling, transportation, storage and disposal are considered. Pertinent activities in managing radioactive waste are schematically given in Fig. 1. This section is focused on introducing different waste management activities with special emphasizes on new waste minimization strategies, importance of quality assurance, risk and performance assessment.

### 4.1 Minimization of waste generation

The objectives of waste minimization strategy are to limit the generation and spread of radioactive contamination and to reduce the volume of the managed wastes in the subsequent storage and disposal activities. The achievement of these objectives will limit the environmental impacts and total costs associated with contaminated material management. The main elements of this strategy can be grouped into four principals: source reduction, prevention of contamination spread, recycle and reuse, and waste management optimization (IAEA 2001 a, 2007). The reduction of the waste generation at the source begins during the planning for any facility that produces radioactive or nuclear wastes. This principal could be achieved by selecting appropriate processes and technologies, the selection of construction and operational material, and the implementation of appropriate procedures during the operational phase. Also, raising the awareness of the importance of waste minimization through training the employees, and the development and application of contamination and quality control procedures represent important tools to implement the waste minimization strategy.

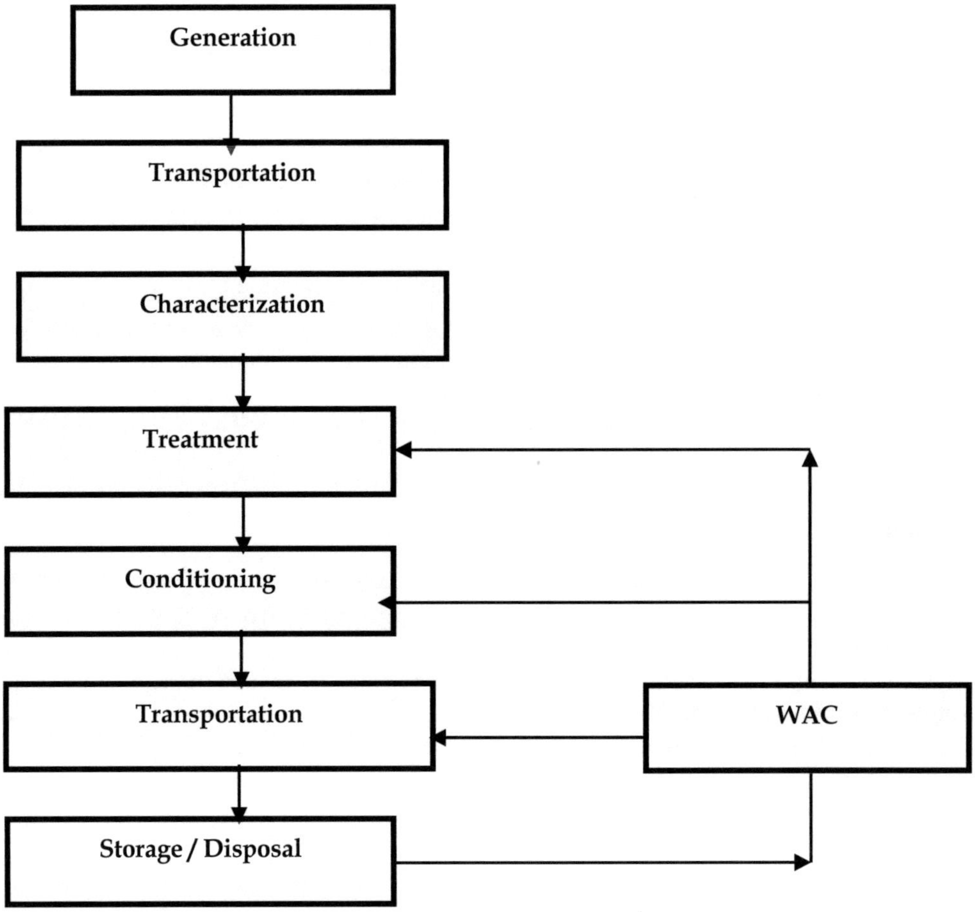


Fig. 1. Radioactive Waste System

Spread of radioactive contamination can lead to creation of secondary wastes, so preventing contamination is consider one of the waste minimization principals. Proper zoning of the facility at the design phase, administrative controls, management initiatives, and selection of decontamination processes are mean keys in reducing the probability of contamination. Finally, the selection of the treatment processes and the utilized chemicals may help in avoiding the production of chemically toxic radioactive wastes.

The recycle and reuse is an attractive method to minimize the generated wastes during the refurbishment and decommissioning of radioactive and nuclear facilities. The decision of selecting this method is dependent on the availability of regulations and criteria, suitable measurement methodology and instrumentation and public acceptance.

The last element in the waste minimization strategies is the optimization of radioactive waste management program that can reduce the volume of the secondary waste. Proper characterization of the generated wastes helps in sorting and segregation of the wastes according to its physical, chemical and radiological characteristics and facilitates the optimization of the treatment option.

### 4.2 Treatment technical options

Treatment is defined as operations intended to benefit safety and/or economy by changing the characteristics of the waste. The basic treatment objectives are volume reduction, removal of radionuclides from the waste and changing the composition of the waste (IAEA 2003 a). There are various commercial volume reduction technologies; the selection of any of these technologies is largely depending on the waste type. To facilitate the selection of the treatment options, the wastes are classified according to their activity limit (e.g. exempt waste, very low level waste, low level waste, intermediate level waste, and high level waste), chemical properties (e.g. aqueous/organic waste, acidity/alkalinity, chemical stability, redox potential, toxicity), physical characteristics (liquid/solid/gas, density, morphology, compactability and level of segregation) and biological properties. Table 2 lists the commercial technical treatment options for managing different waste classes (IAEA 1999, 1994 a ,2009, Ojovan, 2011 ).

| Liquid aqueous waste | Liquid organic waste | Solid wastes | Gaseous |
|---|---|---|---|
| Chemical precipitation (Coagulation/flocculation /separation) | Incineration | Storage for decay (for very low level wastes) | Filtration |
| Ion exchange | Emulsification | Compaction | Sorption, |
| Evaporation | Absorption | Melting, | Scrubbing |
| Reverse osmosis | Phase separation (e.g. distillation) | Fragmentation | |
| Membrane processes | Wet oxidation | Incineration | |
| Evaporation | Alkaline hydrolysis | Encapsulation, | |
| Electrochemical | | | |
| Solvent extraction | | | |

Table 2. Available technical treatment options for different waste categories

### 4.3 Conditioning technical options

The conditioning activity includes the operations that produce a waste package suitable for handling, transport, storage and/or disposal. Conditioning may include the conversion of the waste to a solid waste form (immobilization), enclosure of the waste in containers, and, if necessary, providing an over-pack (IAEA 2003 a). The produced waste form must be structurally stable to ensure that the waste does not degrade and/or promote slumping, collapse or other failure. Chemical and physical immobilizations provide the required structural stability and minimize the contaminant migration. Immobilization techniques consist of entrapping the contaminant within a solid matrix i.e. cement, cement-based material, bitumen, glass, or ceramic (R.O. Abdel Rahman et al. 2007 a).

Cementation of radioactive waste has been practiced for many years basically for immobilization of low and intermediate level radioactive waste. The majority of cementation techniques rely on using Portland Cement as the primary binder. Other binders might be used to improve either the mechanical performance of the final waste matrix or to improve the retention of radionuclides in that matrix, these include fly ash, blast furnace slag, bentonite, zeolite and other materials (R.O. Abdel Rahman & A.A. Zaki 2009 a). The implementation of this technique worldwide is supported by its compatibility with aqueous waste streams, capability of activated several chemical and physical immobilization mechanisms for a wide range of inorganic waste species. Also, cement immobilization possesses good mechanical characteristics, radiation and thermal stability, simple operational conditions, availability, and low cost (R.O. Abdel Rahman et al. 2007 a).

Bituminisation is applied to immobilize the secondary wastes resulting from the treatment of low and intermediate level liquid effluents of very low heat generation (< 40 $TBq/m^3$). The bituminized product has a very low permeability and solubility in water and is compatible with most environmental conditions (IAEA 1998). This kind of immobilization media is restricted for wastes that contain strongly oxidizing components, e.g. nitrates, biodegradable materials and soluble salts. A special care should be given to this waste form during its storage owing to its flammability.

Vitrification is one of the important immobilization techniques which relays on the utilization of glass as immobilizing media, because of the small volume of the resulting waste-form, its high durability and stability in corrosive environments. To ensure the high durability of the produced matrix, the vitrification process should be conducted under very high processing temperatures (>1500 °C), which impose limitations on the immobilized radionuclides and increase the amount of generated secondary wastes. As a result, the most common glasses used in vitrification of nuclear waste are borosilicates and phosphates which use lower processing temperatures (≈1000 °C) while still forming a durable product (M.I. Ojovan & W.E.Lee 2005).

The above-mentioned immobilization technologies are available commercially and have been demonstrated to be viable. The highest degree of volume reduction and safety is achieved through vitrification although this is the most complex and expensive method requiring a relatively high initial capital investment. The potential of using new immobilization matrices were emerged to deal with difficult legacy waste streams. These matrices include crystalline (mineral-like) and composite radionuclide immobilization matrices as well as using thermochemical and in situ immobilization techniques (M.I. Ojovan & W.E.Lee 2005).

### 4.4 Transport of radioactive wastes

The transport of radioactive wastes includes three stage namely; preparation, transfer and emplacement (IAEA 1994 b). The safety of the transport processes could be provided through meeting the provisions of transport regulations, which aim to protect persons, property and the environment from the effects of radiation during the transport of these materials. Transport regulations include requirements on the waste package that ensure its survival under accident conditions. Depending on importance of the shipped wastes from security, safeguards, and safety point of views, the risk assessment of the transport process might include the following (IAEA 2003 b):

1. Shipment information,
2. Radiological, physical, an chemical characteristics of the waste,
3. Physical characteristics of the package and conveyance,
4. Exposure parameters for the transport workers,
5. Routing data and population characteristics,
6. Frequency and severity of accident for a given transport mode, and
7. Estimation of doses to public

### 4.5 Storage technical options

Long-term management of spent fuel is becoming of increasing concern, since few decisions are now available with regard to the implementation of their final disposal. This might be attributed to the public perception towards the final disposal of spent fuel and/or the need to gain better insights into the long-term performance of spent fuel and materials. This class of radioactive wastes is currently stored in different storage types. These include, nuclear power plant pools, wet and dry storage facilities. Figure 2 illustrates the capacity and inventories of different types of spent fuel storage (IAEA 2002).

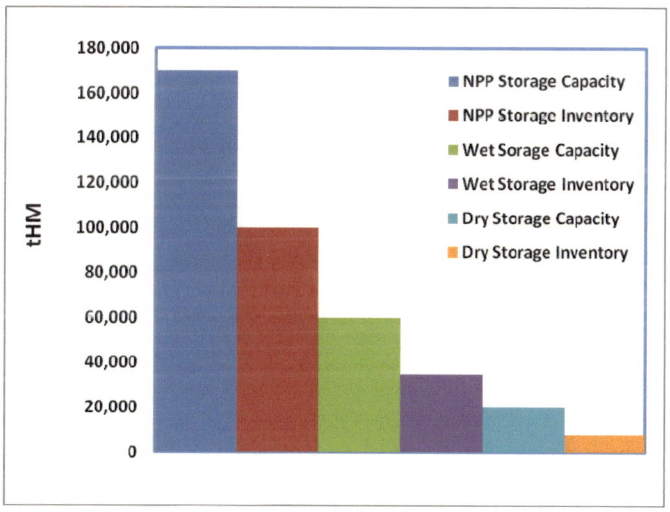


Fig. 2. Comparison of capacities and inventories of different types of spent fuel storage (IAEA 2002)

Interim storage of radioactive waste packages is not only required if the disposal facility is not available but also for wastes those include very short lived radionuclides. The design and operation of storage facilities must comply with the basic safety principles set up on both the national and international scale. To assess the compliance of the storage facility, a licensing process including safety and environmental impact assessments must be part of the waste management system. The main functions of a storage facility for conditioned radioactive waste are to provide safe custody of the waste packages and to protect both operators and the general public from any radiological hazards associated with radioactive wastes. The design of storage facilities should be capable of (IAEA 1998)

1. Maintain the "as-received" integrity of the waste package,
2. Protect the waste from environmental conditions that could degrade it,
3. Keep the external dose rate and contamination limits for waste packages to be accepted by the facility,
4. Minimize the radiation exposure to on-site personnel,
5. Allow control of any contamination from gaseous or liquid releases.

The storage facility may be associated with an area for inspection (including sorting and/or non-destructive examination), certification and labeling of waste packages. The storage facility is usually divided into areas where low contact dose rate packages are stored, areas where packages not meeting waste acceptance criteria (WAC) are stored, and a shielded area where high contact dose rate packages are kept secure (IAEA 1998). The design of the facility usually permits package stacking, sorting and visual inspection. Provision for maintaining a database keeping chain-of-custody for each waste package in storage must be included in the design. Key information about the waste package should include the total radionuclide content, the waste matrix used for immobilization, the treatment and/or conditioning method (as applicable), and the unique package designator. A hard copy file should follow the waste package from conditioning to its final disposal (IAEA 2001 b).

### 4.6 Disposal technical options

Disposal is the last step in the integrated radioactive waste management, it relay on the passive safety concept. The disposal facility includes waste emplacement area, buildings and services for waste receipt. Its design aims to provide isolation of the disposed waste for appropriate period of time taking into account the waste and site characteristics and the safety requirements (Bozkurt 2001, R.O. Abdel Rahman et. al 2005 a, b). To achieve this aim, the multi-barrier concept that relays on using engineered barriers to augment natural barriers has been developed. The use of engineered barriers helps in ensuring that increasingly stringent design aims are satisfied to an appropriate level (IAEA 1997). This concept helps in avoiding over-reliance on the natural barriers to provide the necessary safety (IAEA 1992 a, 1993 a).

Engineered barriers may consist of a number of separate components, including structural walls, buffer or backfill materials, chemical additives, liners, covers, leachate collection and drainage systems, cut-off walls, gas vents and monitoring wells (IAEA 1992 b). The design criteria for each barrier will differ according to the waste class and disposal type, IAEA have define the main function for the engineering barriers in a near surface disposal type. Those functions are listed in Table 3 (IAEA 2001 c).

Disposal facilities could be place in geological formation or near surface. Near-surface disposal includes two main types of disposal systems: shallow facilities located either above or below the ground surface; and underground facilities, usually in rock cavities. Geological disposal refers to disposal at greater depths, typically several hundreds of meters below ground (R.O. Abdel Rahman et. al 2012). Table 4 lists a summary for underground disposal practices.

| Barrier | Function |
| --- | --- |
| Container | Mechanical strength, Limit water ingress, Retain radionuclides |
| Waste form | Mechanical strength, Limit water ingress, Retain radionuclides |
| Backfill | Void filling, Limit water infiltration, Radionuclide sorption<br>Gas control |
| Structural materials | Physical stability containment barrier |
| Cover | Limit water infiltration, Control of gas release, Erosion barrier<br>Intrusion barrier |

Table 3. Function of each engineered barrier.

| Place | Depth (m) | Type of reservoir |
| --- | --- | --- |
| **Czechoslovakia** | | |
| Hostim | 30 | Limestone mine |
| Richard | 70-80 | Limestone mine waste |
| Bratrstvi | -- | Uranium mine |
| **Germany,** | | |
| Asse | 725-750 | Salt mine |
| Morsleben | 400-600 | Potash and salt mine |
| **Swedish** Final Repository | 50 below Baltic Sea | Metamorphic bedrock |
| **Finland** | | |
| Olkiluoto | 60-100 | Crystalline bedrock |
| Loviisa | 70-100 | -------- |
| **USA** WIPP | 655 | Rock salt formation |

Table 4. Summary of some underground disposal

The optimization of the disposal is done by conducting safety assessment studies. These studies are complex due to the dynamic nature of the hydrological and biological subsystems in the host environment that affects the degradation scenarios of the disposal facility. So treating the disposal as one system is not possible, instead these subsystems are decoupled and divided into modules for which the evolution of the disposal is distinguished into step changes rather than continuous time change [NCRP 2005]. Generally, safety assessment relays on specifying assessment context, describing the disposal system, developing and justifying evolution scenarios, formulating and implementing of models; and finally analyzing the assessment results for each module. During the development of safety assessment, all confidence building tools should be utilized and illustrated (R.O. Abdel Rahman et. al 2011 c).

### 4.7 Safety of radioactive waste management.

IAEA recommended that assessment studies have to be developed and well adapted to situations of concern to ensure the protection of human health and the environment (IAEA 1993 b). To apply this recommendation, an initial assessment of the planned waste management practice needs to be performed that identifies the radiological sources, foresees potential exposures, estimates relevant doses and probabilities, and identifies the required radiological protection measures. Various methodologies with varying complexity have been and are being developed to assist in the evaluation of radiological impact of nuclear and radioactive facilities. Despite there are differences in the details of these methodologies to correspond to each facility, the general objective of any radiological assessment is to determine the impact of radioactive material on individuals and their environment (R.O. Abdel Rahman 2010). In 2002, IAEA published procedure for conducting probabilistic safety assessment for non-reactor nuclear facilities (IAEA 2002 b). This procedure is consist six interlinked steps, which include

1. Management and organization,
2. Identification of source of radioactive releases, exposure and accident initiator,
3. Scenario modeling,
4. Sequence quantification,
5. Documentation of the analysis and interpretation of the results, and
6. Quality assurance.

The identification of the source and exposure is done through the consideration of source-pathway- receptor analysis at which different aspects are identified i.e. how radio-contaminants released from the studied facility, the pathways along which they can migrate, and their impacts on human. In developing such analysis, it is important to understand that radio-contaminants are transported by air, soil or water through advective or diffusive processes and that the principal means of human exposure is by direct radiation exposure, inhalation of gases or particulates, and ingestion of contaminated food or water (R.O. Abdel Rahman 2010).

To quantify the sequence of the release there is a need to model the release scenario, this could be performed through the development of conceptual model, mathematical model selection, and development or selection of numerical tools. Generally, a conceptual model describes with words and diagrams the key processes that occur within the studied system (or have a reasonable likelihood of occurring). These models can be formulated at varying levels of complexity and realism to abstract the reality (Environment Agency 1999). The developed conceptual model forms the basis for the selection of mathematical models, which in turn govern the selection and creation of numerical models and computer codes (R.O. Abdel Rahman et. al 2009b).

The planning, development and application of quality assurance program for the safety assessment of the radioactive waste management facilities begin with the identification of quality policy then it associates each step in the assessment. Different quality assurance activities should be performed that include sample control, quality assurance for the documentation. In the scenario modeling step, the range, accuracy and precision of equipment used for input data collection must be verified. The personnel should be suitably trained and qualified to perform the data collection step in accordance with standards. Also the utilized computer software must be verified, validated and documented. Computer software must be placed under configuration control as each baseline element is approved and released. Changes to computer software must be systematically evaluated, co-ordinated and approved to assure that the impact of a change is carefully assessed before updating the baseline (USDOE 2003).

### 4.8 Waste acceptance criteria and quality assurance programs

The waste acceptance is defined as "Quantitative or qualitative criteria specified by the regulatory body or by waste operator and approved by the regulatory body, for radioactive waste to be accepted in a waste management facility "(IAEA 2003). The development of the waste acceptance criteria is carried out in parallel with the development of the waste management facility and is derived from both safety and operational requirements. The compliance with these criteria includes two stages; the first is the definition of the waste characteristics and identification of quality related parameters. This stage is developed by using the results of the safety assessment studies and the operational experience. The second stage is the confirmation of the conformance of the individual waste packages to the WAC, this stage could be checked directly or indirectly by using data sheets that includes information about the preceding waste producer, the waste type, activity, source, description, and radiological characteristics and package identifier number and type if any. The dose and heat rate, surface contamination and the weight are also important parameters that are widely used to confirm conformance with WAC (IAEA 1996). Assurance that a waste package can meet WAC could be provided if the development and design of the management process is carried out under a Quality Assurance Program (QAP).

Inadequate procedure specification and verification of required actions in the selection, design, construction, and operation of individual facilities and processes through the waste management system may lead to a failure in the achievement of waste management goals. The application of a Quality Assurance Program (QAP) to all waste management activities including treatment, conditioning, storage, transport, and disposal is intended to ensure the achievement of the waste management objectives. Within the QAP, there is a need to establish a quality control program that intended to ensure the compliance of the products from the waste management facility with the WAC at the preceding waste management facility and/or meet the regulatory requirements for discharge, transport, condition, store, and or dispose this waste product (R.O. Abdel Rahman 2009 c, 2007 b). The elements of this program are similar to any other program in non-radioactive industry. It includes: organization and responsibilities planning and implementation, personnel training and qualification, existence of procedures and instructions, document control, research and development, procurement, process control, inspection and testing, non-conformance and corrective actions, records, management review and audit.

## 5. Acknowledgment

The author would like to acknowledge her appreciation to Dr M.I. Ojovan, professor at Imperial Colege London, for the time and effort that he spent to review this chapter.

## 6. References

Abdel Rahman R.O., A.M. El Kamash, A.A. Zaki, M.R..El Sourougy, (2005 a) "Disposal: A Last Step Towards an Integrated Waste Management System in Egypt",

International Conference on the Safety of Radioactive Waste Disposal, Tokyo, Japan,, IAEA-CN-135/81, p.p. 317-324

Abdel Rahman, R.O., El-Kamash, A.M, Zaki, A.A., & Abdel-Raouf, M.W. (2005 b). Planning closure safety assessment for the egyptian near surface disposal facility, Presented at the International Conference on the Safety of Radioactive Waste Disposal (pp. 317–324), Tokyo, Japan, IAEA-CN- 135/81

Abdel Rahman R. O., A. A. Zaki, A. M. El-Kamash, (2007 a), Modeling the long-term leaching behavior of (137)Cs, (60)Co, and (152,154) Eu radionuclides from cement-clay matrices, *J. Hazard Mater.*, 145(3) p.p.372-380

Abdel Rahman R. O., A. M. El-Kamash, F. A. Shehata, M. R. El-Sourougy, (2007 b) Planning for a solid waste management quality assurance program in Egypt, *Qual. Assur J.* 11(1) p.p. 53-59

Abdel Rahman R.O., A. A. Zaki, (2009 a), Assessment of the leaching characteristics of incineration ashes in cement matrix, Chem. Eng. J. ,155, p.p. 698-708

Abdel Rahman R. O., (2009 c), Design a quality control system for radioactive aqueous waste treatment facility, *Qual. Assur J*; 12(1)p.p. 31-39

Abdel Rahman R. O., A.M. El Kamash, H. F. Ali, Yung-Tse Hung, (2011 b) Overview on Recent Trends and Developments in Radioactive Liquid Waste Treatment Part 1: Sorption/Ion Exchange Technique, *Int. J. Environ. Eng. Sci.*, 2 (1), PP. 1-16

Abdel Rahman R. O., H. A. Ibrahium, Yung-Tse Hung, (2011 a), Liquid radioactive wastes treatment: A Review , *Water*, 3, P.P.551-565

Abdel Rahman R.O., (2010), Preliminary assessment of continuous atmospheric discharge from the low active waste incinerator, *Int. J. Environ. Sci*, 1, No 2, p.p.111-122.

Abdel Rahman R.O., A.A. Zaki (2011 c), Comparative study of leaching conceptual models: Cs leaching from different ILW cement based matrices, Chem. Eng. J., 173 (2011) p.p. 722- 736.

Abdel Rahman R.O., H. A. Ibrahim, N. M. Abdel Monem, (2009 b), Long-term performance of zeolite Na A-X blend as backfill material in near surface disposal vault. Chem. Eng. J. , 149, 143-152

Abdel Rahman R.O., M. W. Kozak, Yung-Tse Hung, (2012), Radioactive pollution and control, in accepted for publication Handbook of Environmental and Waste Management, Vol 2, Land and Groundwater Pollution Control, chapter 16 World Scientific Publishing Co, Singapore.

Bozkurt, S., Sifvert, M., Moreno, L., & Neretnieks, I. (2001). The long-term evolution of and transport processes in a self-sustained final cover on waste deposits, The Science of the Total Environment, 271, 145–168.

Defra, DTI and the Devolved Administrations, (2007), Policy for the long term management of solid low level radioactive waste in the United Kingdom, Department for environment food and rural affairs.

Department of minerals and energy, (2005), Radioactive waste management policy and strategy for the republic of South Africa, Department of minerals and energy.

Environment Agency, (1999), Guide to Practice for the Development of Conceptual Models and Selection and Application of Mathematical Models of Contaminant Transport Processes in the Subsurface, National Groundwater and Contaminated Land Center Report NC/99/38/2, 1999

IAEA, (1992 a), Review of available options for Low level radioactive waste disposal, iaea-tecdoc-661, International Atomic Energy Agency, Vienna.

IAEA, (1992 b), Performance of engineered Barriers in deep geological repositories, Technical Reports Series no. 342, International Atomic Energy Agency, Vienna.

IAEA, (1993), Report on radioactive waste Disposal, Technical Reports Series no. 349, International Atomic Energy Agency, Vienna.

IAEA, (1993), Use of Probabilistic Safety Assessment for Nuclear Installations with Large Inventory of Radioactive Material, TecDoc-711, International Atomic Energy Agency, Vienna.

IAEA, (1994 a), Advances in Technologies for the Treatment of Low and Intermediate Level Radioactive Liquid Wastes, Technical Reports Series No. 370, International Atomic Energy Agency, Vienna.

IAEA, (1994 b), Interfaces between transport and geological disposal systems for high level radioactive waste and spent nuclear fuel TecDoc-764, International Atomic Energy Agency, Vienna

IAEA, (1995), The Principles of Radioactive Waste Management. Safety Series No. 111-F, International Atomic Energy Agency, Vienna.

IAEA, (1996), Requirements and methods for low and intermediate level waste package acceptability, TecDoc-8 64 International Atomic Energy Agency, Vienna.

IAEA, (1997), Planning and operation of low level waste disposal facilities (proc. Symp. Vienna, 1996), International Atomic Energy Agency, Vienna.

IAEA, (1998), Interim Storage of Radioactive Waste Packages, Technical Reports Series No. 390, International Atomic Energy Agency, Vienna.

IAEA, (1999), Review of the factors affecting the selection and implementation of waste management technologies, TecDoc-1096, International Atomic Energy Agency, Vienna.

IAEA, (2000 a), Regulatory control of radioactive discharges to the environment, IAEA Safety Series Guide No. WS-G-2.3, International Atomic Energy Agency, Vienna

IAEA, (2000 b), Legal and Governmental Infrastructure for Nuclear, Radiation, Radioactive Waste and Transport Safety Requirements, Safety standard series, No. GS-R-1, International Atomic Energy Agency, Vienna.

IAEA, (2001 a), Methods for the minimization of radioactive waste from decontamination and decommissioning of nuclear facilities, TRS 401, International Atomic Energy Agency, Vienna.

IAEA, (2001 b), Waste inventory record keeping systems (WIRKS) for the management and disposal of radioactive waste, TecDoc-1222 International Atomic Energy Agency, Vienna.

IAEA, (2001 c), Performance of engineered barrier materials in near surface disposal facilities for radioactive waste, TecDoc-1255, International Atomic Energy Agency, Vienna.

IAEA, (2002 a), Long term storage of spent nuclear fuel —Survey and recommendations Final report of a co-ordinated research project 1994–1997, TecDoc-1293, International Atomic Energy Agency, Vienna.

IAEA, (2002 b), Procedures for conducting probabilistic safety assessment TecDoc-1267 , International Atomic Energy Agency, Vienna

IAEA, (2003 a), Radioactive waste management glossary, 2003 edition, International Atomic Energy Agency, Vienna.
IAEA, (2003 b) Input data for quantifying risks associated with the transport of radioactive material Final report of a co-ordinated research project 1996–2000, TecDoc-1346, International Atomic Energy Agency, Vienna.
IAEA, (2007), Considerations for waste minimization at the design stage of nuclear facilities, TRS, No 460, International Atomic Energy Agency, Vienna.
IAEA, (2009), Policies and strategies for radioactive waste management, Nuclear energy series NW-G-1.1 International Atomic Energy Agency ,Vienna,
IAEA, (2010) Setting authorized limits for radioactive discharge: practical issues to consider, TecDoc-1638, International Atomic Energy Agency ,Vienna
NCRP, (2005), Performance assessment of near-surface facilities for disposal of low level radioactive waste, NCRP Report No. 152, National Council on Radiation Protection and Measurements.
NEA, (2003), The regulator's evolving role and image in radioactive waste management lessons learnt within the NEA forum on stakeholder confidence, NEA ISBN 92-64-02142-6
NEA, (2005), The regulatory functions and radioactive waste management international overview, NEA No.6041 , ISBN 92-64-01075-0
NEA, (2011), Strategic plan 2011-2016 for the radioactive waste management committee , NEA/RWM(2011)12
Norrby S., S. Wingefors, 1995, formulation of regulatory and licensing requirements, IAEA-TecDoc-853 requirements for the safe management of radioactive waste, p.p.281-286
Ojovan M.I., W.E. Lee, (2005), Introduction to nuclear waste immobilization, Elsevier
Stoiber C, Baer A., Pelzer N., Tonhauser W., (2003), Handbook on nuclear law, International Atomic Energy Agency, Vienna
M.I. Ojovan (Editor) (2011). Handbook of advanced radioactive waste conditioning technologies. ISBN 1 84569 626 3. Woodhead Publishing Series in Energy No. 12, Oxford, 512 p. http://www.woodheadpublishing.com/6269
The three dictionary http://www.thefreedictionary.com/policy
USDOE, (2003), Carlsbad field office quality assurance program document, DOE/CBFO-94-1012 revision 5 2003, United State Department of Energy.

2

# A Controversial Management Process: From the Remnants of the Uranium Mining Industry to Their Qualification as Radioactive Waste – The Case of France

Philippe Brunet
*Université d'Evry, Evry*
*Centre Pierre Naville*
*France*

## 1. Introduction

The analysis of environmental issues inevitably requires the contribution of an array of scientific disciplines. The experimental sciences, whose goal is the knowledge of natural phenomena, cannot aspire, alone, to resolve the problems raised by interactions between human societies and nature. Nor can the social sciences claim any monopoly thereto. This is particularly true of our industrial societies, which ceaselessly produce what Ulrich Beck calls "latent induced effects" (1986) which engender long term environmental and health hazards. Their understanding is always belated. It is very often achieved by expertise in experimental sciences, intersecting with the wisdom of common sense, social mobilisations, and the weight of prevailing social norms (Wynne, 1997). Their extent and lastingness accordingly result from the combination of two factors, one being determined by the other. One factor include the limits, at any time, to the knowledge and predictions that they make possible, in terms of the future trend of a given industrial process ; other factor, the social relationships of production and reproduction whereby this industrial process is implemented by relying on the prior art, but also on prevalent beliefs and ideologies.

These social relationships also produce social values and norms. Under the impact of the rapport between capital and labour, they sustain the subdivisions inherent in any process of industrial production and in its organisation between experts and laymen, producers and consumers, particularly via legitimating arguments (Braverman, 1975). The sociological analysis of environmental issues requires an understanding of the dynamic of these social relationships through the examination of their tensions and conflicts which, very often, are crystallised in these dialectical forms. It must be considered as complementary to the analysis of the nature sciences, without one ever substituting for the other. It is in this perspective that we propose to analyse the production process of the uranium industry, a vital link in the production of nuclear energy. We shall focus particularly on its remnants. We shall show that the qualification of radioactive waste which henceforth attaches to them results from practices of the players within changing configurations, to varying degrees conflictual. The challenge concerns the hegemony of legitimacy to *say* and to *do* with regard to their management. This perspective accordingly implies carrying out a long range analysis to grasp their evolution.

This chapter is divided into two parts. The first describes the emergence of the problematics, in which science, technology, politics and standards are combined in a scheme of specific production relationships. It dwells on the early decades of the atomic complex, to grasp its various structural components and, ultimately, to understand the function of the radioactive waste qualification process. The second part expands the analysis of this mechanism over the long term, based on the case of France. The focus is then directed at the least known productive segment of the nuclear complex, the uranium mining industry, and on its repercussions in terms of waste.

## 2. Science, politics and standards concerned with radioactive waste: A new horizon

By virtue of its history, and its underlying scientific knowledge and techniques, the atomic industry, later called the nuclear industry, is linked with the state of war. This is why no doubt more than any other, this industry has been ambivalent since its inception. It is oriented towards destruction as well as production (Naville, 1977). This attribute is especially pronounced as the structure becomes recursive. Indeed, the earliest large atomic facilities that went on stream in the USA, the USSR, Great Britain and France, were plants simultaneously generating plutonium and electricity for military and civilian uses (Barillot and Davis, 1994). Similarly, the environmental and health hazards associated with the concept of energy generation were precisely the "arguments", amply demonstrated in practice, of its capacity to destroy at a hitherto unsuspected scale. This finding became the background for the many descriptions, popularisations and justifications of the new industry (Ducrocq, 1948; Martin, 1956; Goldschmidt, 1962), giving rise to many consequences that we shall examine in turn, and globally. First, the control of this industry was directly assumed by the States and associations thereof, in peacetime and wartime alike. Second, its technological and strategic sophistication generated an intensive and tight interpenetration of different professional worlds: scientific, military, industrial and political. This tense closed world reflected the elitist, in other words, non-democratic, relationship that became established for decisions pertaining to this industry. And finally, its ambivalence marked an associated process: the qualification of "radioactive waste". The narrow perimeter in which it was long contained caused the slowness of its development, and also, in exchange, the deep democratic penetration that it received.

### 2.1 The atomic and nuclear industry: A matter for States at the planetary scale

After the Second World War, the atomic industry developed essentially in obedience to geostrategic and military objectives. A differentiation set in between States according to whether or not they possessed the atomic weapon and its uranium fuel. This cleavage was not exclusively of a technical or economic nature. It was also political, and had two outcomes. States owning atomic weapons sought to hamper the access of other candidate states to the possession of the industrial process. It hence ordered and crystallised the global ranking of the military powers. This situation still prevails today. For example, the sanctions imposed against Iran since 2006 by the UN Security Council, claimed justification in the fact that these countries had tried or were trying to possess a military nuclear industry (IAEA, 2006). It also fostered a policy of secrecy which was gradually relaxed to facilitate civilian industrial applications.

This unprecedented situation betokened a new relationship between science, industry and politics, with an implication of international controls. The sharing of the world in fact established new geostrategic relations between East and West. Its equilibrium depended on the resources available to each camp to develop the industrial process. Before the war, the Belgian mines of Upper Katanga enjoyed a monopoly of radium production. The importance gained by uranium as a fuel then encouraged the USA to control its production. Despite the discoveries in Canada and Czechoslovakia, uranium was held to be a rare ore (Ducrocq, 1948). Thus, wishing to maintain a lead, which it wrongly believed to be significant in terms of the technology and the uranium raw material, the USA tried to impose its point of view, which only Great Britain and Canada accepted. Faced with the refusal of the USSR, which controlled Czech uranium, the USA decided to maintain its lead by practising a policy of secrecy (Goldschmidt, 1962). Indeed, in late July 1946, a law was passed organising and governing atomic energy in the USA (the *MacMahon Bill*). All the problems of atomic energy, from ore to nuclear fuel, plants included, fell under its authority. Secrecy was maintained, and its violation decreed a capital crime. Finally, the new Bill enshrined isolationism: collaboration with other countries was subject to Congressional approval. This is why from 1946 to July 1954, when the law was first relaxed, even collaboration with English speaking countries was suspended (Goldschmidt, 1962). This policy of secrecy became the international norm. In September 1949, the Russians showed the American that they no longer held exclusive sway. The battle for power and technological sophistication was then joined on a new project based on the thermonuclear reaction, leading to the hydrogen bomb, a thousand times more powerful than the A bomb. At the same time, in 1952, Great Britain broke into the closed club of the atomic countries, followed by France in 1960. This policy of secrecy contained its contradictions. Thus, from the 1950s, the US proposed the Baruch plan to the United Nations (Goldschmidt, 1987). It offered to relinquish atomic secrecy provided that an international agency took charge of the ownership of the uranium mines, atomic materials, and the running of fuel production plants and power reactors. The USSR was opposed and demanded that the USA destroy its arsenal and terminate the arms race. The American proposal was doomed to failure. Certainly, it foreshadowed the various UN regulatory agencies that were progressively set up in the atomic field. This necessity stemmed from the ambivalence of its industry. In terms of destruction, the UN Security Council contained the five foremost historic atomic powers as permanent members[1]. They therefore "monitored" the balance of global forces under the sign of secrecy and mutual mistrust. In terms of production, civilian industrial development could not durably be a subject issue. This conflict was partly resolved at the first international conference in Geneva in 1955, *Atoms for Peace*. The disparateness of its participants and the scheduling of its deliberations (first, states and after Scientists) were symptomatic of its social intricacy and hierarchy, which promoted the existence of the industry, born in the USA in 1943.

---

[1] The list of five permanent members was approved in 1946, long before they became atomic powers. However, the correlation is striking, and the sign of a suite in the state power ranking.

### 2.2 The atomic and nuclear industry, a heterogeneous and closed visage

Its starting point was the *Manhattan Project*. This project was its parent-formula. It associated four different types of social actors, not without some tension: State (for political decisions), Industry (the *Du Pont* company engineers for the practical organisation of the industrial process at Oak Ridge, Tennessee), Scientists (for their investigations), and the Military (for their responsibility in management and control) (N'diaye, 1998). Subsequent industrial developments, each inserted into their specific national frameworks, were differentiated from this initial wartime model. But, with it, they shared the principle of ambivalence between destruction and production, of the disparateness of the social actors, and finally, the closure of this new productive world sustained by the policy of secrecy. The French model was no exception.

Certainly, for no science other than nuclear physics, was the era of its fundamental and theoretical questions and that of its practical applications so intermingled, jump-starting the production of destructive bombs. This unprecedented situation was marked by contradictions and internal tensions, particularly the ambiguous attitudes of the scientists (Martin, 1956). In August 1939, Einstein sent a letter, co-signed by other physicists, to US President Roosevelt, to alert him to the risk of some day finding Nazi Germany in possession of the atom bomb. He decided to move swiftly. This act triggered a process of decisions culminating in the *Manhattan Project* in 1942, in other words, the production of the bomb. It is estimated that 75 000 to 150 000 people were mobilised, particularly in the Oak Ridge plant, until the explosion of the first bomb in New Mexico (Goldschmidt, 1962; N'Diaye, 1998). The scientists, with the army and the industry, were joined under the aegis of the political authority. This created some ambiguity in the attitudes of the scientists in three respects. On the one hand, while nothing in the atomic field could be done without them, its future was beyond their control. On the other, the practical and ideological underpinnings of their professional integrity were denied. This applied to unrestricted access to information and its exchange in the name of priority over the policy of secrecy, and disinterestedness in the name of limited commitment. They tried morally to resolve this contradictory positioning in many ways: through justification, through guilt, or even by engaging in peace movements like *Pugwash* and the *Stockholm Appeal* (Oppenheimer, 1955; Joliot-Curie, 1963; Einstein, 1979).

Similarly, the first international conference in Geneva in 1955, *Atoms for Peace*, which brought together seventy-two countries, tried to resolve the internal contradiction of the atomic complex internationally. It partly relaxed the policy of secrecy and thereby met the desires of the scientists. It made possible the recursiveness of destruction towards production. It timidly addressed the latent induced effects of radioactivity on human health. Its deliberations nevertheless reflected the ambivalence of the atomic complex and the ranking of its players. First, the governments of the atomic countries (USA, Great Britain, USSR and France) held a week-long meeting in July; followed by the scientists and industrialists for twelve days on the civilian applications of the atom. No other social or associative force was invited to the discussion table, confining the issues exclusively in the hands of the experts and political decision makers.

These ingredients of the atomic complex could be found in the French formula, delayed and with specific characteristics. In October 1945, Commissariat à l'Energie Atomique (CEA) was created under the unchallenged authority of Frédéric Joliot-Curie[2]. Its programme was that of atomic science and its civilian applications. The problem of fuel remained to be solved. The CEA had a limited stock of heavy water and uranium in a context of a uranium embargo. France accordingly launched a prospecting programme:

> "Dig everywhere without second thoughts. Have no qualms about your prospecting methods. Besides, if I could, I would send out 2000 prospectors throughout France! They would systematically scour the soil with a Geiger counter, from the Pas-de-Calais to the Pyrenees! Not a single clue of uranium could elude me!"

This was Joliot-Curie's exhortation to the first class of uranium prospectors trained from December 1945 (Paucard, 1994). Until 1950, when Joliot-Curie was dismissed for political and geostrategic reasons, and even beyond, scientists resisted government pressures concerning the assigned objectives. The challenge was the atom bomb and the military presence in the CEA. But, progressively, through the Fifties, the CEA industrialised, militarised and finally escaped the control of the scientists, now more relevant to the initial model promoted by the U.S. So, By government decree in 1951, the CEA was led by a director and no longer by a scientist. In 1955, the government created the consultative commission for the Production of Energy of Nuclear Origin (PEON commission). It is reporting to the government and tasked with supplying justified opinions on decisions to be taken. Also, the government named a military man to direct the CEA's general design office: in 1958 this office became the CEA's Directorate of Military Applications (DAM), charged with setting up France's nuclear weapons programme. After much procrastination, the French government decided to build the atom bomb. The return of General De Gaulle to power in 1958 accelerated the process. Symptomatically, the CEA's director general, P. Guillaumat, was appointed minister of the Armed Forces by de Gaulle in his new Government. Two years later, in 1960, France exploded a bomb in the Algerian Sahara for the first time. In doing so, it joined the club of the four world nuclear powers. It thus marked a crucial step of its scientific, technological and geostrategic history in its quest for international "radiance" (Hecht, 1997).

### 2.3 The atomic and nuclear Industry: Qualifying waste and measuring risks

The concerns that initiated the "radioactive waste" qualification process were present from the outset of the atomic complex, in forms both extensive and unstable. However, they fit into a matrix in which the development of atomic weapons and the corresponding secrecy policies predominated[3]. They were directed towards radioactive materials in use as well as those already used and non-reusable, insofar as they all incurred health hazards. The scientists, engineers and experts, associated with nuclear facilities, investigated and controlled this qualification process. They set up a system of standards and practices to which the governments adhered. Over the long term, this framework stiffened in a context of pressures. This was because a shift in the reference threshold of health and environmental hazards was observed, correlated with a deep public sensitivity, organised or not. In this

---

[2] He was Nobel co-laureate in 1935 with his wife, Irène, for the discovery of artificial radioactivity. During the German Occupation, F. Joliot-Curie secretly joined the Communist party.

[3] An example, among many others, is the circulation of books aimed at the public for protection against the atomic radiation from a bomb. They were generally written by the military. (Gibrin, 1953)

respect, the radioactive waste qualification process was characterised by a democratic penetration that affected the entire nuclear complex.

Without any doubt, the starting point of this process was located in a twofold prolongation. One was the international meetings between experts of the new industry, which became institutionalised, either under the UN or in the form of inter-State treaties in the Fifties (Goldschmidt, 1987)[4]. The aim was to standardize practices to conform to the development of peaceful applications of nuclear energy. It was also a symptom of the public response. In fact, the weight of the military industry and its meshing with the civilian industry limited the quality of available knowledge, the social relationship to this knowledge, and the transparency of the information (Barillot and Davis, 1994). Let us examine these various aspects through three examples in France.

The multiplication of thermonuclear bomb tests came under strong criticism from some of the atomic scientists, who mobilised internationally. Soil contamination by radioactive fallout was condemned with the health hazards associated with the food chain. For example, Linus Pauling, Nobel laureate, in an international conference of conscientious objectors in Germany in June 1959, declared:

> "The government leader who issues the order to explode an experimental atom bomb must realise that it simultaneously condemns 15 000 children yet unborn to suffer serious physical or spiritual handicaps and to have a painful and miserable existence" [press article in *Echo du Centre*, 2 July 1959].

This topic was a pressing concern in the Fifties in France. It was expressed in the political and peaceful battles against atomic weapons. It also raised public awareness about the problems raised by radioactivity. A split accordingly occurred between the good and bad users of the atom, depending on whether they derived respectively from civilian or military applications (Joliot-Curie, 1963). Public attention to the health risks engendered by radioactivity was therefore structured differentially. Notwithstanding this, it forced the CEA to install devices to record the radioactivity produced by this fallout across the country.

Another pressing topic was the dumping at sea of radioactive waste. This method, common to the atomic countries, applied the principle of dilution (Quéneudec, 1965). It was part of an initial presumption of the growth of industrial capitalism. Nature's power of absorption is infinite (Beck, 1986). In October 1960, French press reports that the CEA is planning, in its own words, an experiment to submerge 6 500 drums of low level radioactive waste in the Mediterranean Sea [*Echo du Centre*, October 12, 1960]. In actual fact, from the onset of the Fifties and in secret, the CEA was already implementing the dilution principle by dumping waste into the rivers. The publicity shed on this project sparked a strong reaction from the population concerned: elected officials as well as scientists, biologists and oceanographers in particular, demanded that the Government shelve the experiment. The Minister for Atomic Energy had to explain matters before the Parliament:

---

[4] Without claiming to be complete, examples include UNSCEAR created in 1955 by the UN. Its role was to assess the levels and the effects of exposure to radioactivity. The IAEA, created in 1957 by the UN also, promoted the peaceful uses of nuclear energy. EURATOM, created in 1957 by Europe, was a body that coordinated research programmes on nuclear energy and accompanied the growth of the civilian nuclear industry.

> "We must therefore calm the fears of French opinion by making it understand that, in a century of progress, its vague terrors are no more reasonable than those of our ancestors upon the advent of the railway, of electricity, and of cars. It is a national duty, because it conditions the development of atomic energy in France [...]. We, who are most familiar with the details of the problem, who bear the responsibility, not only to weigh the risks, for ourselves and for our children, with objectivity, but also to inform, have this honour or liberating the men and women of France from their vague and senseless fears, and of restoring their trust" [*Official Bulletin*, Senate, session of 3 November 1960, p. 1435].

Finally, a challenge emerged in France, concerning the health detriment of the radioactivity used in medicine. A teacher, J. Pignero, took the initiative. In 1962 he created the Association against Radiological Hazard (ACDR) to react against the compulsory radiological examinations for schoolchildren. Previously, in 1957, the reading of a popular science magazine alerted him to the risks incurred by the children[5]. The association published a bulletin, *Le danger radiologique* (Radiological hazard) and acted to defend the few teachers who refused the imposition of these examinations on the children. In 1966, the ACDR was converted to the APRI (Association for the Protection of Ionising Radiation) in order to extend the associative battle to the civilian and military industry branches. This appears to have been the first organised opposition to the nuclear industry in France (Prendiville, 1993).

These various forms of public engagement implied an extension and a dissemination of the critical questionings on the subject of the risks of radioactivity, more or less independently of the institutional experts. This extension revealed the instability in identifying the threshold between the benefits and detriments, because science alone could not tell all (Beck, 1986). Thus, a normally positive health use of radioactivity (X-ray examinations) could be challenged for its danger. Moreover, since the diversity of the uses was condensed into a risk bearing aggregate, stretching from the military industry to medical practices, it was not so much the problematics of waste that prevailed in these first challenges, as that of the potential hazard of any radioactive material. It took root in particular in the detachment of some of the scientists from the reassuring and faultless discourse of the atomic institution. Yet its public range remained limited by the small audience of the associative movements that relayed it on, apart from the more political movements focused on the rejection of the bomb. This is why these criticisms did not truly destabilise or delegitimise the power to say and to do of the players of the atomic complex, who generated most of the knowledge and the justifications of this industry.

When the French government decided in 1975 to build a large nuclear power capability to contend with the oil energy crisis, this situation had barely changed. The social criticism of the atomic industry remained very discreet. It is true that the spectre of atomic war had receded and that a political consensus had emerged in France in favour of possession of nuclear weapons. As an emblem of this process, the French communist left, which long argued against nuclear weapons, finally came round to the idea that its possession by France was a guarantee of its independence. This caused a significant weakening of social mobilisation. And it is without any real debate in the parliament that the decision of this

[5] According to testimony obtained by letter dated 25 July 1999.

new nuclear energy plan was taken, because in this case also, the political consensus existed to legitimise the energy independence of France. And yet, it is at the meeting point of the various criticisms that an anti-nuclear movement was taking structural shape in France, with varying strength according to location. This movement was not only heterogeneous in its composition, but also in its arguments and its highly diversified methods of combat. Thus we find three types of critical (Brunet, 2004a; 2006b). First, critics levelled by scientists who organise and popularise for the public the problems raised by the deployment of this energy industry in France. This is the case of Group of Scientists for Information on Nuclear Energy (GSIEN). This group was founded by scientists, particularly nuclear physicists, after the "appeal of the 400" published by the daily *Le Monde* in February 1975. A total of 400 scientists of the CNRS, the College du France and the universities were concerned about the risks incurred by the French nuclear power programme and asked the population to reject the installation of power plants as long as any doubts subsisted. They criticised the secrecy surrounding the nuclear industry. GSIEN published a journal *La Gazette du Nucléaire* which played a considerable role in checks and balances and hence in the democratic penetration of the nuclear industry. This journal was circulated to a nascent antinuclear movement. In this sense, GSIEN was the first independent associative expert. In second, the critics in which the nuclear power programme is assimilated with the installation of a police state ordering an overwhelming consumption disrespectful of nature. This type of critical, more political, was essentially levelled by libertarian and ecological movements as an extension of the criticism of capitalist production and consumption in 1968. Finally, we find the critical type "nimby"[6]. It is truly from this period that the problematics of radioactive waste began to take shape. The inquiry by a journalist among members of the PEON commission in this period was symptomatic of this slow movement. To the question of "waste?" he received the answer: "It is not a current problem. The storage of these wastes today raises no difficulties; they only occupy a few square metres. It will become a substantial problem in the year 2000". (Simonnot, 1978). This issue was essentially centred on the production of industrial facilities qualified as nuclear, in other words, their fuels and wastes and releases. The uranium ore industry remained on the sidelines of this nascent problematics. Accordingly, its associated remnants are difficult to "recognise" as radioactive waste. This is precisely what we shall examine in the second part, covering the long term.

## 3. From the remnants of the uranium mining industry to the qualification of radioactive waste

We have seen that for geostrategic reasons, the context of national reticence and secrecy surrounding the development of the atomic industry internationally after the Second World War compelled France to take steps to assert its independence. Evidence of this is the creation of The CEA in 1945. And in setting up this new industry, uranium procurement became the CEA's top priority. The first class of prospectors was operational in late 1945. From the outset, an ever growing series of survey missions crisscrossed France, focusing on granitic formations. Some were already known from radium mining. This is the case of small deposits known before the war and located on the eastern margin of the Massif Central. As to the remainder, prospecting missions were spread over a vast area forming a V

[6] Nimby: the acronym means "not in my backyard". It is intended to reflect a refusal by future residents, not of this industry as a whole, but of observing the installation of a risky industrial facility nearby.

from Brittany to the Morvan and passing through the Massif Central. Others, like the Limousin, were prospected for the first time. In this region, some twenty kilometres north of the city of Limoges, the richest uranium shows in France and the most promising in terms of quantity were discovered in late 1948 (Paucard, 1992). They allowed the industrial mining of uranium ore and, from the late Fifties, chemical treatment to produce *yellow cake*. The outcome was an industrial configuration which lasted half a century and caused an upheaval in this small rural region, formerly dedicated to agriculture. Like any mining industry, the production of uranium led to the buildup of overburden and tailings. Three successive periods can be distinguished to understand how these tailings were transformed into radioactive waste. They corresponded to different social configurations in which the legitimacy of statement and action tended towards their qualification. Whereas the tailings were treated routinely in the early period, and only raised questions in the second, the analysis of the third period reveals a conflictual context, with growing, permanent, expert and multifaceted vigilance with regard to their management as radioactive waste.

### 3.1 The good old days of uranium: The era of arrangements and convertible industrial remnants

The first configuration, the *good old days of uranium*, lasted about twenty-five years, from 1949 to 1973. It reflected an industrial mining scheme in which the tailings were treated as harmless. As for any mining practice at the time, they were either returned to the environment, or were used for other purposes. The monopoly of knowledge and power over them belonged to the CEA. It alone analysed, guided decisions and set the standards. The knowledge of these tailings was therefore severely restricted by such practices, especially since the mining industry was dissociated from the nuclear industry. Ultimately, risk is intrinsic to mining activity. Only the environment is risk-free.

Because of its duration, this industrial configuration was a structuring factor. It displayed many features. First, it was localised, limited to a few communes of the Monts d'Ambazac. Second, it was closed in on itself. The few kilometres distance from the city of Limoges were a virtual barrier separating the rural and urban worlds. Their links were limited to traditional trading between countryside and city. Besides, this small rural region was the water reservoir of the city of Limoges[7]. This configuration was also dominated by the CEA's mining division, made up of mining engineers and geologists. So, the job organisation of mining production is doubly structured by a geological department, in charge of prospecting, of measuring ore assays, and a mining department which extracts the ore. The Mining Division corresponds to the company which, in addition to these two major technical departments, contains an equipment maintenance and management department and an administrative department. And a significant number of the mine workers were former farmers or their offspring. Moreover, after the discoveries of large uranium deposits, the CEA supplemented the ore mining process with on-site treatment in a plant built in 1957 in the commune of Bessines. For economic reasons relating to the low concentration of the ore, the CEA quickly decided to concentrate the uranium ore chemically on-site. The product obtained was a paste called yellow cake, which had a uranium content of about

---

[7] Since the 19th Century, the City of Limoges has installed reservoirs in the form of ponds, for its water supply.

90%. This cake was then sent to plants in southern France to undergo final treatment to fabricate nuclear fuel by purification and enrichment. Before the Bessines factory went on stream, the CEA transported and processed the ore from Limoges in the Paris area, to the Bouchet factory.

Between this industry and the population, a shared positive vision emerged of its production within a set of arrangements. Uranium ore, an element of an acted nature, became the new wealth of this area and the symbol of its revival. It became a positive heritage. This situation did not discount the "drawbacks" engendered by the proximity of the mine to the villages: collapses of cultivated land, wastewater dumping into the fields, deafening noises and dust clouds from the mine sites. Yet formulated as such, they did not bring into question the mining industry and its vocation for the inhabitants: as the driver of local economic development. Depending on their characteristics, these drawbacks were dealt with under individualised arrangements on an individual case basis, or collectively. Thus, for example, the collective problem of access to water could be solved by its handling by the mining division. Indeed, when the mining division was installed in the Fifties, a collective water supply did not exist, and water was drawn from individual wells. Very often, mining operations intersected the springs and dried up the wells. The mining division then took charge of the collective water supply. In exchange, these arrangements served to reinforce its domination over the area. Thus when the mediation of the mayors was required to address collective drawbacks, the negotiations always took place in the office of the director of the mining division, a venue that was deeply symbolic of the exercise of this unchallenged domination.

As for the treatment of the remains from the industrial process, it was completely unmarked. Two types of waste coexisted: overburden from ore extraction, and mill tailings from chemical treatment. The first were the rock containing the ore, whose economic value was below the assay. Considered as routine and harmless, they were used as backfill for road building projects, or to plug a mine. Part of it was also used in individual arrangements. The latter were present as soon as the Bessines factory became operational. In the form of reddish wet sand, they represented an approximately equivalent mass to the crushed ore. Indeed, in terms of mass, given the very low uranium concentration in the rock, yellow cake represents an infinitesimal part of the total mass of ore treated, about 2 to 5 ppm depending on each case. They were handled in two ways. The first was burial in the excavations of the open pit mines. Thus like the overburden, they were "returned to nature" without any further action. A second, in very small quantities and less frequent was disseminated across the region for masonry. So, examples are not rare of individuals residing on the mining zone who used these residues, some to make a floor slab for a home, build a workshop, obviously unaware of the radioactive hazard associated with the very high radon emanations released by the radioactivity remaining in these residues. This raised no problem: the radioactive composition was considered close to zero because the chemical treatment was supposed to extract all the uranium.

In actual fact, the only recognised risks were contained within the strict limits of the production process. Exposure to these risks concerned the miners, especially the workers. Some of these risks fell into the very broad class of mining operations: cave-ins, silica inhalation. Others were classed as radioactive risks. From the mid-Fifties, a more sophisticated vigilance, taken over by the CEA, was exercised over the work of the miners, with the installation of mine aeration systems and radioactivity measurements, both collective and individual. The chief hazard identified was radon. It was discharged to the exterior by the ventilation systems. A dosimetric measurement system was set up at the same time. In 1951, the CEA formed an inspection body, the Radiation Protection Department (SPR) reporting directly to the High-Commissioner for the mining sector. A methodology and a metrology were then set up (Bernhard & al, 1992).

### 3.2 Nuclear discord: The mining industry, a link of the nuclear industry

The second period, the era of *nuclear discord,* began in the mid-Seventies and ended twenty years later when Compagnie générale des matières nucléaires (COGEMA) announced the indefinite suspension of uranium mining in the Limousin. This period reflected a new industrial configuration, more open and intense than the previous one, with sharp tensions. The construction of a major nuclear power capability in France demanded much higher uranium output than in the past. At the same time, criticism of the government decision roiled across France. It impacted uranium mining, henceforth considered an inseparable component of the nuclear industry. Uranium was no longer acknowledged to be the only positive asset. And tentatively, the issue of waste materialised.

The government decision in 1974 to schedule the construction of nuclear power plants had two major consequences for the CEA's mining sector. One concerned its organisation. To streamline the new energy sector founded on the nuclear industry, the government decided to split off all operations associated with the nuclear fuel cycle from the CEA, ranging from mining to waste reprocessing. It created a subsidiary in 1976 for the purpose, named COGEMA. The second consequence was the transformation of the local industrial configuration in the Limousin. Annual uranium production had to be doubled. This goal implied fresh prospecting and new mine sites. The mining division therefore expanded its operating perimeter and went on a hiring spree. A number of comparative figures can provide an idea of the transformation of the mining division in a few short years. In 1973, it produced 590 tonnes of uranium; in 1980, 1002 tonnes of uranium for 620,300 tonnes of ore extracted. At the same time, its area of occupancy rose from 350 to 1300 hectares, divided into 3300 registered plots. Its workforce also grew from 650 in 1975 to 1000 in 1980. The new mine workers were outsiders. The industrial configuration which, until then, was closed in upon itself, opened up. But the extension of its activities henceforth became a problem for a large segment of the population, farmers and others, who discovered and attempted to legitimise environmental issues with local officials and administrative authorities. The words "pollution", "nature protection", "environmental problems" became current, in opposition to the earlier popularisation. These words come from the new environmentalist vocabulary used by the national associations of conservation and environmental scientists. They are then taken over by the State when it created, in 1973, the first environment ministry in France. (Charvolin, 1997). Local officials passed the word on to COGEMA's mining division and the State authorities. The texture of the individual arrangements which hitherto cemented this configuration disintegrated. Conflicts broke out, essentially collective. Some inhabitants of the mining zone set up owners' and environmental conservation associations. Antinuclear groups were also formed, especially in Limoges. They decried the risks of radioactive pollution of the water catchment basins of their city by mining operations. These conflicts were emblematic of the way in which environmental problems generated by the uranium industry and the solutions made thereto were posed from then on. They also helped to grasp the conditions of the emerging issue of radioactive waste.

A conflict about the definition of the situation broke out on two levels with the mining division. Faced with the industrial breakthrough, these new associative players, pursuing their favourite themes, became spokesmen of a nature and of a living framework that deserved protection, and/or radical critics of a risky energy policy generating very long term waste. In the former case, the associative arguments drew on sensitive past experience: the noise generated by mining, the drying up of the springs, the degradation of the landscape, were criticised. Water and landscaps were defended as common heritage of an abused nature (Dorst, 1965) and as a positive asset whose use was jeopardised by pollution and other industrial detriments. In the latter case, the scientific and critical arguments of the GSIEN (Gsien, 1977) concerning the nuclear industry were mobilised. Appropriated by the antinuclear associations, they were disseminated among the population of Limoges, so that the conflict around the risk of radioactive pollution of drinking water that crystallised in autumn 1979 was acknowledged to be a problem by the municipality and the Prefect, the representative of the State. The confrontation found legitimacy in areas which were no longer those of the mining division but those of the State. In the negotiations initiated in the presence of local officials, while the State obliged COGEMA to protect the water resources of the city from industrial releases, the director of the mining division nevertheless continued to deny the problem. At a joint meeting chaired by the Prefect and attended by the Mayor of Limoges and the director of the mining division, the latter offered an answer:

> "Yes, Prefect, we agree in principle, but provided that it is the prefectural authority that issues the demand, and that it is perfectly clear that it is not a problem of pollution that needs settling, but a problem of psychological damage. In other words, the crowd psychology needs to be corrected." [Prefecture of the Haute-Vienne, "Radioactivity of waters supplying the city of Limoges." - Proceedings of the briefing of 08/10/79. Mimeographed].

At the same time, the State and the nuclear establishment were denounced for their habit of withholding information about radioactivity monitoring measurements. Short of a suspension of the nuclear power programme, which remained its ultimate target - this locally implied suspending the inauguration of new mine operations and the shutdown or slowdown of those incurring a risk for the environment or for the population - the local antinuclear movement defends two others claims. The first is more intensive monitoring of the radioactivity of drinking water and publication of the data. The second is the creation of an enquiry commission to conduct an epidemiological survey by independent bodies.

It is in the tumult of this conflict that the nature protection and antinuclear associations also alerted the public to practices they considered dubious. These included waste dumping at night in an open pit mine by COGEMA. The CEA, which closed its uranium concentration plant in the Paris area to the Bouchet factory, decided to transfer the tailings to the Limousin. The arguments volunteered to justify this practice was their harmlessness and the fact that having originated in the Limousin, they were merely going back to where they came from. The answer offered by the Prefect of the Haute-Vienne in 1979 to a worried local official:

> "[...] At that time, there was no treatment plant on the spot and the ore was sent by lorry to the Paris area, to the Bouchet factory installed by the CEA. The treatment of the ore led to the production of a few thousand tonnes of sterile, just like the sterile at the Bessines factory: these harmless materials are now returning. It is almost like going back home." [press article in *Le Populaire*, 15 février 1979]

The semantics were ingenuous: to qualify mine tailings as "overburden" meant to treat them as any routine form of residue. Their "problem" was nonexistent, and their sole admissible identity was the one attributed by COGEMA. This identity implied an ignorance of the radioactive composition of these residues, and denied the existence of any risk. The associations lacked the means to counter these assertions. Their criticism was limited to inflating the challenge to this industry by arguments targeting waste, which contributed to its disqualification. It sustained a powerful tension between the positive and negative aspects of the industry. Yet too many economic and social challenges precluded its full and unchallenged legitimacy. This is why the issue of radioactive waste from the mining industry was not identified during this period, although emerging details tended in that direction.

### 3.3 Nuclear uncertainty: Managing and qualifying the remnants

The third and final period, of *nuclear uncertainty*, began in 1988 with the announcement of the speedy termination of industrial activity. The social configuration, hitherto centred on industrial operations, and the problems raised by uranium production, were progressively transformed into a solidly environmental configuration that would never end. From uranium as *acted nature*, its remnants were considered *acting nature*. Indeed the radioactive waste qualification process applied to the residues now entailed permanent vigilance and accompaniment. A priori, it's almost impossible to fix the term because the respective time scales for men and radioactivity emitted by these wastes are immeasurable. This situation resulted from the nuclear establishment's progressive loss of hegemony of statement and action with regard to these remnants (Brunet, 2004b). Its weakness had three causes. First, it originates the strengthening of the expertise of the antinuclear movement in France with the emergence of the associative expert, who scientifically challenged the arguments of the nuclear establishment. Second, the local public authorities, no longer anticipating any positive spinoffs from the industry, were vulnerable to the arguments of the associations. Finally, thirdly, the nationwide nuclear establishment was forced into reform, given its relative failure to propose an operationalisation of the comprehensive management of all radioactive waste, high and low level alike. A correlation therefore existed between this weakness, the obvious shift in acceptable standards on radioactive waste management, and the stigmatising image projected by the recognition of the remnants as radioactive waste. Nevertheless, the environmental configuration remained subject to regulatory practice.

With the onset of the Nineties, the mining industry declined and collapsed in late 1995. COGEMA's decision in 1988 to stop any further mining in France had its economic underpinnings: to mine only profitable orebodies. The low price of uranium on the world market made Limousin ore expensive compared to those of Canada and Africa. Despite the very intense but dispersed labour unrest, the mine workers and their unions had to give in, and they quickly disappeared from the social landscape. While the local political officials favoured the resumption of mining, others, urban political officials, promoted the idea of a "green" Limousin for the mining area, oriented towards housing and tourism. The construction of expressways between Limoges and Monts d'Ambazac shortened the distances and many citydwellers came to live there. At the same time, in the late Nineties, the inauguration of a leisure facility on St Pardoux Lake made it a relaxation centre for the population of Limoges and for the tourists. That is why these urban political officials formulated a twofold demand, non-negotiable for them: that COGEMA should finance a conversion plan for the mining territory, and that it clean up the traces left by fifty years of uranium mining.

Yet when mining operations ended, more than twenty million tonnes of residues generated by the industrial process remained on the old mining territory, stored in open pit mines. Added to this mass were wastes of all types, already present or anticipated. The industrial logic of burial for many long years had been fully implemented. Thus, empty drums that previously contained radioactive substances and originated directly from COGEMA and CEA industrial facilities, were regularly dumped in thousands in the mine pits. And in the guise of a conversion project, COGEMA planned to set up an interim storage facility for 200 000 tonnes of depleted uranium produced by the fuel enrichment cycle. All these factors tended to project a negative and stigmatising image on the region, one of a "nuclear dustbin". The unfair tradeoff that triggered it contributed to a dual upheaval symbolic and political. Symbolic, when the uranium converted to residues was no longer considered a positive asset, but became a negative legacy. Political, when the radioactive, economic, environmental and health hazards harboured by this negative legacy, whether real or imaginary, soon spread and tended to convert the officials to the position defended by the environmentalist associations. The industrial configuration blurred and vanished, leaving in its place an environmental configuration in which COGEMA, the elected officials, State representatives, associations and experts were the players. The experts then played a central role in the dynamic of this configuration. They compiled and assessed the controversial knowledge about the remnants, given the turmoil that governed the way in which the questions were asked and answered. This knowledge, no doubt unstable, was nevertheless sufficient to qualify and to manage the remnants.

Two types of expert faced off within this environmental configuration: the establishment expert and the establishing expert (Bonnet, 2006a): the first largely came from the nuclear establishment itself. This period of the late Nineties witnessed the generalised treatment as a problematic issue of all radioactive waste and, as a corollary, a transformation of the institutions which possessed and produced the expertise of the State in this field. We cannot expand further on this issue in this chapter. However, it is clear that the progressive inclusion of the remnants of the uranium industry in the issue of all radioactive waste facilitated the process of its qualification. This transformation stemmed from processes of differentiation and independence. For example, the French National Radioactive Waste Management Agency (ANDRA) was created from a CEA Department. In 1981, with the passage of the first French bill on the nuclear industry specifically and exclusively addressing radioactive waste management, ANDRA became independent of the CEA. This transformation impacted the radioactive waste management policy, the nuclear facility safety policy, and the health and environmental safety policy of the population. The French State did not succeed in resolving the management of high and medium level nuclear waste. Underground storage alternatives were vigorously challenged by the public and the antinuclear groups. From this point of view, the antinuclear groups did not conceal their strategy. One of the routes for securing the shutdown of the nuclear industry in France was to demonstrate the "intestinal blockage" of the system. This strategy consisted in focusing public attention on the waste. Insofar as no transfer solution was accepted by the potential host population (deep burial, underground storage with possible rehandling), the wastes remained where they were generated and ultimately cluttered the area. Added to the establishment expert, arguing by differentiation, were the university laboratories which assumed a role in offering expertise, in which the monopoly of the nuclear establishment disappeared.

During the same period, and at the same time, the associative players transformed themselves into the typical ideal figure of the *associative expert*, with features specific to the Limousin. This figure assumed two forms. On the one hand, it inspired the local environmental defence associations and antinuclear groups, who together created the Limousin anti-waste coordination (CLADE) in 1992. This flexible federative organisation focused exclusively on the "remnants" of the mining industry. Its essential demand was the conduct of a radiological investigation which, independently of the nuclear establishment, could assess the environmental and health hazards incurred by these harmful remnants. The prize was the definitive qualification of these remnants as radioactive wastes. It adjusted its practices to this outcome. Thus, without awaiting the independent investigation that it demanded from the authorities, it concentrated on the burden of proof that bedevilled it and forced it to mobilise science to prove the existence of dangers to health. On the other hand, this associative expert figure also engaged the Commission for Independent Research and Information on Radioactivity (CRII-RAD), an associative radioactivity measurement laboratory. CRII-RAD is a non-profit association created in 1986 to counter the statements of the authorities and experts of the nuclear establishment, who, after Chernobyl Nuclear Power Plant explosion, argued that the radioactive cloud had "stopped" at the French border, and that there had been no significant radioactive fallout on the national territory. This association enjoys the original privilege of having founded a laboratory for measuring the radioactivity that is independent of the state and nuclear authorities, with a nationwide audience. It is therefore at the junction of the activities of the local association, which relentlessly denounced the potentially polluted locations, took samples in suspect places of the old mining zone by observing the procedures recommended by the CRII-RAD and those of the associative laboratory which analysed and interpreted the results, that this associative expert unveiled its reality as a player and, progressively, imposed a new and critical viewpoint on the remnants (Brunet, 2006c). There is therefore an important qualitative change in the practice of associative expertise. Without the GSIEN having disappeared from the associative landscape, the investigation of the antinuclear movement no longer relied exclusively on a critique of the documents produced by the nuclear establishment, but on independent evidence produced in the laboratory. This is one of the reasons why COGEMA, the experts of the nuclear establishment and the authorities were compelled to re-examine the knowledge and management of these "remnants" and to recognise them as radioactive waste.

Firstly, the associative expert forced COGEMA, via the State, to take more restrictive protective measures in its winding-up operations. But above all, with the support of the local authorities, it succeeded in imposing the satisfaction of its central claim: the setting up of an independent investigation of all the mine sites. The Prefect decided in fact to set up a Local Information Commission (CLI). This commission aims to provide information "on the risks incurred by ionising radiation pertaining to the activity of the uranium site"[8]. The composition of the CLI serves to gather together in a single place all the players in this environmental issue: the services of the State, the operator, the eco-environmental and antinuclear associations, local authorities and the experts. This decision is also the translation of many reports produced or under preparation of the State Services and also of the Parliamentary Office for Assessing Scientific and Technological Choices (OPECST) on the mining residues that have supported this qualification (Ministère de l'Environnement, 1991, 1993 ; OPECST, 1992). Thus, mining residues were definitively classed in the category of radioactive waste, which implied a need for management and the establishment of new standards. It is in this new setting of institutionalised consultation that a radiological investigation of the mine zone was decreed.

To allay suspicion, the investigation was funded by the local authorities and took the form of a joint investigation between the CRII-RAD and COGEMA laboratories. The definition of the measurement plan and its implementation lay at the heart of the conflict on the most appropriate definition of the situation. It is therefore not surprising that it was the subject of lengthy and difficult negotiations between the players of the CLI. And when in 1994, the results had to be interpreted by comparing the measurements of the two laboratories, the players of the environmental configuration were unable to agree, leading to the breakup of the CLI. The investigation, all the way to the assessment, clashed on two conflicts of interpretation which prevented settling the argument between the associative movement and COGEMA. First, in this mining area, how the part of the so-called natural radioactivity and that provided by industrial activities? This question evidenced an attempt to establish COGEMA's liability. And besides, is the risk assessed at the source, within the boundaries of the mining sites, or is it, according to the regulations in force, assessed in the environment, outside these boundaries? In the former case, waste monitoring was a public matter; and the second, it remained a private affair, the domain of COGEMA, because access to the source remained prohibited. Thus, while everybody agreed that the results of the two laboratories were identical. But for the associative movement, they offered evidence of the existence of risks which must be neutralised at the source, whereas for COGEMA, they confirmed the absence of any danger to human health and justified his self-inspection.

### 3.4 Continuing tensions between expertise, democracy and social norms

The democratic trial seemed powerless to withstand the ordeal of scientific controversy. Precisely because the scientific controversy extends continuously beyond the narrow issues of Science. These were constantly articulated in terms of social norms: an environmental hazard or a health risk always includes more than just scientific data (Beck, 1986). Both reflect essentially normative points of view drawing not only on scientific reasoning but also on social reasoning. Since it is around this model, initiated by the incomplete experiment of the CLI, that two-track *ad hoc* procedures for consultation and negotiation punctuated every new problem posed by *acting nature*. It brings together the elected officials, State Administration, COGEMA and the experts, which reached the public domain in line with

[8] Prefectoral decision of 7 January 1992.

the action model promoted by the associative expert. Moreover the action model was completed by a legal expertise of the associative movement. All the actions of COGEMA and of the State on radioactive waste management are the subject of closely attentive legal monitoring by the local associative movement. In other words, the recognised legitimacy of the environmental issue by expert knowledge, failed in setting up truly permanent systems for consultation and negotiation, that is to say political. Only *acting nature*, via the spokesmen experts, who claimed to be its interpreters, conditioned its frequency, intensity and scale.

It follows that AREVA Company, formerly COGEMA[9], like any mine producer, wanted to leave Limousin for good after the industrial sites had been redeveloped, and not rehabilitated as this company suggests (Bavoux & Guiollard, 1998). It was forced to remain on the spot. In recent years, the State should take its place for a strictly indeterminate period. Its task was precisely to "contain" this set of remnants now qualified as "radioactive waste" and to meet the standards whose level of acceptability ceaselessly became more restrictive. In other words, while the nuclear establishment was planning to "return these wastes to nature", according to its own terminology, to forget about them, they now became, probably for an unlimited period, the subject of increasingly intensive monitoring, which forced it to remain nearby. This obligation of surveillance and retention was not simply that. AREVA NC and the State are resistant to this because it represents a cost to both economic and symbolic. More the cost of monitoring work grows, less the industry shows that it was profitable. Similarly, more problems appear, less engineering, over the long term, shows its ability to solve them. It's in fact ceaselessly updated by the vigilance of the associative movement. This monitoring is therefore fragile. Indeed, it is largely contingent on the capacity of the movements coming from society to exercise this control which, necessarily, remains discontinuous. The militant capacities of the associative movement are very fragile (Brunet, 2004c; 2006b). More generally, above and beyond the issues of radioactive waste, this situation raises the question of the role of public, the associative movements and experts in a renewed technical democracy. In this context, certainly, the State should reconsider the submission of general interest for the sole benefit of short-term economic, which denies the existence of waste and problems. It needs also to recognize and take account of public engagement in its attention to the commons with their coloration positive or negative, as are the radioactive waste. In the same times, the public must recognize all the commons that are part of the same story. This is certainly one of the most important political challenges of the future of our industrial societies.

## 4. Conclusion

The socio-historical analysis of an industry helps to understand the place that gradually take its waste. In France, in its productive phase, the uranium industry lasted about fifty years. In fact it has no end. Three periods were able to be identified. Each has a very different relationship to his remains. Their succession shows a progressive visibility and legitimacy of his remains to the characterization of radioactive waste. The first period, which lasts nearly twenty five years, shows that the remains do not exist. Either they are "returned to nature," either they are used for other purposes. Uranium is considered by all actors as a common

[9] COGEMA changed its name in 2001 and is now called AREVA NC.

unchallenged. Account only the nature acted. The second period corresponds to a strong growth of the mining industry. This one is disputed because it disrupts the natural environment. Two commons are then in opposition. Water and uranium, respectively, correspond to urban and rural social worlds different. In addition, by the action of anti-nuclear groups, the uranium industry becomes an integral part of the nuclear industry. Their challenge is only to delegitimize the reassuring speech experts from the State and the operator (CEA and COGEMA) about the environmental and health risks associated with radioactivity, affecting the water. However, this challenge is limited because it based solely on the data produced by the nuclear institution itself. Also, in this context of strong activity, industrial remains are hardly questioned. The third period, which has no end, starts when the industrial decline and the operator informs of the imminent closure of the mines. From that moment, the remains are real issues and the problem of radioactive waste emerges. It develops in a context of strong challenges that mobilize elected urban and antinuclear groups against the state and the operator. Uranium as common fades along with its industry. Only exist remnants that become problematic. It then becomes necessary to identify, qualify as radioactive waste, measure and evaluate them in terms of environmental and health risks. Antinuclear associations have acquired a capacity to produce data themselves through the figure of the associative expert. But conflicts over these activities can not diminish for two reasons. The democratic machinery around these radioactive wastes is limited and fragile. And also the actors for the most part unaware of the history of the industrial process in its entirety, including the production of its common, positive and negative. Despite appearances, our society built on the basis of science and technology, is fundamentally a political society.

## 5. References


Bavoux, B., Guiollard, P.-C., *L'Uranium de la Crouzille*, Fichous, Ed. P.-C. Guiollard, 1998

Barillot, B., Davis M., *Les déchets nucléaires militaires français*, Lyon, CDRPC, 1994.

Beck, U., *Risikogesellschaft*, Francfort, Suhrkamp Verlag, 1986.

Bernhard, S., Pradel, J., Tirmarche, M., Zettwoog P., « Bilan et enseignement de la radioprotection dans les mines d'uranium depuis 45 ans (1948-1992), *Revue Générale Nucléaire*, n°6, 1992.

Braverman H., *Travail et capitalisme monopoliste*, Paris, Maspero, 1976, 361 pages.

Brunet, P., *La nature dans tous ses états : Uranium, nucléaire et radioactivité en Limousin*, Limoges, Presses Universitaires de Limoges, 2004a, 353 pages.

Brunet, P., « L'environnement concerté et négocié : un demi-siècle d'exploitation industrielle de l'uranium en Limousin », *Ecologie et Politique*, n°28, 2004b, pp.121-140.

Brunet, P., « L'impossible gouvernance à l'épreuve de la nature agissante » in Scarwell, H.-J., Franchomme, M., (Dir.), *Contraintes environnementales et gouvernance des territoires*, L'Aube, 2004c, pp.147-154

Brunet, P., « L'expert en technosciences : figure « critique » ou « gestionnaire » de la civilisation industrielle contemporaine ? » in Guespin, J., Jacq, A., (Coord.), Le vivant, entre science et marché : une démocratie à inventer, Ed. Syllepse, 2006a, pp. 99-125.

Brunet, P., « Flux et reflux de l'engagement antinucléaire. Entre vigilance et dénonciation », in Roux J. (Coord.), *Etre vigilant – L'opérativité discrète de la société du risque*, Presses Universitaires de Saint-Etienne, 2006b, pp. 189-202

Brunet, P., « La CRII-RAD, un laboratoire « passe-muraille » entre militantisme et professionnalisme », in *Reconversions militantes*, textes réunis par Tissot S., Limoges, Presses Universitaires de Limoges, 2006c, pp. 163-173

Brunet, P., « De l'usage raisonné de la notion de « concernement » : mobilisations locales à propos de l'industrie nucléaire » *Nature, Sciences et Société*, n°4, décembre, 2008

Charvolin F., « L'invention du domaine de l'environnement au tournant de l'année 1970 en France », *STRATES* n°9, pp. 184-196, 1997

Dorst, J, *La nature dénaturée*, Paris, Delachaux et Niestlé, 1965

Ducrocq, A., *Les horizons de l'énergie atomique*, Paris, Calmann-Lévy, 1948

Einstein, A., *Comment je vois le monde*, Paris, Flammarion, 1979

Gibrin, C., *Atomique secours – Etude des effets de l'engin atomique et de la protection familiale et collective contre le danger aérien*, Paris, Charles-Lavauzelle & Cie, 1953, 179 p.

Goldschmidt, B., *L'aventure atomique*, Paris, Fayard, 1962

Goldschmidt, B, *Le complexe atomique*, Paris Fayard, 1980

Goldschmidt, B., *Pionniers de l'atome*, Paris, Stock, 1987

G.S.I.E.N. (Groupement de Scientifique pour l'Information sur l'Energie Nucléaire), *Electro-nucléaire: danger*, Paris, Seuil, 1977.

Hecht G., *The radiance of France : Nuclear Power and National Identity after World War II*, M.I.T. Massachussets, USA, 1998

IAEA, "Implementation of the NPT Safeguards. Agreement in the Islamic Republic of Iran", GOV/2006/14, February, 2006

Joliot-Curie, F., *Textes choisis*, Paris, Editions sociales, 1963

Martin, C.-N., *L'atome maître du monde*, Paris, Le Centurion, 1956

Ministère de l'Environnement, Desgraupes, P., Rapport de la Commission d'Examen des dépôts de Matières radioactives, juillet 1991.

Ministère de l'Environnement, Barthélémy, F., Rapport à Monsieur de Ministre de l'Environnement relatif aux déchets faiblement radioactifs, affaire n°92-282, Conseil Général des Ponts et Chaussées, 14 mai 1993.

Naville, P., *La guerre de tous contre tous*, Paris, Galilée, 1977, 220 p.

N'diaye, P., « Les ingénieurs oubliés de la bombe A », *La Recherche*, n°306, février 1998, p.82-87.

OPECST, Le Déaut, J-Y, Rapport « La gestion des déchets très faiblement radioactifs », Assemblée Nationale n°2624, Sénat n°309, Tome II, avril 1992

Oppenheimer, J.R., *La science et le bon sens*, Paris, Gallimard, 1955.

Paucard, A., *La mine et les mineurs de l'uranium français, I les temps légendaires (1946-1950)*, Cogema, 1992.

Paucard, A., *La mine et les mineurs de l'uranium français, II le temps des conquêtes (1951-1958)*, Cogema, 1994.

Paucard, A., *La mine et les mineurs de l'uranium français, III le temps des grandes aventures (1959-1973)*, Cogema, 1996

Prendiville B., *L'écologie la politique autrement ?*, Paris, L'Harmattan, 1993.

Quéneudec, J.-P., « Le rejet à la mer de déchets radioactifs », *Annuaire français de droit international*, volume 11, 1965. pp. 750-782.

Simonnot, P., *Les nucléocrates*, Grenoble, P.U.G., 1978

Wynne, B., « Le nucléaire au Royaume-Uni » in GODARD, O., (dir.), *Le principe de précaution dans la conduite des affaires humaines,* Paris, Editions de la Maison des sciences de l'homme, INRA Editions, 1997

# 3

# Problems of Uranium Waste and Radioecology in Mountainous Kyrgyzstan Conditions

B. M. Djenbaev, B. K. Kaldybaev and B. T. Zholboldiev
*Institute of Biology and National Academy of Sciences KR, Bishkek*
*Kirghiz Republic*

## 1. Introduction

It is known that uranium industry in the former Soviet Union was a centralized state management. Information flows related to the issues of uranium mining was strictly controlled and is in a vertical subordination of the structures of the Ministry of Medium Machine Building of the USSR. After the USSR collapse, the information about uranium mining and processing were not available in Kyrgyzstan, and all the data related to past uranium production, were in the Russian Federation in the archives of the successor of the former "Minsredmash".

The activity of the regulatory body in the field of radiation safety have been independent of the former USSR. The agency also was part of the "Minsredmash", which was responsible for the nuclear industry. Application of regulatory safety standards ("standards") with respect to exposure and control of emissions of radioactivity in the field of mining and processing was similar in all organizations of the uranium industry, making it easier for their administrative use.

The requirements of radiation safety often disappeared or were not fulfilled, because the task performance of production had priority at the expense of safety. The neglected environmental protection requirements and protection of human health in the process of extraction often the same reason and processing of uranium ores, and recycling. Environmental protection has not been determined as a priority, and have not been identified the relevant criteria of safe operations. While establishing the new mining and uranium ore processing units, the issues of the protection of the environment has been neglected, and the data collection which should become the basis for further evaluation and possible remediation of contaminated areas that make up the heritage of the industry, was not done.

Uranium mining in the country was launched in 1943 year. After Kyrgyzstan gained independence, the uranium tailings are preserved, but without the engineering and technical support outside of Russia and cooperation with other independent countries in the region. Since the early 90's uranium industry of Kyrgyzstan in the region was unexpectedly opened to the world market. A large number of mines in the region during the low profitability of such production has stopped in the 70s of last century. Nowadays, some companies continue to pollute the surrounding areas polluted by dust from uncontrolled waste disposal sites of uranium production, although to a lesser extent than during the current production. The deterioration of the environment as a fact of many experts' associates with a significant economic slowdown in countries faced with serious social problems of local people. This particularly applies to facilities located in Kyrgyzstan, whose economy has suffered more than others in the region. The environmental situation in Kyrgyzstan is exacerbated economic problems, provoking people to predatory use of natural resources (deforestation, poaching, extensive use of arable land, neglect melioration and other measures), which leads, on the basis of the feedback to further environmental degradation.

Thus, the post-war (1941-1945) development of Kyrgyzstan has been closely linked with economic and military policies of the Soviet Union and known that Kyrgyzstan was the largest producer of uranium from 1946 to 1968 for the former Soviet Union. Huge amount of raw materials as a due to inefficient production and wasteful processing of minerals have in the territory of the Republic (747 220 000 $m^3$) with high content of potentially dangerous chemicals stored in waste dumps and tailing. For storage of uranium waste additional waste were also imported from other friendly countries such as Germany, Czech Republic, Slovakia, Bulgaria, China and Tajikistan. The status of these dumps and storage facilities so bad, that radioactive waste, heavy metals and toxic chemicals pollute the environment (soil, air, water) and living organisms. They are involved in biogeochemical cycles in the formation of new biogeochemical provinces (5,6, 14).

In general, the territory of Kyrgyzstan is a large number of radioactive sources (1200). The radioactive sources are stored in premises built storages of primitive methods (overlap of the mountain gorge.) Many of the tailings were formed within settlements (Maili-Suu, Min-Kush, Kaji-Say, Ak-Tuz, Kahn and others) in the mountain valleys and along the river.

Interest in the use of a nuclear facility for peaceful purposes again increased in the early twenty-first century at the decision of the new strategic challenges in the world. For example, at this time (2011) in the country four companies have influenced right to operate at a uranium deposit and 12 companies have licensed right to search for uranium ore. However, it should be noted, after the case in the Japan with nuclear stations (2011), security, use of nuclear energy, require special importance and improvement processes for peaceful purposes.

Thus, in the republic issue of Radioecology and radio biogeochemistry took priority of rare and rare earth elements of the former uranium production (tailing and dumps). The most urgent is to find features radiobiogeochemical enriched uranium and other trace elements and evaluation of reaction areas of organisms in ecosystems of the high content of radionuclide and base metals.

During a long time of economic activities in the Kyrgyz Republic has accumulated a huge amount of industrial and municipal solid wastes containing radionuclide's, heavy metals and toxic substances (cyanide, acids, silicates, nitrates, sulfates, etc.), negatively affecting on the environment and human health. In this regard, the problem of waste management is becoming increasingly important, and some waste has a frontier character.

## 2. Materials and methods

Since 2005 integrated studies for evaluation of radio-ecological features and radio biogeochemical features in uranium tailings and dumps are carried out by us. The survey was carried out according to the modern techniques and methodologies at the territories of radiological, and eco-radio biogeochemical study of the various types of the biosphere(4, 8, 9,11, 13).

The equipment used in research, consists of a set - Dosimeter-radiometer DKS-96, Radiometer PPA-01M-01 with the sampling device POU-4, Photo-electro-colorimeter (SPECOL), liquid scintillation spectrometer, λ - spectrometer (CAMBERRA), radiometer UMF-2000, etc., a satellite instrument to determine the coordinates and a personal computer with data entry module. Distribution and data processing were performed on a personal computer using a special software package. Gamma-ray surveying carried out in accordance with the "Instruction on the ground survey of the radiation situation in the contaminated area" at a height of 0.1 and 1 meter above the ground. According to the technical manuals of dosimeters, at one point was carried out at least three measurements, the log recorded average

Measurements of gross alpha and beta - activity in the mass were performed in the laboratory. For measuring gross alpha and beta - activity in the mass was performed prior to digestion. For that, each sample weighing was carried out separately, and determined their actual weight. They then converted into porcelain crucibles and placed in a cold muffle furnace. Digestion was performed for one hour at 450$^0$C, and then the temperature was raised to 550$^0$C and after three hours muffle furnace turned off. The resulting ash was weighed and ground in a porcelain mortar and homogenized to the state from counting samples were collected weight 0.4 grams for the measurement of alpha and beta - irradiation on radiometer UMF-2000. Volumetric total alpha activity in the sample (Bq/kg) was calculated using the formula:

$$A = (A\alpha \ / \ M) \times (M1 \ / \ m)$$

where Aα - gross alpha-activity of radionuclide in
the counting sample (Bq)
M mass of the original sample (kg)
$M_1$ and $_m$ - mass of the ash samples and aliquots of cell mass (mass of sample countable) (g), respectively.

The total volume of beta activity calculated similarly.

*Determination of the isotopic forms of radionuclide* samples of soil and plants were dried after harvest, soil samples were ground further in a mortar and pestle and sieved through a 2.0 mm diameter, 1.0 mm, 0.25 mm., Plant samples were cut with scissors and prepared at the machine for grinding plant samples. Further sample tests of soil and plants were burnt in a muffle furnace at 400°C, after burning $^{90}$Sr stood by oxalate and antimony-$^{137}$Cs iodine on relevant techniques. Shortchanging the final draft of $^{90}$Sr was carried out on the radiometer UMF-2000, by $^{90}$Sr by instrumental gamma-ray spectrometry. As a model of a radioactive source used a set of solid sources, $^{90}$Sr + It$^{90}$ activity of 50 Bq in the angle 4п and 26 Bq in the angle 2п, with an area of active spot 4 cm$^2$. Cut-off screen for $^{90}$Sr was an aluminum filter with a surface density of 150 mg/cm$^2$, such a filter reduces the effects of $^{90}$Sr in 128 times, and activity It$^{90}$ two times (2, 4).

Satellite device (GPS) with regular frequency automatically recorded the longitude and latitude location, and stores this data in its memory. All coordinate data, indicators of levels of radioactivity, the date, time of measurement later transferred to a computer's memory with the help of the writer. In carrying out studies have been conducted random measure radiation levels in different parts of the tailings piles and indoor as well as selected samples of soil and plants for laboratory analysis.

## 3. Discussion of research results

In connection with the collapse of the USSR on the territory of Kyrgyzstan in derelict condition were 55 of tailings, the total area of 770 hectares, of which more than 132 000 000 m$^3$ of tailings dumps stored, and 85 gained more than 700 m$^3$ of waste, cover an area of over 1,500 hectares. There are 31 tailing dumps and 25 contain the wastes of uranium production volume - 51.830 000 m$^3$, the total radioactivity of more than 90 000 Ci (as of 2010) (1, 6). Since the mid 50s of last century to the present time in the country closed or mothballed 18 mining companies, including 4 for the extraction of uranium (Fig. 1).

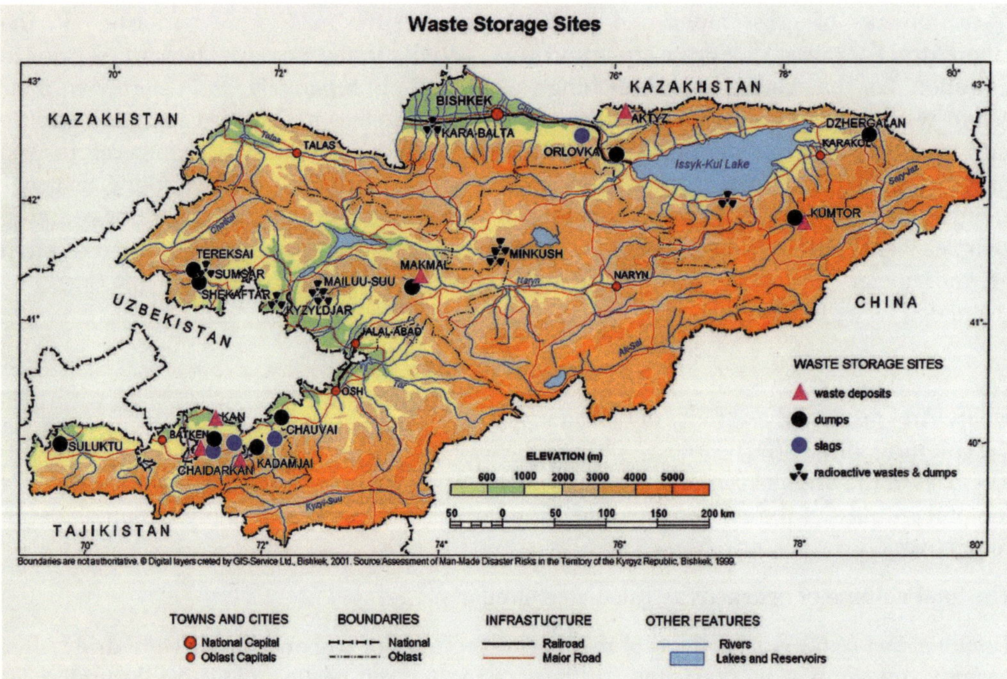


Fig. 1. Layout of the main places of accumulated waste of the former uranium production in Kyrgyzstan

According to the latest data from the National Statistical Committee of Kyrgyz Republic (2010) Most of the toxic waste in the territory of Issyk-Kul (61.4%) and Batken (25.8%) regions. In Issyk-Kul region, the amount of waste has risen sharply since 1997 in connection with the commissioning of the gold processing plant "Kumtor", and in the Batken region of their main sources of formation are Khaidarkan (Hg) and Kadamzhai (Sb) plants.

Toxic chemicals and radionuclide (As, S, Pb, Hg, Sb, U, etc.) in the waste dumps and tailings are found in both soluble and insoluble forms. The most dangerous of them are mobile forms compounds that are primarily involved in the chain: soil, water, vegetation, animals, people. Special problem of waste accumulation (more than 15 million $m^3$) of overburden dumps, tailings and ore-balance, holding large areas near the settlements in the mountains, drainage basin, etc. The greatest threat of contamination remains uranium waste in cross-border areas on the slopes of the Fergana mountain frame and Chui valleys (near Maili-Suu city, settlements Shekaftar, Ak-Tuz and others).

After independence (1991), Kyrgyzstan began to collaborate with many international organizations on this issue, such as the UN, the IAEA, EU, UNESCO, UNDP, IMF and others. The following areas have been designated as priorities for Kyrgyzstan in conclusion with experts of the TC IAEA for the intermediate term period.

### 3.1 Rehabilitation of the effects of uranium mining and processing activities

Kyrgyzstan is facing serious environmental problems associated with uranium mining and processing activities in country. Due to natural disasters such as earthquake, landslide, mudflow and erosion processes increases the threat of further contamination by radioactive substances. As a result of natural processes a number of uranium tailings had been damaged. Most of the tailings storage facilities are in disrepair and poorly controlled.

*The following actions require immediate attention:*

- to develop and confirm the national program of radiating monitoring (at present is not present national the program on radiation monitoring);
- to give radio ecological and radio biogeochemical estimation;
- to estimate and begin rehabilitation works by a priorities;
- to develop correspond uniform regulating infrastructure on the radiating and nuclear safety, capable to operate a situation for the long-term period (till now there is no uniform regulating state structure).

### 3.2 Health: Improved diagnostics and nuclear services radiotherapy

The use of methods based on radiation for the prevention, early detection and treatment of cancer is one of the main priorities of the government in the health sector.

It is known that the use of obsolete equipment in radiotherapy for cancer treatment greatly reduces the chances of survival, and jeopardizes the health of staff. Moreover, the operating costs of equipment, lack of parts and skilled technicians make things worse.

Therefore, the planned improvement of radiotherapy services was an important component of the IAEA TC for the country over the medium term:

- the urgent need to upgrade radiotherapy equipment at the National Center of Oncology, KR;
- modernization of nuclear medicine and diagnostic services through appropriate programs;
- modernization of tomography and diagnostic equipment;

- The need to focus efforts on training of medical staff, as well as the introduction of modern diagnostic techniques.

### 3.3 Knowledge management and rational use of nuclear technology

In 2005, Kyrgyz Republic became a member of the International Nuclear Information System IAEA (INIS). How to create a network of analytical and calibration laboratories.

Kyrgyzstan has received significant assistance through projects of various international organizations such as the World Bank, IAEA, UNDP, IMF, EU and bilateral assistance provided by the governments of Austria, Japan, Netherlands, Sweden, Switzerland and the USA.

By the IAEA in the country, a modern radiology laboratory at the Institute of Biology and Pedology National Academy of Sciences, industry laboratories under the Department of State Sanitary and Epidemiology, Health Ministry of KR and Kara-Balta Environmental Laboratory.

In the framework of national and regional projects of IAEA - agency offers: the expertise, scientific visits, seminars and training courses on various aspects of radiation safety. Kyrgyzstan also has acquired the necessary modern dosimeter and analytical equipment for monitoring and analysis.

**Legal and regulatory framework**

The main basic Law of the KR, which regulates the handling of sources of radiation, is the "Law on Radiation Safety KR" as amended on February 28, 2003 # 48 and August 1, 2003 # 168. This law defines the legal relationship in the field of radiation safety and protection of the environment from the harmful effects of ionizing radiation. The law defines the main concepts, in particular, the term - "contamination" as the presence of radionuclide of technogenic origin in the environment, which may lead to additional exposure in an individual dose of more than 10 μSv year. Additional exposure below this level is negligible and should not be taken into account.

In accordance with the Act in 2005 Kyrgyz Republic, a special representative governing body for radiation protection, regulatory activities with radiation hazardous technologies and sources of radiation under the Ministry of Ecology and Emergency Situations. Since 2006, this Ministry was reorganized into two - "The Ministry for the Protection of natural and forest resources" and "The Ministry of Emergencies." The regulatory role belongs to the Ministry of Health, in particular the Office of the State Sanitary and Epidemiological Surveillance.

The main regulations in the Kyrgyz Republic have been adapted previously developed in the Russian Federation NRB-99 and Sanitary Regulation of Radioactive Waste Management (SRRM-2002). In particular, as the principal dose limit for the staff of the existing enterprises whose activities are related to the practice of radioactive waste management is set at 20 μSv per year, while the limit dose for the population in areas where uranium companies is set at 1 μSv per year. A clear recommendation for establishing intervention levels and regulatory criteria for the study of remediation activities at the former uranium companies has not been established yet.

Till present time establishes the recommendations for remediation of former uranium companies "Sanitary rules of liquidation, and conversion mining of conservation and processing of radioactive ores" (SLCP - 91). "The existing law "On the tailings and dumps" (2001) is a specific document relating to governance and uranium tailings and rock dumps. Earlier as a noted, some of these documents were developed during the former USSR and some are adapted to the Russian Federation, but they must be revised and adapted. These activities are carried out by the Ministry of Emergency Situations and the Agency for Environmental Protection and Forestry of the Kyrgyz Republic.

It should also be noted that the IAEA report ("Radiation and Waste Safety Infrastructure Profile (RWSIP) Kyrgyzstan Part A, 2005), most legal documents in the Kyrgyz Republic of related issues justify rehabilitation, are not available and require development yet.

## 4. Brief description of the major uranium tailings and dumps

### 4.1 Maili-Suu technogenic uranium province

Uranium deposit district in Maili-Suu practiced from 1946 to 1967. Currently, the former enterprise, including in the urban areas are 23 tailings and 13 mining dumps. The total amount of uranium waste, pending in the tailings is approximately 199 000 000 $m^3$ and occupies an area of 432 000 $m^2$. The tailings were conserved in the 1966-1973 years, according to existing regulations. Heaps with a volume of 939 300 $m^3$ and occupied area about 114 700 $m^2$ were not re-cultured (Fig. 2), (1, 3, 5).

For a long time working on repair and maintenance of tailings were sporadic and insufficient. At the present time, the average exposure dose of gamma radiation (gamma-background) on the surface of the tailings is 30-60 mR/h, at some local anomalous areas have greater than 1000 mR/h.

However the science analysts' estimate that extraction from the original rock has been reach up to 90-95% of the uranium, and in the tails is only 5 to 10% or so in today's tails makes great background progeny of the uranium series. In table 1 the structure of original ore and a tail material of the Maili-Suu field are resulted. Elevated levels of Mn and Ca in the tails, as compared with the ore is associated with the use of compounds as a reagent and auxiliary substances in the ore processing and extraction of uranium, and high levels of lead, usually associated with the addition of radiogenic lead, is in the ore.

| Components% | Original ore | Tailings |
|---|---|---|
| Ca | 10-20 | 30 |
| Si | 20 | 6-10 |
| Fe | 2-3 | 0,4-1,0 |
| Pb | 1,5-2,0 | 2,0-3,2 |
| Cr | 4,5-6,0 | 2-3 |
| Mn | - | 50-200 |
| V | 1,0 | 0,4-0,6 |
| Ni | 3-5 | 2 |

Table 1. The average maintenance of separate components in ores and tailings of Maili-Suu

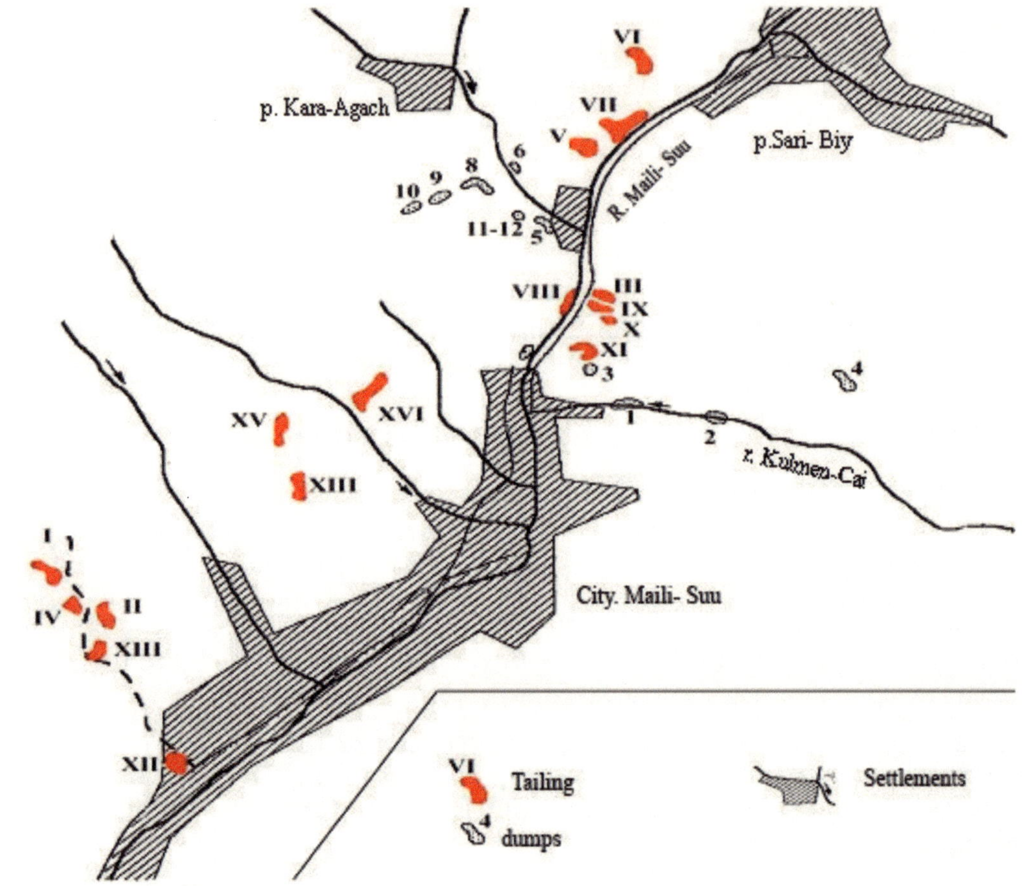


Fig. 2. The scheme of arrangement of tailings and waste dumps in anthropogenic provinces Maili-Suu.

From 1997 to 2003 special rehabilitation work in the country have been done, if they were sporadic. Starting from 2003 to 2007 in the country sharply intensified geomorphologic processes (landslides and floods), and therefore became acutely the question of preservation and rehabilitation of tailings and dumps (Fig.3-4). Upsurge in landslides, mudflows, erosion phenomena on the slopes adjacent to the tailings, the lack of funds for maintenance and repair and maintenance work has created a situation in some of the tailings in which may cause an ecological catastrophe. It should be noted that the destruction of tailings lead to removal of the tail material, not only in to the Maili-Suu river valley, but also into the densely populated Ferghana valley, and further to the basin of Syrdaria river. Fig. 5-6 shows the effect of surface re-vegetation of tailings in the period 1997-2003 compared to 1961. It clear from this scheme that the final completion of the re-cultivation work has far prospective.

The soil cover in the basin area downstream of the river - a typical gray soil, in the middle course - a dark gray soil, and then start mining-brown soil. General characteristics of the soil is as follows: pH 8.2 - 8.8, nitrate - 13.2 - 25 mg/kg of dry matter, chlorides - 25 - 47 mg/kg sulfate - 240 -895 mg/kg and petroleum products - 18 - 128 mg/kg of dry matter. Physical

Fig. 3. Location uranium production Maili-Suu

Fig. 4. The current state of tailings (# 8,9)

and chemical properties of soil cover, the Maili-Suu (except for the area of man-made sites), according to Sanitary and Epidemiology norm (SanEpidN) are in conformity or below the MPC (Maximum Permitted Concentration). More detailed study showed that not all the indicators correspond to the standard level, especially the level of trace elements.

Studies suggest a relatively low level of contamination of the soil cover micronutrients in relation to the background and the MPC. Found a slight increase in the concentration: Al, Mn, Se and U (2 - 3 times) in the autumn and spring, and Zn up to 6 times, the background of U in sub-region in more than 10 times than MPC.

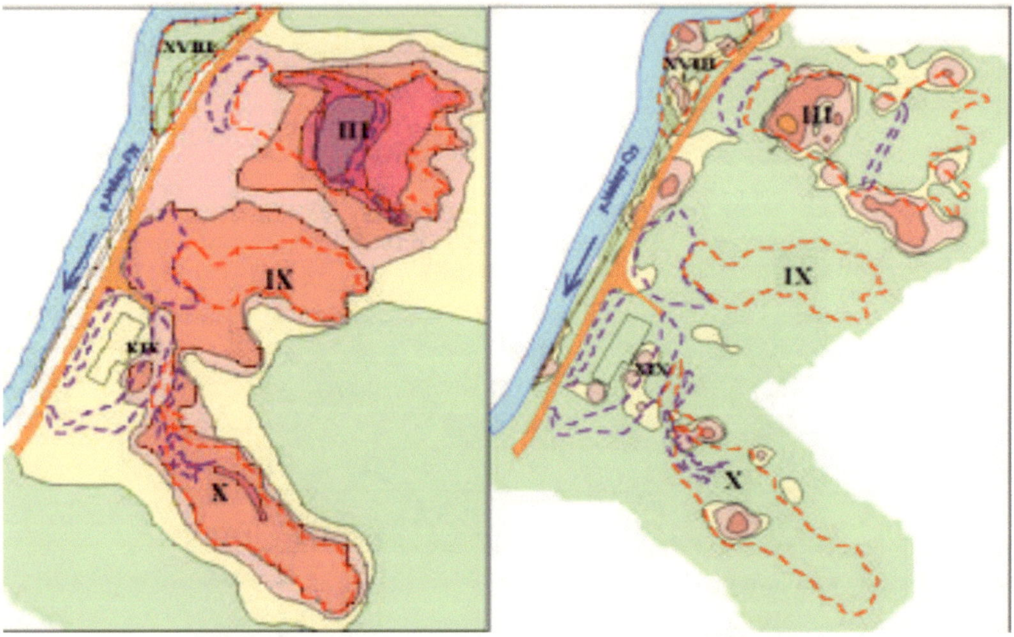


1961 year 2003 year

* Red indicates the areas exceeded MPC (100 mR/$h^{-1}$), green area marked with MPC, complying with local natural background

Fig. 5. and Fig. 6. The effect of surface re-vegetation of tailings in the period 1997-2003.

The vegetation (collection) in the basin Maili-Suu content of most trace elements studied at the level of control areas or slightly higher. Compared with the background sites, content: Al, Ba, Be, Fe, Mn, and Zn 2-times higher; As, Hg, Ni, Pb, Se and U - to 5 times; Mo, Co, Cd - 10-15 times. Increasing concentrations of trace elements observed in the middle and lower reaches of rivers. At some level of background regions is different from the minerals MPC. For example: U, Fe and Co more than a factor of 2, Hg - 10 times higher.

In an average sample of plants in the upper section (conditionally pure), the level of key micronutrients studied is relatively low, except for certain items, such as - Al in the Fergana wormwood (Artemisia ferganensis) - 2.5 times; Cu, Se and V in the astragalus (Astragalus lasiosemius) - 2 - 2.5 times; Ni in astragalus (Astragalus lasiosemius) and Artemisia Fergana (Artemisia ferganensis) - 10 times more compared to the background of other areas of the country.

According to our research water of Maili-Suu r. is not suitable for drinking. In some parts of the river water are found the highest concentrations - Se, exceeding the MPC in 23 times. Fe - concentration exceeds the MPC by 6 times or more, especially in the 2 and 5 points. The content - Cd, Al, Hg, Mn and Pb higher than normal in 2 times. The data obtained by: Ba, Fe, Co, Ni and Zn were not statistically significant (Table 2).

It should be noted that the destruction of tailings lead to removal of the tail material, not only in the valley r.Maili-Suu, but in the densely populated Ferghana valley, then - in the basin r.Syrdaria. In the zone of influence of the tailings of the former enterprise Maili-Suu in

| Elements | MPC | Sampling point (the river) and the mean values | | | | | |
|---|---|---|---|---|---|---|---|
| | | 1 | 2 | 3 | 4 | 5 | Σ |
| 1. Al | 0,5 | 0,55±0,09 | 1,076±0,15 | 0,94±0,031 | 1,026±0,13 | 1,086±0,13 | 0,935±0,22 |
| 2. Ba | 4,0 | 0,068±0,01 | 0,009±0,001 | 0,024±0,003 | 0,088±0,012 | 0,102±0,012 | 0,074±0,068 |
| 3. Co | 1,0 | 0,005±0,002 | 0,0073±0,001 | 0,005±0,001 | 0,006±0,001 | 0,005±0,001 | 0,005±0,001 |
| 4. Cu | 1,0 | 0,004±0,001 | 0,007±0,002 | 0,005±0,001 | 0,008±0,002 | 0,008±0,001 | 0,006±0,001 |
| 5. Fe | 0,5 | 0,248±0,025 | 0,46±0,062 | 0,34±0,025 | 2,54±0,42 | 3,209±0,54 | 2,601±1,01 |
| 6. Hg | 0,005 | 0,01±0,003 | 0,01±0,001 | 0,01±0,002 | 0,01±0,002 | 0,01±0,002 | 0,01±0,001 |
| 7. Mn | 0,1 | 0,07±0,012 | 0,225±0,013 | 0,081±0,012 | 0,181±0,032 | 0,192±0,016 | 0,101±0,03 |
| 8. Co | 0,5 | 0,005±0,001 | 0,004±0,001 | 0,010±0,002 | 0,003±0,001 | 0,010±0,002 | 0,006±0,004 |
| 9. Ni | 0,1 | 0,026±0,006 | 0,032±0,001 | 0,025±0,003 | 0,028±0,004 | 0,025±0,004 | 0,22±0,14 |
| 10. Pb | 0,1 | 0,02±0,001 | 0,035±0,001 | 0,02±0,003 | 0,02±0,003 | 0,02±0,003 | 0,023±0,002 |
| 11. Se | 0,001 | 0,023±0,005 | 0,02±0,003 | 0,023±0,02 | 0,02±0,004 | 0,023±0,005 | 0,021±0,001 |
| 12. V | 0,1 | 0,007±0,001 | 0,011±0,002 | 0,006±0,001 | 0,008±0,001 | 0,006±0,001 | 0,007±0,001 |
| 13. Zn | 1,0 | 0,005±0,001 | 0,011±0,002 | 0,003±0,001 | 0,142±0,023 | 0,077±0,011 | 0,047±0,023 |
| 14. U | 0,037 | 0,004±0,001 | 0,04±0,002 | 0,19±0,021 | 0,04±0,005 | 0,04±0,005 | 0,04±0,01 |
| 15. Cd | 0,001 | 0,002±0,001 | 0,002±0,000 | 0,002±0,001 | 0,002±0,000 | 0,002±0,000 | 0,002±0,0002 |

Table 2. Trace element composition of water in the r. Maili-Suu (average annual mg/kg)

Kyrgyzstan, home to 26 000 people, and Uzbekistan - to 2 400 000, Tajikistan - about 700 000, Kazakhstan - about 900 000 long-term contamination of radionuclides will be subjected to extensive areas Uzbekistan, Kazakhstan, Tajikistan, most of which are in the area of irrigated agriculture. Exposed to infection by rivers and streams, including such major rivers as the Kara-Darya, Syr-Darya. Water supply of drinking water is from rivers and canals, taking them from the beginning. Even if the water supply from groundwater wells may be contaminated with radioactive elements.

It should be noted that the collapse of tailings lead to removal of the tail material, not only in the valley of the Maili-Suu river, but in the densely populated in Fergana Valley, then - in the basin of Syr Darya river. In the zone of tailings influence in Maili-Suu the former enterprise in Kyrgyzstan, lives about 26 thousand people, Uzbekistan - to 2.4 million, Tajikistan - around 0.7 million, Kazakhstan - about 0.9 million. Extensive areas in Uzbekistan, Kazakhstan, and Tajikistan, most of which are in the area of irrigated agriculture, are exposed to long term contamination with radionuclides. The major sources for public exposure are the rivers and streams, including such major rivers as the Kara-Darya and Syr Darya. Water supply of drinking water is from rivers and canals, taking them from the beginning. Even if the water supply from groundwater wells may be contaminated with radioactive elements.

As a whole the soil-vegetative cover near the rivers Majli-Suu according to obtained data is satisfactory. There are no changes revealed of level of the studied elements in a soil-vegetative cover for several years. Naturally, the land covers in the tailings is not suitable for agricultural purposes and require special guidelines for local residents.

Currently, the safe storage of uranium waste in the town of Maili-Suu, has the following problems: disposal facilities are located less than 200 meters from residential city limits, the waste stockpiled near the river bed Maili-Suu. In order to reduce radon load to an acceptable level of sanitary protection zone in the city should be more than 3 km. Tailings dams require constancy of preventive measures in case of catastrophic floods and mud streams.

In the geotechnical investigations and tailings design was not taken into account susceptibility to landslides in the region involving the violation of rock massifs in the development of oil fields. In recent years, large-scale response of the slopes on the mountain of work is expressed in the mass development of landslides in the entire field. They provoke the probability of failure of some tailings. With landslides in the valley may be formed landslide lakes and catastrophic floods. In the flood zone may be tailing located along the river Maili-Suu river, as well as homes and other facilities of the city.

Lack of waterproofing the bottom of the tailings may lead to contamination of groundwater with radionuclide. Studies on the content of radionuclide in ground water and other contaminants have been conducted, as disposal facilities were not equipped with monitoring wells. The situation is complicated by the fact that after the cessation of uranium mining and the collapse of the Soviet Union and tailings dumps were abandoned for a long time in the state. Until 1998, there were only occasional maintenance and repair work. Environmental emergency calls for speedy implementation of measures for rehabilitation of tailings and dumps, to ensure long-term stability and prevent the threat of ecological catastrophe, the consequences of which could cause political complications and also in Central Asia.

Since 2007, the province implemented the project "Prevention of emergency situations", funded by the World Bank, worth 10 950 000 US dollars. The project provides for the identification and prevention of the most significant risks of radioactive tailings in the town of Maili-Suu, hazards of natural origin (landslides) and the improvement of emergency management. Work carried out by VISUTEK (Germany). Earlier, district repeatedly visited various expert missions to the IAEA, the World Bank, ADB, and the Russian Federation and other international organizations. As indicated from the conducted studies, tailing number 3 can impose high risk, so this tailing has been transferred to tailing number 16.

### 4.2 Issyk-Kul province of natural uranium (uranium-technogenic Kadji-Say)

Issyk-Kul province of natural uranium is located on the south shore of the lake Issyk-Kul, in Ton district, at an altitude of 1980 m above sea level. Mining Enterprise of the Ministry of Average Machine Building of USSR for processing uranium ore there was in operation from 1948 to 1969, and was subsequently converted into the electrical engineering plant. The uranium oxide at this site is generated from the ashes of brown coals uraniferous sogutin filed as a by product for the electricity production from coal (5, 15)

Waste and industrial equipment have been buried, forming a tailings pond, with a total volume of uranium waste 400 000 $m^3$, an area of 10 800 $m^2$. Tailings from uranium waste is located 2.5 km east of the residential village, but due to natural factors (rain, groundwater, landslides and mudflows) is an environmental threat to lake Issyk-Kul (1.5 km from the lake) and the nearest towns located on slopes up to 30-45° between the mountains. For 50 years there has been intense uplifting coastal area near the industrial site. A small part of the radioactive ash reached the lake Issyk-Kul.

According to Kovalsky V.V., Vorotnitskaya I.E. and Lekareva V.S. (10), the amount of uranium in the waters of rivers - Ton, Ak- Suu, Issyk-Kul is 5,6 • $10^{-6}$ g\l. According to Kovalsky V.V element content in the river Jergalan varies, depending on season and room selection, from 2,8 • $10^{-6}$ to 1 •$10^{-5}$ g\l. Key water wells and rivers of the Issyk-Kul basin contain 10, in some cases - 100 times more uranium than water areas and non-black earth black earth zone of Russia. Table 3 shows the results of our analysis of natural radionuclides in the water of rivers and tributaries of the lake Issyk-Kul, and the ratio of $^{234}U/^{238}U$. According to scientific estimates of researchers in the lake holds about 100 tons of uranium.

| Location of sampling | Uranium (total) $Bql^{-1}$ | $^{234}U/^{238}U$ | Gross alpha $Bql^{-1}$ | $^{226}Ra$ $Bql^{-1}$ |
|---|---|---|---|---|
| Issyk-Kul lake , Kara Oi v. | 1,79±0,18 | 1,13±0,05 | 1,80 | 0,013 |
| r. Bulan-Sogotu | 0,09±0,01 | - | 0,10 | 0,002 |
| r. Kichi Ak-Su | 0,17±0,02 | - | 0,20 | 0,009 |
| r. Tuip | 0,23±0,02 | - | 0,23 | 0,016 |
| r. Kara-Kol | 0,21±0,02 | - | 0,25 | 0,005 |
| Issyk-Kul lake , Ak-Terek v. | 0,56±0,06 | - | 0,60 | 0,02 |
| Kadji-Say v. a stream number 1 before the rain | 4,21±0,42 | 1,49±0,05 | 4,5 | 0,007 |
| Kadji-Say v. a stream number 2 before the rain | 10,2±1,02 | 1,30±0,05 | 10,0 | 0,005 |
| Issyk-Kul, v. Kadji-Sai river mouth | 1,69±0,17 | 1,52±0,05 | 1,67 | 0,015 |

Table 3. Natural radionuclides in the water of rivers and tributaries of the lake Issyk-Kul

As Table 4 shows, for comparison, the ratio $^{234}U/^{238}U$ at different times and the average content of uranium, this has the same level with slight variations.

| Location of testing | $\gamma = {}^{234}U/^{238}U$ | | Contents Uranium $10^{-6}$ g/l (9) |
|---|---|---|---|
| | 1966-1990 (1-6) | 2003-2004 (12) | |
| r. Toruaygyr | - | 1,49±0,01 | 11,0-19,0 |
| r.Chon-Aksu | 1,39±0,01 | 1,42±0,01 | 6,7-10,7 |
| r. Tup | 1,43±0,01 | 1,34±0,08 | 2,6-8,7 |
| r. Jergalan | 1,23±0,01 | 1,20±0,02 | 4,7-13,0 |
| r.Chon-Kyzylsuu | 1,23±0,01 | 1,20±0,02 | 4,3-11,2 |
| r. Barskaun | 1,14±0,01 | 1,08±0,07 | 7,2-2,7 |
| r. Ak-Terek | 1,23±0,01 | 1,24±0,02 | 0,42-47,0 |
| r. Tamga | 1,22±0,01 | 1,22±0,06 | 15,1-21,6 |
| Spring in the alluvial fan of r.Orukty | 1,51±0,02 | 1,62±0,02 | 2,6 |
| Borehole 3 v. Dzhergalan | 1,20±0,02 | 1,32±0,06 | 0,6-15,6 |

Table 4. Comparison of uranium-isotope data from the test 1966-1970 and 2003-2004 (12)

From radiometric survey we found that radiation levels in the Issyk-Kul basin, and the village itself Kadji-Say and the adjacent territory is relatively low. However, this basin is the natural uranium province, in some areas there is increased radiation background. We found that the beach areas near the southern coast of v. Dzhenish and v. Ak-Terek (placer - Thorium sands) the exposure dose is 30 to 60 mR/h, at least at some points reaches up to 420 mR/h (Table 5, Fig. 7-8).

Background areas were studied by measuring alpha-active isotopes in soils around Lake Issyk-Kul. The level of background radiation on the surface of the industrial zone and the tail short, in a residential area above the 2 time compared with the norms.

On isotopic composition of the soil (Bq/kg), extremely high levels of activity were detected. In the area of the settlement v. Kara-Oy, the content of U238 and Pb210 were found to be 2 - 2.5 times higher in the upper (0-5 cm) soil layers. In the area of the settlements Ak-Terek and Jenish, it was found that for all the thorium (Th) isotopes the level of radiation are higher than any other studied locations by 2 to 10 order of magnitude (Table 6; Fig.9-10).

| **Sampling location** | **T° of water** | **pH** | **Gamma background** | |
|---|---|---|---|---|
| | | | **on the soil surface** | **at a height of 1 m** |
| Kara-Oi v. | 18,5 °C | 8,5 | 150-200 mSv / h | 100 mSv / h |
| Cholpon-Ata t. | 18,8 °C | 8,6 | 200 mSv / h | 150 - 220 mSv / h |
| Bulan-Sogotu v. | 17,5 °C | 8,15 | 150 mSv / h | 100 mSv / h |
| Kichi Ak-Suu r. | 13,2 °C | 7,94 | 160 mSv / h | 150-170 mSv / h |
| Tuip r. | 18,8 °C | 8,12 | 170 mSv / h | 140 mSv / h |
| Kara-kol r. | 15,8 °C | 8,05 | 180 mSv / h | 150-210 mSv / h |
| Ak-Terek v. | 17,5 °C | 8,24 | 470 mSv / h | 420 mSv / h |

Table 5. The level of exposure dose in the Issyk-Kul basin

Fig. 7. Tailing after the rain

Fig. 8. The tailings from the bottom

| Sampling location | Layer cm | Activity of soils by isotope, Bq / kg | | | | | | | | | |
|---|---|---|---|---|---|---|---|---|---|---|---|
| | | U-238 | | Ra-226 | | Pb-210 | | Th-228 | | Ra-228 | |
| | | | +/- | | +/- | | +/- | | +/- | | +/- |
| **Kara-Oi** | 0-5 | 71,8 | 12,7 | 35,1 | 3,9 | 147,4 | 13,0 | 39,5 | 2,2 | 35,2 | 8,8 |
| | 5-10 | 50,8 | 7,3 | 37,7 | 3,4 | 64,6 | 11,4 | 49,0 | 1,9 | 60,1 | 7,5 |
| | 10-15 | 44,0 | 1,7 | 35,1 | 3,2 | 50,1 | 7,2 | 45,6 | 1,8 | 52,3 | 3,5 |
| | 15-20 | 51,7 | 7,4 | 46,1 | 3,5 | 50,2 | 7,7 | 49,9 | 1,9 | 53,6 | 7,7 |
| **Kichi-Ak-suu** | 0-6 | 71,5 | 14,3 | 51,0 | 3,4 | 88,5 | 18,4 | 69,1 | 3,6 | 72,4 | 7,2 |
| | 6 11 | 52,1 | 6,5 | 43,2 | 3,1 | 71,7 | 10,2 | 43,2 | 3,3 | 59,2 | 19,7 |
| | 11-20 | 54,9 | 7,3 | 45,4 | 3,5 | 68,6 | 7,6 | 64,3 | 3,8 | 64,1 | 7,5 |
| **Ak-Terek sand** | 0-5 | 260,0 | 30,0 | 103,0 | 8,0 | 169,0 | 30,0 | 915,0 | 57 | 846,0 | 70,0 |

Table 6. Background values for alpha-active isotopes in soils around Lake. Issyk-Kul and thorium sands

Soil and ground tailings - in the upper layer of soil bulk (0-20 cm) of uranium from 1.1 to 2,6 • $10^{-6}$ g/g, with the depth of the element increases - up to 3,0 • $10^{-6}$ g/g. Most of the uranium concentration was noted in the central zone of tailings: in the upper layer of soil - 4,2 • $10^{-6}$ g/g in the bottom (at depths of 40-60 cm) - 35,0 • $10^{-6}$ g/g, which is 8.3 times more than in the upper horizons.

**The vegetation** is characterized by the province following associations: xerophytic shrub-, sagebrush-efimerovymi deserts, thorny (Akantalimon alatavsky, bindweed tragacanth). The vegetation cover is sparse, the project covering ranges from 5 to 10% and only in some areas

Fig. 9. Dosimeter research.

Fig. 10. Local cattle pastured

up to 50%. The uranium content in different types of wormwood (Artemisia) in the tailings relatively high in relation to the region as a whole - 0,03-0,04 • $10^{-6}$g/g. Representatives of the legume (Salicaceae) - Astragalus (Astragalus) and sweet clover (Melilotus) contain up to 0,09 • $10^{-6}$g/g, while the grass (Poaceae) - a fire roofing (Bromus tectorum) uranium contained in twice to 0,17 • $10^{-6}$g/g. According Bykovchenko J.G. (3) these types of plants can serve as a land-improving plant for reobiletation tailings. According to the results of our studies the percentage of uranium in the province of plants Kaji-Sai is from 0,17 to 4.0 • $10^{-4}$%.

Consequently there is reason to say that most of the plants Kaji-Says region have high uranium content in comparison with other territories in the region. Growth of plants in an environment with high concentrations of uranium is not only accompanied by changes in their biological productivity, but also causes morphological variability in particular: the splitting of Astragalus leaf blade, Peganum garmaly instead of the usual five petals it was noted 6-7 and part of their split, and the black grate observed significant morphological changes - low-growing form with branched inflorescences instead of straight single arrows (5, 10, 13) (Fig.11-12).

Fig. 11. Straight from the top of the tailings

Fig. 12. Color mosaic of plant leaves Iris family (Iridaceae) species-Iris songarica Schrenk

Currently, surface water eroded slopes adjacent to the tailing of relief, ground ash dump, the protective coating surface of the tailings piles and rocks. Diversion of surface water systems tailings are partially destroyed preserved, due to changes in drainage conditions due to existing buildings and structures do not provide normal drainage of surface water. Fences tailings destroyed, the network of groundwater monitoring is absent.

### 4.3 Uranium deposits of settlement

**Min-Kush** (Tura-Kavak) are at an altitude of about 2000 m in the basin of the r. Min-Kush. The population of urban settlement. Min-Kush at present is 4760 persons. In this region there are 4 tailing of radioactive materials - the volume of 1.15 thousand $m^3$, an area of 196.5 thousand $m^2$, and 4 mountain damps (substandard ore, there is no data on the volume) and the whole tail is a flat, land located on slopes up to 25-40° between the mountains. Ore complex operated from 1963 to 1969. After closing all the tailings of the uranium production was inhibited.

Currently, because of the timing of repairs and maintenance, there is a destruction of individual defenses and surface areas. The most dangerous are tailing "Tuyuc-Suu" and "Taldy-Bulak." Tailings "Tuyuc-Suu" is located in line with the same river. The total volume of reclaimed tailings - 450 000 $m^3$, their area - 3,2 hectares. According to the results of radiometric survey the exposure dose at the surface of the tailings - 25-35 mR/h, locally - 150 mR/h. The total radio activity of nuclides in the disposal of the tail material - 1555 Ci.

To skip the reinforced concrete built river bypass channel is now part of ferro-concrete bypass channel structures destroyed, there was differential settlement surface tailings, formed locally closed injury, do not provide a flow of surface waters: a protective coating in some places broken excavation, fences and signs forbidding destroyed. The tailings are located in an area prone to mudslides. Possible violation of the water drainage and destruction of the tailings with the removal of the tail of material in the river Kokomeren and Naryn, then - in Toktogul and the Fergana valley. There has been a movement of an ancient landslide threat of overlap Tuyuk-Suu river and the destruction of the road to tailing (Fig. 13-14).

The radiometric survey of the exposure dose of gamma radiation at various sites of uranium tailings Min-Kush, showed from 27 to 60 mR/h, but at some points is high. For example the tailing Taldy-Bulak - 554 - 662 mR/h (Table 7). In general, the soils of Min-Kush geochemical province largely enriched by uranium, as far as concentration of uranium in them is 5-6 times higher than in other soils of Kyrgyzstan.

| Name of areas | Radiation background in mR/h |
|---|---|
| Min-Kush village | 27,0-28,0 |
| Tailings Tuyuk-Suu gate | 27,5-28,0<br>60,0-61,0 |
| The site- 21(where miners lived) | 32,0-32,5 |
| Tailings Taldy-Bulak | 554 - 662 |
| Water from the tunnels | 61,0-61,5 |
| Hotel Rudnik | 60,0-61,0 |

Table 7. The level of background radiation in a uranium province of Min-Kush

Fig. 13. Arrangement of uranium tailings storage Tuyuk-Suu in the village of Min-Kush

Fig. 14. A landslide in the lower portions of tailing

The soil cover neighborhoods Min-Kush presented, as indicated above, sub-alpine soils of steppe and meadows. The uranium content here, in the middle of the profile ranges - from 3,3 to 17,5 • $10^{-6}$ g/g is relatively high. Moderate pollution (great danger) in the area located above the processing plant where the uranium content in the soil reaches the surface - 30-35 • 10-6 g / g, indicating that the local pollution of this area.

In all soil profiles high concentration of uranium observed in the horizon of 20-40 cm (15-20,0 • $10^{-6}$ g/g). In the adjacent - Kochkor valley where the soils are mountain-valley light brown the uranium content in the range 3,0-5,0 • $10^{-6}$ g/g. Humus to a certain extent helps to perpetuate the uranium in the soil apparently is in the process of sorption of uranium by organic matter of soil and the formation of uranyl humates.

We have also studied the radiation background in some village homes'. V. Min-Kush (Table 8) and measurement results showed that in homes, compared to the MPC, the background radiation slightly increased (2 times) and therefore requires specific measures to reduce. The main reasons for the slight increase in background radiation provided cases of using waste ashes from the local coal for the construction needs.

| **Gamma-ray background: in the attic** | |
|---|---|
| 0,97mcZv/h ± 22%<br>0,88 mcZv/h ± 20% | 0,78 mcZv/h ± 20%<br>0,73 mcZv/h ± 20% |
| Inside appartment 6, in the hall | |
| **Bedroom - - the floor** | **Kitchen floor** |
| 0,76 mcZv/h ± 20%<br>0,65 mcZv/h ± 20%<br>0,75 mcZv/h ± 20% | 0,63 mcZv/h ± 20%<br>0,69 mcZv/h ± 20%0,<br>69 mcZv/h ± 20% |
| **Bedroom - the ceiling** | **Kitchen ceiling** |
| 0,72 м mcZv/h ± 20<br>0,66 mcZv/h ± 22%<br>0,79 mcZv/h ± 22% | 0,57 mcZv/h ± 10%<br>0,80 mcZv/h ± 10%<br>0,71 mcZv/h ± 10% |

Table 8. The level of radiation background in the residential of v. Min-Kush (17 Square, st. Zhusup, Building 10, Apt. 6)

Considered several options for security of stored waste:

- Dismantling and transport the tailings to a safer place;
- Repair of hydraulic structures and constant maintenance of their working condition over a long period of use (thousands of years);
- Conducting sanitation radioecological studies and measures to reduce the exposure dose in dwellings.

### 4.4 Uranium-technogenic provinces Shekaftar

The mine operated from 1946 to 1957 year at this area and also 8 dumps located here. In the dumps warehousing about 700 000 $m^3$ of low-level radioactive rocks and ores substandard.

In the immediate vicinity are houses with gardens. The main pollutants are elements of the uranium series. The average gamma-ray background is 60-100 mR/h on the anomalous areas - up to 300 mR/h. All damps are not re-cultured (Fig. 15).

The material of which is used by local people for household needs. Damp number 5 located on the bank of river Sumsar intense urged by its waters. The lack of vegetation on the surface contributes to the development of wind erosion and surface runoff material stockpiles and distribute them not only to the territory Shekaftar item, but also in adjacent territories of Fergana valley.

A more extensive destruction of stockpiles fall down cross-border contamination of the territory of Uzbekistan and Tajikistan.

Fig. 15. The not re-cultured dumps in the region of Shakaftar

Bringing the dumps in a safe condition requires the following emergency operations:

- strengthening the river banks Sumsar;
- re-culturing of land dumps;
- restoration of fences,
- the installation of warning signs.

### 4.5 Ak-Tuz technogenic provinces of rare and radioactive metals

Ak-Tuz technogenic provinces of rare and radioactive metals are located in the Chui region of KR in the upper part valley river Kichi-Kemin and river basin Chu. The terrain - a complex, mountainous. Absolute altitude exceeds 2000 m above sea level.

The ore field of the region is characterized by an extremely complex structure, and covers about 30 occurrences of lead and rare metals. It is widely developed within a multiplicative and disjunctive offenses manifested repeatedly throughout geological history, ranging from the Precambrian. Within the deposit an oxidized sulfide ores of metals were developed. In industrial concentrations established the presence of: Pd, Zn, Sn, Mn, Cu.

In the region of the Ak-Tuz are 4 tailings. Stored 3900 000 $m^3$ of waste ores, which occupy 117 000 $m^2$, the average gamma-ray background is 60-100 mR/h in the abnormal areas of up to 1000 mR/h (Fig.16-17).

From 1995 to 1999. work to maintain the waterworks were not conducted. In 2000 activities were conducted waterworks tailings number 1 and 3. There is intense erosion of the protective layer tail number 1 and wind erosion surface tailings number 3 with the destruction of the surrounding areas.

Fig. 16. Tailings number 2

Fig. 17. Territory after the development

According to the radiometric measurements the average exposure dose of gamma radiation in part of Ak-Tuz is 21,3 - 33,0 mR/h and around the village within radius of 1 km - 28.8 mR/h. The gamma-ray background in the processing plant is 73,3 mR/h in the sump - 720-740 (in places up to 900) mR/h and near the mines (career) - 50,0-72,0 mR/h. Natural gamma-ray background in the canyons of the Kichi-Kemin is 30,0 mR/h.

The soil cover of the province is typical for middle mountain areas of Kyrgyzstan. Ak-Tuz mining and metallurgical combine mountain-meadow black earth subalpine soil. The texture of the soil has medium and high clay content character. Humus in the upper levels are between 4-8%. Soil reaction (pH) ranges from neutral to slightly acidic and is 6,5-6,8 to 7-7,0. Humus horizon of these soils are rich in potassium (2.2 to 2.6 %.) Contains of 0.35% nitrogen and 0.15-0.30% phosphorus. In the sump pH close to neutral medium (pH = 7) above and below the sump level is the same. Eh - in the region settling tank is moderately increased (210), below the sump decreases.

Results of the analysis of the upper soil layer (up to 0 - 20 cm) are presented in Table 9. The table shows that the maximum concentration of lead found in the area of 500 m below the lagoon (3108,4 ± 415 mg/kg), followed by factories in the area of 1 km (2686,1 ± 287,7 mg/kg) and 4 tailing (1937,0 ± 325,4 mg/kg), which is increased to 10 times compared to other sites, and in relation to the MPC to 200 times.

Zinc concentration increased to 10 times compared to other sites, as compared with up to 15 times MPC. For example, in the factory up to 1 km (720,62 ± 59 mg/kg), the tail region of 3 (818,90 ± 26 mg/kg), 4 tail (756,20 ± 57 mg/kg) and 2 tail (652 70 ± 87,1 mg/kg).

| # | Sampling locations | Pb mg/kg | Zn mg/kg |
|---|---|---|---|
| 1 | 1 km upstream from v.Ak-Tuz (the right bank of the river from the road 60 m) | 621,14±17,82 | 104,83±17,82 |
| 2 | Ak Tuz v. (center, from a point-600 m) | 2057,5±339,4 | 678,79±30,3 |
| 3 | In the area of the factory up to 1 km | 2686,1±287,7 | 720,62±59 |
| 4 | In the area of weight | 398,2±38,2 | 128,3±11,3 |
| 5 | 200 m from the factory (above) | 436,7±45 | 76,83±5,3 |
| 6 | In the area of tank | 453,2 ±37,3 | 631,16±70 |
| 7 | 500 m below the lagoon from the factory | 3108,4±415 | 91,68±33 |
| 8 | In the region of 2 tailings | 370,0±39 | 652,70±87,1 |
| 9 | In the region of 3 tailing | 331,0±34 | 818,90±26 |
| 10 | In the region of 4 tailings from the road above 450-500 m | 1937,0±325,4 | 756,20±57 |

Table 9. The average content of heavy metals in the soil cover Ak-Tuz polymetallic province

Thus we can say that in the village of Ak-Tuz and its surroundings the level of gamma ray background is almost within the natural. Near the mine (quarry) and in the processing plant, where the extraction and processing of ore containing rare earth metals and radioactive thorium, the average exposure dose of gamma radiation exceeds the natural rate of several times, especially Pb and Zn. In unfenced and located near the settlement of the sump in the tens or hundreds of times, which adversely affects of the ecology region. These objects are contained except for radioactive thorium, heavy metal salts. In the event of failure of tailings can take away the tail of material in the basin of Chui river and pollution across national boundaries.

### 4.6 Kara-Balta mining ore plant for production of oxide, oxide of uranium

The plant's capacity to 2,000 tons of uranium a year, in operation since 1955. Tailings mining and metallurgical plant the plain type is located 1,5 km from the town of Kara-Balta, 2380 000 $m^2$ area, maximum height of 35 m. The net capacity 63,5 million $m^3$ is filled with 54.4%. Currently 32,5 million $m^3$ of waste stored, AC power is 84600 Ci.

The main polluting components - uranium series elements. Damp height to 12 m. Completed closed drainage to the production and wells to 35% of the stacked protective plastic film, the rest of the shield consists of loam and clay. There is a technical device abstraction to capture contaminated groundwater from five wells. There is a regular monitoring of groundwater. In the area of tailings maximum contamination up to 3-4 g/l outside the sanitary protection zone at the MPC. In conducting radiometric survey we found that in the tailings level 1 and 2 background radiation is much increased (from 4 to 20 times) compared to the other points. The radiation background at the tail KGRK and surrounding areas from our data is 25 mR/h (the town of Kara-Balta, 200 m from the SPZ) to 550 mR/h (at the base of tail).

## 5. Conclusion

In the complex environmental problems in the country first place put forward the problem of safely storing large quantities of waste mining. The accumulation of significant amounts of radioactive waste resulted from the activities of mining and processing enterprises of the uranium industry 40-70s. Storage in open dumps, tailings and not enough trained squares leads to an intense weathering of toxic substances into the atmosphere, their penetration into the groundwater, soil, surface water and adverse impact on the environment and human health.

Many of the tailings and dumps, radioactive waste disposal in the border areas are in critical condition and cause a risk of contamination and radiation exposure in the territory of Kyrgyzstan, as well as possibly other Central Asian republics. The main causes of environmental stress in the region due to bad choice of storage sites and storage facilities, short-term considerations of economic gain, a low level of geological engineering survey and design lack of foresight and taking into account the effects of technological impacts on the stability of fragile mountain ecosystems. Many of the tailings were formed within settlements.

With the recent surge in industrial and natural catastrophic events, landslides, mudflows, erosion, the threat of radioactive pollution of the environment increases significantly. There is a threat to the health of people living near areas with high levels of radiation and radioactivity in the environment. On many dangerous areas, lack of basic information on the radioactivity content of tailings is not being monitored due to lack of funding and related equipment on the ground.

The main radiological concern in the country is the restoration of plant-soil (gardening) and the bare heaps tailing, protection from the intense erosion of the protective layer of tailings. Thus the uranium tailings are poorly protected and poorly understood features of life in different organisms (biological response to the increased content of radionuclide, the state of microbial complex and human).

Estimated cost of MES KR (approximate) of the reclamation and rehabilitation work only on the tailings will be more than - $ 40 million dollars USA, including:

1. v.Maili-Suu tail. Landslides - 16.8 million USA
2. v. Min-Kush tail. - 4.6 million dollars USA
3. v. Ak-Tuz tail. - 1.6 million USA

4. v. Kaji-Say tail. - 3.6 million USA
5. v. Sumsar tail. - 5.0 million USD
6. v. Shekaftar tail. - 1.5 million USA
7. v. Soviet tail tail. - 2.0 million USA
8. v. Orlovka - 3.0 million USA
9. etc.

Therefore, the Government and the President of the KR pay special attention to these problems and made some steps in this direction. In 2009 the President appealed to the UN Secretary-General, in 2010, the European Union with a request to provide financial and technological support in addressing this issue in Kyrgyzstan and the region.

1. In general, the overall level of external radiation background in Kyrgyzstan is normal except for some man-made and natural areas.
2. Increased radioactive anomaly in the man-made sites marked by three types:
   - Natural radioactivity anomalies associated with layers of loose deposits of radioactive brown coal of Jurassic age.
   - Man-made anomalies, hundreds of times higher than background, are confined to fenced concrete wall piles of gray fine-grained material.
   - Activity of man-made anomalies in the landfill is ten times higher than background.
3. Growth of plants in an environment with high concentrations of uranium is not only accompanied by changes in their biological productivity, but also causes morphological variability - Astragalus borodinii, Peganum garmala, Potentilla argentea
4. It is important to educate the population.

## 6. References


[1] Aitmatov I.T., Torgoev I.A., Aleshin J.G. Geo-environmental problems in the mining complex in Kyrgyzstan //Science and New Technologies.-1997.-#1, P.81-95.

[2] Aleksakhin R.M. Problems of Radioecology: The evolution of ideas. Results. M: Agricultural, 2006. – 880 p.

[3] Bykovchenko J.C., Bykova E.I., BelekovT.B. etc. Man caused uranium contamination of biosphere Kyrgyzstan. - Bishkek, 2005.169 p.

[4] GOST 0.6-90. Radiometric technique express determination by γ-radiation volume and specific activity of radionuclide in water, soil, food, livestock and crop production. - Introduced. 1990-18-06. - M.: Standards Press, 1990. - 35.

[5] Djenbaev B.M. Geochemical ecology of terrestrial organisms.-Bishkek, 2009. 240 p.

[6] Djenbaev B.M Kaldybaev B.K., Zholboldiev B.T. Radiobiogeo-chemical estimate the current state of the biosphere reserve of the Issyk-Kul (the Kirghiz Republic) //International conference "Modern Problems of Geoecology and biodiversity conservation". - Cholpon-Ata. 2009. P 77-81.

[7] Djenbaev B.M., Zholboldiev B.T. The study of the natural uranium isotopes and their relation to uranium biogeochemical provinces of the Issyk-Kul. //Proceedings. # 1. Bishkek. 2010. P. 67-72.

[8] Zyrin N.G .Methodical recommendations on field and laboratory studies of soils and plants under the control of environmental pollution metals/ N.G. Zyrin. - Moscow: Gidrometeoizdat, 1981. – 108 p.

[9] Karpov Y.A., Savostin A.P. Sampling methods and sample preparation. -M.: Bean, lab-knowledge. 2003. - P.68-79.

[10] Kovalsky V.V., Vorotnitskaya I.E., Lekarev V.S,. Nikitina E.V. Uranium biogeochemical food chains in the Issyk-Kul. Proceedings of the Biogeochemical Laboratory. - Moscow "Nauka", 1968, XII. P.25-53.

[11] Mamytov A.M. Soil resources and land registry issues Kyrgyz Republic. - Bishkek: Kyrgyzstan, 1996. -240 p.

[12] Matychenkov V.E., Tuzova E.V. The stability of the isotopic composition of uranium in the waters of Issyk-Kul basin /The study of hydrodynamics of Lake. Issyk-Kul, using isotope techniques. Part 1. 2005. P.133-137.

[13] The vegetation of the Kirghiz SSR (map), M. 1:500000. M. GUGK, 1992 (by Popova L.I., Moldoyarov A., Cheremnykh M.A.).

[14] Torgoev I.A., Aleshin.J.G. Geoecology and waste mining complex in Kyrgyzstan.- Bishkek, Ilim, 2009. 240 p.

[15] Djenbaev B.M., Shamshiev A.B., Jolboldiev B.T., Kaldybaev B.K. Jalilova A. A. The biogeochemistry of uranium in natural-technogenic provinces of the Issik-Kul /Uranium, Mining and Hydrogeology, Technical University «Bergakademie», German, Freiberg, 2008. P.673-680.

4

# Environmental Migration of Radionuclides ($^{90}Sr$, $^{137}Cs$, $^{239}Pu$) in Accidentally Contaminated Areas of the Southern Urals

V. V. Kostyuchenko, A. V. Akleyev, L. M. Peremyslova, I. Ya. Popova, N. N. Kazachonok and V. S. Melnikov
*Urals Research Center for Radiation Medicine, Chelyabinsk, Russian Federation*

## 1. Introduction

In the late 1940s, the facility Mayak Production Association (Mayak PA) for weapon grade plutonium production was put into operation in the vicinity of the town of Kyshtym. The technology used in plutonium production involved generation of high-level waste. A number of accidents that occurred at the plant were associated with inadequate radioactive waste storage techniques. In 1949-1956, radioactive waste with total activity of about 1.8 $10^{17}$ Bq (4.9 MCi) was discharged into the Techa River which resulted in contamination of all river system components. Currently, at late time after the beginning of contamination, $^{90}Sr$ and $^{137}Cs$ still remain essential dose-forming radionuclides on the Techa River. In 1957, the East-Urals Radioactive Trace (EURT), and in 1967 the Karachai Radioactive Trace (KRT), were formed. A distinguishing feature of the radionuclide composition of the releases on the EURT at late time is the prevalence of $^{90}Sr$ and a minimum content of $^{137}Cs$. The composition of radioactive dust on the 1967-Trace is represented primarily by Cs and Sr isotopes in less accessible biological forms compared to those observed on the EURT (Fig. 1). In contaminated areas, measurements of soil contamination levels, analysis of the patterns of radionuclide migration, changes in their biological accessibility, transfer of radionuclides from soil to vegetation, milk and vegetable produce have been conducted on regular basis. Specific activity of $^{90}Sr$ measured in cross sections of the river at most of the riverside villages has decreased to permissible values since the start of observations in 1960. In flooded areas of the bank line, the processes of deepening of radionuclides into soil and a more uniform distribution of radionuclide contents over the soil layer at a depth of 1.5 m were observed. Mean content of $^{90}Sr$ in milk produced in the riverside villages has declined to permissible values. On EURT and KRT, of the total radionuclides contained in the soils, 80% remain deposited in the upper 20-cm layer. Biologically accessible and insoluble forms of $^{90}Sr$ and insoluble forms of $^{137}Cs$ are prevalent. Reduction in radionuclide content in milk has taken place over the first 1-2 years due to deepening of radionuclides into soil and a decrease in their biological accessibility. The main factor that caused cleansing of radionuclides from food chains was radioactive decay and reduced biological accessibility of radionuclides in soils.

### 1.1 Natural-climatic characteristics of the affected territory

As a result of the accidents at the Mayak AP, a number of rivers, water basins and lands of the southern and middle zone of the Trans-Urals region were contaminated. The EURT occupies over 3/4 of the forest and forest-steppe part of the Trans-Urals region where there are numerous lakes, swamps, all kinds of depressions and pits, wood lands and forest outliers which account for non-uniformity of radioactive fallouts. The most common are chernozemic-meadowy and meadowy-chrnozemic soils. The Trans-Urals region has a typical continental climate which is formed by the air masses coming from the Atlantic Ocean. The wind conditions of the region are characterized by prevalence of westerly winds. Whirlwinds are not an infrequent phenomenon. Species of wood prevailing in the forest zone include pines and the main hardwood species - birches and aspens. The floodplain vegetation includes grassy and woody-shrubby species. Birches and willows are encountered in the floodplain. Miscellaneous herbs are characteristic of the Techa floodplain in the middle and lower reaches of the river.

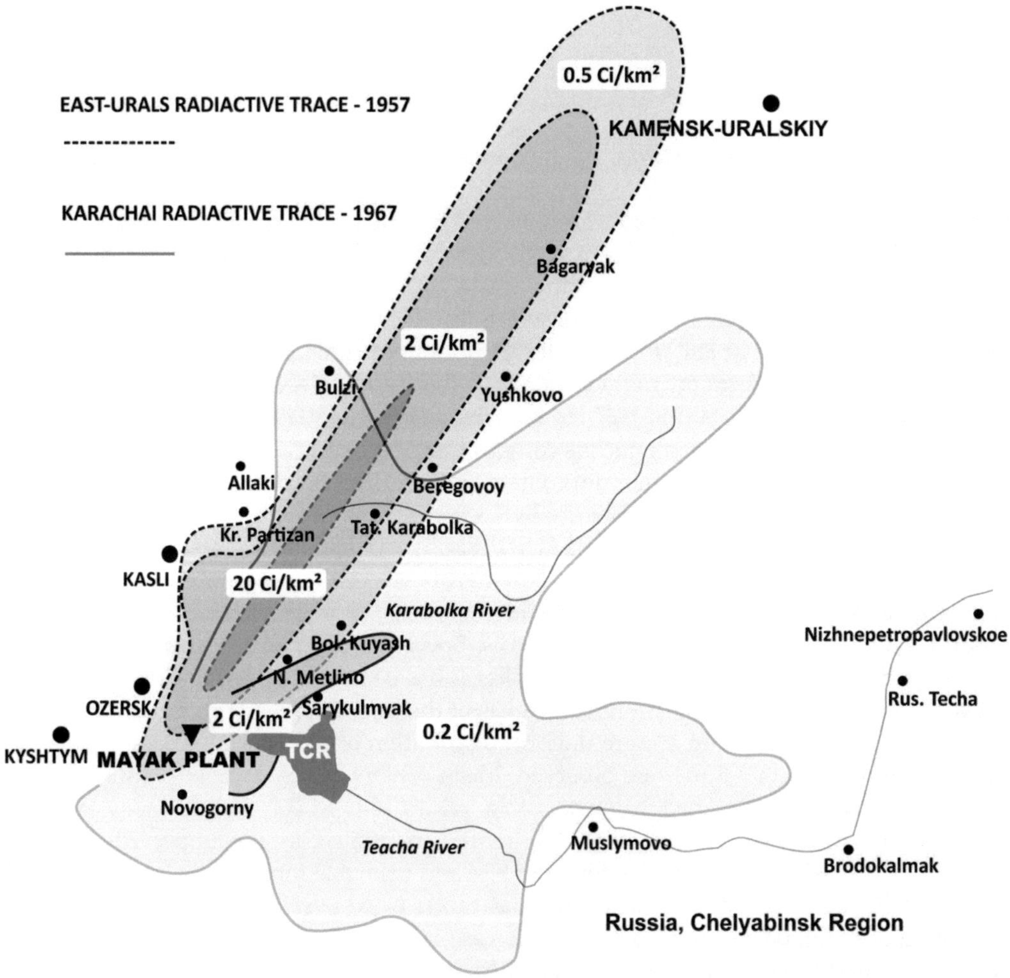


Fig. 1. Schematic map of radiation accidents in the Southern Urals

### 1.2 Methods of the study

The key contaminants of the environment at late phase of the accident are represented by a small number of radionuclides: $^{90}$Sr, $^{137}$Cs, $^{239}$Pu and a few other. Samples of water, bottom sediments, aquatic vegetation, soil and grass were collected at 8 control cross sections on the Techa River in accordance with conventional standard methods of sample drawing. (Basic Requirements…, 1999, in Russian [Основные требования…, 1999]). Samples of bottom sediments were collected at a depth of 60 cm using metallic tubes, the volume of the samples ranged from 20 to 100 cm$^3$. To allow assessment of the distribution and deposits of radionuclides in floodplain soil, samples were drawn at a depth of up to 150 cm. Samples of grass were taken from the area of 0.25 m$^2$. Water was sampled at a depth of 10-50 cm using a sampling device. Measurements of radioactive contamination in the riverside villages and the adjacent areas, contents of radionuclides in food products, migration of activity from soil to food products, horizontal and vertical migration of radionuclides in soils were conducted In the EURT zone. Processing and preparation of food products and other samples to be used in a corresponding assay were conducted using the generally accepted methodology. Measurements of $^{137}$Cs and other gamma-emitting radionuclides were performed using the gamma-spectrometric method. $^{90}$Sr concentration in samples was measured by the radiochemical separation of daughter $^{90}$Y using monoisoochthyl-methyl ether of phosphonic acid (1966) and a subsequent measurement of its activity in a small-background β-metric installation analogous to UMF-2000 based on a flame photometric control of strontium carrier yield. $^{90}$Sr measurement error accounts for 20% at activities of <0.7 Bq/g, and for 10% at higher activity levels. The range of measured values was $0.02\text{-}1{\cdot}10^5$ Bq/dm$^3$.

The method for measuring $^{239}$Pu involves increasing concentration of plutonium ions, and cleaning of isotopes using anion-exchange tar followed up by electro-chemical precipitation on steel disks. Measurements of α-activity were performed in an α-spectrometric installation. Identification and measurements of specific activity of plutonium isotopes was conducted using an indicator mark ($^{236}$Pu or $^{242}$Pu) with a known activity ranging from 1.5 to $1{\cdot}10^5$ Bq/kg, dm$^3$ preliminarily introduced into the sample.

## 2. Techa River

### 2.1 Hydrographic characteristics of the Techa River area

The Techa River basin catchment area is situated between the mountainous Urals region and the Tobol River valley. The water from the Kasli-Irtyash lake system is flowing through the Techa River. After the construction of the Mayak plant was started, a cascade of industrial water reservoirs (IWB) was built for storing low-level liquid waste. In 1956 reservoir B-10, and in 1963-1964 reservoir B-11 were built. Since 1966, the IWRs have been functioning as stagnant water reservoirs. Since 1965, the Techa River was conditionally assumed to rise from the dam of reservoir B-11 (Fig. 2). The drainage of the Kasli-Irtyash lake system takes place through the bypass canals (BC).

From dam P-11 to the village of Muslyumovo, the river flows through a wide valley with numerous swamps. The floodplain is mainly composed of peaty soils. The river bottom is peaty-silty and uneven. Along the river stretch from the village of Muslyumovo and further on, the river flows through a flat even country. The surface of the floodplain is meadowy, formed by sandy-loam soils. The bottom of the river is sandy, miry and loamy at places. In the lower reaches of the river, the width of the valley ranges from 240 m to 2 km. The floodplain is meadowy, loamy, usually flooded during high water. The watercourse is moderately winding. The river is mostly supplied with snow water. The swampy floodplain of the upper reaches retain a considerable amount of thaw water. The river's tributaries are water-short. Floods usually occur in April. Low water lasts till mid-October. Water discharge during the low water periods increases along the river length from 0.84 $m^3/d$ near Muslyumovo to 2.62 $m^3/d$ near the village Klyuchevskoye. The coefficient of ground water supply accounts for 10-30% of the total river drainage.

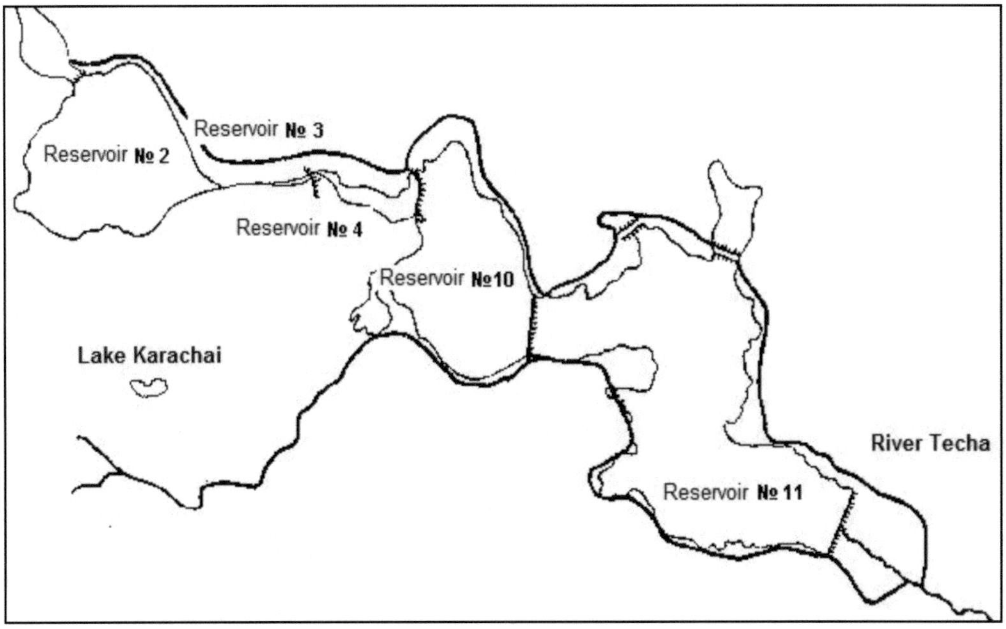


Fig. 2. Schematic map of the Techa cascade of reservoirs

The contamination of the Techa River was caused by the unavailability of reliable technology for reprocessing and storage of liquid radioactive waste (LRW). The approximate total releases into the Techa River over the period 1949-1953 were as follows: $^{89}Sr+^{140}Ba$ 240 kCi, $^{90}Sr$ -320 kCi, $^{137}Cs$ -350 kCi, REE 740 kCi, $^{95}Zr+^{95}Nb$ 37 kCi, $^{103}Ru+^{106}Ru$ kCi (Glagolenko Yu.G., 1966).

### 2.2 Radioactive contamination of water

During the initial period, the studies of radioactive contamination of water were based on measurements of β-emitting nuclide activity. The most well-systematized data were presented in (Marey A.N., 1959). The highest level of β-activity was observed in water in 1951; it was decreasing appreciably with advancing years and increasing distance from the site of releases (Fig. 2). The activity of α-emitters in water was significantly lower. The dependences governing the changes in the concentration of these emitters are similar to those identified for β-emitters. Reduced concentrations of radionuclides in the river water with increasing distance from the release site were accounted for primarily by the dilution processes in the water flow, sedimentation and radionuclide sorption by bottom sediments.

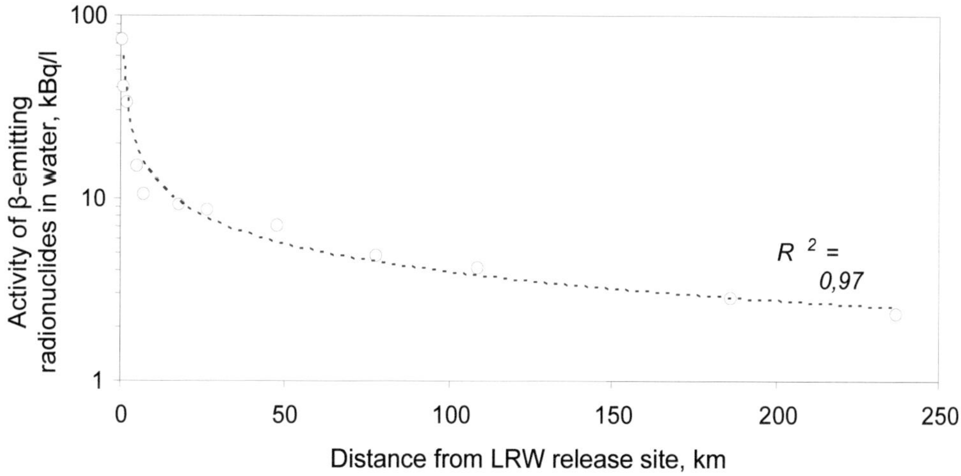


Fig. 3. Changes in volumetric activity of β-emitting radionuclides in Techa River water as a function of distance from LRW release site, 1952 (Marey A.N., 1959).

Nuclide composition of the river water sampled in the middle and lower reaches was for the first time determined in 1951 (Table 1). It was established that a significant proportion of activity of the radionuclides cesium, yttrium, cerium and plutonium is transported by the river stream down the rivercourse on clayey and sandy particles. The same applies to zirconium and niobium. The radionuclides strontium and ruthenium are transferred with river stream mostly in the dissolved state. The basic source of inflow of suspended particles is the surface-slope drainage from the catchment area.

| Village name | β -activity in water, Bq/l | $^{90}Sr$ | $^{137}Cs$ | REE |
|---|---|---|---|---|
| Brodokalmak | $3.1 \cdot 10^4$ | 54.2 | 14.0 | 34.0 |
| Bisserovo | $1.7 \cdot 10^4$ | 50.0 | 32.0 | 19.3 |
| Pershinskoye | $2.1 \cdot 10^4$ | 55.8 | 24.0 | 15.2 |

Table 1. Radiochemical composition of the river water as of 5.08.1951, %

The results of the researches conducted in 1963 showed that small amounts of radionuclides (from 0.001 to 0.014%) (Yu.G. Mokrov, 2002) were carried by the bottom alluvium to the Techa River.

The construction of the Techa cascade of water reservoirs for storing low-level sewage water and re-directing medium-level waste to Karachai Lake resulted in reduced concentrations of radionuclides in river water and bottom sediments. By that time, the radionuclides $^{90}Sr$ and $^{137}Cs$ became the most important contaminants of the Techa River. The long-term dynamics of radionuclide content measured in river water (e.g., at Muslyumovo) up to 1990 was characterized by persistent reduction in $^{90}Sr$ and $^{137}Cs$ concentration. Instability and periodical increases in radionuclide concentrations have been observed in river water (Fig. 4) since 1994. In addition to that, $^{90}Sr$ concentration is stably exceeding the currently permissible level of 4.9 Bq·l$^{1}$. Concentrations of $^{137}Cs$ in river water are less stable along the watercourse (Fig. 5), but the values of the volumetric activity of this radionuclide does not exceed the permissible concentration for drinking water (11 Bq·l$^{-1}$), the role played by $^{137}Cs$ in radiation exposure of the riverside population is not very significant.

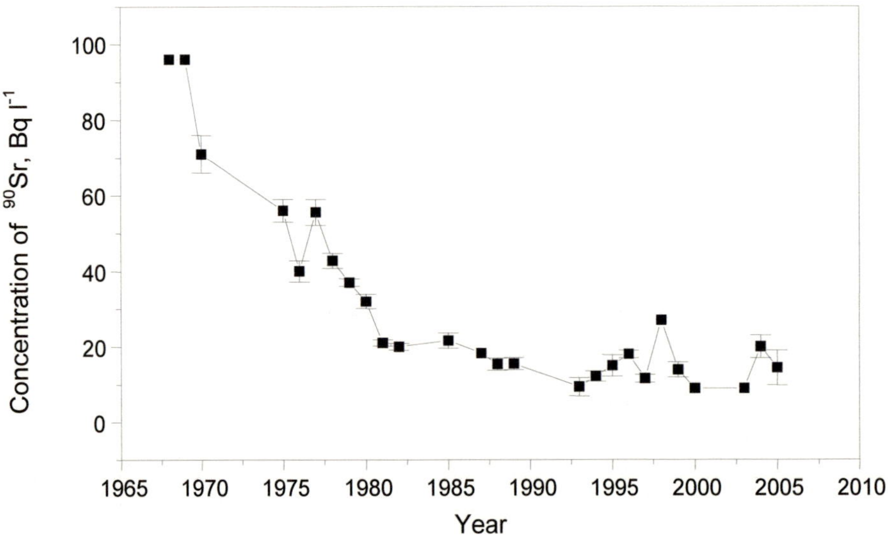


Fig. 4. Specific activity of $^{90}Sr$ in water of the river stretch vis-à-vis Muslyumovo.

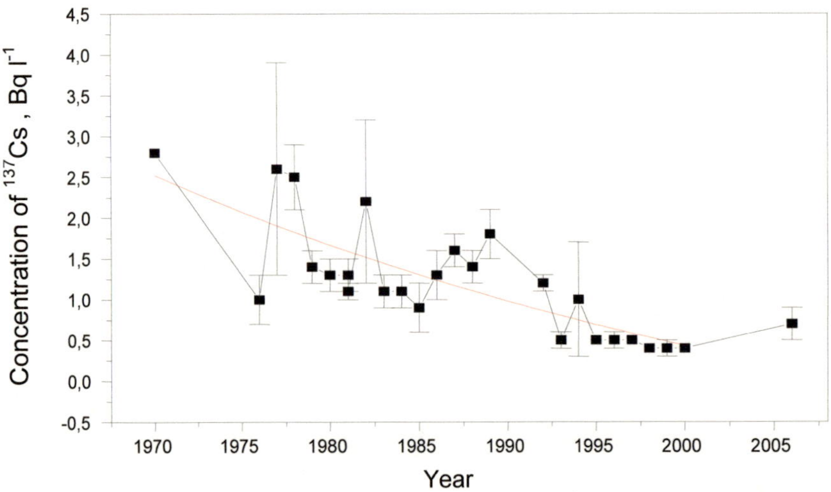


Fig. 5. Specific activity of $^{137}Cs$ in water of the river cross-section opposite Muslyumovo.

Data on contents of plutonium isotopes in Techa River water are scarce. Measurements of specific activity of $^{238,239,240}Pu$ in river water was initiated by the URCRM researchers in 1993

(Table 2). The main source of radioactive contamination of river water is the Techa cascade of reservoirs. Additional contamination is accounted for by desorption of radionuclides from the contaminated floodplain and the river bottom sediments.

In 2009, specific activity of tritium was for the first time determined in water of the Techa river (Fig. 6). Presented in the figure are concentrations of $^{90}Sr$ and tritium in water of the Techa River over its total length down to its confluence with the Isset River measured in samples taken within a week's time in August 2009.

| Sampling site | Distance from dam 11, km | Years | | |
|---|---|---|---|---|
| | | 1994 | 2000 | 2002 |
| Assanov Bridge | 4,5 | 3.2 ± 0.4 (3) | 0.8 (2) | 0.4 (2) |
| Chelyabinsk-Yekaterinburg Bridge | 14 | 3.1 ± 0.8 (3) | 0.46 ± 0.2 (4) | 1.5 (2) |
| Muslyumovo | 40 | 3.1 ± 1 (3) | 0.14 (2) | - |
| Brodokalmak | 71 | - | - | - |
| Russkaya Techa | 95 | 0.3 (2) | - | - |
| Nizhnepetropav-lovskoye | 103 | - | 0.6 ± 0.2 (3) | - |
| Lobanovo | 119 | 1.6 (2) | 0.36±0.2 (3) | - |
| Verkhnyaya | 141 | 2.5 (1) | 1.0 (2) | - |
| Pershinskoye | 170 | 2.7 (1) | - | - |
| Zatechanskoye | 195 | 2.1 (2) | 0.3 ± 0.1 (4) | - |

Note: numbers of samples are given in brackets

Table 2. Specific activity of $^{238, 239, 240}Pu$ in Techa River water in 1994-2002, $Bq \cdot m^{-3}$

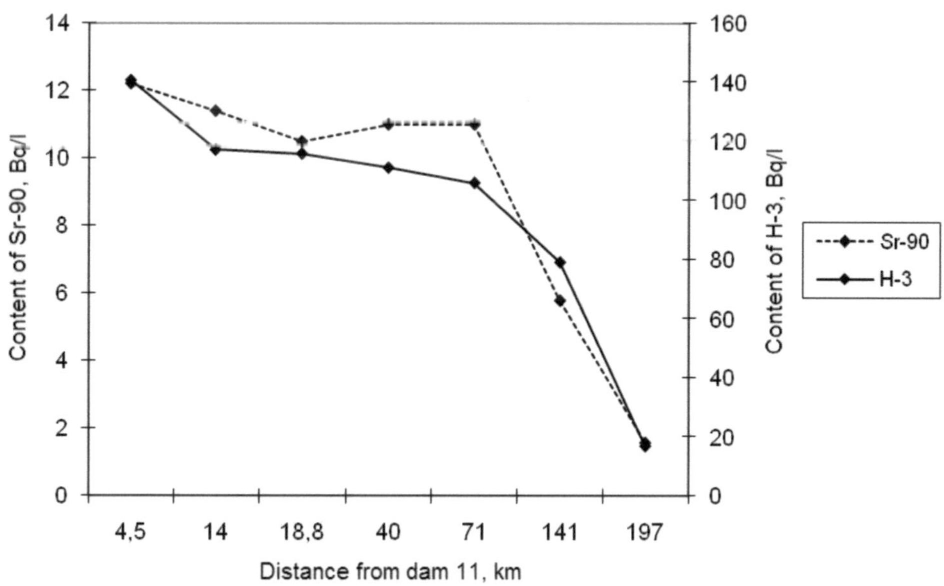


Fig. 6. Content of $^{90}Sr$ and $^{3}H$ in water of the Techa River

Concentrations of $^{90}Sr$ in water measured over the river length has changed from 12 to 1.6 Bq/l, tritium activity ranged from 140 to 17 Bq/l, which is 4-fold higher than the global level. The mean value of $^{3}H/^{90}Sr$ ratio in river water is 10.9±1.2. The total estimated carry-over of $^{90}Sr$ with the Techa river run-off into the Isset River over the period from 1958 through 2001 amounted 2.2×1014 Bq (Yu.A. Izrael, 2000).

| Zone | Volumetric activity (Bq $l^{-1}$) | | | | | | | | |
|---|---|---|---|---|---|---|---|---|---|
| | $^{90}Sr$ | | | $^{3}H$ | | | $^{137}Cs$ | | |
| | Range of values | Mean | Median | Range of values | Mean | Median | Range of values | Mean | Median |
| TCR | 983-5149 | 3007 | 3162 | 763-5935 | 2881 | 3251 | 1.9-914.5 | 102.3 | 32.9 |
| Techa River | 3.8-32.9 | 14.6 | 11.8 | 31-263.2 | 129 | 116 | 0.028-2.4 | 0.39 | 0.17 |
| Drained area | 0.05-173.9 | 27 | 14.1 | 6-15.8 | 10.2 | 9.5 | 0.05-1.51 | 0.55 | 0.49 |
| Tributaries | 0.15-0.18 | 0.15 | 0.15 | 9.5-9.8 | 9.7 | 9.5 | - | - | - |

Table 3. Summarized statistical data on volumetric activities of $^{90}Sr$, $^{3}H$, $^{137}Cs$ in water bodies of different origin

Table 3 shows summarized statistical data on volumetric activities of the radionuclides measured in the Techa River, TCR, surface and ground waters of the catchment area in the upper reaches of the river for the period from 2008 through 2010. In our subsequent calculations, the median values of $^{90}Sr$ and $^{3}H$ were used for TCR water, as well as the $^{3}H/^{90}Sr$ ratio equal to 1, median values of $^{90}Sr$ and $^{3}H$ for Techa River water, $^{3}H/^{90}Sr$ radioactivity ratio equal to 10, rounded median values of volumetric activity of $^{3}H$ equal to 10 Bq $l^{-1}$ in samples of surface and soil water taken in the catchment area of the upper reaches of the river. Concentrations of radionuclides in water of the tributaries corresponded to the background values obtained for contents of radionuclides in surface waterways in the area influenced by the Mayak PA: $^{90}Sr$ - 0.15 Bq $l^{-1}$, $^{3}H$ - 9.7 Bq $l^{-1}$. Concentrations of $^{137}Cs$ were found to be below the detectability level of activity for the volumes sampled. The comparisons of volumetric activity values for $^{90}Sr$ and $^{3}H$ allowed tracing a direct dependence of the concentrations in river as well as in TCR. The only mechanism determining concentrations of $^{90}Sr$ in water is the process of intermix of waters with different initial radionuclide concentrations in different proportions. Also, uranium isotopes ($^{234}U$, $^{238}U$) were used as a radionuclide of reference in relation to $^{90}Sr$ ($^{234}U$, $^{238}U$). Uranium was selected because, unlike $^{3}H$, its isotopes represent ions dissolved in water, as is $^{90}Sr$ (Mokrov Yu.G., 2000); and also because under oxidizing conditions characteristic of surface waters, uranium existing in the form of uranyl-ion ($UO_2^{+2}$) is weakly sorbed by floodplain soils and the river's bottom sediments. The difference between the values of the ratio $^{3}H/^{90}Sr$ for river and TCR waters (10 and 1) is accounted for by dilution of TCR effluent seepage with bypass canal waters in the proportion 1 to 10.

In order to assess the role played by the catchment area of the upper reaches and the river bottom sediments (0-40 km) in contamination of water with $^{90}Sr$, we applied the two-component mixing model ($X_M=X_A \times f + X_B \times (1-f)$, where $X_M$ is the end mixture, $X_A$ and $X_B$ are components, and $f$ is the compound coefficient) using mixture parameters obtained for $^{3}H$. $^{3}H$ and $^{90}Sr$ volumetric activities were measured at cross-sections located at 3.5 km and 40 km from dam 11, respectively. The values of $^{90}Sr$ volumetric activities for waters flowing into the river from the catchment area of the upper reaches range from 4.3 Bq $L^{-1}$ to 19.3 Bq $l^{-1}$, the average value amounting to 9.24 Bq $l^{-1}$. It was concluded based on the calculations that entry of 70% of the total activity of $^{90}Sr$ into the Techa watercourse results from TCR effluent seepage drained through the bypass canal system. The proportion of $^{90}Sr$ activity contributed by washing out of radionuclides from the floodplain and by desorption from bottom sediments accounts for 30%.

## 2.3 Contamination of bottom sediments with $^{90}Sr$ and $^{137}Cs$

According to the data of the first investigations, the highest concentrations of radionuclides in bottom sediments were observed in the reaches close to the release site and in the area of Assanov swamps (30 km from the release site). The lowest level of contamination was registered over the last 40 km stretch down to the outfall. A large amount of activity was accumulated in the surface layer (Table 4) (Glagolenko Yu.G., 1966).

| Distance from the release site, kn | Depth, cm | | |
|---|---|---|---|
| | 0-1 | 5-7 | 10-12 |
| Release site | 103045-486550 | 703-77071 | 262-62678 |
| 7 | 50098 | 8621 | 703 |
| 18 | 47101 | 16539 | 10730 |
| 33 | 6734 | 4625 | 37 |
| 48 | 1961 | 1184 | 59 |
| 78 | 2035 | 292 | 44 |
| 109 | 888 | 540 | 373 |
| 138 | 1221 | 355 | 222 |
| 186 | 255 | 215 | 133 |
| 223 | 307 | 252 | - |

Table 4. Specific β-activity of bottom sediments, August 1952, $kBq \cdot kg^{-1}$ (Glagolenko Yu.G., 1966).

| Distance from the dam 11, km | $^{137}Cs$ | Σ Pu |
|---|---|---|
| 1 | 2 | 3 |
| 1.8 | 27491 | 61 |
| 2.2 | 14874 | 33 |
| 3.8 | 45103 | 133 |
| 5.8 | 23199 | 27 |
| 6.7 | 7104 | 24 |
| 13.0 | 15392 | 57 |
| 14.0 | 17020 | 60 |
| 23.0 | 32190 | 78 |
| 41.0 | 666 | 4 |
| 44.0 | 8325 | 11 |

Table 5. Distribution of contamination density in bottom sediments along the length of the river in 1991-1992, $kBq \cdot m^{-2}$

It was established that radioactivity is best of all accumulated by silt and clay material, while sandy soils manifest a lower rate of accumulation. It was demonstrated that $^{137}Cs$ and plutonium isotopes were intensively sorbed by all varieties of soils. $^{90}Sr$, too, is actively sorbed by soils, but it easily enters into exchange reactions which determines its high migration potential (Saurov M.M., 1968). $^{90}Sr$ and $^{137}Cs$ contamination densities in bottom sediments in the early 1990s are shown in Table 5. During spring floods, silt sediments contaminate the surface of the floodplain which maintains the high levels of contamination of the flooded riverside valley. The levels of silt and water radionuclide contamination are interdependent over the river course. The contents of radionuclides in silt and in water, from dam 11 up to the river outfall are steadily decreasing (Figure 7). Specific activity of $^{137}Cs$ in silts is about 5-fold higher than in water, that of $^{90}Sr$ if 3-fold higher. Compared to $^{90}Sr$, $^{137}Cs$ contamination densities for silts are about 2-fold higher over the total length of the river.

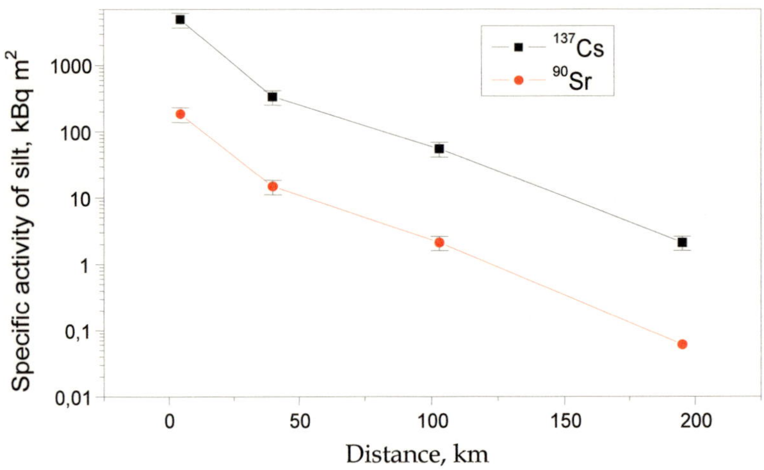


Fig. 7. $^{90}Sr$ and $^{137}Cs$ contamination densities measured in Techa River silts at different distances from dam 11.

Four decades after termination of intensive discharges of radioactive waste into the river, radionuclides deposited in sandy and silty soils migrated to the depth of over 35 cm. The results of the vertical distribution of the radionuclides of interest in the upper reaches of the river are presented in Figures 8 and 9. Compared to $^{137}Cs$, the distribution of $^{90}Sr$ in the bottom soil profiles is more uniform. Maximum values of contamination densities for these radionuclides are in general observed in 0-10 cm layers of soil. $^{137}Cs$ is characterized by a more dramatic decline of contamination density values in lower layers of soil (at the depth of 20-35 cm). In the upper reaches, additional inflow of radionuclides due seepage from TCR and washout of radioactivity from the lands adjacent to the river is observed; in the mid-stream area the contribution of desorption processes from the upper layer is larger.

With distance from the release site, the proportion of exchangeable and mobile forms of $^{90}Sr$ is increasing, on the contrary, the proportion of poorly-accessible forms of $^{90}Sr$ is decreasing.

## 2.4 Contamination of the floodplain with $^{90}$Sr and $^{137}$Cs

Spring overflows of the Techa, and particularly the flood of 1951 contributed to intensive radioactive contamination of the riverside area. The studies of the contents of radioactive substances in floodplain soils started in 1951. The results obtained allowed an insight into the patterns and intensity of the riverside contamination (A.N. Marey, 1959). The width of the floodplain where radioactive contamination was detected did not usually exceed 150-200 m (Table 6). In the upper reaches, in the Assanov swamps area, the overflow reaches 3000 m. A consequent decrease in the levels of floodplain contamination with increasing distance from the release site, and a decrease in the $^{137}$Cs/$^{90}$Sr ratio should also be noted (Table 7).

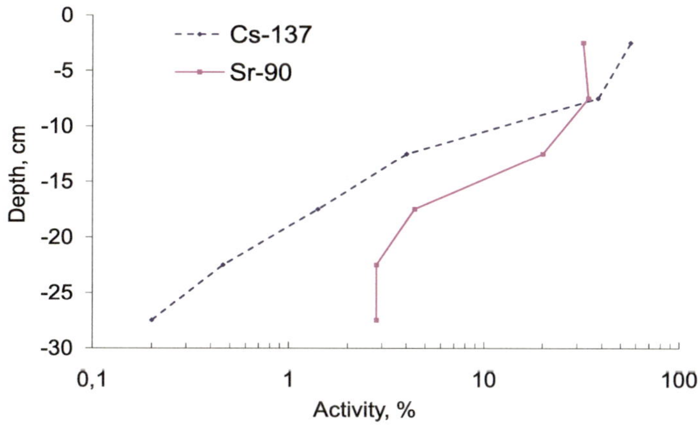


Fig. 8. Distribution of radionuclides along the depth of sandy bottom sediments in the upper reaches of the river.

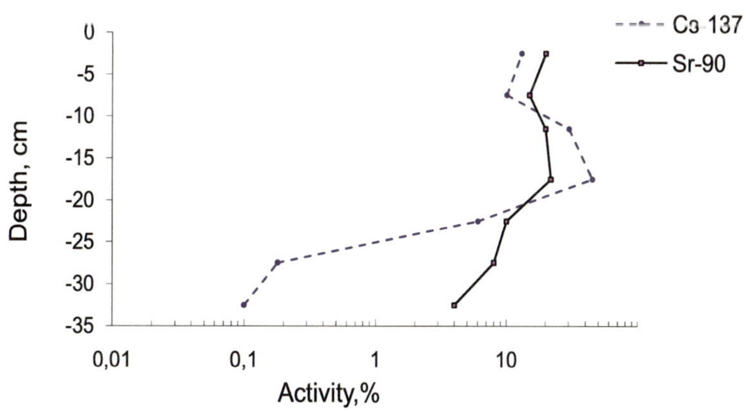


Fig. 9. Distribution of radionuclides along the depth of silty-clayey bottom sediments in the upper reaches of the river.

| Distance from the release site, km | Bank | End of the inundable floodplain |
|---|---|---|
| Release site | 7548 | 141 |
| 18 | 2442 | 18 |
| 33 | 407 | 22 |
| 48 | 74 | 15 |
| 78 | 481 | 4 |
| 105 | 200 | 60 |
| 109 | 63 | 33 |
| 128 | 33 | 4 |
| 138 | 4 | 23 |
| 212 | 3 | 22 |

Table 6. Specific β-activity in the surface layer in 1952 - 1954, $kBq{\cdot}kg^{-1}$ (Marey A.N., 1959)

Changes in the width of the riverside area is, as a rule, determined by soil relief; maximum concentrations of radionuclides are registered in lowland areas.

| Distance from dam11, km | $^{137}Cs$ | $^{90}Sr$ | $^{137}Cs$ / $^{90}Sr$ |
|---|---|---|---|
| 67 | 210 | 150 | 1.4 |
| 102 | 95 | 60 | 1.5 |
| 135 | 80 | 80 | 1.0 |
| 152 | 35 | 100 | 0.3 |
| 237 | 5 | 42 | 0.1 |

Table 7. Changes in contents of $^{137}Cs$ and $^{90}Sr$ in floodplain soils along the river course, $Ci{\cdot}km^{-2}$ (Marey A.N., 1959)

The first maximum level of floodplain contamination density is observed in the area of Assanov swamps: from dam 11 to the distance of 7.5 km. The second peak takes place in lower reaches, at the distance of 37 km from the dam in the area of Muslyumovo swamps. In these areas, the incessant winter run-off of radioactive substances with TCR waters is accumulated. Most thoroughly the floodplain soils are washed during high waters. Surface-downslope waters wash upper layers of the soil, flood waters wash upper layers of the flooded river bank, and ground waters wash deeper layers of the floodplain soils.

With distance along the watercourse, the levels of floodplain contamination with $^{137}Cs$ are appreciably decreasing. In the swamps of the upper reaches, the values of contamination with the radionuclides of interest amount to 150-550 $Ci{\cdot}km^{-2}$, the values for the middle reaches are 20-30 Ci $km^{-2}$, and the respective value for the area close to the estuary is 5 $Ci{\cdot}km^{-2}$. The dynamics of reduction in $^{90}Sr$ contamination density values assumes a more monotonous character. The results of measurements of $^{137}Cs$ and $^{90}Sr$ contamination densities conducted by us in the floodplain in 2005 are presented in figure 10. It can be seen that the level of floodplain contamination with $^{137}Cs$ is higher than the respective value obtained for $^{90}Sr$, with the exclusion of the last locality (village Zatechanskoye).

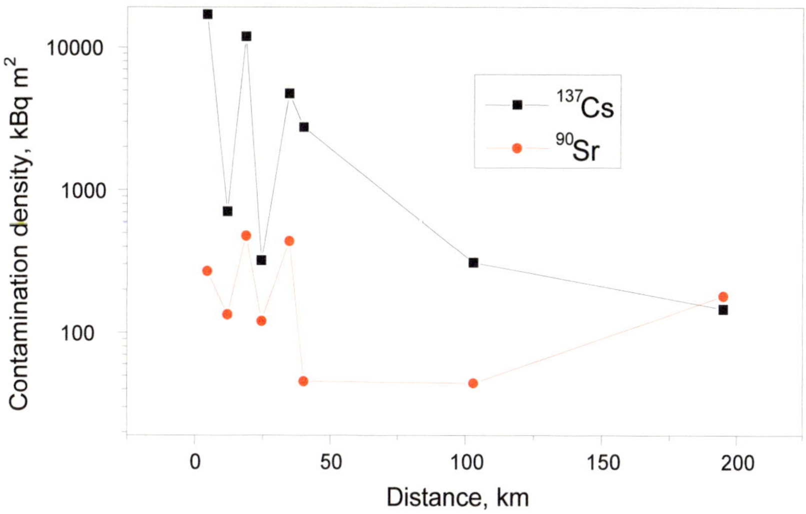


Fig. 10. $^{137}Cs$ and $^{90}Sr$ contamination densities in the floodplain along the watercourse, kBq m$^{-2}$

The study of $^{137}Cs$ distribution by layers of the floodplain along the watercourse (Table 8) has demonstrated that the upper 20 cm layer is contaminated to the highest extent. However, the measurements conducted in the village Zatechenskoye where sandy soils predominate, the concentrations of radionuclides in the lower and surface layers are actually the same. As has been shown by measurements at most of the sampling sites, the upper 0-5 cm soil layer contains a smaller amount of $^{137}Cs$ as compared to the underlying layer. This fact indicates that there has been no intensive contamination of the floodplain with $^{137}Cs$ in the recent years. An extreme non-uniformity in the distribution of $^{90}Sr$ by soil profiles (Table 9) in different reaches of the river can be accounted for by different physical-chemical and morphological properties of the floodplain soils.

| Sampling site | Distance from dam 11, km | Position of layers, cm | | | |
|---|---|---|---|---|---|
| | | 0-5 см | 5-20 см | 20-40 см | 40-60 см |
| Assanov bridge | 4.5 | 1438.7 | 1011.0 | 1595.3 | 1059 |
| Maloye Taskino | 12 | 230.7 | 61.1 | 7.9 | 6.9 |
| Nadyrov bridge | 18.8 | 593.2 | 10084.4 | 1281.3 | 439.0 |
| 6 km below Nadyrov bridge | 24.5 | 129.6 | 28.4 | 53.7 | 44.9 |
| Buslyumovo swamp | 34.7 | 660 | 62.1 | 784.3 | 354 |
| Muslyumovo | 40 | 277.1 | 95.4 | 17.7 | 10.9 |
| Nizhne-Petropavlovskoye | 103 | 8.1 | 65.4 | 19.3 | 6.4 |
| Zatechenskoye | 195 | 11.3 | 99.8 | 31 | 101.7 |

Table 8. Distribution of $^{137}Cs$ by floodplain soil layers over the length of the river, kBq m$^{-2}$

Most of the floodplain soils (Assanov Bridge, Nadyrov Bridge) are characterized by depletion of the upper 0-10 cm layer, presence of maximum $^{90}Sr$ concentrations at the depth of 10-30 cm with a dramatic drop in concentrations measured deeper along the profile. This is caused by washing of the upper layer with the surface waters. In the locality of Muslyumovo, the distribution of $^{90}Sr$ by soil layers is influenced by the continuous run-off from the overlying swampy layers. Further along the watercourse, the processes of cleaning radionuclides from the upper soil layers and a more uniform distribution along the depth of the soil become prevalent. (Nizhnepetropavlovskoye and Zatechenskoye).

| Sampling site | Distance from dam 11, km | Position of layers, cm | | | |
|---|---|---|---|---|---|
| | | 0-5 см | 5-20 см | 20-40 см | 40-60 см |
| Assanov bridge | 4,5 | 20 | 78,5 | 112,6 | 49,4 |
| Maloye Taskino | 12 | 51,8 | 143,8 | 12,6 | 2,1 |
| Nadyrov bridge | 18,8 | 27,2 | 394,1 | 136,9 | 37,6 |
| 6 km below Nadyrov bridge | 24,5 | 46,2 | 39,2 | 26,6 | 8,3 |
| Buslyumovo swamp | 34,7 | 27,7 | 72,1 | 233,9 | 103,7 |
| Muslyumovo | 40 | 10,7 | 33,8 | 6,5 | 2,1 |
| Nizhne-Petropavlovskoye | 103 | 2,5 | 16,4 | 3,6 | 0,74 |
| Zatechenskoye | 195 | 26,3 | 239,8 | 321,1 | 24,5 |

Table 9. Distribution of $^{90}Sr$ by floodplain layers along the watercourse, $kBq \cdot m^{-2}$

From 1991 through 2005, the total content (reserve) of $^{137}Cs$ at the sampling site (Fig. 11) decreased from 574.5 $MBq \cdot m^{-2}$ to 52.8 $MBq \cdot m^{-2}$, accumulating mainly at the depth of 30 cm, and lower, at 40 cm above the aquifer. The total content of $^{90}Sr$ in soil samples taken in the floodplain (Fig. 12) decreased from 14.4 to 8.8 $kBq \cdot m^{-2}$. Over this period, equalization of the contamination was noted at the depth of 55 cm, while in the lower layer an increase in contamination level was observed, with radionuclides moving all the way to the aquiferous layer, obviously due to intensive seasonal washouts of the bog soils.

In 2009, samples were taken in the floodplain area close to Assanov swamps at the depth of 80 cm within 30, 70, 106 and 250 m of the shoreline. Below, an aquifer composed of blue clay was situated. It was shown that $^{137}Cs$ and $^{90}Sr$ contamination densities in the floodplain soil (figure 13) were declining depending on distance from the shoreline. The reserve of $^{137}Cs$ exceeds that of $^{90}Sr$ at any distance, and their ratio changed at different distances from 22-2 times. The distance from 30 to 70 m is characterized by the lowest levels of contamination due to wash-off by ground waters.

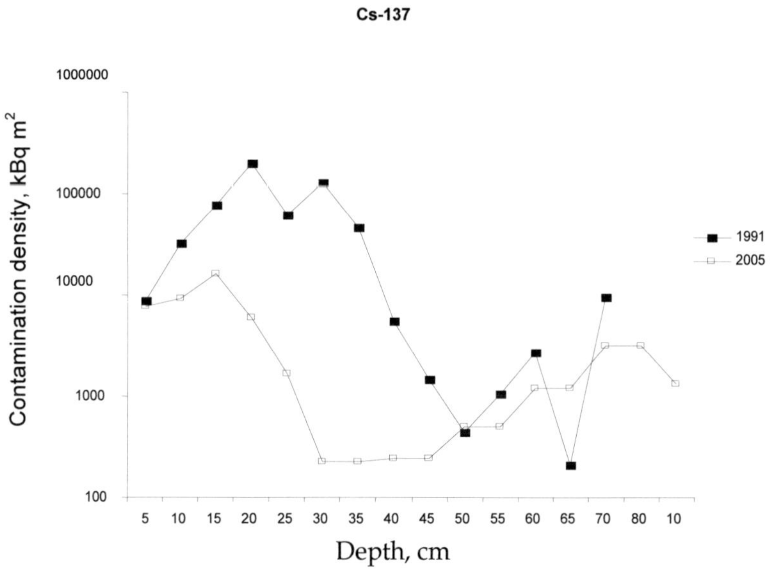


Fig. 11. Contamination densities for $^{137}Cs$ in the Techa floodplain area at sampling site "Assanov swamps" within 6.5 km of dam 11, 10 m from the shoreline.

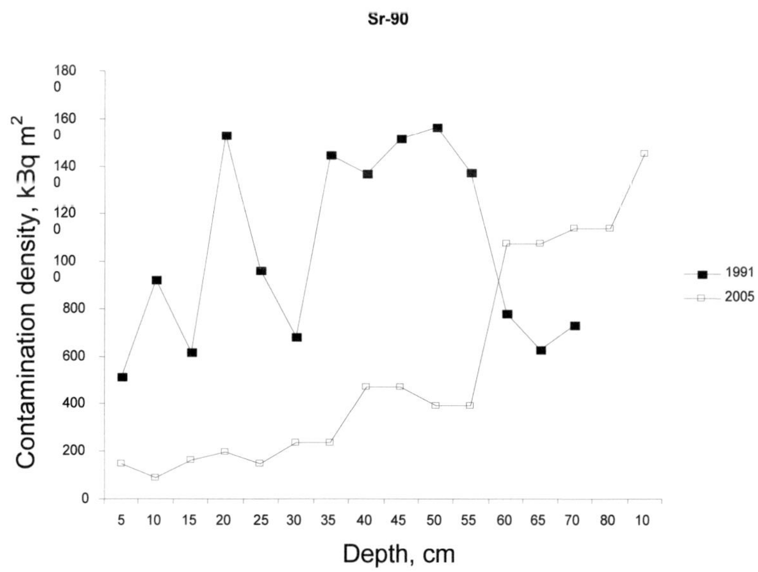


Fig. 12. $^{90}Sr$ contamination densities in the Techa River floodplain at 6.5 km from dam 11, left bank, within 10 of the shoreline.

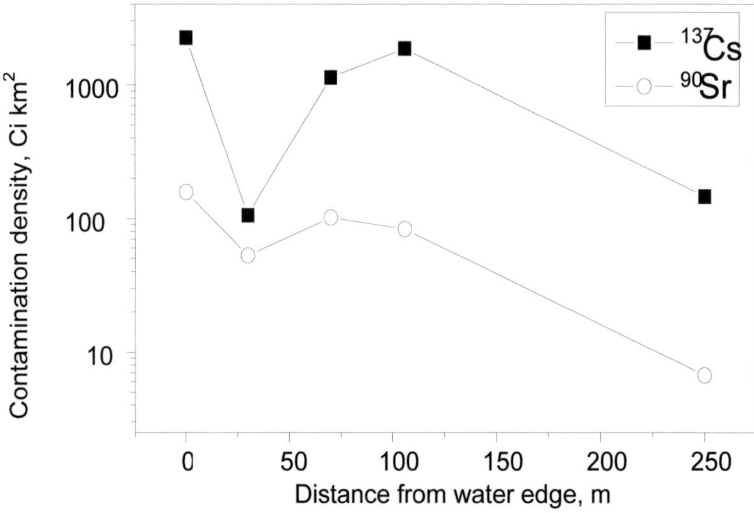


Fig. 13. $^{137}Cs$ and $^{90}Sr$ contamination densities for soils of Assanov swamps, Ci km$^{-2}$

In samples of floodplain taken in the area of Assanov swamps at different depths and sites, prevalent are water-soluble forms of radionuclides deposited in deep layers (70-80 cm) of riverside soil (up to 26%) in localities where they migrate in the direction of the watercourse. A slight increase in exchangeable and mobile forms of $^{137}Cs$ in the lower layer was observed.

In 2005 and 2006, samples of ground and surface waters were taken in the area of Assanov and Muslyumovo swamps with the aim to determine the actual scope of influence exerted by deposited radionuclides on concentrations of radioactivity in swamp water. During cold seasons, when migration processes are at their minimum, concentration of $^{90}Sr$ in flow channels is high, reaching 187 Bq l$^{-1}$, the concentration level registered in ground water is 14-19 Bq l$^{-1}$. In spring, in the portion of the swamp area, where there is no distinctly outlined water flow, $^{90}Sr$ concentrations in water amount to 80-100 Bq l$^{-1}$ in localities far removed from the river banks. In flow channels with a sufficiently dynamic water flow, radionuclide concentrations reach about 40 Bq l$^{-1}$. In Muslyumovo swamps, water sampled in dead channels contains 11-25 Bq l$^{-1}$ of $^{90}Sr$ with radionuclide concentration level in river water of 6.2 Bq l$^{-1}$. The data obtained point out to the fact that it is mainly $^{90}Sr$ which is leached from the floodplain to water, the same applies to $^{137}Cs$, but to lesser extent.

Thus, it was demonstrated that the floodplain and bottom deposits are the key source of secondary contamination of the river water with radionuclides. The level of radioactive contamination of these river components is, in its turn, determined by the radioactive run-offs from the Techa cascade of reservoirs.

## 3. East-Urals Radioactive Trace

### 3.1 Radioactive contamination of soil, vegetation, food products with $^{90}Sr$ and $^{137}Cs$ in the early years after the 1957 accident

A distinguishing feature of the radioactive emission on the EURT is the presence in it of all basic uranium fission products and a minimum content of the $^{137}Cs$ (Table 10).

| Radionuclide | % | Scope of release, PBq |
|---|---|---|
| $^{89}$Sr | traces | - |
| $^{90}$Sr+$^{90}$Y | 5.4 | 2.0 |
| $^{95}$Zr+$^{95}$Nb | 24.8 | 18.4 |
| $^{106}$Ru+$^{106}$Rh | 3.7 | 2.7 |
| $^{137}$Cs | 0.36 | 0.26 |
| $^{144}$Ce+$^{144}$Pr | 66.0 | 48.7 |
| $^{147}$Pm, $^{155}$Eu | traces | - |
| $^{239}$Pu | traces | 0,0014 |

Table 10. Characterization of the radioactive releases and the initial reserve of radionuclides on the EURT outside the Mayak PA industrial site (Avramenko V.I. et al., 1977)

The settling of the radioactive mixture from the cloud which was wind-drawn in the north-eastern direction from the explosion site resulted in the formation of the East-Urals Radioactive Trace (EURT). The Trace encompassed parts of Chelyabinsk, Sverdlovsk and Tyumen oblasts. The length of the EURT is 300 km, its width is 30-50 km. According to refined data, (Korsakov Yu.D. et al., 1996), in 1957 the area with contamination density (η) for $^{90}$Sr >2 Ci km$^{-2}$ was 560 km$^2$, that with η>12 Ci km $^{-2}$ was 230 km$^2$, with η>50 Ci·km$^{-2}$ – 120 km$^2$, with η>200 Ci·km$^{-2}$ –50 km$^2$, with η>800 Ci·km$^{-2}$ – 16 km$^2$, and with η> 2 000 Ci·km$^{-2}$ – 8 km$^2$. There were over 200 populated localities in the Trace area, including several towns and industrial communities. At early time after the accident, the population of the EURT area was exposed to radiation-related hazards, including, in the first place, external exposure to γ-radiation due to prevalence of γ-emitting nuclides in the deposited mixture and, in the second place, internal exposure resulting from intakes of radionuclides contained in food products produced in the localities. As the activity of γ-emitting nuclides (which decayed almost completely 6-7 years after the accident) decreased, the radiation hazards were mostly determined by the radionuclide $^{90}$Sr.

The first measurements conducted on the contaminated territory showed that the γ-radiation dose rate was proportional to the distance from the accident site. According to measurements conducted during the first year after the accident in the area with contamination density amounting to 1 Ci km$^{-2}$ for$^{90}$Sr, the dose in air due to γ-radiation was 1R (G.N. Romanov, 1963). Direct exposure to β-radiation only occurred in areas with contamination density of over 1 500 Ci·km$^{-2}$ for $^{90}$Sr. According to data presented in (A.Ya. Kogotkov, 1968), a decrease in relative contents of $^{90}$Sr in the composition of the mixture depending on distance from the accident site, and, respectively, an increase in the contents of $^{144}$Ce and $^{137}$Cs. Soil contamination densities in some localities may differ by an order of magnitude, or greater.

The coniferous woods were affected most heavily. At the distance of 12.5 km from the contamination source, a total loss of pine woods was registered in the summer of 1958. Mass loss of birch forests were only observed in areas with contamination densities of over 4 000 Ci·km$^{-2}$. Migratory birds were only affected in the spring of 1958, after dose rates in the tree crowns had decreased 10-fold. A reduced number of bird's nests was noted in the areas with contamination level of over 2 000 Ci·km$^{-2}$ for $^{90}$Sr. No loss of animals was registered. In the ensuing years, due to the fact that the contaminated area was made into a sanitary-protection zone, the number of hares, roes and elks increased considerably.

During the first days after the accident, radioactive contamination of grass estimated relative to 1 $Ci/km^2$ of soil contamination was $1.5{\cdot}10^6$ decay $min^{-1}{\cdot}kg^{-1}$. Radioactive contamination of individual food products was very high (Table 11). Since cattle and other domestic animals were fed contaminated forage, contamination of milk and meat was of structural rather than superficial nature. It should be noted that contamination of milk was registered as early as the first 2-3 days after the accident, and that of meat on days 10-12 (R.M. Alexakhin et al., 2001).

During the first days after the accident, radioactive contamination of grass estimated relative to 1 $Ci/km^2$ of soil contamination was $1.5{\cdot}10^6$ decay $min^{-1}{\cdot}kg^{-1}$. Radioactive contamination of individual food products was very high (Table 11). Since cattle and other domestic animals were fed contaminated forage, contamination of milk and meat was of structural rather than superficial nature. It should be noted that contamination of milk was registered as early as in the first 2-3 days after the accident, and that of meat on days 10-12 (R.M. Alexakhin et al., 2001).

| Locality | Soil | Water from water sources | Grass | Milk | Meat | |
|---|---|---|---|---|---|---|
| | | | | | muscles | bones |
| 1 | 2 | 3 | 4 | 5 | 6 | 7 |
| Berdyanish | 21016 | 52 | 360380 | 115 | 58 | 92 |
| Saltykova | 33760 | 104 | 97162 | 218 | - | - |
| Galikayeva | 2405 | 104 | 34262 | - | 6 | 74 |
| Kasli | 18 | - | 32 | - | - | - |
| Russ. Karabolka | 4810 | 42 | 22200 | - | 6 | 137 |
| Yugo-Konevo | 192 | 0.02 | 15207 | - | 1.2 | 10 |
| Yushkovo | 3 | - | - | 3.1 | 3 | 71 |
| Boyevskoye | 266 | - | 755 | - | - | - |
| Bagaryak | 74 | - | 8399 | 4.4 | - | - |
| Kamensk-Uralsky | - | 0.004 | - | - | 2.8 | 1.5 |
| Pozarikha | - | - | 440 | - | - | - |

Note: the results include $^{40}K$ activity

Table 11. β-activity in food products and environmental entities in the EURT area as of 20.10.1957, kBq/kg, l

Food products were the main contributor to radiation exposure of the population. The main cause of the reduction in contents of radionuclides in food products over time is the reduction of radionuclide contents in soils. Table 12 shows the basic composition of radionuclides observed in food products and the dynamics of cleaning the food from radionuclides. Among those radionuclides, $^{144}Ce$+$^{144}Pr$ prevailed during the first 3 years, later on $^{90}Sr$ gained priority.

Beginning from the spring-summer season in 1958, an additional contamination of vegetation and agricultural plants occurred due to a downwind migration of radionuclides from areas with a higher level of contamination density. The proportion of surface contamination of grain crops accounted for 10-15% (P.P. Lyarsky, 1962).

Contamination levels measured for food products of the first post-accident harvest reaped in the fields situated in the Trace zone with contamination densities of 0.2-0.5 $Ci \cdot km^{-2}$ for $^{90}Sr$ or higher, were higher than the permissible limit legally valid at that time: 1300 decay $min^{-1} \cdot kg^{-1}$ (22 $Bq \cdot kg^{-1}$).

| Time after the accident, years | Key radionuclides, % | | | | | Other |
|---|---|---|---|---|---|---|
| | $^{90}Sr+^{90}Y$ | $^{137}Cs$ | $^{144}Ce+^{144}Pr$ | $^{95}Zr+^{95}Nb$ | $^{106}Ru+^{106}Rh$ | |
| 1 (1957) | 10.1 | 1.5 | 76.9 | 10.1 | 1.3 | <1% |
| 2 (1958) | 12.9 | 2.8 | 82.4 | 0.2 | 1.5 | |
| 3 (1959) | 7.0 | 8.0 | 80.6 | - | 4.2 | |
| 5 (1961) | 90.3 | 2.7 | 6.2 | - | 0.6 | |
| 10 (1966) | 98.8 | 1.2 | - | - | - | |

Table 12. Composition of radionuclide observed in food products at different time after the accident

### 3.2 Current levels of soil contamination

The studies conducted in 2006-2009 in the villages Allaki, Bagaryak, Bulzi, Tartar Karabolka and Yushkovo located around the perimeter of the EURT of sanitary-protection zone. The soils of the pasture lands adjacent to these villages are mainly dark-grey clay and leached chernozem.

The mean dose-rate value of gamma-radiation in the localities of Karabolka, Musakayeva and Bagaryak is 0.12 $\mu Sv \cdot hr^{-1}$ which is comparable to natural background values for Chelyabinsk oblast. The mean pasture land contamination density for $Sr^{90}$ in the localities of Tatarskaya Karabolka and Musakayeva is 5.9 $kBq \cdot m^{-2}$, that for Bagaryak is 2-2.5 times higher. Pasture land contamination density for $^{137}C$ is 12.9-24.8 $kBq \cdot m^{-2}$. Mean contamination densities measured in soils of kitchen gardens attached to houses in Bagaryak was 77.5 $kBq \cdot m^{-2}$. Mean specific activity in grass measured in different areas along the EURT axis ranged from 11 to 15 thousand $Bq \cdot kg^{-1}$ for $Sr^{90}$ and from 3 to 60 $Bq \cdot kg^{-1}$ for $^{137}C$. Forms of $Sr^{90}$ and $^{137}C$ encountered in soils sampled in the frontal portion of the Trace (close to Alabuga Lake) have been identified (Table 13). Attention is drawn to the increased number of exchangeable forms of $Sr^{90}$ and mobile forms of $^{137}C$. During the early years after the fallout, the total amount of $Sr^{90}$ was only encountered in soluble state. A decade later, the proportion of exchangeable forms accounted for 65-75%, and the value has not changed since then.

According to data presented by V.V. Martyushov et al. (1996) 36 years after the contamination, the content of exchangeable forms of $^{137}C$ and plutonium did not exceed 3% and 1%, respectively. The proportion of poorly accessible forms of $^{137}C$ and plutonium reached 95-98%. The content of water-soluble forms of $^{137}C$ and plutonium accounts for less than 1%. Water-soluble forms of $Sr^{90}$ are mostly found in cationic compounds (72-76%).

| Radionuclide | Physical-chemical forms of radionuclides , % | | | |
|---|---|---|---|---|
| | Water-soluble | Soluble in 1H $CH_3COONH_4$ (exchangeable) | Soluble in 1H HCl (mobile) | Solid residue (poorly accessible) |
| $^{90}Sr$ | 2.8 ± 0.5 | 73.2 ± 0.4 | 16.3 ± 0.6 | 7.7 ± 0.4 |
| $^{137}Cs$ | 2.7 ± 0.4 | 3.1 ± 1.0 | 19.4 ± 1.5 | 74.8 ± 2.5 |

Table 13. Forms of radionuclides identified in the upper 5-cm layer of soil in the vicinity of Alabuga Lake

In 2009, soil samples were taken at 16 sampling sites at a distance of 20 km from the Mayak PA perpendicular to the EURT axis. Five of those sites are situated in a birch forest. The total contamination density in the soils of the rhizogenic layer and forest litter was, on the average, equal to 737 kBq/m² for $^{90}Sr$, and 41.2 kBqк/m² for $^{137}Cs$. It should be noted that 10.5% of the contamination density for $^{90}Sr$ and 14.1% of contamination density for $^{137}Cs$ are contributed by the debris layer. At 10 points situated in the hayfield extending from the forest up to Alabuga Lake $^{90}Sr$ contamination density ranged from 161 to 350 kBq/m², $^{137}Cs$ contamination density measured in the hayfield ranged from 34 to 93 kBq/m². Contamination densities measured in the wood were higher than those measured in the hayfield.

At a distance of 30 km from the Mayak PA in an area along the Trace axis samples of soils were taken at a depth of 10 cm. $^{90}Sr$ contamination density was found to range from 2.2 kBq/m² to 55.9 kBq/m², that for $^{137}Cs$ ranged from 2.2 to 50.7 kBq/m². At the periphery of the Trace, contamination density for $^{137}Cs$ exceeded that for $^{90}Sr$ 2-5-fold. The contribution made by the debris layer varies significantly. In afforested and steppified areas with well developed steppe debris layer, the contribution of the debris layer to the contamination density for $^{90}Sr$ reaches 37%, for $^{137}Cs$ – 8.8%.

At a distance of about 40 km from the Mayk PA at a right angle to the Trace axis, $^{90}Sr$ and $^{137}Cs$ contamination density distribution was determined for the ploughed layer (0-20 cm). As of the date of the measurements, there were only arable lands that had no sod cover were available for measurements. The highest measured contamination density was 162 kBq/m² for $^{90}Sr$ and 24.4 kBq/m² for $^{137}Cs$

Samples of sod-podzol soil were taken on the Trace axis close to Bolshoi Irtysh Lake (about 55 km from the Mayak PA). Contamination densities in the upper 0-20 cm layer and in the ground litter was 308 kBq/m² for $^{90}Sr$, and 20 Bq/m² for $^{137}Cs$. It is noteworthy that 11.2% of $^{90}Sr$ and 18.0% of $^{137}Cs$ are contributed by the ground litter. The total contamination densities in soil (0-20 cm) and ground litter along the Trace axis are presented in Figure 14.

The studies performed have shown that soil contamination densities close to the EURT axis are still high, even at the present time; besides, in a number of cases a considerable portion of radionuclides is contained in forest debris layer or steppe litter. Contamination levels mostly depend on the distance from the contamination source, i.e. from the Mayak PA. The highest levels of radioactive contamination were identified in the 0-5 cm layer of soil. Below the root-inhabited layer, 2.7%-57.2% of $^{90}$Sr and 28.4%-41.1% of $^{137}$Cs are deposited. In all types of soil, $^{90}$Sr is encountered in biologically accessible forms, and, $^{137}$Cs is contained in poorly-soluble compounds.

### 3.3 Vertical migration of $^{90}$Sr and $^{137}$Cs through soil profiles

To allow assessment of soil contamination levels, two main parameters are used, viz., specific activity of the radionuclide in soil (Bq·kg$^{-1}$) and contamination density (Bq·m$^{-2}$) which takes into account the total contamination density in all *n* soil layers. There is no direct relationship between these two parameters. It was established that over the time period since the 1957 accident the radionuclides had migrated to a significant depth. Actually, samples taken at all the sampling sites showed that the highest level of contamination with $^{137}$Cs and $^{90}$Sr was detected in the upper level of soil and in debris layer (in meadow soils it was found in sod cover and in thick felt of the steppe). The ratio of radionuclide specific activity in the 0-10 cm layer to activity in the 10-20 cm layer did not depend on the summarized contamination density. The value of this ratio is mostly influenced by the type of the ecosystem: the mean value of this ratio for $^{137}$Cs in forest ecosystems is 20.4±4.4, and in meadows it is 2.9±1.6. For $^{90}$Sr, the differences are insignificant: 3.4±0.9 in forests, and 2.8±1.5 in meadows. The ratio of specific activity of $^{137}$Cs in debris cover to activity in the 0-10 cm soil layer amounted on an average to 0.5±0.1, and for $^{90}$Sr to 1.5±0.2. Although the specific activity of $^{137}$Cs and $^{90}$Sr in the litter layer is sufficiently high, it does not exert substantial influence on the summarized contamination density since the volume weight of the litter layer is by two orders of magnitude lower than the volume weight of soil.

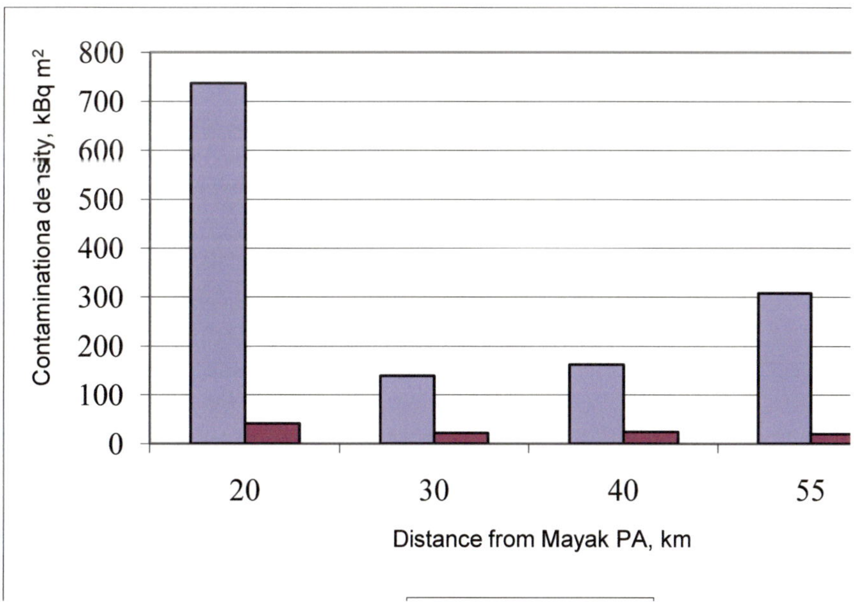


Fig. 14. Contamination density of soil (0-20 cm) and litter layer along the EURT axis as a function of distance from the source of contamination.

Distribution of $^{90}Sr$ and $^{137}Cs$ in soil profiles was studied in 2008-2010 for 3 types of soils: gray forest, sod-podzol and chernozem. Gray forest soils are most prevalent in the forest-steppe zone. They are formed in leaf woods and mixed woods of the Trans-Urals region. Distribution of $^{90}Sr$ and $^{137}Cs$ in gray forest soil profiles was determined at a sampling site within 20 km of the Mayak PA. The highest specific activity of $^{90}Sr$ ($1.6\times10^4$ Bq·kg$^{-1}$) and $^{137}Cs$ ($1.3\times10^3$ Bq·kg$^{-1}$) was measured in the lower layer of the forest litter which contains a half-decayed tree waste. $^{90}Sr$ activity in the upper layer of the forest litter is actually equal to activity found in the 0-5 cm soil layer, $^{137}Cs$ activity is 2.7 times lower. However, the volume weight of the forest litter is low that is why contamination density summarized for its upper and lower layers is lower than that in the underlying 0-5 cm soil layer where the main amounts of $^{90}Sr$ (60%) and $^{137}Cs$ (50%) are deposited. The tree waste of 2008 sampled on October 31 demonstrated an increase in contamination density of the tree waste by 2.4 kBq·kg-m$^{-2}$ for $^{90}Sr$ which accounts for 2.1% of the tree waste contamination measured in spring of 2008. Over the 50 years since the accident, no shift of the maximum values with increasing depth through the soil profiles has taken place. At the same time, both $^{90}Sr$, and $^{137}Cs$, though in small amounts, have at least, reached the depth of 170-175 cm. It should be added that $^{137}Cs$ which is considered to be less capable for vertical migration, is evidently migrating more actively than $^{90}Sr$, 2.7% of $^{90}Sr$, and 28.4% of $^{137}Cs$ are deposited at a depth of over 20 cm.

Sod-podzol soils which most often are found in the northern part of the Trace, are formed under coniferous woods. The distribution of $^{90}Sr$ and $^{137}Cs$ along the profile of sod-podzol soils is comparable to the distribution observed for gray forest soils. The highest specific activity of $^{90}Sr$ ($1.0\times10^4$ Bq·kg$^{-1}$) and $^{137}Cs$ ($1.2\times10^3$ Bq·kg$^{-1}$) was found in the lower part of the forest waste, the highest contamination density for $^{90}Sr$ (39.6%) and $^{137}Cs$ (40.4%) was measured in the upper 0-5 cm soil layer. The forest waste and soil layer up to 20 cm in depth contain 96.2% of $^{90}Sr$ and 58.9% of $^{137}Cs$. At the depth of over 20 cm 3.7% of $^{90}Sr$ and 41.1% of $^{137}Cs$ are deposited below 20 cm. As can be seen from the comparison of the distributions, the mobility of $^{90}Sr$ in the sod-podzol soils differs insignificantly from that registered in gray forest soils, the mobility manifested by $^{137}Cs$ is significantly higher. Already at the depth of 25-30 cm, the specific activities of $^{90}Sr$ and $^{137}Cs$ actually differ one from the other, however, in lower layers $^{137}Cs$ takes the first position.

There occur in steppefied areas of the EURT weak northern or leached chernozems which have been ploughed up for a long time period. It should have been expected, therefore, that the distribution of radionuclides through the 0-20 cm plough-layer would be more uniform and speedy. The site for taking soil samples from the chernozem profile is located close to the EURT axis, however, the contamination density for $^{90}Sr$ turned out to be low, viz., 17.8 kBq·m$^{-2}$ for $^{90}Sr$ and 40.2 kBq·m$^{-2}$ for $^{137}Cs$ over the whole profile. $^{137}Cs$ was found to be significantly prevalent in each layer.

The highest specific activity for $^{90}Sr$ (92.2 Bq·kg$^{-1}$) and $^{137}Cs$ (161 Bq·kg$^{-1}$) was observed in the debris cover which was characteristic of other soil types too. The highest contamination density was measured in the 0-5 cm layer for $^{90}Sr$ (10.4%) and in the 5-10 cm layer (10.9%), and for $^{137}Cs$ in the 0-5 layer (21.4%). At the depth of over 20 cm, 57.2% of $^{90}Sr$, and 37. 4% of $^{137}Cs$ were deposited.

The analysis of $^{90}Sr$ distribution patterns in the EURT soils were made using data of researches conducted in 1963-2008 (figures 15-17). It can be seen that the distribution of $^{90}Sr$ in the profile of the 30 cm layer is well described by the following exponential function: $y=ae^{-bx}$, where $y$ is the content of radionuclides calculated as percentage of the total contamination density in the 30-cm layer, $x$ is reference number of sample taken in the 5-cm soil layer. Using the coefficient $b$ it becomes possible to calculate the depth at which a decrease in contamination density to a preset level takes place. $^{90}Sr$ which settled on the soil surface is slowly migrating to deeper layers, and the coefficient $b$ is decreasing (Table 14).

In gray forest soils, the value of coefficient $b$ correlates with the number of years that have passed since the accident ($r = -0.94$, $p = 0.02$). In sod-podzol soils, especially intensive $^{90}Sr$ migration is going on.

So far, no shift in maximum $^{90}Sr$ and $^{137}Cs$ activities down the soil profile has been observed in any soil types of interest (gray forest, sod-podzol, chernozem). The highest specific activity of both $^{90}Sr$, and $^{137}Cs$ in natural and fallow lands is retained in the lower layers of forest litter and or steppe debris. High specific activity of $^{90}Sr$ and $^{137}Cs$ is also retained in 0-5 soil layer. In deeper soil layers, the activity of these radionuclides is decreasing rapidly and reaches the minimal values in the 25-40 cm layers of eluvial horizons. In illuvial horizons, radionuclide activity is slightly decreased. In general, the specific activity of $^{90}Sr$ and$^{137}Cs$ in natural soils, beginning at a depth 20-25 cm is relatively stable. The calculations have shown that the 180-300 cm soil layer contains about 28% of $^{137}Cs$ and 18% of $^{90}Sr$ of the total contamination density in soil layer 0-300.

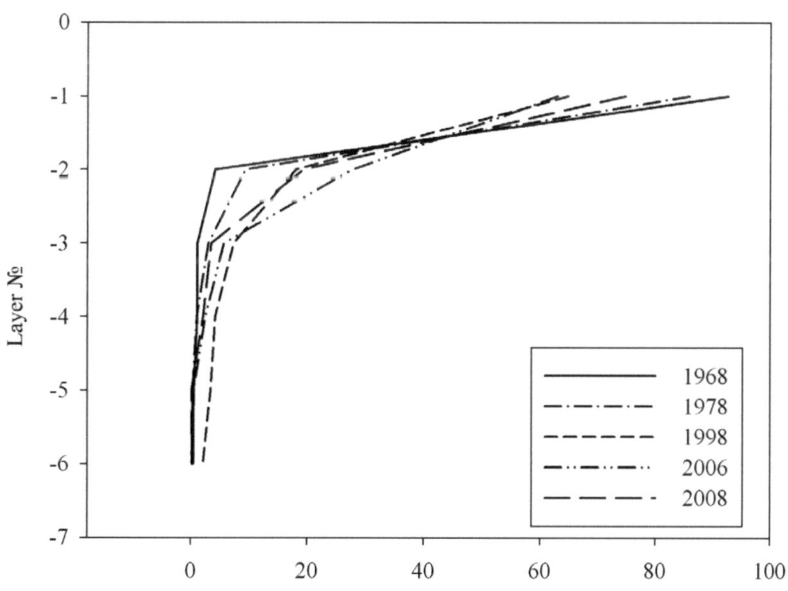


Fig. 15. Dynamics of $^{90}Sr$ distribution in 30-cm layers of gray forest soil

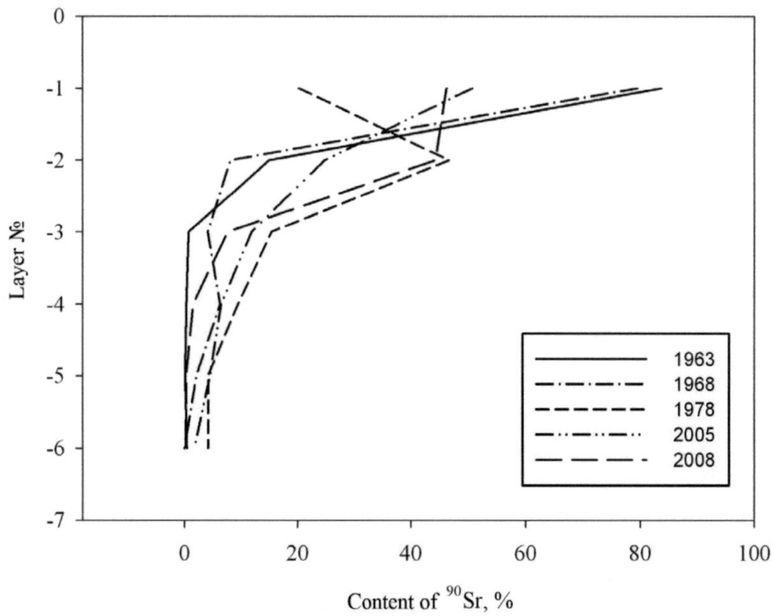


Fig. 16. Dynamics of $^{90}Sr$ distribution in 30-cm layers of sod-podzol soil

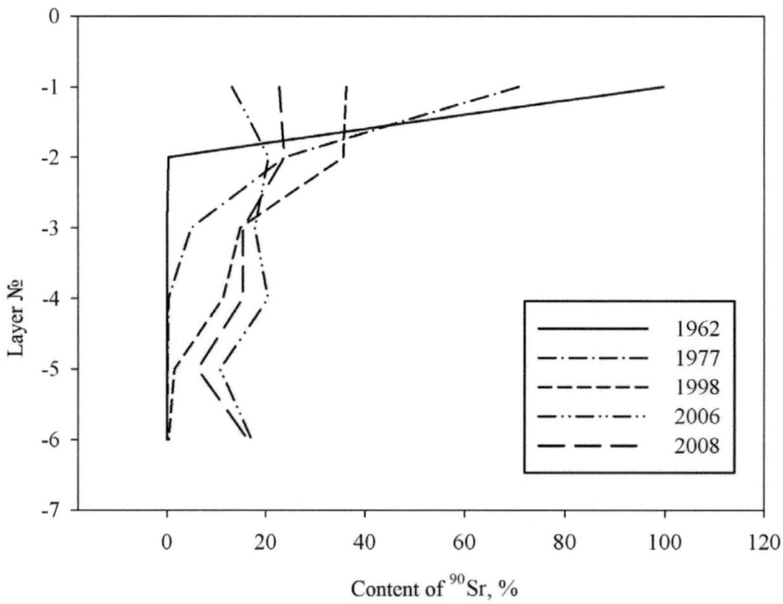


Fig. 17. Dynamics of $^{90}Sr$ distribution in 30-cm layers of chernozem

| Year of sampling | b | R2 |
|---|---|---|
| Gray forest | | |
| 1968 | 3.07 | 0.999 |
| 1978 | 2.15 | 0.999 |
| 1998 | 1.17 | 0.993 |
| 2006 | 0.96 | 0.991 |
| 2008 | 1.38 | 0.999 |
| Sod-podzol | | |
| 1963 | 1.76 | 0.993 |
| 1968 | 2.17 | 0.989 |
| 1978 | 0.30 | 0.435 |
| 2005 | 0.70 | 0.998 |
| 2008 | 0.62 | 0.831 |
| Chernozem | | |
| 1962 | 5.78 | 1.000 |
| 1977 | 1.16 | 0.997 |
| 1998 | 0.45 | 0.874 |
| 2006 | 0.01 | 0.008 |
| 2008 | 0.16 | 0.555 |

Table 14. Parameters of equations describing $^{90}Sr$ distribution in the profile of the 30-cm soil layer

Most of the researchers engaged in studies of radionuclide distribution in soil have noted that in the first years after atmospheric fallouts the highest amounts of $^{90}Sr$ and $^{137}Cs$ settled in sod cover or forest litter. It is known, however, that vegetable waste of herbaceous type decays within one season, while leaf wood waste decays within 3-4 years. It takes longer for waste of coniferous forests to decay, however, over the 50 years since formation of the EURT the forest litter contaminated by atmospheric fallouts should have decayed long ago, while the steppe cover of chernozem did not develop until 1991. Thus, the high specific activity currently observed in the forest litter is due not only to the initial fallouts in 1957 and 1967, but also to a high rate of local radionuclide turnover and continuous fallouts of activity.

The reserve of radionuclides in the vegetation mat and upper soil layers is constantly replenished due to vegetation waste. The waste cover is, in its turn, contaminated due to atmospheric fallouts and transfer of radionuclides to the top by the root system. Since actually the total surface biomass of vegetation is transferred to the waste cover at the end of the year, the yearly carry-over of $^{90}Sr$ to the surface is determined by the annual productivity of the ecosystem.

Waste of woody vegetation accounts for just a portion of the yearly increment, in addition to leaves and grass, the waste composition includes slowly decaying wood, needles and strobiles. That is why radionuclides are deposited not only in soil but also in wood and litter cover. Mineralization of decayed biomass involves an increase in the relative contents of radionuclides in the waste litter. Specific activity of $^{90}Sr$ in the ground litter was 2.2±0.7 times higher than in grass at a distance of 20 km from the contamination source, and 6.4±36 times higher at a distance of 30 km; the respective values for $^{137}Cs$ were 14±8 and 12±8. Table

15 presents specific activity values for radionuclides deposited in litter cover and the upper layer of soil. $^{90}Sr$ activity measured in the upper layer of the litter cover is actually the same as that in the upper layer of the soil, and activity in the lower layer is substantially higher. The difference between levels of $^{137}Cs$ activity in the upper and lower layers of the litter cover is even more substantial than that found for $^{90}Sr$. Ratio of $^{90}Sr$ specific activity to the activity in the 0-10 cm soil layer at a distance of 20 km from the Mayak PA is 0.7±0.3, and that at 30 km is 2.1±1.0. The ratio for $^{137}Cs$ is 0.08±0.04 and 0.4±0.4. Since contamination levels of grass are higher at longer distances, it can be assumed that currently the uptake of $^{90}Sr$ and $^{137}Cs$ by the biomass is going on through the root system. $^{90}Sr$ activity in leaves which contribute the largest portion of the vegetation mat is actually the same as that identified in the 0-5 soil layer. However, as was mentioned above, the weight of the vegetation waste per 1 m$^2$ was not large. On the EURT axis, within 20 km of the Mayak PA, samples of 2008-vegetation waste were taken in a birch forest. Specific activity of $^{90}Sr$ measured in the waste was 5904 Bq·kg$^{-1}$, that of $^{137}Cs$ 54 Bq·kg$^{-1}$. The fall of the aboveground phytomass to the forest floor contributed 2.3 kBq·kg m$^{-2}$ of $^{90}Sr$ and 028 kBq·kg m$^{-2}$ of $^{137}Cs$. This accounted for 2.1% and 3.4% of contamination density in forest litter for $^{137}$ and $^{90}Sr$, respectively, and for 0.36% and 0.8% of the total contamination density in the 0-20 cm layer of forest litter and soil for $^{90}Sr$ and $^{137}Cs$, respectively.

| Distance from Mayak PA, km | Soil | Upper layer of litter cover | | Lower layer of litter cover | | 0-5 cm soil layer | |
|---|---|---|---|---|---|---|---|
| | | $^{90}Sr$ | $^{137}Cs$ | $^{90}Sr$ | $^{137}Cs$ | $^{90}Sr$ | $^{137}Cs$ |
| 20 | gray forest | 7.1×10$^3$ | 0.18×10$^3$ | 16.3×10$^3$ | 1.3×10$^3$ | 7.8×10$^3$ | 0.48×10$^3$ |
| | chernozem | 0.6×10$^3$ | 0.02×10$^3$ | 1.2×10$^3$ | 0.2×10$^3$ | 1.1×10$^3$ | 0.3×10$^3$ |
| 30 | chernozem | 29 | 5 | 92 | 161 | 31 | 143 |
| 55 | Sod-podzol | 2.8×10$^3$ | 0.06×10$^3$ | 10×10$^3$ | 1.2×10$^3$ | 1.9×10$^3$ | 0.2×10$^3$ |

Table 15. Soil and litter contamination levels, Bq ·kg$^{-1}$

Therefore, the plant litter vegetation waste contributes an insignificant proportion of contamination density of the plant litter and upper layer of soil.

It was found out as a result of measurements of the proportions of radionuclides occurring in different forms and having different degrees of accessibility to plants that $^{90}Sr$ is mostly encountered in exchangeable form (64-85%), a larger amount of exchangeable strontium being deposited in sod-podsol soils than in gray forest and chernozem. The largest amount of $^{137}Cs$ is strongly bound together and it occurs in acid-soluble form (4-36%) or as a solid residue (27-82%). In all cases, the content of $^{137}Cs$ accessible to plants in the 5-10 layer proved to be higher, and the content of $^{137}Cs$ in the form of solid residue was significantly lower.

Thus, as of today, the studies of all the soil types have shown that the major part of $^{90}Sr$ and $^{137}Cs$ activity is deposited in the upper layer.

### 3.6 Dependencies governing migration of $^{90}$Sr and $^{137}$Cs in the soil-grass chain and forest products

In 2005-2006, samples of grass and soils taken in the most heavily contaminated areas of the EURT were processed, following which proportionality factors were assessed (PF). Specific activity of $^{90}$Sr in grass ranged from 70 to10940 Bq·kg$^{-1}$, that of $^{137}$Cs from 10 to 997 Bq·kg$^{-1}$. Proportionality factor (Bq/kg in grass/Bq/kg in soil) was within the range 0.2-2.1, the mean value of transfer factor calculated as a ratio of specific activity of radionuclides in grass to soil contamination density, Bq·kg$^{-1}$/ Bq·kg$^{-2}$, was 15.9±8.5 for $^{90}$Sr, while for $^{137}$Cs it was lower: 3.1±3.4. The difference in transfer factor values may depend on soil type and diversity of plant species composing the cover layer. It should be noted that transfer factor values are slightly higher for sod-podzol soils in the EURT zone.

Transfer factor estimated for grass in the EURT zone has changed insignificantly over the recent 10 years, it is 10.3 Bq·kg$^{-1}$ / Bq·kg$^{-2}$ for $^{90}$Sr. The mean transfer factor calculated in 1997 for $^{90}$Sr in pasture soils in the EURT zone was 13.5 Bq·kg$^{-1}$/Bq·kg$^{-2}$. Transfer factor values calculated for $^{90}$Sr relative to standard soil contamination of 1 Ci·km$^{-2}$ (37 kBq·m$^{-2}$) amounted to 230-270 Bq·kg$^{-1}$ / 37 Bq·kg$^{-2}$.

Contamination with $^{90}$Sr of all sampled wild-growing berries exceeds the permissible limit 3.5-13.5 times, $^{90}$Sr proportionality factor is higher than that determined for vegetables and grain (Table 16). Specific activity of $^{137}$Cs in grass and berries does not exceed the permissible limit. It should be taken into consideration that wild-growing berries in the EURT zone present the highest hazard compared to other food products in view of the contribution they can make to dietary intakes of $^{90}$Sr.

| Sampling site | Species | Specific activity of fresh berries,Bq·kg$^{-1}$ | | proportionality factor, (Bq·kg$^{-1}$ in berries)/ (Bq·kg$^{-1}$ in soil) | |
|---|---|---|---|---|---|
| | | $^{90}$Sr | $^{137}$Cs | $^{90}$Sr | $^{137}$Cs |
| 20 km from Mayak PA | Wild strawberry | 813 | 5.2 | 0.140 | 0.0165 |
| | stone bramble | 243 | 2.4 | 0.042 | 0.0076 |
| 30 km from Mayak PA | Wild strawberry | 213 | 1.0 | 0.254 | 0.0054 |

Table 16. Contamination of wild-growing berries with $^{90}$Sr and $^{137}$Cs

Table 17 presents data on levels of contamination of fresh mushrooms. In 2008, specific activity of $^{90}$Sr and $^{137}$Cs in all samples was found to be significantly lower than the permissible limits (50 Bq·kg$^{-1}$ for $^{90}$Sr and 500 Bq·kg$^{-1}$ for $^{137}$Cs). Proportionality factors and transfer factors were significantly lower than those obtained for grass, vegetables and grains sampled in 2007. It can be assumed that mushrooms growing in the EURT zone present no hazards for the population as their contribution to internal dose is insignificant.

| Sampling site | Species | Specific activity of fresh mushrooms, Bq ·kg$^{-1}$ | | Accumulation factor, (Bq ·kg$^{-1}$ in the fruit body)/( Bq ·kg$^{-1}$ in soil) | | Proportionality factor, (Bq ·kg$^{-1}$ in fruit body) / (Bq ·kg$^{-2}$ in soil) | |
|---|---|---|---|---|---|---|---|
| | | $^{90}$Sr | $^{137}$Cs | $^{90}$Sr | $^{137}$Cs | $^{90}$Sr | $^{137}$Cs |
| 50 km from Mayak PA | Birch mushroom | 0.47 | 0.58 | 0.0039 | 0.0113 | 0.0328 | 0.0941 |
| | Coral milky cap | 1.2 | 0.97 | 0.0101 | 0.0189 | 0.0839 | 0.1573 |
| | Mix | 0.7 | 0.9 | 0.0059 | 0.0175 | 0.0489 | 0.1460 |
| | Milk mushroom yellow | 0.34 | 4.6 | 0.0041 | 0.0176 | 0.0305 | 0.1447 |
| | Blewits | 0.59 | 19.9 | 0.0071 | 0.0762 | 0.0529 | 0.6259 |
| | Arachnoid muchroom, excellent | 0.63 | 18.5 | 0.0076 | 0.0709 | 0.0565 | 0.5819 |
| | Mix | 1.4 | 8.2 | 0.0169 | 0.0314 | 0.1255 | 0.2579 |
| | Blewits white | 0.7 | 3.9 | 0.0092 | 0.0169 | 0.0749 | 0.1383 |
| | Mix | 1.7 | 18.8 | 0.0531 | 0.0450 | 0.2798 | 0.3705 |
| 20 km from Mayak PA | Honey agaric summer mushroom | 9.8 | 39.0 | 0.0017 | 0.124 | 0.0133 | 0.9476 |
| 30 km from Mayak PA | Yellow boletus | 5.5 | 10.9 | 0.0065 | 0.0586 | 0.0396 | 0.4950 |
| | Mix of the lamellar | 8.9 | 2.0 | 0.0106 | 0.0108 | 0.0641 | 0.0908 |

Table 17. Contamination of mushrooms with $^{90}$Sr and $^{137}$Cs

### 3.7 Migration of radionuclides along food chains

Levels and dynamics of $^{90}$Sr specific activity in milk produced in villages at the periphery of the EURT in 1958-2006 are presented in figure 24. A dramatic decrease in contents of $^{90}$Sr in milk had occurred before 1963 while later on a slow decrease was observed overtime. Since 1960 the period of a 2-fold decrease in $^{90}$Sr contents in milk ($T_{1/2\ eff.}$) was 23 years. The content of $^{137}$Cs in milk produced in the villages studied was 2 times lower as compared to that of $^{90}$Sr. Currently, this value is1.1 ±0.4 Bq/l for $^{137}$Cs.

In 1960, the transfer factor in the soil-milk chain was 340 μBq·l$^{-1}$/Bq·m$^{-2}$, in 2006 it was 45 μBq·l$^{-1}$/Bq·m$^{-2}$. A reduction in the transfer factor resulted from a decrease in biological accessibility of the radionuclide in the soil-pasture grass chain.

Potatoes take the second position after milk in terms of $^{90}Sr$ contributed by food produce. The dynamics of reduction in specific activity of $^{90}Sr$ in potatoes has been the sane over the total period of observations, viz., from 7.0 to 0.9 Bq·kg$^{-1}$, the respective value for $^{137}Cs$ is from 1.5 to 0.7 Bq·kg$^{-1}$. Within a number of years, stable values of transfer factor have been established for agricultural products (Table 18).

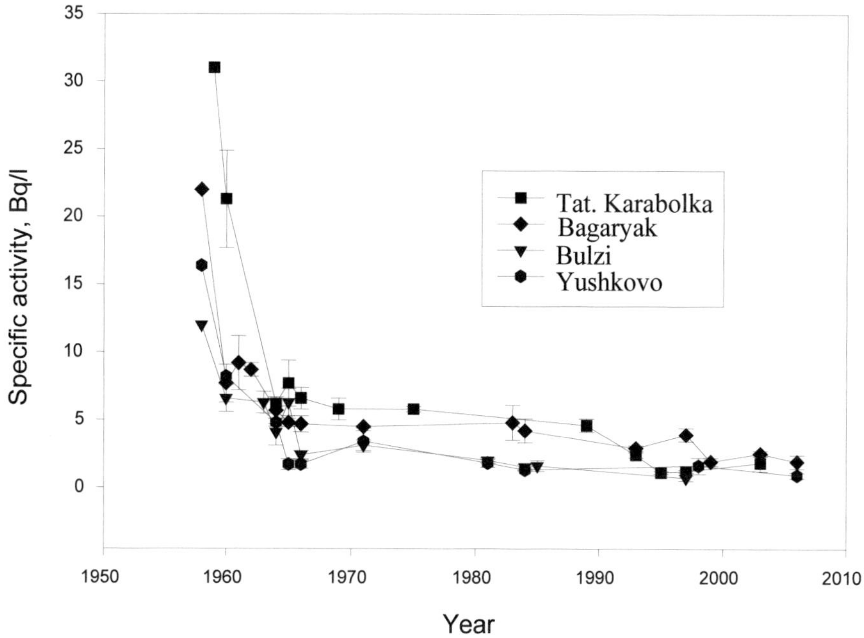


Fig. 24. Dynamics of specific activity of $^{90}Sr$ in milk produced in EURT villages in the period 1958-2006

| Crops | Products | Proportionality factors | |
|---|---|---|---|
| | | $^{90}Sr$ | $^{137}Cs$ |
| Peas | Grain | 11.4±6.6 | 15.2±5.2 |
| Wheat | Grain | 14.4±9.1 | 6.6±1.9 |
| Oats | Grain | 5.7±2.5 | 9.9±4.6 |
| Barley | Grain | 6.5±2.1 | 4.4±1.3 |
| Cabbage | Head of cabbage | 4.7±1.6 | 3.0±1.4 |
| Carrots | Root crop | 5.2±1.1 | 2.2±0.8 |
| Beet | Root crop | 43.3±44.7 | 16.5±15.7 |
| Potato | Tubers | 4.8±1.3 | 2.3±0.5 |
| Corn | Paste (dry) | 1.4±0.6 | 3.6±0.5 |
| Sudanese grass | Paste (dry) | 9.0±7.2 | 46.0±36.7 |

Table 18. Mean values of proportionality factors for $^{137}Cs$ and $^{90}Sr$ measured in the portion of cash crop grown in 0-20 cm layer of gray forest soils

Values of proportionality factor decrease overtime and, as a result, it becomes possible to grow agricultural product with admissible level of conatimation in soils with a higher contaminated level.

## 4. Karachai Radioactive Trace

In 1997 the area covered by the Karachai Radioactive Trace (KRT) delineated by $^{90}$Sr contamination isoline of 7.4 kBq/m$^2$ (0.2 Ci/km$^2$) amounted to 1660 km$^2$. The total amount of the radionuclides deposited on this territory was 800 Ci. Radinuclide composition of dust fallouts was as follows: 32 % of $^{90}$Sr+$^{90}$Y, 47% of $^{137}$Cs, 21% of $^{144}$Ce+$^{144}$Pr. Biological accessibility to plants was 90 % for $^{90}$Sr, and 12% for $^{137}$Cs (Yu.D. Korsakov et al., 1996).

In 1967 about 97% of the total $^{137}$Cs deposited in pasture soils settled in the upper 0-3 cm layer. Currently, 38 years after the fallout, $^{137}$Cs is accumulated in the soil layer at a depth of 13 cm, and $^{90}$Sr is mostly deposited in the soil layer to a depth of 0-20 cm (89.5%). A small portion of radionuclides which settled in the layer at a depth of 0-70 cm migrated to a depth of 70 cm.

In 1967, the transfer factor for contents of $^{90}$Sr in grass and soil was 0.09 Bq·kg$^{-1}$ / kBq·m$^{-2}$. A sharp decline in transfer factor values for $^{137}$Cs within the soil-grass chain occurred one year later after it had been cleaned of surface contamination, and in the subsequent years no changes in transfer factor values were noted. During early years, the uptakes of the radionuclide were going on from sod cover and soil, and transfer factor values fluctuated between 0.0025 and 0.005 Bq/kg : Bq/m$^2$.

Long-term studies of $^{90}$Sr and $^{137}$Cs transfer from soil to grass were conducted in the grazing land of Sarykulmak village where dairy cattle was grazing (Figure 25).

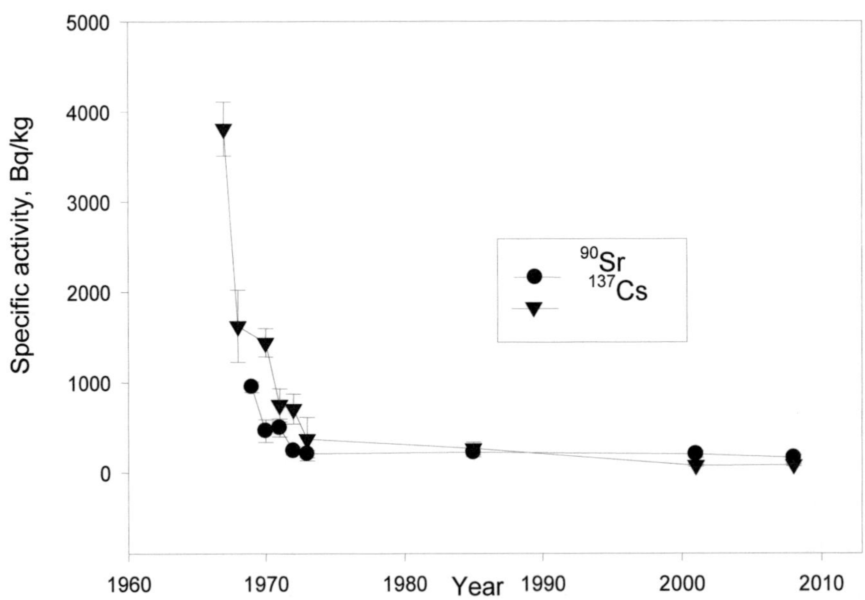


Fig. 25. Levels and dynamics of decrease in radionuclide contents measured in pasture grass

A reduction in specific activity of $^{137}Cs$ in the pasture grass over the period from 1967 through 2009 is described by the two exponential dependences: $T_{1/2}$ was 1.8 years in the initial period after deposition of radionuclides, and in the subsequent period it was 3.8 years. The first exponent describes the cleaning of grass from surface contamination, the second describes the reduction in uptakes of radionuclides by grass from soil. Over the period from 1967 through 2009, the specific activity of $^{137}Cs$ in grass decreased 100-fold , that of $^{90}Sr$ 10-fold.

The initial specific activity of $^{90}Sr$ in milk late in April, 1967, was 140 Bq/l, that of $^{137}Cs$ - 237 Bq/l. From 1967 through 2009, the average values of $^{90}Sr$ and $^{137}Cs$ specific activity in milk decreased 30-fold. The decrease in $^{90}Sr$ and $^{137}Cs$ levels in milk was determined based on the decrease in grass contamination. Beginning from 1970, the period of half-cleaning of milk from $^{90}Sr$ was 20 years, and from $^{137}Cs$ 10 years. Proportionality factors for $^{90}Sr$ in the soil-milk chain estimated on the Karachai Radioactive Trace were on the average 5-fold lower than the values estimated for the EURT for the same time periods elapsed after the radioactive fallouts on the soil. Evidently, the long period for which $^{90}Sr$ and $^{137}Cs$ remained deposited in silt and soil on the shores of Karachai led to a reduction in biological accessibility of radionuclides to plants and, as a result, to a reduced uptake of radioactivity by milk. Specific activities of $^{90}Sr$ in milk are more correctly described by the log-normal distribution. After the accident, the contents of $^{137}Cs$ in potatoes was insignificant, and overtime it decreased about 10-fold, while the contents of $^{90}Sr$ decreased 3-fold.

## 5. Conclusion

1. Major radiation accidents that took place in the Southern Urals during the period 1949-1967 brought about contamination of vast territories with radionuclides and exposure of the local population. The situation on the Techa River involved contamination of river water, bottom sediments and the floodplain with $^{137}Cs$ and $^{90}Sr$. On the East-Urals Radioactive Trace $^{90}Sr$ was the prevalent contaminant, while on the Karachai Trace biologically poorly-accessible compounds of $^{137}Cs$ and $^{90}Sr$ prevailed.
2. The key mechanisms by which contamination of the environment can be eliminated include radioactive decay, reduction in biological accessibility and deepening of radioactive substances in soil. Pronounced sorption capabilities and poor solubility hamper migration of $^{137}Cs$ in the environment, while $^{90}Sr$ is more mobile.
3. Due to a number of protection measures implemented from 1965 through 2004 on the Techa River, concentrations of $^{90}Sr$ and $^{137}Cs$ in river water have decreased, and currently the specific activity of $^{90}Sr$ amounts to 10-15 Bq·$l^{-1}$, and that of $^{137}Cs$ to 0.5-1.5 Bq·$l^{-1}$. Radionuclides deposited in bottom sediments migrated to the depth of over 35 cm. Contamination density in the 0-10 cm layer depends on the concentration of radionuclides in the watercourse. $^{90}Sr$ and $^{137}Cs$ are revealed in floodplain soil at a depth of over 70 cm where a decrease and averaging of concentrations overtime has been observed. At the present time, it is expedient to retain the constraints on the use of river water.

   After the use of river water was banned in 1956, the major pathway for contribution of radioactivity to dietary intakes has been made by milk (87-95 %) and vegetables. Since 1967, the content of $^{90}Sr$ in milk, with rare exception, has not exceeded the permissible limit (25 Bq·$l^{-1}$).

4. The rate of vertical migration of $^{90}Sr$ in EURT soils ranges from 0.25 to 0.35 cm year$^{-1}$, the largest amount of radionuclides which remains deposited in the upper part of the soil profile (0-20 cm) is decreasing with increasing depth. Nonmobile $^{137}Cs$ is mosty retained in the upper 10-cm layer. Biological accessibility of $^{90}Sr$ has decreased over the past period 7-10 times and it has not actually changed over the recent years. Currently, 40 years after the accident, the content of fixed forms of $^{90}Sr$ in soil has reached 34%, that of $^{137}Cs$ and plutonium 95-98%.
   The period of half-cleaning of milk from $^{90}Sr$ was 2-3 years during the first years after the accident, and during the subsequent period it amounted to 15 years.
5. On the Karachai Trace, due to the prevalence of $^{137}Cs$ and $^{90}Sr$ with poor biological accessibility, the reduction in contamination of soils, grass and food products was going on more speedily (than that registered on the EURT). Specific activity of $^{137}Cs$ and $^{90}Sr$ in milk exceeded the permissible limits only during the first month after the accident.
6. The prognosis for the further development of the radiation situation on the Techa River is determined predominantly by the radioactive runoff from the Techa cascade of reservoirs. With the lapse of time, the part of the EURT territory where the use of agricultural lands is restricted, will diminish. There are no such restrictions in the Karachai Trace area.

## 6. References

M.I. Avramenko, A.N. Averin., Yu.V. Glagolenko, et al.. 1957 Accident. Assessment of the parameters of the explosion and analysis of the character of radioactive contamination of the territory // Radiation Safety Issues. – 1997. - № 3. - P. 18-28.

R.M. Aleksakhin, L.A. Buldakov, V.A. Gubanov et al. Major radiation accidents: consequences and protection measures. Under the editoship of L.A. Ilyn and V.A. Gubanov / - M., IzdAT, 2001 – 752 p.

Yu.G. Glagolenko, E.E. Dzekun, E.G. Drozhko, G.M. Medvedev, S.I. Rovny, A.P. Suslov. Strategy for radioactive waste management at the Mayak PA// Radiation Safety Issues. №2.-1966.-p 3-10.

Yu.A. Izrael, E.M. Artemov, V.N. Vasilenko, I.N. Nazarov, A.I. Nakhutin, A.A. Uspin, A.M.Kamkin. Radioactive contamination of the Urals region by the Mayak Production Association // Radioactivity in case of nuclear explosions and accidents. (Proceedings of the International Conference, Moscow, April 24-26, 2000, C.-P., 2000.- P. 411-424.

A.Ya. Kogotkov. Behavior of radionuclides in soils of the Mid-Transuralian region: Doctoral thesis (Biological Sciences). – M., 1968. – 556 p.

Yu.D. Korsakov, E.A. Fedorov, G.N. Romanov, L.I. Panteleyev. Assessment of radiation conditions on the territory contaminated due to downwind transfer of radioactive aerosols in the area around the facility in 1967: Abstract // Radiation Safety Issues. - 1996. - № 4. - P. 50-59.

P.P. Lyarsky. Sanitary consequences of contamination of the territory with long-lived products of uranium fission and arrangement for sanitary-prophylactic measures: Doctoral Thesis (Medical sciences). - M., 1962. – 928 p.

A.N. Marey. Sanitary consequences of discharges of radioactive waste from atomic facility into water basins: Doctoral thesis (madical sciences) / Biophysics Instituteof the Ministry of Health, USSR. - M. - 1959. - 441 p.

V.V. Martyushov, D.A. Spirin, G.N. Romanov, V.V. Bazylev. Dynamics of the state and migration of Strontium-90 in soils of the East-Urals Radioactive Trace // Radiation Safety Issues. -1996.-№3.-p. 28-35.

Yu.G. Mokrov. Reconstruction and prognosis of radioactive contamiation of the Techa River. Part I. Role of weighted particles in the processes of radioactive contamination of the Techa Rivert in 1949- 1951 – Ozyorsk: Editorial-Publishing Center VRB, 2002.- 176 p.

Basic requirements of samplig (MU 2.6.1.715-98 and MUK 2.6.1.016-99).

M.M. Saurov. Radiation-hygienic assessment of the natural migration of the population chronically exposed to uranium fission products: Doctoral thesis (Madical sciences).-M.,1968.-663 p.

Compedium of methods for measurements of radioactivity in the environment. Methods for radiochemical analysis. Editor: G.A. Sereda, Z.S. Shulenko. GidrometeoIzdat, M., 1966, 51 p.

Consequences of anthropogenic radiation exposure and rehabilitation issues of the Urals region. Editor: S.K. Shoigu. M.,2002, 287 p.

Nature of Chelyabinsk Oblast. – Chelyabinsk, 1964.

V.P. Yulanov. Pine growth peculiarities (Pinus silvestris) in areas contaminated with radioactive substances: Candidate,s thesis (Biological sciences). - Chelyabinsk, 1965. – 154 p.

### 6.1 Original references in Russian language

Авраменко М.И., Аверин А.Н., Глаголенко Ю.В. и др. Авария 1957 г. Оценка параметров взрыва и анализ характеристик радиационного загрязнения территории // Вопросы радиационной безопасности. – 1997. - № 3. - С. 18-28.

Алексахин Р.М., Булдаков Л.А., Губанов В.А. и др. Крупные радиационные аварии: последствия и защитные меры. Под общей редакцией Л.А.Ильина и В.А.Губанова/ - М., ИздАТ, 2001 – 752 с.

Глаголенко Ю.Г., Дзекун Е.Г., Дрожко Е.Г., Медведев Г.М., Ровный С.И., Суслов А.П. Стратегия обращения с радиоактивными отходами на ПО «Маяк»//Вопросы радиационной безопасности №2.-1966.-с 3-10.

Израэль Ю.А., Артемов Е.М., Василенко В.Н., Назаров И.Н., Нахутин А.И., Успин А.А., Кямкин А.М. Радиоактивное загрязнение Уральского региона производственным объединением "Маяк" // Радиоактивность при ядерных взрывах и авариях. (Труды Международной конференции, Москва, 24-26 апреля 2000, С.-П., 2000.- С. 411-424.

Коготков А.Я. Поведение радионуклидов в почвах среднего Зауралья: Дис. …докт. биол. наук. – М., 1968. – 556 с.

Корсаков Ю.Д., Федоров Е.А., Романов Г.Н., Пантелеев Л.И. Оценка радиационной обстановки на территории, загрязненной в результате ветрового переноса радиоактивных аэрозолей в районе предприятия в 1967 г.: Реферат // Вопр. радиационной безопасности. - 1996. - № 4. - С. 50-59.

Лярский П.П. Санитарные последствия загрязнения территории долгоживущими продуктами деления и организация на ней санитарно-профилактических мероприятий: Дис…докт. мед. наук. - М., 1962. – 928 с.

Марей А.Н. Санитарные последствия удаления в водоемы радиоактивных отходов предприятия атомной промышленности: Дисс. на соискание степени докт. мед. наук. / Институт биофизики МЗ СССР. - М. - 1959. - 441 с.

Мартюшов В.В., Спирин Д.А., Романов Г.Н., Базылев В.В. Динамика состояния и миграции стронция-90 в почвах Восточно-Уральского радиоактивного следа // Вопросы радиационной безопасности. -1996.-№3.-с. 28-35.

Мокров Ю.Г. Реконструкция и прогноз радиоактивного загрязнения реки Теча. Часть I. Роль взвешенных частиц в процессе формирования радиоактивного загрязнения реки Теча в 1949- 1951 гг. – Озерск: Редакционно-издательский центр ВРБ, 2002.- 176 с.

Основные требования, лежащие в основе отбора (МУ 2.6.1.715-98 и МУК 2.6.1.016-99).

Сауров М.М. Радиационно-гигиеническая оценка естественного движения населения, подвергшегося хроническому воздействию продуктов деления урана: Дис. …докт. Мед.наук.-М.,1968.-663 с.

Сборник методик по определению радиоактивности окружающей среды. Методики радиохимического анализа. Под редакцией Середы, Г. А., Шулепко, З. С. Гидрометеоиздат, М., 1966, 51 с.

Последствия техногенного радиационного воздействия и проблемы реабилитации Уральского региона. Под.ред. С.К.Шойгу. М.,2002, 287 с.

Природа Челябинской области. - Челябинск, 1964.

Юланов В.П. Особенности роста сосны (Pinus silvestris) на территории, загрязненной радиоактивными веществами: Дис. …канд.биол.наук. - Челябинск, 1965. - 154с.

# Section 2

## Pre-Disposal Activities

5

# Radioactive Waste Assay for Reclassification

Timothy Miller
*Atomic Weapons Establishment*
*United Kingdom*

## 1. Introduction

Following the ban on deep sea disposal of radioactive wastes in 1983 the Atomic Weapons Establishment (AWE) in the United Kingdom (UK) has stored plutonium (Pu) and uranium (U) contaminated wastes on site in 200 l steel drums or as wrapped packages containing filters or other materials. This was because of difficulty in demonstrating compliance with the with the 100 Bq/g Pu alpha activity limit for waste disposal at the national Low Level Waste (LLW) repository at Drigg in Cumbria (UK) and the absence of a national Intermediate Level Waste (ILW) repository. Wastes that were potentially only lightly contaminated were consigned to Drigg as LLW because of difficulties in demonstrating that they could be disposed of as either Very Low Level Waste (VLLW) or Exempt Waste (EW). Table 1 summarises the current situation in the UK regarding disposal route and costs for each waste category. This UK classification system is different from that recommended by the IAEA or that used in the USA.

| Category | Activity range ($Bq\ g^{-1}$) | Disposal route | Disposal cost (£ per 200l drum) |
|---|---|---|---|
| ILW | > 4,000 Pu, U<br>> 12,000 beta | Indefinite storage at AWE until a national ILW repository is available | 40,000 |
| LLW | < 4,000 U<br>< 12,000 beta | LLW repository at Drigg | 250 |
| LLWD | < 100 Pu alpha | LLW repository at Drigg | 250 |
| VLLW | < 4 Pu, U | As authorised | 50 |
| EW | < 1 U<br>< 0.15 Pu | As AWE policy dictates | 20 |

Table 1. Waste disposal routes and costs

In recent years AWE has successfully developed robust techniques for reducing both the quantity and category of its Pu and U contaminated wastes. The objective was to reduce costs and ensure that wastes provisionally classified as ILW were reclassified to LLW for off-site disposal to Drigg where possible and that the limited space available at Drigg was used for genuine LLW and not VLLW or EW. This chapter describes the waste reclassification techniques, recently developed at AWE, for ILW reclassification to LLWD (i.e. LLW acceptable to Drigg) and for VLLW reclassification to EW that is not subject to regulatory control in the UK. The techniques were applied to almost all waste streams encountered at AWE and around 50 % of provisionally classified ILW was downgraded to LLW and 70- 90 % of provisionally classified VLLW to EW. However, further development is required in order to address the problems posed by the most challenging waste streams. These problems are discussed and suggestions are made for future work.

## 2. Reclassification techniques

Measurement of 200l steel waste drums, at the 100 Bq/g Pu alpha threshold, was beyond the capabilities of most assay techniques because the low yields of neutron or photon emissions from plutonium isotopes gave detection limits that were several hundred or thousand Bq/g (Miller, 2002). Hence it was necessary to develop techniques that measure the relatively low energy (60 keV), but high yield (36 %) photon from Am-241 (Pu-241 daughter) and apply the Pu alpha/Am-241 activity ratio to calculate the total Pu alpha activity. The main drawback with this strategy is the potential for underestimating Pu in scenarios where photon attenuation is severe due to the presence of high density and high atomic number (Z) materials in the waste. However, this was not found to be a problem for waste drums encountered at AWE because activity was well distributed within the waste (Miller, 2009b). Three techniques were developed. Firstly, High Resolution Gamma Spectrometry (HRGS), using a High Purity Germanium (HPGe) crystal detector and Spectral Non-destructive Assay Platform (SNAP) analytical software to calculate isotopic activities together with associated errors and Detection Limits (DL). Secondly, Low Resolution Gamma Spectrometry (LRGS), using a thin NaI(Tl) crystal detector calibrated using a traceable Am-241 source and simulated waste. Thirdly, Gross Counting (GC) using plastic scintillation photon detectors.

### 2.1 HRGS/SNAP

The SNAP assay system consists of standard HRGS hardware coupled with SNAP analytical software. This allows the photon detector calibration to be corrected for both counting geometry (e.g. drum dimensions and detector location) and gamma ray attenuation (e.g. waste matrix density and composition. Figure 1 shows the standard counting geometry with the detector end cap located at 60 cm from the centre/middle of the drum wall.

Fig. 1. SNAP assay system monitoring a 200l waste drum on a rotating turntable

The gamma spectrum, for the waste drum, was stored in the form of a Region of Interest (ROI) report (.Rpt) file containing the counting data for the photons of interest (e.g. Am-241 @ 60 keV) and imported into SNAP for radionuclide identification, modelling, assay calculation and reporting. Radionuclide identification can be performed using a full radionuclide library or a sub-library pertaining to the waste-stream of interest (e.g. plutonium). Each ROI in the spectrum was selected, by the analyst, from the library and the spectrum saved as a RPu file.

The RPu files can then be imported into the SNAP modelling. Early versions of the SNAP software have two models: a cylinder or a box. More recent versions have additional models, such as a disk. The dimensions of the cylinder (height and diameter) or the box (height, width and depth) pertain to the dimensions of the waste material within the waste package. For a completely filled waste drum this would be the internal dimensions of the drum that define the size of the item being assayed.

The detector location, relative to the waste, was defined by three measurements: the detector to item distance was measured from the detector end cap to the surface of the waste material; the detector height was measured from the base of the waste to the detector axis and the left of centre was measured from the centre of the waste to the detector axis. The detector calibration was selected together with the collimator position (e.g. flush with the detector end cap).

The waste material was defined by three or four variables: the matrix mass, primary matrix percentage by volume, primary matrix material and (if applicable) secondary matrix material. Three layers of shielding material and thickness may be applied, but for a 200 l drum 0.11 cm of iron was used with the other two layers set to 0 cm of 'none'.

Modelling was completed by entering the count time, height above sea level, detection limit required (i.e. Critical Level or Minimum Detectable Amount) and number of sides of the waste package counted (2 or 4) for computation of the Geometric Attenuation (GA) error. This was the percentage difference in activity between a uniformly distributed matrix and activity compared to a single point source of activity at a 'worst case' location.

Assay calculations were performed once all of the above modelling information had been entered. This gave radionuclide activities, based on uniform activity and matrix distribution, for each photon measured. All photon energies, for a given radionuclide, should yield consistent results. If not the modelling was adjusted to get the best agreement. For example, if the 60 keV signature from Am-241 was underestimating by a factor of two, compared to the 662 keV Am-241 photon, increasing the steel shielding thickness by 1 mm would give better agreement.

Another feature is the lump correction routine for U and Pu waste-streams. For example, the main photons from Pu are at 129 and 414 keV. Underestimation at 129 keV compared to 414 keV is often an indication of photon self-absorption within Pu. The software allows the analyst to progressively increase the size of the Pu until consistent results are obtained at 129 and 414 keV.

When the analyst is satisfied with the modelling the software can be used to generate an htm report file. This summarises all of the sample details, modelling and results for presentation to the customer as an electronic or paper copy.

The software is easy to use and a complete assay, from spectrum acquisition to report generation, takes only a few minutes. This can be further streamlined by using the software to generate calibration curves when the only significant variable is the matrix mass . As the mass increases the calculated activity increases and the response factor is yielded from the ratio of the count rate to calculated activity. In this situation the observed net count rate, divided by the response factor, yields the radionuclide activity or mass as required.

The main limitation of the software is that it is less accurate for objects counted very close to the detector due to solid angle and inverse square law effects. Normal counting geometry is with the detector at one drum diameter from the drum wall.

There are a number of other software packages available that can be used as an alternative to SNAP. However, SNAP has a proven record at both Los Alamos National Laboratory (LANL) in the USA and AWE. Also inter-comparison studies, against benchmark techniques like the Segmented Gamma Scanner (SGS), have shown that SNAP gives similar results with the advantages of simplicity, speed, lower costs and improved detection limits (Miller, 2008). Eberline Services have reported similar results (Lasher, 2009).

## 2.2 LRGS

The IS 610 LRGS was originally developed by AWE for the detection of low energy photons from low level Pu and U ground contamination. It can be used as a hand held monitor or mounted on its tripod (figure 2).

Fig. 2. LRGS positioned for HEPA filter monitoring

The Pu version of the IS610 uses a 75 mm diameter by 1 mm thick NaI(Tl) crystal and three ROIs at 10-24, 47-72 and 10-72 keV for detection of Am-241 @ 60 keV and L x-rays @ 17 keV . The U version has a thicker (10 mm ) crystal with ROIs at 10-28, 48-74 and 161-237 keV for detection of the higher energy U-235 photon @ 186 keV.

The IS610 is cheaper and easier to use than SNAP, but is much less versatile and has poorer detection levels. It was therefore only used in less challenging scenarios, where it could be calibrated using a traceable source and simulated waste, such as Pu contaminated HEPA filters where the only significant photon yield is at 60 keV (Miller, 2003).

### 2.3 GC

Better detection levels were achieved using a GC system. This detects all photons from 50 keV up to 2 MeV and has a 350 l counting chamber (63.5 cm high, 63.5 cm wide, 87 cm deep). All six sides of the chamber are surrounded with plastic scintillation photon detectors (50 mm thick) and lead shielding (25 mm thick). The aluminium base plate of the chamber is linked to a load cell in order to provide a measure of the waste mass. Waste items are introduced and removed using doors at the front and rear of the monitor (figure 3).

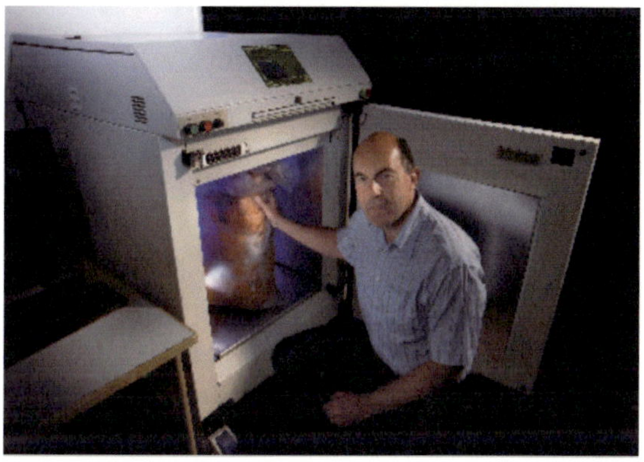

Fig. 3. GC system being loaded with bagged waste

The GC system must be well shielded and sited in an area with a low and non-fluctuating photon background. Furthermore wastes must be carefully segregated to provide materials with low photon attenuating properties, low Naturally Occurring Radioactive Material (NORM) content and an isotopic fingerprint that is consistent with the calibration (Miller, 2010).

## 3. Pu waste reclassification

All three techniques described above have been applied to low activity Pu waste streams to give an Am-241 activity or detection limit that was subsequently converted to a Pu activity or detection limit using the known Pu/Am-241 activity ratio for the item measured. Equipment calibrations were checked, using waste package standards prepared by the National Physical Laboratory (NPL) and measured Am-241 activities were found to be within a few percent of the true values (Miller, 2011). The following sections review, for each application, the calibration and performance (i.e. detection levels and uncertainties) checks done and subsequent results achieved with real wastes where monitored.

### 3.1 Reclassification of soft drummed Pu ILW using SNAP

Detector response factors (cps/Bq) were measured for an Am-241 source at various locations within 200 l waste drum calibration standards containing low Z (soft), low bulk density paper and PVC matrices over the range 16.6-72.5 kg net (34.6-90.5 kg gross). These factors were weighted, according to the relative volumes which they represent, in order to derive a uniform response factor and quantify the systematic error for non-uniform activity distributions. Detection levels and random errors were also derived from the counting data (Miller, 2002). Figure 4 shows how the maximum, minimum and uniform response factors reduced as gross drum mass increased. The results were fitted with polynomial curves.

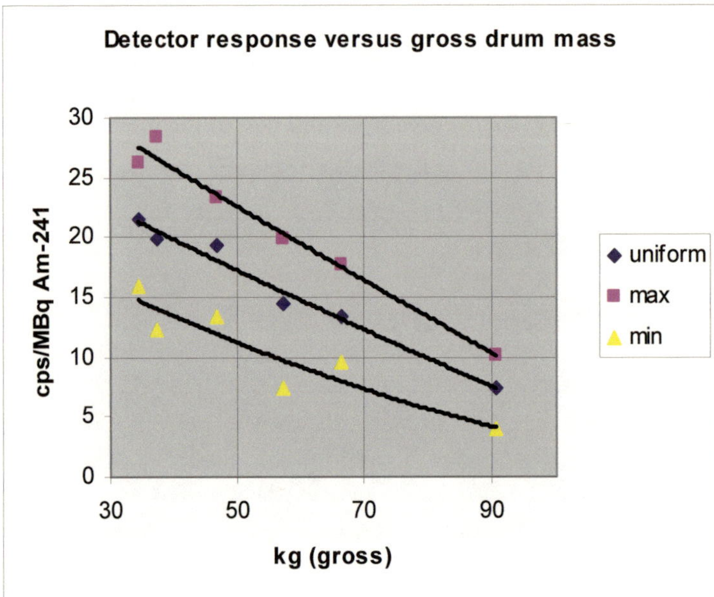


Fig. 4. SNAP response factors

The calibration curve, generated using SNAP software, was virtually identical to the uniform response curve in figure 4 that was achieved using waste drum standards (Miller, 2009b). Maximum and minimum response was within +/- 30 % of uniform for the lighter drums and within +/- 50 % for the heavier drums.

Repeat measurements, on blank drums, indicated that the background standard deviation ($\sigma$) was similar to the square root of the background counts ($\sqrt{B}$). Hence, equation 1 gave a Pu DL ($2\sigma$) of 1.2 Bq/g Pu, given: B = 30 counts, T = 100s count time, F = 15 cps/MBq, G = 60 kg and R = a Pu alpha/Am-241 activity ratio of 10.

$$DL = (2\sqrt{B}/TFG)R \quad (1)$$

Over the past nine years (December 2002 til June 2011) 4071 legacy drums, provisionally classified as ILW for long term storage at AWE, were assayed using SNAP. Only 535 (13.1 %) were confirmed ILW, with 1287 LLW (31.6 %) and 2252 LLWD (55.3 %). Similar figures were obtained for recently generated drums from decommissioning operations: 535 ILW (16.7 %), 1636 LLW (51 %) and 1037 LLWD (32.3 %). These figures include the results for both soft and hard drummed wastes.

### 3.2 Reclassification of hard drummed Pu ILW using SNAP

Drum radiography indicated that a number of drums contained high Z (hard) wastes, such as metals, so it was necessary to verify that the calibration curves in figure 4 were also applicable to these waste streams. This was achieved by comparing SNAP results (Pu g) with Passive Neutron Multiplicity Counting (PNMC) results (Pu g) for drums having progressively more metal content and Pu content (Miller, 2009b), but under 100 kg gross. Figure 5 shows that consistent results were obtained with the 60 keV Am-241 signature and more penetrating emissions from Pu-239 (129 & 414 keV) and fast neutrons from Pu-240 for drums containing 100 % metal waste. The SNAP results at the three photon energies were plotted against the PNMC results and fitted using linear trend-lines (figure 5).

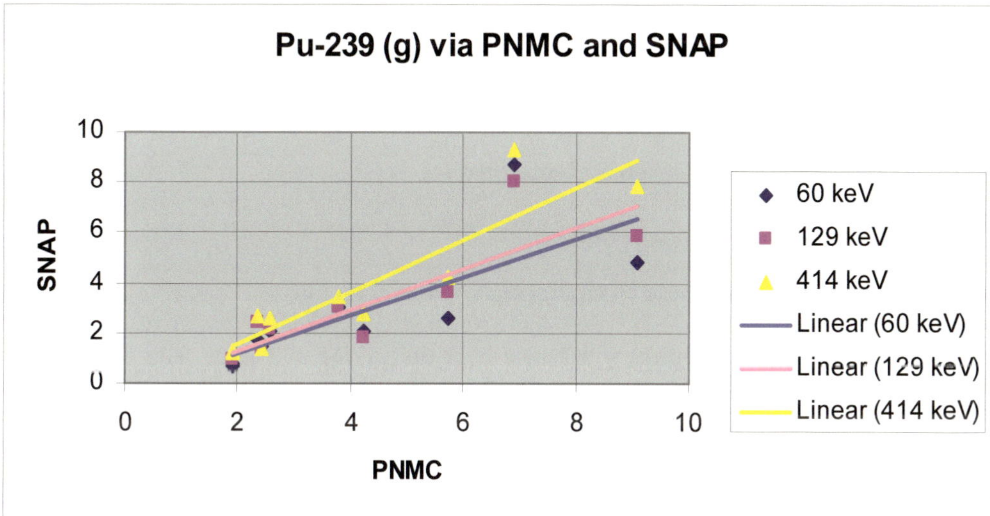


Fig. 5. Comparative results for drums containing metallic wastes

### 3.3 Reclassification of Pu ILW HEPA filters using SNAP and LRGS

Uniform, maximum and minimum response factors, for Am-241 in HEPA filters, were derived for the SNAP (HRGS) and IS610 (LRGS) detectors (Miller 2003). Table 2 summarises the DLs (2σ) calculated using equation 1, for a filter counted at 60 cm on two sides, given: T = 100s; G = 20 kg; R = 10; B = 0.3 cps for SNAP and 10 cps for LRGS; F = maximum, minimum and uniform response factors for SNAP and LRGS.

| Technique | Uniform | Minimum | Maximum |
|---|---|---|---|
| SNAP | 0.48 | 0.43 | 0.59 |
| LRGS | 2.14 | 1.64 | 3.09 |

Table 2. Pu DLs (Bq/g) for HEPA filter assay

Both techniques gave adequate performance for the 100 Bq/g Pu alpha Drigg LLWD threshold and around 2,000 filters have been downgraded from ILW to LLWD. Further development is needed to meet the latest EW threshold of 0.15 Bq/g total Pu alpha and beta activity when R was > 10. This might be achieved by counting for longer within a fully shielded room as used for In-Vivo Monitoring (IVM) of Am-241 in lungs.

### 3.4 Reclassification of soft bagged Pu VLLW using SNAP

Exempt clearance levels, in the UK, have recently reduced from < 11.1 to < 1 Bq/g for U wastes and from < 0.4 to < 0.15 Bq/g for Pu wastes. It can be seen, from table 1, that VLLW categorization avoids relatively costly LLW disposal charges and utilization of limited space at the UK national LLW repository at Drigg. The lower EW categorization gives a relatively smaller cost saving compared to VLLW.

Hence, the principal objective of this application was to develop portable HRGS (SNAP) for the best detection levels and lowest measurement uncertainties for low density soft wastes generated at AWE. Studies have focused on a typical 10-11 kg bag of waste that can be conveniently contained in a reproducible counting geometry by placement inside a standard shortened 200 l plastic waste drum liner. This was monitored as a rotating cylinder and as a disk counted on each broad side in order to determine the counting geometry with the best combination of low detection levels and uncertainties.

#### 3.4.1 Measurements on rotating cylinder

The plastic drum liner was cut to 50 cm in height and had an internal diameter of 55 cm and external diameter of 56.4 cm. Soft waste was represented by 11 kg of paper rolls, with a fill height of 38 cm, giving a typical soft waste bulk density of 0.12 g/cc. The cylinder was placed on a rotating turntable and a general purpose, flush collimated, HRGS detector (HPGe, N-type, crystal: 6.14 cm diameter x 8 cm thick) located at 10, 20, and 25 cm offsets from the centre/middle of the lead brick shielded drum liner (figure 6).

Fig. 6. Counting Geometry for cylinder

A traceable Am-241 source (259 kBq encapsulated in thin plastic) was placed inside the waste material and the detector response (cps/Bq) measured at different locations. Each detector response factor was weighted, according to the volume element represented by the source position, in order to derive the detector response for uniform distribution of Am-241. This was plotted against detector offset, together with the maximum and minimum detector response (figure 7). Data points in figures 7 and 8 were fitted using polynomial (poly) curves.

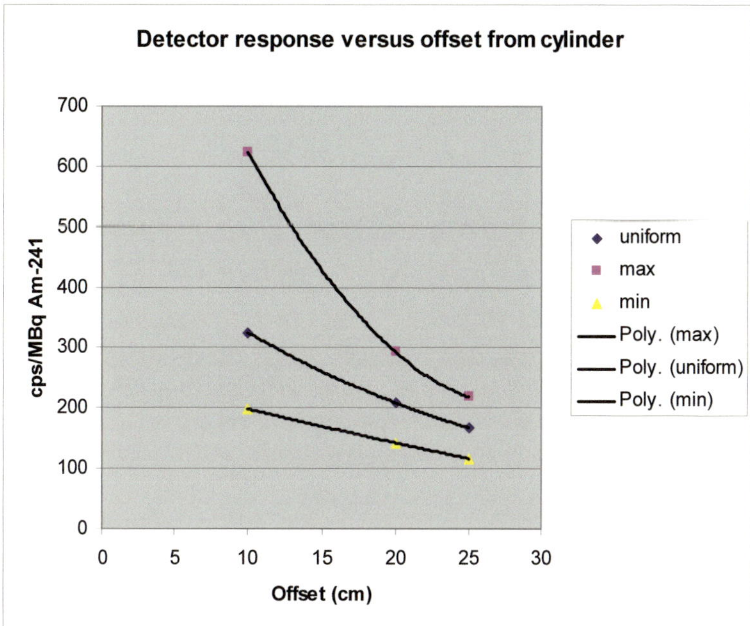


Fig. 7. Detector response factors for cylinder

As the offset reduced detector response increased, but uncertainties increased more sharply. SNAP software was used to generate comparative uniform response factors that were similar to those measured (table 3).

| Detector offset (cm) | Measured | SNAP |
|---|---|---|
| 10 | 324.7 | 283.5 |
| 20 | 208.7 | 194.3 |
| 25 | 167.5 | 161 |

Table 3. Response factors (cps/MBq) for cylinder assay

### 3.4.2 Measurements on waste disk

The Am-241 source was measured at 5 cm intervals along the detector axis and at 5 cm intervals at 7.9, 15.7, 23.6 and 27.5 cm off axis. All response factors were adjusted for attenuation by 0.1 g/cc density soft waste by using tables of mass attenuation coefficients and path-lengths from the source positions to the detector. The uniform response factor, for a 10 kg disk of waste measuring 55 cm in diameter by 40 cm depth, was then calculated by weighting each source location according to the volume element that it represented for detector locations at 10, 20 and 25 cm from the centre of the broadside of the disk. For two sided counting (i.e. inverting the bag halfway through the count), the maximum response was calculated as the average of the maximum and minimum on-axis response factors. The minimum response was achieved for a source location at the mid edge of the disk. Figure 8 shows that all response factors increased as detector offset reduced, but uncertainties rose even more sharply than noted for the cylinder (figure 7).

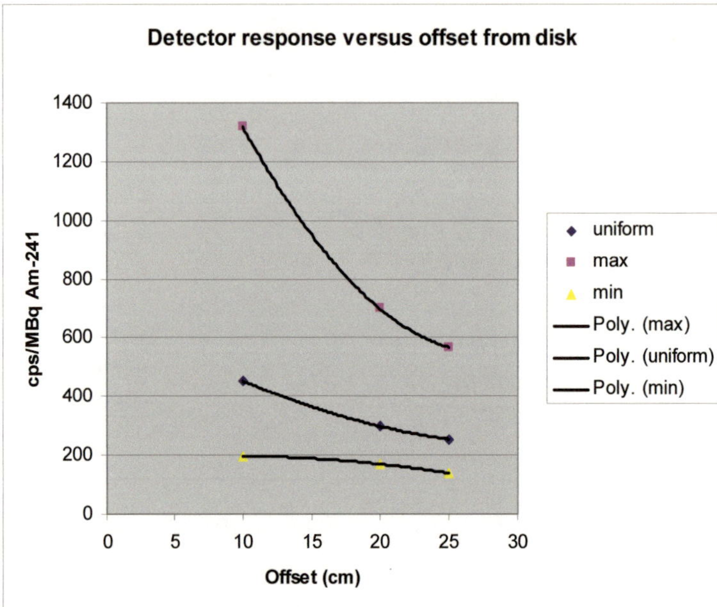


Fig. 8. Detector response factors for disk

Table 4 summarises the close agreement between measured response factors for the disk geometry with those calculated using SNAP software.

| Detector offset (cm) | Measured | SNAP |
|---|---|---|
| 10 | 453 | 446 |
| 20 | 297.9 | 308.4 |
| 25 | 253.9 | 255.1 |

Table 4. Response factors (cps/MBq) for disk assay

### 3.4.3 DL calculations

The DLs (2σ) in table 5 were calculated using equation 1 with: B = 0.2 cps; T = 1000s; G = 10 kg; R = 10 and the response factors in tables 3 and 4 and figures 7 and 8 at 10 cm detector offset.

| Geometry | Uniform | Minimum | Maximum |
|---|---|---|---|
| Disk | 0.063 | 0.021 | 0.14 |
| Cylinder | 0.087 | 0.045 | 0.14 |

Table 5. Pu DLs (Bq/g) for bagged waste assay

Better DLs were achieved with the disk geometry. However, the 0.15 Bq/g clearance level for Pu compositions was only just achieved for activity located at the minimum response position within the waste bag. Furthermore, Pu compositions with higher Pu/Am-241 activity ratios could not be confidently placed in the exempt category.

The performance of the technique could be enhanced by using a larger diameter HPGe crystal since Am-241 response factors are proportional to the detector frontal surface area. For example an 85 mm diameter probe should improve DLs by a factor of two.

Compressing the disk and increasing the waste mass, was also estimated to give a factor of two improvement in DLs. However, spectral summing could give a factor of ten improvement if the spectra from 100 bags were added together (equation 1).

A further factor of 3 improvement could be achieved by monitoring within a shielded room, such as that used for in-vivo monitoring (IVM), where backgrounds are a factor of 10 lower. At present this technique is still being developed and not yet being routinely used to reclassify wastes.

### 3.5 Reclassification of soft bagged Pu VLLW using GC

GC had the potential to assay wastes with consistent isotopic fingerprints, at the new exempt thresholds of < 1 Bq/g U and < 0.15 Bq/g Pu, when sited in areas with low/non-fluctuating photon background. However, careful segregation of wastes was required because the background counts were significantly modified by waste characteristics, such as: NORM content, density, composition and distribution in the counting chamber.

#### 3.5.1 Measurements on uncontaminated materials

Only low bulk density, low Z materials were examined because of the difficulties in GC calibration for high density, high Z objects (Miller, 2010). Approximately 35 kg of each material type was spread evenly throughout the 350 l counting chamber to give a bulk density of 0.1 g/cc. Repeat 600 s counts were done to generate ten measurements for each material. Table 6 summarises the mean and σ values achieved for the net counts (i.e. gross counts minus counts for the continuously updated GC background).

| Material | Net cps | σ | Net cps + 2σ |
|---|---|---|---|
| Empty chamber | 1.8 | 9.7 | 21.2 |
| PVC | 3.4 | 6.1 | 15.6 |
| Tyvek | 25.0 | 3.8 | 32.6 |
| Paper | 41.1 | 9.0 | 59.1 |
| Marigold gloves | 61.5 | 16.3 | 94.1 |
| Overshoes | 66.1 | 5.9 | 77.9 |
| Supertex coveralls | 68.0 | 15.1 | 98.2 |
| Orange coveralls | 138.4 | 6.9 | 152.2 |

Table 6. GC results for uncontaminated materials

The results in table 6 highlight the importance of segregating waste into material types having similar net cps above background. Detection levels (2σ) of 0.09 Bq/g Pu were calculated using equation 2, where: σ = 10 cps; F = 0.0606 cps/Bq Am-241; G = 35 kg; R = 10. Variations in F (DL) with source location were around +/- 25 % (Miller, 2010).

$$DL = (2\sigma/FG)R \tag{2}$$

### 3.5.2 Measurements on Pu wastes

Table 7 summarises the results for eight bags of waste produced by a Pu facility. The sample activity was calculated by dividing the net cps by the Pu counting efficiency and the sample mass. The blank activity (i.e. blank cps/FG) was then subtracted to give the Pu activity measured above the apparent Pu activity indicated for the blank.

| Material | kg | Net cps | Sample (Bq/g) | Blank (Bq/g) | Pu (Bq/g) |
|---|---|---|---|---|---|
| PVC | 29.9 | -8.6 | -0.047 | 0.016 | -0.063 |
| PVC | 30.2 | -5.4 | -0.030 | 0.016 | -0.046 |
| PVC | 33.7 | 25.4 | 0.124 | 0.016 | 0.108 |
| PVC | 28 | 10.2 | 0.060 | 0.016 | 0.044 |
| PVC | 35.3 | 9.8 | 0.046 | 0.016 | 0.030 |
| Orange coveralls | 17.2 | 152.5 | 1.463 | 0.653 | 0.810 |
| Tyvek | 13.1 | 1944 | 24.488 | 0.118 | 24.370 |
| Cotton gloves | 30.8 | 35.1 | 0.188 | 0.112 | 0.076 |

Table 7. GC results for Pu contaminated materials

The tyvek bag indicated a total activity of 24.37 x 13.1 = 319.2 kBq Pu which was equivalent to 31.92 kBq Am-241 since the GC was calibrated using an Am-241 source and a Pu/Am-241 activity ratio of 10/1 because Pu makes no significant contribution to the photon count rate.

The tyvek bag was placed in a standard 200 l waste drum and assayed using SNAP. This indicated 38.8 kBq Am-241 which was 22 % higher than the GC, but within the +/- 25 % uncertainty established for both techniques.

### 3.5.3 Measurements on NPL standard

GC calibration has been achieved using a volume weighted technique (Miller, 2010). Verification has been done using a certified NPL volume source that fills the GC counting chamber. This consists of a corrugated carton (62x62x84 cm) containing a series of nine filter papers (each 58x58 cm), each spiked uniformly with a standard solution of Am-241 and each separately laminated. The filter papers were interspersed within the carton between a series of ten polythene inactive low density inserts (each 60x60x8 cm). The total Am-241 activity was 2857 +/- 12 Bq (reference time 01/01/10) in 29 kg (around 0.1 Bq/g and 0.1 g/cc).The mean cps/Bq Am-241 counting efficiency for the source (6.14 %) was similar to that achieved using volume weighted Am-241 point source measurements in soft waste (6.06 %). A blank NPL source, with no added Am-241, gave no net counts above background.

## 4. U waste reclassification

The lower specific activity of U compositions, compared to Pu compositions, results in a relatively low proportion of U ILW. For example, at the 4,000 Bq/g ILW threshold, a typical 60 kg 200 l drum would contain: 80 mg Pu (3 GBq/g); 80 g Enriched Uranium (EU, 3 MBq/g) and 12 kg Depleted Uranium (DU, 20 kBq/g). Studies have shown that most drums contain < 1g of Pu or U and underestimation of activity, due to self absorption within lumps of Pu or U, was relatively rare at low activities and was be addressed using the SNAP lump correction routine (Miller, 2007, 2008, 2009a).

### 4.1 Reclassification of drummed U ILW using SNAP

Around 100 of the most active EU drums were assayed using SNAP and the SGS. Figure 9 shows that the results correlated reasonably well, but the SNAP results were higher because small lump corrections were required to obtain consistent results at all photon energies. The need for a lump correction was flagged by abnormally low 143/205 kev peak ratios, but was not performed by the SGS.

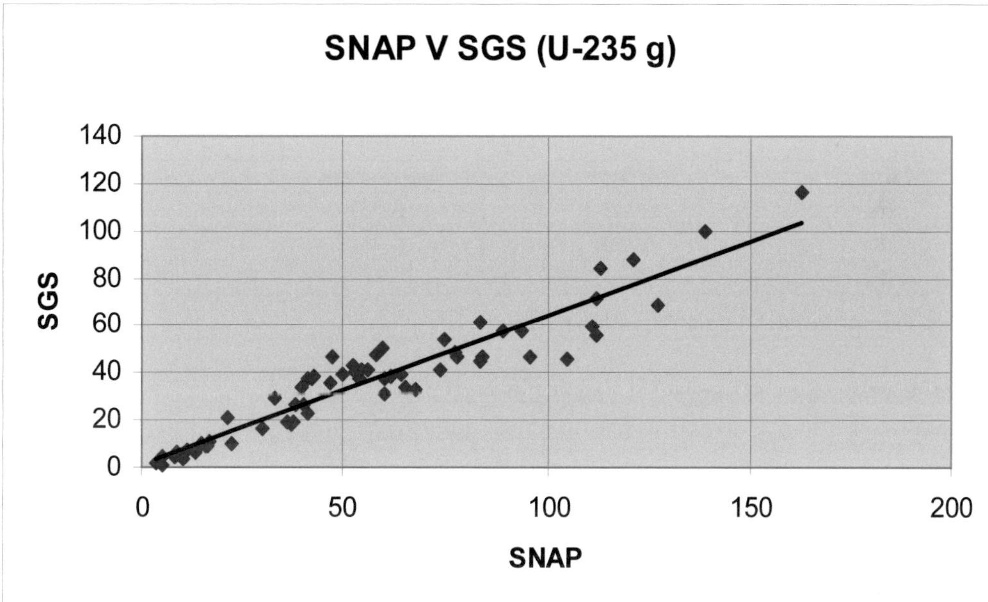


Fig. 9. Comparative results for SNAP and SGS

### 4.2 Reclassification of bagged U VLLW using SNAP

SNAP software was used to calculate detector response factors, for bagged U VLLW in a disk geometry, like those presented in table 4 for Am-241 @ 60 keV. For DU the 93 keV photon from Th-234 was assumed to be in secular equilibrium with U-238. For EU the 186 keV photon was a direct measurement for U-235. Table 8 summarises the comparative uniform response factors.

The DLs (2σ) in table 9 were calculated using a background count rate of 0.2 cps for each photon, 1000s count time, 10 kg waste, the response factors in table 8 and figure 8 at 10 cm and typical isotopic multipliers (i.e. Am-241 x 10 = Pu; U-235 x 40 = EU; Th-234 x 1.55 = DU).

Although not measured, the maximum DLs for U compositions are expected to be well within the 1 Bq/g clearance level for U wastes based on comparison with maximum an minimum Am-241 (Pu) source measurements (figure 8). This recently developed technique has not yet been used for routine measurements.

| Detector offset (cm) | Nuclide | keV | Measured | SNAP |
|---|---|---|---|---|
| 10 | Am-241 | 60 | 453 | 446 |
| 20 | Am-241 | 60 | 297.9 | 308.4 |
| 25 | Am-241 | 60 | 253.9 | 255.1 |
| 10 | Th-234 | 93 | - | 67.3 |
| 20 | Th-234 | 93 | - | 46.5 |
| 25 | Th-234 | 93 | - | 38.5 |
| 10 | U-235 | 186 | - | 550.3 |
| 20 | U-235 | 186 | - | 378.9 |
| 25 | U-235 | 186 | - | 313.4 |

Table 8. Comparative detector response factors for VLLW disk (cps/MBq)

| Material | Uniform | Minimum | Maximum |
|---|---|---|---|
| Pu | 0.063 | 0.021 | 0.14 |
| DU | 0.065 | - | - |
| HEU | 0.21 | - | - |

Table 9. Comparative DLs for VLLW disk (Bq/g)

### 4.3 Reclassification of bagged U VLLW using GC

Thirty waste bags, from a enriched uranium facility, were assayed using Cronos. The waste was mixed material (PVC, paper, coveralls, gloves) and so it was not possible to subtract a blank activity for the apparent EU Bq/g from the waste material itself. Hence the net cps was divided by the EU counting efficiency and the sample mass to give a total EU activity which included the apparent EU activity from the waste material in addition to any EU present. The two most active bags were checked using SNAP and this gave consistent results (table 10). Around 70 % of the bags indicated < 1 Bq/g HEU (table 11). DLs, given a 30s count time and $\sigma$ = 16 cps, were around 0.1 Bq/g for the heavier bags and 1 Bq/g for the lighter bags.

| Bag | GC | SNAP | GC/SNAP |
|---|---|---|---|
| 1 | 23.9 | 19 | 1.26 |
| 8 | 49.4 | 41.8 | 1.18 |

Table 10. Comparative EU mass (U-235 mg)

The background $\sigma$, achieved for 600 s counting (3.8-16.3 cps), was much greater than the square root of the background ($\sqrt{B} = (\sqrt{1800 \times 600})/600 = 1.7$ cps). Also, increasing count time from 60s to 600s gave little reduction in $\sigma$ compared to the $\sqrt{10} = 3.2$ x reduction in $\sqrt{B}$ cps. This highlights the importance of locating GC in an area of low and non-fluctuating photon background. High energy photons from NORM (e.g. K-40 in building materials) and cosmic radiation can penetrate the GC lead shielding and generate lower energy photons that increase the background $\sigma$. Locating GC within a shielded room, as used for in-vivo monitoring equipment, would be a costly, but potentially very beneficial option.

| Bag mass (kg) | Net cps | Bq/g |
|---|---|---|
| 5.9 | 482 | 12.9 |
| 7.1 | 8.0 | 0.2 |
| 9.4 | 30.5 | 0.5 |
| 33.7 | 52.2 | 0.2 |
| 4.4 | 6.1 | 0.2 |
| 7.3 | 15.1 | 0.3 |
| 3.8 | -9.3 | -0.4 |
| 3.4 | 996.1 | 46.1 |
| 10.1 | 14.2 | 0.2 |
| 6.8 | 12.0 | 0.3 |
| 7.8 | 30.4 | 0.6 |
| 3.1 | 22.2 | 1.1 |
| 2.7 | -6.6 | -0.4 |
| 15.6 | -0.7 | 0.0 |
| 6.0 | 110.1 | 2.9 |
| 4.1 | 10.4 | 0.4 |
| 11.0 | 132.7 | 1.9 |
| 5.9 | 12.5 | 0.3 |
| 6.1 | 22.8 | 0.6 |
| 7.5 | 8.4 | 0.2 |
| 10.3 | 7.3 | 0.1 |
| 11.7 | 49.7 | 0.7 |
| 4.2 | 20.8 | 0.8 |
| 2.2 | 13.6 | 1.0 |
| 10.3 | 41.9 | 0.6 |
| 5.4 | 61.0 | 1.8 |
| 5.5 | 16.8 | 0.5 |
| 4.4 | 21.5 | 0.8 |
| 3.9 | 150.9 | 6.1 |
| 6.0 | 59.1 | 1.6 |

Table 11. GC results for EU contaminated waste bags

## 5. Conclusions

The techniques presented have been successfully applied to the majority of waste streams encountered at AWE. However there are two main areas of weakness. Firstly, the potential for underestimation with high density, high Z waste streams. Secondly, achieving adequate DLs for Pu compositions with high Pu/Am-241 activity ratios.

The NPL Measurement Good Practice Guide, for radiometric non-destructive assay, indicates that underestimation caused by even the most severely attenuating wastes reduces to factors of 2 or 3 for distributed contamination that is typical of AWE wastes. Hence, the techniques can be applied, with caution, to the denser waste streams.

Improvement in DL performance may be achieved by a combination of improved shielding, more efficient detectors and spectral summing for similar waste packages.

## 6. References

Miller, T. (2002). Assay of Low Level Plutonium in Soft Drummed Waste, *Proceedings of INMM 2002 43rd Annual Meeting on Nuclear Materials Management*, paper 118, Orlando, Florida, USA, June 23-27, 2002.

Miller, T. (2003). Plutonium Assay in HEPA Filters at the Drigg Threshold, *Proceedings of INMM 2003 44th Annual Meeting on Nuclear Materials Management*, paper 12, Phoenix, Arizona, USA, July 13-77, 2003.

Miller, T. (2007). Depleted Uranium Waste Assay at AWE, *Proceedings of Waste Management Symposia 33rd Annual Meeting*, paper 7017, Tucson, Arizona, USA, February 25-March 1, 2007.

Miller, T. (2008). Applications Where SNAP is Best Practical Means for Radioactive Waste Assay, *Proceedings of Waste Management Symposia 34th Annual Meeting*, paper 8033, Phoenix, Arizona, USA, February 24-28, 2008.

Lasher, D. (2009). Use of High Resolution Gamma Technology During Decontamination and Decommissioning of Nuclear Facilities, *Proceedings of Waste Management Symposia 35th Annual Meeting*, paper 9220, Phoenix, Arizona, USA, March 1-5, 2009.

Miller, T. (2009a). Enriched Uranium Waste Assay at AWE, *Proceedings of Waste Management Symposia 35th Annual Meeting*, paper 9034, Phoenix, Arizona, USA, March 1-5, 2009.

Miller, T. (2009b). Developments in Plutonium waste Assay at AWE, *Journal of Radiological Protection*, Vol.29, No.2, (May 2009), pp. 201-210.

Miller, T. (2010). Waste Assay at the Free Release Threshold Using a Box Monitor, *Proceedings of Waste Management Symposia 36th Annual Meeting*, paper 10007, Phoenix, Arizona, USA, March 7-11, 2010.

Miller, T. (2011). Application of NPL Radioactive Waste Package Standards at AWE, *Proceedings of Waste Management Symposia 37th Annual Meeting*, paper 11004, Phoenix, Arizona, USA, February 27-March 3, 2011.

6

# Estimation of Induced Activity in an ADSS Facility

Nandy Maitreyee[1,*] and C. Sunil[2]
[1]*Saha Institute of Nuclear Physics, Bidhannagar, Kolkata,*
[2]*ARSS, H.P. Division, Bhabha Atomic Research Centre, Mumbai,*
*India*

## 1. Introduction

In order to meet the ever growing global demand for electricity, accelerator driven subcritical systems (ADSS) are emerging as one of the preferred choices. This is due to the fact that in ADSS, the neutrons required to sustain the fission chain reaction in a reactor is supplied from spallation reaction induced by high energy protons from an accelerator, on a heavy target. The spallation target is placed in the core of the reactor. The neutrons produced are used to drive the chain reaction as well as for transmutation of radioactive waste. When the accelerator beam is turned off the supply of neutrons is stopped and any criticality accident may be averted. So ensuring safety against any type of nuclear accident is easier in these facilities as compared to the conventional power reactors. In view of this, concentrated efforts in the field of nuclear engineering are being directed towards the development of ADSS facilities. Production of high flux of fast neutrons through high energy nuclear (spallation) reactions is the main aim of the booster accelerator in an ADSS. This is achieved through the interaction of high current high energy proton beams on suitable targets. Running a high energy accelerator incurs a considerable expenditure and the accelerator parameters are decided for optimizing the cost-benefit of the system (maximum neutron yield for certain beam energy and beam current). It has been found that the neutron economy is optimized around 1 GeV proton energy [1, 2]. However, various structural parameters, target stability, heat generation profile and other logistics are studied at much lower beam energies [3], which are easily available. Running a high energy machine at high currents would lead to the generation of considerable amount of radioactive waste depending on the target-projectile combination. So, in these high current machines one of the factors constraining the beam current and irradiation time is the production of induced radio- and chemical toxicity.

Choice of the target in an ADSS facility is governed by several important factors like high neutron yield, hard neutron spectrum, low radio- and chemical toxicity, low probability of fire hazard, easy cooling of the system, high target stability, reduced cost, etc. Requirement of high neutron yield mandates a heavy element as the target material for the proton induced reaction besides demanding high current beam. Several targets were studied in this

* Corresponding Author

context. Lead and lead-bismuth alloys possess certain desirable properties which have made them to be some of the most suitable candidates for the ADSS target-coolant system [4]. Both Pb and Bi targets produce hard neutron spectrum for high energy proton induced reactions. One of the critical operating conditions of ADSS is high volumetric heat deposition rate. In this regard, Pb and Bi offer low possibility of boiling and loss of cooling, less fire-hazard (compared to liquid Na), and thereby a lower cost. Hence lead-bismuth eutectic (LBE) is one of the suitable targets for an ADSS facility. Some of the other heavy mass targets are W, U. Of these probable ADSS targets, W has similar neutron yield, while radiotoxicity of U target will be a cause of concern. In ADSS, in order to fully utilize the high energy projectile beam and to achieve the maximum neutron yield thick targets are used. In an ADSS, the projectile beam is completely stopped inside the target and maximum number of neutrons is produced from secondary interactions. So, the thickness of the target is kept much larger than the range of the projectile in the target. The amount of undesirable radionuclides produced in an ADSS giving rise to induced radioactivity depends on the target material employed, the projectile current, the transport properties of the projectile and the neutrons produced. This induced activity in ADSS target constitutes the major part of radioactive waste in the facility. The qualitative and quantitative character of this radioactive waste should be known for safe hands-on maintenance of the facility and final disposal of the used target and other materials. The nature and amount of radioactive waste that may be generated can be assessed either through experimental measurements or with the help of theoretical model based calculations. Experimental measurement is always a better choice for precision and accuracy in results, but sometimes we need to have 'a priori' estimate of the induced activity that may build up in an ADSS facility. Secondly, it is not always feasible to experimentally measure the induced activity at high energies and high currents. But, even then, the reaction models used to calculate the activity are not capable of incorporating the characteristics of all physical processes involved in the reaction. Hence the results need to be validated against experimental data.

Currently two types of target configurations are adopted for ADSS: i) windowless target and ii) target separated from the beam pipe by a thin window. Different types of stainless steel are used for the window material of which T91 and D9 are most important. Through the window the intense beam passes through and a large amount of radioactivity is induced in the window material [5]. In this paper we have estimated the radioactivity induced in an LBE target by 1 mA proton beam in the energy range of 400 MeV to 1.2 GeV. As has been mentioned earlier in this section that the neutron cost (estimated from a correlation between neutron yield and cost of running the accelerator) is minimized around 1 GeV proton energy. But the neutron yield saturates at ~2 GeV. So we have chosen the upper limit of our study at 1.2 GeV which is well suited for running an ADSS and at the same time is often affordable for relatively smaller laboratories. But the neutron yield, most often, will be higher at 2 GeV which has prompted other workers to study this energy range for characterization of ADSS parameters. It has also been mentioned earlier in this section that target stability and other studies can be carried out at a much lower energy [3]. Moreover, spallation reaction starts to predominate at proton energy of around 300 MeV. Hence we have taken the lower limit of projectile energy in this study as 400 MeV. Thus the three energies selected in our study, 400 MeV, 800 MeV and 1.2 GeV will help to understand the energy dependence of different variables studied for the ADSS employing a LBE target.

We have also estimated the activity induced in a window material of T91 or D9 stainless steel. In section 2 we describe the method of calculation with the specification of target materials, description of nuclear reaction model codes and method of calculating thick target yield of the radionuclide. In section 3 we discuss the results of our calculations.

## 2. Method of calculation

### 2.1 Specification of the target and window materials

LBE target contains 55% Bi and 45% Pb by weight. The two types of stainless steel D9 and T91 contains the naturally occurring isotopes of several elements like Fe, Cr, Mn, Mo, Ni and others. The composition of LBE, T91 and D9 are given in Table 1.

| Material | Constituting Element | Weight % | Isotope (Atomic %) |
|---|---|---|---|
| | | | |
| LBE | Bi | 55 | $^{209}$Bi (100) |
| | Pb | 45 | $^{204}$Pb (1.4), $^{206}$Pb (24.1), $^{207}$Pb (22.1), $^{208}$Pb (52.4) |
| | | | |
| T91 | Fe | 88.82 | $^{54}$Fe (5.85%), $^{56}$Fe (91.75%), $^{57}$Fe (2.12%), $^{58}$Fe (0.28%) |
| | Cr | 9.0 | $^{50}$Cr (4.34%), $^{52}$Cr (83.79%), $^{53}$Cr (9.50%), $^{54}$Cr (2.37%) |
| | Mo | 1.0 | $^{92}$Mo (14.84%), $^{94}$Mo (9.25%), $^{95}$Mo (15.92%), $^{96}$Mo (16.68%), $^{97}$Mo (9.55%), $^{98}$Mo (24.13%), $^{100}$Mo (9.63%) |
| | Mn | 0.45 | $^{25}$Mn (100) |
| | Ni | 0.4 | $^{58}$Ni (68.08%), $^{60}$Ni (26.22%), $^{61}$Ni (1.14%), $^{62}$Ni (3.63%), $^{64}$Ni (0.93%) |
| | V | 0.2 | $^{50}$V (0.25%), $^{51}$V (99.75) |
| | C | 0.1 | $^{12}$C (98.89), $^{13}$C (1.11) |
| | P | 0.02 | $^{31}$P (100) |
| | S | 0.01 | $^{32}$S (95.02),$^{33}$S (0.75),$^{34}$S (4.21),$^{36}$S (0.02) |
| | | | |
| D9 | Fe | 63.782 | $^{54}$Fe (5.85%), $^{56}$Fe (91.75%), $^{57}$Fe (2.12%), $^{58}$Fe (0.28%) |
| | Ni | 15.5 | $^{58}$Ni (68.08%), $^{60}$Ni (26.22%), $^{61}$Ni (1.14%), $^{62}$Ni (3.63%), $^{64}$Ni (0.93%) |
| | Cr | 14.5 | $^{50}$Cr (4.34%), $^{52}$Cr (83.79%), $^{53}$Cr (9.50%), $^{54}$Cr (2.37%) |
| | Mo | 2.5 | $^{92}$Mo (14.84%), $^{94}$Mo (9.25%), $^{95}$Mo (15.92%), $^{96}$Mo (16.68%), $^{97}$Mo (9.55%), $^{98}$Mo (24.13%), $^{100}$Mo (9.63%) |

| Material | Constituting Element | Weight % | Isotope (Atomic %) |
|---|---|---|---|
| | Mn | 2.35 | $^{25}$Mn (100) |
| | Si | 0.75 | $^{28}$Si (92.23), $^{29}$Si (4.67), $^{30}$Si (3.10) |
| | Ti | 0.25 | |
| | V | 0.05 | $^{50}$V (0.25%), $^{51}$V (99.75) |
| | Al | 0.05 | $^{27}$Al (100) |
| D9 | Nb | 0.05 | $^{93}$Nb (100) |
| | Co | 0.05 | $^{59}$Co (100) |
| | Cu | 0.04 | $^{63}$Cu (69.17), $^{65}$Cu (30.83) |
| | As | 0.03 | $^{75}$As (100) |
| | P | 0.02 | $^{31}$P (100) |
| | Ta | 0.02 | $^{181}$Ta (99.988) |
| | S | 0.01 | $^{32}$S (95.02), $^{33}$S (0.75), $^{34}$S (4.21), $^{36}$S (0.02) |
| | N | 0.005 | $^{14}$N (99.634), $^{15}$N (0.366) |
| | B | 10-20 ppm | $^{10}$B (19.9), $^{11}$B (80.1) |

Table 1. Isotopic composition of LBE, T91 and D9 [5]

### 2.2 Source term

Radioactivity induced in the window and the target material by nuclear interaction of the high energy proton beam leads to the generation of radioactive waste. We have calculated the source term for the production of different radionuclides using two Monte Carlo nuclear reaction model codes QMD (Quantum Molecular Dynamics) [6] and FLUKA [7]. QMD calculates the production cross section for different radionuclides. From these calculated cross sections we have determined the total yield for each radionuclide in the entire target volume considering the gradual and continuous degradation of projectile energy and flux inside the target. The radiation transport code FLUKA gives the total yield of some of the radionuclides. The results obtained from the two model calculations are compared for some of the cases. The calculated results are also compared with a few experimental data.

### 2.3 Nuclear reaction models

#### 2.3.1 QMD

The QMD model simulates the nuclear reaction in an event-by-event basis. Pauli exclusion principle is taken into account by forbidding collisions among the nucleons which lead to transition to already occupied or partially occupied states. The nucleons are represented by Wigner densities of Gaussian wave packets of width $L$ of the form [6]

$$f_i(\vec{r}_i,\vec{p}_i,t)=\frac{1}{\pi h^3}\exp\left(-\frac{[\vec{r}_i-\vec{r}_{i0}(t)]^2}{2L}-[\vec{p}_i-\vec{p}_{i0}(t)]^2\frac{2L}{h^2}\right) \tag{1}$$

$r_{i0}$ and $p_{i0}$ are the centroids of position and momentum of the $i$th nucleon. The distribution function for the total system is given by

$$f_i(\vec{r},\vec{p},t)=\sum_i f_i(\vec{r}_i,\vec{p}_i,t)$$

The time evolution of $r_i$ and $p_i$ is described by,

$$\frac{d}{dt}(\vec{r}_i)=\{H_i,\vec{r}_i\}$$
$$\frac{d}{dt}(\vec{p}_i)=\{H_i,\vec{p}_i\} \tag{2}$$

where $H$ is the Hamiltonian for the system. $H$ consists of kinetic energy, Skryme, Coulomb, Yukawa interaction part and the symmetry energy. Transition between equilibrium, pre-equilibrium and spallation mechanisms including fast multi-particle emissions is effected by changing the relative importance of mean field effects and nucleon-nucleon collisions. Non-equilibrium emissions which play an important role in high energy reaction have been taken into account [2].

The QMD theory considers realistic momentum distribution of the nucleons inside the nucleus, multistep process, multiple pre-equilibrium emission process and variation of the mean field potential due to excitation. Time evolution of $r_i$ and $p_i$ are determined from eqs (2). Particle trajectory is followed through the nuclear volume and once a particle escapes the nuclear boundary, emission is considered. If the positions and momenta of some particles lie within a previously defined range, they are considered to form a cluster. A statistical decay model (SDM) is employed to estimate evaporation from residual nuclei following fast particle emissions as estimated by the QMD model.

QMD calculates the production cross section of different nuclides. The reaction cross section of the projectile in the target calculated by the code is not explicitly expressed. In absence of such information we have used the value of the maximum impact parameter ($b_{max}$) to approximate the reaction cross section and determine the degradation of projectile flux inside the thick target. The reaction cross section ($\sigma_r$) is calculated by,

$$\sigma_r=\pi b_{\max}^2\left(1-\frac{B_C}{E_P}\right) \tag{3}$$

where $B_C$ is the Coulomb barrier for the reaction and $E_p$ is the projectile energy

### 2.3.2 FLUKA

FLUKA [7] is a general purpose tool for calculation of particle transport and interactions with matter, covering an extended range of applications spanning from proton and electron accelerator shielding to target design, calorimetry, activation, dosimetry, detector design, Accelerator Driven Systems, cosmic rays, neutrino physics, radiotherapy etc. It can simulate the interaction and propagation of 60 different types of particles from electrons, photons, neutrons to optical photons, neutrinos, muons etc., most of them of energy from as low as few keV to several TeV. The code has the latest physics models incorporated and is constantly being updated with the latest available data. It can handle even very complex geometries, using an improved version of the well-known Combinatorial Geometry (CG) package. Several user routines are available for very advanced scoring and tracking of particles of interest.

FLUKA(2011) Monte Carlo code was used to calculate the induced activity production in the LBE column. The irradiation profile card (IRRPROFI) allows the flexibility of switching on and off constant beam currents for several intervals and can simulate the irradiation condition with in-between temporary shut offs. This card was used to simulate a constant irradiation of the LBE by a 1 mA beam current for a period of one month. A cylindrical LBE target was chosen as the geometry for irradiation. The length of the cylinder was kept more than the range of the proton (calculated using the SRIM [8] code) while the diameter was fixed at 30 cm to take care of the lateral hadronic cascade. RESNUCLE card was used to calculate the residual nuclei production. For accurate results the new evaporation, heavy fragment evaporation and coalescence modules were activated with the PHYSICS card. Heavy recoils too, were transported using the IONTRANS option as stipulated in the manual. Radioactive decay was requested through the RADDECAY card while induced activity was calculated at different cooling times defined by the DCYTIMES and associating one RESNUCLE cards for every cooling time using the DCYSCORE card. The results were obtained in the units of Bq $cm^{-3}$ and were converted using the volume of the target.

### 2.4 Thick target calculation

The transport code FLUKA gives the total yield of radionuclides in a thick target while the nuclear reaction model code QMD calculates the production cross section of the radioisotopes. From the production cross sections generated by QMD, total yields of different radioisotopes were calculated by adopting a conservative approach. The total thickness of the target is such that the projectile is completely stopped in the target. It has been considered that the thick target is made of a number of thin slabs, such that the projectile loses equal amount of energy, ΔE, in each of these slabs. Energy lost by the projectile per unit path length or the stopping power is calculated using the code SRIM [8]. The total yield of a radionuclide produced is the sum of yields of that particular nuclide from all of the thin slabs at gradually degrading projectile energy weighted by the properly attenuated projectile flux. Moreover it is assumed that the scattering and absorption of the neutrons produced by the primary interaction of the proton beam is negligible. The neutrons produced also contribute to the production of some of the radioisotopes responsible for augmentation of the final repository of the radionuclides.

The kinetic energy $E^i_p$ of the projectile incident on the *i*-th slab and the slab thickness $x_i$ are, respectively, given by [9,10],

$$E_p^i = E_p^0 - (i-1)\Delta E$$

$$x_i = \int_{E_p^i}^{E_p^{i+1}} \frac{dE}{-dE/dx}$$

where $E^0_p$ is the incident energy. Using this formalism we estimated the yield and activity of radionuclides from calculated formation cross sections in the QMD code.

### 3. Results and discussion

We have used reaction model codes to estimate the activity induced in LBE, T91 and D9 targets. Since the reaction model codes use several approximations and cannot incorporate all the detailed features of nuclear reactions, these codes need to be validated against experimental data. We have compared a few experimentally measured cross sections for formation of nuclides by proton induced reaction on Pb and Bi target with those obtained from QMD calculations. This comparison is shown in Table 2. From Table 2 we see that the measured cross sections are fairly reproduced by the calculated values in the energy range considered. We have not compared the cross section calculated by FLUKA with measured data in this work, but the code has already been well validated.

| Reaction | Energy (MeV) | Cross section (mb) | |
|---|---|---|---|
| | | Experiment | QMD |
| $^{nat}Pb$ (p, X)$^{200}Tl$ | 660 | 28.0 ± 8 [11] | 17.0 |
| $^{nat}Pb$ (p, X)$^{198}Au$ | 660 | 1.8 ± 0.3 [11] | 2.7 |
| $^{209}Bi$(p,7n5p)$^{198}Au$ | 1500 | 0.4 ± 0.1 [12] | 1.0 |

Table 2. Comparison of experimentally measured [11,12] and QMD calculated cross sections for a few nuclides

| Z | A | Nuclide | Half-life | Radiation | γ energy (MeV) |
|---|---|---|---|---|---|
| | | | | | |
| 1 | 3 | $^{3}H$ | 12.33 y | β– | |
| 6 | 14 | $^{14}C$ | 5730 y | β– | |
| 11 | 22 | $^{22}Na$ | 2.6 y | ε | 0.511, 1.27 |
| 11 | 24 | $^{24}Na$ | 14.96 y | β– | |
| 19 | 40 | $^{40}K$ | 1.277e+9 y | β–, γ | 1.46 |
| 20 | 45 | $^{45}Ca$ | 162.61 d | β– | |
| 27 | 60 | $^{60}Co$ | 1925.1 d | γ | 1.17, 1.33 |
| 53 | 125 | $^{125}I$ | 59.408 d | ε | |
| 53 | 126 | $^{126}I$ | 13.11 d | ε, β– | |
| 53 | 131 | $^{131}I$ | 8.02 d | β– | 0.364 |
| 54 | 127 | $^{127}Xe$ | 36.34 d | ε | |
| 55 | 131 | $^{131}Cs$ | 9.689 d | ε | |
| 55 | 132 | $^{132}Cs$ | 6.479 d | ε, β– | |
| 55 | 134 | $^{134}Cs$ | 2.0648 y | β– | 0.136, 0.475 |
| 56 | 131 | $^{131}Ba$ | 11.5 d | ε | 0.123, 0.216, 0.496 |
| 56 | 133 | $^{133}Ba$ | 3854 d | ε | 0.303, 0.356, 0.383 |
| 57 | 140 | $^{140}La$ | 1.678 d | β– | 0.487 |
| 78 | 188 | $^{188}Pt$ | 10.2 d | ε, α | |

| Z | A | Nuclide | Half-life | Radiation | γ energy (MeV) |
|---|---|---|---|---|---|
| 78 | 190 | $^{190}Pt$ | 6.5e+11 y | α | |
| 78 | 191 | $^{191}Pt$ | 2.96 d | ε | |
| 79 | 195 | $^{195}Au$ | 186.1 d | ε | 0.261 |
| 79 | 196 | $^{196}Au$ | 6.183 d | ε, β– | 0.333, 0.355 |
| 80 | 194 | $^{194}Hg$ | 520 y | ε | |
| 80 | 197 | $^{197}Hg$ | 64.14 h | ε | 0.077 |
| 80 | 203 | $^{203}Hg$ | 46.612 d | β– | 0.279 |
| 81 | 200 | $^{200}Tl$ | 26.1 h | ε | |
| 81 | 201 | $^{201}Tl$ | 72.91 h | ε | 0.167 |
| 81 | 202 | $^{202}Tl$ | 12.23 d | ε | |
| 81 | 204 | $^{204}Tl$ | 3.78 y | β–, ε | 0.439 |
| 82 | 202 | $^{202}Pb$ | 52.5e3 y | ε | 0.422, 0.787, 0.96 |
| 82 | 203 | $^{203}Pb$ | 51.873 h | ε | 0.279, 0.401 |
| 82 | 205 | $^{205}Pb$ | 1.53e7 y | ε | |
| 83 | 205 | $^{205}Bi$ | 15.31 d | ε | |
| 83 | 206 | $^{206}Bi$ | 6.243 d | ε | |
| 83 | 207 | $^{207}Bi$ | 31.55 y | ε | 0.569, 0.894, 1.43, |
| 83 | 208 | $^{208}Bi$ | 3.68e+5 y | | 0.51, 0.65, 0.921 |

Table 3. Half-life and radiation type of radionuclides for which activity is estimated after 30 days of irradiation [13]

In this work, we have studied the activities of various radionuclides produced, for 30 days of irradiation, due to proton induced reactions on LBE target in the projectile energy range of 400 MeV to 1.2 GeV at a beam current of 1 mA. Induced activities of nuclides calculated using QMD and FLUKA are reported. The list of radionuclides along with their half-lives and type of radiation emitted are given in Table 3. The neutrons produced in the primary interaction also contribute to activity build-up for nuclides near the target mass range. But this has not been taken into account in the QMD calculations. Activity of different radionuclides is estimated for continuous irradiation ranging from 1 day to a maximum period of 30 days. In calculating the induced activity in LBE target we have considered the target thickness to be sufficiently larger than the range of the proton in LBE at the incident projectile energy. As a result the radioactivity induced by the primary projectile has become independent of the target volume.

In figures 1 to 4 we have shown the activity of radionuclides in the mass range of 3 to 208 formed due to proton induced reactions on LBE for 400 MeV, 800 MeV and 1.2 GeV calculated using the code QMD. The radionuclides considered are those having half-lives of more than several days. In some of the cases the product radionuclide may be formed either in the ground state or in a metastable state. Since the codes do not distinguish between formation in the ground or metastable state, in the case of such radionuclides, we have considered the half-life of the ground state unless the metastable state is much longer lived than the ground state. Figure 1 shows the activity build-up of $^{3}H$, $^{14}C$, $^{22,24}Na$, $^{45}Ca$ and $^{60}Co$.

$^{3}H$, $^{45}Ca$ and $^{60}Co$ are nuclides of concern for radiotoxicity of the environment while $^{14}C$ and $^{22}Na$ are biologically important. It has been observed that for 400 MeV incident proton energy the activity build-up for 30 days of irradiation is in the range of $10^{10}$ to $2x10^{11}$ Bq for all the species except for $^{14}C$. Maximum activity of the order of $2.3x10^{11}$ Bq is produced for $^{45}Ca$. $^{14}C$ is a much longer lived isotope than the other product nuclides shown here. The reaction system considered in this plot produces an activity of ~$2x10^{7}$ Bq for $^{14}C$.

As we go to higher projectile energies the variation in the activity build-up pattern with nuclide species changes for the same mass range. In figure 1b we see that for 800 MeV proton energy maximum activity of the order of ~ $8.8x10^{11}$ Bq is obtained for $^{3}H$ while $^{45}Ca$ activity is slightly less (~ $7.0x10^{11}$ Bq). At the highest energy 1.2 GeV of the entire energy range considered $^{3}H$ still have maximum activity of $2.3x10^{12}$ Bq while $^{45}Ca$ has a slightly lower activity of $6.3x10^{11}$ Bq than that at 800 MeV. This observation is explained from the fact that as we go to higher energies, production of high energy neutrons and projectile-like fragments increase [14, 15], competition between a larger number of evaporation channels also come into play [5]. It has been observed from figure 1 that for the nuclide species and the energy range considered, saturation activity is not reached for any of the products for 30 days of irradiation.

Figures 2 (a), (b) and (c) show the activity build-up of different radionuclides in the mass range of 125 to 140. $^{125,131}I$ are biologically important radionuclide as they are used for diagnosis and therapy of thyroid malfunction, Cs is a bone seeking element, radioactive Xe adds to the radiotoxicity of the environment while Ba and La are important lanthanides. From figure 2 we observe that at 400 MeV beam energy maximum activity is obtained for $^{132}Cs$ which is of the order of $1.9x10^{12}$ Bq while minimum activity (~$1.2x10^{10}$ Bq) is obtained for $^{134}Cs$ which is a very long lived isotope. At 800 MeV projectile energy the trend of the activity build-up changes and maximum activity is produced for $^{131}Ba$ while $^{132}Cs$ is produced with an activity less than that by 7.7%. Activity of $^{134}Cs$ is higher (~$3.6x10^{10}$ Bq) than the minimum activity of the order of $1.6x10^{10}$ Bq obtained for $^{133}Ba$. At 1.2 GeV, variation of induced activity formation with nuclide species again changes and $^{140}La$ is produced with maximum activity of ~ $3.0x10^{12}$ Bq. Minimum activity at this energy is obtained with $^{133}Ba$. It has also been observed that for the mass range considered in figure 2 and for 30 days of irradiation, saturation activity is reached only $^{140}La$ at 800 MeV and 1.2 GeV beam energy, while at 400 MeV no significant activity of $^{140}La$ is produced.

In figures 3 (a), (b) and (c) we have plotted the induced activity of some radionuclides in the noble metal group and Hg for 400, 800 MeV and 1.2 GeV proton induced reaction on LBE. From these figures we see that for the isotopes of Pt, Au and Hg shown, the magnitude of maximum activity obtained is much larger than the other two mass groups studied. Maximum activities of $3.2x10^{13}$ Bq, $4.0x10^{13}$ Bq, $3.4x10^{13}$ Bq, respectively, are produced for $^{197}Hg$ for the three beam energies considered while minimum activity is achieved for $^{194}Hg$. In this mass region saturation activity is obtained for $^{194}Au$ and $^{197}Hg$. In figures 4 (a), (b) and (c) we have plotted the activities produced for nuclides close to the target nuclides. From this figure it has been observed that for the entire mass range of product nuclide considered maximum activity is obtained for $^{203}Pb$ and $^{200}Tl$. These activities attain the highest value at 800 MeV and are of the order of $5.5x10^{13}$ Bq and $6.5x10^{13}$ Bq, respectively. Activity produced for $^{201}Tl$ in 30 days' irradiation is similar to that for $^{200}Tl$ at 400 MeV, but at 800 MeV and 1.2 GeV proton energy, these are less than the induced activity of $^{200}Tl$ by 27% and 15% respectively.

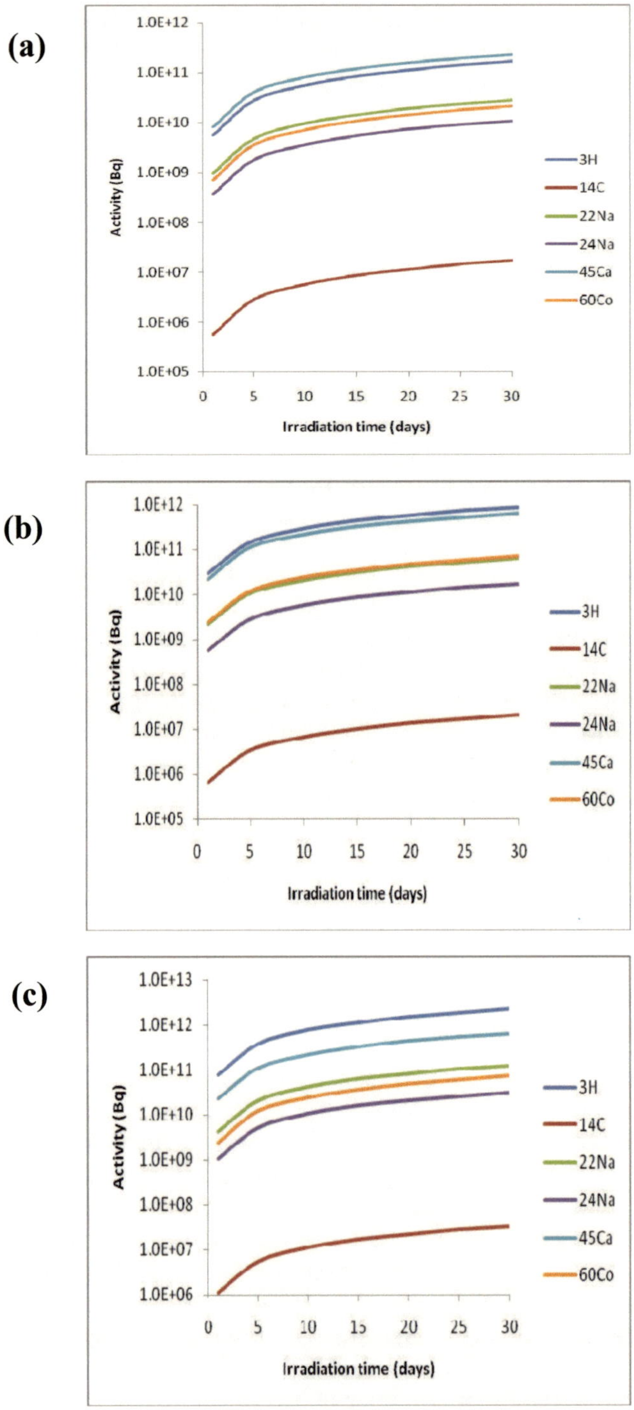


Fig. 1. Activity build-up of some radionuclides in the mass range 3 – 60 for 1 mA proton beam induced reaction on LBE target at beam energy of (a) 400 MeV, (b) 800 MeV and (c) 1.2 GeV

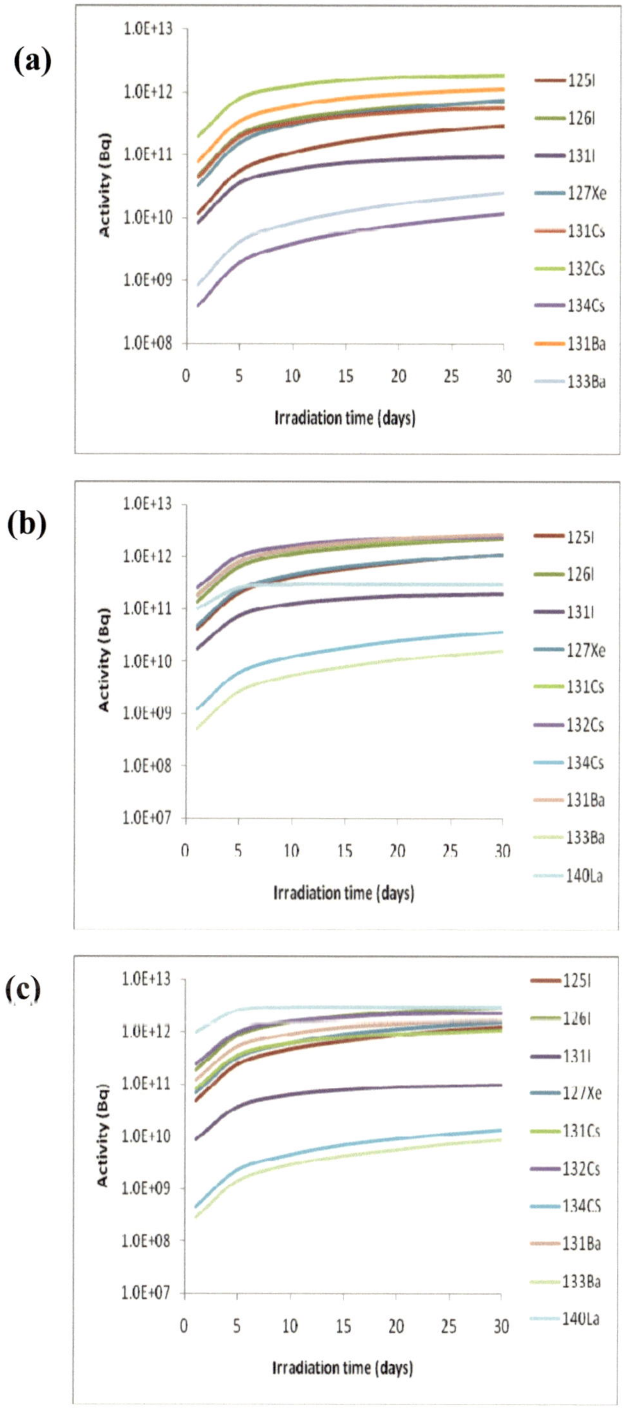


Fig. 2. Activity build-up of some radionuclides in the mass range 125 –140 for 1 mA proton beam induced reaction on LBE target at beam energy of (a) 400 MeV, (b) 800 MeV and (c) 1.2 GeV

**(a)**

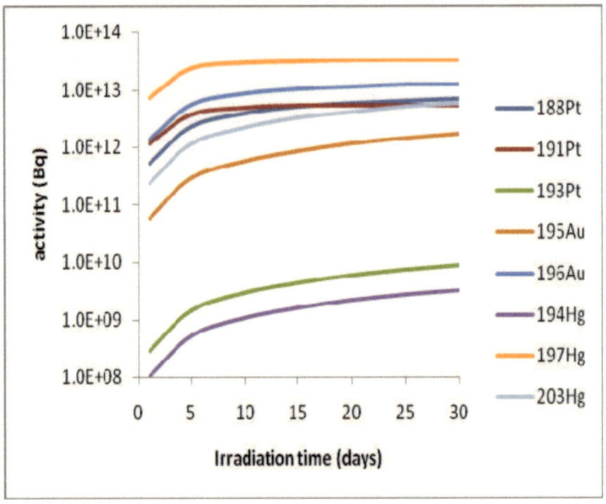


**(b)**

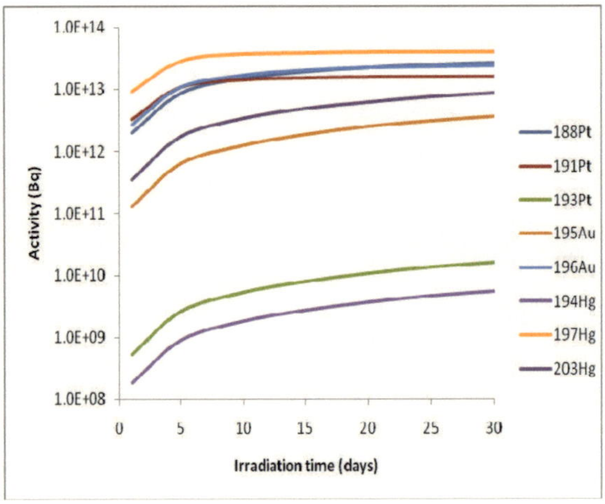


**(c)**

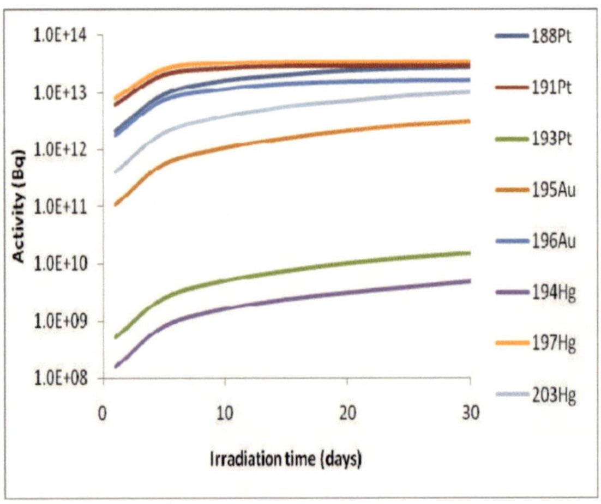


Fig. 3. Activity build-up of some radionuclides in the mass range 188 – 203 for 1 mA proton beam induced reaction on LBE target at beam energy of (a) 400 MeV, (b) 800 MeV and (c) 1.2 GeV

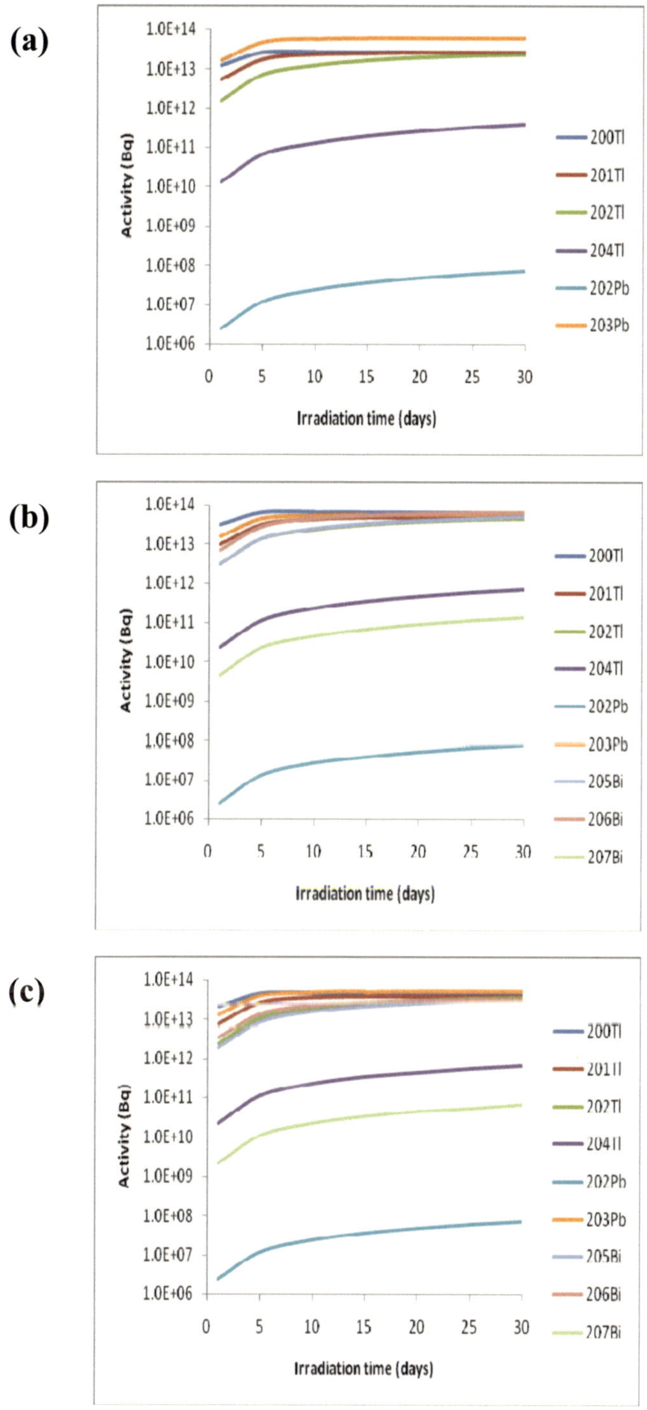


Fig. 4. Activity build-up of some radionuclides in the mass range 200 - 207 for 1 mA proton beam induced reaction on LBE target at beam energy of (a) 400 MeV, (b) 800 MeV and (c) 1.2 GeV

We have also calculated the activity of various radionuclides produced in proton induced reaction on LBE target in the energy range from 400 MeV to 1.2 GeV using the code FLUKA. In Table 4 we have given these calculated activities of several radionuclides for proton energies of 600 MeV and 1.0 GeV. From the table we see that for both the projectile energies considered maximum activities of the order of $6.0 \times 10^{14}$ Bq and $1.0 \times 10^{15}$ Bq, respectively are obtained for $^{201}$Tl while minimum activity is produced for $^{205}$Pb. This minimum activity is of the order of $3.6 \times 10^{6}$ Bq and $7.7 \times 10^{6}$ Bq respectively. It has also been observed that saturation activities are reached for $^{191}$Pt, $^{196}$Au, $^{197}$Hg, $^{200,201}$Tl, $^{203}$Pb. Our analysis revealed that the activity induced for various radionuclides after 30 days of irradiation as calculated by FLUKA is underpredicted to a small extent by the QMD calculation. This is explained from the fact that FLUKA simulations also take into account the activation of the target nuclei by the neutrons produced in primary interaction besides activation by the primary projectile. But in QMD calculation we have not included the production of radionuclides by neutrons.

In figures 5 and 6 we have shown the yield of different nuclides plotted against atomic number Z and mass number A for proton energy 600 MeV and 1.0 GeV, respectively. In fig. 5 we see that in the mass range upto A=150, yield of most of the nuclides are in the range of $10^{11}$ atoms. Some of the nuclides in the range Z= 40 - 50 and A = 100 - 125 are produced with a yield of $10^{14}$ atoms. But as we approach nearer to the target mass region the yield of nuclides increases to ~ $10^{15}$ - $10^{16}$ atoms. At projectile energy of 1.0 GeV this general trend of variation of yield for different species of nuclides remain the same, but the absolute yield of the nuclides increase, particularly for higher mass nuclides. The maximum yield of the target like nuclides is of the order of $10^{16}$.

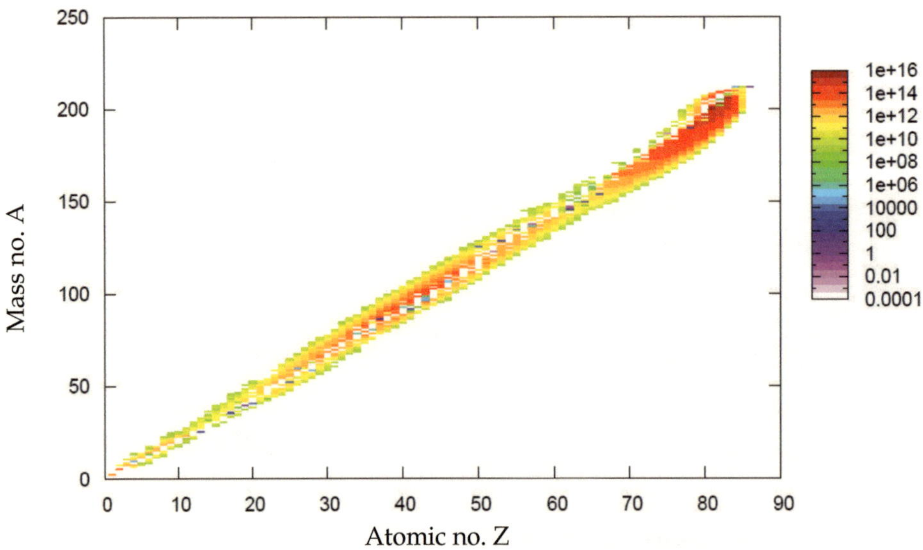


Fig. 5. Yield of nuclides for 1 mA, 600 MeV proton irradiation of LBE target for 30 days, plotted against atomic no. Z and mass no. A

| A | Z | Nuclide | Activity (Bq) for proton energy (MeV) | |
|---|---|---|---|---|
| | | | 600 MeV | 1.0 GeV |
| | | | | |
| 208 | 83 | $^{208}$Bi | 2.5E+08 | 5.8E+08 |
| 205 | 82 | $^{205}$Pb | 3.62E+06 | 7.70E+06 |
| 203 | 82 | $^{203}$Pb | 7.7E+14 | 1.4E+15 |
| 202 | 82 | $^{202}$Pb | 7E+08 | 1.2E+09 |
| 204 | 81 | $^{204}$Tl | 1.1E+12 | 2.2E+12 |
| 202 | 81 | $^{202}$Tl | 8E+13 | 1.6E+14 |
| 201 | 81 | $^{201}$Tl | 6E+14 | 1E+15 |
| 200 | 81 | $^{200}$Tl | 5.3E+14 | 8.8E+14 |
| 203 | 80 | $^{203}$Hg | 1.9E+12 | 4.4E+12 |
| 197 | 80 | $^{197}$Hg | 3E+14 | 4.8E+14 |
| 194 | 80 | $^{194}$Hg | 2.3E+10 | 3.6E+10 |
| 196 | 79 | $^{196}$Au | 2.3E+12 | 6E+12 |
| 195 | 79 | $^{195}$Au | 2.4E+13 | 3.7E+13 |
| 193 | 78 | $^{193}$Pt | 2.4E+11 | 3.9E+11 |
| 191 | 78 | $^{191}$Pt | 9.8E+13 | 1.8E+14 |
| 188 | 78 | $^{188}$Pt | 6.2E+13 | 1.3E+14 |
| 60 | 27 | $^{60}$Co | 6.9E+09 | 2E+10 |
| 45 | 20 | $^{45}$Ca | 1.8E+10 | 7.6E+10 |
| 24 | 11 | $^{24}$Na | 4.5E+11 | 1.8E+12 |
| 22 | 11 | $^{22}$Na | 3.7E+08 | 2E+09 |
| 14 | 6 | $^{14}$C | 2E+07 | 9E+07 |
| 3 | 1 | $^{3}$H | 3.3E+12 | 8E+12 |

Table 4. Activity of radionuclides after 30 days of irradiation as calculated by FLUKA for 600 MeV and 1.0 GeV proton induced reaction on LBE target.

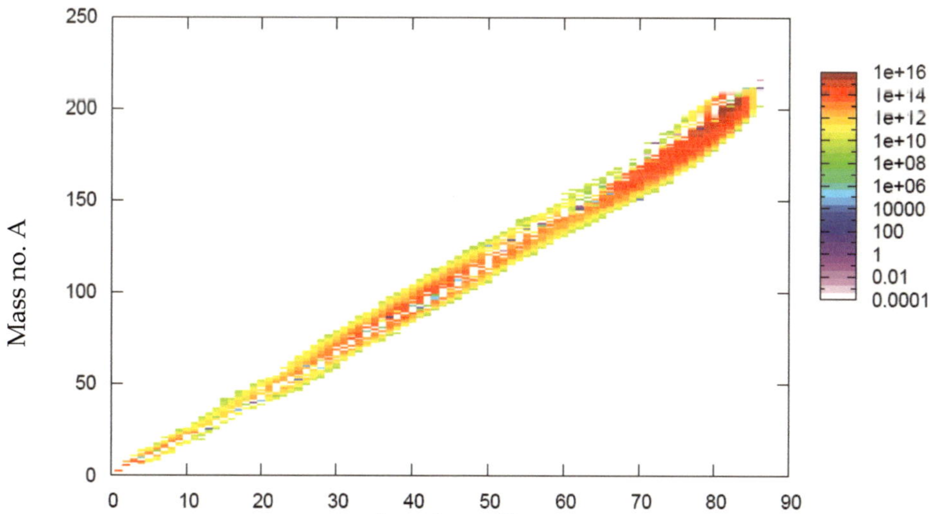


Fig. 6. Yield of nuclides for 1 mA, 1000 MeV proton irradiation of LBE target for 30 days, plotted against atomic no. Z and mass no. A

In the case of LBE target irradiated by a proton beam, one of the major concern on radiotoxicity is the production of $^{210}Po$ via the production of β– emitter $^{210}Bi$ by neutron induced reaction. $^{210}Po$ is an α emitter and pose a potential hazard for internal contamination. We have estimated the production and decay of $^{210}Po$ using the code FLUKA. Figure 7 shows the activity of $^{210}Po$ over cooling times of $10^2$ sec to $10^7$ sec after beam shutdown at the end of 30 days of irradiation.

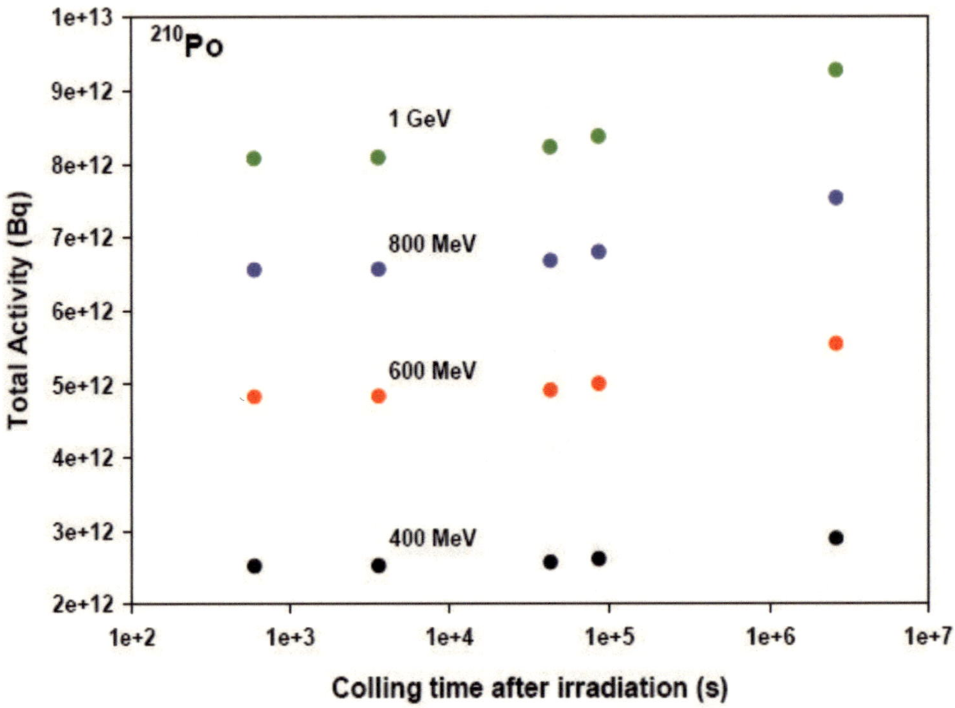


Fig. 7. Activity of $^{210}Po$, produced from 1 mA proton beam irradiation of LBE target at different projectile energies, for 30 days, over different cooling times.

In the next part we report the induced activity in the window material made of T91 and D9 stainless steel. T91 and D9 are metal alloys used as the beam window in an ADSS facility. Isotopic composition of the window materials are given in section 2.1, Table 1. The window is not used as a stopping material, but only attenuates the beam. We have considered a thickness of 2.5 cm for the beam window. The build-up of radioactivity in T91 windows by proton induced reactions at beam energies of 400 MeV, 800 MeV and 1.2 GeV is shown in figures 8 (a), (b) and (c). From these figures we see that maximum activity is reached for $^{51}Cr$ of the order of $4.6 \times 10^{13}$ Bq - $6.0 \times 10^{13}$ Bq. The radioactivities produced for most of the other nuclides are in the range of $10^{11}$ Bq - $10^{12}$ Bq. Smallest activity is produced for $^{22}Na$ and $^{39}Ar$ in the range of $10^9$ Bq - $10^{10}$ Bq. From these figures we see that for 30 days of irradiation, saturation activities are reached for $^{24}Na$ and $^{41}Ar$. Gaseous radiotoxicity due to $^3H$ builds up from an activity of ~ $10^{11}$ - $10^{12}$ Bq and is much lower than the maximum activity produced.

(a)

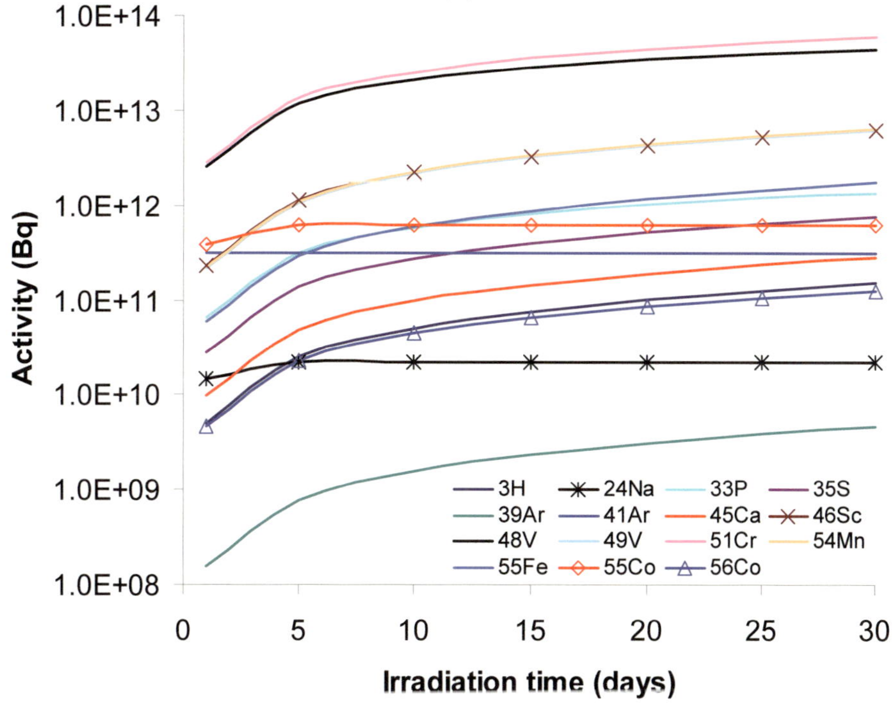
1.0E+14
1.0E+13
1.0E+12
1.0E+11
1.0E+10
1.0E+09
1.0E+08
Activity (Bq)
0
5
10
15
20
25
30
Irradiation time (days)
3H
24Na
33P
35S
39Ar
41Ar
45Ca
46Sc
48V
49V
51Cr
54Mn
55Fe
55Co
56Co

(b)

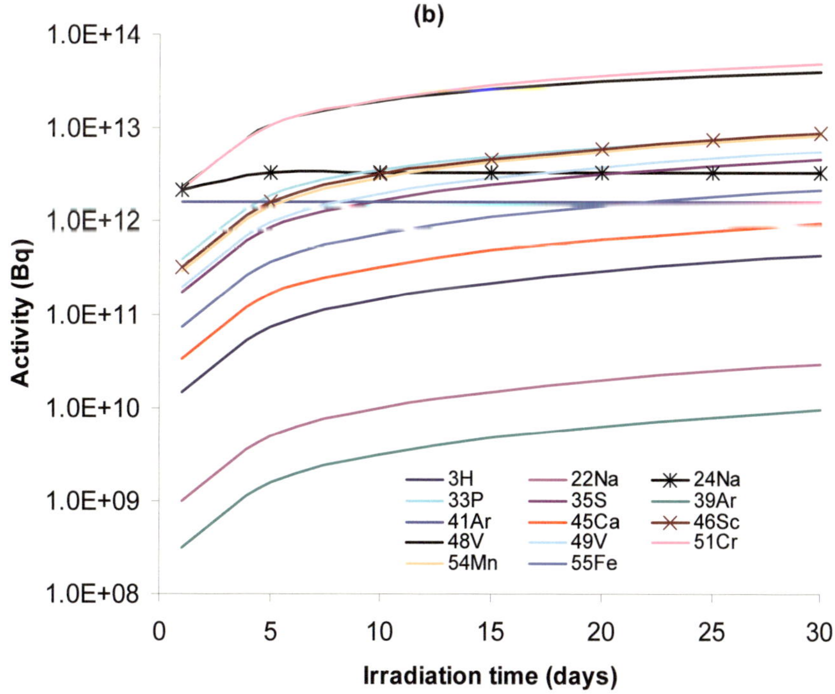
1.0E+14
1.0E+13
1.0E+12
1.0E+11
1.0E+10
1.0E+09
1.0E+08
Activity (Bq)
0
5
10
15
20
25
30
Irradiation time (days)
3H
22Na
24Na
33P
35S
39Ar
41Ar
45Ca
46Sc
48V
49V
51Cr
54Mn
55Fe

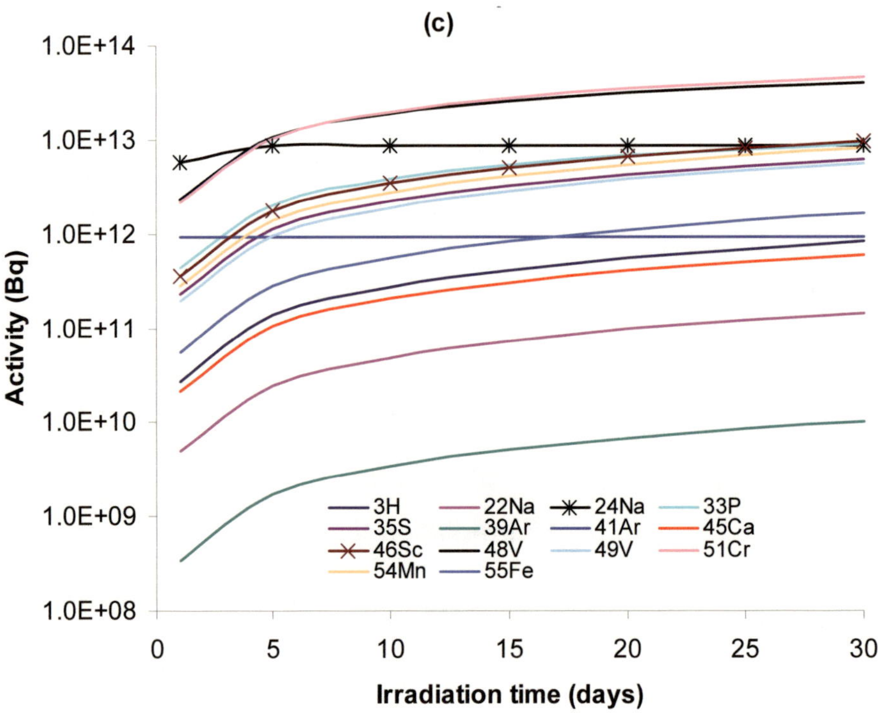


Fig. 8. Activity build-up of radionuclides in T91 window material for 1 mA proton beam induced reaction at beam energy of (a) 400 MeV, (b) 800 MeV and (c) 1.2 GeV

In figures 9 (a), (b) and (c) we have shown the build-up of radioactivity for the same radionuclides as in the case of T91 along with some additional nuclides produced in proton induced reactions on D9 at projectile energies of 400 MeV, 800 MeV and 1.2 GeV. The activity build-up pattern for different radionuclide species in this case is similar to that in the case of T91 target. In the case of D9 window we observe production of $^{60}Co$ for beam energies of 800 MeV and 1.2 GeV. Among the various radioisotopes studied for D9 window, $^{60}Co$ is produced with minimum activity, has a half-life of 5.27 years, has two strong characteristic γ-rays and hence is a source of potential hazard. Activity of $^{60}Co$ is produced in the range of $5.0 \times 10^8$ Bq – $1.2 \times 10^9$ Bq. Proper precautionary measure need be taken for handling these window materials.

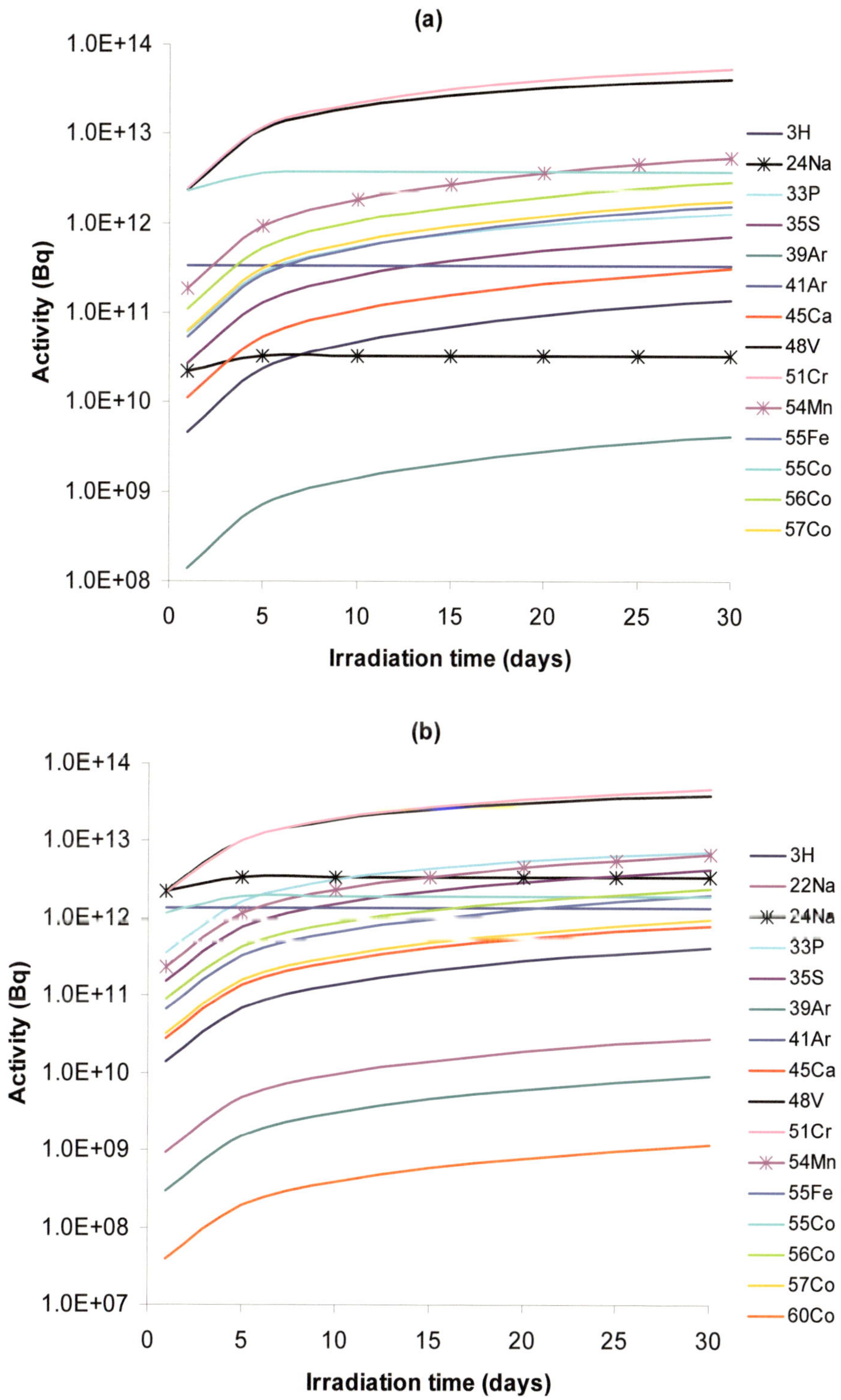
(a)
Activity (Bq)
Irradiation time (days)
1.0E+14
1.0E+13
1.0E+12
1.0E+11
1.0E+10
1.0E+09
1.0E+08
0
5
10
15
20
25
30
3H
24Na
33P
35S
39Ar
41Ar
45Ca
48V
51Cr
54Mn
55Fe
55Co
56Co
57Co
(b)
Activity (Bq)
Irradiation time (days)
1.0E+14
1.0E+13
1.0E+12
1.0E+11
1.0E+10
1.0E+09
1.0E+08
1.0E+07
0
5
10
15
20
25
30
3H
22Na
24Na
33P
35S
39Ar
41Ar
45Ca
48V
51Cr
54Mn
55Fe
55Co
56Co
57Co
60Co

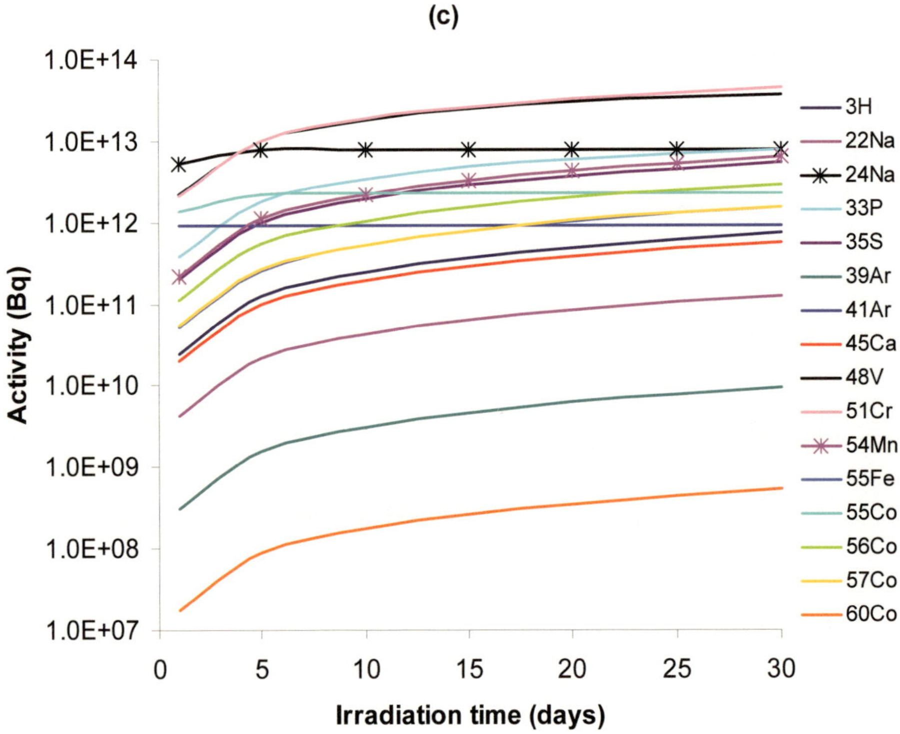


Fig. 9. Activity build-up of radionuclides in D9 window material for 1 mA proton beam induced reaction at beam energy of (a) 400 MeV, (b) 800 MeV and (c) 1.2 GeV

## 4. Conclusion

It has been observed from the analysis of the present work that a significant inventory of radiotoxic isotopes are produced due to irradiation of LBE target by protons in the energy range of 400 MeV to 1.2 GeV at a beam current of 1 mA. Highest yield of radioisotopes with high radiotoxicity like $^{3}H$, $^{14}C$, $^{22}Na$, $^{60}Co$, $^{203}Hg$, $^{210}Po$ is obtained at 1.2 GeV proton energy and it has been observed that production of radiotoxic nuclides increases with increasing proton energy. This is undesirable, but on the other hand neutron yield increases with increasing proton energy, which is desirable. Thus a compromise has to be struck between the two conflicting requirements. QMD calculations for 0.4, 1.2 and 2.0 GeV proton induced reactions on LBE target show that the ratio total neutron yield/total mass yield (which is an indicator of the neutron economy against radioactive waste generation) increases from ~8.0 at 0.4 GeV to 9.0 at 1.2 GeV and then decreases to ~8.36 at 2.0 GeV.

In an ADSS, a window is used to separate the beam transport pipe from the liquid metal target. T91 target window is known to perform better under severe stress to which the ADSS target window is subjected to and in handling the large amount of heat developed. Our study shows that the induced radioactivity in T91 window due to 1.2 GeV 1 mA proton beam is higher than that in D9 window by ~10% but is still preferred for its better performance under severe conditions [16].

Many of the radioisotopes produced are β– or β+ active and should be properly shielded against internal contamination along with the α emitters. Maximum activity is produced for $^{200,201}$Tl and $^{203}$Pb of the order of 6x$10^7$ MBq. Half-lives of these isotopes are of the order of 1-3 day and proper decay time may be allowed for these activities to cool down to the acceptable limit. $^{207}$Bi, having a half-life of 31.55 years and characteristic γ-ray emissions with energies 0.569, 0.894 and 1.43 MeV, is produced with an activity of $10^4$ MBq. Precautionary measure should be taken for handling of target assembly and disposal of used target with such long-lived activities. Total containment of induced activity during and after operation and proper waste disposal facility should be in place for efficient running of the ADS system. $^{210}$Po activity, which is a potential source of α-contamination, is generated to the extent of ~ 8 - 9 x$10^6$ MBq. $^3$H, which poses radiation hazard threat to the environment by adding to ground water contamination, is produced with an activity of $10^5$ MBq from the irradiation of T91 and D9 beam windows. Proper containment for both $^3$H and $^{210}$Po should be ensured to protect plant personnel and environment from unwanted exposure to radiation while operating a target assembly containing a beam window.

In a high energy accelerator like one in ADSS, neutrons from the primary interaction contribute a significant fraction of the total radioisotope inventory. This should be taken into account while calculating the production cross section of the nuclides in the framework of QMD model. The inventory of radioisotopes presented in this work should provide an estimate of the total induced activity in such an ADSS facility using LBE target either with a beam window of stainless steel or a windowless target. Calculated values of induced activity will be scaled up by the ratio of currents if the system need be run at higher currents.

## 5. Acknowledgement

The authors gratefully acknowledge the support provided by Ms. C. Lahiri, presently at Physics Department, University of Calcutta, Kolkata, by assisting in a part of the calculations." should be replaced by "....part of the calculations and by CBAUNP project, SINP.

## 6. References

[1] A Letourneau *et al*, *Nucl. Instrum. Methods* B70, (2000) 299.
[2] Maitreyee Nandy and P.K. Sarkar, PRAMANA-Journal of Physics 61 (4) (2003) 675

[3] P. Vladimirov, A. Mslang, J. Nucl. Mater Part 1 329–333 (2004) 233.
[4] T. Obara, T. Miura, H. Sekimoto, Prog. Nucl. Energy 47 (1–4) (2005) 577; J. Zhang, N. Li, J. Nucl. Mater 326 (2–3) (2004) 201.
[5] Maitreyee Nandy and P.K. Sarkar, Nucl. Instr. Meth. Phys. Res. A 583 (2007) 248
[6] K Niita, S Chiba, T Maruyama, T Maruyama, H Takada, T Fukahori, Y Nakahara and A Iwamoto, *Phys. Rev.* C52, 2620 (1995); K Niita, T Maruyama, T Maruyama and A Iwamoto, *Prog. Theor. Phys.* 98, 87 (1997)
[7] A Fasso, et al FLUKA: a multi-particle transport code, (2005), CERN-2005-10.
[8] J.F. Ziegler, J.P. Biersack, U. Littmark, "The Stopping and Range of Ions in Solids", Pergamon Press, New York, 1985.
[9] M. Nandy, T. Bandyopadhyay, P.K. Sarkar, Phys. Rev. C 63 (2001) 034610.
[10] Sunil C., Maitreyee Nandy and P.K. Sarkar, Phys. Rev. C 78 (2008) 064607.
[11] http//www.nndc.bnl.gov, C,96MOSCOW, 221,96.
[12] http//www.nndc.bnl.gov, C,96SAROV,184,(1996).
[13] R.B. Firestone et al, "Table of Isotopes", ed. S.Y. Frank Chu, Coral M. Baglin, 8th edition, Lawrence Berkeley National Laboratory, University of California (1999)
[14] Anna Kowalczyk, "Proton induced spallation reactions in the energy range 0.1 - 10 GeV" arXiv:0801.0700v1 [nucl-th] 4 Jan 2008
[15] Y. Kadi * and J.P. Revol, "Design of an Accelerator-Driven System for the Destruction of Nuclear Waste", Workshop on Hybrid Nuclear Systems for Energy Production, Utilisation of Actinides & Transmutation of Long-Lived Radioactive Waste, Trieste, LNS0212002 3 - 7 September (2001)
[16] V. Mantha, A. K. Mohanty and P Satyamurthy, PRAMANA-Journal of Physics 68 (2) (2007) 355

# Low-Waste and Proliferation-Free Production of Medical Radioisotopes in Solution and Molten-Salt Reactors

D. Yu. Chuvilin, V. E. Khvostionov,
D. V. Markovskij, V. A. Pavshouk and V. A. Zagryadsky
*NRC "Kurchatov Institute"*
*Russian Federation*

## 1. Introduction

There is an old dispute what is more valuable in work of atomic power stations - the electric power or by-products of its operation - produced stable and radioactive isotopes. Among the experts in reactors physics and techniques, an opinion prevails that nuclear reactors make profit exclusively from electric power production on atomic power stations. However some specialists consider this opinion rather far from a real state of things.

In the early nineties on the initiative of the American Council for Energy Awareness, USCEA, detailed research of influence of nuclear and radiating technologies (except for defensive) on national economy of the USA (Rabotnov, 1999) was undertaken. It consisted of two parts. In one part applications of radioactive materials in the industry, medicine and scientific researches were only considered, in the other - actually nuclear power, that is electric power production at atomic power stations. Results have surprised both authors of the report, and customers. First, the scale of figures. It appeared, that the total annual volume of the business connected with radioisotopes application in the United States made 257 billion dollars in 1991. Secondly, this sum appeared three and a half times more than the full cost of the nuclear electric power (73 billion dollars). Radioisotopes are used in eight various branches, 3.7 million persons of almost five hundred specialties are employed in this activity that makes about three percent of full employment in the USA. The total product of 330 billion dollars means that the nuclear-power and radiation complex of the USA looks as the eleventh on size a world industrial state.

Actually nuclear reactors are "generators" of a large amount of artificial radioisotopes, basic of which are fission products and actinium series. It is assumed to consider as fission products not only the radioisotopes got directly as a result of heavy nuclei fissions, but also the radioisotopes formed as a result of radioactive transformations and nuclear reactions of type (n, γ), (n, 2n), (n, p), etc. on radioactive and stable nuclei of fission products.

The chapter presents low-waste technologies of medical fragment radioisotopes production. Schemes of fragment radioisotopes production in reactors with fuel in the form of uranium salts solution or with molten-salt fuel on the basis of metal fluorides LiF, $BeF_2$, $UF_4$ are considered. The aggregate state of liquid fuel opens the possibility of selective extraction of fragment radioisotopes $^{99}Mo/^{99m}Tc$, $^{89}Sr$, $^{131,133}I$, $^{133}Xe$, etc. from fuel not involving both $^{235}U$ and the basic group of fragment elements. Radiochemical processing of the irradiated uranium typical for solid fuel systems is as a result excluded, that finally should lead to decrease in an exit of a highly active waste and reduction of the fission materials flow in technological process.

## 2. Application of fragment radioisotopes in medicine

One of the most scale consumers of radioisotopes is nuclear medicine where they are applied in diagnostics and therapy of heavy diseases. Production of medical radioisotopes has turned to an important branch of the industry sharing more than 50 % of annual radioisotopes production all over the world. Today more than 160 radioisotopes of 80 chemical elements are produced by means of nuclear reactors and charged particles accelerators. The nomenclature of preparations labeled with radioisotopes permanently extends, new diagnostic instruments are developed. High efficiency of radioisotopes use in medicine is confirmed by long-term clinical practice. Radioisotope methods are used in oncology, cardiology, hematology, urology, nephrology, etc.

The simplest and cheapest way of scale radioisotopes production is based on fission reaction (n,f). The basic medicine radioisotopes formed as a result of uranium-235 fission by neutrons are $^{137}Cs$, $^{131}I$, $^{133}Xe$, $^{90}Sr$ and $^{99}Mo$.

$^{99}Mo$ is one of the most demanded radioisotopes in nuclear medicine. It is used in $^{99}Mo/^{99m}Tc$ generators widely applied in the world at early diagnostics of oncologic, cardiovascular and a number of other diseases. More than 80 % of radiodiagnostic procedures in the world are carried out by preparations labeled with $^{99m}Tc$ (IAEA, 1999). From the IAEA data, approximately 15 millions of diagnostic procedures were carried out with $^{99m}Tc$ annually in the world ten years ago, from them 7 millions in Europe and 8 millions in the USA. The total world consumption of $^{99}Mo/^{99m}Tc$ for that period was estimated at a level of 6 000 Ci a week (6 day pre-calibration). The prediction of $^{99}Mo$ demand till 2006 made in works (Ball, 1999; Ball, Nordyke & Brown, 2002) is illustrated in the diagram in figure 1. About 30 millions of diagnostic procedures are carried out with $^{99m}Tc$ in the world annually today (Hansell, 2008). The world consumption of $^{99}Mo/^{99m}Tc$ has exceeded 12 000 Ci a week, and annual volume of the world market of $^{99m}Tc$ consumption is estimated as US$ 3.7 billion (including cost of medical services).

The basic producers of $^{99}Mo$ are Canada, EU and South Africa, which supply about 95 % of this radioisotope in the world. The technology of $^{99}Mo$ production is based on irradiation in research reactors and subsequent radiochemical processing of the targets from highly-enriched uranium. Table 1 provides further information on the major $^{99}Mo$ producing reactors (Westmacott, 2010).

Recent series of the not planned stops of the reactors producing $^{99}Mo$, mainly due to significant age of these reactors, have shown to the world imperfection of the existing system of supplying this radioisotope and have caused growing concern for the future of nuclear medicine. The faults in $^{99}Mo$ production had negative effect on treatment of oncologic patients in Europe and Northern America. For example, during these stops, supply of $^{99}Mo/^{99m}Tc$ generators in some British hospitals has reduced to 30 % from normal. The doctors had to choose, who from the patients more than others requires the procedures with $^{99m}Tc$.

| Reactor name | Location | Annual operating days | Normal production per week[1] | Weekly % of world demand | Fuel/ Targets[2] | Date of first commis-sioning |
|---|---|---|---|---|---|---|
| BR-2 | Belgium | 140 | 5 200[3] | 25-65 | HEU/HEU | 1961 |
| HFR | Netherlands | 300 | 4 680 | 35-70 | LEU/HEU | 1961 |
| LVR-15[4] | Czech Rep. | - | >600 | - | HEU[5]/HEU | 1957 |
| MARIA[4] | Poland | - | 700-1 500 | - | HEU/HEU | 1974 |
| NRU | Canada | 300 | 4 680 | 35-70 | LEU/HEU | 1957 |
| OPAL | Australia | 290 | 1 000-1 500 | -[6] | LEU/LEU | 2007 |
| OSIRIS | France | 180 | 1 200 | 10-20 | LEU/HEU | 1966 |
| SAFARI-1 | South Africa | 305 | 2 500 | 10-30 | LEU/HEU[7] | 1965 |
| RA-3 | Argentina | 230 | 200 | <2 | LEU/LEU | 1967 |

1. Six-day curies end of processing. In some cases, the maximum production can be substantially higher that the values listed here for normal production.
2. Fuel elements and targets are classified as either LEU, containing less than 20% of $^{235}U$, or HEU, which contains greater than 20% $^{235}U$ (in some cases greater than 95%).
3. Does not account for increase in capacity since April 2010 with the installation of additional irradiation capacity. This increases BR-2 available capacity to approximately 7 800 six-day curies EOP: however it is not yet clear what "normal" production will be at the facility.
4. These reactors started production in 2010, - so some data is not available yet.
5. The LVR-15 reactor uses fuel elements that are enriched to 36% $^{235}U$.
6. The OPAL reactor started production in 2007 for domestic use but has not yet exported significant amounts.
7. SAFARI-1 is in the process of converting to using LEU targets and expects to have completed conversion in 2010.

Table 1. Major $^{99}Mo$ producing reactors

An attempt to solve the problem of $^{99}Mo$ production by building two specialized reactors MAPLE in Canada was not successful - May 16, 2008 these reactors were finally stopped because of the made design mistakes. Against steady ageing of the research reactors which operation time sometimes exceeds 50 years (reactor NRU), this problem becomes sharper.

Besides, the isotope crisis has drawn attention to one more acute problem - production technology. The fission fragment method is a basis of current $^{99}Mo$ production. A highly-enriched uranium target is irradiated in a reactor, and then processed by one of the known radiochemical ways. $^{99}Mo$, which specific activity reaches several tens thousand Ci per gram of molybdenum is separated from fission products (Gerasimov, Kiselev & Lantcov, 1989). More than 95 % of $^{99}Mo$ are produced using highly-enriched uranium (HEU), ~90 % of $^{235}U$. Commercial producers of $^{99}Mo$ spend about 50 kg of HEU annually.

The weakest feature of this way of $^{99}Mo$ production is extremely inefficient use of uranium. Only ≈0.4 % of $^{235}U$ is used for production of $^{99}Mo$, and the other part is directed in waste. Moreover, at fission of uranium a wide spectrum of fragments is formed, which total activity at the moment of irradiation end exceeds activity of $^{99}Mo$ by two orders of value. That results in necessity of contamination of plenty radioactive waste, including long-lived ones.

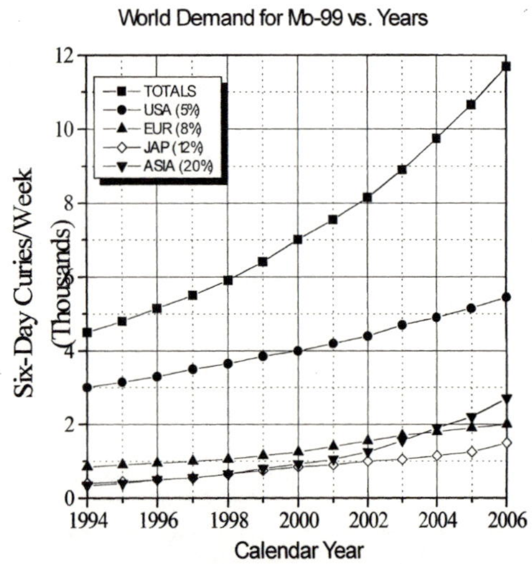


Fig. 1. Prediction of growth of demand for $^{99}Mo$ in the world (Ball, 1999; Ball, Nordyke & Brown, 2002)

In Fig. 2 the basic stages of the existing $^{99}Mo$ production technology are shown - from manufacturing of a uranium target to delivering the final product to the consumer.

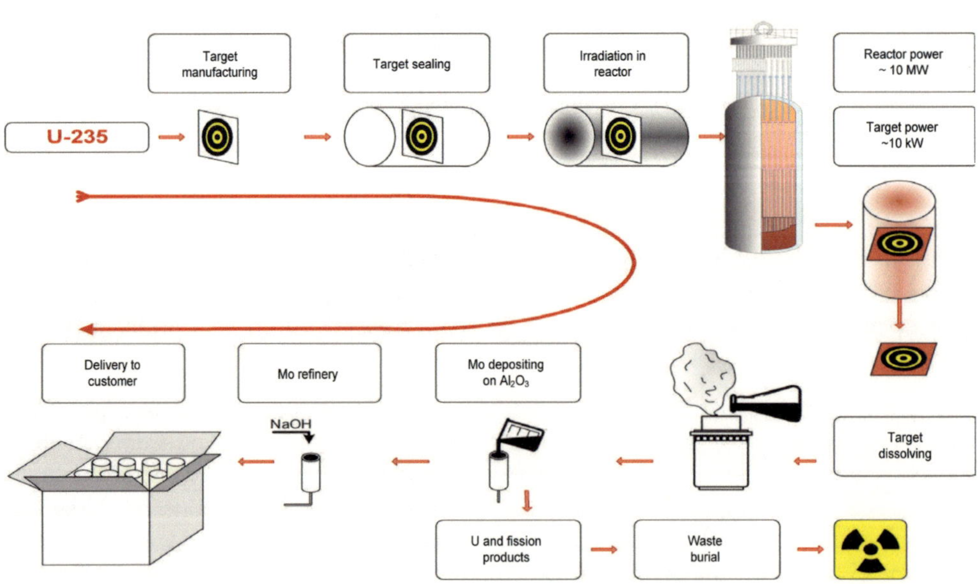


Fig. 2. Basic stages of the existing $^{99}Mo$ production technology

The third problem of the traditional $^{99}Mo$ production is that its target waste can be easily transformed by known chemical methods to metal HEU - material used in manufacture of nuclear military loads. Orientation of modern $^{99}Mo$ production on the use of highly-enriched uranium, against gradual deducing of HEU from a civil turn according to the IAEA "non-proliferation" concept, creates additional risks for $^{99}Mo$ consumers. Thereupon expected rapid development of the nuclear medicine using $^{235}U$ fission products can lead the next years to serious problems with ensuring the efficiency of the international monitoring system for non-proliferation of nuclear materials.

In the international community there is a consensus about necessity to limit or even to bring to naught HEU turn in the civil sphere. Programs, in particular, the program RERTR (Reduced Enrichment for Research and Test Reactors) on reduction of HEU turn in peace sectors of economy are accepted. The research reactors used for production of $^{99}Mo$ will be gradually converted to low-enriched uranium fuel (LEU) and LEU-targets. Risks from use of uranium of high enrichment will start to constrain sooner or later, and then can even stop this production completely.

The world crisis of $^{99}Mo$ production gives an exclusive chance for realization of alternative methods of this radioisotope production, free from disadvantages of the modern technologies and ensuring large-scale low-waste production of $^{99}Mo$ based on the use of low-enriched uranium.

What could be alternative? The choice, in effect, is limited: building of a new multi-purpose research reactor, development of accelerating methods or building of inexpensive and simple in operation solution reactors. Taking into account inertia of the licensing system and time for designing and building the facilities, the choice needs to be made the next years.

It would seem, the most real alternative to HEU for today is conversion of a research reactor core for work with LEU-fuel and introduction of new types of targets of low-enriched uranium. However the "target" technology characterized by the lowest efficiency of use of uranium, most likely will not give competitive advantages as the use of LEU will inevitably lead to involving for manufacture of a considerable amount of irradiated uranium. It will result in additional loadings on radiochemical processing and in considerable amount of radioactive waste that, obviously, will raise production cost price. The most research reactors now producing the basic amount of $^{99}Mo$ are depreciated and for their maintenance in operating and safe condition more and more considerable efforts and expenses are required. The next years these reactors will demand capital reconstruction or will be deduced from operation. Other alternative technologies based on neutron activation of molybdenum isotope $^{98}Mo$ or on uranium photo-fission at accelerators, are not effective for large-scale production and can be applied only for satisfying regional needs.

Renunciation of the traditional technology of uranium targets irradiation and transition to technologies based on specific physical-chemical properties of liquid nuclear fuel, ensuring an opportunity to extract $^{99}Mo$ directly from the reactor fuel leaving uranium and fission fragments in it, allow to exclude a problem of highly-enriched uranium contamination and to simplify the problem of handling the radiochemical processing waste radically. Water solutions of uranium salts or melts of fluoride salts can be used as liquid nuclear fuel.

The aggregate state of liquid fuel opens the possibility of selective extraction of fragment radioisotopes $^{99}Mo$/$^{99m}Tc$, $^{89}Sr$, $^{131,133}I$, $^{133}Xe$, etc. from fuel not involving both $^{235}U$ and the basic group of fragment elements. It will lead to sharp decrease in the flow of nuclear materials in the technological process and in the yield of highly active waste. Estimations show, for example, that a solution reactor of 200 kW power is capable to produce more than 1000 Ci/w of $^{99}$Mo with calibration for 6-days from the moment of delivery.

The technology based on use of homogeneous liquid nuclear fuel possesses a number of obvious advantages in comparison with a "target" method:

- All fuel of such a reactor is a "target";
- There is no necessity for creation of special facility for manufacturing and radiochemical processing of targets;
- $^{99}$Mo is selectively taken from solution fuel by sorption on special filters or spontaneously leaves the molten salt fuel in the form of flying fluorides;
- Uranium and the basic highly active fission fragments remain in fuel;
- Reactors with homogeneous fuel possess immanent nuclear safety (negative thermal reactivity factor);
- Research mini-reactor "Argus" with solution fuel in the form of $UO_2SO_4$ has operated in Russia for about 30 years;
- Research reactor MSRE of power 8 MW with molten salt fluoride fuel $LiF\text{-}BeF_2\text{-}UF_4$ successfully operated in the USA within four years, when the effect of spontaneous exit of a part of fission fragments in the gas phase was observed (including $^{99}$Mo);
- There are no problems with reactor operation on LEU.

Now development and creation of a specialized nuclear-chemical complex for commercial production of fragment radioisotopes $^{99}$Mo, $^{89}$Sr, $^{133}$Xe, $^{131}$I, $^{133}$I, etc. for satisfaction of growing requirements of nuclear medicine look actual.

## 3. Production of fragment radioisotopes with help of molten salt fluoride nuclear fuel

Transition from the traditional technology of uranium targets irradiation to the technology based on an unique effect of spontaneous exit of fragment $^{99}$Mo from molten salt fluoride fuel to the gas phase (Grimes, 1970), observed in the late sixties in the experimental reactor MSRE (ORNL) allows to simplify a problem of waste disposal at production of fragment radioisotopes considerably. According to the results of measurements, more than 50 % of all molybdenum formed at fission of uranium leaves melts. The most of other fragment elements, such as Zr, Ba, Sr, Cs, Br, I, rare-earth elements and all uranium having steady well soluble in fuel salt compounds remain in melts.

The idea of use of molten salt nuclear fuel in the reactor technology was proposed at the beginning of the 50-ties at creation of small power autonomous energy sources. A distinctive feature of this kind of fuel is the possibility of continuous correction of the chemical structure and control of nuclear-physical, chemical and thermal processes at its work in the nuclear reactor.

The molten salt reactor concept was realized at creation of experimental reactors ARE (Aircraft Reactor Experiment) by power of 2.5 MW and MSRE (Molten-Salt Reactor Experiment) of 7.3 MW. The first experimental molten salt reactor, ARE, constructed in the USA in 1954, worked with salt composition $NaF\text{-}ZrF_4\text{-}UF_4$ at temperature 860$^{o}$C (Bettis e.a., 1957). The basic task of the experiment was confirmation of serviceability and stability of the reactor with circulating fuel. MSRE operated in Oak Ridge National Laboratory (ORNL) within four years in the late 60s (Haubenreich & Engel, 1970) with fuel $^7LiF\text{-}BeF_2\text{-}ZrF_4\text{-}UF_4$ at temperature 650°C. The purposes of MSRE construction consisted in working out the molten salt nuclear fuel technology, checking the serviceability of some structure units, study of neutronics characteristics of this reactor type. Corrosion steady nickel alloy Hastelloy-N was developed, some questions of processing and cleaning the salt were solved during reactor operation, and its nuclear safety was shown.

Significant researches of molten salt coolants were performed in the Soviet Union (Novikov et al., 1990). The structural materials on the basis of nickel, corrosion stable in fluorides are created, the pump, heat exchanging devices of various purposes etc are developed and checked in conditions of stand tests. Now these researches proceed, mainly within the framework of the international project "Generation-IV" on creation of new generation nuclear reactors.

### 3.1 Behavior of fission fragments in molten salt fuel

In MSRE, the basic group of fragment elements, such as Zr, Ba, Sr, Cs, Br, I and rare earth elements, having well soluble in salt compounds remained in fuel. Insoluble compounds deposited on the surface of structural materials of the reactor facility (graphite, Hastelloy-N). Krypton and xenon, having low solubility in salt, spontaneously or under bubbling of fuel by inert gas left the salt melt. Moreover it was found, that a small group of elements, so-called "noble" metals Mo, Nb, Te, Ru in the form of volatile fluorides or aerosols entered gas covering free surfaces of salt. About half of these elements deposited on surfaces of reactor structural materials and less than 1% remained in fuel salt.

In a series of ORNL 1966-1970 reports the results of numerous measurements of fuel structure, researches of structural material samples and gas tests from MSRE contour are published. An analysis of these experimental data has allowed us to reveal certain laws of distribution of some fragment elements in the reactor facility, to put forward a hypothesis of the mechanism of "noble" metals exit to gas volume above the salt melt and on this basis to propose a new way of fragment $^{99}Mo$ production.

In particular, in (Rosenthal et al., 1969) (see tab. 2) the balance of radionuclides in reactor MSRE is shown. Significant fraction of $^{99}Mo$, $^{132}Te$, $^{103}Ru$ entered gas covering free surfaces of salt. The detection of significant amount of $^{89}Sr$ in the gas volume above the salt surface is connected with its predecessor $^{89}Kr$ entering into gas.

| Nuclide | Contents (% from calculated value) | | | | |
|---|---|---|---|---|---|
| | Fuel | Graphite | Hastelloy-N | Evacuated gases | Total yield |
| $^{99}Mo$ | 0.17 | 9.0 | 28 | 50 | 87 |
| $^{132}Te$ | 0.47 | 5.1 | 14 | 74 | 94 |
| $^{129}Te$ | 0.40 | 5.6 | 17 | 31 | 54 |
| $^{103}Ru$ | 0.033 | 3.5 | 3.7 | 49 | 56 |
| $^{95}Nb$ | 0.001 | 41 | 18 | 11 | 70 |
| $^{89}Sr$ | 83 | 17 | 8.5 | 0.11 | 109 |

Table 2. Distribution of fission products in reactor MSRE

The assumption was made, that these fission fragments arising in extremely electron-scarce states, in the process of their thermalization catch electrons, passing a number of unstable, but volatile valent states (Kirslis & Blankenship, 1967). If the reactor structural materials are rather inert, and the fuel does not contain significant amount of other absorbers of radiolytic fluoride, some part of Mo, Te and Ru will oxide up to maximal fluorides and pass to the gas phase.

In work (Rosenthal et al., 1968), 50 g salt samples were located in hermetic reactionary volumes connected to a gas contour, from which helium or mix of helium with hydrogen could be made to fuel melt. For melting of fuel the reactionary volume was heated up to 600°C. The gas moved in the tube at the flow rate 10 - 15 $cm^3 \cdot min^{-1}$, allowing both blowing the salt melt mirror and bubbling the fuel. After passage of reactionary volume the gas was evacuated through a system of filters including few sections of fission fragments catching – metal-ceramic filter, sorbent on the basis of NaF and chemical sorbent. After reactor operation of 32 000 MW×h the filters were removed, each section was separately dissolved in acid and the solutions were analyzed on the contents of fission fragments and $^{235}U$. Some measurement results of the salt radioisotope structure are given in table 3.

Calculated by us total activities of radioisotopes collected in all elements of filtration system, normalized to initial activity in salt test are shown in the same table. The cumulative yields of radioisotopes at $^{235}U$ fission by thermal neutrons, their half-decay periods, chemical formulas of maximal fluorides of elements and boiling temperatures of these chemical compounds are also shown here.

| Parameter | $^{103}$Ru | $^{132}$Te | $^{95}$Nb | $^{99}$Mo | $^{95}$Zr | $^{144}$Ce | $^{140}$Ba |
|---|---|---|---|---|---|---|---|
| $T_{1/2}$ | 39.36 days | 78.2 h | 34.97 days | 66.0 h | 64.02 days | 284.4 days | 12.75 days |
| Cumulative yield of nuclide at fission of $^{235}U$,% | 3.04 | 4.28 | 6.5 | 6.15 | 6.29 | 5.47 | 6.08 |
| Cross section of the reaction (n,γ), barn | 7.71 | | <7 | 1.73 | 0.49 | 2.6 | 1.6 |
| Nuclide specific activity in melt, $decay \cdot min^{-1} \cdot g^{-1}$ | $7.5 \cdot 10^6$ | $1.7 \cdot 10^8$ | $10^9$ | $1.2 \cdot 10^9$ | $1.2 \cdot 10^{11}$ | $7.3 \cdot 10^{10}$ | $1.6 \cdot 10^{11}$ |
| Nuclide concentration in melt, $g^{-1}$ | $6.1 \cdot 10^{11}$ | $1.2 \cdot 10^{12}$ | $7.3 \cdot 10^{13}$ | $6.8 \cdot 10^{12}$ | $1.6 \cdot 10^{16}$ | $4.3 \cdot 10^{16}$ | $4.2 \cdot 10^{15}$ |
| Ratio of total nuclide activity in filters to its activity in 1 g of initial melt, rel. un. | 5.6 | 0.7 | ≈0.02 | 1.1 | $1.1 \cdot 10^{-4}$ | $2.0 \cdot 10^{-5}$ | $1.2 \cdot 10^{-5}$ |
| Maximal fluorides of nuclides | $RuF_5$ | $TeF_6$ | $NbF_5$ | $MoF_6$ | $ZrF_4$ | $CeF_3$ | $BaF_2$ |
| Boiling temperature of maximal fluorides at saturated vapor pressure 760 mm, °C | 272 | -38.9 | 233 | 36.2 | 903 | 2327 | 2260 |

Table 3. Radioisotope structure of MSRE fuel ("blowing" of a salt mirror with helium)

First of all one may indicate a wide scatter of fission fragments concentrations in the initial salt sample and a possibility to divide them into two groups by this parameter. The concentrations of $^{99}Mo$, $^{132}Te$, $^{95}Nb$, $^{103}Ru$ are lower by two - four orders of value than the concentrations of $^{89}Sr$, $^{141,144}Ce$, $^{95}Zr$, $^{140}Ba$, though cumulative yields of these fragments differ not more than twice. It is impossible to explain the so large difference by burning up of fragments in the reactor, at the given cross sections. The most probable reason is connected to individual behavior of chemical compounds of these elements in molten salt fuel.

In table 3 there are two radioisotopes $^{95}Nb$ and $^{95}Zr$ presenting the generically connected pair in the decay $^{95}Kr \rightarrow ^{95}Rb \rightarrow ^{95}Sr \rightarrow ^{95}Y \rightarrow ^{95}Zr \rightarrow ^{95}Nb \rightarrow ^{95}Mo$. At long irradiation of fuel, these radioisotopes should be in balance, i.e. their activities should be equal. The difference of $^{95}Nb$ and $^{95}Zr$ activities by more than two orders of value means, that $^{95}Nb$ mainly leaves melt, and $^{95}Zr$ remains in fuel.

The observable difference in concentrations of the two groups of radioisotopes specified in table 3 can be explained by exit of $^{99}Mo$, $^{132}Te$, $^{95}Nb$, $^{103}Ru$ to the gas phase in form of volatile fluorides, boiling temperature of which is much lower than the temperature of salt melt. Thus refractory fluorides $^{95}Zr$, $^{141,144}Ce$, $^{140}Ba$ mainly remain in salt.

Molybdenum has a few stable fluorides, which can be formed under γ-irradiation of the nuclear reactor - $MoF_3$, $MoF_4$, $MoF_5$, $MoF_6$. At the fuel melt temperature ≈600$^0$C, significant fractions of tetra- and pentafluoride of molybdenum $MoF_4$, $MoF_5$ and also molybdenum hexafluoride $MoF_6$ (boiling temperature of 36.2$^0$C (Nikolsky, 1998) are in the gas phase. In case these compounds arise near to borders with the gas phase, they can leave the melt.

In work (Kirslis & Blankenship, 1969), the degree of fission fragments penetration in graphite of MSRE core was investigated. As well as for the previous experiment, we were interested by penetration into graphite of generically connected pair $^{95}Zr \rightarrow ^{95}Nb$. It follows from experimental data that $^{95}Nb$ concentration in graphite has appeared approximately by three orders of value higher, than $^{95}Zr$. Their distribution on depth of the graphite sample is shown in figure 3.

It is known, that the radioactive atoms formed as a result of nuclear transformations, partially appear in chemical compound, at which they were formed, as so-called "parent" compound (Nesmeyanov, 1978). Taking into account that zirconium has sole stable fluoride - $ZrF_4$, we estimated the possibility of preservation of the initial chemical form of a molecule as tetra fluoride after decay of a parent nucleus $^{95}Zr$ and formation on its place of $^{95}Nb$ nucleus.

The energy of the exited molecule, which can be spent on break of the chemical binding, is equal:

$$E_B = E \frac{M_R}{M + M_R}$$

where E is recoil energy, M is mass of the recoil atom, $M_R$ - mass of the rest of molecule. 0.44E will be spent on break of the chemical binding for a molecule $ZrF_4$, i.e. about a half of $^{95}Zr$ recoil energy.

At β⁻-decay of $^{95}Zr$ basically two groups of β⁻-particles with average energies of 121 and 110 keV (Burrows, 1993) are formed. Recoil energy of $^{95}Nb$ for β⁻-particles with energy 121 keV makes 0.77 eV and for 110 keV - 0.69 eV. Besides at decay of $^{95}Zr$ the γ-quanta with energy about 750 keV are emitted. Thus the recoil energy of $^{95}Nb$ will make 3.2 eV.

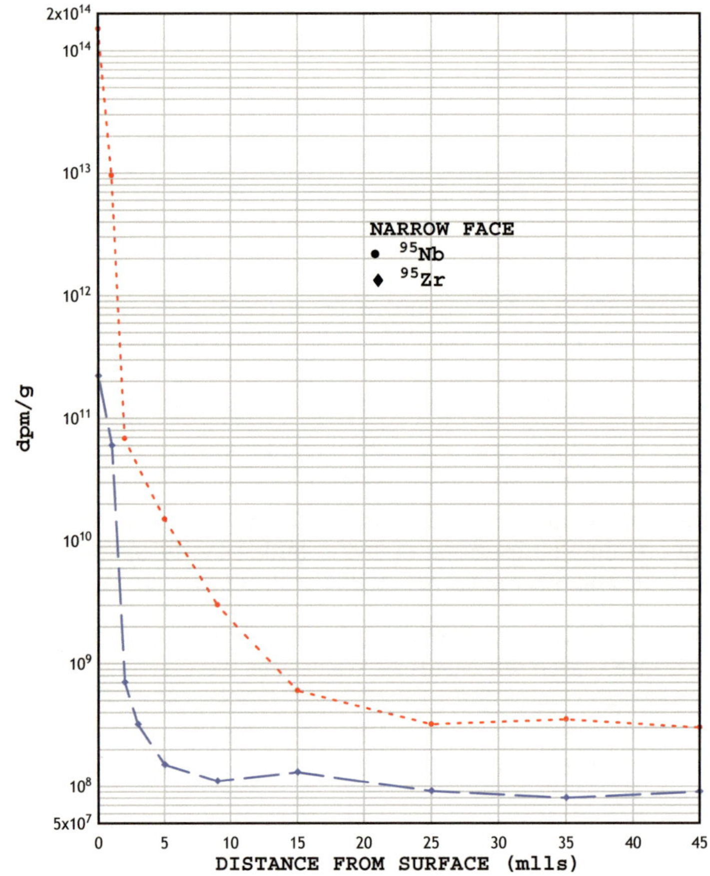


Fig. 3. Distribution of $^{95}Zr$ and $^{95}Nb$ activities on depth of graphite sample contacted with salt melt in the reactor MSRE core (Kirslis & Blankenship, 1969)

As the binding energy of atoms in molecules is 2-5 eV (Nesmeyanov, 1978), in most cases of $^{95}Zr$ decay in a molecule $^{95}ZrF_4$ the formed new atom $^{95}Nb$ will be kept in the chemical form of the parent compound, i.e. as a molecule $^{95}NbF_4$. The further fate of niobium tetra fluoride $^{95}NbF_4$ is determined by thermodynamics and radiation-chemical processes in molten salt fuel. In the reactor γ-field formation of niobium pent fluoride $^{95}NbF_5$ is possible, which at temperature ≈600°C will be in gas form ($T_{boiling}$ = 234°C), and at presence of the phase border "liquid - gas" it will leave fuel melt, penetrating, including, into open gas pores of graphite, which is in contact with molten salt fuel. The boiling temperature of zirconium fluoride is $T_{boiling}$ = 912°C, that is much higher than the fuel temperature. Therefore $^{95}ZrF_4$ does not enter gas bubbles and, accordingly, does not penetrate into pores of graphite that explains significant difference of $^{95}Zr$ and $^{95}Nb$ contents in graphite.

In our opinion, the observed experimental results can be explained by formation and exit of easily volatile fluorides of noble metals Mo, Nb, Te from molten salt fuel to the gas phase. In detail the mechanism of $^{99}Mo$ exit from molten salt fuel is considered in (Zagryadsky et al., 2008; Menshikov & Chuvilin, 2008).

The proposed radiochemical model of "noble" metals exit from fluoride fuel in a gas phase assumes, that chemical processes in fuel $LiF-BeF_2-UF_4$ vary sharply in the field of an ionizing radiation and are determined by a number of active particles - a complex ion $BeF_4^2$, radicals $BeF_4^-$, $Be_2F_8^{3-}$ and a number of other ions. The fragment molybdenum formed in molten-salt fuel at uranium fission exists in the melt mainly in the form of complex ions, such, for example, as $MoF_5^{2-}$. Reacting with ions $Be_2F_8^{3-}$ within characteristic time $\tau \approx 3 \cdot 10^{-4}$ s, molybdenum ions can form at temperature ≈ 600°C volatile fluorides $MoF_4$, $MoF_5$, $MoF_6$ which, being near to borders with a gas phase will leave the melt. Due to presence in a gas phase of radiolitic fluorine, molecules $^{99}MoF_4$, $^{99}MoF_5$ are oxidized to a higher fluoride $^{99}MoF_6$ which is delivered by helium flow into catching system along transport communications.

### 3.2 Advantages and possible variants of realization of fragment $^{99}$Mo production on the basis of molten salt nuclear fuel

Basing on the results of experiments performed within the framework of the research program at the reactor MSRE and the hypothesis of the mechanism of $^{99}Mo$ exit from molten salt fuel, we proposed a new way of fragment $^{99}Mo$ production, using a spontaneous (or stimulated by bubbling of fuel with inert gas) exit of fragment $^{99}Mo$ from molten salt fuel to the gas phase. For the first time this idea was mentioned in (Chuvilin & Zagryadsky, 1998).

The aggregate condition of fuel on the basis of Li, Be and U fluorides allows to take $^{99}Mo$, $^{89}Sr$, $^{133}Xe$ from salt melt selectively, leaving $^{235}U$, and the basic group of fragment elements. As a result inherent for solid fuel systems radiochemical processing of the irradiated uranium is excluded, what at the end should result in reduction of highly active waste and of the flow of fission materials in the technological process.

Principally the process of $^{99}Mo$ production in a research reactor can be organized as follows (Chuvilin & Zagryadskiy, 1997). One of fuel assemblies of the reactor core is replaced with a loop facility filled with molten salt fuel, attached to a gas contour. As a result of $^{235}U$ fission, in the salt melt $^{99}Mo$ is generated, which moves to the salt - gas border and passes to the gas phase above the salt melt surface in form of fluorides and aerosols. $^{99}Mo$ removal can be intensified by bubbling of helium or argon through salt.

The collected in free volume of the loop system gas-aerosol fraction is removed by purge of the cavity with inert gas. The gas - carrier containing aerosols and volatile fluorides of molybdenum enters the catching system, which can be realized as a set of filters, chemical sorbents or freezing traps. Cleaned from fragment elements, inert gas recirculated in the loop system. For $^{99}Mo$ removal, the filters periodically are directed to radiochemical processing. Accompanying molybdenum fragment elements deposited in the catching system are utilized after an appropriate procedure of extraction. The basic scheme of fragment $^{99}Mo$ production in the molten salt reactor is shown in Fig. 4.

The possible design solutions of the loop system will be determined by the technology of $^{99}Mo$ removal from molten salt fuel. At the first stage of its introduction it is expedient to use elementary devices that will allow checking up the basic technical decisions, investigating radionuclide purity of the product, estimating economic parameters of the proposed way of $^{99}Mo$ production.

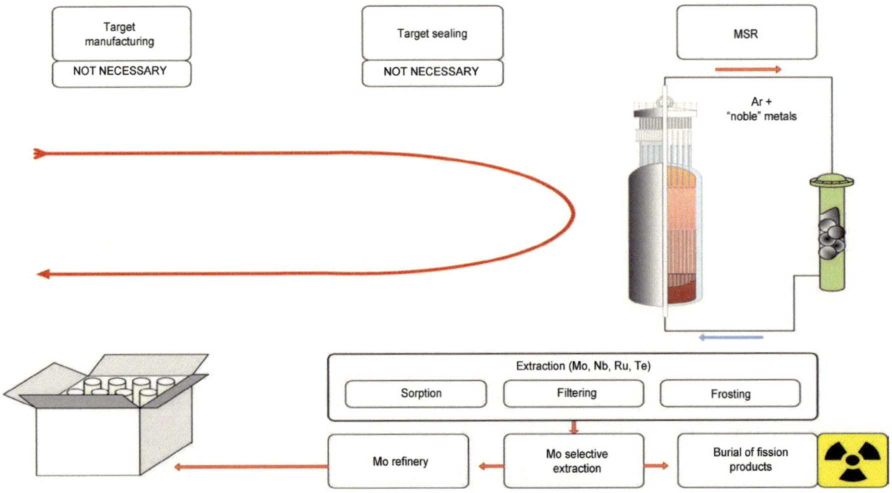


Fig. 4. Scheme of fragment $^{99}Mo$ production in a molten salt reactor

### 3.3 Experimental ampoule loop system "RAMUS"

For demonstration of physical practicability of the proposed new technology of $^{99}Mo$ production, in the research reactor IR-8 of the NRC "Kurchatov Institute" experimental ampoule loop system "RAMUS" with molten salt fluoride fuel is created, where the process of $^{99}Mo$ production will be realized according to (Chuvilin & Zagryadskiy, 1997). It is planned to determine the efficiency of $^{99}Mo$ extraction from fuel, possibility of transporting molybdenum fluoride compounds from the reactor core to the accumulating zone for extraction from the gas - carrier flow, and other parameters. Besides that it is planned to produce as well two other accompanying radioisotopes of medical purpose: $^{89}Sr$, $^{133}Xe$ in the loop system "RAMUS".

The loop system "RAMUS" will be placed in the beryllium reflector of IR-8 having neutron flux $\approx 10^{13}$ $n \cdot cm^{-2} \cdot s^{-1}$. The loop system has the following characteristics:

| | |
|---|---|
| Molar structure of fuel composition, % | $66LiF\text{-}33.9BeF_2\text{-}0.15UF_4$ |
| Enrichment by isotope $^{235}U$, % | 90 |
| Volume of salt, $cm^3$ | 450 |
| Power of reactor IR-8, MW | 8 |
| Heat generation in fuel salt, kW | 3.2 |
| Range of salt working temperatures, $^{o}C$ | 613 - 680 |

The basic scheme of the ampoule part of the loop system "RAMUS" is shown in Fig. 5. The radioisotopes $^{99}Mo$, $^{89}Sr$, $^{133}Xe$ will be evacuated from molten salt fuel by bubbling of melt with helium. After passage of gas-carrier along the communications and filters, in which decay and depositing of short-lived fission products take place, the target radioisotopes are caught in depositing devices, from which they are extracted subsequently.

During design of the loop system "RAMUS" some technical decisions have required to be experimentally substantiated. In particular, the data on formation and carrying over of aerosol particles in the process of molten salt bubbling by inert gases and spontaneous thermocondensation of salt component vapors are practically absent in the literature. At bubbling the large interphase surface on "liquid-gas" border is created that promotes an intensification of mass-exchange processes, and also a more complete chemical interaction of gases with liquids.

The experiments on formation and carrying over of aerosol particles by helium in the process of molten salt fuel bubbling were performed to make more precise the parameters of the unit "RAMUS". The regime of the vertical bubble flow in molten salt 66LiF-34$BeF_2$ was investigated at temperature 530-630°C. The experiments are carried out on a model installation reproducing the dimensions of the ampoule "RAMUS". Weight and height of the salt melt made 900 g and 450 mm, respectively. For creation of the gas bubble flow, a tube was immersed in the salt melt, on which bottom end cylindrical sprayers with face or lateral holes were installed. 6 types of sprayers were tested. The stabilized helium flow rate varied in the range of 1.5-4.5 $cm^3 s^{-1}$.

The rate of carrying over of aerosol particles was determined with help of the photometer FAN-A. Besides that the aerosols were taken out on filters and then the deposit was weighted and analyzed by ICP-AES and DTA-TDA methods.

Foam was observed on the liquid surface in the experiments with bubbling of the salt melt by helium. It was possible to get the minimal height of the foam layer - 15-18 mm with the sprayer of three holes Ø=0.5 mm, L = 0.5 mm located uniformly on the circle. The mass concentration of aerosol particles did not exceed 100 $mg \cdot m^{-3}$, and the rate of their carrying over - 0.1 $\mu g \cdot s^{-1}$. The size distribution of aerosol particles has appeared bimodal. The first mode (superfine aerosol, average particle diameter ≈0.05 microns) is caused by spontaneous thermocondensation of salt components vapors at their cooling from 630°C to 60°C. 90÷95 percents of aerosol particles mass consist of beryllium difluoride. That is because $BeF_2$ vapors pressure at temperature 630°C more than 50 times exceeds LiF vapors pressure.

The second mode of submicron aerosol is caused by collapsing of bubbles on the boundary surface "foam - gas" and corresponding dispersion of liquid. The diameter of particles taken out from an ampoule does not exceed 4 microns, at average diameter 0.5 microns. Their chemical structure coincides with the structure of the salt melt. It means that the carrying over of salt components in form of aerosols at bubbling of melt is determined by dispersion of liquid at collapsing of bubbles on the surface of two phase boundary, and the contribution of $BeF_2$ superfine particles generated at vapors spontaneous thermocondensation makes less than 5% of weight.

The received experimental results have allowed making a conclusion that for the planned term of the loop system "RAMUS" operation in the reactor IR-8 channel the convective carrying over of salt components in form of vapors and aerosols will be negligible small and will not influence gas-dynamic characteristics of communications and the gas flow filtration system of the loop system.

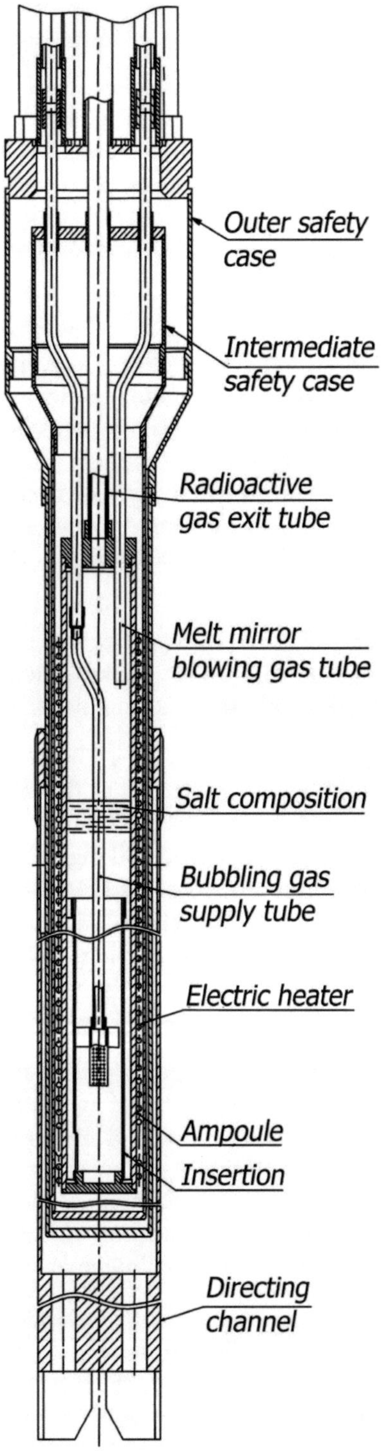


Fig. 5. General scheme of the ampoule part of the loop system "RAMUS"

In other series of the experiments, the possibility of $MoF_6$ loss at its transport along technological communications of the loop system was studied depending on temperature. Sorption of $MoF_6$ from the gas mixture flow (He + $MoF_6$) with the contents of molybdenum hexafluoride in the range $10^{12}$ - $10^{13}$ $cm^{-3}$ on granulated NaF was also investigated. The content of $MoF_6$ in the gas mixture corresponds to the calculated concentration of fragment molybdenum in the loop system "RAMUS". The infra-red Fourier-spectrometer FSM-1202 is used for $MoF_6$ measurement in the gas mixture. The experiments are carried out for the middle and near infra-red areas of the spectrum, with use of multirunning flask with the optical way 4.8 m long, which has allowed measuring $MoF_6$ concentration down to $10^{12}$ $cm^{-3}$. The measurements have shown that deep extraction of $MoF_6$ from the gas flow takes place even at high flow rates of the mixture up to 10 $cm^3 \cdot s^{-1}$.

The process of $MoF_6$ desorption from the sorbent surface becomes noticeable at temperature 100°C. However, as in reactor conditions the temperature at location of the column with sorbent will not exceed 60°C, the effect of $MoF_6$ desorption in the loop system "RAMUS" could be neglected. The measurements of $MoF_6$ losses during its transport along metal communications specially prepared in fluoride atmosphere have shown that they are within uncertainty limits of the spectrometer FSM-1202, and the loss of target product could be neglected.

Thus, the carried out pre-reactor experiments have allowed estimating a number of technological parameters of the loop system "RAMUS", choosing the optimal design of pneumatic sprayers and the regime of molten salt bubbling, predicting possible losses of $MoF_6$ at its transport along gas communications of the loop system.

## 4. Production of $^{99}$Mo in solution reactors

The reactor is named "solution" if its fuel is in the form of water solution of uranyl-sulphate $UO_2SO_4$ or uranyl-nitrate [$UO_2(NO_3)_2$] (Kolesov, 1999). At production of $^{99}$Mo in the solution reactor, the necessity of target manufacturing and their radiochemical processing disappears. All the reactor fuel is a target. $^{99}$Mo is taken from solution by sorption on special sorbent.

Russell Ball (companies Babcock & Wilcox and Ball Systems, the USA) has proposed a design of the liquid homogeneous reactor MIPR (Medical Isotope Production Reactor) for production of radioisotopes, in particular $^{99}$Mo (Ball, 1999). This reactor can work on uranylnitrate solution in water with low-enriched $^{235}$U (to 19 %) as fuel composition. At power of 200 kW it is capable to produce by estimations up to 2000 Ci of $^{99}$Mo a day. The proposed design of a reactor and its characteristics allows receiving the permission to its placing even in the centre of big cities. The power of a solution reactor is more than 10 times less than of a traditional research reactor and demands hundreds times smaller amount of $^{235}$U for production of equal amount of $^{99}$Mo. According to estimations, building and physical start-up of a reactor will take 36 months with the total cost from 6 to 26 million US dollars.

### 4.1 Production of fragment $^{99}$Mo on basis of solution reactor ARGUS

"Argus" is a thermal homogeneous solution reactor, operating at stationary power up to 20 kW in the NRC "Kurchatov Institute". Water solution of uranylsulphate by volume 21.1 l with uranium concentration 81.3 g/l and enrichment by $^{235}U$ 90 % (Pavshouk & Chuvilin, 2005; Afanasev et al., 1986) is used as fuel.

The fuel solution is located in the case - a welded cylinder with a hemispherical bottom and a flat cover in which a cooling coil is located. Vertical "dry" channels are placed in the case: central and two symmetric peripheral channels in which control and safety rods are located. The reactor case is surrounded by a lateral and bottom face graphite reflector. Gaseous radiolysis products of fuel solution are regenerated by means of the system including a catalytic recombiner, which together with the case forms a tight system excluding leak of fission products to the environment. The recombination system is based on the principle of natural circulation of the gas mix along the contour. The solution type of the reactor provides conditions of the maximum safety which is guaranteed by the big negative reactivity effect and the optimum concentration of uranium in solution, leading to self-regulation of the reactor facility.

Since the middle of the 90s years the National Research Centre "Kurchatov Institute" has been working out a technology and creating a demonstration nuclear-technological complex of fragment $^{99}Mo$ production on the solution reactor "Argus". The aggregate state of the reactor "Argus" fuel allows $^{99}Mo$ extraction directly from fuel solution without influence of structure and characteristics of the reactor core. For that, after reactor operation on power solution is pumped through special sorbent on the basis of titanium oxide which provides the first stage of $^{99}Mo$ extraction from all mass of fission products. Thus uranium is not absorbed, and fuel solution completely comes back to the reactor case. All amount of the uranium which has undergone fission reaction, necessary for maintenance of reactor power, is spent for production of the target isotope.

This newest technology of $^{99}Mo$ production allows lowering the necessary reactor power 100 and more times as compared to the traditional technology that not only results in reduction of radioactive waste production, but also supposes reactor locating in the occupied areas and promotes solving the licensing problems. Thus efficiency of $^{235}U$ use comes practically to 100 % while in target technologies only 0.4 % of $^{235}U$ are spent for getting the product, and its other part after target processing, as a rule, goes to waste (Starkov, 1996). The necessity of uranium targets manufacturing is, besides, excluded. The developed technology allows to solve an actual problem of uranium enrichment reduction in production of $^{99}Mo$ and to eliminate public concern in connection with non-proliferation problem. The reactor which core is water uranylsulphate solution is used for accumulation of $^{235}U$ fission fragments. Thus the yield of fragment $^{99}Mo$ per fission of $^{235}U$ makes 6 %. Accumulation of $^{99}Mo$ in the core during the reactor operation on power of 20 kW is shown in Fig. 6.

For demonstrating the practicability of $^{99}Mo$ extraction technology directly from fuel solution, the demonstration experimental complex of the reactor "Argus" presented in Fig. 7 and including a reactor loop, transportation means and hot cells equipment for purifying $^{99}Mo$ from fission products has been designed and put into operation.

As a result of reactor experiments the possibility of $^{99}Mo$ extraction from reactor solution fuel has been shown for the first time, and also the following technological process has been developed. The reactor "Argus" operates at power within 5 days. After some endurance, the fuel solution is pumped through sorbent within 3-6 hours. For returning of the rests of fuel solution into the reactor case the reactor loop together with the sorption column is washed out by the condensate which has been accumulated in a system of radiolysis products regeneration during power operating time of the reactor. Then the column is remotely placed in a transport container and goes to hot cells, where after sorbent washing by acid solutions radionuclide $^{99}Mo$ is desorbed with alkali. For getting demanded radionuclide purity two stages of purification with help of refinery columns filled with the same sorbent of type "Thermoxid" are taken.

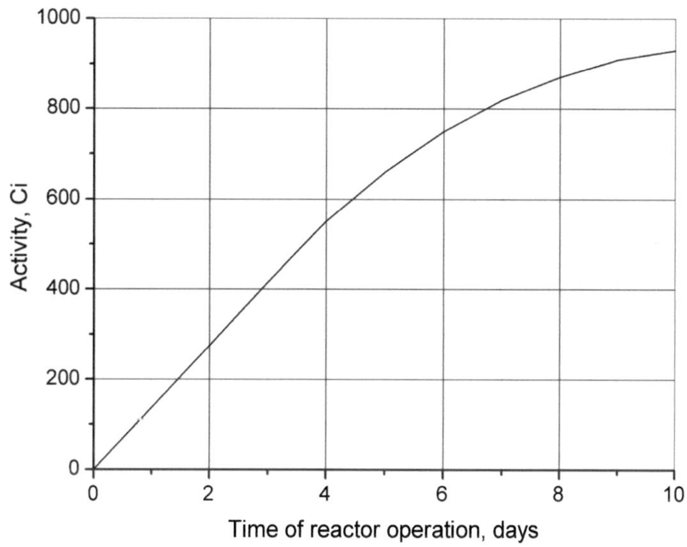


Fig. 6. Activity of $^{99}Mo$ in the core depending on reactor operating time

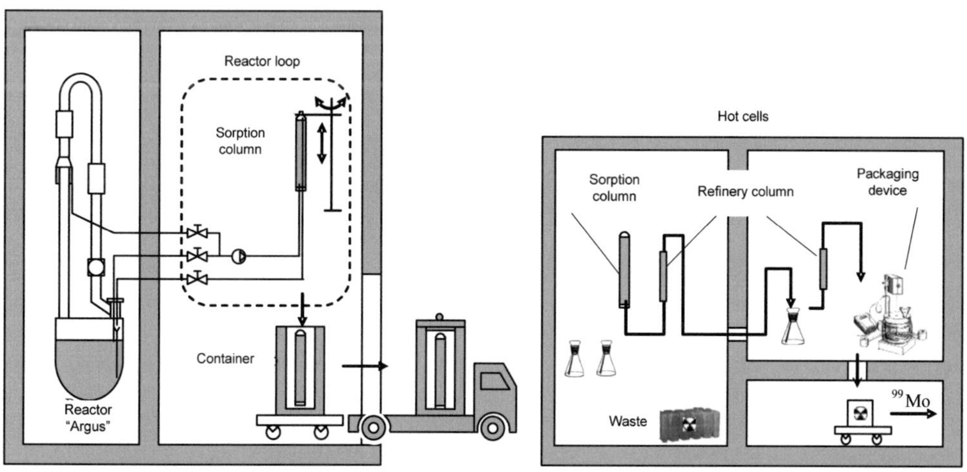


Fig. 7. Transport-technological scheme of demonstration complex at reactor "Argus"

Using the created demonstration equipment reactor experiments are performed on production of $^{99}Mo$ samples and working out of solution technologies. Quality certification of the samples was performed in Europe (Institute of radioactive elements, Belgium), and in the USA (Argonne National Laboratory) where the impurity content in the sample more than 10 times less than in the international requirements (see table 4) was confirmed. The maximum activity of samples made ~ 10 Ci.

On the basis of $^{99}Mo$ produced in the solution reactor "Argus", generators of $^{99m}Tc$ with the standard column filled with sorbent of type "Thermoxid" have been made. Eluted $^{99m}Tc$ has been investigated on injection suitability in State Scientific Center "Institute of Biophysics" of the Russian Federation and has received a good rating as for radionuclide purity, and absence of macro impurity of stable elements (manganese).

| Nuclide | European requirements | Results of the samples analyses | |
|---|---|---|---|
| | | IRE (Belgium) | ANL (USA) |
| α-activity<br>$\alpha/^{99}Mo$ | $<1.0\cdot10^{-9}$ | $1.4\cdot10^{-10}$ | $< 7.1\cdot10^{-10}$ |
| β-activity<br>$^{89}Sr/^{99}Mo$<br>$^{90}Sr/^{99}Mo$ | <br>$<6.0\cdot10^{-7}$<br>$<6.0\cdot10^{-8}$ | <br>$6.4\cdot10^{-9}$<br>$6.4\cdot10^{-10}$ | <br>$1.2\cdot10^{-8}$<br>$7.8\cdot10^{-11}$ |
| γ- activity<br>$^{131}I/^{99}Mo$<br>$^{103}Ru/^{99}Mo$ | <br>$<5.0\cdot10^{-5}$<br>$<5.0\cdot10^{-5}$ | <br>$1.1\cdot10^{-5}$<br>$0.9\cdot10^{-6}$ | <br>$4.2\cdot10^{-7}$<br>$1.4\cdot10^{-7}$ |
| Others γ-, β-<br>$(\beta\text{-},\gamma\text{-})_{\Sigma}/^{99}Mo$<br>$^{132}Te/^{99}Mo$<br>$^{133}I/^{99}Mo$<br>$^{125}Sb/^{99}Mo$ | <br><br><br>$<1.0\cdot10^{-4}$<br> | <br>$<1.0\cdot10^{-4}$<br>$4.4\cdot10^{-5}$<br>$3.4\cdot10^{-7}$<br>$2.7\cdot10^{-7}$ | <br>$<3.8\cdot10^{-7}$<br>$8.9\cdot10^{-8}$<br>-<br>$2.1\cdot10^{-7}$ |

Table 4. Results of $^{99}Mo$ samples quality analysis

### 4.2 Technology of radionuclide $^{89}Sr$ production in the solution reactor

One of the most effective ways is production of $^{89}Sr$ from uranium fission products, but formation of impurity $^{90}Sr$ makes impossible practical application of this way when uranium targets are used. The solution reactor type gives a unique possibility of solving this problem.

The method of fragment radionuclide $^{89}Sr$ extraction is based on a possibility during fission reaction in nuclear fuel of solution reactor to influence the genetic predecessors of a target radioisotope resulted in nuclear transformations of fission fragments in a decay chain of elements with the mass number 89: $^{89}Se\rightarrow^{89}Br\rightarrow^{89}Kr\rightarrow^{89}Rb\rightarrow^{89}Sr$. The earlier experimental researches have shown that practically all long-living isotopes of krypton and xenon leave in the gas phase of solution reactors (Loboda et al., 1989). The basic decay chains of the fission products, leading to formation of strontium radioisotopes with a half-life period of gaseous predecessors more than 1 second, are presented in Fig. 8.

Apparently from consideration of the basic decay chains leading to formation of radioactive strontium isotopes, the half-life period of the gaseous predecessor ($^{89}Kr$) of a target radionuclide $^{89}Sr$ makes 3.2 minutes, and the half-life period of a gaseous predecessor ($^{90}Kr$) of the main impurity radionuclide $Sr^{90}$ is essentially less and makes 33 s. Using this circumstance, in the isolated volume of a gas after corresponding endurance it is possible to reach demanded value of the activity ratio of target gaseous predecessors and impurity strontium radionuclides, and then to pump over the gas mix through filters in accumulating volume in which after decay of predecessors $^{89}Sr$ will accumulate (Loboda et al., 1989; Abalin et al., 2000).

An experimental loop system consisting of a closed technological loop with rotation pumps, valves and vacuum-tight connecting units was used for taking gas samples from the reactor free volume. The hydraulic scheme of the experimental installation is shown in Figure 9.

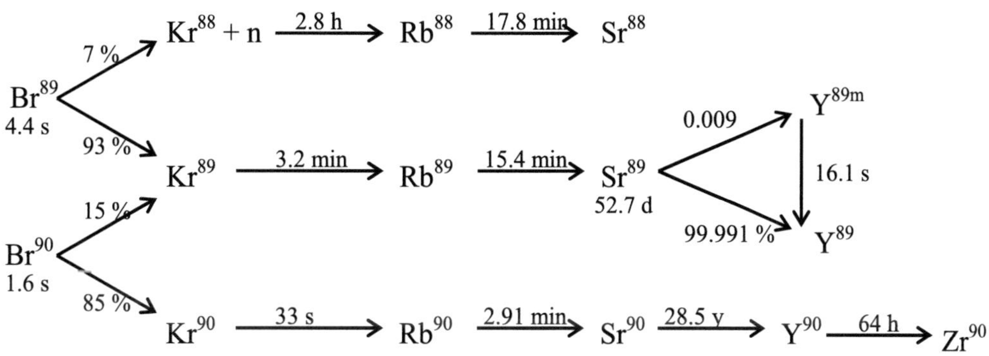


Fig. 8. Decay chains of fission products with atom masses 89 and 90

The experiments confirming the possibility of $^{89}Sr$ extraction from fuel solution of the reactor "Argus" were performed under the following scheme. The reactor operated at power from 5 to 20 kW within 20 minutes. Then after endurance within 5-6 minutes the gas-air mix was pumped over through a column filled with Rashig rings. The reactor loop mentioned above was used. Pumping time of the gas mix was 5 minutes, which is enough for filling the column. The columns were removed after 2-3 days of gaseous fission products decay. In the laboratory the aggregate taken from the column was washed out with solution of hydrochloric acid for extraction of $^{89}Sr$ and some accompanying fission products.

A series of experiments for confirmation of fragment $^{89}Sr$ production using exit of the gaseous predecessor $^{89}Kr$ from fuel solution was performed. Samples of $^{89}Sr$ chloride of high radionuclide purity are got in which $^{90}Sr$ impurity content has made less than $5 \cdot 10^{-4}$ %. The typical content of the produced $^{89}Sr$ chloride solution is given in Table 5.

With the obtained experimental estimations of $^{89}Sr$ yield in a solution reactor we can compare the specific productivities of reactors BR-10 (Russia, Obninsk) (Zvonarev et al., 1997), BOR-60, BR2 (Koonen, 1999).) with "Argus" reactor, normalized to the target mass and reactor power. For the sake of comparison, we accept the mass of $^{235}U$ in the solution equal to 1600 g, as a target. Table 6 shows specific yields of $^{89}Sr$ in various reactor installations.

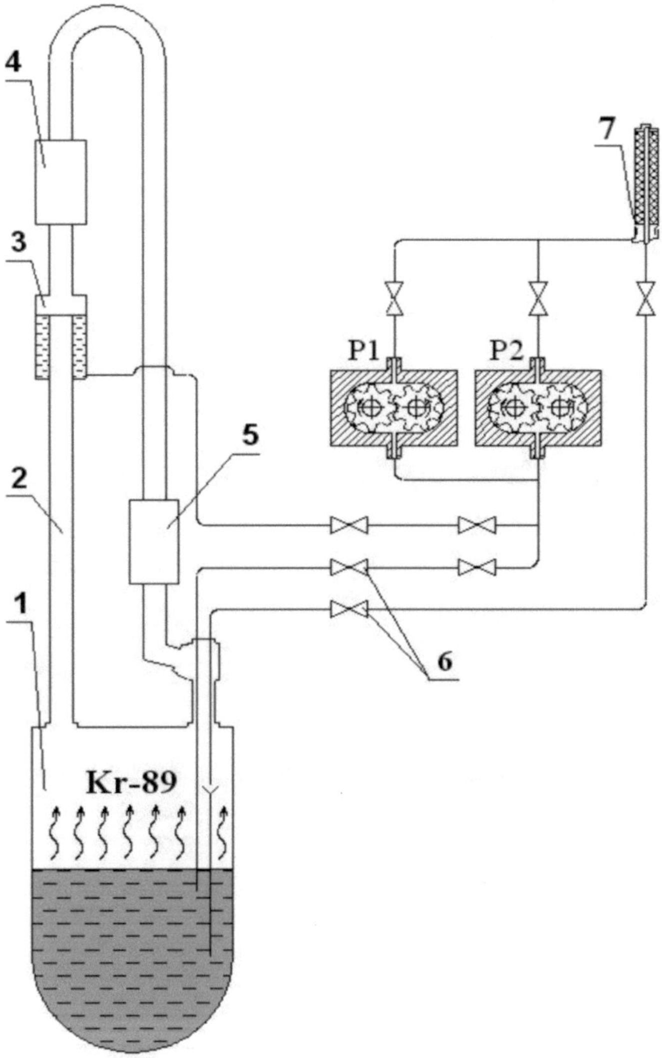


Fig. 9. Scheme of reactor "Argus" with experimental loop. 1 – reactor core; 2 – system of catalytic regeneration; 3 – accumulator of condensate; 4 - heat exchanger; 5 – catalytic recombiner; 6 – valves; 7 – sorption column; P1, P2 – pumps.

| Radionuclide | Activity, relative units. | | |
|---|---|---|---|
| | HEPA | MKF-F | MKF-U |
| Sr-89 | 1 | 1 | 1 |
| Cs-137 | $1.2 \cdot 10^{-2}$ | - | - |
| Ba-140 | $2.3 \cdot 10^{-2}$ | $4.5 \cdot 10^{-4}$ | $4.5 \cdot 10^{-6}$ |
| Ce-141 | $5.4 \cdot 10^{-4}$ | - | - |

Table 5. Typical content of the produced $^{89}Sr$ chloride solution

| Specific yield of $^{89}Sr$, MBq/g/kW | | | |
|---|---|---|---|
| BR-10 | BOR-60 | BR2 | "Argus" |
| $7.4 \cdot 10^{-3}$ | 0.01 | 0.4 | 45.0 |

Table 6. Specific yields of $^{89}Sr$ in various reactor facilities

The advantage of the 20 kW solution reactor is obvious and undeniable. The expected "Argus" reactor productivity of (1.5-1.8) $\cdot 10^3$ GBq of $^{89}Sr$ a year is comparable with that of the 60-MW BOR-60 Reactor ($\approx 4 \cdot 10^3$ GBq/y). However, it is worth noting that $^{89}Sr$ in a solution reactor is a byproduct from its "main" production of medical $^{99}Mo/^{99}Tc$. The target radionuclide separation occurs spontaneously; a majority of fission fragments remains in the fuel, and the buildup of radioactive waste with the new technology is very low. The separated impurity radionuclides $^{133}Xe$ and $^{135}Xe$ can also be used in medicine, while $^{137}Cs$ is a valuable radionuclide for technical applications.

The unique properties of the solution reactor fuel open the doors for this type of nuclear installations to the nuclear medicine industry. Separation of radionuclides directly from fuel solution allows using the low power (50 kW) reactor for production of $^{99}Mo$ with the minimum yield of $1.85 \cdot 10^4$ GBq/day, 700 GBq/week of $^{133}Xe$, as well as a number of other valuable radionuclides (Burrows, 1993; Zagryadsky et al., 2008).

So, the new method for medical $^{89}Sr$ production in the reactor with solution fuel is proposed which is characterized by simplicity, high production efficiency and low buildup of radioactive waste. The main advantages of the new technology were validated by numerous experiments. Basic features of the method are as follows:

- the mechanism for $^{89}Sr$ delivery to the sorption volume of the "Argus" reactor experimental loop is based on transport of gaseous $^{89}Sr$ predecessor - radionuclide $^{89}Kr$;
- the radionuclide impurity composition of $^{89}SrCl_2$ solution includes elements with nuclear masses 137, 140 and 141, which have gaseous elements in their decay lines;
- filtration of the gas flow with mechanical aerosol filters reduces the content of radionuclide impurities $10^2$-$10^3$ times;
- cleaning of $^{89}Sr$ chloride solution in chromatographic columns with DOWEX-50×8 or Sr-Resin ensures full removal of $^{137}Cs$ from the solution and significant reduction of $^{140}Ba/^{140}La$ impurity;
- $^{90}Sr$ impurity in $^{89}SrCl_2$ solution was not detected with the measuring instruments within the sensitivity limits of ($\approx 5 \cdot 10^{-4}$ %).

## 5. Conclusions

The key problem of modern medical radionuclide production in research nuclear reactors - considerable amount of a highly active waste and use in technology of highly-enriched uranium - can be solved by transition to use of homogeneous liquid nuclear fuel enriched by isotope of uranium $^{235}U$ less than 20 %. It is known (Basmanov, E. et al., 1998), that conversion traditional «target» technology of fragment radioisotopes production from HEU to LEU in a target cycle not only does not solve a problem of a radioactive waste (the yield of fission products per unit $^{99}Mo$ activity is the same), but also aggravates it: the amount of the irradiated uranium, the radioactive waste accompanying processing of more uranium for the same $^{99}Mo$ yield, both accumulation of $^{239}Pu$ and other transuranic elements increase in times. That increases the product cost and difficulty of disposal. A target turnover cycle of fission materials is absent in the proposed concept of medical isotopes production. Therefore, on the one hand, the overwhelming part of the waste connected with fission of uranium appears "locked" in a reactor core or a loop installation, and, on the other hand, there is no problem of fission materials proliferation in a target cycle. The question of conversion of the considered reactors with liquid fuel from HEU to LEU does not concern actually technology of medical isotopes production, and should be considered within the general strategy of nuclear reactors safety. Technical possibility of such conversion does not call doubts.

The accumulated operational experience with solution and molten salt fuel, presence of the developed experimental base provide a possibility for performing a broad spectrum of technological researches directed on creation and realization of new methods for production of radionuclides and radiopharmaceuticals for nuclear medicine.

Carried out in the NRC "Kurchatov Institute" model experiments at the solution reactor "Argus" have confirmed basic provisions of the new law-waste technologies of $^{99}Mo$ and $^{89}Sr$ production.

This newest technology of the most important medical radionuclide production allows to lower necessary power of a reactor 100 and more times comparing to traditional technology that not only leads to decrease in total radioactive wastes of production, but also supposes placing the reactor in the occupied areas and promotes solving the licensing problems. The efficiency of $^{235}U$ use comes practically to 100 % while traditional target technologies use only 0.4 % of $^{235}U$ for reception of the final product, and other part of uranium after target processing, as a rule, goes to utilization.

## 6. References

Abalin, S. et al. (2000). Method of strontium-89 radionuclide reception. *Patent of Russian Federation # 2155399*, 2000

Abalin, S. et al. (2002). Method of Strontium-89 Radioisotope Production. *US Patent 6,456,680*. September 24, 2002.

Afanasev, N., Benevolenskij, A., Ventsel, O. et al. (1986). Reactor "Argus" for laboratories of nuclear-physical methods of the analysis and the control, *Atomic energy* (Russian), 1986, v. 61, iss. 1, pp. 7-9.

Ball, R. (1999). *IAEA-TECDOC-1065*, IAEA, Vienna, 1999, pp.5-17.

Ball, R., Nordyke, H. & Brown, R. (2002). Considerations in the Design of a High Power Medical Isotope Production Reactor, *Abstracts at the 24th International Meeting RERTR*, Argentina on November 3-8, 2002.

Basmanov, E. et al. (1998). Management of radioactive waste from $^{99}Mo$ production. *IAEA-TECDOC-1051*, Vienna, November 1998, p. 39.

Bettis, E., Schroeder, R., A.Cristy G. et al. (1957). The Aircraft Reactor Experiment. Design and Construction. *Nucl. Sci. Engng*. 1957, v.2, N 6, pp. 804-826.

Burrows, T. (1993). Nuclear Data Sheets Update for A = 95. *Nuclear Data Sheet,* 68, 1993.
Chuvilin, D. & Zagryadskiy V. (1997). The low radioactive waste method of a fragment Mo-99 production. *Proc. of the Third European Workshop on Chemistry, Energy and the Environment,* Estoril, Portugal, 25-28 May 1997, pp. 381-388.
Chuvilin, D. & Zagryadsky, V. (1998). Method of reception of isotope $^{99}$Mo. *Patent of Russian Federation #2102807.* 1998
Gerasimov, A., Kiselev, G. & Lantcov M. (1989). Reception of $^{99}$Mo in nuclear reactors. *Atomic Energy* (Russian), v.67, iss. 2, 1989, pp. 104-108 (in Russian).
Grimes, W. (1970). Molten-salt reactor chemistry. *Nucl. Appl. Technol.* 1970, v.8, N 2, pp. 137-155.
Hansell, C. (2008). Nuclear medicine's double hazard. Imperiled Treatment and the Risk of Terrorism. *Nonproliferation Review,* v.15, No. 2, July 2008.
Haubenreich, P. & Engel, J. (1970). Experience with the Molten-Salt Reactor Experiment. *Nucl.Appl. Technol.* 1970, v.8, N 2, pp.107-140.
IAEA (1999). Production Technologies for Molybdenum-99 and Technetium-99m, *IAEA-TECDOC-1065,* 1999.
Kirslis, S. & Blankenship, F. (1967). *Report ORNL-4076,* March 1967, p. 51.
Kirslis, S. & Blankenship, F. (1969). *Report ORNL-4344,* February 1969, p. 119.
Koonen, E. (1999). The BR2 Reactor, a Major European Radioisotope Producer. *Int. Conf. GLOBAL'99 "Nuclear Technology - Bridging the Millennia",* August 29 - September 3, 1999.
Kolesov, V. (1999). Aperiodic pulse reactors. Sarov, RFNC VNIIEF, 1999 (in Russian).
Loboda, S. et al. (1989). Exit of fission products from a fuel of solution reactor. *Atomic energy* (Russian), v. 67, iss. 6, December 1989, pp. 432-433 (in Russian).
Menshikov, L. & Chuvilin, D. (2008). *Preprint IAE-6526/12,* (2008) (in Russian).
Nesmeyanov, A. (1978). Radiochemistry. M., Chemistry, 1978 (in Russian).
Nikolsky, B. (Ed(s).). (1998). *Directory of the chemist.* Vol. 1. Ed. Chemistry, Leningrad branch, 1971, (in Russian).
Novikov, V. et al. (1990). Molten salt NEU: prospects and problems. M., Energoatomizdat, 1990 (in Russian).
Pavshouk, V. & Chuvilin, D. (2005). Reception of radionuclides – fission fragments of nuclear fuel. In: *Book of Isotopes.* V. 1. Properties. Reception. Application. M, Phyzmatlit, 2005, p. 524 (in Russian).
Rabotnov, N. (1999). Radiation pharmacology - revolution in public health services. *Nuclear society* № 2-3, (September 1999) pp. 35-38 (in Russian).
Rosenthal, M., Briggs, R. & Kasten, P. (1968). *Report ORNL-4254,* August 1968, pp.101-107.
Rosenthal, M., Briggs, R. & Kasten, P. (1969). *Report ORNL-4344,* February 1969, p.136.
Starkov, O. (1996). Production of radioisotopes for medicine and scientific researches. SSC-RF PhEI:50 Obninsk, 1996, pp. 359-363 (in Russian).
Westmacott, C. (2010). The Supply of Medical Radioisotopes. An Economic Study of the Molybden-99 Supply Chain, *OECD,* 2010, NEA No6967
Zagryadsky, V., Menshikov, L. & Chuvilin, D. (2008) *Preprint IAE-6518/4,* 2008 (in Russian).

Zvonarev , A. et al. (1997). Reception of $^{89}Sr$ in fast reactors. *Atomic Energy* (Russian), v.82, iss.5, May 1997, pp. 396-399 (in Russian).

# 8

# Substantial Reduction of High Level Radioactive Waste by Effective Transmutation of Minor Actinides in Fast Reactors Using Innovative Targets

Michio Yamawaki[1], Kenji Konashi[2], Koji Fujimura[1,3] and Toshikazu Takeda[1]
[1]*Research Institute of Nuclear Engineering, University of Fukui*
[2]*Institute for Materials Research, Tohoku University*
[3]*Hitachi Research Laboratory, Hitachi, Ltd.,*
*Japan*

## 1. Introduction

The problem of managing high-level long-lived radioactive wastes is one of the difficult issues associated with fission reactors. Long-term radiotoxicity is dominated primarily by minor actinides (MAs: Np, Am, Cm) and long-lived fission products (FPs: $^{99}Tc$, $^{129}I$, and so on). The potential radiotoxicity of an isotope is defined as the ratio of its radiotoxicity to the annual limit on intake (ALI) by ingestion. Calculations have shown that the potential radiotoxicity level of the waste would be reduced to that of natural uranium ore after 1,000 years provided that the MAs and long-lived FPs are removed from the waste and transmuted in reactors (Kondo & Takizuka, 1994). Extensive core design studies have been performed to assess the fast reactor (FR) capability for transmuting the MAs (Wakabayahi et al., 1995), (Kawashima et al., 1995). It was pointed out that while the MAs are to be charged to the core in FRs and an annual transmutation rate of more than 10% is possible, significant problems would be encountered in the core safety characteristics, such as the sodium void reactivity and the Doppler coefficient. Many concepts for transmutation of the MAs have been proposed using light water reactors (LWRs) (Takano et al., 1990), (Masumi et al., 1995) as well as FRs. In LWRs, the neutron fluxes are lower than those in FRs, but neutron spectra are so softer that the neutron cross sections of the MAs are larger. Thus, LWRs provide similar MA transmutation performance to FRs according to core analysis. However, more of the higher actinides are produced by MA recycling and the reactivity penalty for MA loading is larger than that in FRs.

A study on burning long-lived actinides, especially Am, was done in FRs using moderated targets in the periphery of the core or in the core region (Rome et al., 1996). The target consisted of $AmO_2$ surrounded by $ZrH_2$ or $CaH_2$ or mixtures thereof. On the other hand, uranium zirconium hydride (U-Zr-H) fuel has been used in TRIGA (Training Research Isotopes and General Atomics) research reactors for about forty years (Simnad, 1981) and the hydrogen absorption properties of U-Th-Zr and U-Th-Ti-Zr fuels have been examined for their possible development as a new U-Th mixed hydride fuel (Yamamoto et al., 1995), (Yamawaki et al., 1998). The hydride fuel has several advantages such as high hydrogen atom concentration as well as inherent safety, low release of fission products, and high thermal conductivity.

In this chapter, we propose effective MA transmutation core concepts using an FR. Moderated regions containing Np and Am are constructed in a driver core region because Np and Am will be effectively transmuted into other actinides of shorter half-lives or effective Pu fuels in the high neutron flux with soft spectra. The MA-containing hydride fuel (U-MAs-Zr-H) is loaded in the target assembly of the moderated region, from the viewpoints of high hydrogen and MA concentration and MAs-H homogeneity, which mean good transmutation characteristics. The resulting Cm is partitioned and stored to decay for over 100 years, and the resulting Pu is recovered and recycled in the driver core region. Systematic parameter survey has been carried out to investigate the fundamental characteristics of MA transmutation and the core safety parameters such as sodium void reactivity in a 1,000 MWe-class fast reactor core.

In the second part of this chapter, the amount of MAs loaded in the MA burning core was increased by loading MA targets even in the radial blanket region, which increased the transmutation amount of the MA. And transmutation rate and incineration rate of MA in the MA once-through core were increased by being lengthened the irradiation period of MA targets. Based on the enhanced MA transmutation characteristics, a scenario for introducing the concept was investigated.

In the third part of this chapter, feasibility of the actinide hydride containing Np and Am as a transmutation target fuel to reduce the amount of long-lived actinides in the high level nuclear waste was studied by employing $UTh_4Zr_{10}H_x$ as a simulated actinide hydride fuel where Th is a surrogate for minor actinides. Irradiation tests of the simulated actinide hydride fuel target have been carried out in Japan Material Testing Reactor (JMTR) of Japan atomic Energy Agency (JAEA). After irradiations, both non-destructive and destructive examinations were carried out for each test. The integrity of the hydride fuel pellet was confirmed through irradiation, supporting the feasibility of the concept of hydride fuel targets.

## 2. Fast reactor core concepts for MA transmutation using hydride fuel targets

### 2.1 Actinide-hydride target and its features

Special target assembly, which contains actinide-hydrides, is considered to achieve high transmutation rate in fast reactor. The actinide-hydride target assemblies are loaded in core region of fast reactor containing mixed oxide fuels. Fast neutrons generated in the core fuel region are moderated in the actinide-hydride target assemblies and then produce high flux of thermal neutrons, which have large cross section of nuclear reaction with actinides. The target contains $^{237}Np$, $^{241}Am$ and $^{243}Am$ with ratio of 77.4, 5.0 and 17.6, which corresponds to that for LWR $UO_2$ fuel irradiated with burnup of 45GWd/t. The Cm was excluded to be stored for over 100 years.

The U-(Np, Am)-Zr hydride composition has the atomic ratio of U:(Np, Am):Zr:H=1:4:10:27, which was determined based on the experimental study on the U-Th-Zr-H system

(Yamawaki et al., 1998). The Use of actinide-hydride target enables to increase amount of actinides loaded in the target assemblies. Fig.1 compares loading amount of actinides in the case of the actinide-hydride target with that in the case of mixed actinide oxide pins and $ZrH_{1.6}$ pins. The latter loading arrangement was discussed in previous report. In the condition of same densities of hydrogen atoms, the amount of loaded actinides by the former method is about six times larger than that in the latter method.

The actinide-hydride has very high hydrogen atom density, which is equivalent to that of liquid water. The moderation of fast neutrons in the actinide-hydride target has been simulated by the Monte Carlo code, MCNP (Briesmeister, 1993) with the JENDL-3.2 library. The calculations have been done for a slab of actinide-hydride with the length of 2m side and 20cm thick (Konashi et al., 2001). The source neutron with fission energy spectrum is incident at a point in center of a face of the actinide-hydride slab. The neutron spectra were averaged over the volume of each small cell (20cm side and 1cm thick), which is placed at the center of the slab. Fig.2 shows the results of the Monte Carlo code calculations. Incident fast neutrons are adequately moderated within 5cm depth of actinide-hydride target. Total flux decrease with increase of distance from the target surface, since the thermalized neutrons are effectively captured by actinide nuclides. The Monte Carlo calculation results show that the actinide-hydride target is an excellent integral target-modulator system.

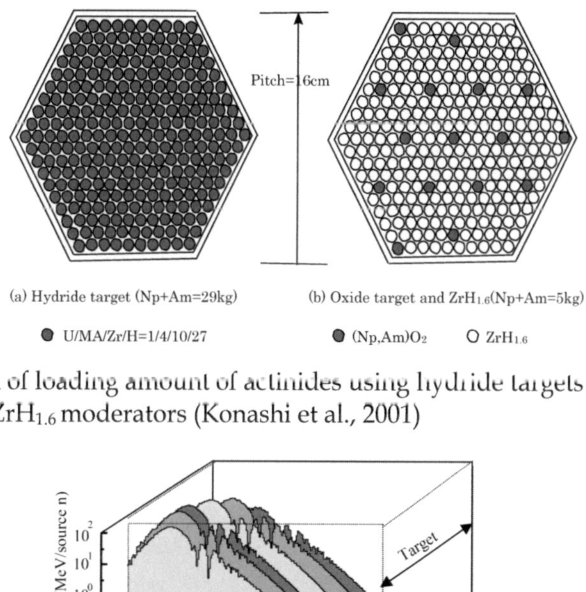


Fig. 1. Comparison of loading amount of actinides using hydride targets with that using oxide targets and $ZrH_{1.6}$ moderators (Konashi et al., 2001)

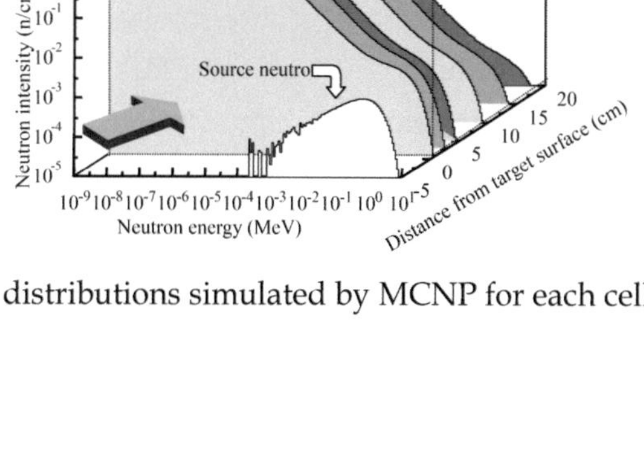


Fig. 2. Neutron energy distributions simulated by MCNP for each cell of target (Konashi et al., 2001)

## 2.2 Calculation conditions

Systematic survey calculations were implemented to investigate the basic characteristics of MA transmutation based on a 1000 MWe-class fast reactor core with mixed oxide fuel using hydride fuel targets. The design parameters are shown in Table 1. The core layout with 36 target assemblies is shown in Fig.3.

Core analyses were performed with the SRAC (Tsuchihashi et al., 1986) and CITATION codes using the JENDL-3.2 library. The 107-group effective microscopic cross sections in the target assembly, inner core, outer core, and blanket regions were calculated using the SRAC code. The effective cross section of the target assembly was calculated with a super cell model, which is a one-dimensional cylinder model consisting of the homogeneous target assembly surrounded by the inner core region. The effective cross sections of the other regions were calculated with the homogeneous cell model. Burnup calculations were made with the two-dimensional RZ model using the CITATION code with the 107-group microscopic cross sections. The burnup period was divided into some steps of about 180 days. The actinide transmutation chain was considered up to $^{245}$Cm as shown in Fig.4. The MA transmutation characteristics such as the transmutation rate in the driver core were calculated with this model. Those in the central target assembly were calculated with this model and those in the off-center target assembly were evaluated by the following procedure. The RZ model is suitable for calculation of the MA transmutation characteristics for the central target assembly, but not the off-center target assembly. So, transmutation characteristics of the off-center target assembly were estimated based on those of the central target assembly by using the relative total neutron fluences, because the transmutation characteristics of the target are almost proportional to the total neutron fluences in the target. The total neutron fluences were calculated in the clean core without the target assembly at the corresponding positions of the center and off-center target assemblies. Strictly, these estimations need 3-dimensional calculations because the radial flux distribution varies with the irradiation. However, the total transmutation rate of all target assemblies could be estimated to a certain extent in the model described above, because the target assemblies are loaded throughout the driver core.

| Item | spec. |
|---|---|
| Reactor electric power | 1,000 MW |
| Reactor thermal power | 2,600MW |
| Cycle length | 1year (3 batch refueling) |
| Core concept | 2 region-homogeneous |
| Core fuel element | $PuO_2$+$UO_2$ |
| Blanket fuel element | Depleted-$UO_2$ |
| Coolant material | Sodium |
| Core diameter/Height | 3.00/1.00m |
| Pu enrichment (IC/OC) | 15.8/20.2wt% |

Table 1. Main design parameters of 1,000 MWe fast reactors

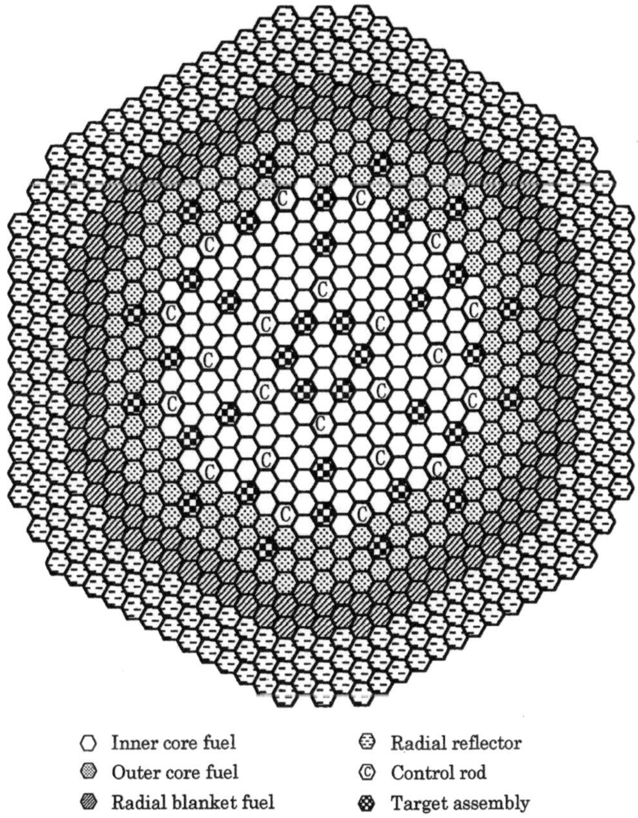


Fig. 3. Core layout of 1,000 MWe fast reactor with 36 target assemblies (Sanda et al., 2000)

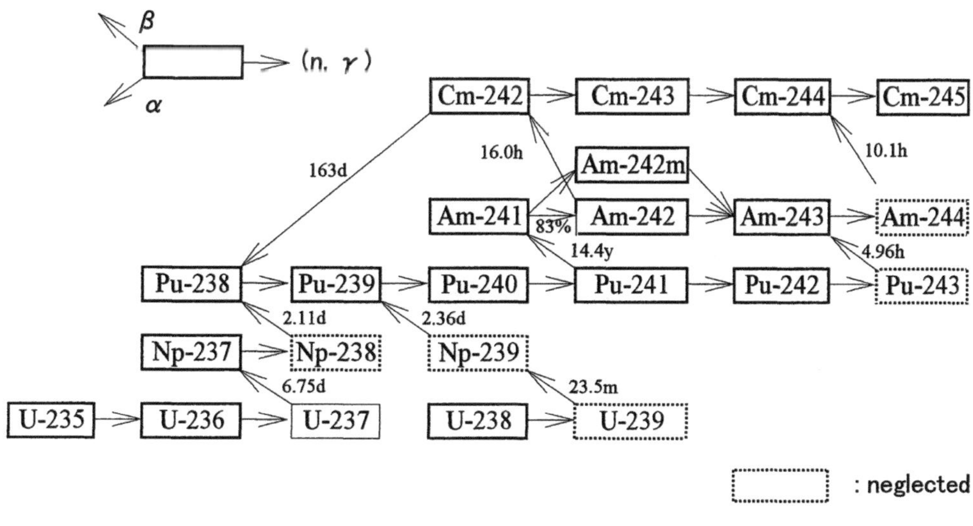


Fig. 4. Actinide transmutation chain employed in this chapter

In this chapter, both the MA transmutation and MA incineration were estimated. The MA transmutation is defined as the initial MAs minus the residual MAs, and the MA incineration as the incineration by fission of MAs and their daughter nuclides during the irradiation (Takeda & Yokoyama, 1997). So, transmuted MAs include incinerated MAs and the resulting Pu. The reactivity coefficients were directly calculated from the difference in multiplication factors between the reference and perturbed cores at BOC (Beginning of cycle) with the two-dimensional RZ model using the CITATION code. The target assemblies were modeled by some rings according to the number of target assemblies. Doppler coefficients in the core, blanket, and target assembly regions, and the sodium void reactivity in the core and target assembly regions were calculated.

The composition U/MAs/Zr/H in the hydride fuel was assumed to be 1/4/10/27 based on earlier study results (Yamamoto et al., 1995) for the case of the highest hydrogen and MA concentration, and the compositions were changed in order to evaluate the effect on MA transmutation characteristics. The density of U-MAs-Zr-H fuel was set at 7.15 g/cm$^3$ based on the preliminary measured data of U-Th-Zr-H fuel with the composition of 1/4/10/27. The amount of loaded MAs for the target assembly in this case could be about 7.6 times larger than that in the mixed $MAO_2$ and $ZrH_{1.6}$ target studied previously with the same numbers of hydrogen atoms. The hydride fuel target assemblies containing Np and Am were loaded in the core region. Cm was excluded, being instead stored to decay for over 100 years. The isotopic content $^{237}$Np/$^{241}$Am/$^{243}$Am of the MAs in the target used for the reference analysis was 77.4/5.0/17.6. These values were based on the content without the cooling for the discharged LWR $UO_2$ fuel of about 45 GWd/t burnup, in which the Cm was assumed to be partitioned in the reprocessing process. The effect of the isotopic content of the discharged fuel with some cooling time on the transmutation was studied later.

## 2.3 Nuclear performance and core concepts

### 2.3.1 Parameter survey

#### (a) MA transmutation

To evaluate the fundamental MA transmutation characteristics, one MA hydride fuel target assembly was loaded in the center of the1000 MWe-class FR core described above. Figure 5 shows the dependency of MA transmutation rate on hydrogen density in the hydride fuel target assembly loaded with 29.1kg of MAs. In this figure, 100 % hydrogen density fuel means U-MAs-Zr-H fuel with the composition of 1/4/10/27. The target assembly was irradiated in the core center for 3 years. The MA transmutation rate of U-MAs-Zr-H fuel is about two times larger than that of U-MAs-Zr fuel without hydrogen. Figure 6 compares the averaged neutron energy spectra in the inner core region and hydride fuel target assembly. Higher fluxes in the thermal and epithermal spectra are obtained in the target assembly, where the MA transmutation is better mainly because some MAs have the large resonance of fission cross sections in the epithermal energy region. Figure 7 shows the neutron flux radial distributions at typical energies near the target assembly loaded in the core center at BOC. Thermal and epithermal fluxes increase rapidly near the target assembly and have a flat distribution in the target assembly. The heavy nuclei composition and the transmutation characteristics in the target assembly with and without hydrogen loaded in the core center are listed in Table 2, where the transmutation rate means the transmuted MAs divided by the initial MAs. Since Cm remaining in a MA-burning target is assumed to be partitioned and stored for decay for over 100 years, and the resulting Pu and the remaining Cm (mainly fissile $^{245}$Cm) can be recovered and recycled in the driver core, Cm is excluded from the residual MAs in this table. The MA incineration rate is given in this table, which means the incinerated MAs by fission divided by the initial MAs. The ratio of the incineration rate to the transmutation rate as well as transmutation and incineration rates increase with the irradiation period (the operation cycle). As the ratio increases, the contribution of the MA incineration increases in the transmutation. The ratio in the hydride fuel is 0.51 for the 3-year irradiation. This is much larger than that (0.32) of the metal fuel without hydrogen, but the MLHGR (Maximum Linear Heat Generation Rate) is too large for the hydride fuel after the 3-year irradiation. In this study, the MLHGR limit of the hydride fuel target was assumed to be 500W/cm based on the metal fuel core design (Yokoo et al., 1995) in a sodium cooled FR. The MLHGR of the target in the middle of the second operation cycle exceeds the limit as shown in Table 2. This is due to the increase of fissile isotopes ($^{239}$Pu, $^{241}$Pu, $^{242m}$Am, $^{245}$Cm) produced by neutron capture reactions of MAs during the irradiation and the large fission cross sections of these isotopes in the epithermal energy regions. Fast neutron fluence ($\geqq$0.1MeV) in the target is about 40% smaller than that in the driver core even after the 3-year irradiation, so cladding materials have significant design margins for neutron irradiation in the target. In the core design using the target, it is desirable that the MLHGR of the target be as large as possible within the limit for the higher MA transmutation, because higher neutron fluence will be obtained in the target. So, we studied shuffling of these target assemblies in each cycle and/or adjusting the composition of the MAs and hydrogen in the targets. The results are discussed in Section 2.3.2.

Next, the amount of MAs loaded in the target fuel was decreased by changing the MAs/Zr/H composition to study increase of the incineration rate, where U-free fuels were applied so as not to produce new MAs. By using these fuels, neutron fission reactions will increase because a large epithermal neutron flux is obtained in the target due to the little neutron flux depression at BOC . Table 3 shows MA transmutation and incineration rates, and the MLHGR for 3-year irradiation for various targets loaded in the core center. The density of MAs Zr H fuel was set at 7.15 g/cm$^3$ for all cases. The MA incineration rate reaches over 90% for many cases. A core concept for the once-through option using these targets is discussed in Section 2.3.2.

The isotopic content of the MAs in the target was based on the content for the discharged LWR fuel without the cooling for the reference analysis. In order to get the effect of the isotopic content, transmutation characteristics were estimated for the central target assembly with different isotopic contents. The contents $^{237}$Np/$^{241}$Am/$^{243}$Am were set at 59.6/28.2/12.2 and 41.9/49.6/8.5 for the discharged LWR fuel with 3-year and 10-year cooling times, respectively (At. Energy Soc. Jpn. Ed., 1994). As the cooling time increases, $^{237}$Np decreases and $^{241}$Am increases. Table 4 shows the MA transmutation and incineration rates of U-MAs-Zr-H fuel with the compositions of 1/4/10/27 and 1/4/10/0 for different isotopic contents of the MAs. This table shows the effect of isotopic contents is small for the transmutation rates, though the transmutation rates decreases a little with the cooling time. Also, the MA incineration rates increase remarkably as the cooling time increases. This is due to the fact that the fission cross section of $^{242m}$Am transmuted by $^{241}$Am is much larger than that of $^{237}$Np in the epithermal energy region.

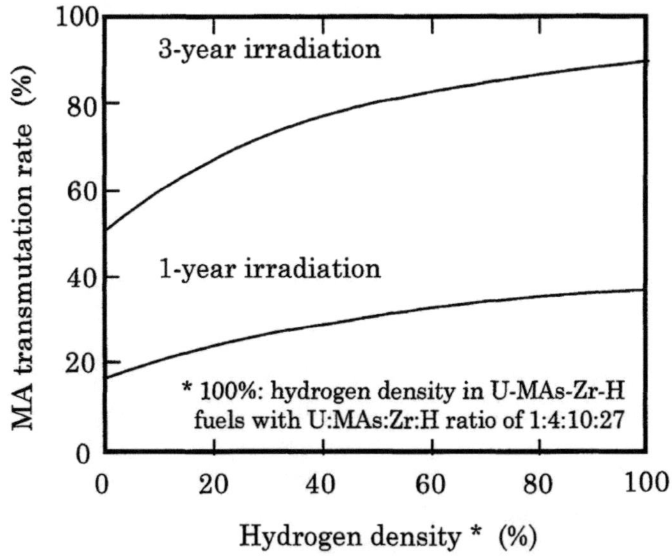


Fig. 5. Dependency of MA (Np+Am) transmutation rate on hydrogen density in a hydride fuel target assembly loaded in core center (Sanda et al., 2000)

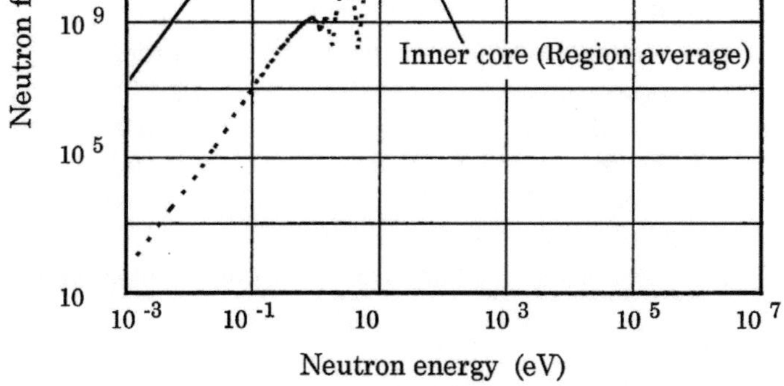


Fig. 6. Comparison of neutron energy spectra between inner core and hydride fuel target assembly (Sanda et al., 2000)

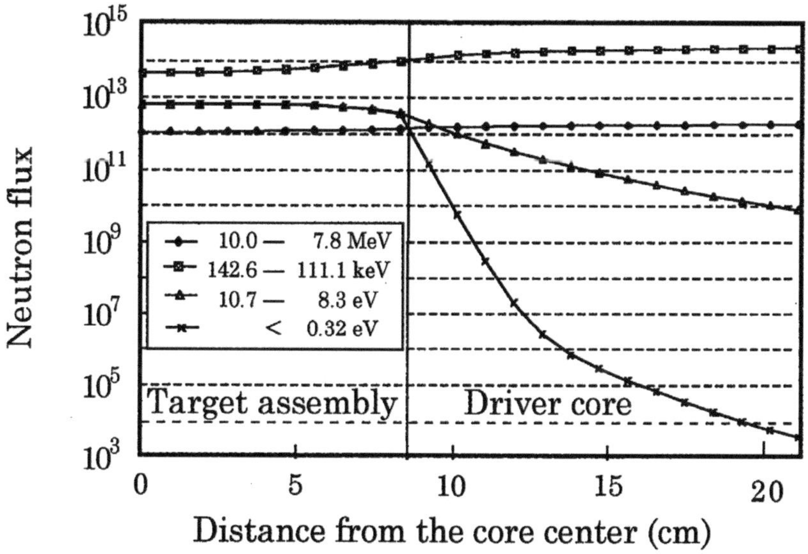


Fig. 7. Neutron flux radial distribution at typical energies near the target assembly loaded in core center
(Sanda et al., 2000)

| | With hydrogen | | | | Without hydrogen | | | |
|---|---|---|---|---|---|---|---|---|
| Irradiation period (yr) | 0 | 1 | 2 | 3 | 0 | 1 | 2 | 3 |
| Heavy nuclei mass (kg) | | | | | | | | |
| U | 7.27 | 6.35 | 5.23 | 3.94 | 7.27 | 6.81 | 6.69 | 6.35 |
| $^{237}$Np | 22.44 | 14.20 | 7.16 | 2.55 | 22.44 | 18.14 | 14.12 | 10.59 |
| $^{239}$Np | 0.00 | 0.01 | 0.01 | 0.01 | 0.00 | 0.00 | 0.00 | 0.00 |
| Pu | 0.00 | 8.08 | 13.04 | 12.97 | 0.00 | 3.51 | 6.25 | 8.03 |
| $^{241}$Am | 1.47 | 0.71 | 0.26 | 0.07 | 1.47 | 1.18 | 0.91 | 0.68 |
| $^{242}$Am | 0.00 | 0.00 | 0.00 | 0.00 | 0.00 | 0.00 | 0.00 | 0.00 |
| $^{242m}$Am | 0.00 | 0.04 | 0.02 | 0.01 | 0.00 | 0.03 | 0.04 | 0.04 |
| $^{243}$Am | 5.22 | 2.93 | 1.16 | 0.30 | 5.22 | 4.32 | 3.45 | 2.68 |
| Cm | 0.0 | 2.4 | 3.3 | 2.6 | 0.0 | 0.9 | 1.5 | 2.0 |
| MAs (Np+Am) mass (kg) | 29.1 | 17.9 | 8.6 | 2.9 | 29.1 | 23.7 | 18.5 | 14.0 |
| MA transmutation rate $TR$ (%)$^{\dagger1}$ | 0.0 | 38.6 | 70.4 | 89.9 | 0.0 | 18.7 | 36.4 | 52.0 |
| MA incineration rate $IR$ (%)$^{\dagger2}$ | 0.0 | 5.4 | 20.3 | 45.5 | 0.0 | 4.2 | 9.4 | 16.6 |
| $IR/TR$ | — | 0.14 | 0.29 | 0.51 | — | 0.22 | 0.26 | 0.32 |
| MLHGR$^{\dagger3}$ of target (W/cm) | 67 | 327 | 796 | 1,096 | 118 | 190 | 259 | 309 |

$^{\dagger1}$(Initial MAs-Residual MAs)/Initial MAs
$^{\dagger2}$MAs and their daughters incinerated by fission/Initial MAs
$^{\dagger3}$Maximum Linear Heat Generation Rate

Table 2. Heavy nuclei mass and MA transmutation characteristics for two types of targets with and without hydrogen (U/MAs/Zr/H=1/4/10/27 or 0) loaded in core (Sanda et al., 2000)

These results show the incineration rates of the targets in the reference analysis are conservative when the isotopic content of the MAs in the target was based on the content for the discharged LWR fuel without cooling.

| Case | Composition U/MAs/Zr/H | MAs loaded (kg/assembly) | MA transmutation rate (%) | MA incineration rate (%) | MLHGR of target (W/cm) | MLHGR† of core (W/cm) |
|---|---|---|---|---|---|---|
| A | 1/4/10/27 | 29.1 | 89.9 | 45.5 | 1,096 | 558 |
| B | 0/1/10/15 | 13.6 | 98.6 | 79.3 | 728 | 710 |
| C | 0/0.7/10/15 | 9.5 | 99.3 | 91.7 | 600 | 790 |
| D | 0/0.5/10/15 | 6.8 | 99.7 | 97.6 | 535 | 851 |
| E | 0/0.5/10/10 | 6.8 | 99.3 | 90.4 | 411 | 700 |
| F | 0/0.1/10/15 | 1.4 | ~100.0 | 99.8 | 125 | 1,023 |

† Thermal power peak in fuel assembly adjacent to target assembly

Table 3. Comparison of MA transmutation characteristics for various targets loaded in core center for 3-year irradiation (Sanda et al., 2000)

| | With hydrogen | | | | Without hydrogen | | | |
|---|---|---|---|---|---|---|---|---|
| Irradiation period (yr) | 0 | 1 | 2 | 3 | 0 | 1 | 2 | 3 |
| MA transmuration rate (%) | | | | | | | | |
| Case 1 (Cooling time=0 yr)† | 0.0 | 38.6 | 70.4 | 89.9 | 0.0 | 18.7 | 36.4 | 52.0 |
| Case 2 (Cooling time=3 yr)† | 0.0 | 38.6 | 71.9 | 91.3 | 0.0 | 18.5 | 36.0 | 51.6 |
| Case 3 (Cooling time=10 yr)† | 0.0 | 38.6 | 72.9 | 92.7 | 0.0 | 18.2 | 35.7 | 51.2 |
| MA incineration rate (%) | | | | | | | | |
| Case 1 (Cooling time=0 yr)† | 0.0 | 5.4 | 20.3 | 45.5 | 0.0 | 4.2 | 9.4 | 16.6 |
| Case 2 (Cooling time=3 yr)† | 0.0 | 10.0 | 30.6 | 57.0 | 0.0 | 5.7 | 12.8 | 21.5 |
| Case 3 (Cooling time=10 yr)† | 0.0 | 13.6 | 40.0 | 67.5 | 0.0 | 7.0 | 15.8 | 25.8 |

† Isotopic content $^{237}Np/^{241}Am/^{243}Am$ of MAs in targets
Case 1: 77.4/5.0/17.6 (reference), Case 2: 59.6/28.2/12.2, Case 3: 41.9/49.6/8.5

Table 4. MA transmutation and incineration rates for two types of targets (U/MAs/Zr/H=1/4/10/27 or 0) with different isotopic contents loaded in core center (Sanda et al., 2000)

**(b) Reactivity coefficient**

As the MA enrichment increases in an FR, the neutron energy spectrum gets harder and the fast fission rate increases. Therefore, safety characteristics deteriorate; i.e. the value of the sodium void reactivity becomes more positive and the absolute value of the Doppler coefficient decreases. For our core concepts, the spectrum gets softer due to the hydride fuel targets and safety characteristics are improved. Figures 8 and 9 show the relations between reactivity coefficients and the number of target assemblies composed of U/MAs/Zr/H fuel with a composition of 1/4/10/27. In the figures, the corresponding reactivity coefficients are described for an FR core loaded with the same amount of MAs homogeneously. For a reference core without MAs, sodium void reactivity is 5.0 $ in the core region, and Doppler coefficient is $-1.0\times10^{-2}$ Tdk/dT in the core and blanket regions. In the proposed cores, the positive sodium void coefficient decreases with the number of target assemblies. In contrast, the negative Doppler coefficient shows little dependency on the number of target assemblies and the relative value slightly increases with the number. This little dependency is due to compensation from the positive effect of spectrum softening and the negative effect of higher Pu enrichment in the driver core by loading the target assemblies, because higher Pu enrichment means less $^{238}U$, the main contributor of Doppler coefficient. However, these figures show safety characteristics can be improved by loading MA hydride targets, compared with an FR core loaded with MAs homogeneously.

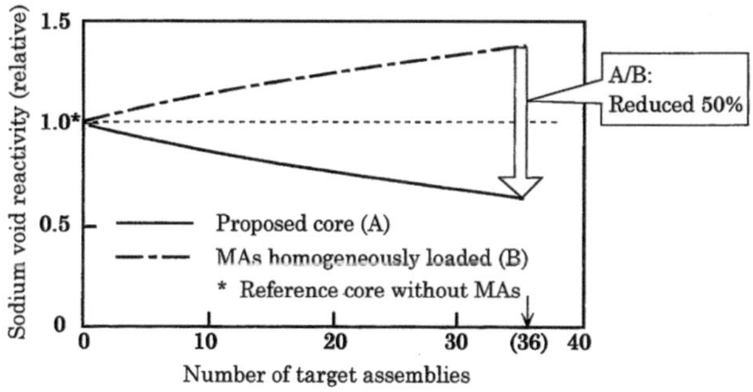


Fig. 8. Relation between sodium void reactivity target assemblies (Sanda et al., 2000)

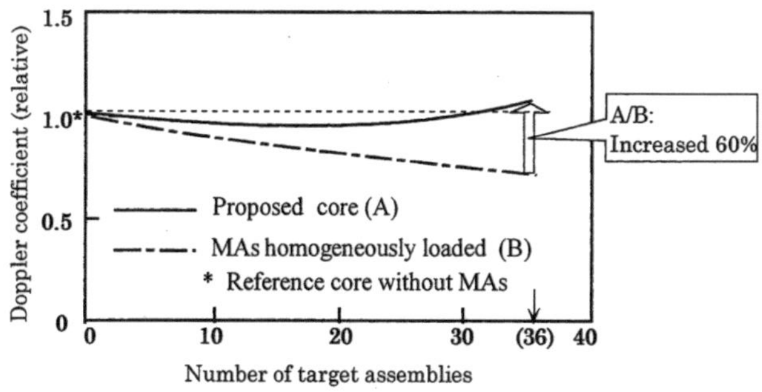


Fig. 9. Relation between Doppler coefficient and number of and number of target assemblies (Sanda et al., 2000)

**(c) Thermal power peaking for fuel assembly adjacent to target assembly**

Thermal power peaking is generated in the fuel assemblies adjacent to the target assembly by epithermal neutrons coming from the target. When the amount of MAs loaded is small in a target, the thermal power peaking causes a problem which must be considered in the core design as neutron flux depression is small in the target assembly due to fewer MA neutron capture reactions. In order to decrease the thermal peak, we studied an assembly composed of hydride fuels surrounded by $^{99}$Tc, which is one of the long-lived FPs and makes a large contribution to long-term radiotoxicity. Figure 10 shows the radial power distribution near the target with and without $^{99}$Tc after the 3-year irradiation in the case of target assembly composed of MAs/Zr/H fuel with the composition of 0.7/10/15. The density of this fuel was set at 6.0 g/cm$^3$ based on TRIGA type fuel data of U-Zr-H fuel with the composition of 0.7/10/27 (Simnad, 1981). The density of $^{99}$Tc was set at the theoretical density of 11.5 g/cm$^3$. While the big thermal peak in the adjacent fuel assemblies is generated for the target assembly without $^{99}$Tc rods, the MLHGR of the adjacent fuel assemblies comes within the limit by replacing two layers of hydride fuel rods with $^{99}$Tc rods as shown in Fig. 10. The MLHGR decreases with the incineration rate of MAs for the target assembly. This is due to decreasing the amount of the fissile nuclei by fission after the 3-year irradiation with a larger fission incineration rate. Figure 11 shows the radial power distribution near the target with 2 layers of $^{99}$Tc rods for different irradiation periods. In the target assembly, the LHGR (Linear Heat Generation Rate) increases in the central region up to the 2-year irradiation. This is due to the increase of fissile nuclei such as $^{239}Pu$, $^{241}Pu$, $^{242m}Am$, and $^{245}Cm$, which are produced by neutron capture reactions of MAs with irradiation and have very large fission cross sections in the thermal and epithermal energy region. After the 2-year irradiation the LHGR decreases in the target assembly. This is due to the decrease of the fissile nuclei by the fission reaction. The transmutation and incineration rates in the discharged target decrease to 98.4 % and 82.4 % from 99.4 % and 92.6 %, respectively, by using 2 layers of $^{99}Tc$ rods. These target assemblies can be applied to the core concept for the MA once-through option due to the high incineration rate described below.

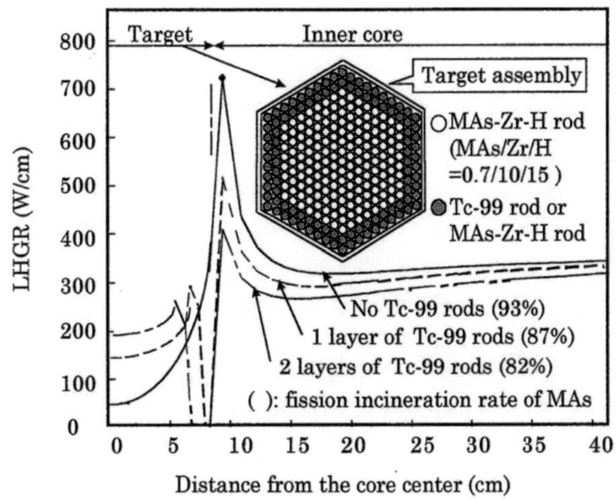


Fig. 10. Radial power distribution near target fuel assembly with and without $^{99}Tc$ rods after the 3-year irradiation (Sanda et al., 2000)

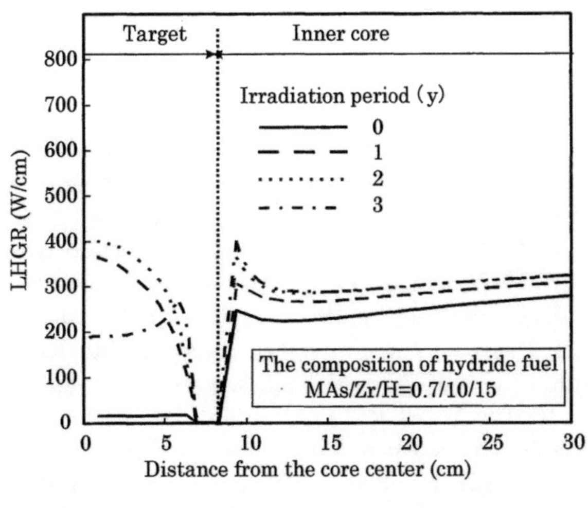


Fig. 11. Radial power distribution near target fuel assembly with 2 layers of $^{99}Tc$ rods for different irradiation periods (Sanda et al., 2000)

### 2.3.2 Proposed core concepts

We proposed two core concepts using MA-containing hydride fuel targets based on the survey results described above. One is for the MA burner option with Pu multi-recycling to transmute a larger amount of MAs, and the other is for the MA once-through option. The merit of the MA once-through option is that it can avoid some problems such as spent targets transporting and reprocessing, and new targets fabrication.

**(a) MA burner option with Pu multi-recycling**

An MA burner core was studied using target assemblies composed of U/MAs/Zr/H fuel with the composition of 1/4/10/27 without Tc. The core layout was shown in Fig.3, in which 36 target assemblies were loaded on 4 rings. The MA transmutation and the MLHGR of the targets were evaluated for two cases. The target assemblies were discharged after one and two cycle irradiations for the first and second cases, respectively. The MLHGR is much smaller than the limit for the first case, and nearly reaches the limit for the second. More of the MAs in the target assemblies on the inner ring are transmuted than those on the outer ring. The target assemblies were shuffled every irradiation cycle for the second case; the assemblies on the first and second rings were moved to the fourth ring at the beginning of the second cycle to obtain similar MA transmutation every irradiation cycle. This contributes to flattening the radial power distribution for each cycle. Table 5 shows heavy nuclei mass and MA transmutation of target assemblies for the proposed burner core concept. The transmutation and incineration rates are 30.6 and 3.7% in the discharged target for the first case, and 57.3 and 14.1% for the second case. For this core concept, the MAs produced in about 14 and13 LWRs can be transmuted every year in the target, respectively, for the first and second cases. Table 6 compares MA transmutation and reactivity coefficients in various core concepts. The MA transmutation and safety characteristics of the proposed cores are improved in comparison with conventional FR cores loaded the same amount of MAs.

Figure 12 shows a fuel cycle including the proposed MA burner and the heavy nuclei flow. In the MA burner core, Np and Am of the hydride fuel target are provided from reprocessing of the spent fuel of LWRs (FRs in future) and the MA burner core, and the discharged target. The reprocessed Pu is recovered and recycled in the driver core. As an option, long-lived FPs such as $^{129}I$ and $^{99}Tc$ are partitioned and mixed with a hydride such as $ZrH_{1.6}$. These are loaded in the radial blanket region in order to be transmuted into stable nuclides. When the radial blanket assemblies adjacent to the driver core region were replaced by $^{99}Tc$ target assemblies composed of $^{99}Tc$ and zirconium hydride pins with the same volume ratio, the annual transmutation mass is about 75kg /y (0.9%/y) which is equivalent to the amount produced by about 2 LWRs and a self-generated one.

**(b) MA once-through option**

An MA incineration core for the MA once-through option was studied using target assemblies composed of MAs/Zr/H fuel with a composition of 0.7/10/15 surrounded by two layers of $^{99}Tc$ as shown in Fig.10. The core layout is shown in Fig.3, and was the same as in the MA burner option. These target assemblies were shuffled every irradiation cycle and discharged after a 3-cycle irradiation within the limit of MLHGR. During shuffling the target assemblies on the inner ring were moved to the outer ring at each cycle to obtain a similar

MA transmutation for each target. Table 5 also includes heavy nuclei mass and MA transmutation characteristics of target assemblies for this proposed core concept. The transmutation rate and incineration rate are 93.0% and 63.9% in the discharged targets, respectively. For this core concept, the MAs produced in about 2 LWRs can be incinerated every year in the targets. For $^{99}Tc$ pins in the target assemblies, the annual transmutation mass is about 60kg / y (4.1%/y) which is equivalent to the amount produced by about one LWR and a self-generated one.

Similar core concepts are realized using the mixed MA and $ZrH_{1.6}$ target assemblies, but the MA transmutation characteristics would be inferior to those for the hydride fuels due to lower hydrogen atom concentration in the assembly as described in sec. 2.2, especially in the MA burner option.

| | Burner core | | | | Incineration core | |
|---|---|---|---|---|---|---|
| Discharged cycle | 1 (1-year irradiation) | | 2 (2-year irradiation) | | 3 (3-year irradiation) | |
| | Charged | Discharged | Charged | Discharged | Charged | Discharged |
| Heavy nuclei mass (kg) | | | | | | |
| U | 261.68 | 235.51 | 130.84 | 102.53 | 0.00 | 0.00 |
| $^{237}Np$ | 807.92 | 572.69 | 403.96 | 181.36 | 45.23 | 3.64 |
| $^{239}Np$ | 0.00 | 0.24 | 0.00 | 0.17 | 0.00 | 0.00 |
| Pu | 0.00 | 230.63 | 0.00 | 197.80 | 0.00 | 18.61 |
| $^{241}Am$ | 52.88 | 31.13 | 26.40 | 8.00 | 2.97 | 0.06 |
| $^{242}Am$ | 0.00 | 0.04 | 0.00 | 0.02 | 0.00 | 0.00 |
| $^{242m}Am$ | 0.00 | 1.12 | 0.00 | 0.54 | 0.00 | 0.01 |
| $^{243}Am$ | 187.80 | 122.45 | 93.90 | 34.01 | 10.55 | 0.41 |
| Cm | 0.00 | 68.23 | 0.00 | 52.46 | 0.00 | 3.10 |
| MAs (Np+Am) (kg) | 1,048.6 | 727.7 | 524.3 | 224.1 | 58.8 | 4.1 |
| MA transmutation (kg) | — | 320.9 | — | 300.2 | — | 54.7 |
| MA incineration (kg) | — | 38.6 | — | 74.1 | — | 37.6 |
| MA transmutation rate (%) | — | 30.6 | — | 57.3 | — | 93.0 |
| MA incineration rate (%) | — | 3.7 | — | 14.1 | — | 63.9 |
| MLHGR of target (W/cm) | 65 | 227 | 65 | 500 | 16 | 286 |

Table 5. Heavy nuclei mass and MA transmutation characteristics of targets for burner and incineration core concepts (Sanda et al., 2000)

| | Conventional fast reactor core | | | Burner core | | Incineration core |
|---|---|---|---|---|---|---|
| Type of MA loading | Reference (No MAs) | Homo-geneous 3.8% MAs | Hetero-geneous MA target | 36 MA hydride targets (U/MAs/Zr/H =1/4/10/27) | | 36 MA hydride targets (MAs/Zr/H =0.7/10/15, with 2 layers of $^{99}Tc$) |
| MA loaded (kg) | 0 | 1,050 | 1,050 | 1,050 | | 176 |
| Annual transmutation rate (%/yr) | (—) | $14^{*1}$ | $11^{*1}$ | $28^{*2,\dagger}$ $(31)^{\dagger\dagger}$ | $26^{*3,\dagger}$ $(29)^{\dagger\dagger}$ | $16^{*1,\dagger}$ $(31)^{\dagger\dagger}$ |
| Sodium void reactivity (relative) | 1.0 | 1.4 | 1.2 | 0.7 | | 0.7 |
| Doppler coefficient (relative) | 1.0 | 0.7 | 0.8 | 1.1 | | 0.9 |

$^{*1}$: 3-year irradiation, $^{*2}$: 1-year irradiation, $^{*3}$: 2-year irradiation, $^{\dagger}$: Driver core+Targets, ( )$^{\dagger\dagger}$: Targets

Table 6. Comparison of MA transmutation characteristics and reactivity coefficient (Sanda et al., 2000)

### 2.4 Summary for section 2

Fast reactor core concepts were studied to reduce the long-term radiotoxicity of nuclear waste by using minor actinides (MAs) in the form of hydride fuel targets. The hydride fuel target assemblies containing Np and Am were loaded in the core region, and the Cm was partitioned and stored to decay for over 100 years. A systematic parameter survey was carried out to investigate the fundamental characteristics of MA transmutation and reactivity coefficients in a 1,000 MWe-class FR core. Results showed safety characteristics could be improved by loading MA hydride targets, as compared with the FR core loaded with MAs homogeneously. Two core concepts were proposed using 36 target assemblies. One was the MA burner core to transmute a larger amount of MAs by neutron capture and fission combined with Pu multi-recycling in FRs, where the MAs produced in about 13 LWRs could be transmuted every year with 58% transmutation rate ( 14% incineration rate) in discharged targets. The other was the MA once-through core to incinerate a small amount of MAs by fission, the MAs produced in about 2 LWRs could be incinerated every year with about 64% fission incineration rate (93% transmutation rate) in discharged targets. This study showed that the proposed core concepts using MA-containing hydride fuel targets have great potential to achieve good transmutation of MAs while providing the improved safety characteristics of an FR core.

MA transmutation characteristics using hydride fuel targets was studied from the viewpoint of reactor physics. However, future attention should be given to the following.

i. Hydrogen is dissociated from hydride fuels at higher temperatures. When the fuel temperature rises under the transients, positive reactivity would be inserted by hydrogen dissociation in an FR. We estimated its effect roughly. The hydrogen dissociation pressures are about 0.1 and 16 atm at 700 and 1,000 deg. for U-$ZrH_{1.6}$ fuel, respectively Simnad, 1981. As a result, the hydrogen composition decreases from 1.6 to 1.5969 (0.2 % decrease) at 1,000 deg. under the assumption that the volume and temperature of the fuel are the same as those of the gas plenum in a fuel pin. And, the 0.2 % hydrogen decrease corresponds to about an inserted reactivity of 2 ¢ for the proposed core with 36 target assemblies composed of U/MAs/Zr/H fuel with a composition of 1/4/10/27. So, the hydrogen dissociation effect under the transients is expected to be little.
ii. Hydrogen penetration increases in the stainless steel cladding material of the fuel pin at higher temperatures. Countermeasures, such as increasing the coolant flow rate for target assemblies so as to decrease the cladding temperatures and internal coating of the cladding material would be effective. Silicon carbide has a very good hydrogen retention capability. It is being used very successfully in HTGR (High Temperature Gas-cooled Reactor) fuel as the cladding against hydrogen (tritium) penetration even up to temperatures exceeding 1,000 deg. (Greenspan, 1997).
iii. It is necessary to develop new MA burning target materials, which are stable under high temperatures and irradiation conditions. The U-$ZrH_{1.6}$ fuel for the TRIGA reactor was reported by General Atomics to be stable at steady-state temperatures up to 700 deg. Under transient conditions the U-$ZrH_{1.6}$ fuel can withstand temperatures at high as 1,000 deg. (Simnad, 1980). The U-Th-Zr-H fuels have been fabricated and examined for thermal and mechanical properties through a preliminary irradiation examination (Yamawaki et al., 1998). These experiences are expected to contribute to development of Np and Am hydride fuels.

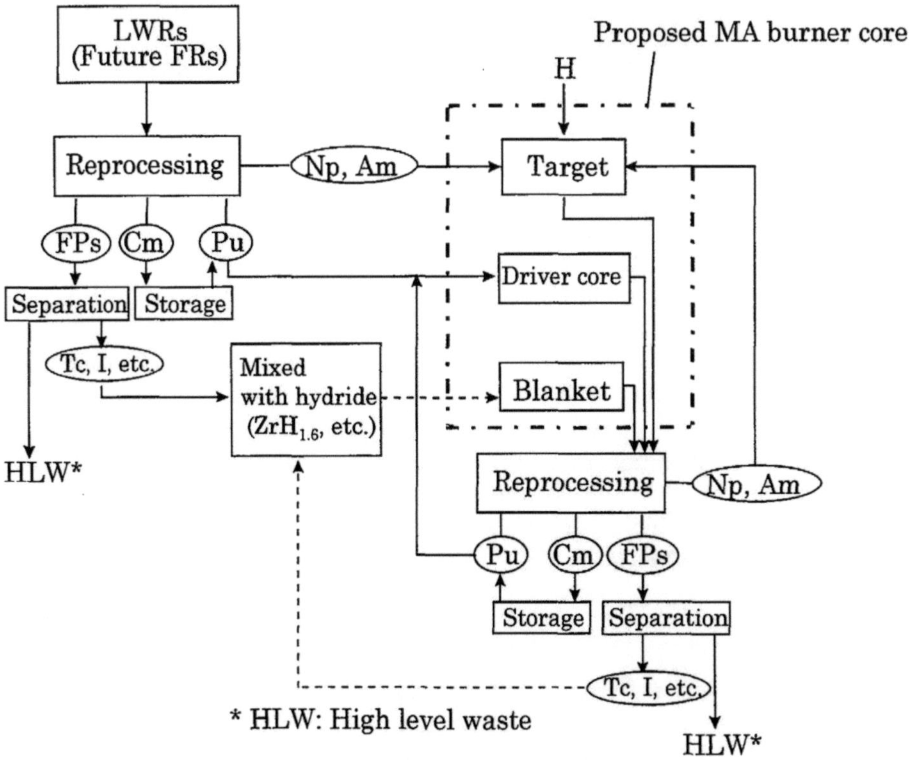


Fig. 12. Fuel cycle including the proposed MA burner core and heavy nuclei flow (Sanda et al., 2000)

## 3. Enhancement of transmutation characteristics of the MA transmutation fast reactor core concept using hydride fuel targets and its introduction scenario

### 3.1 Enhancement of transmutation characteristics

Two types of core concepts for the MA burning FR core using hydride fuel targets were proposed in Sec. 2. One is for an MA burner option to transmute a larger amount of MAs by neutron capture and fission combined with Pu multi-recycling in FRs. The other is for an MA once-through option to completely incinerate a smaller amount of MAs by fission. These concepts have great potential to achieve good transmutation of the MAs with improved of safety characteristics in the FR core.

In this study we try to enhance the MAs transmutation characteristics of both types of MA burning FR core using hydride fuel targets.

#### 3.1.1 MA burner option with Pu multi-recycling

The U/MAs/Zr/H composition in the hydride fuel is assumed to be 1/4/10/27. In this option, a large amount of Pu is produced by the neutron capture reaction of MAs. In sec. 2, the MLHGR limit of the hydride fuel target was assumed to be 500W/cm based on the metal fuel core design. Therefore the target assembly loaded in the core fuel region should be discharged after a 2-year irradiation. The MAs produced in about 13 LWRs could be transmuted every year by loading 36 target assemblies in the 1,000MWe-class FR core. The average transmutation rate in discharged targets was 58%. The main design parameters of 1,000 MWe-class FR were shown in Table 1.

First irradiation test with simulated actinide-hydride target has been done in JMTR. The U/Th/Zr/H type target material was used for the irradiation test (Konashi et al., 2000), since the hydrogen chemical potential of Th hydride is close to that of Np and Am hydrides. The MLHGR limit was set less than 200 W/cm in order that an enough thermal margin should be ensured to the actinide-hydride fuel. It is important to show the potential of the actinide-hydride as the target for MA transmutation reactor within a feasible condition. For that sake, a new loading scheme was proposed as follows.

All target assembly which are irradiated in the core fuel region within a year, are shuffled to the first row of the radial blanket (RB) region. As 72 target assemblies can be loaded in the first row of the RB region, 36 target assemblies can be irradiated within 2 years. As a result, 36 target assemblies will be discharged after a 3-year irradiation. Increase in number of the target brings the decrease in number of the core fuel, which will increase MLHGR of the core fuel within the condition shown in Table 1. Therefore, the number of target loaded in the core fuel region was limited to 36 according to sec. 2. Figure 13 shows the dependency on irradiation period of MA transmutation and MLHGR in a typical target assembly. In the RB region, the MA transmutation rate becomes small because the neutron flux level is smaller than in the core region. The average MA transmutation rate is 47%. On the other hand, the MLHGR becomes very small, which increases the margin of hydride fuel temperature. The total number of targets loaded in the core is 3 times more than the previous case, which increases the MA transmutation amount. As a result, the MA produced in about 21LWRs can be transmuted in the 1,000MWe-class FR core (Table 7).

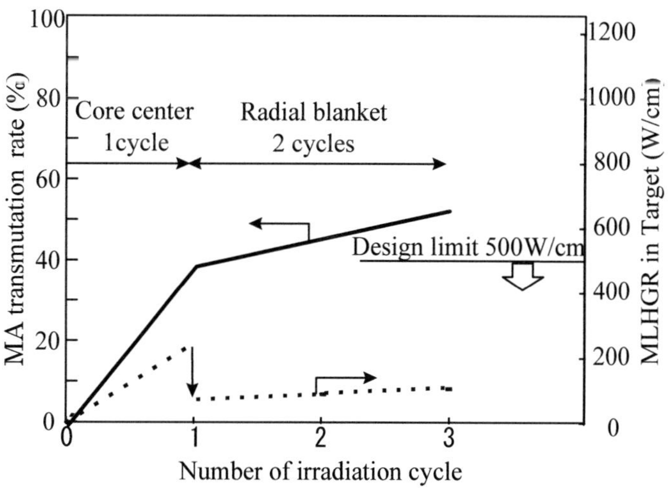


Fig. 13. Burnup characteristics of hydride fuel target of MA burner with Pu multi-recycling core (Fujimura et al., 2001)

| Item | Unit | LWR | FBR*1 | | |
|---|---|---|---|---|---|
| | | | Without Targets | MA burner | MA once-through |
| Output | GWe | 1 | ← | ← | ← |
| RB | — | — | 2 layers | None | 2 layers |
| BR | — | — | 1.2 | 0.9 | ~ 1.0 |
| Pu loaded | t/y | 0 | 1.7 | 2.7 | 2.7 |
| Pu produced | t/y | 0.2 | 0.3 | -0.3 | ~ 0 |
| MA loaded | t/y | 0 | 0.2 | 1.1 | 0.03 |
| MA transmuted | t/y | -0.02 | 0.0 | 0.50 (47%)*2 | -0.03 (~100%)*3 |
| | *4 LWR/y | -1 | ~ 0 | 21 | 1.3 |

*1 MAs self-generated in core fuel are recycled to core fuel,
*2 Including transformation of nuclei, *3 Incineration rate by fission,
*4 Divided by annual MAs produced in a LWR of same electric output

Table 7. Mass balance data of each type of core (Fujimura et al., 2001)

### 3.1.2 MA once-through option

In this option, the amount of MAs loaded in a target is small and the aim is for almost all of the MAs to be incinerated completely, including Pu which is produced by MAs transformation with neutron capture reaction. This option is free from reprocessing of spent target fuel. The target assembly is composed of MAs/Zr/H fuel with a composition of 0.7/10/15 according to sec. 2. The reason for no U is to minimize production of MAs. Hydride fuels are surrounded by two layers of $^{99}$Tc rods to prevent thermal peaking at adjacent fuel rods. Figure 14 shows the dependency on irradiation period of MA transmutation rate and incineration rate in a typical target assembly. In the previous section, all target assembly in the MA once-through core was discharged after 3-year irradiation with core fuel assembles. Core average MA transmutation rate was 93% and incineration rate was 64%.

In this section, we let the irradiation period of a target assembly be twice as that of the core fuel to increase the MA transmutation rate and incineration rate. Both the MA transmutation rate and incineration rate is almost 100% after a 6-year irradiation as shown in Fig.14. In this option, MLHGR of a target will not be over its design limit because the inventory of MAs and Pu produced from transformation of MAs is very small. The merit of this option is that it avoids such problems as spent target transport, reprocessing and new target fabrication. Figure 15 shows the two options of the MA burning FR core using hydride fuel targets which we used to investigate an introduction scenario.

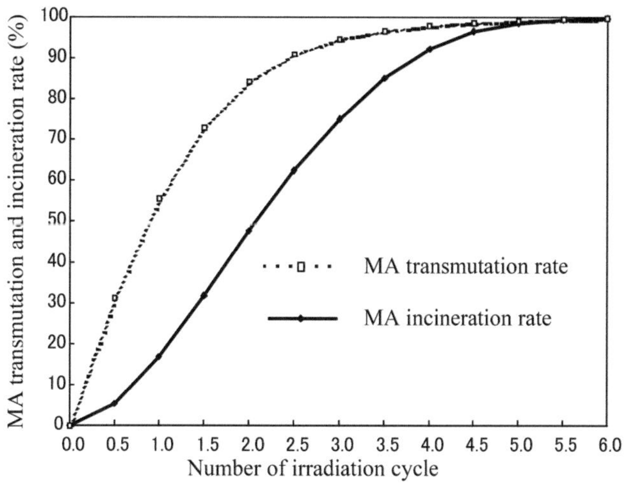


Fig. 14. Burnup characteristics of hydride fuel target of MA once-through core (Fujimura et al., 2001)

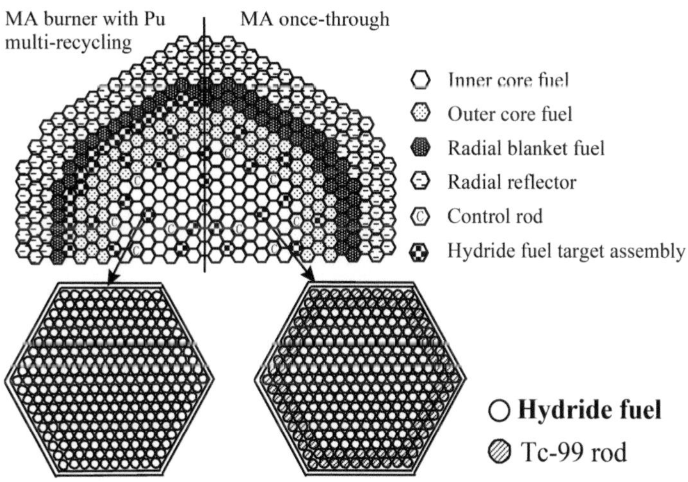


| Option | Hydride | Target position | Feature |
|---|---|---|---|
| MA burner | U/MA/Zr/H 1/4/10/27 | Core(1 cycle)→ RB (2 cycles) | Relatively large mass of MAs is loaded |
| MA once-through | MA/Zr/H 0.7/10/15 | Core (6 cycles) | Small amount of MAs completely incinerated →Tc-99 transmutation rate (~4%/y) |

Fig. 15. Two options of MA burning fast reactor core using hydride fuel targets (Fujimura et al., 2001)

### 3.2 Investigation of introduction scenario

By introducing two options of the MA burning FR core using hydride fuel targets in the early stage of the fast breeder reactor period, the following scenario is feasible. The MA burning FR core using hydride fuel targets will be replaced by a self-generated MA transmutation FR core without the target after most of the MAs produced in the LWR are transmuted. In the reprocessing of the spent fuel of the self-generated MA transmutation FR core, MAs are not separated from Pu.

#### 3.2.1 Nuclear power capacity and installation scenario of FBR

Nuclear power capacity and installation scenario of the FBR (Fast Breeder Reactor) are set with the high projection case as the reference (Wajima et al., 1996). Nuclear power capacity between the years 2010 and 2,100 is extrapolated based on its value of 45GWe in 2,000 and seventy GWe in 2010 which were given in the 1994's Long-Term Program for Development and Utilization of Nuclear Energy of Japan. Capacity growth rate between 2011 and 2050 is fixed at 14GWe per year which is equivalent to the minimum growth rate per 10 years between 1971 and 2010. The growth rate between 2051 and 2100 is fixed at 10 GWe/yr. Seventy GWe in 2010 is consistent with the value shown by the Electricity Industry Council's report for Japan given in July 1998. Above nuclear power capacity well coincides with that supposed in the paper (Hamamoto et al., 2001) which discussed the MOX cores for effective use of Pu in LWRs. The following installation projection of a large FBR is assumed. Installation of the FBR plant will start in 2010 when commercial LWRs will be replaced and installation of two commercial sized FBR plants should be completed prior to 2030. The FBR will be commercialized after 2030. The FBR plant installation rate is assumed to be relatively large in which the utilized amount of natural uranium ore in Japan will not be over 10 % of the world's ultimate resources (about 1.7 Mt). Figure 16 shows the yearly dependency of the nuclear power capacity for high projection case in Japan.

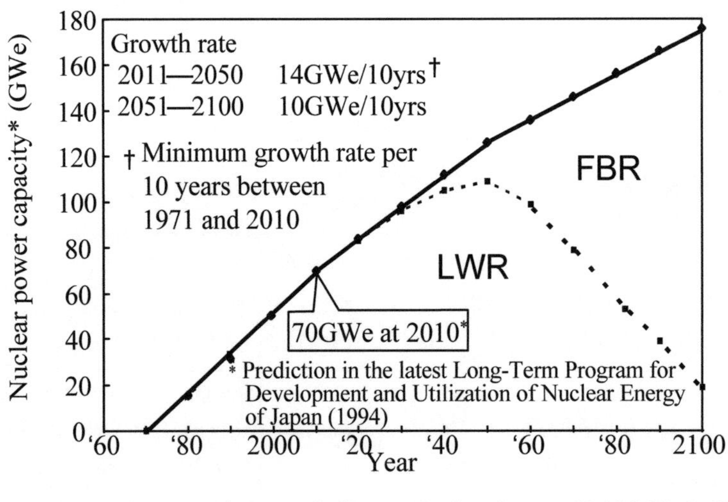


*High projection case - T. Wajima et al.: Journal of Japan Nuclear Society, Vol.38, No.2, 133(1996).

Fig. 16. Yearly dependency of the nuclear power capacity in Japan (Low projection case) (Fujimura et al., 2001)

Nuclear power capacity and installation scenario of the FBR with the low projection case as the above reference was also set. In this scenario, capacity growth rate after 2070 will be fixed at zero. The FBR will be installed after 2030 and commercialized at 2050. Figure 17 shows the yearly dependency of the nuclear power capacity for low projection case in Japan.

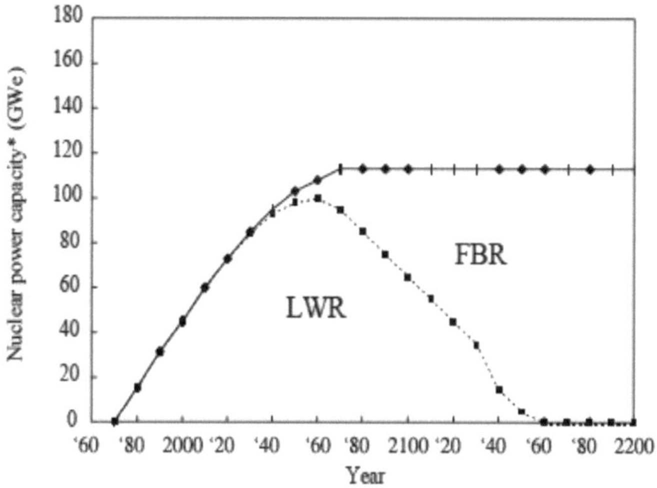


* * Low projection case - T. Wajima, et. al.: Journal of Japan Nuclear Society, Vol.38, No.2, 133 (1996).

Fig. 17. Yearly dependency of the nuclear power capacity in Japan (High projection case) (Fujimura et al., 2001)

### 3.2.2 Fuel cycle flow

Fuel cycle flow of each reactor being considered in this capter is shown in Fig. 18. All the Pu and MAs being extracted from reprocessing of LWR spent fuel are assumed to be utilized and transmuted by the FBR. No utilization of Pu in the LWR is assumed. In this study, all FBRs which will be installed after the residual amount of MAs from LWR spent fuel are transmuted, are assumed to have a self-generated MA transmutation FR core. MAs included in the spent fuel are not separated from Pu . They are utilized as fuel and transmuted. Concentration of the MAs in heavy metal is about 1-2 % in the equilibrium cycle core which does not affect nuclear performance of the core so much. On the other hand, it was shown that a 1,300MWe commercial size FBR can transmute self-generated six long-lived FPs ($^{79}Se$, $^{99}Tc$, $^{107}Pd$, $^{129}I$, $^{135}Cs$, $^{151}Sm$) in the radial blanket and part of the axial blanket (Kobayashi et al., 1997). Fuel cycle flow of the core fuel of the MA burning FR using hydride fuel targets is the same as that of the self-generated MA transmutation FR core. And all MAs from the LWR spent fuel are loaded in hydride fuel targets. About half of the MAs will survive in a spent target fuel in the MA burner option core and they will be reloaded in the reactor as targets. In the once-through option core, most of the actinides are incinerated by fission when being discharged. Therefore a simple reprocessing will be needed for extracting FPs.

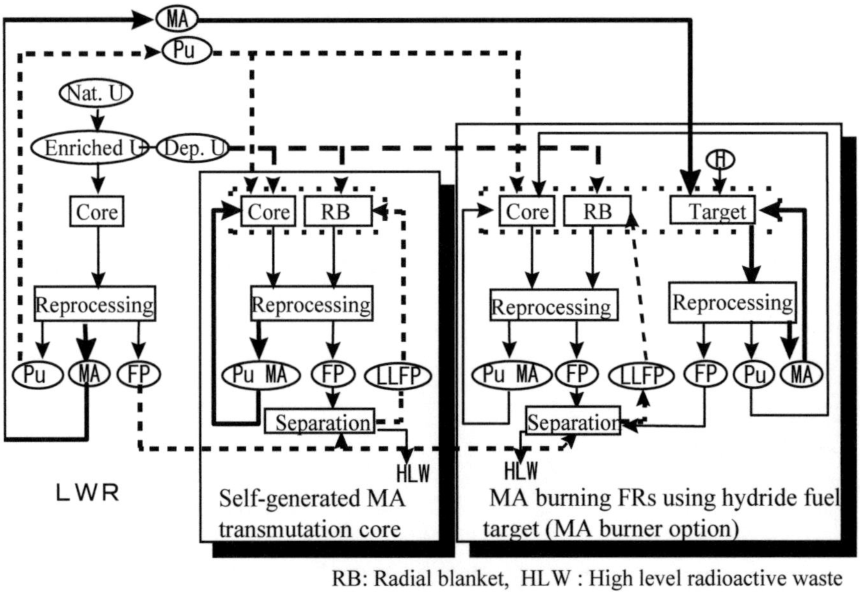


Fig. 18. Fuel cycle flow (Fujimura et al., 2001)

### 3.2.3 Mass balance data for nuclear material

Table 7 shows the mass balance data for investigating the introduction scenario. Electric output of each core is 1GWe.

**(a) Mass of MAs produced and transmutated**

Twenty three kg of MAs are generated in an LWR per year. The FR without targets in table 7 is the self-generated MA transmutation FR core. Where transmuting no MAs, 28 kg of MAs are generated in an FBR per year. In the MA burner option core, 36 targets in the core region and 72 in the radial blanket region, for a total of 108 targets are loaded. Targets in the core region are shuffled every irradiation cycle and moved to the first row of the radial blanket region. They will be discharged after two more years irradiation. The MAs transmutation rate is 47% in the discharged target and the annual transmutation mass is about 500kg /yr. In the once-through option core, the target assemblies are discharged after six years of irradiation and shuffling in the core region. The core fuel assemblies are discharged after three years. Both MA transmutation and incineration rates are about 100% in the discharged targets. The mass being transmuted and incinerated is about 30kg /yr.

**(b) Amount of MAs produced and transmuted**

The produced mass of Pu in an LWR is about 20 kg/yr. Because some assemblies in the radial blankets will be exchanged with target assemblies, the breeding ratio will be less than 1 and Pu will be consumed. The consumed mass is about 300 kg/yr. Although depleted uranium is loaded in the radial blanket region of the once-through option core, high Pu enrichment due to the moderating effect of hydride in targets makes its breeding ratio about 1. Therefore annual Pu production and consumption becomes about zero. Target assemblies are discharged after six years of irradiation and shuffling. Core fuel assemblies are discharged after three years irradiation.

### 3.2.4 Introduction scenario

Transmutation characteristics of the MA burning FR core using hydride fuel targets were enhanced in sec. 3.1. In this section, the effectiveness of thr MA accumulation and the total number of spent target ass being reprocessed for the following cases.

**(a) High projection case**

**(1) Without MA transmutation**

MA accumulation was evaluated when an LWR and FBR without MA transmutation function were assumed to be installed according to the nuclear power capacity shown in Fig. 16. MA accumulation in 2100 is about 320 t as shown in Fig. 20.

**(2) With MA transmutation**

**Case-0 LWR→the self-generated MA transmutation FR core**

In an FBR, the self-generated MAs are transmuted. This does not contribute to an increase of MA accumulation out of reactors. Therefore MA accumulation in 2100 is about 190 t as shown in Fig. 20. The decreasing rate of MA accumulation is about 40% for Case-0.

**Case-1 LWR→MA burner option→the self-generated MAs transmutation FR core**

All FBRs being installed after the 2010s were assumed to be the MA burner with Pu recycling option core. New FBR plants, which will be installed after the 2050s when the accumulation mass of MAs out of reactors becomes about zero, will be replaced by the self-generated MAs transmutation FR core as shown in Fig. 19. In this scenario, MA accumulation out of reactors will be kept at zero after 2060 as shown in Fig. 20. As mentioned above, the spent target assemblies should be reprocessed. In this case, the total number of spent target assemblies to be reprocessed is about 14,000.

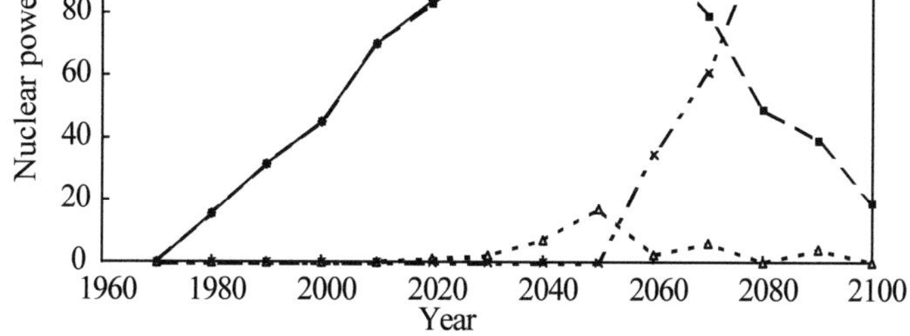


Fig. 19. Projection of installed nuclear capacity (Case-1) (Fujimura et al., 2001)

The number of installation plants for each type of FBR (the MA burner with Pu recycling option core, the self-generated MA transmutation FR core) are set so that the amount of Pu and MAs needed to start up their initial core will not be larger than their accumulation out of reactors. The same applies for Case-2.

If MAs would not be transmuted, more than 300 t of MAs should be disposed as HLW. In this scenario of Case-1, MA accumulation out of reactors will be zero at 2060 which reduces potential radiotoxicity so much in relatively short period and cost for managing the HLW.

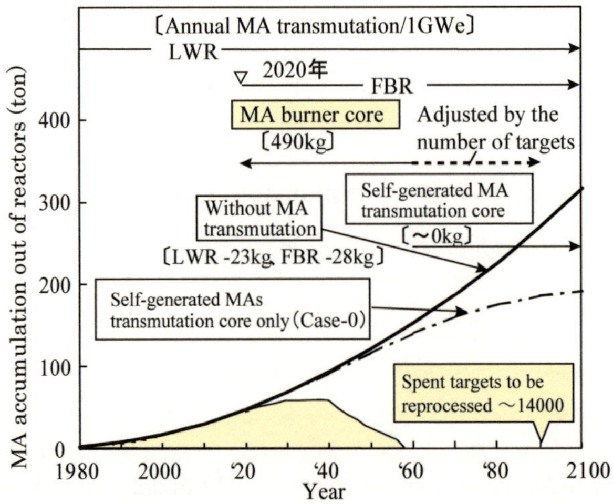


Fig. 20. MA accumulation out of reactors (Case-1) (Fujimura et al., 2001)

**Case-2 LWR→MA burner option→Once-through option→the self-generated MA transmutation FR core**

All FBRs being installed after the 2010s were assumed to be the MA burner with Pu recycling option core until the ratio of FBR plant numbers to the LWR number is 1 to 21 as shown in Fig. 21.

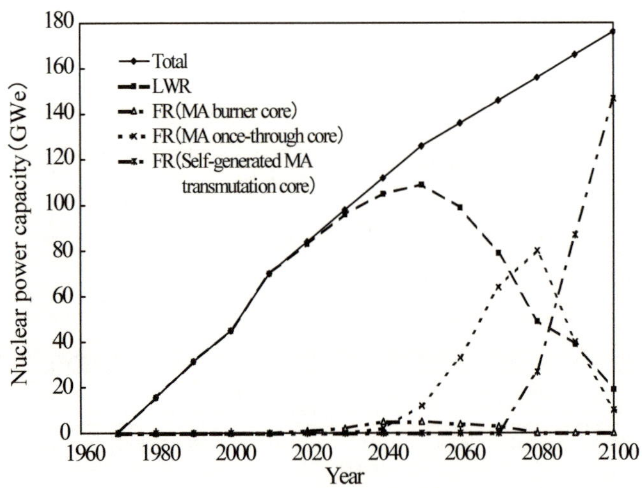


Fig. 21. Projection of installed nuclear capacity (Case-2) (Fujimura et al., 2001)

The once-through option core will be installed if the accumulation of MAs newly produced by LWRs would not increase after 2040. After the 2070s, the number of target assemblies loaded in the once- through option will be decreased in accordance with the decrease of the increasing rate of cumulative amount of MAs due to the decrease in number of LWRs plants. The once- through option core will be replaced by the self-generated MA transmutation FR core. MA accumulation out of reactors will be kept zero in the year 2090 as shown in Fig. 22. Total number of MA burner option cores with Pu multi-recycling is less than that of Case-1 and also the total number of spent target assemblies being reprocessed is about 7000 which is half of that in Case-1. In this scenario of Case-2, the cost for reprocessing the spent fuel will be decreased.

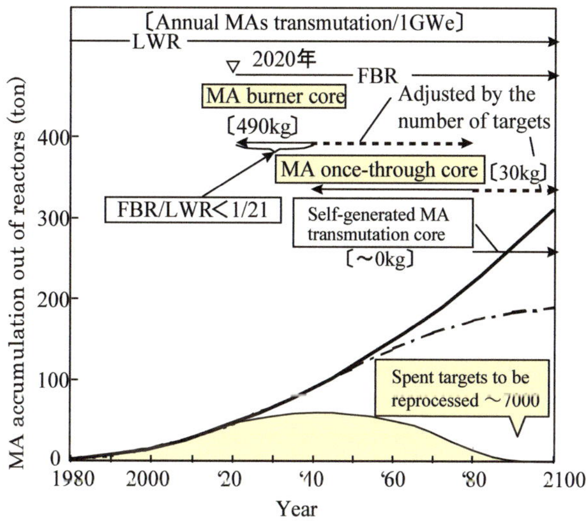


Fig. 22. MA accumulation out of reactors (Case-2) (Fujimura et al., 2001)

**(b) Low Projection Case LWR→FR (No target)→MA burner option→the self-generated MA transmutation FR core**

Projection of installed nuclear capacity is shown in Fig. 23. In this figure, FR with no target includes conventional type FBR core without MA transmutation and the self-generated MAs transmutation FR core. Installation of the FBR will be started at the 2030s as conventional type core without MA transmutation. The capacity growth of FBR is set to be about 1 GWe or 3 GWe per 10 years between 20231 and 2060, and 10 GWe per 10 years between 2061 and 2160. Installation of the MA burner with Pu recycling option core will be started at the 2040s. The capacity growth of the core is set to be about 1 GWe or 3 GWe per 10 years between 2041 and 2070. Its maximum capacity will be 18 GWe at 2080 as shown in Fig. 23. The accumulation mass of MAs out of reactors becomes about zero at 2080 as shown in Fig. 24. After this year, the number of MA targets loaded in the MA burner with Pu recycling option core will be reduced and some of the new FBR plants will be replaced by the conventional type core without MA transmutation to keep the MA accumulation out of reactors zero. Even though the number og spent target assemblies to be reprocessed was increased being compared with that of Case-1, it is feasible that MA accumulation out of reactors would be held to about zero within the 21st century.

As shown above, the scenario with MA accumulation out of reactors as zero is feasible within the 21st century by introducing the MA burning FR core using hydride fuel targets.

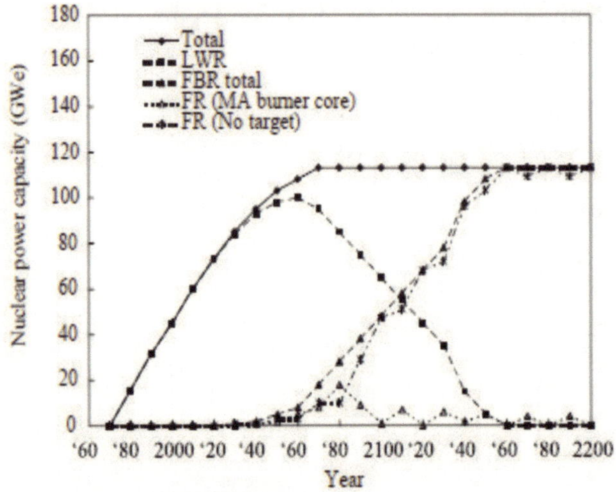


Fig. 23. Projection of installed nuclear capacity (Case-3) (Fujimura et al., 2001)

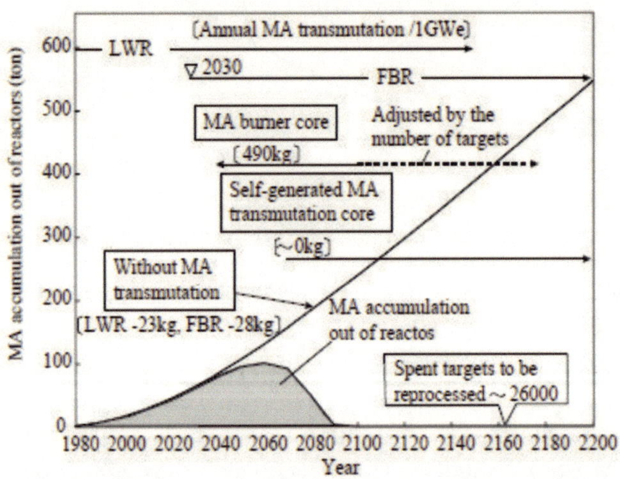


Fig. 24. MA accumulation out of reactors (Case-3) (Fujimura et al., 2001)

### 3.3 Summary for section 3

Transmutation characteristics of MA burning FR core using hydride fuel targets were enhanced and an introduction scenario for the core was investigated.

1. The MA burner core with Pu multi-recycling could transmute the MAs produced in about 21 LWRs each year. The targets were shuffled after the 1-year irradiated in the core region and further irradiated for 2 years in the first row of the radial blanket region. MA transmutation amount could be increased compared with that in the sec. 2 while being increased thermal margine of MA hydride fuel.

2. The MA once-through core could incinerate almost all MAs loaded in targets by fission during a 6-year irradiation in the core region. The merit of this option is that it could avoid such problems as reprocessing of spent target.
3. By introducing all FRs with the MA burner core after the year of 2020, the following scenario would be feasible. The residual amount of MAs from LWR spent fuel would be held to about zero within the 21st century and all FRs would be changed to the self-generated MA transmutation core without targets.
   When the MA once-through cores are introduced with MA burner cores, the total number of targets to be reprocessed could be reduced by about 50%.
4. Even in the low projection case while the MA burner cores are installed at the 2030s after conventional fast breeder reactor without MA transmutation, MA accumulation out of reactors would be held to about zero within the 21st century.

## 4. Feasibility of the actinide hydride containing Np and Am as a transmutation target fuel

### 4.1 Feasibility study using a simulated actinide hydride fuel

Feasibility of the actinide-hydride containing $^{237}Np$, $^{241}Am$ and $^{243}Am$ as a transmutation target fuel to reduce the amount of long-lived actinides in the high level nuclear waste was studied by employing $UTh_4Zr_{10}H_x$, as a simulated actinide hydride fuel. Th was used as a surrogate for minor actinides from the viewpoint of handling radioactive material as well as its similar thermodynamic stability as those of minor actinide hydrides.

### 4.2 Fabrication and property measurement of the simulated actinide hydride fuel

The pellets of the simulated actinide hydride fuel were successfully fabricated through alloying and hydrogenation within expected diameter errors. Figure 25 shows that the U-Th-Zr hydride consists of three phases: U-metal, $ThZr_2H_x$ and $ZrH_x$. As shown in Fig. 26, the U-Th-Zr hydride can hold more hydrogen at temperatures above 1173K than the U-Zr hydride (Yamamoto et al., 1994). This is realized due to the higher thermodynamic stability of $ThZr_2H_x$ phase formed in the U-Th-Zr hydride than the U-Zr hydride used in the TRIGA reactors. This finding led to the present concept of the hydride fuel target containing $^{237}Np$, $^{241}Am$ and $^{243}Am$ for effective transmutation (Yamawaki et al., 1997).

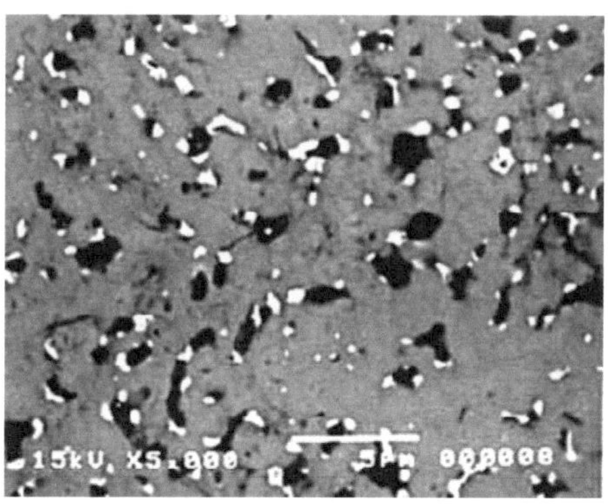


Fig. 25. Back-scattered electron image of $UTh_4Zr_{10}H_{20}$. The black areas consist of Zr hydride; the gray region of $ThZr_2H_x$; the white areas of U metal (Yamamoto et al., 1994)

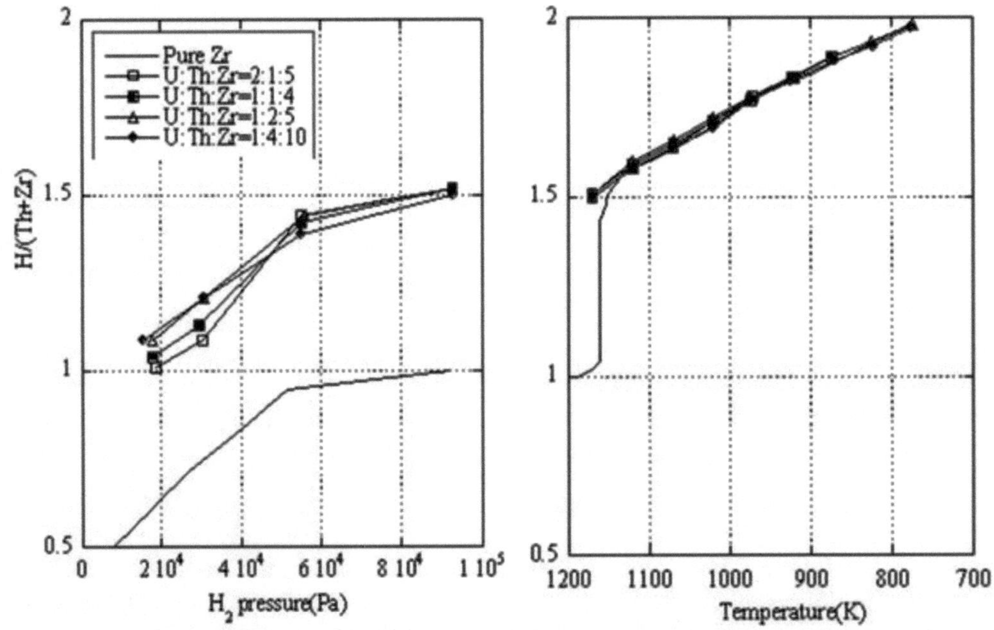


Fig. 26. Equilibrium hydrogen concentration in U-Th-Zr alloys under various hydrogen pressures at 1173K (left figure) and that at various temperatures under $10^5$ Pa (right figure). Those of unalloyed Zr are shown as solid lines without symbols attached (Yamamoto et al., 1994)

Fundamental properties such as thermal diffusivity (Fig.27) and thermal expansion (Fig.28) have been measured for the hydride fuel pin design. The thermal diffusivities, α, of $UTh_4Zr_{10}H_{18\text{-}27}$ were measured using a laser-flash method (Tsuchiya et al., 2002) in the temperature range from room temperature to 950 K as shown in Fig.27. The thermal diffusivities have been measured both during increase and decrease of the temperature. The results of the respective values were in good agreement. This indicates that the hydrogen release from the specimens was negligible during the measurement. The thermal diffusivity is described as the sum of the lattice contribution and the electronic contribution. The defects due to hydrogen losses in the crystal structure of the hydride increase with decrease of hydrogen content. The marked decrease of the thermal diffusivity at temperatures lower than about 650K seems to be attributed to the effect of such hydrogen defects on the lattice contribution. The thermal conductivity, λ, of $UTh_4Zr_{10}H_x$ was calculated from the following relation of the measured α, the literature data of the density, ρ, and the estimated value of specific heat Cp:

$$\lambda = \alpha \times Cp \times \rho \ (\mathrm{W/cm{\cdot}K}), \tag{1}$$

where

$Cp = -0.110 + 6.87\times10^{-4}T + 6.36\times10^{-3}x \quad (\mathrm{J/g{\cdot}K})$,

$\alpha = (1.11x-21.2)/T + 2.29\times10^{-2} + (-3.18\times10^{-6}x + 7.59\times10^{-5})T \ (\mathrm{cm^2/sec})$,

$\rho = 8.4-2.99\times10^{-2}x \ (\mathrm{g/cm^3})$.

The thermal conductivity evaluated in this study as shown in Fig. 29 and the thermal expansion measured in this study are essentially important to estimate the temperature and the mechanical integrity of the hydride fuel during irradiation. The hydride fuel decomposes at high temperature, so that the temperature evaluation and mechanical behavior estimation are especially important for this fuel.

### 4.3 Irradiation tests

Irradiation tests of the simulated actinide hydride target have been conducted in Japan Material Testing Reactor (JMTR) of JAEA. Two irradiation tests of the U-Th-Zr hydride fuel were carried out. The irradiation conditions of the first test were burnup of 0.2% FIMA (Fraction per Initial Metal Atom), linear heat rate of 140 W/cm , fast neutron dose of $1.10\times10^{19}$ n/cm$^2$ and thermal neutron dose of $1.23\times10^{20}$ n/cm$^2$. The irradiation conditions of the second test were changed to burnup of 1.1% FIMA, linear heat rate of 178 W/cm, fast neutron dose of $4.66\times10^{19}$ n/cm$^2$ and thermal neutron dose of $6.43\times10^{20}$ n/cm$^2$. After irradiation, non-destructive and destructive examinations were performed in each test. It was confirmed that the integrity of the hydride fuel was kept intact through irradiation, supporting the feasibility of the present concept for the hydride fuel target.

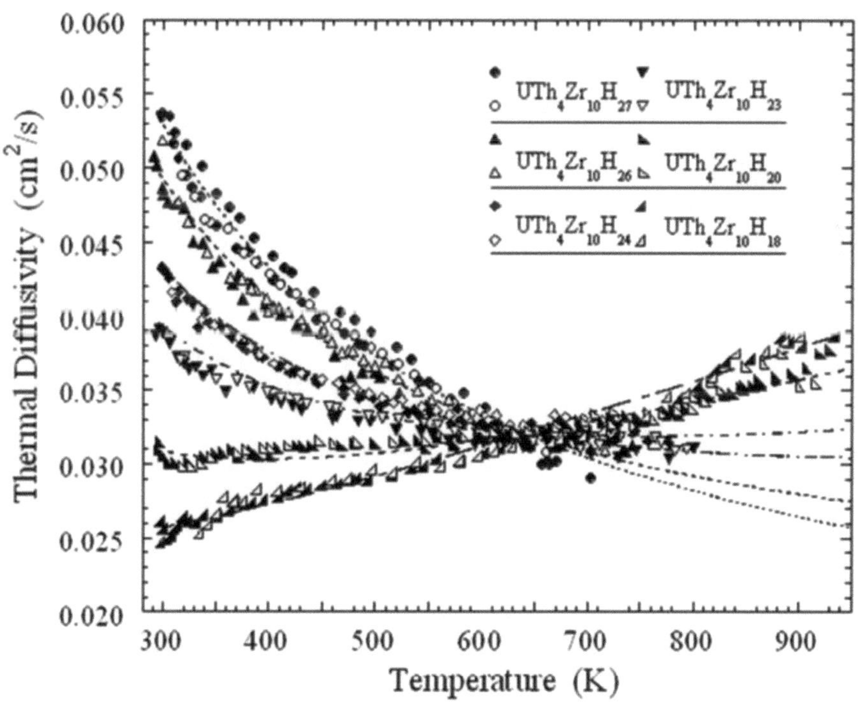


Fig. 27. Thermal diffusivity of $UTh_4Zr_{10}H_{18-27}$ (Open symbols: increasing temperature; Solid symbols: decreasing temperature) (Tsuchiya et al., 2002)

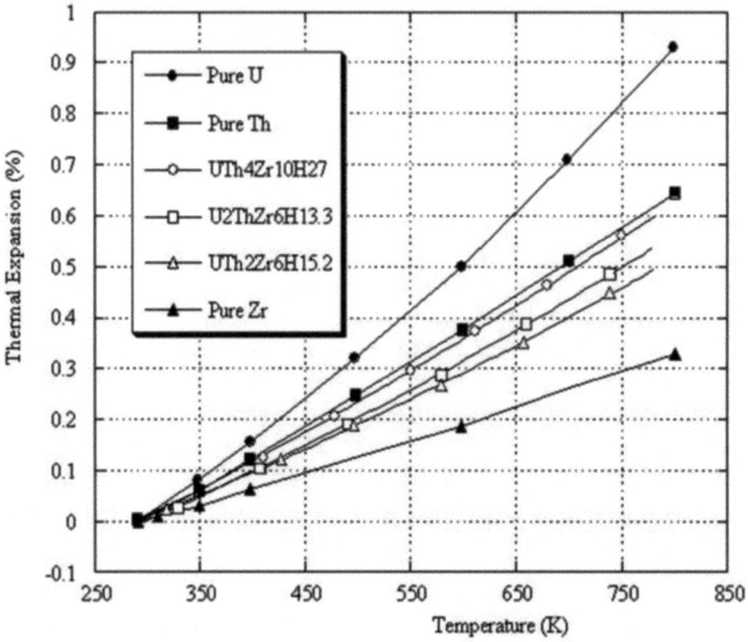


Fig. 28. Thermal expansions of U-Th-Zr hydrides

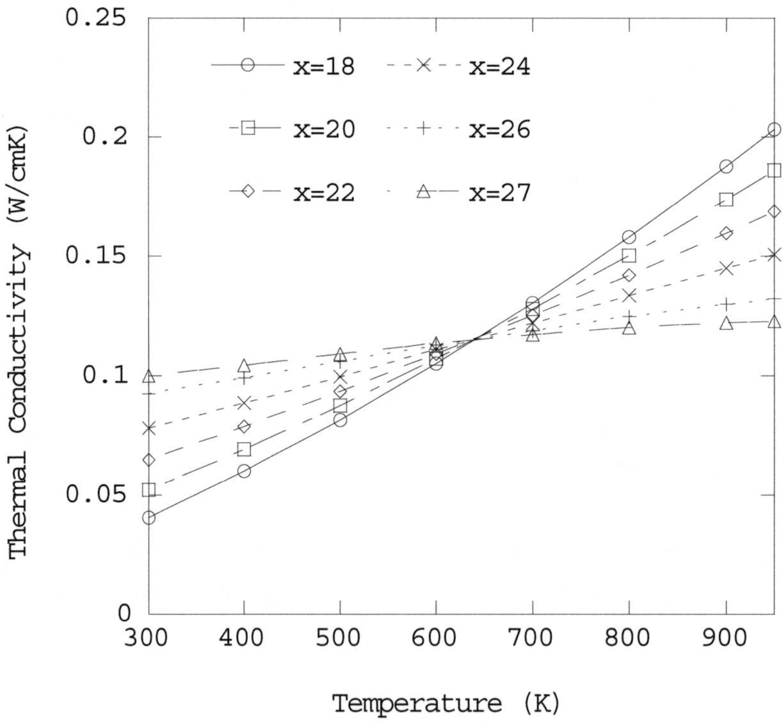


Fig. 29. Thermal conductivity of $UTh_4Zr_{10}H_x$ (x=18-27)

### 4.4 Summary for section 4

Feasibility of the actinide hydride containing $^{237}Np$, $^{241}Am$ and $^{243}Am$ as a transmutation target fuel to be used to reduce the amount of long-lived actinides in the high level nuclear waste has been studied by employing $UTh_4Zr_{10}H_x$ as a simulated actinide hydride fuel, where Th is a surrogate for minor actinides. The pellets of the simulated actinide hydride fuel were successfully fabricated through alloying and hydrogenation within expected diameter errors. It was shown that the U-Th-Zr hydride fuel has higher stability at high temperature than U-Zr hydride. Fundamental properties of $UTh_4Zr_{10}H_x$ such as thermal diffusivity and thermal expansion have been measured and then thermal conductivity was evaluated. These properties are important to evaluate the temperature and mechanical integrity of the hydride fuel during irradiation. Irradiation tests of the simulated hydride fuel were carried out in JMTR of JAEA. It was found that the integrity of the simulated hydride fuel was kept intact through irradiation. This suggests the integrity of the actual hydride fuel to be kept through irradiation, supporting the feasibility of the present concept of using the hydride as transmutation target.

## 5. Conclusion

High-level radioactive wastes are generated after reprocessing of spent fuels from nuclear reactors, which include long-lived radioactive nuclides of actinides and fission products. The currently available method for final disposal of the high-level wastes is to vitrify them, to store them and to dispose them underground. To reduce the work needed to carry out this type isolation job, a number of transmutation methods have been proposed and studied. This chapter reports an innovative efficient method for transmutation of minor actinides (MAs) by applying hydride targets to fast reactors.

Fast rector core concepts have been studied to reduce long-life radiotoxicity of nuclear waste by applying MA-containing zirconium hydride fuel targets. Systematic parameter survey has been carried out to investigate the fundamental characteristics of MA transmutation and the core safety parameters such as sodium void reactivity in a 1,000 MWe-class fast reactor core. Two core concepts were proposed, using 36 target assemblies, by adjusting the composition of hydride fuels. One is the MA burner core to transmute a large amount of MAs in a short time combined with Pu multi-recycling in fast reactors, whereby the MAs produced in about 13 light water reactors (LWRs) can be transmuted every year with a 58% MA transmutation rate in discharged targets. The other is the MA once-through core to incinerate a small amount of MAs by fission, whereby the MAs produced in about 2 LWRs can be incinerated every year with a 64% MA incineration rate due to 93% transmutation rate in discharged targets. These concepts have been shown to have great potential to achieve good transmutation characteristics of MAs while providing the improved safety characteristics of a fast reactor core.

A scenario to reduce the long-life radiotoxicity of nuclear waste by applying fast reactor cores loaded with hydride fuel targets have been investigated. The MA burner core with Pu multi-recycling can transmute MAs as much as produced in 21 LWRs a year. The targets are shuffled after irradiated 1 year in the core region, then further irradiated for 2 years in the radial blanket region. The MA once-through core can incinerate almost all of the MAs in targets by fission during 6-year irradiation in the core region. Introduction of the MA burner core for all fast reactors (FRs) after the year 2020 allows the following scenario. The residual amount of MAs from LWR spent fuel can be held nearly zero within the 21st century and all FRs will be changed to the self-generated MA transmutation core without the targets. When the MA once-through cores are introduced with MA burner cores, the total number of targets to be reprocessed can be reduced by 50%. Even in the low projection case while the MA burner cores are installed after conventional FRs, MA accumulation would also be held to nearly zero within the 21st century.

Feasibility of the actinide hydride containing Np and Am as a transmutation target fuel to reduce the amount of long-lived actinides in the high level nuclear waste was studied by employing $UTh_4Zr_{10}H_x$ as a simulated actinide hydride fuel where Th is a surrogate for minor actinides. The pellets of this simulated actinide hydride fuel were successfully fabricated through alloying and hydrogenation within expected diameter error. The U-Th-Zr hydride can hold a larger amount of hydrogen above 1173K than the U-Zr hydride due to higher thermodynamic stability of the former than the latter. Thermal diffusivity and thermal expansion coefficient of $UTh_4Zr_{10}H_x$ were measured, from which thermal conductivity was evaluated.

Irradiation tests of the simulated actinide hydride fuel target have been carried out in Japan Material Testing Reactor (JMTR) of JAEA. Two tests were conducted up to the burnup of 0.2% and 1.1% FIMA (Fission per initial metal atom) each. After irradiations, both non-destructive and destructive examinations were carried out, which showed that the integrity of the hydride fuel was kept intact through irradiation. This result supports the feasibility of the concept of the hydride fuel as a transmutation target.

## 6. References

Kondo, Y. & Takizuka, T.(1994). Technology Assesment of Partitioning Process (1) – Status of the Partitioning Technology-, *JAERI-M-94-067* [in Japanese].

Wakabayashi, T., et al., (1995). Feasibility Studies of an Optimized Fast Reactor Core for MA and FP Transmutation, *Proceeding of the International Conference on Evaluation of Emerging Nuclear Fuel Cycle Systems (GLOBAL'95)*, Versailles.

Kawashima, K., et al., (1997). Utilization of Fast Reactor Excess Neutrons for Burning Minor Actinides and Long Lived FPs, *Proceeding of the International Conference on Future Nuclear Systems (GLOBAL'97)*, Yokohama.

Takano, H., et al., (1990). Higher Actinide Confinement Transmutation Fuel Cycles in Fission Reactors, *Proc. PHYSOR 90*, Marseilles, P III-145.

Masumi, R., et al., (1995). Minor Actinide Transmutation in BWR Cores for Multi-recycle Operation with Less Minor Actinide-to-Fissile Plutonium Amount Ratio, *J. Nucl. Sci. Technol.*, 32[10], 965.

Rome, M., et al., (1996). Use of Fast Reactors to Burn Long-Life Actinides, Especially Am, Produced by Current Reactors, *Proc. PHYSOR 96* , Mito, M-52.

Simnad, M.T., (1981). The U-ZrHx Alloy: Its Properties and Use in TRIGA Fuel, *Nuclear Engineering and Design*, 64, 403-422.

Yamamoto, T., et al., (1995). Development of New Reactor Fuel Materials, Hydrogen Absorption Properties of U-Th, U-Th-Zr and U-Th-Ti-Zr Alloys, *J. Nucl. Sci. Technol*, 32[3], 260.

Yamawaki, M., et al., (1998). Concept of Hydride Fuel Target Subassemblies in a Fast Reactor Core for Effective Transmutation of MA, J. Alloys Comp. Vol. 530, pp.271-273.

Sanda, T., et al., (2000). Fast Reactor Core Concepts for Minor Actinide Transmutation Using Hydride Fuel Targets, *J. Nucl. Sci. Technol.*,37[4],335.

Konashi, K., et al., (2001). Development of Actinide-Hydride Target for Transmutation of Nuclear Waste, *Proceeding of the International Conference on Future Nuclear Systems (GLOBAL'01)*, Paris.

Briesmeister, (Ed.) J., (1993). MCNP Đ A General Monte Carlo N-Particle Transport Code, Version 4A, *Report LA-12625-M*, Los Alamos, USA.

Tsuchihashi, K., et al., (1986). Revised SRAC code system, *JAERI-1302*.

Takeda, T., & Yokoyama, K., (1997). Study on Neutron Spectrum for Effective Transmutation of Minor Actinides in Thermal Reactors, *Ann. Nucl. Energy*, 24[9], 705.

Yokoo, T., et al., (1995). Performances of Minor Actinides Recycling LMRs, *Proceeding of the International Conference on Evaluation of Emerging Nuclear Fuel Cycle Systems (GLOBAL'95)*, Versailles, 1570.

At. Energy Soc. Jpn. Ed., (1994). Present Status of Transmutation Research and Development, p.1-10, [in Japanese].

Greenspan, E., (1997). Private Communication.

Simnad, M.T., (1980). GA-A16029.

Fujimura, K., et al., (2001). Enhancement of Transmutation Characteristics of the Minor Actinide Burning Fast Reactor Core Concept Using Hydride Fuel Targets and Its Introduction Scenario, *J. Nucl. Sci. Technol.*,38[10],879.

Konashi, K., et al., (2000). Transmutation Method Using Neutron Irradiation Targets of Actinide-Hydrides, *Proc. ICENES 2000* , Petten, the Netherlands, 244.

Wajima, T., et. al., (1996). A Strategy Analysis of the Fast Breeder reactor Introduction and Nuclear Fuel Cycle Systems Development, *Nihon-Genshiryoku-Gakkai Shi* ( At. Energy Soc. Jpn.), 38[2], 133, [in Japanese].

Hamamoto, K., et. al., (2001). High Moderation MOX Cores for Effective Use of Plutonium in LWRs, *Nihon-Genshiryoku-Gakkai Shi* ( At. Energy Soc. Jpn.), 43[5], 503, [in Japanese].

Kobayashi, K., et al., (1997). Applicability Evaluation of MOX Fuelled Fast Breeder Reactor to the Self-Consistent Nuclear Energy System, *Proceeding of the International Conference on Future Nuclear Systems (GLOBAL'97)*, Yokohama, 1062.

Yamamoto, T., Kayano, T., Suwarno, H. and Yamawaki, M., (1994). Hydrogen Absorptin Properties of U-Th-Zr and U-Th-Ti-Zr Alloys, *Sci. Rep. RITU*, A40, 13-16.

Yamawaki, M., et al., (1997). Development of MA-Containing Hydride Fuel as MA Burning Target Material, *Proceeding of the International Conference on Future Nuclear Systems (GLOBAL'97)*, Yokohama..

Tsuchiya, B., Konashi, K., Yamawaki, M. and Nakajima, Y., (2002). Thermal Diffusivity of U-Th-Zr Hydride at High Temperature, *J. Nucl. Sci. Technol. Supplement 3*, p.855.

# 9

# Clean-Up and Decontamination of Hot-Cells From the IFIN-HH VVR-S Research Reactor

A. O. Pavelescu and M. Dragusin
*Horia Hulubei National Institute of Physics and Nuclear Engineering (IFIN-HH), Bucharest-Măgurele, Romania*

## 1. Introduction

The "Horia Hulubei" Institute of Physics and Nuclear Engineering VVR-S type research reactor from Magurele-Bucharest, Romania was shut down 13 years ago, in 1997, after 40 years of operation.

Radioisotope production generated a significant contamination in the reactor main building, ventilation system and radioactive leakage drainage, overflow and collecting system. Major radioactive contaminants generated by this activity, with the half life higher than one year, are: Co-60, Cs-134, Cs-137, Sr-90, Eu-152, U-238 and Am-241 [1].

Four hot cells, a transfer room, a chemical laboratory and a decontamination solution preparation room are placed in the basement of the reactor hall and were used for processing the radioactive materials coming from the reactor. The hot cells and the transfer room are disposed one next to another and the communication between them was performed through a transfer channel in which a transfer truck, carried materials inside the hot-cells. The VVR-S hot cells are highly contaminated and contain a lot of radioactive sources and activated materials. For this reason, this area is inaccessible and contamination measurements are not possible. The total activity of materials abandoned in these rooms it is not known, but it is expected to be of about 15 Ci (0.55 TBq). During the hot-cells decommissioning, these radioactive sources and materials will be evacuated.

The cleanup of the hot cells will be carried out under preservation license on the basis of an Activity Plan approved by Romanian National Commission of Nuclear Activities (CNCAN). IFIN-HH has an organizational structure in accordance with the Quality Management Program provisions for the activities developed inside the Institute. The reactor decommissioning department (DDR) includes a Quality Assurance Compartment (AC). The AC staff verifies the observance of the quality assurance conditions and the radiological security. The AC compartment reports to the IFIN-HH top management all the nonconformities related to the quality management systems and regarding the radiological security requirements. All the members of the AC compartment are assigned with the CNCAN approval and are given responsibilities in accordance with the legislation in force.

## 2. General data necessary for the waste/material clean-up

### 2.1 Layout of the hot cells

The layout of the VVR-S research reactor hot cells is presented in the Fig. 1 bellow.

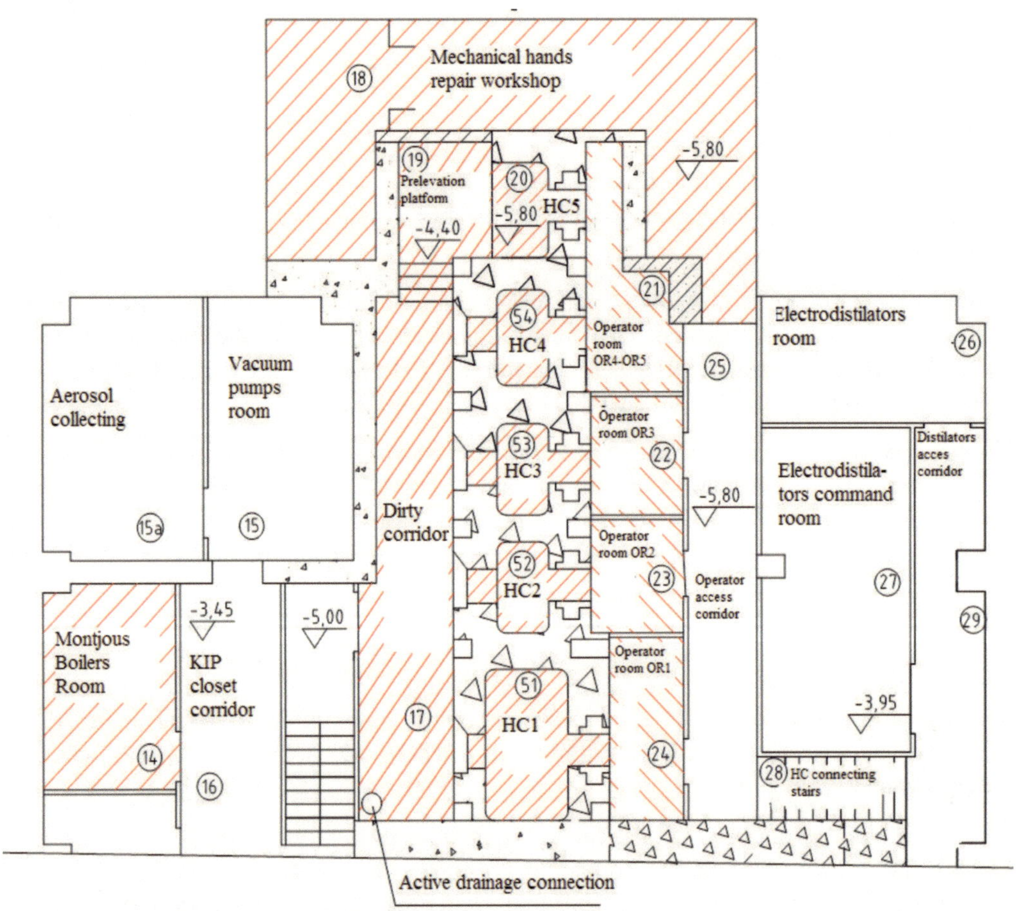


Fig. 1. The hot cells complex (HC1-HC5) [2]

Access to hot cells 1, 2, 3, 4 and 5 is made from the "dirty corridor" 17. Hot cell No. 1 (HC 1) has the following dimensions: length 3200 mm, width 2000 mm, height 3400 mm (2800 compartment without lamps). HC 1 can be accessed through a locked access corridors with a metal door (cast iron) having a thickness of 350 mm. Access corridors dimensions are: length 750 mm, width 900 mm, height 1600 mm. Hot cells 2, 3, 4 are identical in size and access, and have the dimensions of: length 2000 mm, width 1200 mm, height 2600 mm. (2000 mm without lamps compartment);

Access to HC 2, is made through a locked access corridors with a metal door (cast iron) having a thickness of 350 mm towards room 17 and a thin door (protection against contamination) of steel to hot room . Access corridors dimensions are: 1250 mm length mm, width 600 mm, height 1200 mm.

Access to HC3 or HC4 is made through identical in size corridors with the corridor for accessing the described HC2. Hot cells are fully clad with stainless steel sheet. For handling hot objects in the hot-cells mechanical arms are used, two for each room. Visibility is provided by the visor of Pb glass and lighting by lamps placed in separated niches in the ceiling.

Hot cells are ventilated on the inside (to ensure a depression of 5 mm $H_2O$ from the outside). Hot air is discharged from the room through HEPA filters.

The HC 1 represented in Fig. 1 above, has the following dimensions: 2000 mm length, 1200 mm wide, 2600 mm height (2000 mm without the light compartment) [2]. The access in hot cell 4 (HC 4), is made by a closed metallic door from cast iron with a thickness of 350 mm towards room 17 and a thin door (for protection against contamination) made from stainless steel towards the hot cell. The access corridor dimensions are: 1250 mm length; 600 mm wide; 1200 mm height. The hot cells are entirely covered with stainless steel plate.

To calculate the biological shielding of the hot cells, it has been considered that the highest activity which one works with is 2.69 x $10^4$ Ci (approx. 1 PBq) and the gamma radiation energy is 1.65 MeV. [2].

The biological shielding has the following characteristics [1]:

- The biological shield in the upper side is achieved through a 2 m concrete layer of 2.3 g/$cm^3$ density and laterally through a concrete layer of 75 cm and a density of 4.2 g/$cm^3$;
- Lead-glass eye-sleds, with 0.72 m of thickness;
- Between the hot cells, there are heavy concrete walls with 0.65 cm of thickness.

The radiological situation presented below is based on the hot-cell operating history described in an internal report made by the IFIN-HH Centre of Radioisotopes Production in 2003.

## 2.2 Hot-cells operating history

The VVR-S hot-cells operation history was used to approximate the contaminants in the HC interior and is presented in Table 1 [3].

| Period | Operations | Sources Inventories |
|---|---|---|
| 1957 - 1990 | Production of radioisotopes for medical purposes, radiochemicals and various industrial applications (furnaces, measuring, etc) | Medical products: $^{131}I$ approx. 100 Ci/yr; $^{99}Mo$ approx. 50 Ci/yr; $^{198}Au$ approx. 50 Ci/yr; $^{192}Ir$ sources of SRC type approx. 1000 pc/yr x 50 mCi/pc = 50 Ci/an; $^{90}Y$ approx. 2Ci/yr; Radiochemicals: $^{82}Br$ approx. 2Ci/yr; $^{42}K$ approx. 4 Ci/yr; $^{32}P$ approx.3 Ci/yr; $^{35}S$ approx.300 mCi/yr, various radiochemicals: activities max. 5 Ci/yr; $^{60}Co$ sources for furnaces and other industrial applications, approx. 5 Ci/an |
| 1970- 1983 | Additional $^{192}Ir$ sources were produced for gammagraphy | $^{192}Ir$ sources with activities of approx. 15 Ci - 20 Ci/source and total activities between 1000 Ci/yr and 8500 Ci/yr |
| 1980-1985 | Small uranium quantities have been irradiated for research purposes, while trying to perform fission molybdenum separation for obtaining $^{99}Mo$-$^{99m}Tc$ generators | |
| 01.01.1990 - 01.01.1998 | Radionuclides were produced for medical and industrial use by irradiation of targets embedded in non-radioactive nuclear grade aluminum blocks with dimensions of 37 x 140 mm and 22 x 140 mm in wet canals. | Medical products: $^{131}I$ approx. 100 Ci/year; $^{99}Mo$ approx. 50 Ci/yr; $^{198}Au$ approx. 50 Ci/yr; sources $^{192}Ir$ type SRC approx.1000 pc/yr x 50 mCi/pc = 50 Ci/yr; Irradiated silicon for the electronics industry: between 10 and 30 blocks; Various radiochemicals, activity max. 5 Ci / year; $^{60}Co$ sources for furnaces and other industrial applications, activity approx. 2 Ci/yr. |
| 01.01.1998 - 01.01.2003 | Operations for the production of radioisotopes were not conducted. | |

Table 1. Hot-cells operation history

### 2.3 Radiological situation

Existing estimated sources inventory in the VVR-S reactor hot-cells are given in Table 2 below [3]:

| Hot cells | Inventory | Activity |
|---|---|---|
| Hot cell 1 | In the existing deposit under the room worktop low activity sources of $^{60}Co$ may be found (this deposit has not been accessed since 1985);<br>$^{60}Co$ small spheres, activity 1 Ci in 1997, respectively 0,5 Ci at 1.01.2003;<br>$^{134}Cs$ 50 mCi in 1994, respectively 1.6 mCi in 2003. | |
| Hot cell 2 | Approximately 1000 bars containing sources of $^{192}Ir$, produced after 1983. | |
| Hot cell 3 | Bottles with residual solutions of $^{60}Co$, $^{134}Cs$, $^{133}Ba$, $^{63}Ni$. Total activity may be around 50 mCi;<br>Bottles containing fission production prepared in the period 1980-1982: $^{90}Sr$, $^{137}Cs$, $^{134}Cs$. Total activity may be around 50 mCi; | Estimated activities are:<br>$^{60}Co$ - 30 mCi in irradiation boxes;<br>30 mCi in taps;<br>4 mCi in bars;<br>$^{65}Zn$ - 7mCi in irradiation boxes;<br>7 mCi in caps;<br>1 mCi in irradiation bars.<br>The rest of generated radionuclides: $^{51}Cr$, $^{59}Fe$, $^{46}Sc$, $^{140}La$, $^{24}Na$ have most certainly decayed. Resulted residues from cutting the irradiation boxes contain the same radionuclides in quantities approx. 10 times smaller than the irradiation boxes. [3] |
| Hot cell 4 | Plastic bag containing textile material used in decontamination,<br>Plastic bag containing cellulosic material - filter paper,<br>Irradiation box lead lest,<br>Cotton mop, cylinder (filled) ø 37 x 60 - possibly of Pb - counterweight;<br>Empty can box without cap 800 ml,<br>Stainless steel cylinders without cap ø 25 x 70 empty,<br>Plastic bag containing 2 bars probably with sources at their end;<br>Mechanical hand tweezers | |
| Hot cell 5 | HC 5 is empty and presents no radiological risk. | |
| Hot cells 1-4 | In all the HC there are also irradiation boxes as follows:<br>Maximum no. of irradiation boxes 1500 boxes distributed in HC 1,2,3 and 4;<br>1500 taps and approximately 600 irradiations bars;<br>An undetermined number of irradiation boxes and irradiation bars with uncut caps.<br>Approx. 80 kg of Pb pellets of unknown isotopic composition are distributed in HC 1, 2 and 4 as well. [3] | |

Table 2. VVR-S hot-cell sources inventory and activity

In HC 3 there are approximately 500 small paper baskets, approx.10 kg of glass, and approx. 2 kg of paper, as it can be seen in Fig. 2.

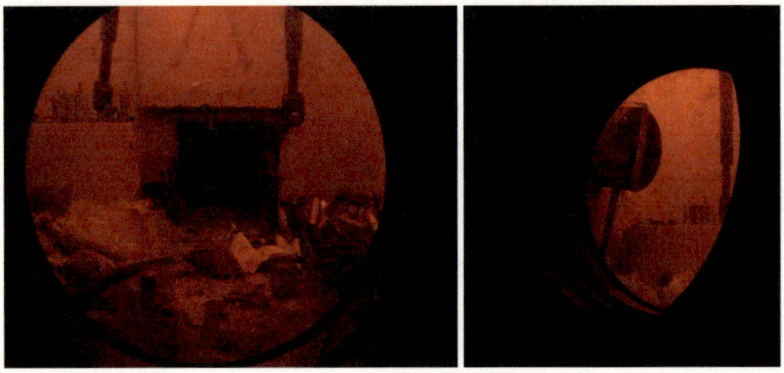

Fig. 2. Images through HC 3 visor. [4]

Due to the fact that it was open in 2006, HC 4 is the best characterized hot cell (Fig. 3 and Fig. 4).

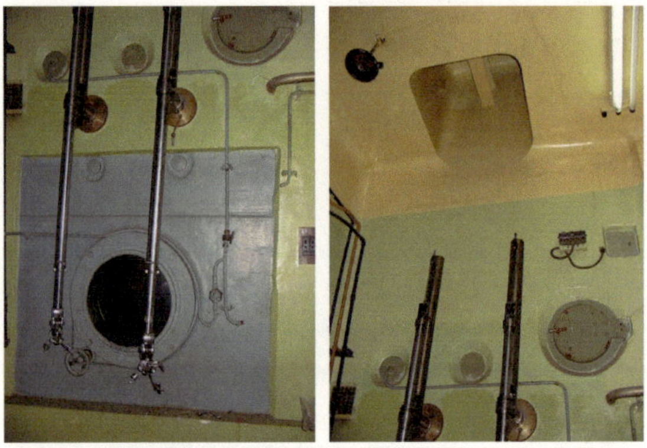

Fig. 3. Images from the Operator Room No. 4 (OR 4) serving the HC 4 [4]

Fig. 4. Interior of the hot cell no. 4 (HC 4), opened in July 2006 [4]

The HC 4 measured dose rates are presented in Table 3

| No. | Device | Detector | Measuring Location | Measuring Value |
|---|---|---|---|---|
| 1 | Eberline | Γ | Margins of opened exterior iron-cast protection door | 42 μSv/h |
| 2 | Eberline | Γ | Margins of plated interior door | 280 μSv/h |
| 3 | Eberline | Γ | Protection metal plate of the visitation window from transfer truck tunnel | 200 μSv/h |
| 4 | Eberline | γ | HC4 Centre at 0,5 m above the floor | 1.4 mSv/h |
| 5 | Eberline | γ | HEPA filter level (near the filter) | 0.7 mSv/h |
| 6 | Eberline | γ | Room Centre, floor level | 9.2 mSv/h |
| 7 | Eberline | γ | Room centre, 1,5 m above floor level | 0.3 mSv/h |
| 8 | Eberline | γ | Near floor drainage | 4.4 mSv/h |
| 9 | Eberline | γ | Near the wall, left of the entrance | 1.4 mSv/h |
| 10 | Eberline | γ | Middle waste tray. Probably the 2 bars from plastic bag contain sources | 18 mSv/h |

Table 3. Dose rate measurements in HC 4 in 24.07.2006 according to dosimetric service procedure AC-PO-DDR-501-01 [4]

The presumed total radionuclide inventory in 2003 is presented below in Table 4:

| Radionuclide | $T_{1/2}$ | Observations | Radionuclide | $T_{1/2}$ | Observations |
|---|---|---|---|---|---|
| **I-131** | 8.04 d | Fission product | Sr-90 | 29.1 y | Fission product |
| **Mo-99** | 2.75 d | Fission product | Cs-137 | 30 y | Fission product |
| **Au-198** | 2.69 d | Radioisotope | Zn-65 | 244 d | Radioisotope |
| **Ir-192** | 74 d | Radioisotope | Cr-51 | 27.7d | Radioisotope |
| **Y-90** | 2.62 d | Radioisotope | Fe-59 | 45.1 d | Radioisotope |
| **Br-82** | 1.47 d | Radioisotope | Sc-46 | 83.8 d | Radioisotope |
| **K-42** | 12.44 h | Radioisotope | La-140 | 1.68 d | Radioisotope |
| **P-32** | 14.3 d | Radioisotope | Na-24 | 15 h | Radioisotope |
| **Tc-99** | 6.02 h | Fission product | Co-60 | 5.27 y | Radioisotope |
| **Cs-134** | 2.06 y | Fission product | Mo-99 | 2.75 d | Radioisotope |
| **Ba-133** | 10.7 y | Radioisotope | Ni-63 | 96 y | Radioisotope |

Table 4. Approximate contamination nuclide inventory and their half-life time in the hot-cells, reported in 2003, for the working activity cease in 1998 [4]

Dose rate in hot cells during and after their clean-up is currently unknown and an evaluation is presented in section 4.

## 3. Clean-up and decontamination activities for hot cell no. 4

### 3.1 Material transfer

The following activities will be accomplished in the following sequence [5]:

1. **Waste transfer from HC 4 to HC 5 and then to the conditioning containers or intermediate disposal.**

**Transfer route:** HV 4 → (direct connecting conduct between HC 4–HC 5, or HC 4 → connecting tunnel → DC 17 → Room 19 → HC 5 → HC 5 (preliminary measurement and containing) → lift-up in Room 102 (auto lock), containing in adequate barrels → Room 101 (NR hall quota 0,00 m) → Room 103 (radiological characterization of contained wastes) → Room 101 (NR hall quota 0,00 m) → room 102 (auto lock), expedition in to the treatment plant in view of conditioning or intermediate disposal.

2. **Identification/inventory of the evacuated wastes**: Wastes from HC4 are identified using photos obtained while opening HC4 from 24.07.2007 for measuring the dose rate.
3. **Measures in HC4** (in 24.07.2006 according to the dosimetry service AC-PO-DDR-501-01) [4]
4. **Evacuation of materials from HC 4** (for measuring the dose rate)

Using the direct connecting conduct from HC 4 and HC 5, the following operations are performed:

a. The access door to the direct connecting conduct is opened from HC5 (room 20);
b. Utilizing the mechanical hands from HC 4, the small wastes can be grasped then pushed through the inclined conduct towards HC5 in the following order as follows: plastic bag containing textile material used in decontamination, plastic bag containing cellulosic material - filter paper, irradiation box lead lest penal, cotton mop, cylinder (filled) ø 37 x 60 - possibly of Pb - counterweight; empty can box without cap 800 ml, stainless steel cylinders without cap ø 25 x 70 empty, plastic bag containing 2 bars probably with sources at their end; mechanical hand tweezers.

Wastes extracted one by one are measured with a dosimeter probe already installed in HC 5 in view of transfer in the adequate container.

**NOTE:** From the analysis of the dose rate measurement of 18 mSv on the waste tray, it is presumed that the greater dose is given by the plastic bag containing 2 bars probably with sources at the end on the respective tray. Before the evacuation of this bag, in HC 5, a lead container is let down from Room 102, which is adequate for disposal of this waste. The manipulation of the transferred wastes from HC 5 is made according to the operation procedure AC-PL-DDR-03. [6]

5. **Opening the HC 5 doors**

After the opening of doors from behind HC 5 the following operation take place:

a. Opening of the biological protection door from HC 4 as follows:
- The dose rate is measured and noted in the centre of the biological protection door opening, the centre of connecting tunnel with the HC4 contour and in the centre of interior protection door of the HC 4 (closed)

- The dose rate is measured at the surface in the centre of the protection plate of the carriage tunnel (on the floor)
- Samples are taken and they are contamination of the walls is analyzed, as well as the ceiling and the floor and of the ceiling of the HC4 connecting tunnel;
- The HC 4 connecting tunnel interior surfaces are decontaminated. Non-fixed contamination is removed to prevent its transfers on the protection wearing or clothing. The decontamination is made by vacuum cleaner fitted with a HEPA filter, and the by cleaning with a wet cloths using organic solvents or with detergent and finally with water wet cloths for cleansing.

b. Interior protection door is opened in HC 4 by sliding to the left as follows:
- Dose rate is measured and noted in the centre of the opened door, in the centre of the room, in the centre of the room, near various objects in the room.
- Working time are calculated for manipulation operations of the radioactive wastes (the most active object)
- Manipulation times are established in function of route distances, as well as the layout of the objects in the containers;
- The operation will begin with the object having the greatest dose rate

c. Transfer/evacuation of the wastes from HC 4
- From the operation room, with the aid of the mechanical hands the remaining objects are pushed one at a time, towards the centre of the room and in the access door of the HC 4, in such a way that the operator will not be forced to expose his body to the HC 4 interior;
- The most active object is grabbed with the manipulator having the length of 1 meter. The object is taken out of the HC 4 and it is put in the prepared container, situated either in the connecting tunnel with HC 4, either in the corridor 17 near the biological protection door of the HC 4. The contained cover is mounted, the interior protection HC 4 door is closed and it is assured by sealing.
- The container is transported in HC 5. From the HC 5 operator room, the object activity is measured. The manipulation of the devices found in the hot cells of the VVR-S nuclear reactor is made according to the AC-PL-DDR-03 procedure [6]

d. The container is transported in the room 102, there it is loaded in an adequate barrel and it is transferred through hall 101 in room 103 in view of characterization and later transfer towards the Department of Radioactive Wastes Management, according to the operating procedure PO-DDR-827 [7].

e. The steps from c), d) and e) are retaken until reaching the complete emptiness of the HC 4.

**NOTE**: Special working obligatory equipment is required as follows: sealed protection costume, resistant to water drops, with prepared air feeding and telephonic communication line. The operations are performed only with manipulators, pincers and homeostatic pincers for griping, manipulation of the HC 4 objects. Operator will not expose his full body by entering the hot cell, in order to retrieve and process the objects inside.

### 3.2 Decontamination of the HC 4 interior

The following activities will be accomplished in the following sequence [5]:

- The dose rate in the completely empty HC 4 is measured;
- The working times are calculated for the scenario in which the operator exposes his full body by completely enter the room;
- The interior of the HC is decontaminated by washing with water and detergent jet, and then only with water jet utilizing the pressure equipment. Small fine water droplets are produced which can contaminate the protection costume and the HC 4 connecting tunnel;
- The decontamination solution is evacuated from HC 4 by removing the lid from the room drainage system (utilizing the manipulator). If the drainage is stuck, the decontamination solution is recovered by peristaltic pump (80 l/h) and it is spilled in the active drainage connection from corridor 17;
- When attaining an acceptable dose rate of 60-100 μSv/day measured in the centre of the room, the operator enters inside the HC 4 and the walls are decontaminated with a wet cloth with organic solvents, collecting the cloths in recipients situated at the beginning of the connecting tunnel.

### 3.3 Closing of the HC 4

The procedure is similar with the one for opening if HC 4, but in reverse order [5]:

- The HC 4 connecting tunnel door is closed;
- The radiation field is measured and recorded in the connecting tunnel from corridor 17 and HC 4;
- The biological protection door towards the corridor 17 is closed and sealed;
- Secondary wastes resulted from corridors 17, 19 and 20 (floors foils);
- The radiation field from the corridors17, 19 and 20 is measured.

**NOTE**: After closing the HC 4 becomes clean and fully operational.

## 4. Dose/Risk modeling using RESRAD computer codes package

### 4.1 RESRAD Build 3.5 Code

RESRAD is a computer code designed to estimate radiation doses and risks from RESidual RADioactive materials (Fig. 5) [8]. The only code designated by Department Of Energy for the evaluation of radioactively contaminated sites.

United States National Regulatory Commission (NRC) has approved the use of RESRAD for dose evaluation by licensees involved in decommissioning, NRC staff evaluation of waste disposal requests and dose evaluation of sites being reviewed by NRC.

RESRAD has been applied to over 300 sites in the U.S. and other countries. Environmental Protection Agency (EPA) Science Advisory Board reviewed the RESRAD model. EPA used RESRAD in rule-making on radiation site cleanup regulations.

RESRAD code has been verified and has undergone several benchmarking analyses, and has been included in the IAEA's VAMP and BIOMOVS II codes to compare environmental transport models. RESRAD training workshops have been held at DOE, NRC, and EPA headquarters. Around 800 people have been trained at these workshops and RESRAD has been used by several universities as a teaching tool as well.

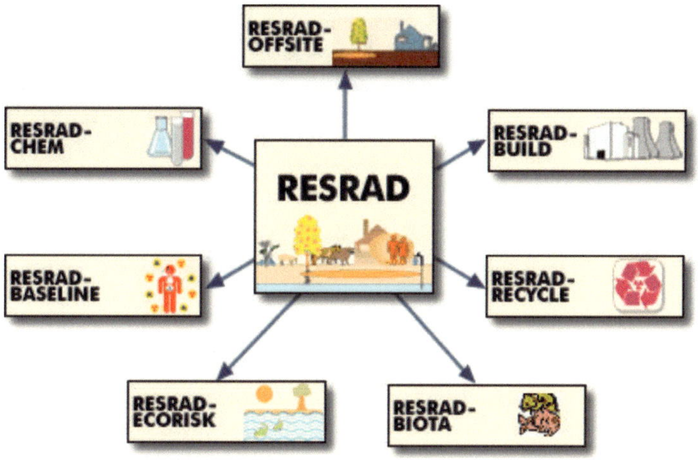


Fig. 5. RESRAD computer code package [8]

The RESRAD Build Code is a model for analyzing the radiological doses resulting from the remediation and occupancy of buildings contaminated with radioactive material with the following features (Fig. 6) [9]:

- Considers external exposure, inhalation of dust and radon, and ingestion of soil/dust.
- Up to 10 sources and 10 receptors can be modeled.
- Sources geometry can be point, line, plane, or volume.
- Building can be any structure composed of up to three compartments.
- Radioactive contamination can be on surface or in building material.
- Exposure scenarios considered include building occupancy (residential use and office worker) and building remediation (decontamination worker and building renovation worker).

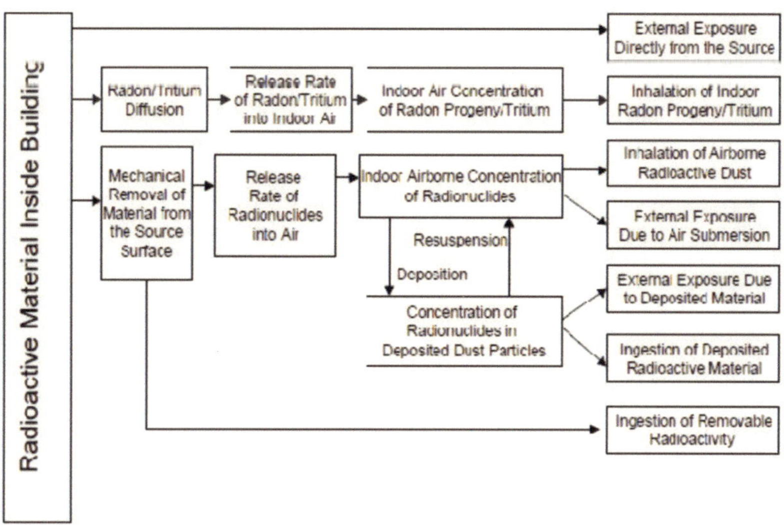


Fig. 6. Exposure Pathways Incorporated into the RESRAD-BUILD Computer Code [9]

| Parameter | Unit | Building Occupancy[a] | Building Renovation[b] | Remarks |
|---|---|---|---|---|
| Exposure duration | days (d) | 365.25 | 179.00 | To match the occupancy period of 365.25 days in NUREG/CR-5512 building occupancy scenario (Beyeler et al. 1999 [8]) and renovation period of 179 days in NUREG/CR-5512 building renovation scenario (Wernig et al. 1999 [9]). |
| Indoor fraction | _[c] | 0.267 | 0.351 | To match the 97.5 d/yr time in building in NUREG/CR-5512 building occupancy scenario (Beyeler et al. 1999) and 62.83 days spent in the building during renovation period in NUREG/CR-5512 building renovation scenario (Wernig et al. 1999 [9]). |
| Receptor location | m | 0, 0, 1 | 0, 0, 1 | At 1-m from the center of the source. |
| Receptor inhalation rate | $m^3/d$ | 33.6 | 38.4 | For building occupancy scenario it matches with 1.4 $m^3/h$ breathing rates in NUREG/CR-5512 (Beyeler et al. 1999 [8]) and for building renovation scenario it matches with 1.6 m3/h breathing rate of moderate activity given in the EPA Exposure Factor Handbook (EPA 1997 [9]). |
| Receptor indirect ingestion rate | $m^2/h$ | $1.12 \times 10^{-4}$ | 0 | Value for the building occupancy scenario is the mean value from the distribution and for the building renovation scenario it is assumed the ingestion is only from the direct contact with the source. |
| Source type - | _ | Area | Volume | For building occupancy scenario it is assumed that contamination is only on the surfaces, whereas for the building renovation scenario is volumetric |

| Parameter | Unit | Building Occupancy[a] | Building Renovation[b] | Remarks |
|---|---|---|---|---|
| Direct ingestion rate | 1/h (area) g/h (volume) | $3.06 \times 10^{-6}$ | 0.052 | Calculated from the default ingestion rate of $1.1 \times 10^{-4}$ m$^2$/h in NUREG/CR-5512 building occupancy scenario (Beyeler et al. 1999 [8]). The effective transfer rate from NUREG/CR-5512 building renovation scenario for ingestion of loose dust to the hands and mouth during building renovation (Wernig et al. 1999 [9]). |
| Air release fraction | – | 0.357 | 0.1 | For the building occupancy scenario, it is the mean value from the parameter distribution. For the building renovation scenario, a smaller fraction is breathable. |
| Removable fraction | – | 0.1 | NR[d] | 10% of the contamination is removable (NUREG/CR-5512 building occupancy scenario default [10]). The parameter is not required for the volume source. |
| Time for source removal or source lifetime | d | 10,000 | NR | Value for the building occupancy scenario is the most likely value from the parameter distribution. The parameter is not required for the volume source. |
| Source erosion rate | cm/d | NR | $4.1 \times 10^{-4}$ | For the building renovation scenario, it is assumed that the total source thickness of 15 cm can be removed in 100 years of building life. |

[a] Parameter values used in the building occupancy scenario.

[b] Parameter values used in the building renovation scenario.

[c] A dash indicates that the parameter is dimensionless.

[d] NR = parameter not required for the analysis.

Table 5. Key parameters used in the building occupancy and building renovation scenarios [10]

### 4.2 Calculation of the intake rates and time integrated cancer risks

RESRAD is able to calculate lifetime cancer risks resulting from radiation exposure. The radiation risk can be computed by using the U.S. Environmental Protection Agency (EPA) risk coefficients with the exposure rate (for the external radiation pathways) or the total intake amount (for internal exposure pathways).

The EPA risk coefficients are estimates of risk per unit of exposure to radiation or intake of radionuclides that use age- and sex-specific coefficients for individual organs, along with organ-specific dose conversion factors (DCF). The EPA risk coefficients are characterized as best-estimate values of the age-averaged lifetime excess cancer incidence risk or cancer fatality risk per unit of intake or exposure for the radionuclide of concern. Detailed information on the derivation of EPA risk coefficients and their application can be found in several EPA documents [11] [12]. The methodology used in the RESRAD code for estimating cancer risks follows the EPA risk assessment guidance (EPA 1997) and is presented very briefly in the following section.

Intake rates calculated by the RESRAD code are listed by radionuclide and pathway and correspond to specific times. Intake rates for inhalation and ingestion pathways are calculated first for all of the principal radionuclides and then multiplied by the risk coefficients to estimate cancer risks.

For inhalation and soil ingestion pathways (p = 2 and 8, respectively), the intake rates (Bq/yr or pCi/yr) can be calculated by using the following Equation (1) [13]:

$$(Intake)_{j,p}(t) = \sum_{i=1}^{M} ETF_{j,p}(t) \times SF_{i,j}(t) \times S_i(O) \times BRF_{i,j}, \quad (1)$$

where:

$(Intake)_{j,p}$ $(t)$ = intake rate of radionuclide j at time t (Bq/yr or pCi/yr),
$M$ = the number of initially existent radionuclides,
$ETF_{j,p}$ $(t)$ = environmental transport factor for radionuclide j at time t (g/yr),
$p$ = primary index of pathway,
$SF_{ij}(t)$ = source factor,
$i,j$ = index of radionuclide (i for the initially existent radionuclide and j for the radionuclides in the decay chain of radionuclide i),
$S_i(0)$= initial soil concentration of radionuclide i at time 0,

The cancer risk at a certain time point from external exposure can be estimated directly by using the risk coefficients, which are the excess cancer risks per year of exposure per unit of soil concentration. Because the risk coefficients are derived on the basis of the assumptions that the contamination source is infinite both in depth and lateral extension and that there is no cover material on top of the contaminated soil, it is necessary to modify the risk coefficients with the cover and depth, shape, and area factors to reflect the actual conditions. Non-continuous exposure throughout a year also requires that the occupancy factor be considered when calculating the cancer risks.

Consequently, the RESRAD code uses the environmental transport factor for the external radiation pathway, along with the risk coefficient and exposure duration, to calculate the excess cancer risks as seen in Equation (2) [13]:

$$(Cancerrisk)_{j,1}(t) = \sum_{i=1}^{M} ETF_{j,1}(t) \times SF_{ij}(t) \times S_i(O) \times BRF_{i,j} \times RC_{j,1} \times ED, \quad (2)$$

where:

$RC_{j,1}$ = risk coefficient for external radiation (risk/yr)/(pCi/g),
$ED$ = exposure duration (30 yr).

For the inhalation and ingestion pathways, the cancer risks at a certain time point are calculated as the products of intake rates, risk coefficients, and exposure duration. Unlike the intake rates, cancer risks from inhalation of radon and its decay progeny are reported as total risks that include radon and progeny contributions. Therefore, the radon risks are the sums of the products of the individual radon or progeny intake rates and their risk coefficients.

### 4.3 Modeling results

In a previous study [14] the dose rate and the associated risk related to the clean-up and decontamination of HC 4 was evaluated. Using the measured doses, the activity of these sources and the associated risk was estimated, due to the difficulty of the spectroscopy analysis in order to determine the exact nuclide sources and their activity. The presumed time for clean-up and decontamination operation is 1 month for each hot cell, and therefore the estimation was be made for this period. The dose rate and the associated risk related to the clean-up and decontamination of hot cell no. 4 were calculated using RESRAD Build 3.5 code.

In the model 1 receptor was considered and 3 nuclide sources were taken into account: Co-60, Cs-137, Sr-90, because these nuclides are most probable in the hot cells. The activity of these sources was estimated by using Rad Pro 3.26 software models [15], starting from the measured doses. This was done, due to the lack of a difficult to perform spectroscopy analysis (inside the hot cell) in order to determine the exact nuclide sources and their activity.

In Fig. 7, a graphic is presented, showing the evolution of modeled hourly dose rate at the beginning of the operations, then after 1, 2, 3 and 4 weeks of clean-up operations when only the external pathways are taken into account. At the beginning of clean-up operations, the calculation revealed an initial equivalent hourly dose rate of 7 mSv (700 mrem). The relative high dose rate is due to the fact that receptor was modeled without a biological lead protection. The dose decreases in time towards insignificant values, as the hot cell no. 4 is cleared and decontaminated [14].

As it can be seen in Fig. 8, in the unlikely scenario in which the exposure pathways are summed (inhalation, ingestion, immersion, external, deposition and radon) the equivalent dose rate becomes extremely high. The corresponding cancer risk, presented in Fig. 9, is higher than 100 for this particular scenario. Due to high dose/risks involved, complying with ALARA principle, the utilization of a robot is proposed, instead of a human operator.

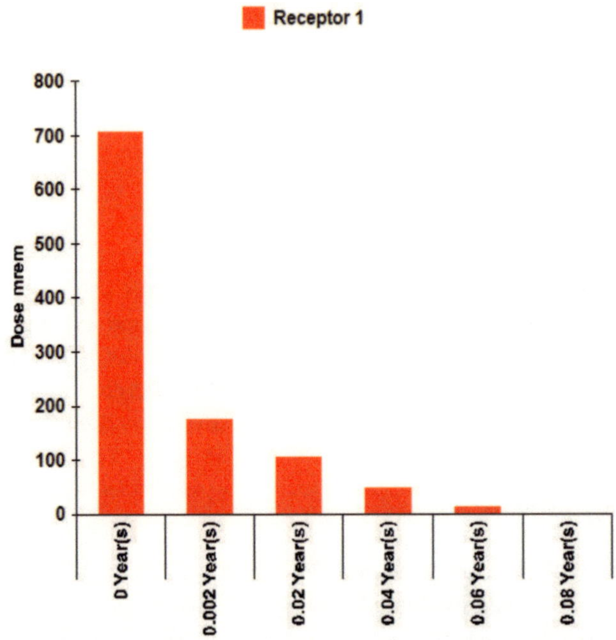


Fig. 7. Dose received by time for summed nuclides and external pathways [14]

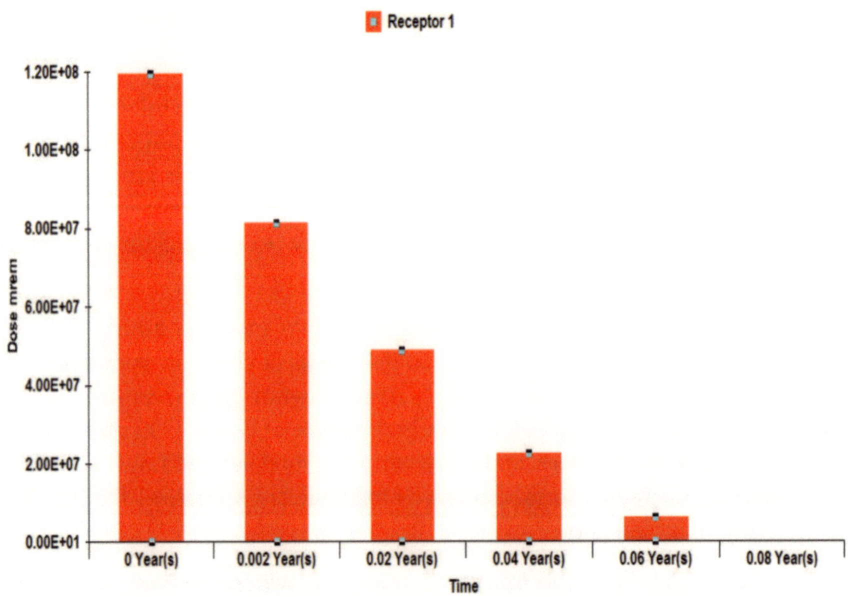


Fig. 8. Dose by time for summed nuclides, summed sources and summed pathways [14]

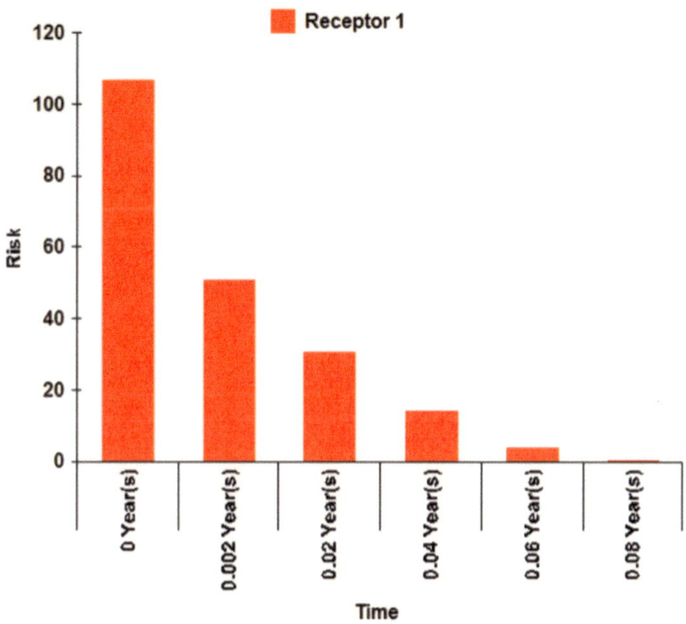


Fig. 9. Risk by time for summed nuclides, summed source and summed pathways. [14]

Furthermore, the modeling revealed each nuclide contribution to the overall risk. In the proposed reference scenario, the greatest risk is presented by Sr-90 followed by Cs-137 and Co-60 (see Figure 10) [14].

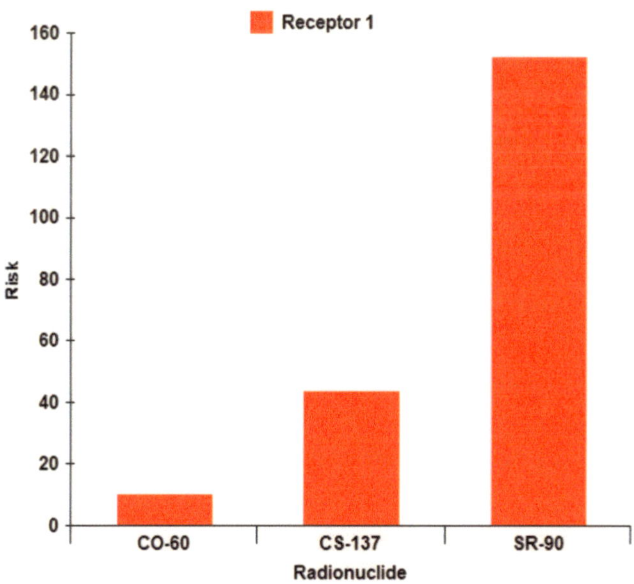


Fig. 10. Risk by each nuclide for the reference scenario [14]

In Fig. 11, the risk for all the exposure pathways is presented. At the beginning at the operations, both the risks from aerosols inhalation and ingestion are presented, then only the risk from ingestion remains high. This is due to the fact that the modeling did not take into account any breathing equipment for the operator. The risks from external exposure, deposition, immersion and radon are considerable smaller.

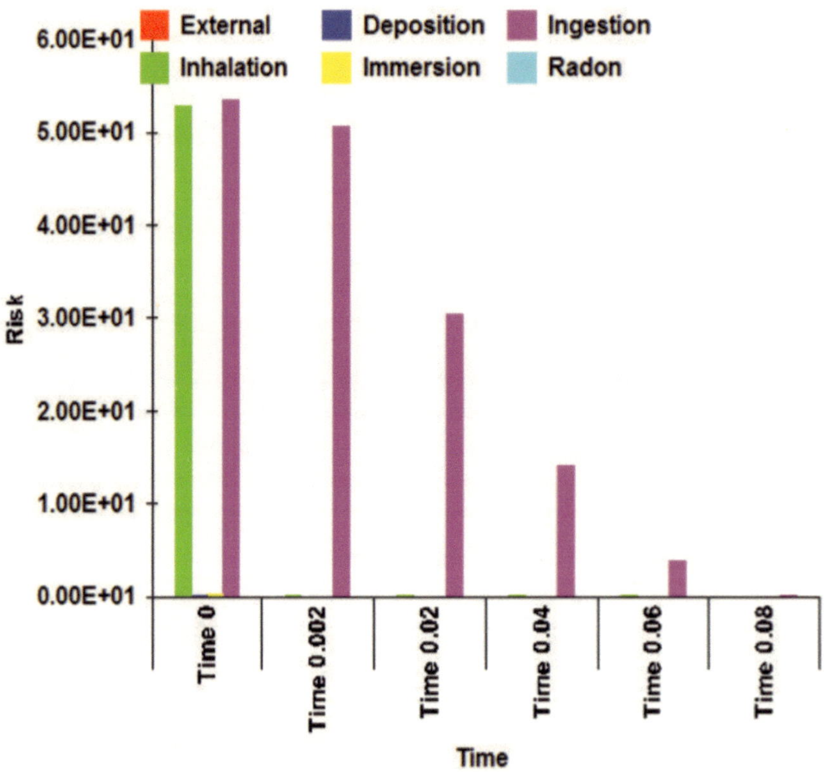


Fig. 11. Risk by time and pathway for summed nuclides and summed sources [14]

## 5. Remote operation alternative for clean-up and decontamination operations

Due to high dose/risks involved, complying with the ALARA principle, the utilization of a robot is proposed instead of a human operator. Using this robot, endowed with a high versatility arm (Fig. 12), the room will be cleared of remaining objects.

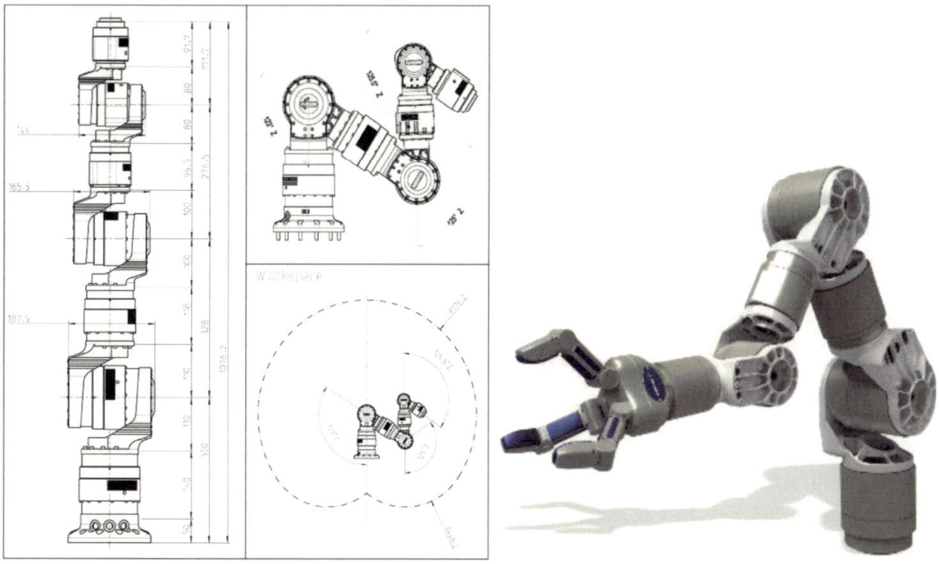

Fig. 12. Schunk robotic arm (a) [16]

The LWA 3 arm has the following features:

- Standard assembly of servo-electric swivel modules (PRL) and standard connecting elements to one 7-axis structure and 6 degree of freedom (2x PRL 120, 2x PRL 100, 2x PRL 80, 2x PRL 60);
- The arm can be configured in a reduced version with 5 modules and 4 degree of freedom;
- Nominal payload : 5 kg in standard configuration;
- Maximum payload: 10 kg in the reduced version;
- Arm length: 1073 mm (1.07 m) in standard configuration, but may be configured for longer reach by adding new actuator modules;
  When mounted on the MP S500 Neobotix mobile platform (with a height of 0,59 m), the total height of the assembly, arm + platform, is approximately 1,40 m in normal functioning, but the arm can be folded in order to pass through the hot cell access tunnels (1.2 m of height);
- The IP protection class against penetration of water and foreign objects is 65. This represents o good protection and allows the decontamination with water jet, although the manufactured does not recommend its utilization in extremely dirty environments;
- Completely interconnected cable routing through the hollow shaft integrated into the PRL-modules;
- Interconnected aluminum structure for weight optimization;
- Open software architecture for controlling the axes;
- Precision: 1 mm;
- Power supply 24 VDC / 20 A;
- Battery operation possible;
- Brushless servomotors

In Fig. 13 below, the Neobotix platform is shown and its dimensions are emphasized.

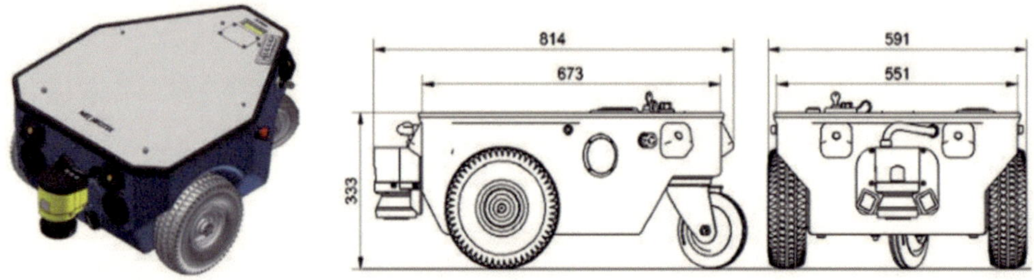


Fig. 13. Neobotix Platfrom [16]

Technical data for the Neobotix MP-S500 Platform:

- Dimensions (mm): 814 x 591 x 333(L x l x H);
- Weight: 80 kg;
- Payload: 80 kg;
- Moving speed: 1.5 m/s;
- Battery capacity: 10 h;
- Sensors: 1 x 2D laser sensor laser and 5x ultrasonic sensor;
- Control: external PC by wireless communication and onboard computer;
- The platform satisfies the maximum width criterion for accessing the hot cells doors of 600 mm, having a lateral dimension of 591 mm.

Fig. 14. Assembley of Schunk robotic arm mounted on the Neobotix Platform

In Fig, 14 above, the robot assembly is shown in standard configuration. Given its technical specifications and presented characteristcs , the robot will be capable to perform all the decontamination operations, thus eliminating the need of a human operator.

## 6. Conclusion

The presumed time for clean-up and decontamination operation of Hot Cell 4 is 1 month. The modeling of the dose rate revealed an initial equivalent dose rate of 7 mSv and an associated cancer risk of 120, due to external exposure only, which subsequently decreases in time towards insignificant values, as the hot HC 4 is cleared and decontaminated. Furthermore, in the unlikely scenario that the exposure pathways are summed, the equivalent dose rate is extremely high. Nevertheless, in the proposed reference scenario the greatest risk is presented by Sr-90 followed by Cs-137 and Co-60. Due to high dose/risks involved, complying with ALARA principle, the utilization of a robot is proposed.

## 7. References

[1] M. Dragusin, et al., *IFIN-HH VVR-S Research Reactor Decommissioning Plan, Revision 10,* Institute of Physics and Nuclear Engineering - Horia Hulubei (IFIN-HH), Centre of Decommissioning and Radioactive Waste Management (CDMR), February 2010;

[2] M. Dragusin, A.O. Pavelescu, I. Iorga, *Good Practices In Decommissioning Planning And Pre-Decommissioning Activities For The Magurele VVR-S Nuclear Research Reactor,* Nuclear Technology & Radiation Protection Journal, ISSN 1452-8185, 2011;

*[3]* IFIN-HH M. Sahagia et al, *VVR-S Research Reactor Hot-Cells Radiological Situation Report,* Institute of Physics and Nuclear Engineering - Horia Hulubei (IFIN-HH), Centre of Radioisotopes Production Report, 2003.

[4] D. Stanga et. al, *Dose Rate Measurements in VVR-S Research Reactor Hot Cell No. 4,* Dosimetric Service Procedure AC PO DDR-501-01, 24.07.2006;

[5] V. Popa, A.O. Pavelescu, et al., *IFIN-HH VVR-S Reactor Hot Cells Clean-up Plan (draft version),* Institute of Physics and Nuclear Engineering - Horia Hulubei (IFIN-HH), 2011

[6] V. Popa et al, *Manipulation of the Transferred Wastes from the VVR-S Reactor Radioactive Wastes Treatment Facility,* Institute of Physics and Nuclear Engineering - Horia Hulubei (IFIN-HH), Operation Procedure AC-PL-DDR-03, Revision 1/June 2006;

[7] V. Popa, *Interface Between Department of Decommissioning of the Reactor and Department of Management of the Radioactive Wastes for the transfer of the radioactive wastes,* Institute of Physics and Nuclear Engineering - Horia Hulubei (IFIN-HH) Operating Procedure PO-DDR-827, Revision 2/October 2007;

[8] C. Yu, A.J. Zielen, J.J. Cheng, D.J. LePoire, E. Gnanapragasam, S. Kamboj, J. Arnish, A. Wallo III,W.A. Williams, H. Peterson, *User's Manual for RESRAD Version 6,* Environmental Assessment Division, Argonne National Laboratory (ANL), ANL/EAD-4, July 2001;

[9] C. Yu, D.J. LePoire, J.J. Cheng, E. Gnanapragasam, S. Kamboj, J. Arnish, B.M. Biwer, A.J. Zielen, W.A. Williams, A. Wallo III, H.T. Peterson, Jr., *User's Manual for RESRAD-BUILD Version 3,* Environmental Assessment Division, Argonne National Laboratory (ANL), ANL/EAD/03-1, June 2003;

[10] B.M Biwer, et al., *Technical Basis for Calculating Radiation Doses for the Building Occupancy Scenario Using the Probabilistic RESRAD-BUILD 3.0 Code,* NUREG/CR-6755, ANL/EAD/TM/02-1, prepared by Argonne National Laboratory, Argonne, Ill., for Division of Systems Analysis and Regulatory Effectiveness, Office of Nuclear Regulatory Research, 2002;

[11] U.S. Environmental Protection Agency, *Exposure Factor Handbook,* EPA/600/P-95/002Fa, Office of Research and Development, National Center for Environmental Assessment, 1997;

[12] K.F. Eckerman, et al., *Cancer Risk Coefficients for Environmental Exposure to Radionuclides,* EPA 402-R-99-001, Federal Guidance Report No. 13, prepared by Oak Ridge National Laboratory, Oak Ridge, Tenn., for U.S. Environmental Protection Agency, Office of Radiation and Indoor Air, 1999;

[13] S. Kamboj, et al., *Probabilistic Dose Analysis Using Parameter Distributions Developed for RESRAD and RESRAD-BUILD Codes,* NUREG/CR-6676, ANL/EAD/TM-89, prepared by Argonne National Laboratory, Argonne, Ill., for Division of Risk Analysis and Applications, Office of Nuclear Regulatory Research, U.S. Nuclear Regulatory Commission, May 2000;

[14] A. O. Pavelescu, V. Popa, M. Drăguşin, *Modelling of the Dose Rates and Risks Arising from Hot-cells Clean-up Activities in the Decommissioning of the VVR-S Research Reactor,* Romanian Reports in Physics, Bucharest, Romania, No. 2/2012 (in press);

[15] R. McGinnis, *Rad Pro Calculator 3.26 Software Site Online,* Massachusetts Institute of Technology, http://www.radprocalculator.com/Index.aspx, 2011;

[16] Schunk GmbH & Co. Catalog Automation Highlights, *LWA3 Lightweight Arm Modular Robotic Specification Sheet,* 2009.

# 10

# Decontamination of Radioactive Contaminants Using Liquid and Supercritical $CO_2$

Kwangheon Park[1], Jinhyun Sung[2],
Moonsung Koh[3], Hongdu Kim[4] and Hakwon Kim[5]
[1]*Department of Nuclear Engineering, Kyung Hee University, Yongin,*
[2]*Radiation Engineering Center, Hankook Jungsoo Industries Co. Shiheung,*
[3]*Nuclear Security Center, Korea Institute of Nuclear Nonproliferation and Control, Daejeon,*
[4]*Department of Advanced Materials Engineering for Information and Electronics, Kyung Hee University, Yongin,*
[5]*Department of Chemistry, Kyung Hee University, Yongin,*
*South Korea*

## 1. Introduction

Nuclear power plants have been used as one of major sources of energy all over the world [1]. Nuclear energy is intrinsically clean in environmental viewpoints since it is a highly-concentrated energy source generating much less amount of wastes compared to chemical energy sources such as coals and oils. However, nuclear energy produces inevitable dangerous radioactive wastes. Volume reduction of radioactive wastes is desirable, and additionally should be done in environmentally favorable ways. We introduce $CO_2$ as a cleaning solvent for the decontamination of radioactive contaminants. The subjects for decontamination are diverse – clothes, parts, equipments, tools, solid/liquid wastes, and soil. Main target elements are Co, Sr, Cs, Tc and actinides with chemical forms of metal spikes, oxides, hydroxides and metallic salts [2,3].

Decontamination technology should assist in minimizing waste volume by concentrating the radioactivity of the wastes. However, generation of secondary radioactive waste during decontamination processes produces another problem. If we apply carbon dioxide as a cleaning medium, the secondary waste can be minimized because of the ease of $CO_2$ recycle (simply depressurize and pressurize again). This is why $CO_2$ is gaining attention as an alternative solvent for decontamination of radioactive contaminants. However, $CO_2$ is non-polar, and allows very limited solubility of polar materials such as metallic ions. To make it possible to dissolve polar and ionic materials in $CO_2$, we developed water-in-$CO_2$ micro- and macroemulsions. Water-in-$CO_2$ microemulsion can dissolve most of the polar and ionic substances as well as non-polar substances because aqueous and organic phases can be dispersed in $CO_2$ uniformly at the given sufficient stabilization of the interface.

This chapter shows the surface decontamination of radioactive specimens based on $CO_2$ micro- and macroemulsions. Since most radioactive contaminants contain a variety of metal oxides or metal salts, it is needed that acid solution instead of water be added to the emulsions. We briefly reviewed past works regarding the microemulsion of water in $CO_2$ and its applications. Then, the formation regions of microemulsions with small amounts of water (or acid solution) were mentioned. And, ultrasound effects on the stability of micro- and macroemulsions were explained. Then, we showed the applications of micro- and macroemulsions of acid in $CO_2$ in the dissolution test of Cu-coated parts (nuts) and real contaminated radioactive samples. And the applicability of this technique was discussed.

## 2. Reviews on the application of micro- and macroemulsions in $CO_2$

In 1990s, much experimentation attempted to identify surfactants capable of forming water in liquid/supercritical $CO_2$ microemulsions. K. A. Consani and R. D. Smith [4] reported that the solubilities of over 130 surfactants were tested in supercritical $CO_2$ at 50°C and pressures of 100~500 bar. Consequently, most ionic surfactants (e.g., salts, acids, quaternary ammonium compounds and alkyl phosphate salts) were found to be relatively insoluble in $CO_2$. This is mainly due to the fact that the majority of the amphiphiles evaluated were found to have minimal solubility in $CO_2$. However, nonionic surfactants with suitable hydrocarbon chain length appeared to be able to exhibit reasonable solubility in liquid/ supercritical $CO_2$ at moderate pressure.

On the other hand, owing to favorable solubility parameters, certain fluorocarbons and to a lesser extent silicon, a few surfactants dissolve in $CO_2$. Based on the fact that fluorocarbons and $CO_2$ are compatible, Beckman's group designed the first effective fluoro-surfactants for $CO_2$ [5]. Next, the fluorinated surfactants variously synthesized by Johnston [6-10] and Desimone [10, 11] seem to have high solubility in supercritical $CO_2$.

A micelle is an aggregation (or cluster) of surfactant molecules. The molecule must have a polar 'head' and a non-polar hydrocarbon chain 'tail'. When this type of molecule is added to water, the non-polar tails of the molecules clump into the center of a ball like structure called a micelle, because they are hydrophobic. In a reverse micelle, the polar groups of the surfactants are concentrated in the interior and the lipophilic groups extend towards and into the non-polar solvent (Fig. 1). At certain water-to-surfactant ratios (W), water will be incorporated into the core of the micelle, generating a nano-droplet of water in a $CO_2$ solution (i.e., microemulsion). These droplets are extremely small, ranging from 5 to 100 nm in diameter. The nano-sized water droplet is carried to surfaces by the supercritical $CO_2$, allowing extraction of metals in a way not possible with liquid/ supercritical $CO_2$ only. Therefore, the reverse micelle may be used as an environmentally benign solvent in synthesis of nanometer-sized metal catalysts, cleaning of precision parts and extraction of polar species. Generally, microemulsions are clear, thermodynamically stable solutions.

The macroemulsions are a dispersion of droplets of one liquid in another with which it is incompletely miscible. As shown in Fig 2, macroemulsions of droplets of an organic liquid (a '$CO_2$' like oil) in an aqueous solution are indicated by the symbol $CO_2$-in-water (C/W) and macroemulsions of aqueous droplets in an organic liquid as water-in-$CO_2$ (W/C). In macroemulsions the droplets exceed 0.1μm [12]. The relative sizes are shown in detail in Fig. 3.

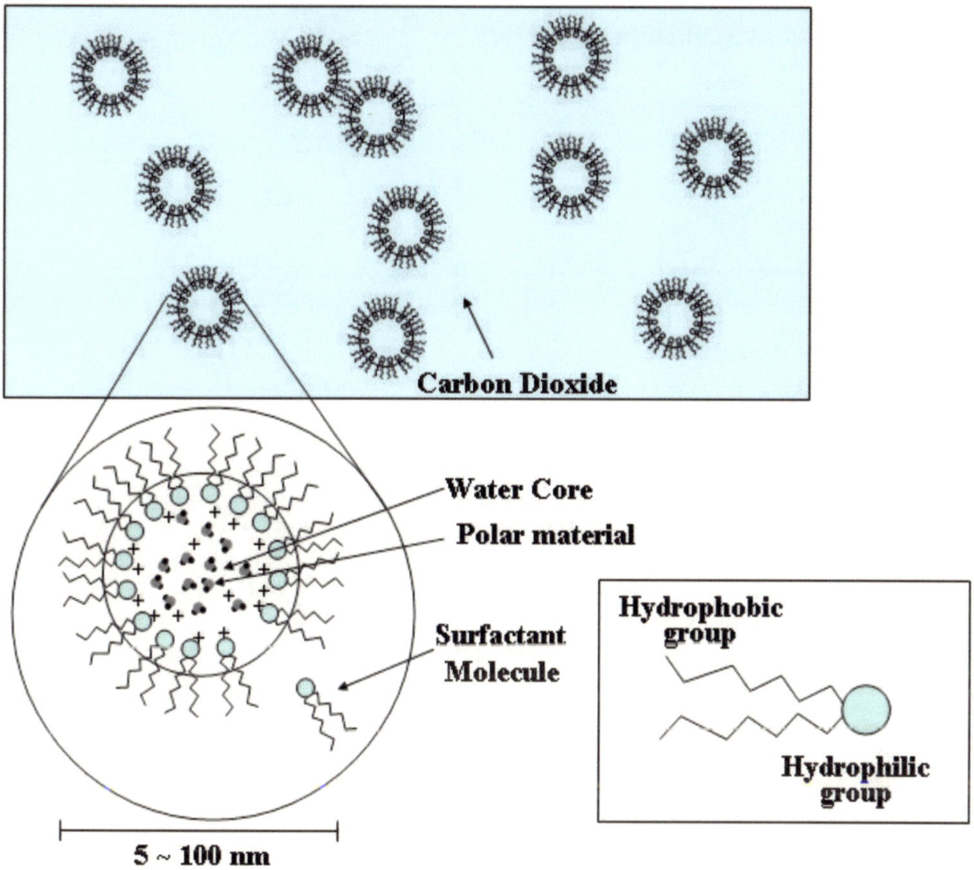


Fig. 1. Structure of reverse micelles and formation of microemulsions in $CO_2$.

Microemulsions are thermodynamically stable, and typically consist of dispersed phase droplets of 5 to 100 nm in diameter. In contrast to microemulsions, macroemulsions are thermodynamically unstable, often requiring considerable energy input to induce their formation, and existence. Furthermore, macroemulsions may be formed with higher interfacial tensions between water and oil (or $CO_2$) than in the case of microemulsions, and thus have lower values of surfactant adsorption at the interface. Therefore, emulsions may be formed for a wide range of surfactant concentrations [13].

There is increasing interest in using reverse micelle and microemulsions for many applications such as solutions for enhanced oil recovery, for the separation of proteins from aqueous solutions, as media for catalytic or enzymatic reactions, and as mobile phases in chromatographic separations.

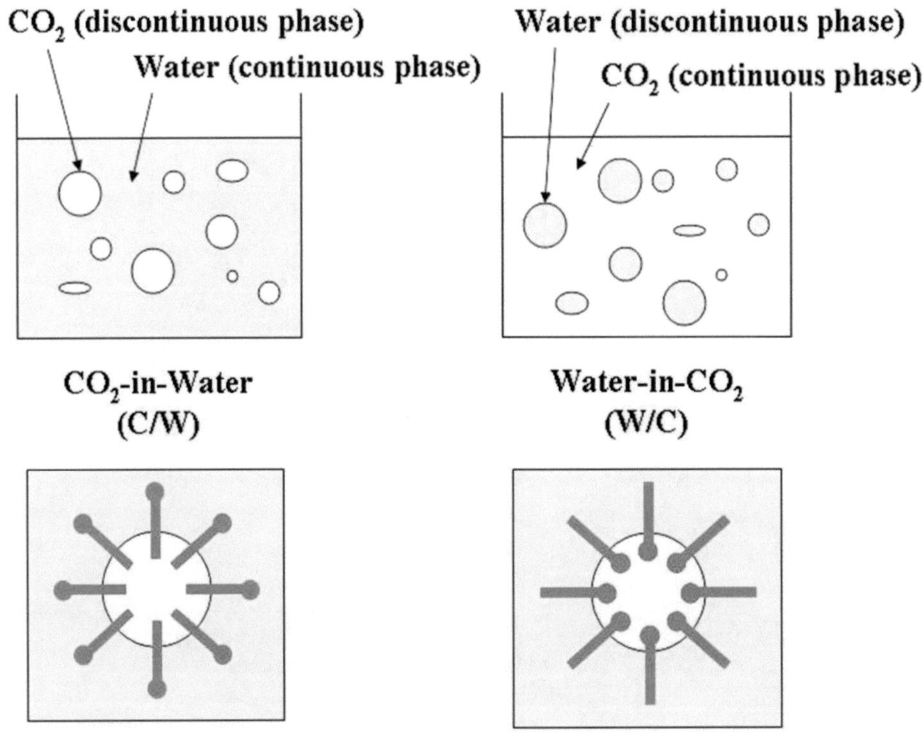


Fig. 2. Structures of C/W and W/C emulsions in $CO_2$.

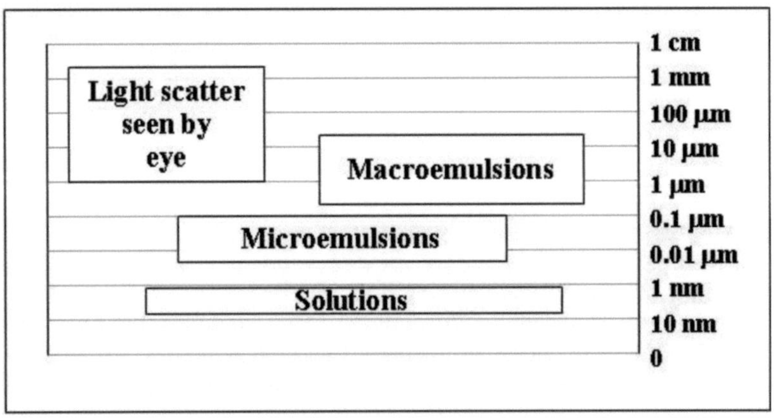


Fig. 3. Size ranges of micro- and macroemulsions [12].

In 1987, Fulton and co-workers reported the first observations of microemulsions in supercritical fluid [14]. The formation of reverse micelles and water-in-oil (W/O) microemulsions in supercritical fluids using the surfactant sodium bis(2-ethylhexyl) sulfosuccinate (AOT). Fluids that are supercritical at moderate temperatures and pressures include ethane ($T_C$=32.4°C, $P_C$=34.8bar), propane ($T_C$=97°C, $P_C$=43.3bar), and xenon

($T_C$=16.6 °C, $P_C$=58.4bar) [14-16]. Also, the properties of microemulsions such as structure, microemulsion size, and the solvent environment of AOT were measured by using dynamic light scattering, small angle X-ray scattering (SAXS) and view cell determinations of phase behavior [17,18].

Johnston and co-workers initially studied the phase behavior of hydrophilic substances such as proteins and amino acids and described this as microemulsions with AOT. They reported large effects on the droplet size and phase behavior of microemulsions in compressible liquids with changes in four principal variables such as pressure, temperature, salinity, and molecular volume of the alkane solvent [19-21]. Later, to favor bending of the interface around water, and to prevent liquid crystal formation, hybrid surfactant with a fluorocarbon and hydrocarbon tail, $C_7F_{15}CH(C_7H_{15})OSO_3^-Na^+$, was newly synthesized. The formation of microemulsions with a fluorocarbon-hydrocarbon hybrid surfactant must have been due primarily to the presence of the fluorocarbon tail, given the lack of success of many nonfluorinated surfactants [22]. In 1996, his group developed a new fluoroether surfactant, ammonium carboxylic perfluoropolyether (PFPE), to form stable reverse micelles in supercritical $CO_2$ [23]. The water cores within these microemulsions were probed by elegant techniques including X-band electron paramagnetic resonance (EPR), time-resolved fluorescence depolarization [24], UV-VIS spectroscopy, and FTIR [25].

Water-in-$CO_2$ macroemulsions may be substituted for toxic organic solvents in chemical processing. Macroemulsions have been used for phase-transfer reactions between $CO_2$-soluble substrates and hydrophilic nucleophiles. Further studies of the mechanism of emulsion stabilization are needed for the rational design of surfactants for this newly emerging field. However, the formation of macroemulsions in $CO_2$ has been challenging due to its weak van der Waals forces, as reflected by its low polarizability per unit volume. To date, a majority of the investigation into the formation of macroemulsions in $CO_2$ has focused on the development of surfactants. Johnston found that water-in-supercritical $CO_2$ (W/C) emulsions were formed with a variety of surfactants such as fluorinated [26, 27], block copolymer [28-30], nonionic polymer [31], and hydrocarbon [32].

Micro- and macroemulsions have been used to solubilize hydrophilic substances in $CO_2$ (e.g., salts, water soluble catalysts, metal nano-particles, electroplating, metal ions, electroactive probes, proteins, and enzymes). McCleskey [33, 34] demonstrated the extraction of metal ions from a variety of solid substrates using microemulsions in supercritical $CO_2$. Microemulsions are especially advantageous for the extraction of metals, because the amount of water required is only proportional to the amount of metal to be extracted, not to the amount of waste to be cleaned. Liu [35] also used microemulsions with nonionic surfactant (X-100) to extract copper ions. More than 99% of the copper is extracted from filter paper surface, and 81% of the used surfactant was recovered and generated. Water-in-microemulsions were presented as a new strategy for promising method of metal ion extraction.

Sawada applied microemulsions for dyeing fiber in supercritical $CO_2$. Pentanethylene glycol *n*-octyl ether and a co-surfactant were used for forming the microemulsions [36]. Co-surfactants generally make the chemical process complicated. Sawada also used one type of surfactant such as PFPE to make microemulsions in $CO_2$[37].

Yonker [38] has demonstrated membrane separations with reverse micelles in near- and supercritical fluid solvents. Reverse micelles were formed and used for the separation of polar, water-soluble macromolecules in solvents. This technique directly extends the capabilities of membrane separations in supercritical fluids to include both non-polar and polar molecules.

Wai [39] reported that metal nano-particles were synthesized within the core of microemulsion droplets. Microemulsions containing $AgNO_3$ were stabilized by a mixture of AOT and PFPE-$PO_4$, acting as surfactant and co-surfactant, respectively. On addition of a reducing agent to the microemulsion system, silver particles were formed inside the emulsion droplets and the particle dimensions were controlled by the size of the water cores. The metal nano-particles formed in this way can be used as in situ catalysts for chemical reactions in the fluid phase, for example the hydrogenation of olefins [40]. Also, new nanomaterials such as CdS and ZnS [41] were obtained using the mixture of two microemulsions containing different metal ions in their micelle water cores.

Sone [42-44] found that electroplating using nickel could be performed in a macroemulsion of supercritical $CO_2$ with nonionic or anionic surfactants. Generally, conventional wet plating methods produce lots of liquid toxic waste, which need expensive waste treatment systems. And the products usually have surface defects such as pinholes and cracks by hydrogenation. By comparison, better quality nickel films have been produced by electroplating in macroemulsion in supercritical $CO_2$. The electroplated film has a good and uniform with a smaller nano-grain size, and a significantly higher Vickers hardness.

Our group has also found that Ni electroplating could be carried out in the emulsion of supercritical $CO_2$ formed by ultrasound. Emulsion of nickel plating solution in $CO_2$ was formed by the agitation from an ultrasonic horn. Electroplating of Ni on the iron sheet was successfully made. The coated surface was very uniform without any pinholes or cracks due to hydrogenation. Only 5-10% of electroplating solution was needed for Ni plating compared to general wet plating. The used electroplating solution could be reused after a recovery process [45].

## 3. Experimental section

### 3.1 Chemicals and specimens

We selected the two types of surfactants. One is sodium bis(2,2,3,3,4,4,5,5-octafluoro-1-pentyl)-2-sulfosuccinate (fluorinated-AOT), a fluorinated ionic surfactant synthesized by our group [41]. The other is the ethoxylated nonyl phenol series (NP-series), a non-ionic surfactant which is commercially available. NP-series was supplied by Nicca Korea Company in Korea. The surfactants were NP-2, 3, 4, 6, 8, 10, and 16. Nitric acid and organic solvents (HPLC grade) were obtained from Ducksan Pure Chemical Co. Ltd. Inorganic acids such as oxalic and citric acids were obtained from Aldrich Chemical Co.

Cu-coated parts (nuts) plated by electroplating were prepared. The conditions for electroplating were as follows; 2.5 $mA/dm^2$ for 7minutes on a nut (surface area: 4.4±0.5 $cm^2$) under Cu electroplating solution. Real radioactive parts (bolts, nuts, connectors), contaminated by radioactive nuclides (such as Co-60, Cs-137, Mn-54, etc), were obtained from Unit 1 of the Wolsung and Unit 2 of the Kori Nuclear Power Plants during overhaul periods.

### 3.2 Apparatuses for measurements

A variable volume cell (from 4.2mL to 22.4mL, Hanwoul Eng.) was used to measure the solubility of a subject in a high pressure media. The known amounts of a surfactant was placed into the cell, and heated to a goal temperature. $CO_2$ was introduced by a syringe pump (260D, ISCO, USA). The solubility or cloud point could be determined by a direct visual observation through sapphire windows placed on both sides. The deviation of temperature was ±0.5 °C and that of pressure was ±1bar.

Two agitation methods were used to form a stable microemulsion - stirring by a magnetic bar and direct agitation by an ultrasonic horn. Stirring agitation was obtained by putting a magnetic bar into the reactor cell which was located on a magnetic rotator. We made a high-pressure cell containing an ultrasonic horn that was connected to a sonar vibrator outside (Fig. 4). The frequency of the horn was 20kHz. The amount of energy dissipation by the horn inside the cell is not clearly known, however, we guess approximately 10~20 W, which is the energy efficiency of 5~10 % of 200W of the total energy consumption of the sonar [46, 47].

The decontamination efficiency of actual radioactive contaminated parts was analyzed by the gamma-spectrum using a Ge-detector (HPGE P-type, EURISYS, France).

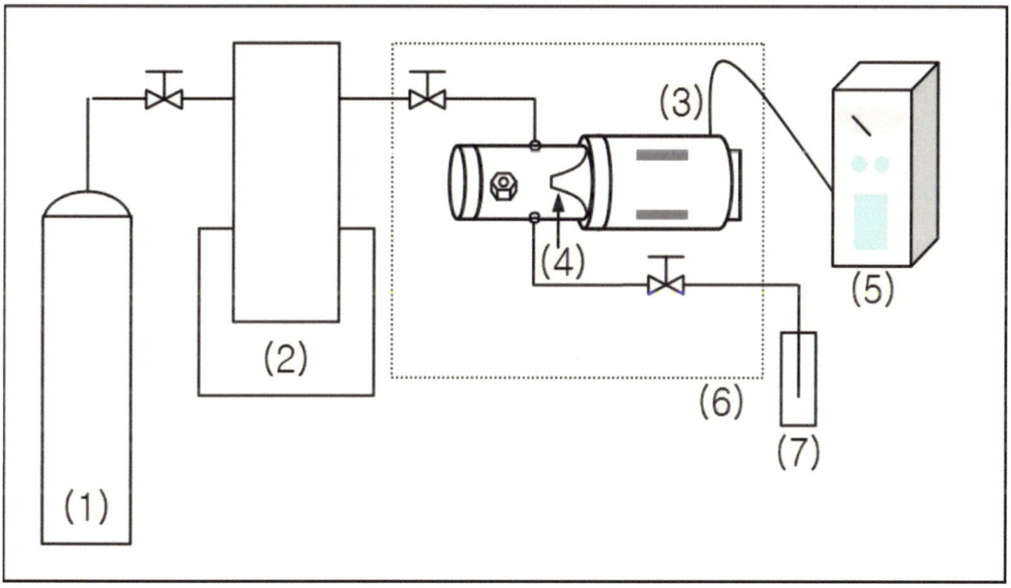


Fig. 4. Decontamination apparatus with ultrasound. (1)$CO_2$ cylinder, (2)syringe pump, (3)high pressure cell (87mL), (4)ultrasonic horn, (5)ultrasonic power supply, (6)oven, (7)collecting vial.

## 4. Results and discussion

### 4.1 Microemulsion formation in supercritical $CO_2$

The basic structure of NP-series is shown in Fig. 5. It has both a hydrophilic and a $CO_2$-philic group. The length of the hydrophilic functional group becomes longer as n value increases, and thereby polar solubility increases. The NP-series surfactants are commercially available with economic price. Since it is electrically neutral, it is less sensitive to the presence of electrolyte, and less affected by pH value. Solubilities of NP-series surfactants in supercritical $CO_2$ were measured using the solubility apparatus, and the results are shown in Fig. 6. Due to the $CO_2$-philic property of the alkyl chain, NP-series are quite soluble in $CO_2$. The ethylene oxide chain is a hydrophilic functional group, and as its length increases, solubility within $CO_2$ becomes lower. For example, $1.0 \times 10^{-3}$ mol fraction of NP-2 was clearly dissolved at (or above) 95bar. On the contrary, 300 bar or higher pressure was needed to dissolve the same mol fraction of NP-16. From the solubility measurements of surfactants, we concluded that NP-series surfactants were soluble to $CO_2$ enough to form a microemulsion in $CO_2$.

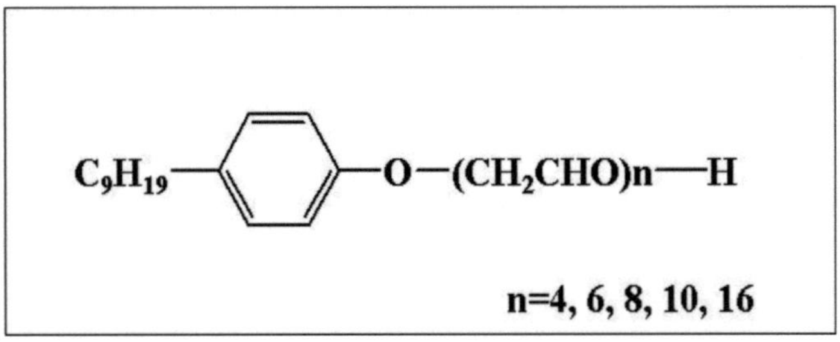


Fig. 5. Basic structure of the NP-series.

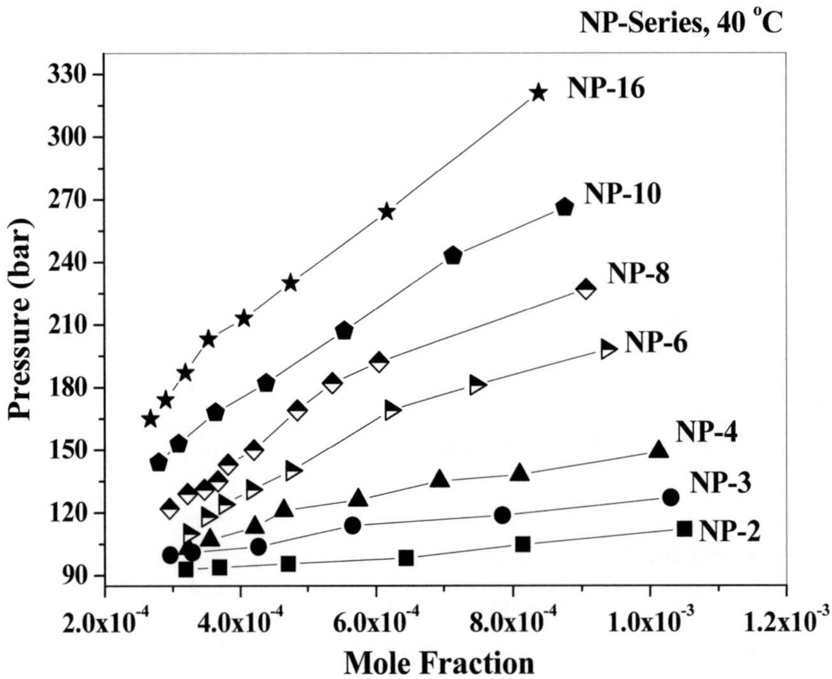


Fig. 6. Solubilities of the NP-series in supercritical $CO_2$.

The formation pressures of microemulsions of water in supercritical $CO_2$ by NP-series surfactants are shown in Fig. 7. The W value for each surfactant was taken as 20, and temperature was maintained at 40 °C. The formation pressures were measured to be in the order of NP- 4 > 6 > 3 > 2 > 16 from the lowest. The stability of a microemulsion depends on the stability of interface between water and $CO_2$. The stability of the interface depends on the types of surfactants. Microscopically speaking, the water cores are surrounded by the surfactants and the surfactants should make the water cores stable to be a reverse-micelle. Both hydrophilic and $CO_2$-philic parts of the surfactant should be mechanically balanced to maintain stable water cores in $CO_2$ matrix. NP-4 seems to make mechanical equilibrium easily in the interface between the water droplets and the surrounding $CO_2$. We chose NP-4 as an optimal surfactant for the microemulsion formation.

The formation pressures of a microemulsion with different W values were also measured using NP-4. The cloud points at different W values are shown in Fig. 8. In each case, a greater pressure is required to make a microemulsion as the concentration of surfactants (and water the amount of which was set by W value) increases in $CO_2$. In other words, a greater pressure is required to dissolve a large amount of water at a given mole fraction of surfactants.

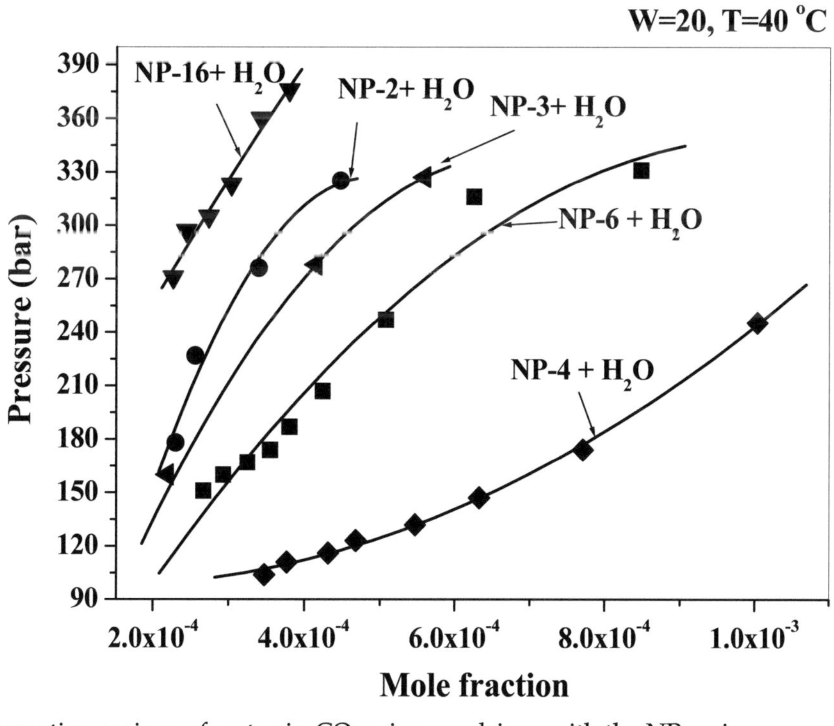


Fig. 7. Formation regions of water-in-$CO_2$ microemulsions with the NP-series.

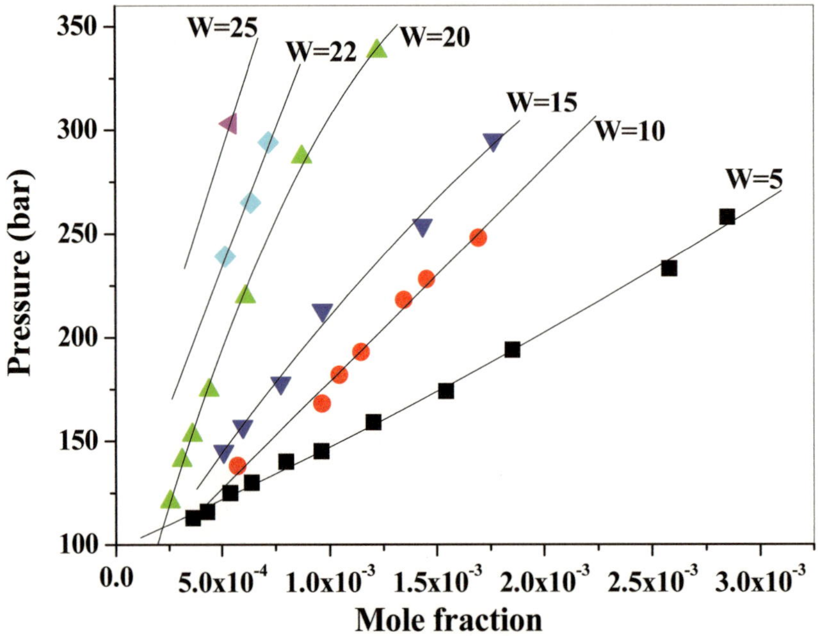


Fig. 8. Formation regions of water in supercritical $CO_2$ microemulsions with the NP-4 mole fraction in $CO_2$. NP-4:20mM, temperature:40 °C.

Acid microemulsions in supercritical $CO_2$ were made and their formation pressures were measured using ionic (F-AOT) and non-ionic (NP-4) surfactants. The results are shown in Fig.9 and Fig.10, respectively. The formation pressure of a microemulsion by F-AOT was higher than its solubility pressure line (Fig.9). The formation pressure of the microemulsion of water in $CO_2$ by F-AOT increased continuously as the mol fraction of surfactant (and also that of water) increased. It seemed that higher pressure should be applied to stabilize the increased amount of aqueous droplets in $CO_2$ matrix. However, in the case of the microemulsion of acid, we could not obtain a microemulsion state when the mol fraction of surfactants exceeded a certain critical value (~ $10^{-2}$ of mol fraction in the case of 3M nitric acid). The acid cores in a microemulsion seemed unstable above this critical mol fraction. The exchange reaction of Na in the hydrophilic head with protons in the acid cores might be a reason for the destabilization.

On the contrary, the NP-4 maintained the stability of microemulsions of acid cores with the increase of mol fraction of surfactant and acid (Fig.10). The formation pressure of acid microemulsion in $CO_2$ was measured to be higher than that of water in $CO_2$. It is believed that the non-ionic surfactant, NP-4 forms a stable microemulsion of acidic solution in $CO_2$ enough to be used in decontamination.

Organic acids are more commonly used in surface treatment of metals. The formation pressures of the microemulsion containing oxalic and citric acids by NP-4 were measured respectively. As shown in Fig. 11, stable microemulsions of organic acid were possible with non-ionic surfactant, NP-4. The formation pressure of a microemulsion containing citric acid in the core turned out to be lower than that of oxalic acid. When the two acids were mixed together, the formation pressure increased. Organic acids needed higher pressure to form microemulsions in $CO_2$ than inorganic acid.

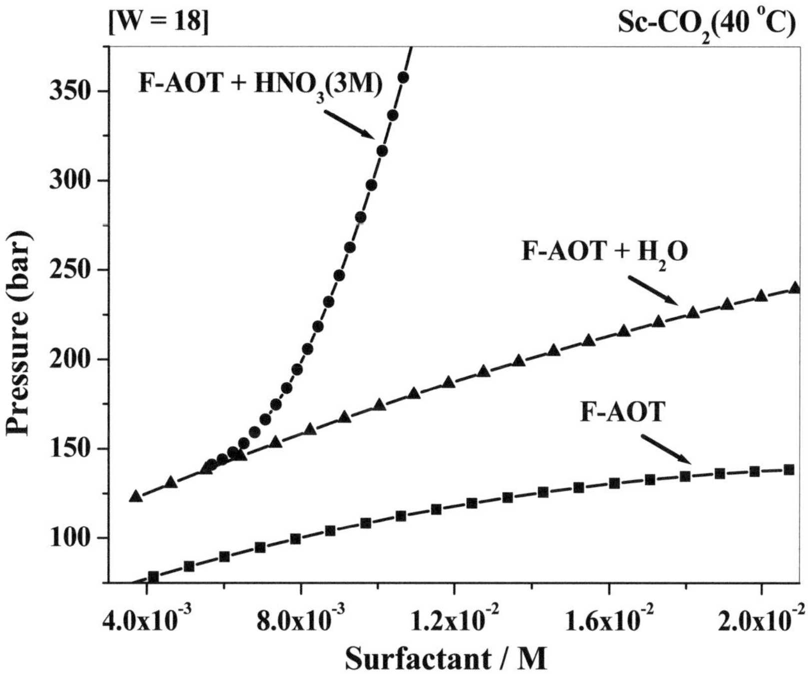


Fig. 9. Formation regions of microemulsions with F-AOT as acidity increases.

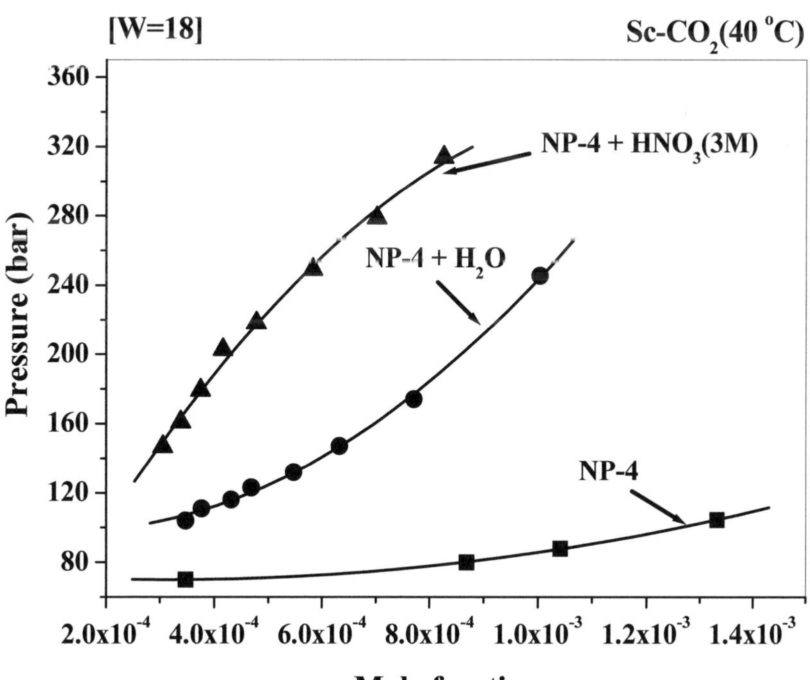


Fig. 10. Formation regions of microemulsions of acid cores and those of water cores with NP-4.

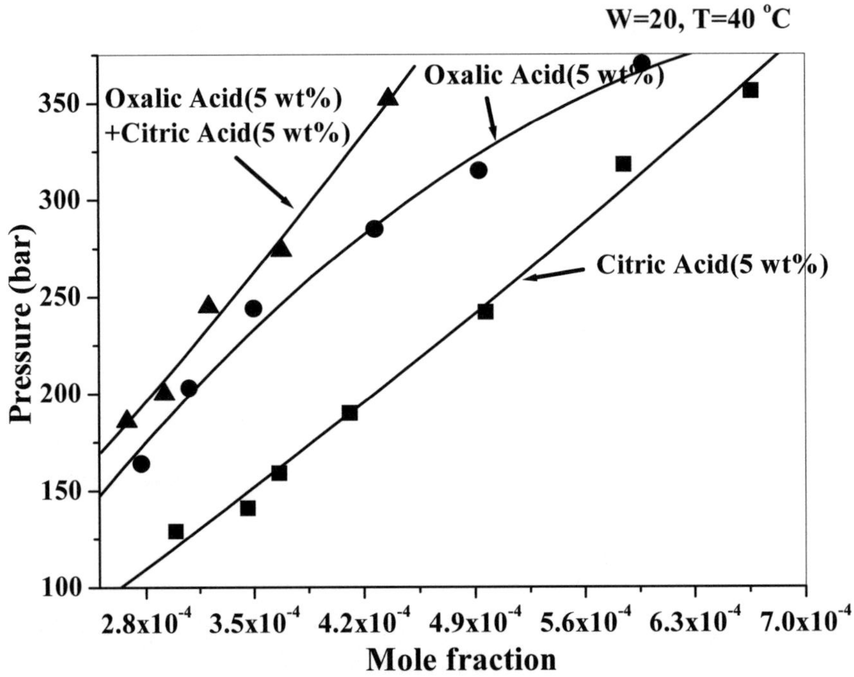


Fig. 11. Formation regions of microemulsions with organic acids (citric acid, oxalic acid).

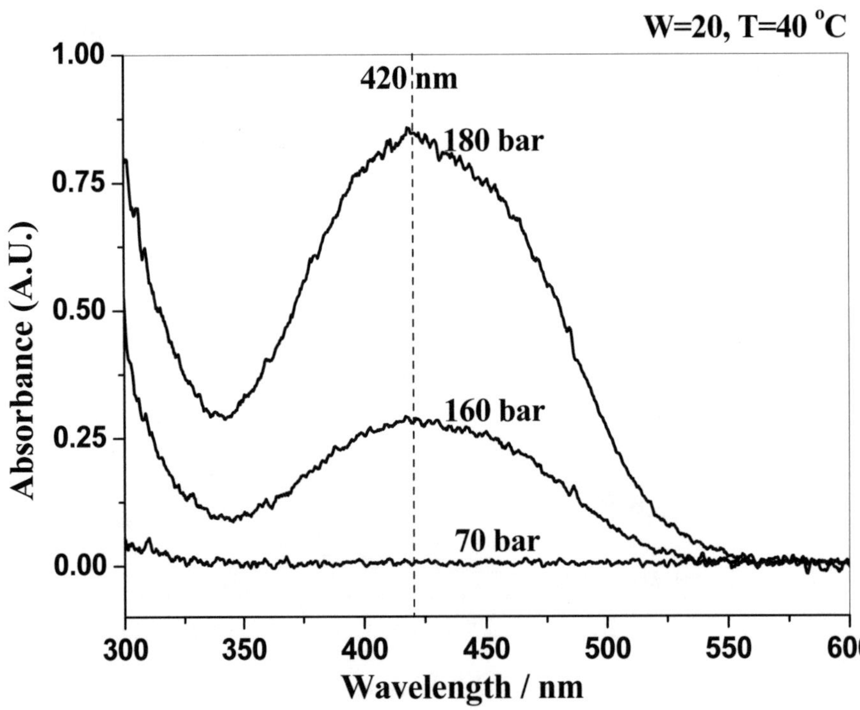


Fig. 12. UV-VIS spectrum of methyl orange in supercritical $CO_2$ microemulsions with NP-4.

The formation of a microemulsion can be checked by the absorption spectrum of a UV-VIS spectrometer. For a clear observation, methyl orange was injected into the cell. We measured the variation of the UV-VIS spectrum with the increase of $CO_2$ pressure (Fig.12). A clear peak (420nm) of methyl orange appeared at the pressure of 160bar at 40 °C. And, the fact of the peak growth with applied $CO_2$ pressure indicated that more stabilized microemulsion was formed under the higher pressure.

### 4.2 Enhanced stabilization of micro- and macroemulsions by ultrasound

Two types of agitating methods - a stirrer and an ultrasound horn- were used for comparing the stability of microemulsions. In the stirrer system, it takes about 20 minutes to form a stable microemulsion. The dispersed behavior of inner contents (nitric acid, surfactant and $CO_2$) and a magnetic bar were observed through a view cell. We photographed the formation process with time, as shown in Fig.13.

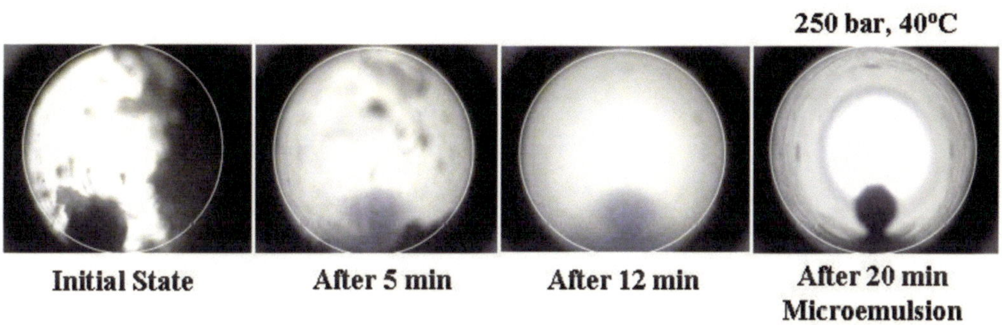


Fig. 13. Microemulsions formation using a stirrer (400rpm) with time at 250bar and 40 °C (to form the microemulsions, nitric acid (1M, 24.5μL) and surfactant (NP-4, 24.8μL) were used inside the pressure cell (10mL)).

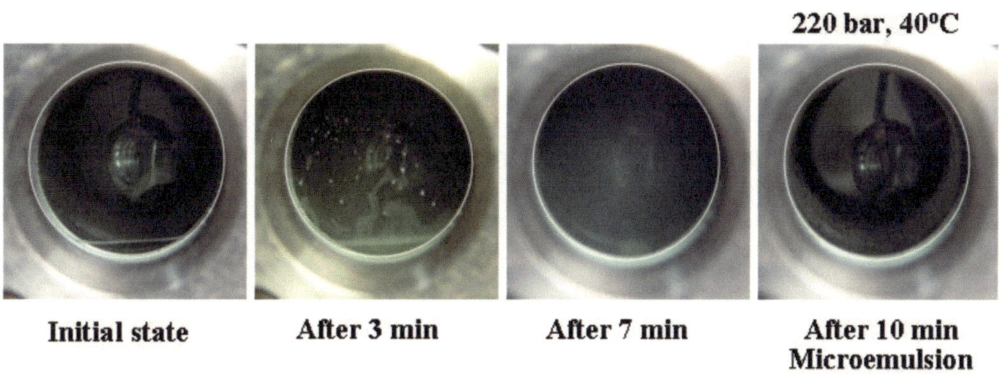


Fig. 14. Microemulsions formation using the ultrasound over time at 220bar and 40 °C.
(to form the microemulsions, nitric acid (1M, 214μL) and surfactant (NP-4, 216μL) were used inside a pressure cell (87mL)).

In the ultrasound system, a microemulsion was formed in about 10 minutes as shown in Fig. 14. As seen in the photos, nitric acid was partially dissolved at 3 min. and a cloud state was seen at 7 min. And finally a clear microemulsion appeared after 10 min. Microemulsions are thermodynamically stable and their dispersed phase is at the nano-scale level. The ultrasound horn applied intense dispersing, stirring and emulsifying effects to the fluid [48], which provide additional power to mix water (or acid) with $CO_2$. All of these effects from the ultrasound horn enhance the stabilization of emulsion.

We conducted experiments about the formation of macroemulsions using an ultrasonic horn. The formation of macroemulsion can be observed visually through the sapphire windows of the high pressure vessel. After inserting the aqueous solution (10% of vessel volume, 5mL) and the surfactant (6v/o of the aqueous solution, 900μL) into the vessel (50mL), it was heated to 55 °C and pressurized to 120bar by $CO_2$. These mixtures existed in a completely separated state initially (Fig.15(a)). Then localized macroemulsions were formed after 10 seconds after agitation (Fig.15(b)), and the macroemulsion was uniformly formed in total in less than 30 seconds (Fig.15(c)). The macroemulsion existed as long as the ultrasound horn agitated the fluid in the cell. After the shut-down of power, the macroemulsion returned back to the initial state, i.e., separated mixtures.

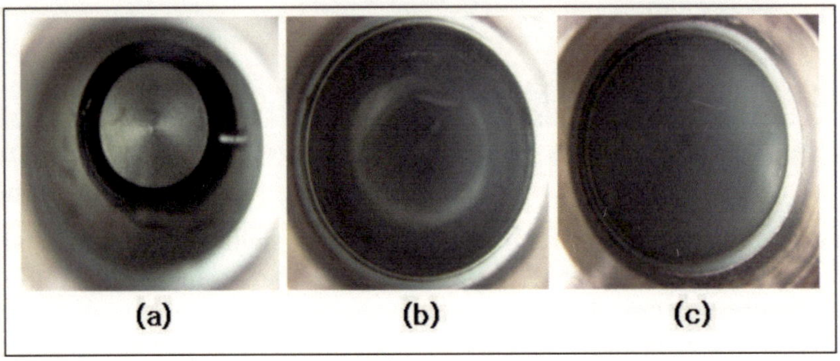


Fig. 15. Photos of emulsion forming processes by ultrasound. (a)state not subjected to ultrasound, (b)state in which an emulsion was partially formed after agitation for 10 seconds, (c)state in which an emulsion was uniformly formed after generation of ultrasound for 30 seconds.

### 4.3 Decontamination using Micro- and Macroemulsions in $CO_2$

#### 4.3.1 Dissolution test of Cu from Cu-coated substitute

As a mock-up test for surface decontamination of radioactive components, Cu-coated nuts were prepared. We put a Cu-coated nut into the reactor cell and let the nut contacted the acid-containing microemulsion. The Cu coating was removed under the microemulsions of acids (1M or 6M-$HNO_3$) in $CO_2$. After one hour of the removal process, most of the Cu-coating was removed and dissolved into the microemulsions. Fig. 16 shows the specimen before and after the experiments. Surface damages of the specimens (corrosion) were observed. We examined the removed surface via an optical microscope to see the surface damage by concentrated nitric acid (Fig. 17). 1M of $HNO_3$ removed almost all of the Cu-coating with small surface damage. However, 6M of $HNO_3$ not only removed the Cu-coating but also damaged heavily some portions of the surface. Anyway, with a very small amount of acid in the microemulsion, we could dissolve and remove the Cu-coating of the specimens very effectively.

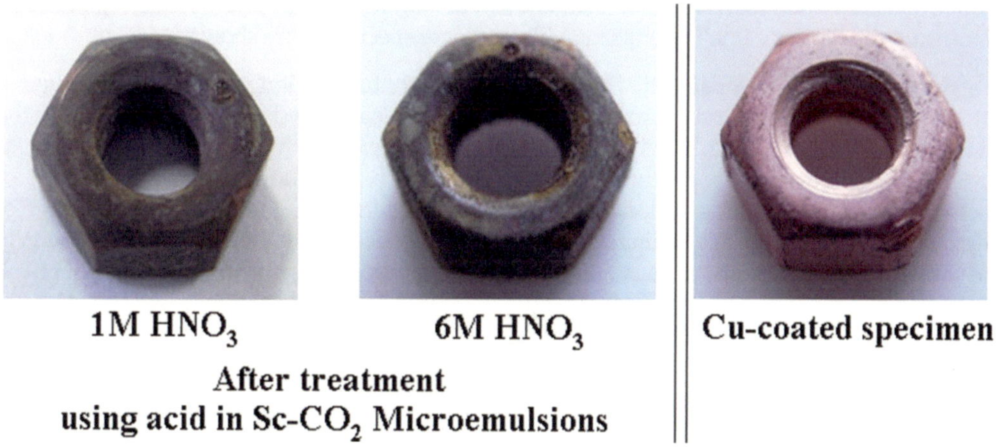


Fig. 16. The photos of specimen before and after acid in Supercritical $CO_2$ microemulsion treatment.

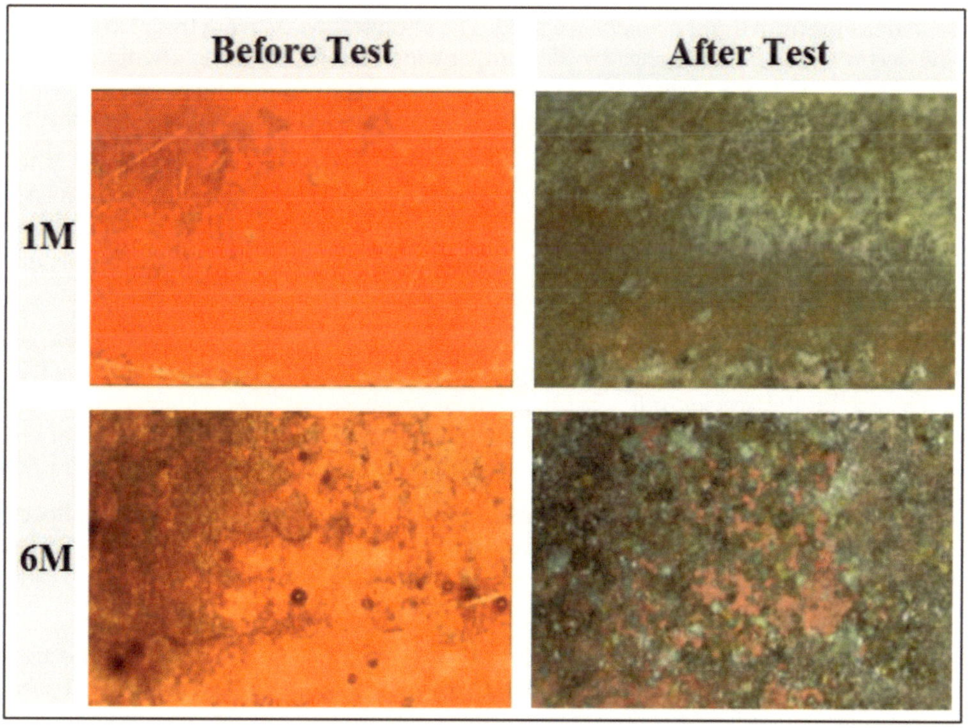


Fig. 17. Microscopic images of a Cu-plated nut using the acid-in-$CO_2$ microemulsions.

### 4.3.2 Decontamination of radioactive components

We obtained radioactive components (bolts, nuts, small pieces of pipes) from nuclear power plants (Wolsung and Kori) in Korea. The radioactivity of the obtained radioactive components were measured by a Ge detector, and classified. We selected radioactive specimens based on the level of contamination that should be mild enough to handle in the laboratory, and high enough to measure the decontamination efficiency. The selected specimens are shown in Fig.18.

For the purpose of comparison, three different decontamination methods were tried - cleaning by a conventional acid solution, microemulsion, and macroemulsion. In the conventional acid cleaning, the specimens were placed into a nitric acid (250mL, 0.1M) at 50°C for an hour with stirring by a magnetic bar. And additional experiments of decontamination by acid solution were done under ultrasonic agitation. In the case of microemulsion cleaning, the radioactive samples were decontaminated for about an hour in the microemulsions with a very small amount (~30μL) of nitric acid (1M, 6M) in supercritical $CO_2$ at 60 °C. Agitations by magnetic-bar stirring and by ultrasonic-horn vibration were done separately for comparison purposes.

A macroemulsion was made by ultrasonic horn agitation. In the reactor cell, 10% of total volume was filled by acid (0.1M or 5% oxalic acid) and the rest volume was by pressurized $CO_2$. The reactor was maintained at the temperature of 50°C and pressure of 120bar during the cleaning process. The cleaning process lasted for an hour.

We measured the radioactivity of the specimen before and after the experiment, respectively; then the cleaning efficiency was calculated by the comparison of the activities of these two. A few of gamma radiation peaks (Co-60, Sb-125, Mn-54, Zr-95) were observed by a Ge-detector from the specimens we used. The strongest peaks were from Co-60, and we set the decontamination efficiency by the comparison of Co-60 activities. The uncertainty of gamma radiation detection using the Ge-detector was about 5%. Based on this uncertainty, we assume that 10% be the uncertainly limit in the value of decontamination efficiency.

The results of decontamination experiments using radioactive specimens are shown in Table 1. The conventional cleaning using 0.1M nitric acid with a stirrer gave the value of about 50% of decontamination efficiency. When ultrasonic agitation was applied, the efficiency increased up to 78%. Additional cleaning (2$^{nd}$ time) barely increased the efficiency (up to 84%). This might be the limit of cleaning efficiency when we used 0.1M nitric acid as a cleaning solution.

The microemulsion with nitric acid clearly worked well in decontamination of radioactive metal components. The volume fraction of acid used in the microemulsion cleaning was only 0.25%, and the size of the reactor cell was about 1/20 compared to the container used in the conventional acid cleaning. The decontamination efficiency was about 63% if 0.1M nitric acid was used. When strong nitric acids were used, the decontamination efficiency increased up to 100%.

A macroemulsion containing acid was also effective in decontamination of radioactive specimens. Due to strong agitation by an ultrasonic horn, the decontamination efficiency went to up to 89% when 0.1M nitric acid was used. Interestingly, oxalic acid (5%) was very effective in decontamination of metallic parts (100% elimination).

The cleaning methods using microemulsion and macroemulsion seemed very effective in decontamination of metallic parts contaminated on the surface. The amounts of acid used in these cleaning methods were very small (0.25%~10% in volume). Because it contains the removed radioactive contaminants, the aqueous acid used in cleaning becomes a secondary waste. We can reduce the amount of the secondary waste revolutionarily if micro- or macroemulsion of acid in $CO_2$ is used in decontamination.

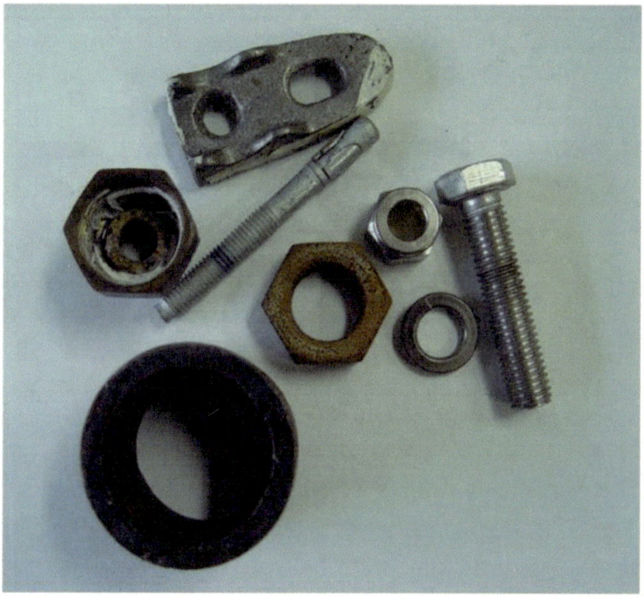

Fig. 18. Radioactive contaminants supplied from Unit 1 of the Wolsung and Unit 2 of the Kori Nuclear Power Plants during overhaul periods.

| Decontamination methods | Acid types | Agitation methods | Acid in vol % | Efficiency / % | |
|---|---|---|---|---|---|
| | | | | 1st | 2nd |
| Conventional acid cleaning | 0.1M $HNO_3$ | Stirrer | 100% | 50±10 | - |
| | | Ultrasound | | 78±10 | 84±10 |
| Supercritical $CO_2$ microemulsions | 0.1M $HNO_3$ | Stirrer | 0.25% | 63±5 | - |
| | 1M $HNO_3$ | Ultrasound | | 87±10 | |
| | 6M $HNO_3$ | | | 100 ±10 | - |
| Supercritical $CO_2$ macroemulsions | 0.1M $HNO_3$ | Ultrasound | 10% | 89±10 | - |
| | Oxalic Acid (5%) | Ultrasound | | 100±10 | - |

Table 1. The decontamination efficiencies of radioactive components (nut, bolt, etc) with respect to different cleaning methods.

## 5. Conclusion

New surface decontamination techniques using acid-$CO_2$ micro- and macroemulsions were developed to decontaminate radioactive components. NP-series, commercially available surfactants, were applied to form the micro- and macroemulsions in supercritical $CO_2$. We found that micro- and macroemulsion could be formed if nonionic surfactants (NP-series) were used. And the formation points were measured with respect to the concentrations of surfactants in $CO_2$. The results showed that the solubility of surfactant increased gradually as the hydrophilic group became shorter. In the formation of the microemulsion with water, it was confirmed that the NP-4 surfactant formed the most stable microemulsion, because the hydrophilic and the $CO_2$-philic parts of the surfactant were mechanically well balanced to maintain stable water cores in $CO_2$ matrix.

An ultrasonic horn was used to enhance the formation of micro- and macroemulsions. The agitation by an ultrasound horn was superior to that by a stirrer in obtaining a stable microemulsion. When the ultrasound was used, the microemulsion was formed quickly even at the milder condition.

We tested Cu-coated parts to see whether the microemulsion containing nitric acid could remove the surface metal layer. Almost all of the Cu-coating was removed under the microemulsion of 1M $HNO_3$ in $CO_2$. Both microemulsion and macroemulsion containing nitric acid clearly worked well in decontamination of radioactive metallic parts. The decontamination efficiency increased as the acidity of nitric acid increased. And oxalic acid in macroemulsion was very effective in the surface decontamination of metallic parts. The cleaning methods of micro- and macroemulsion seemed better in decontamination of metallic parts than conventional cleaning. Moreover, these methods produce very small amount of secondary wastes.

## 6. References


[1] International Atomic Energy Agency, "Energy, Electricity and Nuclear Power Estimates for the Period up to 2020 July 2002", IAEA-RDS-1/22, Vienna, (2002).

[2] Korea Atomic Energy Research Institute, "Decontamination and Decommissioning Technology Development of Nuclear Facilities", KAERI/RR-798/88, (1988)

[3] Korea Atomic Energy Research Institute, "Nuclear Fuel Cycle Waste Recycling Technology Development", KAERI/RR-1830/97, (1997)

[4] K. A. Consani and R. D. Smith, "Observations on the Solubility of Surfactants and Related Molecules in Carbon Dioxide at 50 °C", J. of Supercritical Fluids, 3 (1990) 51.

[5] T. A. Hoefling, R. M. Enick, and E. J. Beckman, "Microemulsions in Near-critical and Supercritical $CO_2$", Journal of Physical Chemistry, 95 (1991) 7127.

[6] K. Harrison, J. Goveas, and K. P. Johnston, "Water-in-Carbon Dioxide Microemulsions with a Fluorocarbon-Hydrocarbon Hybrid Surfactant", Langmuir, 10 (1994) 3536.

[7] M. J. Clarke, K. L. Harrison, K. P. Johnston, and S. M. Howdle, "Water in Supercritical Carbon Dioxide Microemulsions: Spectroscopic Investigation of a New Environment for Aqueous Inorganic Chemistry", J. Am. Chem. Soc., 119 (1997) 6399.

[8] M. P. Heitz, C. Carlier, J. deGrazia, K. L. Harrison, K. P. Johnston, T. W. Randolph, and F. V. Bright, "Water Core Within Perfluoropolyether-based Microemulsions Formed in Supercritical Carbon Dioxide", J. Phys. Chem. B, 101 (1997) 6707.

[9] C. T. Lee, Jr., P. A. Psathas, K. J. Ziegler, K. P. Johnston, H. J. Dai, H. D. Cochran, Y. B. Melnichenko, and G. D. Wignall, "Formation of Water-in-Carbon Dioxide Microemulsions with a Cationic Surfactant: a Small-Angle Neutron Scattering Study", J. Phys. Chem. B, 104 (2000) 11094.

[10] J. S. Keiper, R. Simhan, and J. M. DeSimone, "New Phosphate Fluorosurfactants for Carbon Dioxide", J. Am. Chem. Soc., 124 (2002) 9.

[11] J. L. Kendall, D. A. Canelas, J. L. Young, and J. M. DeSimone, "Polymerizations in Supercritical Carbon Dioxide", Chem. Rev., 99 (1999) 543.

[12] K. Shinoda and S. Friberg, "Emulsions and Solubilization", John Wiley & Sons, (1986) 3.

[13] C. T. Lee, Jr., P. A. Psathas, and K. P. Johnston, "Water-in-Carbon Dioxide Emulsions: Formation and Stability", Langmuir, 15 (1999) 6781.

[14] R. W. Gale, J. L. Fulton, and R. D. Smith, "Organized Molecular Assemblies in the Gas Phase: Reverse Micelles and Microemulsions in Supercritical Fluids", J. Am. Chem. Soc, 109 (1987) 920.
[15] R. W. Gale, J. L. Fulton, and R. D. Smith, "Reverse Micelle Supercritical Fluid Chromatography", Anal. Chem., 59 (1987) 1977.
[16] J. L. Fulton and R. D. Smith, "Reverse Micelle and Microemulsion Phases in Supercritical Fluids", J. Phys. Chem., 92 (1988) 2903.
[17] J. L. Fulton, J. P. Blitz, J. M. Ti ngey, and R. D. Smith, "Reverse Micelle and Microemulsion Phases in Supercritical Xenon and Ethane: Light Scattering and Spectroscopic Probe Studies", J. Phys. Chem., 93 (1989) 4198.
[18] J. L. Fulton, D. M. Pfund, J. M. DeSimone, and M. Capel, "Aggregation of Amphiphilic Molecules in Supercritical Carbon Dioxide: A Small Angle X-ray Scattering Study", Langmuir, 11 (1995) 4241.
[19] R. M. Lemert, R. A. Fuller, and K. P. Johnston, "Reverse Micelles in Supercritical Fluids. 3. Amino Acid Solubilization in Ethane and Propane", J. Phys. Chem., 94 (1990) 6021.
[20] G. J. McFann and K. P. Johnston, "Phase Behavior of AOT Microemulsions in Compressible Liquids", J. Phys. Chem., 95 (1991) 4889.
[21] D. G. Peck and K. P. Johnston, "Theory of the Pressure Effect on the Curvature and Phase Behavior of AOT/Propane/Brine Water-in-Oil Microemulsions", J. Phys. Chem., 95 (1991) 9549.
[22] K. Harrison, J. Goveas, and K. P. Johnston, "Water-in-Carbon Dioxide Microemulsions with a Fluorocarbon-Hydrocarbon Hybrid Surfactant", Langmuir, 10 (1994) 3536.
[23] K. P. Johnston, K. L. Harrison, M. J. Clarke, S. M. Howdle, M. P. Heitz, F. V. Brigth, C. Carlier, and T. W. Randolph, "Water-in-Carbon Dioxide Microemulsions: An Environment for Hydrophiles Including Proteins", Science, 271 (1996) 624.
[24] M. P. Heitz, C. Carlier, J. deGrazia, K. L. Harrison, K. P. Johnston, T. W. Randolph, and F. V. Bright, "Water Core within Perfluoropolyether-Based Microemulsions Formed in Supercritical Carbon Dioxide", J. Phys. Chem. B, 101 (1997) 6707.
[25] M. J. Clarke, K. L. Harrison, K. P. Johnston, and S. M. Howdle, "Water in Supercritical Carbon Dioxide Microemulsion: Spectroscopic Investigation of a New Environment for Aqueous Inorganic Chemistry", J. Am. Chem. Soc., 119 (1997) 6399.
[26] C. T. Lee, Jr., P. A. Psathas, K. P. Johnston, J. deGrazia, and T. W. Randolph, "Water-in-Carbon Dioxide Emulsions: Formation and Stability", Langmuir, 15 (1999) 6781.
[27] G. B. Jacobson, C. T. Lee, Jr., K. P. Johnston, and W. Tumas, "Enhanced Catalyst Reactivity and Separations using Water/Carbon Dioxide Emulsions", J. Am. Chem. Soc., 121 (1999) 11902.
[28] K. P. Johnston, "Block Copolymers as Stabilizers in Supercritical Fluids", Current Opinion in Colloid & Interface Science, 5 (2000) 351.
[29] P. A. Psathas, S. R. P. DaRocha, C. T. Lee, Jr., and K. P. Johnston, "Water-in-Carbon Dioxide Emulsions with Poly(dimethylsiloxane)-Based Block Copolymer Ionomers", Ind. Eng. Chem. Res., 39 (2000) 2655.
[30] J. L. Dickson, C. O. Estrada, J. F. J. Alvarado, H. S. Hwang, I. C. Sanchez, G. L. Barcenas, K. T. Lim, and K. P. Johnston, "Critical Flocculation Density of Dilute Water-in-$CO_2$ Emulsions Stabilized with Block Copolymers", Journal of Colloid and Interface Science, 272 (2004) 444.
[31] S. R. P. DaRocha, P. A. Psathas, E. Klein, and K. P. Johnston, "Concentrated $CO_2$-in-Water Emulsions with Nonionic Polymeric Surfactants", Journal of Colloid and Interface Science, 239 (2001) 241.

[32] K. P. Johnston, D. Cho, S. R. P. DaRocha, P. A. Psathas, W. Ryoo, S. E. Webber, J. Eastoe, A. Dupont, and D. C. Steytler, "Water in Carbon Dioxide Macroemulsions and Miniemulsions with a Hydrocarbon Surfactant", Langmuir, 17 (2001) 7191.

[33] M. Z. Yates, D. L. Apodaca, M. L. Campbell, E. R. Birnbaum, and T. M. McCleskey, "Metal Extractions using Water in Carbon Dioxide Microemulsions", Chem. Commun., (2001) 25.

[34] M. L. Campbell, D. L. Apodaca, M. Z. Yates, T. M. McCleskey, and E. R. Birnbaum, "Metal Extraction from Heterogeneous Surfaces using Carbon Dioxide Microemulsions", Langmuir, 17 (2001) 5458.

[35] J. Liu and W. Wang, G. Li, "A New Strategy for Supercritical Fluid Extraction of Copper Ions", Talanta, 53 (2001) 1149.

[36] K. Sawada, T. Takagi, and M. Ueda, "Solubilization of Ionic Dyes in Supercritical Carbon Dioxide: A Basic Study for Dyeing Fiber in Non-Aqueous Media", Dyes and Pigments, 60 (2004) 129.

[37] J. H. Jun, K. Sawada, and M. Ueda, "Application of Perfluoropolyether Reverse Micelles in Supercritical $CO_2$ to Dyeing Process", Dyes and Pigments, 61 (2004) 17.

[38] C. R. Yonker, J. L. Fulton, M. R. Phelps, and L. E. Bowman, "Membrane Separations using Reverse Micelles in Nearcritical and Supercritical Fluid Solvents", J. of Supercritical Fluids, 25 (2003) 225.

[39] M. Ji, X. Chen, C. M. Wai, and J. L. Fulton, "Synthesizing and Dispersing Silver Nanoparticles in a Water-in-Supercritical Carbon Dioxide Microemulsion", J. Am. Chem. Soc., 121 (1999) 2631.

[40] H. Ohde, C. M. Wai, H. Kim, J. Kim, and M. Ohde, "Hydrogenation of Olefins in Supercritical $CO_2$ Catalyzed by Palladium Nanoparticles in a Water-in-$CO_2$ Microemulsion", J. Am. Chem. Soc., 124 (2002) 4540.

[41] H. Ohde, M. Ohde, F. Bailey, H. Kim, and C. M. Wai, "Water-in-$CO_2$ Microemulsions as Nanoreactors for Synthesizing CdS and ZnS Nanoparticles in Supercritical $CO_2$", Nano Letters, 2 (2002) 721.

[42] H. Yoshida, M. Sone, A. Mizushima, H. Yan, H. Wakabayashi, K. Abe, X. T. Tao, S. Ichihara, and S. Miyata, "Application of Emulsion of Dense Carbon Dioxide in Electroplating Solution with Nonionic Surfactants for Nickel Electroplating", Surface and Coating Technology, 173 (2003) 285.

[43] H. Yan, M. Sone, A. Mizushima, T. Nagai, K. Abe, S. Ichihara, and S. Miyata, "Electroplating in $CO_2$-in-Water and Water-in-$CO_2$ Emulsions using a Nickel Electroplating Solution with Anionic Fluorinated Surfactant", Surface and Coating Technology, 187 (2004) 86.

[44] H. Wakabayashi, N. Sato, M. Sone, Y. Takada, H. Yan, K. Abe, K. Mizumoto, S. Ichihara, and S. Miyata, "Nano-Grain Structure of Nickel Films Prepared by Emulsion Plating using Dense Carbon Dioxide", Surface and Coating Technology, 190 (2005) 200.

[45] M. Koh, M. Joo, K. Park, H. Kim, H. Kim, S. Han, and N. Sato, "Ni Electroplating in the Emulsions of Sc-$CO_2$ Formed by Ultrasonar", J. Kor, Inst, Surf. Eng., 37 (2004) 344.

[46] K. Park, M. Koh, C. Yoon, H. D. Kim, and H. W. Kim, "Solubilization Study by QCM in Liquid and Supercritical $CO_2$ under Ultrasonar", 2003 American Chemical Society, 860 (2003) 207.

[47] K. Park, M. Koh, C. Yoon, H. Kim, H. Kim, "The Behavior of Quartz Crystal Microbalance in High Pressure $CO_2$", J. of Supercritical Fluids, 29 (2004) 203.

[48] D. Ensminger, "Ultrasonics - Fundamentals, Technology, Applications", 2$^{nd}$ edition, Marcel Dekker, INC., (1988) 391.

# 11

# Radionuclide and Contaminant Immobilization in the Fluidized Bed Steam Reforming Waste Product

James J. Neeway[1], Nikolla P. Qafoku[1], Joseph H. Westsik Jr.[1], Christopher F. Brown[1], Carol M. Jantzen[2] and Eric M. Pierce[3]
*[1]Pacific Northwest National Laboratory, Richland, WA,*
*[2]Savannah River National Laboratory, Aiken, SC,*
*[3]Oak Ridge National Laboratory, Oak Ridge, TN,*
*USA*

## 1. Introduction

The goal of this chapter is to introduce the reader to the Fluidized Bed Steam Reforming (FBSR) process and resulting waste form. The first section of the chapter gives an overview of the potential need for FBSR processing in nuclear waste remediation followed by an overview of the engineering involved in the process itself. This is followed by a description of waste form production at a chemical level followed by a section describing different process streams that have undergone the FBSR process. The third section describes the resulting mineral product in terms of phases that are present and the ability of the waste form to encapsulate hazardous and radioactive wastes from several sources. Following this description is a presentation of the physical properties of the granular and monolith waste form product including and contaminant release mechanisms. The last section gives a brief summary of this chapter and includes an overview of strengths associated with this waste form and needs for additional data. The reader is directed elsewhere for more information on other waste forms such as Cast Stone (Lockrem, 2005), Ceramicrete (Singh et al., 1997, Wagh et al., 1999) and geopolymers (Kyritsis et al., 2009; Russell et al., 2006).

Classical steam reforming is a versatile process that decomposes organic materials through reactions with steam (Olson et al., 2004a). Steam reforming has been used on a large scale by the petrochemical industry to produce hydrogen for at least 65 years. If the material being reformed contains halogens, phosphorus, or sulfur, mineral acids are also formed (e.g., hydrochloric acid, phosphorous acid, phosphoric acid, and hydrogen sulfide) unless inorganic materials capable of scavenging these species are present in the waste or additives (Nimlos & Milne, 1992; Olson et al., 2004a). Organic nitrogen is converted to $N_2$, and organic oxygen is converted to CO or $CO_2$ (Olson et al., 2004a). In the presence of a reducing agent such as organic carbon, nitrates and nitrites are converted to nitrogen gas (Vora et al., 2009). The waste feed may be either basic or acidic (Lorier et al., 2005). Alkali elements, including sodium, potassium, and cesium in the wastes, "alkali activate" the unstable $Al^{3+}$ in the clay to form new mineral phases. The other waste component cations and anions are captured in the cage structures of the sodium aluminosilicate minerals.

In 1999, the Studsvik facility in Erwin, Tennessee demonstrated the ability to commercialize the FBSR process (Mason et al., 2003). The facility uses steam reforming based on a process known as THermal Organic Reduction (THOR®). The THOR® FBSR process is being used commercially to process liquid radioactive waste waste streams, including ion exchange resins, charcoal, graphite, sludge, oil, and solvents that contain up to ~4.5 $\times 10^5$ Sv/hr (Mason et al., 2003). Steam reforming thermally treats wastes at temperatures ranging from 625 to 750°C using a fluidized bed reformer (Vora et al., 2009). During mineralization with superheated steam, organic matter is converted to carbon dioxide and steam, while nitrates and nitrites are reduced to nitrogen. The non-volatile solids in the residue are converted to water-insoluble stable crystalline minerals that incorporate contaminants.

Other THOR Treatment Technologies, LLC (TTT) designed FBSR testing platforms have also been built and used to demonstrate steam reforming technologies for the immobilization of radioactive and simulant wastes at the bench-, pilot-, and engineering-scale. An FBSR facility is being designed and constructed at Idaho National Engineering and Environmental Laboratory (INEEL) for the treatment of sodium-bearing waste (SBW) to be sent for disposal at the Waste Isolation Pilot Plant (WIPP) in Carlsbad, New Mexico (Marshall et al., 2003; Olson et al., 2004b; Soelberg et al., 2004a). Another such facility is being considered for converting Savannah River Site (SRS) salt supernate waste (Tank 48), containing nitrates, nitrites, and cesium tetraphenyl borate (CsTPB) to carbonate or silicate minerals which are compatible with subsequent vitrification (Soelberg et al., 2004b). Pilot-scale testing has also been performed for other DOE wastes including Waste Treatment Plant-Secondary Waste (WTP-SW) and Hanford Low-Activity Waste (LAW).

Another system called the Bench-Scale Reformer (BSR) has been developed at Savannah River National Laboratory (SRNL) to help assess the suitability and effectiveness of the FBSR process for the treatment of Hanford LAW (Burkett et al., 2008). The major difference between the FBSR and BSR designs is that the BSR bed is not completely fluidized due to its containment in shielded cells, restricting the height necessary to allow for this process disengagement (Burket et al., 2008). Mineralization is still created by the BSR steam reforming reactions and homogenization occurs by turbulent mixing after sample formation.

### 1.1 Fluidized bed steam reformer process description

A typical THOR® mineralizing FBSR process can use either a single reformer or dual reformer. The dual reformer flowsheet is only needed if the waste being mineralized contains organics that need to be destroyed. Currently, there is no flowsheet for encapsulating the FBSR granular product in a binder to produce a monolithic waste form although this has been proven at the bench-scale. The dual reformer consists of the following primary subsystems (Olson et al., 2004b):

1. A feed for gases, liquids, slurries, and co-additives such as clay and denitration catalysts;
2. The fluidized bed reactor vessel known as the Denitration and Mineralization Reformer (DMR)
3. A high temperature filter (HTF) to catch fines and recycle them to the DMR bed to act as seeds for particle size growth;
4. The solid and product collection from the DMR and HTF;
5. The off-gas treatment which can include the second reformer known as the Carbon Reduction Reformer (CRR);
6. The monitoring and control of the system.

The FBSR process used for mineralizing high sodium wastes (nitrate and/or hydroxide) can use either a single reformer flowsheet, the denitration DMR, or a dual reformer flowsheet, including a CRR, to handle organics, as shown in Figure 1.

The DMR operating temperature is maintained at 700 to 750°C for generating the sodium aluminosilicate (Na-Al-Si or NAS) end product. The flow diagram shows the feed preparation, denitration and mineralizing, and off-gas portions of the FBSR process. All mineralization reactions take place in the DMR. Granular products are removed from the bottom of the DMR and finer product solids are separated from the process outlet gases by the HTF. The finer HTF mineral solids can be recycled back to the DMR bed as seed material to the DMR bed, as shown in Figure 1, or the HTF mineral solids can be combined with the granular mineral solids from the DMR to make a monolith. The process outlet gases are treated in the off-gas treatment system to meet air permit emission limits. In the application of FBSR to Hanford wastes, a clay mineralizing agent, any reductants, and any catalysts are co-added to the DMR along with the waste(s) in the feed tank as shown in Figure 1. The bed is fluidized with superheated steam and near-ambient pressure. A carbon source, such as coal, wood product, or sucrose, is injected into the bed as a fuel source and reducing/denitration agent. Within the fluidized bed, the waste-feed droplets coat the bed particles and rapidly dry. Nitrates, nitrites, and organics are destroyed (TTT, 2009; Vora et al., 2009). In the steam environment, the clay mineralizing agent injected with the wastes becomes unstable as hydroxyl groups are driven out of the clay structure (Jantzen, 2008). The clays become amorphous, and the silicon and aluminum atoms become very reactive.

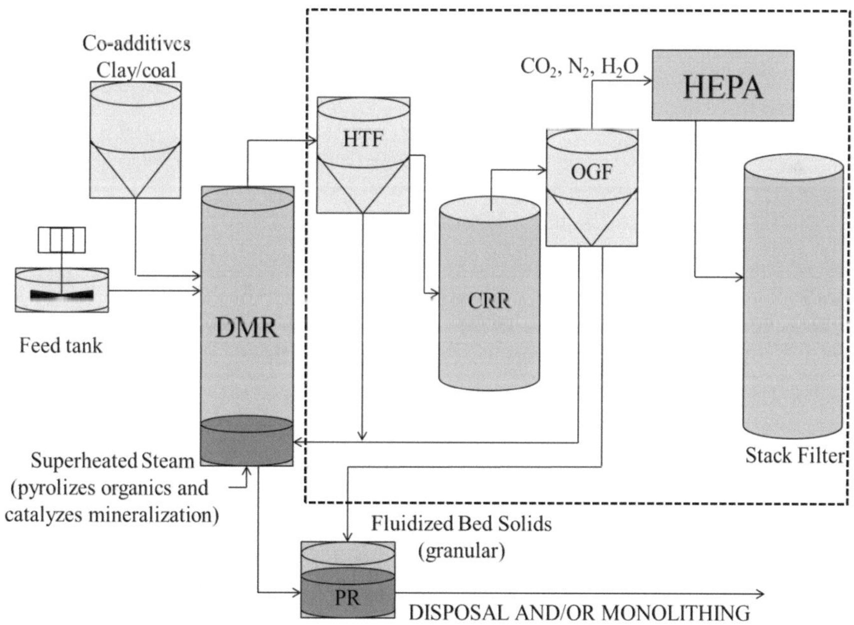


Fig. 1. FBSR Sodium Aluminosilicate (NAS) Waste Form Dual Processing Flowsheet (DMR = Denitration & Mineralizing Reformer; PR = Product Receiver; HTF = High Temperature Filter (material recycled to DMR); CRR = Carbon Reduction Reformer (treats gases only); OGF = Off-Gas Filter; HEPA = High-Efficiency Particulate Air filter) (Adapted from Jantzen, 2008)

### 1.1.1 Denitration and Mineralization Reformer

The DMR is a fluidized bed reformer with the bed media being fluidized using superheated steam injected by a distributor at the bottom of the vessel. Granular carbon is also fed directly into the bed to provide energy for the process. The reaction of solid carbon with steam generates carbon monoxide and hydrogen, via the water/gas reaction (see Olson et al., 2004b for a list of reactions). A small amount of oxygen is bled in to react with any excess hydrogen to create more steam and heat because the reaction of oxygen and hydrogen is exothermic. A metered flow of atomizing compressed air is used to atomize the waste/clay slurry introduced to the bed. As the waste/clay slurry is injected into the bed by the waste feed injectors, mineral products are formed by the hydrothermal reaction of the alkali metals (sodium, cesium, and potassium) with the added clay (aluminosilicate). The reaction may also involve clay and alkaline earths or other inorganics in the waste. DMR bed materials consist of accumulated mineral stable, leach-resistant product granules that are fluidized by the low-pressure steam.

When the waste feed is introduced into the fluidized bed as fine spray, the waste feed reacts to form new minerals after contacting the heated fluidized bed. Nitrates and nitrites in the feed react with reductive gases to produce mainly nitrogen gas with traces of $NO_x$. The non-volatile contaminant constituents, such as metals and radionuclides, are mineralized by being incorporated into the final mineral species in the bed product. The process gases exit the DMR through the HTF and consist mainly of steam, $N_2$, CO, $CO_2$, and $H_2$. Some low levels of $NO_X$, acid gases, and short-chained organics may also be present and these can be destroyed in the CRR.

### 1.1.2 Off-gas treatment system

The off-gas treatment system provides high-efficiency filtration and oxidation of any residual volatile organics and small amounts of carbon monoxide and hydrogen from the DMR. The process off-gas from the DMR is routed into a HTF to trap small mineral product particles called fines. The gases pass through the carbon reduction reformer (CRR) to reduce residual $NO_X$ to $N_2$ in the lower reducing zone and to oxidize CO, $H_2$, and the residual hydrocarbons into $CO_2$ and water vapor in the upper oxidizing stage of the CRR. The CRR has a semi-permanent bed media composed of alumina. No additional $NO_X$ abatement or acid gas removal is required because the nitrates and nitrites are converted into nitrogen gas inside the DMR and CRR with a very high efficiency (TTT, 2009). Acid gases are minimized as the S, Cl, and F are incorporated into the bed product mineral structures such that no wet scrubber is required to remove acid gases in the off-gas treatment system. The off-gas from the CRR is cooled and then passes a through a Off-Gas Filter (OGF) and then through a high-efficiency particulate air (HEPA) filter to remove any further particulates. Further cooling occurs in an off-gas cooler and the off-gas is then passed through another set of HEPA filters and a mercury adsorber (if needed) before being exhausted out of a stack. The FBSR process outlet gases are compliant with Maximum Achievable Control Technology (MACT) limits for metals, $HCl/Cl_2$, particulates, dioxins/furans, volatile and semi-volatile organic compounds, total hydrocarbon, and carbon monoxide as well as site discharge limits for $NO_X$ and $SO_X$ (TTT, 2009; Vora et al., 2009).

### 1.2 Process control

Waste form process monitoring and control are necessary to obtain the desired mineral product. The proper amount of additives and operation in the proper REDuction/OXidation (REDOX) range ensure autocatalytic heating and pyrolysis. Mineralization control produces the desired NAS phases and is based on the MINCALC™ process control system (See Section 2.1.1). This strategy favors the formation of nepheline and sodalite.

To assist in finding the process REDOX potentials, an Electromotive Force (EMF) series developed by Schreiber (2007) is used to determine what the oxygen fugacity in the DMR is during mineralization. Iron, in the form of $Fe^{3+}(NO_3)$ is added as a REDOX indicator and the final product REDOX is measured spectrochemically using the Baumann method (Baumann, 1992). By measuring the ratio of $Fe^{+2}/\Sigma Fe$ in the resulting mineral product, the EMF is used to back calculate the oxidation of the remaining multivalent species of interest, e.g. S, Re, Cr. This calculation is more accurate if the unreacted coal in the bed product is less than 10 wt%. All control boundaries lead to a complete denitration of the product. Typically the FBSR process is run with a reducing environment with a log oxygen fugacity of -15 to -16 Pa.

## 2. FBSR waste form process description

Kaolin clay, the aqueous waste stream, steam, and a carbon or other heat source, are the only ingredients for the FBSR NAS granular product. This mix of liquids, solids, and gases is provided through feed subsystems in order to obtain a desired FBSR process material and conditions. The granular product is removed from the FBSR either as product from the bottom of the bed or as particulates removed from the fluidizing gases by the HTF. Unless the HTF particulates are recycled to the DMR bed as seed particles, these two materials are then combined and encapsulated in a binder for final disposal. This section gives a more detailed description of the additives used in the FBSR process.

### 2.1 Waste form ingredients

#### 2.1.1 Kaolin clay

Kaolin clay is a key ingredient in the FBSR process. In kaolin clay, the aluminum atom is octahedrally coordinated with bonds to two oxygen atoms and four hydroxyl ions. During processing at 700-750°C, the kaolin clay is destabilized and the four OH- atoms are vaporized which leaves the clay Al atoms unstable and amorphous at the nanoscale. Alkalis in the waste react with the unstable Al atom and rearrange to form a crystalline mineral species with a low free energy tetrahedral configuration such as $NaAlSiO_4$.

Several sources of kaolin clay have been used in the FBSR pilot-scale testing program with Table 1 listing some of their properties. SnoBrite and OptiKasT kaolin clays were used for tests with Hanford LAW (Olson et al., 2004b) while Troy and K-T Sagger XX kaolin clays were used for tests with INEEL SBW (Olson et al., 2004a). All of these kaolin clays were investigated for use as a mineralizing additive in the INEEL pilot-scale experiments. All the clays were evaluated based on XRD, particle-size analyses, whole element chemistry, and rheological properties (Olson et al., 2004b).

| Clay | Major Phases | Minor phases | Si:Al atom ratio | Total moisture (wt %) | Particle size (wt% less than 10% - 50% - 90%) | Particle density (g/cm3) |
|---|---|---|---|---|---|---|
| SnoBrite | Kaolinite[(a)] | Muscovite[(b)]; Rutile ($TiO_2$) possible | 1.02 | 14.20% | 0.82µm - 5.00µm - 20.8µm | 2.77 |
| OptiKasT | Kaolinite[(a)] | Muscovite [(b)] | 1.04 | 15.15% | 0.74µm - 4.22µm - 15.9µm | 2.69 |
| Troy | Kaolinite[(a)] | Muscovite [(c)]; Quartz ($SiO_2$) possible | 01:01.2 | 14.65% | 1.83µm - 14.83µm - 57.1µm | 2.74 |
| K-T Sagger XX | Kaolinite[(d)] | Muscovite[(b)] Rutile ($TiO_2$) | 01:01.7 | 10.60% | 1.34µm - 6.55µm - 21.5µm | 2.73 |

(a) Kaolinite (PDF#75-1593) ($Al_2O_3 \cdot 2SiO_2 \cdot 2H_2O$)
(b) Muscovite (PDF#07-0042) (K, Na)(Al, Mg, Fe)$_2$-($Si_{3.1}Al_{0.9}$)$O_{10}$(OH)$_2$
(c) Muscovite (PDF#86-1385) (($K_{0.86}Al_{1.94}$)($Al_{0.965}Si_{2.895}O_{10}$)- ((OH)$_{1.744}F_{0.256}$)
(d) Kaolinite (PDF#78-1996)

Table 1. Properties of Kaolin Clays Used in FBSR Pilot Scale Tests (from Olson et al., 2004b, 2004b)

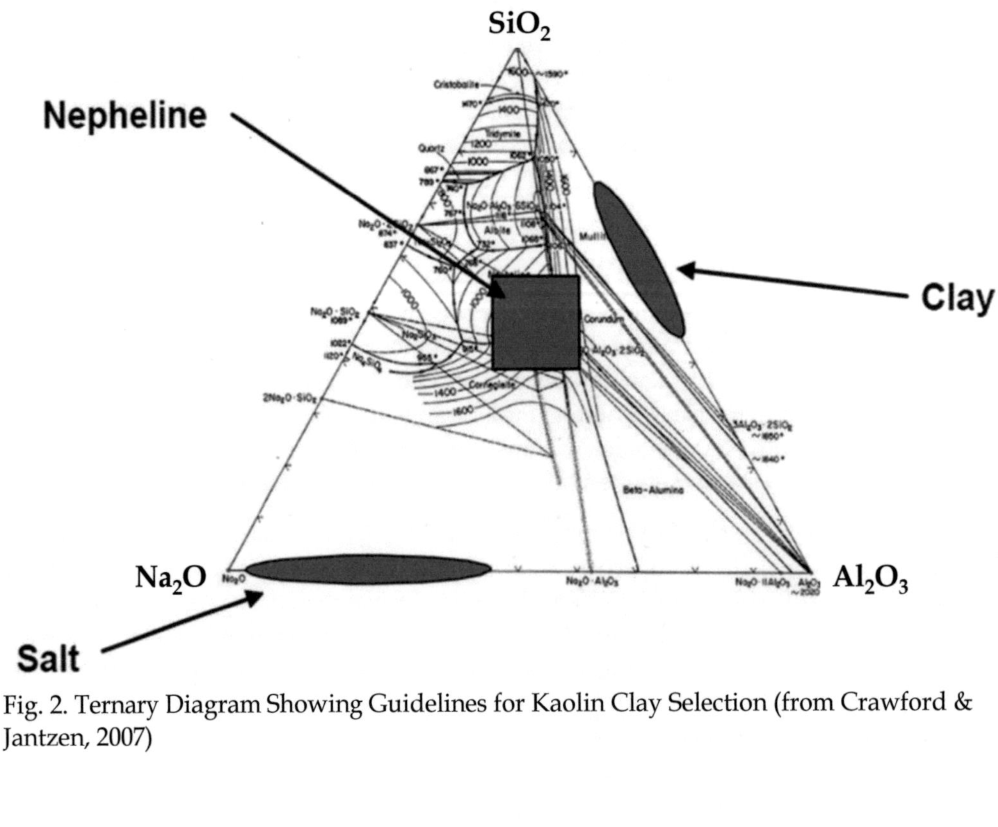


Fig. 2. Ternary Diagram Showing Guidelines for Kaolin Clay Selection (from Crawford & Jantzen, 2007)

The mineralizing agent needs to have a Si/Al that will react with the liquid waste and produce desired minerals. The kaolin clay type was selected with the appropriate Al:Si mole ratio that would suitably react with the Na and anions in the waste liquids. Usually, a ternary phase diagram (such as the one depicted in Figure 2) shows the target region of compositions that is thought to be the most favorable for producing the desired mineral products (Crawford & Jantzen, 2007). Similar diagrams may be seen in Olson et al. (2004a) and TTT (2009) for Hanford LAW and WTP-SW wastes, respectively. The most favorable atomic ratios that would produce the desired nepheline and sodalite products are thought to be $1 < M/Si < 1.33$, $1 < M/Al < 1.33$, $1 \leq Al/Si$, and $M/(Al+Si) = 0.5\text{-}0.67$, where M represents an alkali metal, mostly Na in this case. Atomic ratios provide guidelines due to the possibility of significant substitution of different alkali and alkaline earths, and some Fe for Al, in these feldspathoid minerals (Olson et al.c 2004a). SRNL has developed a spreadsheet called MINCALC™ that can be used as a tool to select the clay formulation and carbon addition and to make adjustments, such as accounting for extra aluminum and potassium in the wastes (Crawford & Jantzen, 2007; Pareizs et al., 2005). Additional considerations for choosing the optimum mineralizing additive include particle-size distribution that is optimized to make sure that as much clay as possible reacts with the liquid waste as well as eliminating clays that contain additional elements that will not react to form the desired mineral phases.

### 2.1.2 Carbon source

Different reductants (e.g., charcoal, carbon, or sugar) are used in the FBSR process to aid in pyrolyzing organics, and removing nitrate. Charcoal or carbon plus steam creates $H_2$ for autocatalytic heating, while sugar is not a good heat source and can only be used for the denitration and pyrolysis. For example, different types of carbon were considered to be used as reductants in the INEEL pilot-scale experiment (Olson et al., 2004a, 2004b). Carbon was evaluated based on reactivity, particle size, particle fracturing (attrition) resistance, moisture content, loss on ignition, and the ash composition. A wood-based carbon was chosen based on its efficient performance

### 2.1.3 Other properties of the starting bed

The primary solid feed material in the fluidized bed reactor consists of granular carbon and an initial starting bed media. The chosen starting material must be dense, inert, and have a high heat capacity due to high processing temperatures (Olson et al., 2004b). Bed materials that meet these criteria that were considered for an INEEL pilot-scale steam reformer were alumina, dolomite, sintered bauxite beads, nepheline syenite, and sintered calcium silicate (Olson et al., 2004b). Choice of starting bed material was based on composition, melting point, resistance and durability, particle size, and availability of the material. 70-grit alumina was finally chosen for the test. Alumina facilitates processing through its high heat capacity which transfers heat to the atomized feed and prevents over-quenching in the feed zone. Also, hot alumina does not seem to be coated by the product and was chosen because of its attrition resistance and inertness (Olson et al., 2004a). The 2008 Hanford LAW and WTP-SW pilot-scale tests at the Hazen Research Inc. (HRI) facility also used alumina as the starting bed material.

### 2.2 Process details

After startup, the FBSR process is straightforward. The feeds for the clay and tank wastes are fed into the FBSR. Products exiting through the DMR are passed through the HTF or recycled back into the DMR. The granular product is collected and allowed to cool before sending to the binder station for encapsulation into a monolith.

A last step in FBSR, as related to radioactive waste management, is the product must be encapsulated to meet disposal system requirements for compressive strength and to minimize the dispersability of the material. Several binders have been tested (Jantzen, 2006; TTT, 2009) and further testing and development is anticipated before a final form is selected. Simple, inexpensive binders, such as a cementitious material, a geopolymer, or Ceramicrete, are likely candidates. The geopolymer appears to be the most promising.

### 2.3 FBSR process chemistry

Two major processes occur simultaneously in the steam reforming of a nitrate salt: the mineralization process and the denitration process. Mineralization is the reaction of activated clay with estranged cations (Na, Cs, Tc, etc.) and other species present in the salt waste (Cl, F, I, $SO_4$, etc.). Stable crystalline clays become reactive amorphous clays at FBSR processing temperatures because clays lose their hydroxyl groups above 550°C. The waste species react with the reactive amorphous meta-kaolin-clay to form new stable crystalline mineral structures allowing formation and templating of mineral structures at moderate temperatures. The resulting stable crystalline structures leave the process as a granular solid product. Iron-bearing co-reactants can be added during processing to stabilize multivalent hazardous species present in the waste in durable spinel phases, i.e. Cr, Ni, Pb iron oxide minerals. Denitration, in the presence of a carbon source, is the reformation of the $NO_3$ and $NO_2$ anions to $N_2$, $H_2O$ (steam) and $CO_2$.

## 3. FBSR waste form description

The FBSR waste form is composed of two main components: a granular product and an encapsulating binder material. The primary granular product made by processing wastes in the FBSR is formed of geophases (minerals). These phases may provide leach resistant (durable) waste forms for immobilizing the contaminants that are present in different waste liquids (Olson et al., 2004a). This granular product is then encapsulated in a binder material to form a monolith which limits dispersability and provides some structural integrity for subsidence prevention in the disposal facility. The existing FBSR waste form data are derived from a number of pilot-scale FBSR tests conducted with INEEL SBW and Hanford LAW and secondary waste simulants. Table 2 summarizes these tests.

### 3.1 Sodium aluminosilicate (NAS) primary waste form

#### 3.1.1 Major mineral form attributes

The primary product from the FBSR process with kaolin clay is a granular product composed of NAS minerals which bond with radionuclides and contaminants of concern (COCs). The NAS FBSR granular product is a multiphase mineral assemblage of NAS feldspathoid group minerals (sodalites, nosean, and nepheline) with cage and ring structures that sequester anions and cations (Jantzen et al., 2007). The nomenclature of this series of mineral species is governed by the species that occupies the cavities in the aluminosilicate framework as well as the type of crystal structure, either cubic or hexagonal. Nepheline is the basic NAS mineral with the formula $Na_2O$-$Al_2O_3$-$2SiO_2$. When sulfates are captured within the cage structure, nosean forms with the formula $3Na_2O$-$3Al_2O_3$-$6SiO_2 \cdot Na_2SO_4$. When chlorides are captured within the cage structure, sodalite forms with the formula $3Na_2O$-$3Al_2O_3$-$6SiO_2 \cdot 2NaCl$. Depending on the waste compositions, process additives such as magnetite are added to retain Cr as $FeCr_2O_4$ (Jantzen et al., 2007). The retention of anions and cations within the mineral structures of the nepheline, sodalite, and nosean phases, as well as the role of magnetite, will be discussed here. The most comprehensive work on the subject has been performed by Jantzen et al. (2005) and Jantzen (2008) and this section summarizes the work presented there.

| Waste | Pilot-Scale Facility | Date | Sample ID | Monolith |
|---|---|---|---|---|
| **Hanford Wastes** | | | | |
| LAW AN-107, Envelope C [(a), (b)] | Hazen Research Facility, 6-inch FBSR | Dec. 2001 | SCT02-098-FM, Fines PR-01 | No |
| LAW Saltcake blend [(c)] | SAIC STAR, 6-inch FBSR | Aug. 2004 | Bed 1103, Bed 1104, Fines 1123 | Blend[(i)] |
| LAW Saltcake blend [(d)] | Hazen Research Facility, 15-inch FBSR | 2008 | P1 PR bed, HTF fines | Yes |
| WTP-SW LAW melter off-gas recycle [(d)] | Hazen Research Facility, 15-inch FBSR | 2008 | P2 PR bed, HTF fines | Yes |
| LAW Saltcake blend [(e)] | SRNL BSR, 2.75-inch | 2010 | Module B samples | No |
| WTP-SW LAW melter off-gas recycle [(e)] | SRNL BSR, 2.75-inch | 2010 | Module A samples | No |
| **Idaho Wastes** | | | | |
| SBW [(f)] | SAIC STAR, 6-inch FBSR | Jul-03 | Bed 260, Bed 272, Bed 277 | Blend[(i)] |
| SBW [(g)] | SAIC STAR, 6-inch FBSR | 2004 | Bed 1173 | Blend[(i)] |
| SBW [(h)] | Hazen Research Facility, 15-inch FBSR | 2004 | DMR4xxx, HTF4xxx | No |

(a) Jantzen, 2002
(b) Pareizs et al., 2005
(c) Olson et al., 2004a
(d) TTT, 2009
(e) Jantzen et al., 2011
(f) Marshall et al., 2003
(g) Olson et al., 2004b
(h) Ryan et al., 2008
(i) Blends are samples used in these studies that combined 20% LAW, 32% SBW and 45% starting bed material

Table 2. Summary of FBSR Pilot-Scale Sodium Aluminosilicate Waste Form Preparation Tests

The NAS mineral waste forms are comprised mostly of nepheline (nominally $NaAlSiO_4$ or $Na_3KAl_4SiO_{16}$) which is a hexagonal structured feldspathoid mineral, sodalite and nosean following the reactions:

$$\underbrace{2NaOH}_{\text{waste}} + \underbrace{Al_2O_3 \bullet 2SiO_2}_{\text{kaolinclayadditive}} \rightarrow 2 \underbrace{NaAlSiO_4}_{\text{Nephelineproduct}} + H_2O \tag{1}$$

$$\underbrace{8NaOH + 2Cl}_{\text{waste}} + \underbrace{3(Al_2O_3 \bullet 2SiO_2)}_{\text{kaolinclayadditive}} \rightarrow \underbrace{Na_6Al_6Si_6O_{24}(2NaCl)}_{\text{Sodaliteproduct}} + 3H_2O + 2OH^- \tag{2}$$

$$\underbrace{8NaOH + SO_4}_{\text{waste}} + \underbrace{3(Al_2O_3 \bullet 2SiO_2)}_{\text{kaolinclayadditive}} \rightarrow \underbrace{Na_6Al_6Si_6O_{24}(Na_2SO_4)}_{\text{Nosean product}} + 3H_2O + 2OH^- \tag{3}$$

$$\underbrace{8NaOH + 2ReO_4^-}_{\text{waste}} + \underbrace{3(Al_2O_3 \bullet 2SiO_2)}_{\text{kaolinclayadditive}} \rightarrow \underbrace{Na_6Al_6Si_6O_{24}(2NaReO_4)}_{\text{Sodaliteproduct}} + 3H_2O + 2OH^- \tag{4}$$

$$\underbrace{6NaAlSiO_4}_{\text{nephelineproduct}} + \underbrace{2NaReO_4}_{\text{waste}} \rightarrow \underbrace{Na_6Al_6Si_6O_{24}(2NaReO_4)}_{\text{Sodalite}} \tag{5}$$

The ring-structured aluminosilicate framework of nepheline forms cavities which can be 8- or 9-coordinated. There are eight large (9-fold oxygen) coordination sites and six smaller (8-fold oxygen) sites. The larger 9-fold coordination sites ionically bond with Ca, K, and Cs and the smaller 8-fold sites ionically bond with Na (Deer et al., 1963). When K is the cation it is known as kalsilite ($KAlSiO_4$). In nature, the nepheline structure is known to accommodate Fe, Ti and Mg, as well (Deer et al., 1963). In addition, rare earth nephelines are known, e.g. $NaYSiO_4$, $Ca_{0.5}YSiO_4$, $NaLaSiO_4$, $KLaSiO_4$, $NaNdSiO_4$, $KNdSiO_4$, and $Ca_{0.5}NdSiO_4$, where the rare earth substitutes for Al in the structure (Barrer, 1982). A sodium-rich cubic structured nepheline with excess Na is also known, e.g. $(Na_2O)_{0.33}$ $Na[AlSiO_4]$ and was found in a FBSR mineralized product (Jantzen, 2002, Pareizs et al, 2004). This nepheline structure has large cage-like voids in the structure where Na can bond ionically to 12 framework oxygens (Klingenberg & Felsche, 1986). This cage-structured nepheline has not been shown to occur in nature. Despite this, the large cage-like voids should be capable of retaining large radionuclides, especially monovalent anions such as $(ReO_4)^{-1}$ (Olson et al., 2004a). Likewise, $Na_2O$ deficient nepheline structures have been found in other FBSR mineralizing campaigns for INEEL alumina-rich SBW (Olson et al., 2005).

Structurally related to the aluminosilicate framework of nepheline is the sodalite group of minerals. Sodalite minerals have cage-like structures formed of aluminosilicate tetrahedra. The cage structures retain anions and/or radionuclides that are ionically bonded to the aluminosilicate tetrahedral and to sodium. The cage in sodalite is occupied by two sodium and two chlorine ions can be written as $Na_6[Al_6Si_6O_{24}] \cdot 2NaCl$ to highlight that the NaCl molecules are chemically bonded in the cavities. If the NaCl molecules are replaced by $Na_2SO_4$, $Na_2CO_3$ and NaOH the mineral names are known as nosean, natrodavyne, and basic sodalite, respectively. Sodalite minerals are known to accommodate Be in place of Al and $S_2$ in the cage structure, along with Fe, Mn, and Zn, e.g. danalite ($Fe_4[Be_3Si_3O_{12}]S$), helvite ($Mn_4[Be_3Si_3O_{12}]S$), and genthelvite ($Zn_4[Be_3Si_3O_{12}]S$) (Deer et al, 1963). These cage-structured sodalites are also found to retain Mo, Cs, Sr, B, Ge, I and Br (Buhl et al., 1989;

Deer et al., 1963; Fleet, 1989). Regardless of the oxidation state of sulfur during FBSR processing, the feldspathoid minerals can accommodate sulfur as either sulfate or sulfide. Although neither Cs nor Rb sodalites have been identified as phase pure end members, Cs and Rb are tolerated in the sodalite structure (Deer et al, 1963; Deer et al., 2004). In addition, non-naturally occurring Zeolite-A structures are known to form from the reaction of CsOH and RbOH with kaolin clay (Barrer et al., 1968).

Due to the flexibility of the sodalite, monovalent species such as $Cs^+$, $K^+$, $Ca_{0.5}$, $Sr_{0.5}$, etc. can substitute for $Na^+$, while $(SO_4)^{-2}$ , $(MoO_4)^{-2}$, $(AsO_4)^{-2}$, $(MnO_4)^{-1}$, and $(ReO_4)^{-1}$ [and thus presumably $(TcO_4)^{-1}$], can all substitute for the $Cl^-$ in the structure. In addition, $I^-$, $Br^-$, $OH^-$, and $NO_3^{-2}$ can all substitute for the $Cl^-$ atoms. In the above-listed oxyanions, the oxygens in the tetrahedral polyhedral around the metal provide the oxygen bonds for the tetrahedral $XO_4$ groups. These oxygen come from four of the six tetrahedra forming a ring along the body diagonal of the cubic unit cell (Hassan & Gruncy, 1984). Boron and beryllium can substitute for Al in a tetrahedral polyhedra in the sodalite structures as can Ti. Elements such as Fe and Zn can substitute for $Na^+$ (Deer et al., 2004; Hassan & Gruncy, 1984). Figure 3 shows the flexibility of sodalite to incorporate various species in its structure.

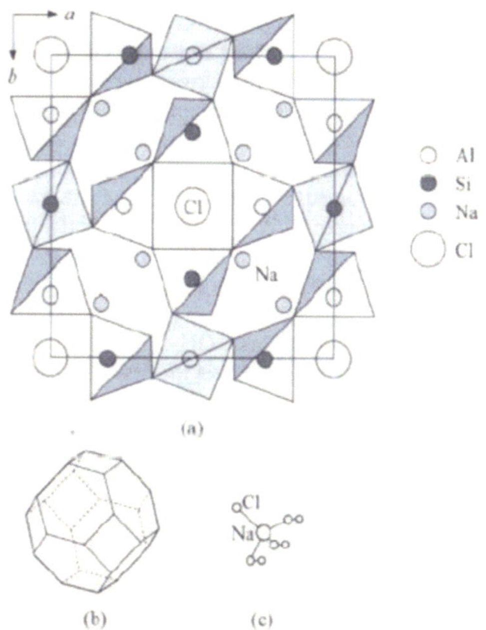


Fig. 3. Structure of Sodalite showing (a) two-dimensional projection of the (b) three-dimensional structure and (c) the four fold ionic coordination of the Na site to the $Cl^-$ ion and three framework oxygen bonds. (Deer et al., 2004)

The last of the mentioned feldspathoid cage-structured minerals is nosean, $Na_6[Al_6Si_6O_{24}](Na_2SO_4)$. Nosean has $Na_2SO_4$ bonded in the sodalite cage like structure. All bonding in the sodalite/nosean single unit cell is ionic and the atoms are regularly arranged. This is similar to the manner of ionic bonding in glass, but more highly ordered than the atomic arrangements in glasses which have no long-range order (Jantzen, 2008).

Another possible additive in FBSR processing to catalyze denitration is an iron oxide co-reactant containing the spinel mineral magnetite (Jantzen, 2002). In this situation, and potentially for waste streams rich in iron, spinels form that retain cations such as Cr within the structure as $FeCr_2O_4$. This incorporation is made possible due to the reduction of $Cr^{6+}$ to $Cr^{3+}$ at the elevated temperature and reducing atmosphere of the FBSR process (Jantzen et al., 2007). The spinels such as $Fe_3O_4$ ( $Fe^{+2}Fe_2^{+3}$ + $O_4$) are known to take $Cr^{+3}$ and $Ti^{+3}$ into their lattice in place of $Fe^{+3}$, as well as many of the divalent transition metals like $Ni^{2+}$, $Mn^{2+}$, $Zn^{2+}$, and $Mg^{2+}$ (Deer et al., 1963). Spinels have both tetrahedral and octahedral coordination spheres with oxygen. The trivalent ions reside in the four-fold coordination positions and the divalent ions reside in the six-fold coordination positions. All the trivalent and divalent ions are ionically bonded to oxygen (Jantzen, 2008).

### 3.1.2 FBSR product morphology

The FBSR granular product is composed of two fractions from the FBSR process. Solids collected from the bottom of the fluidized bed are captured in the PR. Figure 4 shows a photograph of the PR material from the 2008 Hazen pilot-scale test with the Hanford LAW simulant. The PR material includes residual carbon from coal or wood products used in the FBSR as an energy source and as a reductant. Unreacted carbon can be removed by roasting at 525°C (Bullock et al., 2002). Carbon is often removed to ensure that it does not contribute to the measured surface area used for durability testing (Jantzen & Crawford, 2010). This temperature is high enough to remove carbon in an oxidizing atmosphere but not too high to change the composition of the phase assemblages of the product. The PR material may also include residual alumina used as an initial seed material when the FBSR is first started up. Solids leaving the FBSR entrained in the fluidizing gases are captured in the high-temperature filter (HTF). The morphology of the finer particles, which tend to clump together but are relatively easy to collect are also shown in Figure 4.

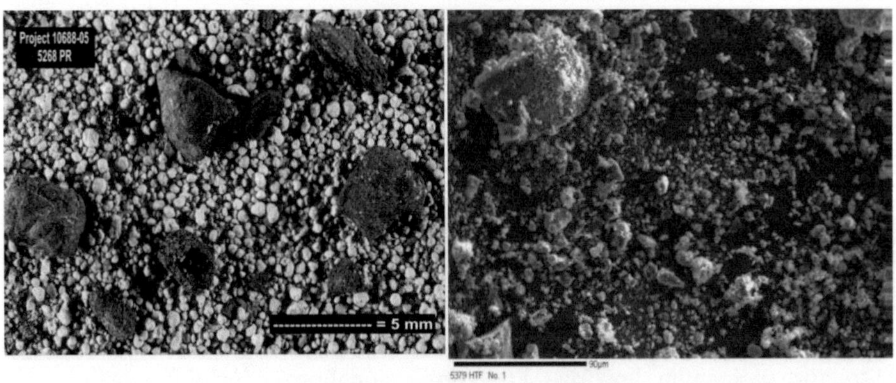


Fig. 4. Example of FBSR Granular Product from the Product Receiver (left). Microprobe Photographs of High Temperature Filter (HTF) Fines (right) (from TTT, 2009)

Some scanning electron microscopy (SEM) micrographs of the FBSR material from the 2001 Hazen Research facility test with Hanford LAW are presented in Figure 5. The micrograph on the left clearly shows FBSR materials to be highly porous. It is also worth noticing the black dots in the micrograph on the right which indicate the presence of magnetite ($Fe_3O_4$). Iron oxides are also good hosts for contaminants, e.g. forming spinels with Cr and Ni in the wastes.

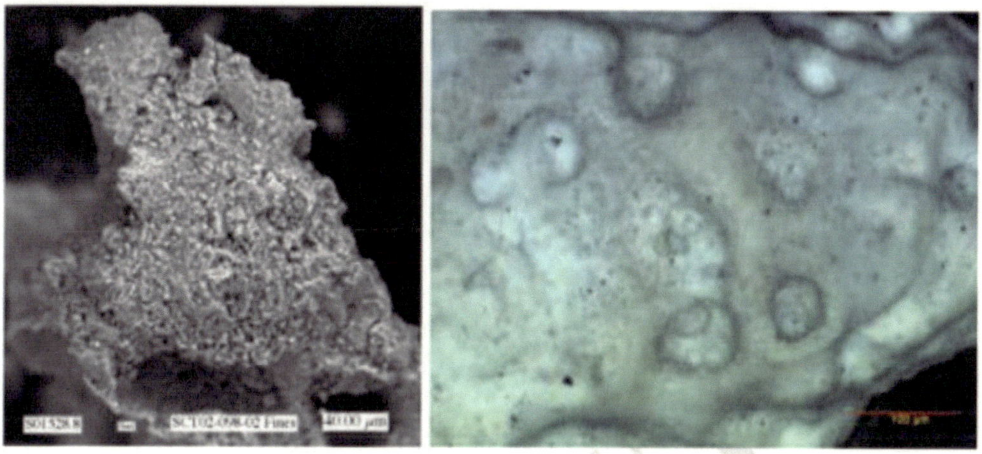

Fig. 5. SEM Micrographs of Typical FBSR Product Grain (left). Optical photograph of SCT02-098 particle (FBSR material); the black particles are magnetite ($Fe_3O_4$) (McGrail et al., 2003a).

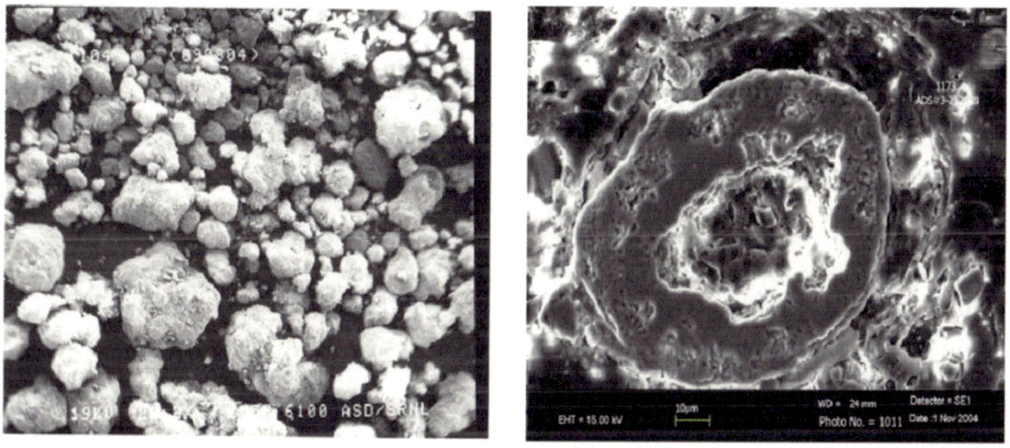

**(a)** 1123 Bed product **(b)** 1173 Bed product (sectioned)

Fig. 6. SEM Photomicrographs of FBSR Bed Product Showing the Surface Topography and Porosity (Lorier et al., 2005). Also see Table 2 for sample origins.

Figure 6 shows SEM micrographs in as received (left) and in as embedded and sectioned (right) by Lorier et al. (2005) of FBSR materials from tests in 2004 with Hanford LAW saltcake simulant (Bed 1123) and INEEL SBW simulant (Bed 1173), respectively. The LAW saltcake FBSR product shows the particle shape and irregular surface topography of the granular product. The SBW micrograph shows the internal porosity of a granule in cross-section. Large uncertainties associated with the porosity of the FBSR granular product gives a large discrepancy between a calculated geometric surface area and the measured surface area of the material using Brunauer, Emmett, and Teller (BET) (Brunauer et al., 1938) surface area measurements. Using the smaller surface area from the geometrical value would lead to an underestimation in dissolution rates used for product durability studies therefore it has been recommended to use the surface area obtained from BET measurements (McGrail et al., 2003a).

### 3.2 Phase composition and mineralogy

Studies have been conducted with different FBSR materials to determine the minerals that are present. Most of these minerals are believed to belong to the groups of nepheline, sodalite, and carnegieite, all feldspathoids with a one molar ratio of Si:Al:Na. Carnegeite is a metastable form of nepheline with the same chemical composition but with less atomic order. It readily transforms to nepheline upon heating (Jantzen & Crawford, 2010). Both nepheline and carnegeite have ring structures. In Table 3 one may see a summary of the mineral phases identified by Jantzen and Crawford (2010) in the Hanford LAW and WTP-SW pilot-scale tests as well as the INEEL SBW tests. Nepheline ($NaAlSiO_4$) is the primary phase formed. The fines captured in the HTF have a shorter residence time in the FBSR and contain low-carnegieite. Data in the table are from an XRD method that gives information on the specific crystalline phases present by comparison to reference library spectra. No internal standards were used to allow for quantitative measurements of the phases. Information on 'major,' 'minor' and 'trace' phases present are given by intercomparison of the main peaks of each crystalline pattern within a sample.

In addition to the NAS phases, other minor phases have been identified using XRD of the FBSR granular product. These include quartz ($SiO_2$) and anatase ($TiO_2$), which are impurities in the clays used as mineralizing agents in the FBSR. Corundum (crystalline alumina, $Al_2O_3$) results from the seed material for starting up the FBSR process. Its concentration decreases over time as the starting FBSR bed material leaves through the PR. Hematite ($Fe_2O_3$) and magnetite ($Fe_3O_4$) were identified in the Hanford Envelope C product, most likely due to the iron oxide additive used as a denitration catalyst (Jantzen, 2002). Jantzen et al. (2006b) also report that amorphous metakaolin was identified by SEM in the early LAW Envelope C and Envelope A FBSR tests as unreacted cores in the mineralized granules.

Nosean and sodalite have been identified as minor NAS phases in the FBSR granular product. Nosean and sodalite have cage structures that can retain anions and radionuclides that bond ionically within the structure. Table 4 shows how various elements within Hanford tank wastes may substitute in the nepheline, sodalite, and nosean structures (Jantzen et al., 2011).

The oxidation state can impact how and where contaminants are captured in the FBSR product (Jantzen, 2008). The FBSR process is run under reducing conditions with a log oxygen fugacity of -15 to -16 Pa. Under the designed REDOX conditions, a REDOX-sensitive species, such as chromium, is predicted to be 50 to 70% reduced to $Cr^{3+}$ and would be sequestered in a spinel (hematite or magnetite) phase. Sulfur is predicted to be only 1 to 19% reduced to $S^{2+}$ and would enter the sodalite phase as $SO_4$ in the +6 oxidation state.

| FBSR Product | Low-Carnegieite[a] | Nepheline | Nosean and/or Sodalite | Other Minor Components |
|---|---|---|---|---|
| **Hanford Envelope "C" LAW Wastes (2002) $Fe^{+2}/\Sigma Fe$ of Bed = 0.15** | | | | |
| SCT02-098-FM | | X | Y | $Al_2O_3$, $Fe_2O_3$, $Fe_3O_4$ |
| Fines PR-01 | X | X | Y | $Al_2O_3$, $Fe_2O_3$, $Fe_3O_4$ |
| **Hanford Envelope "A" LAW Wastes (2004) $Fe^{+2}/\Sigma Fe$ of Bed = 0.28-0.81** | | | | |
| Bed 1103 | X | X | Y | $TiO_2$ |
| Bed 1104 | X | X | Y | $TiO_2$ |
| Fines 1125 | X | Y | | $TiO_2$ |
| **INEEL SBW Wastes (2003-2004) $Fe^{+2}/\Sigma Fe$ of Bed = 0.51-0.61** | | | | |
| Bed 260 | Y | X | TR | $Al_2O_3$ and $TiO_2$ |
| Bed 272 | Y | X | TR | $TiO_2$ |
| Bed 277 | Y | X | TR | $TiO_2$ |
| Bed 1173 | | X | TR | $Al_2O_3$, $SiO_2$, $NaAl_{11}O_{17}$, $(Ca,Na)SiO_3$ |
| **Hanford Rassat LAW Wastes (2008) $Fe^{+2}/\Sigma Fe$ of Bed = 0.41-0.90** | | | | |
| PR Bed Product 5274 (P1A) | Y | X | X | $Al_2O_3$ |
| PR Bed Product 5316 (P1A) | Y | X | X | Pyrophyllite* |
| HTF Fines 5280 (P1A) | X | Y | | $NaAl_{11}O_{17}$ (Diaoyudaoite),$TiO_2$ |
| HTF Fines 5297 (P1A) | X | Y | X | $SiO_2$ |
| PR Bed Product 5359 (P1B) | Y | X | X | Pyrophyllite* |
| PR Bed Product 5372 (P1B) | Y | X | X | Pyrophyllite* |
| HTF Fines 5351 (P1B) | X | Y | Y | $SiO_2$ |
| HTF Fines 5357 (P1B) | X | Y | Y | $TiO_2$ |
| Composite (P1A) | X | Y | Y | $SiO_2$ and $TiO_2$ |
| Composite (P1B) | X | Y | Y | $SiO_2$ and $TiO_2$ |
| **Hanford Melter Off-Gas Recycle (WTP SW) Wastes (2008) $Fe^{+2}/\Sigma Fe$ =0.41-0.90** | | | | |
| PR 5475 (P2A) | Y | Y | X | Pyrophyllite* |
| HTF Fines 5471 (P2A) | X | X | X | $SiO_2$ |
| PR 5522 (P2B) | Y | Y | X | Pyrophyllite*, $TiO_2$ |
| HTF Fines 5520 (P2B) | X | X | X | $SiO_2$ and $TiO_2$ |
| Composite (P2B) | Y | X | X | $SiO_2$ |

X = Major constituent; Y = Minor constituent; TR = trace constituent; a = the PDF for this phase states it is orthorhombic nepheline and possibly low-carnegeite (PDF 052-1342). Note low-carnegeite also has ring structures that are oval for sequestration of K, Cs, etc; *$Al_{1.333}Si_{2.667}O_{6.667}(OH)_{1.333}$

Table 3. Mineral Phases Analyzed in FBSR Products (from Jantzen & Crawford, 2010).

Rhenium, often used as a non-radioactive surrogate for technetium, is predicted to be only 2 to 6% reduced to the +4 oxidation state at the nominal operating conditions. At the +7 oxidation state, rhenium, and by association, technetium, are predicted to enter the sodalite phase. Mattigod et al. (2006) were able to synthesize sodalite [$Na_8(AlSiO_4)_6(ReO_4)_2$] that contained Re(VII). Its crystal structure was determined from Rietveld refinement of experimental XRD data. This study showed that Re(VII) can be incorporated into NAS solids. REDOX control is important for making certain that the contaminants enter the desired FBSR mineral phases. Only 2.5% of the Re was in the +7 state and 1% of the S was in the +6 state in the HTF product compared to 94 to 95% Re(VII) and 86 to 89% $S^{6+}$ in the PR product.

| Nepheline—Kalsilite Structures (a) | Sodalite Structures (b) | Nosean Structures |
|---|---|---|
| $Na_xAl_ySi_zO_4$ (c)<br>where x=1-1.33<br>y and z = 0.55-1.1 | $[Na_6Al_6Si_6O_{24}](NaCl)_2$ (c) | $[Na_6Al_6Si_6O_{24}](Na_2SO_4)$ (c, d) |
| $KAlSiO_4$ | $[Na_6Al_6Si_6O_{24}](NaF)_2$ (c) | $[Na_6Al_6Si_6O_{24}](Na_2Mo_4)$ (c, e) |
| $K_{0.25}Na0._{75}AlSiO_4$ (c) | $[Na_6Al_6Si_6O_{24}](NaI)_2$ (d) | $[Na_6Al_6Si_6O_{24}]((Ca,Na)SO_4)_{1-2}$ (f) |
| $CsAlSiO_4$ (c) | $[Na_6Al_6Si_6O_{24}](NaReO_4)_2$ (g) | $[(Ca,Na)_6Al_6Si_6O_{24}]((Ca,Na)S,SO_4,Cl)$ |
| $RbAlSiO_4$ (c) | $[Na_6Al_6Si_6O_{24}](NaMnO_4)_2$ (h) | |
| $(Ca_{0.5},Sr_{0.5})AlSiO_4$ (c) | $(NaAlSiO_4)_6(NaBO_4)_2$ (i, j) | |
| $(Sr,Ba)Al_2O_4$ (c) | $Mn_4[Be_3Si_3O_{12}]S$ (d) | |
| $KFeSiO_4$ (c) | $Fe_4[Be_3Si_3O_{12}]S$ (d) | |
| $(Na,Ca_{0.5})YSiO_4$ (h) | $Zn_4[Be_3Si_3O_{12}]S$ (d) | |
| $(Na,K)LaSiO_4$ (h) | | |
| $(Na,K,Ca_{0.5})NdSiO_4$ (h) | | |

(a) Iron, $Ti^{3+}$, Mn, Mg, Ba, Li, Rb, Sr, Zr, Ga, Cu, V, and Yb all substitute in trace amounts in nepheline (Deer et al., 2004).
(b) Higher valent anionic groups such as $AsO_4^{3-}$ and $CrO_4^{2-}$ form $Na_2XO_4$ groups in the cage structure where X= Cr, Se, W, P, V, and As (Barrer, 1982).
(c) Deer et al., 2004
(d) Deer et al., 1963
(e) Brookins 1984
(f) Dana, 1931
(g) Mattigod et al., 2006
(h) Barrer 1982
(i) Buhl et al. 1989
(j) Tobbens and Buhl 2000

Table 4. Cation and Anion Substitution in Feldspathoid Mineral Structures (from Jantzen et al., 2011)

### 3.3 Encapsulating materials

The FBSR granular product will need to be encapsulated in a binder or be contained within high-integrity containers to meet Hanford IDF requirements for compressive strength of 3.4 MPa (500 psi). The compressive strength requirement is driven by the need to prevent subsidence, or sinking, of the disposal facility to maintain surface cap and barrier functionality. Encapsulating the granular product also helps reduce the impact of the dispersible materials in human intrusion scenarios.

Several works have studied the encapsulation of the FBSR granular products with various binders (Jantzen, 2006, 2007; TTT, 2009). Different matrices that have been evaluated as potential binders for FBSR granular product encapsulation have been ordinary Portland cement (OPC), Ceramicrete phosphate bonded ceramic, hydroceramic cements, and several geopolymers. The cement monoliths were prepared with Type II Portland cement and Portland cement plus precipitated silica as a chemically pure representative of a fly ash pozzolanic material. Ceramicrete is a phosphate-based cement developed at Argonne National Laboratory (ANL) (Singh et al., 1997, 1998; Wagh et al., 1997, 1999). It is made from an acidic solution of potassium dihydrogen phosphate ($KH_2PO_4$) and magnesium oxide (MgO). Hydroceramics are prepared through the reaction of a sodium hydroxide solution with metakaolin clay. Under controlled curing conditions, the clay and caustic react to form zeolite mineral phases. Geopolymers are amorphous ceramic-like, inorganic polymers made from the cross-linked three-dimensional structure of aluminosilicates. No zeolite phases form because of the insufficient presence of water. All binders are formed at room temperature. The monoliths must be able to include the FBSR granular bed product material, the starting bed product, HTF fines, and unreacted carbon. Comparisons of some different binder materials can be seen in Table 5.

| Binder Type | Monolith | Waste Loading | Compressive Strength (psi) | Cure Time (days) | Density (g/cc) | BET Surface area (m2/g) |
|---|---|---|---|---|---|---|
| OPC | OPC-1 | 80% | 1,630 | 12 | 1.64 | 31.5 |
| OPC | OPC-2 | 87% | 573 | 28 | 1.61 | 21.3 |
| High Al Cement | FON-1 | 68% | 770 | 7 | 1.77 | 20 |
| High Al Cement | FON-2 | 74.16% | 490 | 7 | 1.75 | 15.5 |
| High Al Cement | S41-1 | 68.60% | 672 | 7 | 1.75 | 10.7 |
| High Al Cement | S41-2 | 74.16% | 340 | 15 | 1.7 | 10.7 |
| High Al Cement | S71-1 | 68.60% | 1,120 | 7 | 1.7 | 13.1 |
| High Al Cement | S71-2 | 74.16% | 550 | 15 | 1.65 | 9.2 |
| Geopolymer | GEO-1 | 67% | 1,510 | 11 | 1.87 | 15.2 |
| Geopolymer | GEO-2 | 72% | 860 | 14 | 1.87 | 17.3 |
| Geopolymer | GEO-3 | 67% | 1,270 | 11 | 1.81 | 10.9 |
| Geopolymer | GEO-4 | 71% | 410 | 11 | 1.84 | 6.2 |
| Geopolymer | GEO-5 | 63% | 950 | 7 | 1.88 | 10.6 |
| Geopolymer | GEO-6 | 66% | 1,080 | 7 | 1.82 | 10 |
| Ceramicrete | CER-1 | 67% | 520 | 8 | 1.81 | 32.2 |
| Ceramicrete | CER-2 | 73% | 550 | 28 | 1.81 | 27.7 |

Table 5. A summary of 2-inch Monolith Cubes (TTT, 2009).

## 4. Physical properties

Table 6 gives data on the density, particle size, and surface area of the FBSR granular product from the early FBSR campaigns with INEEL SBW and the Hanford LAW campaign (Jantzen et al., 2006). The measured BET surface areas were measured in this study on a carbon-free basis (coal removed by roasting), and the value measured by McGrail et al. (2003a) (coal removed manually) was obtained for comparison. The geometric surface area is from Lorier et al. (2005).

| | Density by Pycnometry (g/cm³) | Surface Area by BET (m²/g) | Surface Area Geometric (m²/g) |
|---|---|---|---|
| **SBW SAIC STAR (Marshall et al., 2003)** | | | |
| Bed 260, 272, 277 | 3.30, 3.13, 2.73 | 6.03 | |
| **LAW Saltcake (Rassat blend) SAIC STAR (Olson et al. , 2004a)** | | | |
| Bed 1103 | 2.53 | 4.53 | |
| Bed 1123 | 2.53 | 4.43 | 0.0212[(a)] |
| Fines 1125 | 2.46 | 4.41 | |
| **SBW Hazen (Olson et al., 2004b)** | | | |
| Bed 1173 | 2.76 | 2.36 | 0.0194[(a)] |
| **LAW Envelope C (AN-107) Hazen (Jantzen, 2002; Pareizs et al., 2005)** | | | |
| Bed SCT-02 | 2.66, 2.764[(b)] | 2.37[(b)] | 0.0193 |
| PR-01 Fines | 2.5 | 5.15 | |

(a) Lorier et al., 2005
(b) McGrail et al., 2003b

Table 6. Density and Surface Area of FBSR Granular Products

## 5. Contaminant release mechanisms

A current summary of knowledge regarding the mineralization of radioactive wastes by the FBSR process and a comparison of the durability glass and the FBSR mineral phases has been provided in a previous publication by Jantzen (2008). A basic understanding of FBSR contaminant release mechanisms starts with an understanding of the crystalline product. Mineral waste assemblages formed by FBSR possess short-range order (SRO). SRO in the NAS mineral structures allows contaminants to be trapped in the cage-shaped structures while those external to the cage-shaped structures are bound ionically to oxygen atoms. NAS cage-structured feldspathoid minerals present in FBSR material (mostly sodalite, nosean, and nepheline) are formed by SRO. SRO becomes medium-range order (MRO) through the structures of $(SiO_4)^{-4}$ and $(AlO_4)^{-5}$ tetrahedra, which are joined by sharing one or more of the four oxygen atoms with another tetrahedra. The tetrahedra are arranged to form a cage (sodalite, nosean) or rings (nepheline) via one or two of the tetrahedral oxygen atoms (bridging oxygens), while the other tetrahedral oxygen atoms (non-bridging oxygens) are available to bond ionically with the cations inside or outside the cage. These cations may be alkali or alkaline earths which may be hazardous or radioactive. The cage and/or ring structures are repeated at regular periodicity, which is the long-range order (LRO) characteristic of mineral/crystalline structures. The LRO provides shorter and more regular oxygen-cation (ionic) bonding and a periodic ordering. Glasses do not possess LRO, but they do possess SRO and MRO. Sometimes glasses have more highly ordered regions, referred to as clusters or quasicrystals that have atomic arrangements that approach those of crystals. The ordered regions are metastable compared to crystalline minerals because crystalline species are at their lowest thermodynamic free energy. Therefore, the NAS FBSR mineral structure waste forms are inherently stable.

An explanation of the enhanced durability has been given by Jantzen (2008). According to this author, the dissolution mechanisms (contaminant release mechanisms) of the SRO and MRO in mineral (ceramic) and vitreous waste forms are similar. Mineral waste forms can afford better retention of cationic species compared to glass waste forms due to the LRO of the mineral structure and the regularity of the coordination and bonding of a given coordination polyhedra in which a cation or radionuclide resides. While the activation energy required to break an Si-O, Al-O, B-O bond may be similar in a glass and a ceramic/mineral, due to the SRO, the $(SiO_4)^{-4}$, $(BO_4)^{-5}$, $(BO_3)^{-3}$ and some $(AlO_4)^{-5}$ are more rigidly retained in a mineral structure due to the LRO and periodicity (repeated pattern) of the polyhedra. This author also asserts that in mineral waste forms, as in glass, the molecular structure controls contaminant release by establishing the distribution of ion exchange sites, hydrolysis sites, and the access of water to those sites. It has been demonstrated experimentally that ion exchange in glass occurs along percolation channels that exist in glass. The cations in the percolation channels are ionically bonded to the non-bridging oxygen (NBO) bonds, just as they are in the more ordered crystalline mineral species. In the mineral waste forms there are no percolation channels and dissolution with water must attack the ionically bonded lattice from the surface. The basic difference is that there may be fewer bonds around a given cation in a glass or the bonds may have varying lengths compared to those in a crystalline or mineral waste form.

## 6. Concluding remarks

The primary product from the FBSR process is a granular product composed of NAS minerals. The NAS FBSR granular product is a multiphase mineral assemblage of Na-Al-Si feldspathoid minerals (sodalite, nosean, and nepheline) with cage and ring structures that sequester anions and cations (Jantzen et al., 2007). Nepheline is the basic NAS mineral with the formula $Na_2O$-$Al_2O_3$-$2SiO_2$. When sulfates are captured within the cage structure, nosean forms with the formula $3Na_2O$-$3Al_2O_3$-$6SiO_2 \cdot Na_2SO_4$. When chlorides are captured within the cage structure, sodalite forms with the formula $3Na_2O$-$3Al_2O_3$-$6SiO_2 \cdot 2NaCl$. Depending on the waste compositions, process additives such as magnetite are included to form iron-bearing spinel minerals to sequester Cr and Ni in the waste.

The FBSR waste form may be encapsulated in a binder to minimize dispersability. A number of binders have been evaluated including ordinary Portland cement, high-alumina cements, geopolymers prepared with either kaolin clay or fly ash, various hydroceramic cements and an advanced silicone geopolymer composite material. Characterization data are available on the FBSR granular product prepared with the LAW, WTP-SW, and SBW simulants. This includes some contaminant release studies to support risk assessments and LAW waste form down selections in the early 2000's. Some characterization data is also available on the various binders being evaluated.

### 6.1 Strengths associated with this waste form

Fluidized Bed Steam Reforming (FBSR) offers a moderate temperature continuous method by which LAW and/or WTP-SW wastes can be processed irrespective of whether they contain organics, nitrates, sulfates/sulfides, chlorides, fluorides, volatile radionuclides or other aqueous components. The FBSR technology can process these wastes into a crystalline ceramic (mineral) waste form. The FBSR process also differs from glass or ceramic waste form production in that it is carried out at temperatures ranging from 700 to 750°C whereas glass are ceramics are made from temperatures ranging from 1000 to 1500°C (Jantzen, 2008; Vora et al., 2009; Williams et al., 2010). This makes FBSR an attractive option in secondary liquid waste treatment especially for encapsulation of volatile contaminants of concern.

Monolithing of the granular FBSR product is being investigated to prevent dispersion during transport or burial/storage, which is a regulatory driver and not necessary for the durability and performance of this waste form. The mineral product degrades by the breaking of atomic bonds in the mineral structure in the same fashion as atomic bonds are broken in vitreous waste forms. Therefore, monolith formulation versus durability is considered supplementary since monolith selection is based on the scenario that the monolith will not compromise the mineral product durability.

The FBSR process has been demonstrated at a non-radioactive scale using simulants for Hanford LAW wastes and Idaho SBW. The FBSR process has been demonstrated with a secondary waste stream (WTP-SW) from the WTP based on an early LAW scenario in which the LAW melter submerged bed scrubber and wet electrostatic precipitator condensates are sent from the WTP as a secondary liquid stream for treatment. A testing program is currently underway using a bench-scale steam reformer and actual tank wastes.

Research teams from SRNL and PNNL are currently working to perform different tests with the FBSR materials. Some of the major findings and results collected so far and are summarized below:

- Data indicates $^{99}$Tc, Re, Cs, and I (all isotopes) report to the mineral product and not to the off-gas
- $^{99}$Tc and Re show similar behavior in partitioning between product and off-gas
- $^{99}$Tc, Re, $SO_4$ and Cr behavior are controlled by the oxygen fugacity in the FBSR/BSR process, i.e. control of the REDuciton/OXidation (REDOX) equilibrium
- XAS results show that Re is in the 7+ oxidation state and contained in the sodalite structure
- Re is a good surrogate for $^{99}$Tc during leaching experimentation proving that the current radioactive and simulant BSR campaign products using Re and $^{99}$Tc match the historic and engineering scale data that used Re only proving the "tie back" strategy
- All monoliths made from radioactive and non-radioactive granular products pass compression testing at >3.4 MPa (500 psi), maintain PCT leach rates <2 g/m$^2$, and perform well in ASTM C1308 (ASTM, 2008b) and ANSI 16.1 (ANS, 2008) testing indicating that the binder material is not degrading the granular product durability response.
- In order to match the Bench-Scale Reformer (BSR) REDOX to the Engineering Scale Technology Demonstration (ESTD) REDOX, the addition of reductants such as coal and control of gas inputs were adjusted during production: a more rigorous REDOX control strategy needs to be developed to ensure the COCs are in the correct oxidation states.

### 6.2 Sparse data and unresolved issues

Though the FBSR products have been studied over the last years, there are still some areas that remain with sparse data and unresolved issues. This is in contrast with glasses that have been studied as confinement materials for more than 30 years. More durability testing of all FBSR products using both the Single Pass Flow-Through (SPFT) and Product Consistency Test (PCT) methods to compare to various glasses are needed to better understand contaminant release mechanisms. The physical characteristics of the FBSR material must also be studied. This includes a better standing of the effective surface area available for leaching by water. Further research needs to be conducted to determine the preferred host mineral phase for different radioactive materials and COCs under oxidizing and reducing conditions. Data must be obtained regarding the impacts of radiation, biodegradation, and water immersion on the compressive strength of the FBSR granular product encapsulated in any of the binder materials being considered for compression tests. Methods to consolidate the FBSR granular product into monolith waste forms are being investigated (Jantzen, 2007; TTT, 2009). The porosity and void volume of these materials on an engineering or production scale are, as yet, unknown. More research also needs to be conducted concerning the fabrication of the binder.

## 7. References

ANS—American Nuclear Society (2008). Measurement of the Leachability of Solidified Low-Level Radioactive Wastes by a Short-Term Test Procedure. *ANSI/ANS-16.1-2003; R2008,* American Nuclear Society, La Grange Park, Illinois.

ASTM—American Society for Testing and Materials (2008a). Standard Test Method for Determining the Chemical Durability of Nuclear Waste Glasses: The Product Consistency Test (PCT). *ASTM C1285 - 02(2008),* American Society for Testing and Materials, West Conshohocken, Pennsylvania.

ASTM—American Society for Testing and Materials (2008b). Standard Test Method for Accelerated Leach Test for Diffusive Releases from Solidified Waste and a Computer Program to Model Diffusive, Fractional Leaching from Cylindrical Waste Forms. *ASTM C1308-08,* American Society for Testing and Materials, West Conshohocken, Pennsylvania.

Barrer, R.M. (1982). *Hydrothermal Chemistry of Zeolites.* Academic Press, ISBN: 0120793601, New York

Barrer, R.M, Cole, J.F., & Sticher. H. (1968). Chemistry of Soil Minerals. Part V, Low Temperature Hydrothermal Transformations of Kaolinite. *Journal of the Chemical Society A: Inorganic, Physical, Theoretical,* pp. 2475-2485

Baumann, E.W. (1992). Colorimetric Determination of Iron(II) and Iron(III) in Glass. *Analyst,* Vol. 117, No. 5, pp. 913-916

Brunauer, S., Emmett, P.H., & Teller, E. (1938). Adsorption of Gases in Multimolecular Layers. *Journal of the American Chemical Society,* Vol. 60, pp. 309-319

Buhl, J.C, Engelhardt, G., & Felsche, J. (1989). Synthesis, X-Ray Diffraction, and MAS n.m.r. Characteristics of Tetrahydroxoborate Sodalite, $Na_8[AlSiO_4]_6[B(OH)_4]_2$. *Zeolites,* Vol. 9, No. 1, pp. 40-44

Bullock Jr, J.H., Cathcart, J.D., & Betterton. W.J. (2002). Analytical methods utilized by the United States Geological Survey for the analysis of coal and coal combustion by-products. *Open-File Report 02-389.* U. S. Geological Survey, Denver, Colorado.

Burket, P.R., Daniel, W.E., Nash, C.A., Jantzen, C.M. & Williams, M.R. (2008). Bench-Scale Steam Reforming of Actual Tank 48H Waste. *SRNS-STI-2008-00105,* Savannah River National Laboratory, Aiken, South Carolina.

Crawford, C.L. & Jantzen, C.M. (2007). Durability Testing of Fluidized Bed Steam Reformer (FBSR) Waste Forms for Sodium Bearing Waste (SBW) at Idaho National Laboratory (INL). *WSRC-STI-2007-00319,* Savannah River National Laboratory, Aiken, South Carolina.

Dana, E.S. (1932). *A Textbook of Mineralogy,* ISBN: 9780471193050, John Wiley & Sons, Inc., New York

Deer, W.A., Howie, R.A., & Zussman, J. (1963). *Rock-Forming Minerals.* ISBN: 9781862392595, John Wiley & Sons, Inc., New York

Deer, W.A., Howie, R.A., Wise, W.S., & Zussman, J. (2004). *Rock-Forming Minerals, Framework Silicates: Silica Minerals, Feldspathoids and the Zeolites.* ISBN: 9781862391444, The Geological Society, London

Fleet, M.E. (1989). Structures of sodium alumino-germanate sodalites. *Acta Crystallographica C: Crystal Structure Communications C,* Vol. 45, No. 6, pp. 843-847

Hassan, I. & Gruncy, H.D. (1984). The Crystal Structures of Sodalite-Group Minerals. *Acta Crystallography Section B,* Vol. 40, No. 1, pp. 6-13

Jantzen, C.M. (2002). Engineering Study of the Hanford Low Activity Waste (LAW) Steam Reforming Process (U). *WSRC-TR-2002-00317, Rev. 1,* Savannah River National Laboratory, Aiken, South Carolina.

Jantzen, C.M. (2006). Fluidized Bed Steam Reformer (FBSR) Product: Monolith Formation and Characterization. *WSRC-STI-2006-00033,* Savannah River National Laboratory, Aiken, South Carolina.

Jantzen, C.M. (2007). Fluidized Bed Steam Reformer (FBSR) Monolith Formation, *Proceedings of Waste Management 2007,* Tucson, Arizona, March 2007.

Jantzen, C.M. (2008). Mineralization of Radioactive Wastes by Fluidized Bed Steam Reforming (FBSR): Comparisons to Vitreous Waste Forms, and Pertinent Durability Testing. *WSRC-STI-2008-00268,* Savannah River National Laboratory, Aiken, South Carolina.

Jantzen, C.M & Crawford, C.L. (2010). Mineralization of Radioactive Waste Wastes by Fluidized Bed Steam Reforming (FBSR): Radionuclide Incorporation, Monolith Formation, and Durability Testing. *Proceedings of Waste Management 2010,* Phoenix, Arizona, March 2010.

Jantzen, C.M., Pareizs, J.M., Lorier, T.H., & Marra, J.C. (2005). Durability Testing of Fluidized Bed Steam Reforming (FBSR) Products (U). *Proceedings of Waste Management Technologies in Ceramic and Nuclear Industries,* American Ceramic Society, Westerville, Ohio.

Jantzen, C.M., Lorier, T.H., Marra, J.C., & Pareizs, J.M. (2006). Durability Testing of Fluidized Bed Steam Reforming (FBSR) Waste Forms. *Proceedings of Waste Management 2006.* Tucson, Arizona, February 2006

Jantzen, C.M., Lorier, T.H., Pareizs, J.M., & Marra, J.C. (2007). Fluidized Bed Steam Reformed (FBSR) Mineral Waste Forms: Characterization and Durability Testing. In: *Scientific Basis for Nuclear Waste Management XXX,* eds. Begg, B., Dunn, D. S., Poinssot, C., 985: pp. 379-386. Materials Research Society, Warrendale, Pennsylvania.

Jantzen, C.M., Crawford, C.L., Burket, P.R., Daniel, W.G. Cozzi, A.D., & Bannochie, C.J. (2011). Radioactive Demonstrations of Fluidized Bed Steam Reforming (FBSR) as a Supplemental Treatment for Hanford's Low-Activity Waste (LAW) and Secondary Wastes (SW). *Proceedings of Waste Management 2011,* Phoenix, Arizona, February 2011

Klingenberg, R. & Felsche, J. (1986). Interstitial Cristobalite-Type Compounds $(Na_2O)_{0.33}Na[AlSiO_4]$). *Journal of Solid State Chemistry,* Vol. 61, No. 1, pp. 40-44.

Kyritsis, K., Meller, N. & Hall, C. (2009). Chemistry and Morphology of Hydrogarnets formed in Cement-Based CASH Hydroceramics Cured at 200° to 350°C. *Journal of the American Ceramic Society*, Vol. 92, No. 5, pp. 1105-1111

Lockrem, L. L. (2005). Hanford Containerized Cast Stone Facility Task 1 - Process Testing and Development Final Test Report, *RPP-RPT-26742, Rev. 0,* CH2M HILL Hanford Group, Inc., Richland, Washington.

Lorier, T.H., Pareizs, J.M., & Jantzen. C.M. (2005). Single-Pass Flow Through (SPFT) Testing of Fluidized-Bed Steam Reforming (FBSR) Waste Forms. *WSRC-TR-2005-00124,* Savannah River National Laboratory, Aiken, South Carolina.

Marshall, D.W., Soelberg, N.R., & Shaber, K.M. (2003). THOR® Bench-Scale Steam Reforming Demonstration. *INEEL/EXT-03-00437,* Idaho National Engineering and Environmental Laboratory, Idaho Falls, Idaho.

Mason, J.B., McKibbin, J., Ryan, K., & Schmoker, D. (2003). Steam Reforming Technology for Denitration and Immobilization of DOE Tank Wastes. *Proceedings of Waste Management 2003,* Tucson, Arizona, February 2003

Mattigod, S.V., McGrail, B.P., McCready, D.E., Wang, L.-Q., Parker, K.E., & Young, J.S. (2006). Synthesis and Structure of Perrhenate Sodalite. *Microporous and Mesoporous Materials,* Vol. 91, No. 1-3, pp. 139-144

McGrail, B.P., Pierce, E.M., Schaef, H.T., Rodriguez, E.A., Steele, J.L., Owen, A.T., & Wellman, D.M. (2003a). Laboratory Testing of Bulk Vitrified and Steam Reformed Low-Activity Waste Forms to Support a Preliminary Risk Assessment for an Integrated Disposal Facility. *PNNL-14414,* Pacific Northwest National Laboratory, Richland, Washington.

McGrail, B.P., Schaef, H.T., Martin, P.F., Bacon, D.H., Rodriguez, E.A., McCready, D.E., Primak, A.N,. & Orr, R.D. (2003b). Initial Evaluation of Steam-Reformed Low Activity Waste for Direct Land Disposal. *PNWD-3288,* Pacific Northwest Division, Richland, Washington.

Nimlos, M.R. & Milne, T.A. (1992). Direct Mass-Spectrometric Studies of the Destruction of Hazardous Wastes. 1. Catalytic Steam Re-Forming of Chlorinated Hydrocarbons. *Environmental Science & Technology,* Vol. 26, No. 3, pp. 545-552

Olson, A.L., Soelberg, N.R., Marshall, D.W., & Anderson, G.L. (2004a). Fluidized Bed Steam Reforming of Hanford LAW Using THOR® Mineralizing Technology. *INEEL/EXT-04-02492,* Idaho National Engineering and Environmental Laboratory, Idaho Falls, Idaho.

Olson, A.L., Soelberg, N.R., Marshall, D.W., & Anderson, G.L. (2004b). Fluidized Bed Steam Reforming of INEEL SBW Using THOR® Mineralizing Technology, *INEEL/EXT-04-02564,* Idaho National Engineering and Environmental Laboratory, Idaho Falls, Idaho.

Olson, A.L., Soelberg, N.R., Marshall, D.W., & Anderson, G.L. (2005). Mineralizing, Steam Reforming Treatment of Hanford Low-Activity Waste. *Proceedings of Waste Management 2005,* Tucson, Arizona, February 2005

Pareizs, J.M., Jantzen, C.M., & Lorier, T.H. (2005). Durability Testing of Fluidized Bed Steam Reformer (FBSR) Waste Forms for High Sodium Wastes at Hanford and Idaho (U). *WSRC-TR-2005-00102,* Savannah River National Laboratory, Aiken, South Carolina.

Russell, R.L., Schweiger, M.J., Westsik Jr, J.H., Hrma, P.R., Smith, D.E., Gallegos, A.B., Telander, M.R., & Pitman, S.G. (2006). Low Temperature Waste Immobilization

Testing. *PNNL-16052 Rev 1*, Pacific Northwest National Laboratory, Richland, Washington.

Ryan, K., Mason, J.B., Evans, B., Vora, V., & Olson, A. (2008). Steam Reforming Technology Demonstration for Conversion of DOE Sodium-Bearing Tank Waste at Idaho National Laboratory into a Leach-Resistant Alkali Aluminosilicate Waste Form. *Proceedings of Waste Management 2008*, Phoenix, Arizona, February 2008

Schreiber, H.D. (2007). Redox of Model Fluidized Bed Steam Reforming Systems Final Report Report Subcontract AC59529T, VMI Research Laboratories, Lexington, Virginia.

Singh, D., Wagh, A.S., Cunnane, J.C., & Mayberry, J.L. (1997). Chemically Bonded Phosphate Ceramics for Low-Level Mixed-Waste Stabilization. *Journal of Environmental Science and Health Part A-Environmental Science and Engineering & Toxic and Hazardous Substance Control*, Vol. 32, No. 2, pp. 527-541

Singh, D., Wagh, A.S., Tlustochowicz, M., & Jeong, S.Y. (1998). Phosphate Ceramic Process for Macroencapsulation and Stabilization of Low-Level Debris Wastes. *Waste Management*, Vol. 18, No. 2, pp. 135-143

Soelberg, N.R., Marshall, D.W., Bates, S.O., & Taylor, D.D. (2004a). Phase 2 THOR® Steam Reforming Tests for Sodium-Bearing Waste Treatment. *INEEL*/EXT-04-01493, Idaho National Engineering and Environmental Laboratory, Idaho Falls, Idaho.

Soelberg, N.R., Marshall, D.W., Bates, S.O., & Siemer, D.D. (2004b) SRS Tank 48H Steam Reforming Proof-of-Concept Test Report, *INEEL/EXT-03-01118, Rev 1*, Idaho National Engineering and Environmental Laboratory, Idaho Falls, Idaho.

Spence, R.D., & Shi, C. (2005). *Stabilization and Solidification of Radioactive and Mixed Waste*. CRC Press, ISBN: 9781566704441, Boca Raton, Florida

TTT. (2009). Report for Treating Hanford LAW and WTP SW Simulants: Pilot Plant Mineralizing Flowsheet. *RT-21-002*, THOR Treatment Technologies, LLC, Denver, Colorado.

Vora, V., Olson, A., Mason, B., Evans, B., & Ryan, K. (2009). Steam Reforming Technology Demonstration for Conversion of Hanford LAW Tank Waste and LAW Recycle Waste into a Leach Resistant Alkali Aluminosilicate Waste Form. *Proceedings of Waste Management 2009*, Phoenix, Arizona, March 2009

Wagh, A.S., Strain, R. Jeong, S.Y., Reed, D., Krause, T., & Singh, D. (1997). Stabilization of Rocky Flats Pu-Contaminated Ash within Chemically Bonded Phosphate Ceramics. *Journal of Nuclear Materials*, Vol. 265, No. 3, pp. 295-307

Wagh, A.S., Jeong, S.Y., & Singh, D. (1997). High Strength Phosphate Cement Using Industrial Byproduct Ashes. In: *Proceedings of First International Conference*, ed. A. Azizinamini et al., American Society of Civil Engineers. pp. 542-553

Williams, M.R., Jantzen, C.M., Burket, P.R., Crawford, C.L., Daniel, W.E., Aponte, C., & Johnson, C. (2010). 2009 Pilot Scale Fluidized Bed Steam Reforming Testing Using the THOR® (THermal Organic Reduction) Process: Analytical Results for Tank 48H Organic Destruction. *Proceedings of Waste Management 2010*, Phoenix, Arizona, October 2010

12

# Experimental Verification of Solidification Stress Theory

Charles Solbrig, Matthew Morrison and Kenneth Bateman
*Idaho National Laboratory*
*USA*

## 1. Introduction

Two Ceramic Waste Forms (CWFs) have been formed in this work without the benefit of an analysis tool that could stop large scale cracking that occurs during solidification. Both showed severe cracking. This chapter describes a new theory for a stress not modeled before, termed here, the solidification stress. This stress is set-into a glass or ceramic cylinder being formed during the time period of solidification. It is due to the temperature gradient existing during solidification. This stress is in addition to the normal thermal stress calculated in a solid due to a temperature gradient.

This chapter describes 1) how this stress can be controlled to prevent damage, 2) the methods available to measure this stress, and 3) the significant damage which occurred during the formation of the two large ceramic cylinder prototype high level nuclear waste forms, and 4) measurements made during CWF2 that verified the theory can predict the conditions under which cracking occurs.

This research program is being conducted to develop a crack-free ceramic waste form (CWF) to be used for long term encasement of fission products and actinides resulting from electrorefining of spent nuclear fuel. A crack-free waste form should have more resistance to leaching than one with many cracks. The fission products are deposited in electrofiner electrolyte salt as a byproduct of the removal of uranium and actinides process during reprocessing of spent nuclear fuel. The encasement is accomplished by absorbing the radioactive salts into zeolite, mixing the zeolite-salt mixture with glass frit in a stainless steel cylindrical can, heating to a temperature range (600 °C) where consolidating and melting take place, then further heating up to a completely molten state at 915 °C causing the zeolite to convert to sodalite glass matrix, and then the sodalite glass matrix is solidified by cooling it through solidification (~625 C), then further cooling to near ambient temperature. If cracking occurs which is usually the case, it will occur during the cooldown phase which is a detriment to long term encasement.

In this research, a model was developed that proposes a permanent stress develops, called solidification stress, when the melt solidifies and that this stress, if large enough, will cause failure as the CWF nears room temperature. This stress is proportional to the rate of cooling during solidification. This stress is in addition to the thermal stress which develops during rapid cooldown of a solid. Two CWF's have been formed in this research, both of which encountered severe cracking. Sufficient data was recorded on the second to test the theory developed. Recording of temperatures and the cracking sounds during CWF2 cooldown shows that the cracking is from this newly recognized stress and not the thermal stress. The theory was provisionally verified on two small scale experiments and were reported on in Solbrig and Bateman, 2010. A third CWF formation is planned that is predicted by the theory to be crack free if the specification is followed.

The solidification stress is of opposite sign of the thermal stress and remains constant after solidification. Its derivation is reported on in Solbrig and Bateman (2010) and summarized in this chapter. The theory predicts that cracking of the CWF would occur at low temperatures if caused by solidification stress but at high temperatures (somewhat below the solidification temperature) if caused by thermal stress. To reduce solidification stress, the cooldown rate during solidification should be reduced. Recording cracking sounds confirm the existence of this solidification stress since cracking occurred during the low temperature phase of the cooldown. A cooldown rate history is proposed that should eliminate cracking in the next CWF formed.

CWF2 is a prototype vertical ceramic waste cylinder formed over a period of 10 days by heating a mixture of 75% zeolite, 25% glass frit in an argon atmosphere furnace through melting to 925 C and then cooling through solidification to room temperature. It is approximately 1 m high, 0.5 m in diameter, weighs about 400 kg, and is formed in a stainless steel can 0.5 cm thick. This cylinder developed many cracks on cooldown. At least 15 loud cracks were recorded over a period of 4 days at the end of cooldown when the temperatures were below 400 C, the last being after the CWF was removed from the furnace when the surface temperature was below 100 °C.

The CWF2 surface and centerline temperatures at mid height were recorded which allowed the stresses to be calculated. The timing of the cracks was compared to the time the calculated total stress exceeded the tensile stress limit and verified that the cause of the cracking was solidification stress and not thermal stress. Since the CWF is encased in a can in a furnace, the cracks cannot be easily observed but can be detected with sound measurements. Similarly, the stress cannot be measured but only estimated with analysis. Destructive examination of the CWF after cooldown was used to show the large amount of the cracking which occurred. It appeared to be initiated mainly in the inner region which is further evidence the cracking is due to solidification stress since solidification stress is tensile in the inner region and thermal stress is compressive in the inner region.

## 2. Theoretical background

The first part of this section describes some of the significant experimental work which has been reported on in the literature. The second part of this section summarizes the theory developed in this work which describes why the solidification stress is formed during solidification and how the solidification stress causes the CWF to form significant cracks near room temperature as they cool down.

## 2.1 Background experimental work

Faletti and Ethridge (1986) summarized the work done by several investigators who formed full-size waste form cylinders and supplied enough information for analysis of the results. The cylinder sizes were 50 to 60 cm in diameter and 150 to 300 cm in length and were either ceramic or glass. This is the size of the cylinders scheduled to be stored in U.S. Department of Energy (DOE) waste storage containers. Glass or ceramics have the characteristic of failing at their elastic limit. Due to nature of ceramics and glass, they crack to relieve tension instead of yield. Even though they are contained in a stainless steel can 0.5 cm thick, it is assumed that the can will rust away in a hundred years or so that the CWF will be exposed to the environment after this.

The Pacific Northwest Laboratories (PNL) report shows the cross section of a 60-cm-diam cylinder that illustrated considerable stress damage that occurred during cooling. The report goes on to state that all cylinders show similar damage during formation. One of the few known cases of crack-free waste forms were 15 cm (6 in.) formed by the continuous melt process, as indicated by Slate et al.(1978). Faletti and Ethridge (1986) state that cooling can proceed at any practical rate until the glass reaches its annealing point of 500 to 550 °C. Further cooling would have to proceed on the order of 1 °C/h implying nearly 3 weeks to cool a 60-cm-diam (24 in.) canister.

The formation of the cylinders reported on here is different than the PNL cylinders since the zeolite glass mixture consolidates as it heats up. A heavy weight is placed on the CWF as it heats up to compress it and keep voids from forming during consolidation. By the time it is ready for cooling, the zeolite has reacted to sodalite and has consolidated by a factor of more than 2.

## 2.2 Solidification stress model

For the convenience of the reader, the equation development from Solbrig and Bateman (2010) is summarized in this section. Only the axial stresses are modeled. The circumferential and axial stresses are equal on the circumference at the-plane so considering one of them is approximately equivalent to considering both. The radial stresses are small.

**Thermal Stress:** To understand the solidification stress model, it is first necessary to review how thermal stresses are modeled that are induced in a solid by a temperature distribution in a non prestressed cylindrical solid. This development is consistent with Timoshenko and Goodier (1970). Non prestressed means that there is no residual stress when the solid is at a uniform temperature. The cylinder is modeled as a series of concentric annuli. Each cylindrical annulus is intrinsically attached to the ones on either side of it. Each annulus temperature is uniform in the axial direction but the temperature of each is different in the radial direction. The length of the cylinder at any time is determined by length predicted using the Coefficient of thermal Expansion (CTE) based on the average temperature of the cylinder. Due to the intrinsic attachment of the annuli, the length of each is forced to be equal to the average length. Thus those annuli with temperatures above ($T_{avg}$) are forced to be shorter in length than their unattached thermally expanded length should be so they are in compression and those below $T_{avg}$ are forced to be longer so are in tension.

Thus, the amount that annulus i must be elongated to reach the average length, L, is

$$\begin{aligned} &\Delta L_i = L\alpha(T_{avg} - T_i) \\ &where\ \alpha = thermal\ \text{expansion}\ coefficient \\ &\quad T_i = Temperature\ of\ annulus\ \ i \end{aligned} \tag{1}$$

The stress, $\sigma_i$, induced in an annulus elongated by a length, $\Delta L_i$, where a positive number signifies tension, is

$$\begin{aligned} &\sigma_i = \frac{E}{1-\mu}\frac{\Delta L_i}{L} \\ &where\ \ E = Young's\ Modulus \\ &\quad \mu = Poisson's\ ratio \end{aligned} \tag{2}$$

These two equations are combined to get the axial thermal stress at the mid-plane at any r as

$$\sigma_i = \frac{E\alpha}{1-\mu}(T_{avg} - T_i) \tag{3}$$

This equation indicates that tensile stress in a cooling cylinder is largest at the surface. It should be noted that the circumferential surface stress is the same as the axial. So to get the magnitude of the total stress on the surface, this equation should be multiplied by the square root of two. For simplicity, in this chapter, the equivalent stress limit is applied in the axial direction only.

**Solidification Stress is a Set-in Stress:** A pre-stress is defined as a stress that exists when the temperature is uniform. A pre-stress induced by solidification is referred to here as a set-in stress. The total stress at any location is equal to the set-in stress plus the thermal stress. The stress which will cause the CWF to crack occurs during the cooldown. The cooling process starts from the highest temperature attained in the formation process at 925 °C where the CWF is all liquid and has no stress. As it is cooled, it eventually begins to effectively solidify at $T_s$ (around 625 °C) where the viscosity becomes very large. During the cooling, the surface temperature is the lowest T because it is being cooled from the outside. Therefore, in terms of the model, the outside annulus solidifies first, then the second one in solidifies, followed by the third, etc. Each annulus is the same length when it solidifies $L(T_s)$. When the second annulus inward solidifies, the outer is shorter due to thermal contraction of the outside annulus as is has cooled below $T_s$. The two annuli are intrinsically connected along length $L(T_0)$ where $T_0$ is the temperature of the outer cylinder when the second one in inward solidifies. Therefore, these two cylinders are intimately connected along the length $L(T_0)$. The length not connected between these two annuli is defined here as the Length deficit, $L(T_s)$-$L(T_0)$.

As each successive cylinder solidifies, the solid one outward from it is shorter than it is. The length deficit for each annulus is the length not connected to the next outer annulus. When the whole cylinder reaches room temperature, the length deficits add up to form a dome shape. The dome height is larger with higher cooling rates. The length deficit causes a set-in stress when the temperature becomes uniform because all of the annuli are forced to be the same length. This stress is defined here as the solidification stress.

Figure 1 shows an exaggerated drawing of the total length deficit as a function of radius where the centerline total length deficit is shown at the left and the surface at the right.. The length deficits are small and would be difficult to see if the total annulus length were drawn instead of just the deficit. The total length deficit at any radius, r, is the sum of the individual length deficits from the outside and summing inward to radius, r. Each annulus is attached to the annulus on its inside but not along the length deficit.

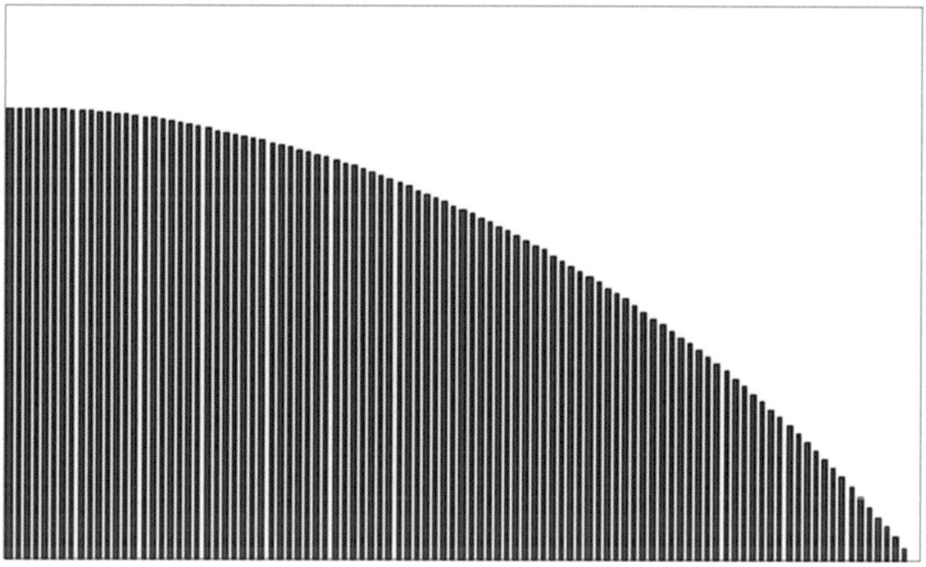

Fig. 1. Length Deficits.

The total length deficit at any r is

$$L_d = \int_{R_O}^{r} \frac{dL}{dr}\Big|_{S^+} dr \tag{4}$$

where the derivative is evaluated on the outer side of solidification front, $s^+$, as the solidification front moves inward.

In a manner similar to the length deficit, a temperature deficit is defined as

$$\Delta T_i = T_s - T_{i+1} \tag{5}$$

where the annulus at $r_i$ has just solidified, and $T_{i+1}$ is the Temperature of the annulus at $r_{i+1}$ at that time.

The total temperature deficit at any r is defined as

$$T_d = \int_{R_O}^{r} \frac{dT}{dr}\Big|_{S^+} dr \tag{6}$$

where the derivative is evaluated on the outer side of the solidification front, $s^+$ as the solidification from moves inward.

The temperature deficit is related to the length deficit by coefficient of thermal expansion as presented in equation 1. Substituting the length deficit into the equation for axial stress results in an equation for the set-in stress in terms of the temperature deficit distribution as:

$$\sigma_{\Delta i} = -\frac{E\alpha}{1-\mu}(\Delta T_{avg} - \Delta T_i) \tag{7}$$

The negative sign in front of the term on the right side of this equation is due to the fact that the change in the connective length is the negative of the change in the length deficit.

Note that both the thermal and solidification stresses are proportional to the CTE. Since the tensile stress at the failure limit is also calculated using the CTE, the stress limit is proportional to CTE as well.

## 3. Large scale experimental results – Verification of theory

CWF2 is a prototype vertical ceramic waste cylinder formed over a period of 10 days by heating a mixture of 75% zeolite, 25% glass frit in an argon atmosphere furnace through melting to 925 C and then cooling through solidification to room temperature. It is approximately 1 m high, 0.5 m in diameter, weighs about 400 kg, and is formed in a stainless steel can 0.5 cm thick. This cylinder developed many cracks on cooldown. At least 15 loud cracks were recorded over a period of 4 days at the end of cooldown when the temperatures had decreased below 400 C, the last occurred after the CWF was cool enough that it had been removed from the furnace. This section describes the results of this test.

### 3.1 Methods used to estimate stress

Stress cannot be measured during or after cooldown. Cracks cannot be visually observed as they occurr because the CWF is encased in a stainless steel can and is in a furnace. Two methods used here to assess the stress: sound recordings and destructive examination. Although the cracking cannot be seen, the timing cracks can be detected with sound measurements. The CWF2 surface and centerline temperatures at mid height were measured which allowed the stresses caused by these temperature histories to be calculated using the theory. The timing of the cracks was compared to the time the calculated total stress exceeded the tensile stress limit and verified that the cause of the cracking was solidification stress and not thermal stress. So although the stress in the hot CWF cannot be measured, it can be estimated with the theory presented in this chapter. In order to know if the stress calculated will cause damage, it is necessary to know the failure limit. Subsection 3.2 discusses the tensile failure stress for the INL CWF and the method of measuring it. Subsection 3.3 presents the data obtained for the cracking sound recordings which were used to verify the theory. The next section (3.4) describes the temperature data that was used to estimate the stresses. Section 3.5 then presents the stresses calculated from these temperatures and determines the time of failure and compares these times to the times at which cracking sounds were recorded. Destructive examination of the CWF after cooldown (described in Subsection 3.6) determined the location of cracking. It was initiated mainly in the inner region which is evidence the cracking is due to solidification stress.

### 3.2 Tensile stress limit measurements in CWF2

The tensile failure stress and the coefficient of thermal expansion (CTE) values in the above analysis have been estimated at 82.4 mpa (12000 psi) and $45x10^{-6}/°C$. The stress limit used is based experiments done on glass cylinders, Bateman and Solbrig (2008)a&b. The CTE was measured on CWF surrogates, Bateman and Capson (2003). The work described here is the first attempt to measure the tensile failure stress of the INL CWF and was made on CWF2 formed material. It was cut out of a piece removed from CWF2 at about the 1 foot high level. This region had cracked into many pieces during formation. The removed piece was a pie shaped wedge of 75 degrees, a radius of 25.4 cm (10 in), and a depth of about 0.94 cm (2.375 in.). A rectangular beam which was 0.34 cm (0.875 in) thick and 0.94 cm (2.375 in) wide was cut out of the wedge by the "A Core" Company. Due to the difficulty of cutting the material without breaking it, the beam was only about 8 inches long with uneven ends.

The method of determining the failure stress was to install the beam as a cantilever and apply a sufficient load on the cantilever to cause failure. In order to apply enough moment to reach the failure stress a metal extension was attached to the beam as shown in Figure 2.

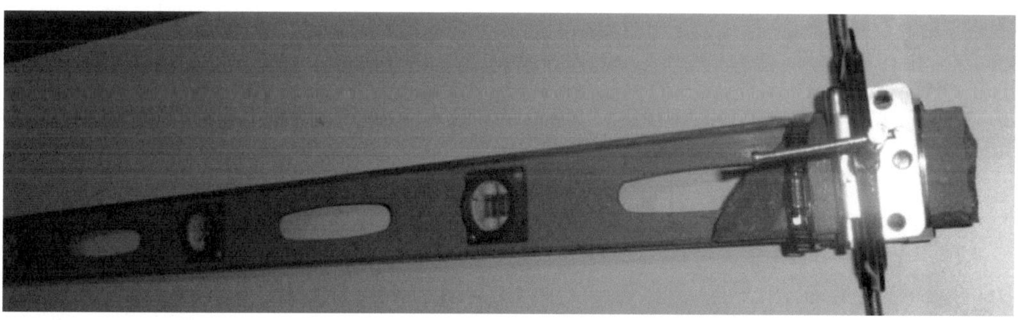

Fig. 2. Metal Extension attached to the CWF2 Beam

The exposed part of the CWF beam (about 2.54 cm ,1 in) was inserted into a vertically wall mounted vise as shown in Figure 3. The metal extension was 1.81 m (6 ft) long with a center of gravity of 0.9 m (3 ft) and clamped the CWF material to provide a cantilever. Weights were placed on the beam at the center of gravity (Figure 3). Weights were added until failure.

Fig. 3. Weight placed on the beam 0.9 m (3 ft) from the vise

The first two pieces of the beam that broke off in the first two tests are shown in Figure 4. The longer one on the right was the first one broken off. The break planes were at about 5 degrees less than a 90 degree break.

Fig. 4. Second (left) and First (right) Failed Sections

The maximum horizontal tensile stress for a given load can be calculated with an equation derived in Timoshenko and Mcaugh (1949). The stress calculated with the load at failure is the tensile failure stress limit. The maximum occurs at the top surface at the cantilever connection with the beam loaded at the end. The maximum tensile stress at the wall with the beam anchored as a cantilever is

$$\begin{aligned}&Stress = \frac{Mc}{I}\\&where\ M = Total\ Moment\ about\ the\ cantilever\ support = L * P\\&L = Length\ of\ the\ beam,\ P = total\ force\ on\ the\ end\ of\ the\ beam\\&c = half\ the\ height\ of\ the\ beam = h/2,\ h = height\ of\ the\ beam\\&I = Vertical\ moment\ of\ inertia = \frac{b * h^3}{12},\ b = width\ of\ beam\end{aligned} \tag{8}$$

Substituting the moment of inertia and moment into stress equation A.1 yields

$$Stress = \frac{6PL}{bh^2} \tag{9}$$

Three tests were run. The first three were successful and showed a range of values for failure tensile stress. The fourth may have been too short to get a realistic test. It seemed to crumble rather than break. The measurements are summarized in Table.1. The second and third tests are deemed to be the most accurate with the load reported for Test 2 being somewhat less than that which caused failure and the load for Test 3 being somewhat greater than the failure load. The accuracy of the load causing failure was limited by the use of discrete weights for the loading rather than a continuous loading device and the lack of additional samples to measure.

| Date | Test | Failure load kg (lb) | Beam Weight kg (lb) | Total Load kg (lb) | Lever arm m (in) | Comment |
|---|---|---|---|---|---|---|
| 6/21/10 | 1 | 4 (8.8) | 1.36 (3) | 5.36(11.8) | 0.9(36) | First test. |
| 6/22/10 | 2 | 6 (13.2) | 1.36 (3) | 7.36(16.2) | 0.9(36) | Load <than failure |
| 6/23/10 | 3 | 8 (17.6) | 1.36 (3) | 9.36(20.6) | 0.9(36) | Load >than failure |
| 6/24/10 | 4 | 2 (4.4) | 1.36 (3) | 3.36 (7.4) | 0.9(36) | Crumbled, short |

Table 1. Results of CWF2 Failure Measurements

The failure stress was calculated for the above experiments (Table.2). Neglecting the fourth test, the results show a failure yield stress of between between 9.66 mpa (1402 psi) and 16.9 mpa (2447 psi). The first test is a bit questionable, so a value of 15.2 mpa (2200 psi is reasonable. The last line was added to the table which shows that the beam would have had to support over 100 lbs to have a yield stress of 12000 psi. This is over 5 times as great as the loads which actually caused failure.

| Test | Total Load kg (lb) | Failure Stress Mpa (psi) |
|---|---|---|
| 1 | 5.36 (11.8) | 9.6 (1402) |
| 2 | 7.36 (16.2) | 13.2 (1924) |
| 3 | 9.36 (20.6) | 16.9 (2447) |
| 4 | 3.36 (7.4) | 6.1 (879) |
| 12 k Load | 45.9 (101) | 82.4 (12000) |

Table 2. Failure Stresses Calculated for the Above Failure Loads

Thus, the testing has shown that the tensile yield stress of the sample removed from CWF2 is about 15.2 mpa (2200 psi). This is much lower than the 82.4 mpa (12000 psi) used in the model. However, the calculation of the stress is proportional to the Coefficient of Thermal Expansion. The CTE value used in the model is quite high for a ceramic, 45 x $10^{-6}$ /C, and was chosen in order to agree with measured data on contraction during cooldown, Bateman and Capson (2003). . It is the ratio of the CTE and the stress limit that is the actual criterion in determining failure If the CTE is actually lower on the order of what is usual for a glass or ceramic of about 9 x $10^{-6}$ /C, then CWF2 would have failed when the total stress reached 16.5 mpa (2400 psi) which is close to the above measured value.

One other factor which should be considered is the actual state of the sample tested. It was taken from the outside region of the CWF which would have been in compression during cooldown before failure. There may have been fine cracks in the tested material because of all the damage which occurred. That is, the test may have been conducted on flawed material and perhaps a higher value would be obtained with a specimen from a CWF that does not crack during cooldown.

### 3.3 Sound recordings and timing of the cracking

A recording of the sound from the first crack recorded is shown in Figure 5. Time is plotted on the horizontal axis and proceeds from left to right. Sound power is plotted on the vertical axis. Background noise is recorded until the crack occurs evidenced by a large increase in power. The power then decreases exponentially back to the background level.

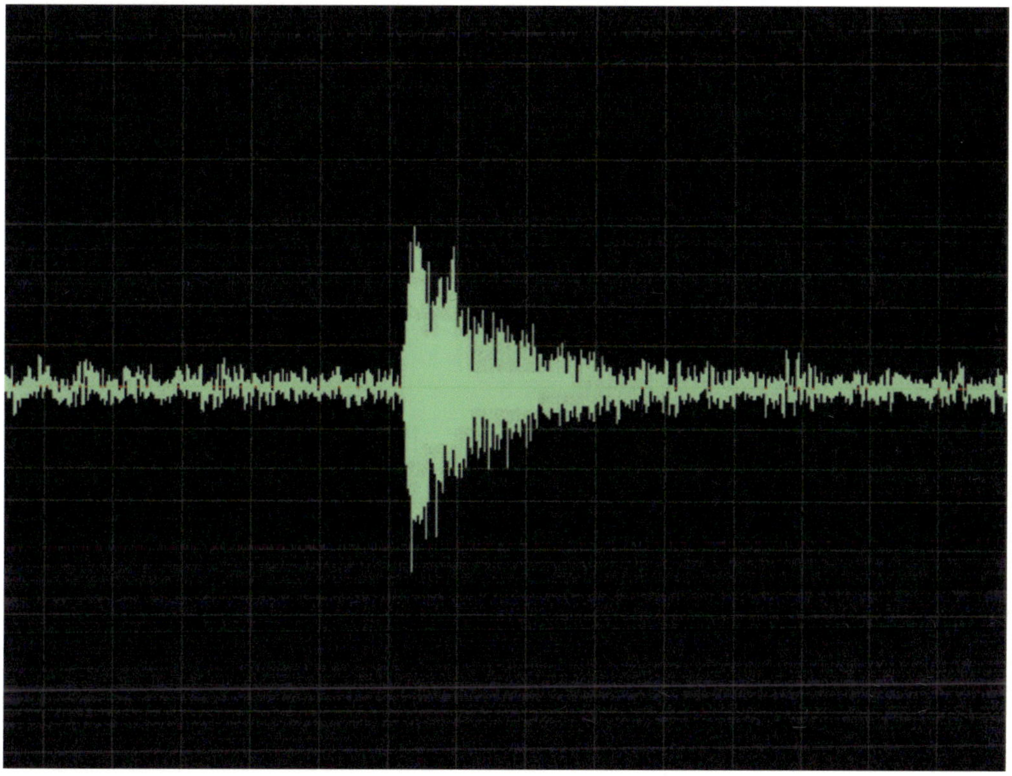

Fig. 5. Sound Pressure of the First Crack Recorded.

All of the recordings looked similar but the power of the crack sound varied from crack to crack relative to the background noise. In general, the sound level decreased with each succeeding crack. The crack sounded like a loud gunshot even though the CWF was encased in a furnace with a insulating wall one foot thick. The total time duration in Figure 5 is about 2 seconds. The first crack occurred when the centerline temperature was 420 C and the surface temperature was about 400 C. Cracking continued down to near room temperature. The last crack occurred after the CWF had been removed from the furnace.

The timing of the cracks is included in Table B.1 along with the temperature data. The cracks do not occur at evenly spaced intervals. They usually occur several hours apart. The cracking analysis presented in this chapter applies to an uncracked cylinder and the cracking is probably relieving some of the stress. However, the theory predicts that the total stress continues to increase as the temperature decreases which explains why the cracking continues. The decrease of the sound power as cracking proceeds seems to indicate that the stress in the remaining pieces is actually decreasing. The last crack occurred almost two days after the previous one and occurred after the CWF was removed from the furnace.

| | Month/Day | Time | Hours since start. | Time between cracks | Sound level (db)* | Center-line Temperature of CWF (°C)** |
|---|---|---|---|---|---|---|
| Heat up | 3/22 | 2:20 PM | 0 | | NA | 70 |
| Cooldown | 3/28 | 7:20 AM | 137 | | NA | 920 |
| Crack 1 | 3/30 | 10:58 PM | 200.6 | | 91 | 420 |
| Crack 2 | 3/31 | 1:28 AM | 202.1 | 1.5 | 88 | 405 |
| Crack 3 | 3/31 | 6:31 AM | 208.2 | 6.1 | 91 | 382 |
| Crack 4 | 3/31 | 3:11 PM | 216.9 | 8.7 | 87 | 335 |
| Crack 5 | 3/31 | 4:12 PM | 217.9 | 1.0 | 85 | 325 |
| Crack 6 | 4/01 | 12:55 AM | 226.6 | 8.7 | 89 | 300 |
| Crack 7 | 4/01 | 6:39 PM | 244.3 | 17.7 | 81 | 220 |
| Crack 8 | 4/01 | 7:49 PM | 245.5 | 1.2 | 83 | 210 |
| Crack 9 | 4/02 | 12:47 AM | 250.5 | 5.0 | | 190 |
| Crack 10 | 4/02 | 2:11 AM | 251.9 | 1.4 | | 170 |
| Crack 11 | 4/02 | 7:56 AM | 257.6 | 5.8 | | 150 |
| Crack 12 | 4/02 | 10:39 AM | 260.3 | 2.7 | | 130 |
| Crack 13 | 4/02 | 6:01 PM | 267.7 | 7.4 | | 110 |
| Crack 14 | 4/03 | 10:57 AM | 284.6 | 16.9 | | 90 |
| Crack 15 | 4/05 | 9:29 AM | 331.2 | 46.5 | | 70 |

* Sound level relative to a reference of 100 dB

**Temperatures estimated for Cracks 9 to 15

Table 3. Timing and Sound Levels of Cracking

### 3.4 Temperature data for CWF2

A cooling rate program was specified for the formation of CWF2 which was designed to eliminate cracking. This program was overridden by the protection program for the coolant blower so that severe cracking did occur. In the following section, the stress versus time in the CWF from the temperature data is estimated using the stress theory developed above. Then in a later section, the stress developed is compared to the stress limit and to the timing at which cracking occurred to demonstrate that not only does solidification stress exist but that it and not the thermal stress is responsible for cracking the CWF.

Surface and centerline temperatures were measured at the mid-plane. The surface mid-plane temperature history was then matched (by adjusting the heat transfer coefficients throughout the cooldown) in the CWF heat transfer, densification model developed in this work. Once this temperature history was matched with the code output, the temperature distributions in the solid were known. Then the stresses, thermal, solidification, and total, determined by the above equation could be calculated for the CWF. Then the cracking times were included on the stress plots. The cracking sounds all occurred when the calculated total stresses were above the stress limit. Although the thermal stress component exceeded the stress limit early in the cooldown, cracking was not recorded during this time. When cracking occurred, the thermal stress was below the limit.

The mid-plane temperature measurements versus time made during the cooldown of CWF2 are shown in Figure 6. The centerline and surface temperatures are shown in this figure as well as the furnace wall temperature. In addition to the measured temperatures, the temperature history prescribed for the cooldown which should have kept the stress below the limit are shown. The measured rate of temperature drop (EXPerimental) through solidification is much more rapid than the desired (SPECification) curves. This results in a much larger temperature drop from the center to the surface than desired.

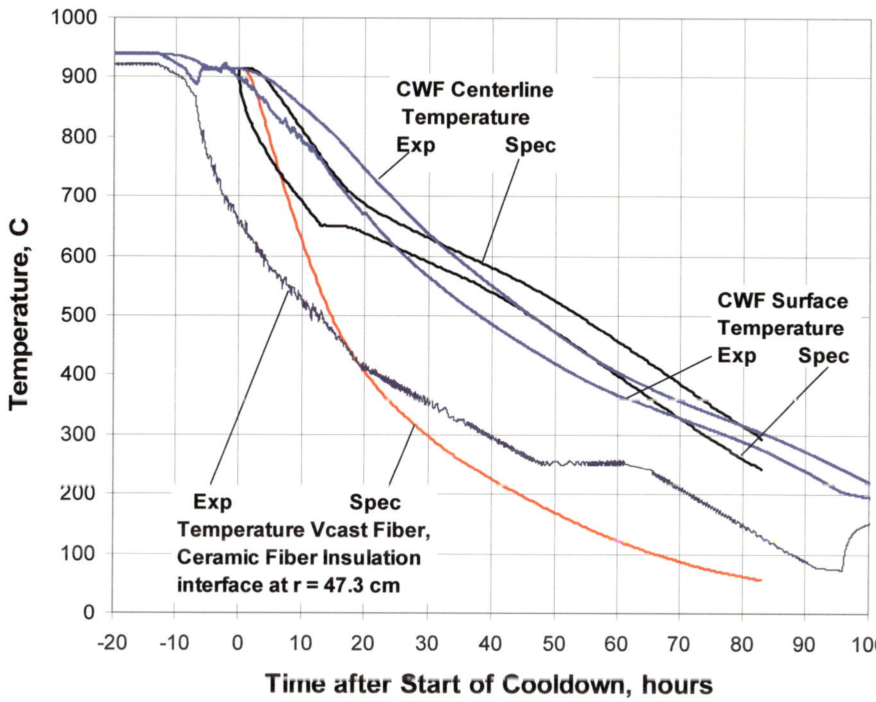


Fig. 6. Experimental and Specification Temperatures at Mid-Plane

Note that the centerline temperature may be distinguished from the surface temperature since it is always greater than the surface temperature. The time scale in Figure 6 was adjusted so that zero time corresponds to the time that cooldown starts. It is 143 hours after the start of the initial heatup. The plan was to start cooldown 12 hours earlier. From Figure 6, it is seen that cooldown did start 12 hours earlier but after 4 hours the CWF cooldown and heat transfer from the CWF aborted, resulting in the surface and center temperatures equilibrating over the next hour. This was caused by the coolant blower tripping off due to a temperature limit (100 °C) being exceeded. It stayed off for about 8 hours but then the fan restarted so CWF cooldown started again (at zero time in Figure 6). The furnace wall continued cooling during the time that the fan was off because the control system kept the heating coils off. As temperatures decreased, the pump stayed on for longer periods of time causing the average coolant flow to increase resulting in more heat being removed from the CWF. The largest heat removal occurred during solidification causing a large set-in stress.

To more clearly show the cooling rate problem encountered in CWF2, the temperature difference between the center and the surface is plotted in Figure 7. This difference during solidification is proportional to the solidification stress which is set-in.. The calculation specified that the temperature difference should decrease in the solidification range (about 625 C). The data shows the CWF2 temperature difference actually increased significantly.

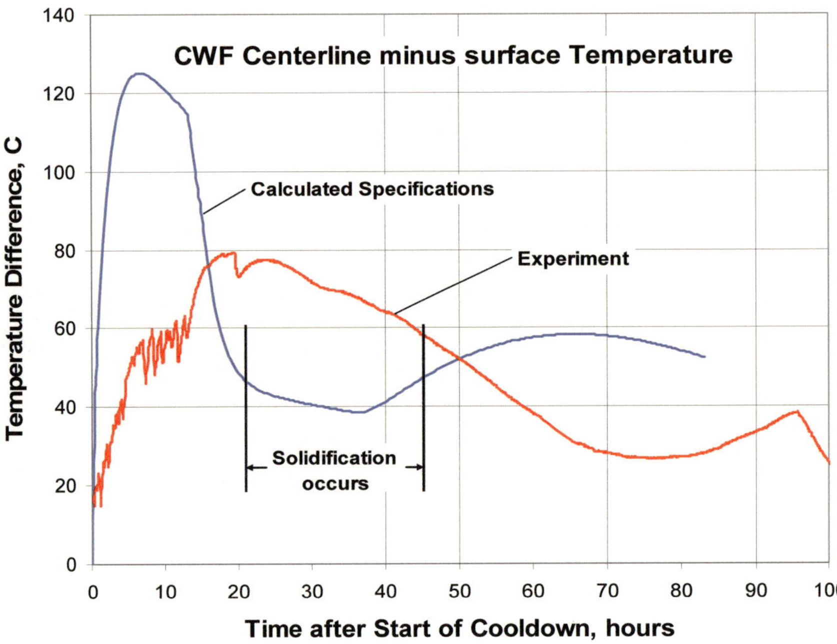


Fig. 7. Temperature difference at Mid-Plane

In fact, the largest differences occur during solidification. The specification required the temperature difference to be less than 40 C but the data show it was nearly 80 C. This temperature difference resulted in CWF2 cracking. With the successful match of the surface temperature the other quanties of interest were available from the code. In particular, the centerline temperature, the radial temperature profile, and stresses developed, both solidification and thermal as well as total were available as code output. Figure 8 shows the agreement between the calculated temperatures on the surface and center of the CWF and the data. The calculations are a bit high in the CWF center (the higher temperature curve in Figure 8) at high temperature and a bit low at low temperature but the agreement is good in the important solidification region (between 725 C and 525 C). Below 300 C, the discrepancy between measured and calculated is not important because the thermal stress will continue to decrease and eventually the total stress will be just the solidification stress.

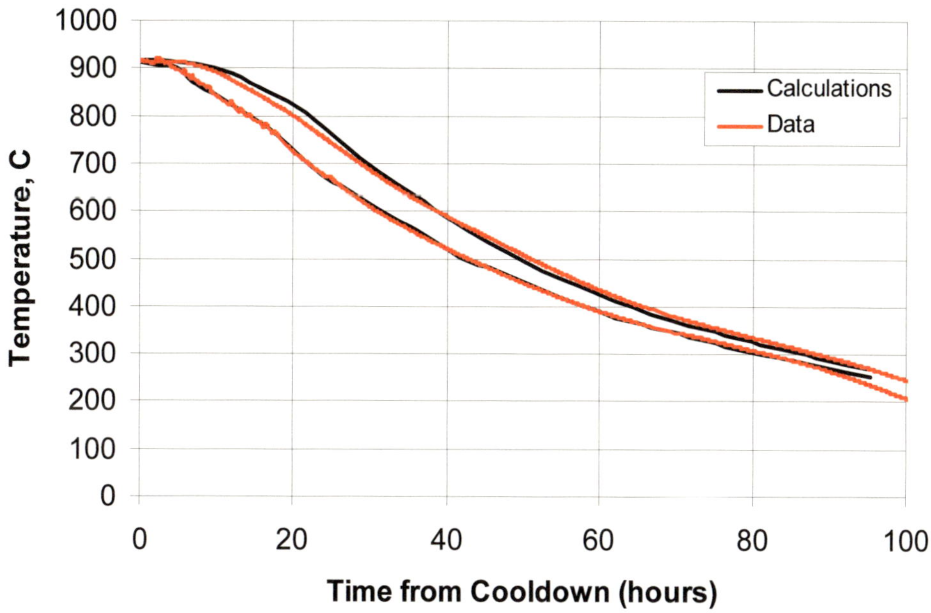


Fig. 8. Comparison of Calculated Center and Surface Temperatures to Data (CWFformBig P01-CWF2.xls)

### 3.5 Comparison of calculated stress to the times cracking occurred

The next three figures show the thermal stress, the solidification stress and the total stress respectively output from the above mentioned code after the surface temperatures were matched. In each figure, the estimated tensile limit is included 82.4 mpa (12000 psi).

The thermal stress calculated through the mid-plane for this cooldown transient is shown in Figure 9. The stresses are shown at ten evenly spaced radial increments (spaced about 2.5 cm apart) with the center line stress being the largest negative value or compression and the outer surface being the largest positive value or tension. Thus, the thermal stress predicts that the stress will be in tension in the outer radial region of the CWF so that cracking would be expected to start in the outer radial region if solidification stress did not exist. In fact, the thermal stress at the outer radius is seen to exceed the tensile limit from 32 hours to 48 hours. The thermal stresses slowly decrease after that. If thermal stress were the cause of cracking in the CWF, it would be expected that cracking would occur during this time period. As mentioned sound measurements recorded 15 loud cracks during the course of the cooldown. The time of the first six cracks is indicated in the figure with numbers from 1 to 6. The first one occurred at 61 hours, much later than the 32 to 48 hours time interval when the thermal stress was above the tensile limit. Since all the cracks occurred significantly after the thermal stress reduced below the tensile limit, it cannot be responsible for the CWF damage. As mentioned previously only axial stresses are discussed in this chapter. Circumferential stresses are the same magnitude. Radial stresses are very small.

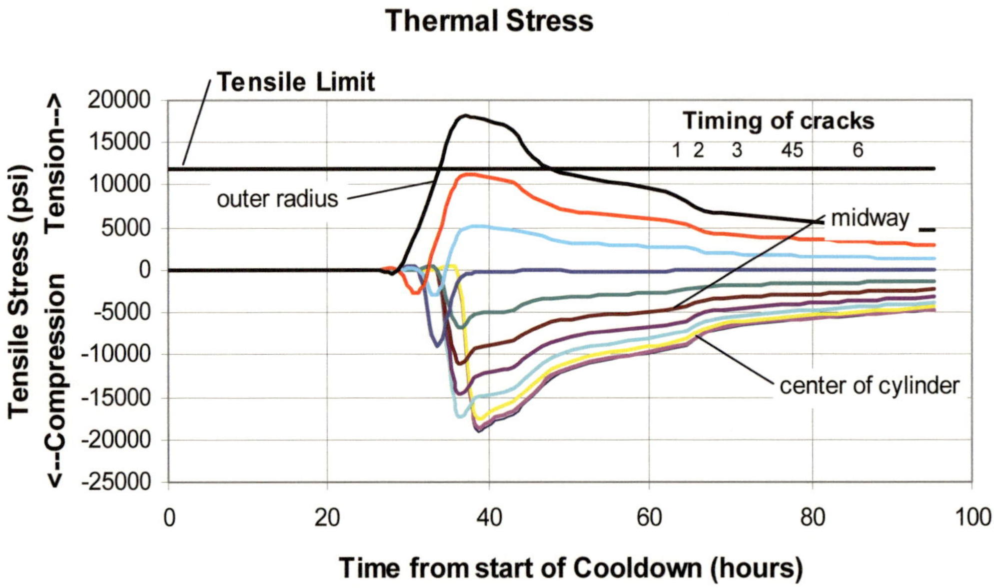


Fig. 9. Thermal Stress at the mid-plane in the CWF (Note: 10000 psi=68.9 mpa)

Figure 10 shows the solidification stress calculated using the method developed in Solbrig and Bateman (2010) and summarized in the previous theory section. It develops while the CWF is solidifying and occurs because while one layer of the CWF is solidifying, it attaches itself to an adjacent solidified layer of a shorter length. As the solid then is cooled down to room temperature, all these different lengths are forced to the same length causing stresses.

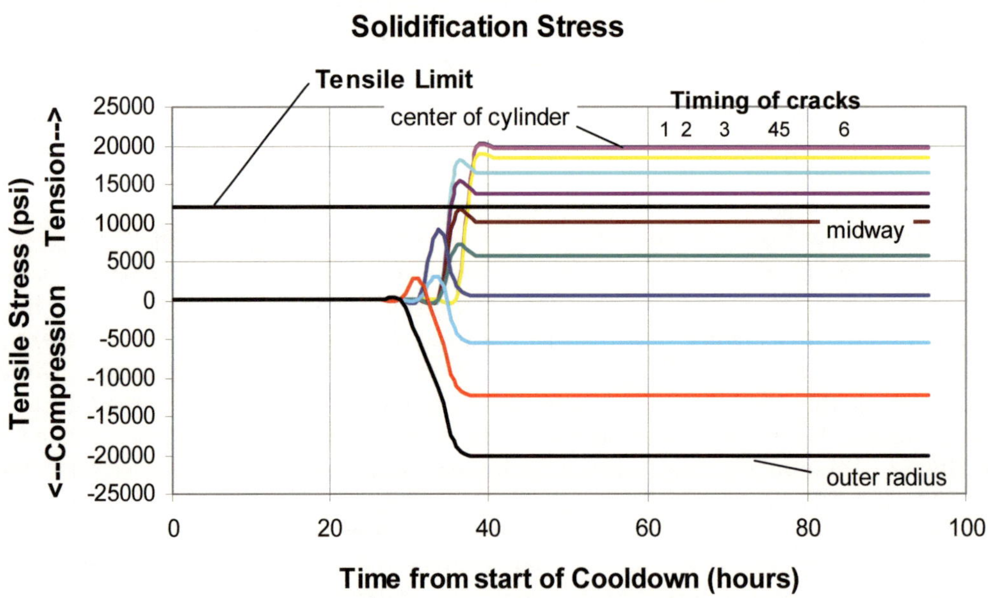


Fig. 10. Solidification Stress of the CWF (Note: 10000 psi=68.9 mpa)

The solidification stress developed is dependent on the temperature profile during solidification but is independent of temperature profile during the remainder of the cooldown. Note that the solidification stress is of the opposite sign than the thermal stess and solidification stress and thermal stress subtract from each other.

Once the solidification stress develops, it is constant. The maximum solidification stress is about 137.9 mpa (20000 psi) which is well over the tensile stress limit of 82.4 mpa (12000 psi). The total stress is the sum of the thermal and the solidification stress, and since they are of opposite sign, the thermal stress partially cancels out the solidification stress especially during the early portion of cooldown in the solid phase. The thermal stress during the early period is responsible for keeping the total stress less than the tensile limit. But the thermal stress decreases as the temperature decreases, it cancels out less and less of the solidification stress as the temperature decreases and the temperature profile flattens out. When the temperature becomes uniform, the thermal stress is zero so the total stress is then equal to the 137.9 mpa (20000 psi) solidification stress at the cylinder surface at the mid-plane of the CWF. This explains why the CWF cracks at low temperature instead of high temperature where the thermal stress is high. That is, the combination of the solidification stress and the thermal stress results in the total stress continuing to increase as the average temperature decreases and the temperature profile flattens out and is highest when it is flat..

Both the solidification stress and the thermal stress are added together to obtain the total stress shown in Figure 11. The tensile total stress is zero out to 34 hours after the start of cooldown and then increases as the temperature decreases to room temperature.

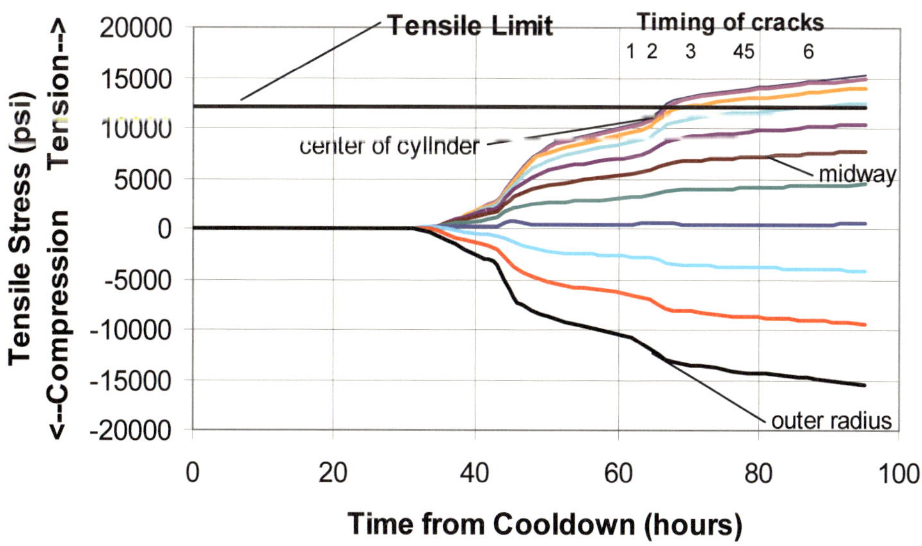


Fig. 11. Total Stress of the CWF (Note: 10000 psi=68.9 mpa)

During the initial solidification period, the two stresses almost cancel each other out. The solidification stress is always greater than the thermal stress because the solidification temperature profile is always steeper than the temperature profile. Since the solidification stress is always in tension in the inner region, the total stress is also in tension in the inner portion of the cylinder. The total stress at the centerline eventually exceeds the tensile limit at 68 hours. The timing of the first six cracks is indicated by the upper numbers in Figure 11. The first cracking sound that was recorded occurred at 61 hours. This figure shows that the total surface stress is slightly less than the stress limit for the first two cracks but greater than the stress limit after that. The tensile limit shown is only an estimate and could be as low as 68.9 mpa (10000 psi). The damage continued all the way down to room temperature. In all, 15 loud cracks were heard including the last one which occurred after the CWF was at a low enough temperature that it was removed it from the furnace. Post test destructive examination of CWF2 confirmed the considerable damage which occurred.

It should be noted that if the temperature profile were to be flattened out at a high temperature, say 500 °C to ostensibly relieve stress, it would cause the CWF to crack at that temperature since the thermal stress would have been removed leaving the solidification stress to exceed the stress limit..

### 3.6 Visual confirmation of stress

Since it is not possible to measure the stress in the ceramic as it is forming, other means must be used to determine the stress which occurs during the formation. The timing of the sounds of cracking versus the calculated stresses is one method of estimating the stresses. Destructive examination of the resulting CWF is another. Since the CWF is formed in a steel canister, damage cannot be observed without destructive examination. Consequently, three one inch thick slices were cut out of both CWF1 and CWF2 with a 4 foot diameter diamond tipped saw blades. One slice was cut out at the mid section, another near the top, and a third near the bottom. In addition, an axial slice was cut through the center of the bottom half of CWF2. The cutting of the CWF did not seem to cause any of the cracking but the cracking occurred before the cutting. CWF2 was the first experiment that had sound recordings so similar measurements are not available for CWF1 but cracking seemed to occur in a similar manner as in CWF2..

Since CWF1 and CWF2 were run with the same cooldown cycle, damage to both were similar. Visual observation of CWF2 showed as much or more damage than CWF1. In addition, the axial cut of the lower half of CWF2 showed considerably more damage with parts of it appearing to be almost rubble.

The most egregious cracking which occurred in CWF1 is shown in Figure 12. This is a picture of the one inch thick slice cut out of the midsection of CWF1. Both axial and circumferential stress damage are observed in this picture. The pieces that have fallen out of the slice are caused by axial stress exceeding the stress limit in the axial direction. The radial cracks are due to the circumferential stress. Most of the damage occurs in the inner portion of the CWF confirming that the damage was caused solidification stress which is tensile in the inner region rather than thermal stress which is compressive in the inner region. Cracks in the circumferential direction are indications of radial stress. These appear in the outer region and may be caused by radial thermal stress.

Fig. 12. Cracking in the mid-plane in CWF1

There is less damage to the one inch slices cut out of the top sections and bottom sections. This may be due to axial temperature gradients causing the highest solidification stresses to occur in axial middle portions of the CWF.

## 4. Conclusions

A theory has been developed to model a stress which was posited to develop when a ceramic solidifies due to the temperature gradient which exists during the solidification process. An experiment was run which verifies the existence of this stress. Thermal stress alone would have predicted cracking to occur while temperatures are high but when the solidification stress is added, the total stress calculation predicts cracking of the CWF will occur at low temperatures. Cracking sounds were recorded in this experiment and are used in this chapter to show that the existence of this stress is probable since cracking occurred during the low temperature phase of the cooldown. Confirmation of this model provides confidence in the ability of the model to predict a cooldown history for the next CWF formation which will eliminate cracking. Without including the solidification stress in the calculation, the low cooling rate needed to prevent cracking would be prescribed when thermal stress is high instead of during solidification and cracking would not be prevented with such a prescription.

## 5. Acknowledgement

Work supported by the U.S. Department of Energy, Office of Nuclear Energy (NE) under DOE Idaho Operations Office Contract DE-AC07-05ID14517.

## 6. References

Bateman, K. J. and Capson, D. D. (2003). "Consolidating Electrorefined Spent Nuclear Fuel Waste: Analysis and Experiment," *Proc. ASME Int. Mechanical Engineering Congress andExposition (IMECE 2003)*, Washington, D.C., November 16–21, 2003. Figure 12.

Bateman, K. J., and Solbrig, C.W. (2008)a. "Use of Similarity Analysis on Experiments of Different Size to Predict Critical Cooling Rates for Large Ceramic Waste Forms," *Proc. 16th Int. Conf. Nuclear Engineering (ICONE16)*, Orlando, Florida, May 11–15, 2008, American Society of Mechanical Engineers (2008).

Bateman, K. J., and Solbrig, C.W. (2008)b. "Stabilizing Glass Bonded Waste Forms Containing Fission Products Separated from Spent Nuclear Fuel," *Sep. Sci. Technol.*, 43, *9*, 2722 (July, 2008).

Faletti, D. W., and Ethridge, L. J. (1986). "A Method for Predicting Cracking in Waste Glass Canisters," PNL 5947, Pacific Northwest Laboratories (Aug. 1986)

Slate, S. C., Bunnell, L. R., Ross, W. A., Simonen, F. A., and Westsik, J. H. (1978). "Stress and Cracking in High-Level Waste Glass," PNL SA 7369, Pacific Northwest Laboratories, Dec. 1978.

Solbrig, C. W., and Bateman, K. J. (2010). Modeling Solidification-Induced Stress in Ceramic Waste Forms Containing Nuclear Wastes, Nuclear Technology, Vol. 172, P 189-203, Nov, 2010.

Timoshenko, S., and Goodier, J. N. (1970). *Theory of Elasticity*, Third Edition, McGraw-Hill, New York, NY (1970).

Timoshenko, S., and MacCullough, G. H. (1949). "Elements of Strength of Materials," Van Nostrand, 3rd Ed., 1949, (P.122, Eq 49)

# 13

# Cadmium Personnel Doses in an Electrorefiner Tipping Accident

Clinton Wilson, Chad Pope and Charles Solbrig
*Idaho National Laboratory*
*USA*

## 1. Introduction

This chapter estimates airborne cadmium concentrations caused by the facility design base earthquake (DBE) at the INL Fuel Conditioning Facility which damages the MK-IV electrorefiner (ER) vessel so that cadmium spills out onto the floor. In addition, the seismically qualified safety exhaust system (SES) is assumed to fail. The SES is a safety grade system that is large enough to keep the flow through any DBE caused breach into the cell. But with SES inoperative, failure of non -seismically qualified cell boundary penetrations allows release of cadmium vapor to the facility workers, site workers, and general public. Consequence categories are designated by estimating airborne concentrations at specific personnel locations and comparing them to applicable exposure guidelines. Without the failure of the SES, there would be negligible doses to all workers and the general public.

This scenario involves a spill of 587.4 kg of 500°C liquid cadmium from the MK-IV electrorefiner to the argon cell floor. Based on the methods and assumptions described in this chapter, the maximum potential chemical toxicity consequences of this beyond DBA release scenario are: Moderate for facility workers, low for collocated workers, low for bus staging area, and negligible for the off-site public. This safety assessment is concerned only with the cadmium release. Bounding calculations are used as required in safety assessments to assure that exposures would not be greater than calculated. Consequences of potential releases of the radiological materials caused by the DBE alone are analyzed elsewhere.

The main metabolic feature of cadmium is a long biological half-life, resulting in a virtually irreversible accumulation of this metal in the body throughout life. In non-occupationally exposed subjects, ingested and inhaled cadmium content in the body increases continuously with age. Cadmium poses a chemical toxicity hazard to kidneys, lungs, and/or liver if ingested or inhaled. Repeated or prolonged exposure to cadmium can damage target organs. Severe over-exposure can result in death.

## 2. Description of facility

The Idaho National Laboratory (INL) Fuel Conditioning Facility (FCF) uses an engineering scale pyrometallurgical process developed by Argonne National Laboratory (Till and Chang, 1989 and Till, et. al., 1997) to reprocess metallic fast reactor fuel. It is currently being used to treat fast reactor spent fuel from the Experimental Breeder Reactor II. Operations in the facility began June 7, 1996. The process is necessary because the Experimental Breeder Reactor II fuel design uses sodium metal bonding between the fuel pin and stainless steel cladding. The use of sodium prevents the spent fuel from being suitable for direct geologic disposal. Since the Experimental Breeder Reactor II fuel is metallic rather than oxide, a pyrometallurgical process can be used for treatment rather than the traditional solvent extraction method. Since this is a nuclear facility, safety analyses of possible incidents and highly unlikely accidents are required in order to license and run such a facility. The design basis accident (DBA) for this facility is an earthquake which causes failure of pipes that go through the walls to supply argon, cell atmosphere purification, and remote operations. An exhaust system was designed and installed to mitigate this accident.

FCF is pictured in Figure 1. FCF is in the foreground with the Experimental Breeder Reactor II, located in the domed containment building, in the background. The two facilities are connected, allowing direct spent fuel transfers from the reactor to the processing facility so that spent fast reactor fuel could be processed and returned to the reactor for further power generation. Although the mission has changed and the Experimental Breeder Reactor II has been shut down and decommissioned, the spent fuel from this reactor must still be treated for long term storage.

Fig. 1. Fuel Conditioning Facility.

The entire pyrometallurgical process is conducted remotely in a hot cell environment. Figure 2 shows the FCF hot cell layout. The Air Cell, the rectangular portion, contains an air atmosphere and is used for spent fuel storage, fuel assembly disassembly into individual rods, and product storage. An operating corridor surrounds the whole facility. The Argon Cell, the annular portion, contains an inert argon atmosphere which allows metallic fuel rods to be disassembled and processed. The release of radioactive gases when the fuel rods are chopped and the exposure of heavy metal to the Argon Cell atmosphere in the form of chopped fuel or metal extracted in the refining process presents the possibility of radiological releases due to cell breaches caused by an earthquake. The vessel of the MK-IV electorefiner shown at the top of Figure 2 is hypothesized to crack open due to the design basis earthquake causing the cadmium to spill on the floor.

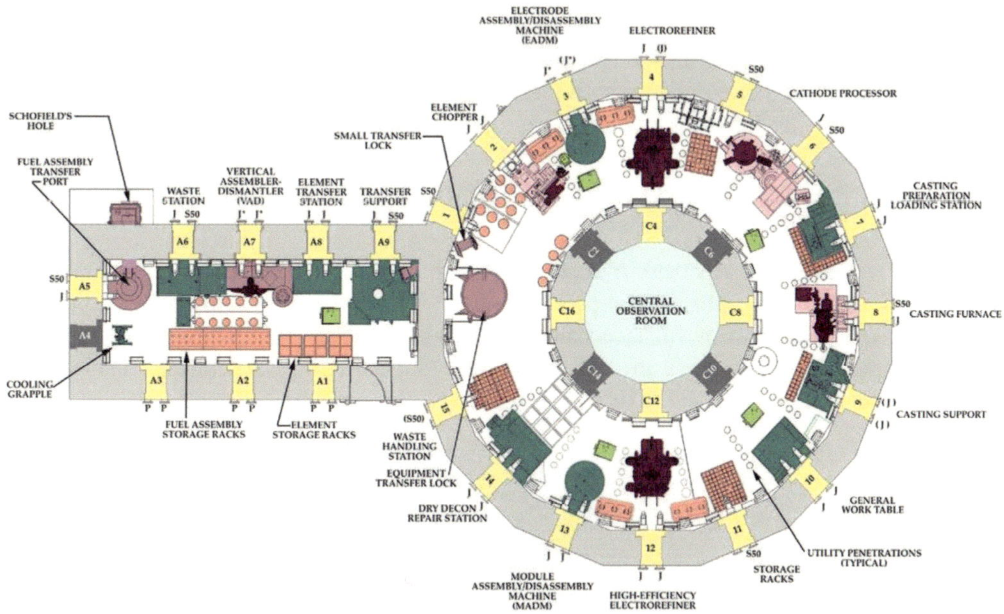


Fig. 2. Facility Layout.

The process starts with chopping spent fuel elements and loading them into baskets. Once filled, a basket (anode) is lowered into an electrorefining vessel (Figure 3) which contains molten salt (LiCl-KCl) electrolyte and in the case of the MK-IV electrorefiner, a pool of cadmium below the salt. A stainless steel mandrel (cathode) is also lowered into the salt. The basket walls are perforated allowing molten salt to contact the element segments.

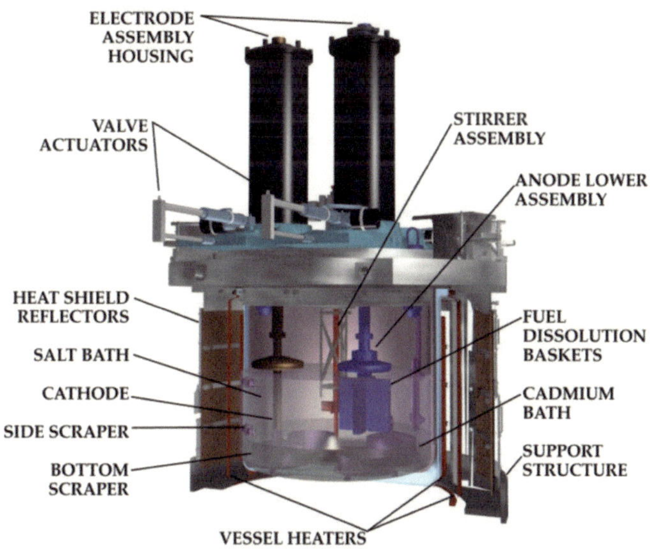


Fig. 3. Electrorefiner.

In the electrorefiner, active metal fission products, transuranic metals, and sodium metal undergo chemical oxidation and form chlorides. Voltage is applied between the basket, which serves as an anode, and a rod (mandrel), which serves as a cathode. This causes metallic uranium in the spent fuel to undergo electro-chemical oxidation, thereby forming uranium chloride and simultaneously, uranium chloride to undergoes electro-chemical reduction at the cathode depositing uranium metal dendrite onto the mandrel (Figure 4). Fission products, plutonium, and other transuranic metals remain as chlorides in the salt during this step but are removed later on.

Upon removal from the electrorefiner, the uranium metal dendrites are stripped from the mandrel. These and adhering fission product laden salt are placed in a crucible. The crucible is transferred to a distillation furnace where the uranium metal melts, forming an ingot, and the salt vaporizes under vacuum conditions and subsequently condenses in a separate crucible.

Fig. 4. Cathode Deposit.

Pyroprocessing spent metallic nuclear fuel requires the use of an inert atmosphere to maintain product purity and because pyrophoric materials, such as the cathode deposits, are produced at certain points in the process. In addition to providing an inert atmosphere, radioactive contamination and special nuclear material security concerns can be satisfied by thick concrete cell walls (1.5 m thick), a subatmospheric inert (i.e., argon) environment, remote operations, and specialized windows (1.5 m thick) to enable operations with the remote controls.

The argon cell has airborne and surface radioactive contamination and is therefore of concern for possibly releasing radioactivity. More importantly, much of the pyrophoric material is laden with radioactive fission products that can become airborne if the material begins burning. In addition, the electrorefiner contains cadmium and radioactive fission product compounds in the electrolyte salt . During normal operation, the cell pressure is kept at 745 Pa below atmospheric pressure to ensure that no radioactive material leaks out to the operating corridor which surrounds the cell or the environment.

Vertical cross sections of the circular argon cell and below grade support are shown in Figure 5. A dashed line shows the confinement boundary of the argon cell (that portion of the boundary which is not considered to fail). The boundary between the cooling cubicles and argon cell is sheet metal and is assumed to fail when the earthquake occurs if that cooling cubicle is assumed to have a breach. The failed cooling cubicles initially contain air, which is assumed to mix with the argon in the cell. The locations of the three potential breach areas are also indicated. The SES, which initiates when the cell pressure increases to -248 Pa g, draws on the argon cell and both cooling cubicles. (Note 248 Pa = 1 iwg.)

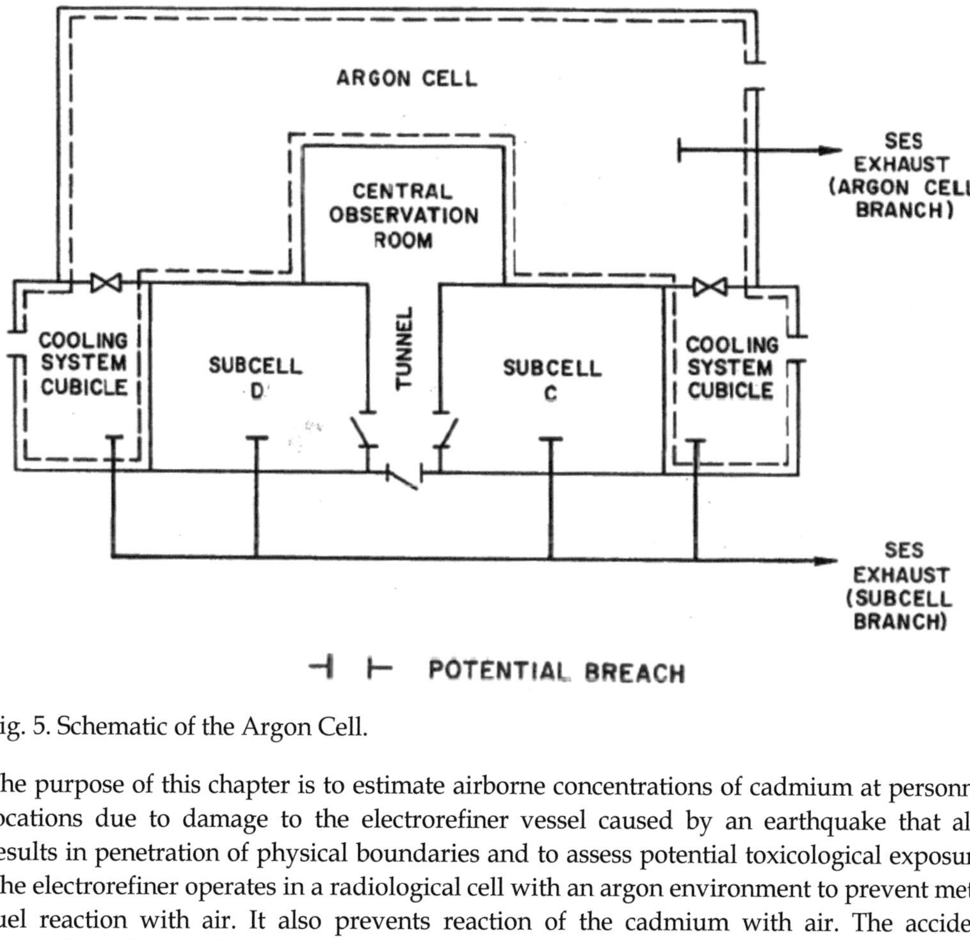


Fig. 5. Schematic of the Argon Cell.

The purpose of this chapter is to estimate airborne concentrations of cadmium at personnel locations due to damage to the electrorefiner vessel caused by an earthquake that also results in penetration of physical boundaries and to assess potential toxicological exposure. The electrorefiner operates in a radiological cell with an argon environment to prevent metal fuel reaction with air. It also prevents reaction of the cadmium with air. The accident proceeds with the electrorefiner splitting open or tipping over due to the design base earthquake. Thirty centimeters of liquid salt are on top of 15 centimeters of cadmium in a one meter diameter vessel, at 500 C. The cadmium spills out on the floor of the argon filled reprocessing cell and is assumed to be uncovered by the salt. The mass transfer of cadmium to the argon atmosphere takes place in two phases. The first phase occurs as the cadmium is spilling on the cell floor with the cadmium splashing back up and spreading on floor. The second phase occurs after the cadmium spreads out over the floor. The floor is concrete covered by a steel liner. Cadmium evaporates from the surface until it solidifies. This chapter evaluates the mass transfer in each phase and estimates the transport and health effects to personnel in the facility, to personnel on the site, and to the general public from exposure to the cadmium.

The cooling system is assumed to trip off due to a power trip which occurs when an earthquake is sensed. Combustion cannot occur in this facility unless enough oxygen is introduced into the cell to bring the oxygen concentration above the 4% spontaneous combustion limit. The facility has a gas purification system which limits the oxygen concentration to below 50 ppm during normal operation which is far below the 4% combustion limit of metal fuel exposed to a mixture of oxygen and nitrogen or argon. A breach of the cell boundary followed by the SES initiation can cause sufficient oxygen to be introduced into the cell to cause a fire. The flow rate of the SES is large enough to limit or inhibit the unfiltered outflow of cell atmosphere. It causes the pressure to become subatmospheric by removing cell atmosphere from the cell. The SES flow is all filtered through HEPA filters before being exhausted to the atmosphere. It has the adverse effect of causing significant inflow of air into the cell in sufficient quantity to cause combustion. In order for people to be exposed to the cadmium from the rupture of the electrorefiner vessel, the SES must fail to allow outflow of the argon cell atmosphere to the operating corridor or to the outside of FCF.

## 3. Cadmium physical, chemical, and toxicity properties

Cadmium is a solid at normal atmospheric temperatures and pressures. Abundance earth's crust: 150 parts per billion by weight, 30 parts per billion by moles. Cadmium and its compounds are highly toxic. Cadmium is a soft, malleable, ductile, bluish-white metal, which is easily cut with a knife. It is an excellent electrical conductor and shows good resistance to corrosion and attack by chemicals. It is similar in many respects to zinc in its chemical properties. Cadmium tarnishes in air and is soluble in acids but not in alkalis. The metal burns in air to form brown cadmium oxide (CdO). Cadmium most often occurs in small quantities associated with zinc ores, such as sphalerite (ZnS). Greenockite (CdS) is the only cadmium mineral of any consequence. Almost all cadmium is obtained as a by-product of zinc, copper, and lead ore refining operations.

Most cadmium is used in batteries (especially rechargeable nickel-cadmium, NiCad, batteries). As a result of its low coefficient of friction and its high fatigue resistance, cadmium is used in alloys for bearings. Cadmium is used in low melting alloys and is a component of many kinds of solder. It is also is used in electroplating. Compounds containing cadmium are used in black and white television phosphors, and in the blue and green phosphors for color television picture tubes. Cadmium sulfide is used as a yellow pigment, and cadmium selenide is used as a red pigment, often called cadmium red. (Chemicool, 2011).

Cadmium and tellurium can be compounded into CdTe thin-film photovoltaic modules whose physical characteristics are ideal for solar cell production. They are relatively low cost and have an almost perfect bandgap for solar energy harvesting. (Chemicool, 2011).

| Substance Name | Cadmium | Cadmium Oxide | Cadmium Chloride |
|---|---|---|---|
| Molecular formula | Cd | CdO | $CdCl_2$ |
| Molecular weight | 112.41 | 128.41 | 183.32 |
| Density [ g/cm³ ] | 8.646 | 8.15 | 4.048 |
| Boiling point | 765 - 767°C<br>1,410 - 1,413°F | 1,385 - 1,559°C<br>2,525 - 2,838°F | 956.9 - 969.6°C<br>1,754 - 1,777°F |
| Melting point | 321°C<br>610°F | 1,430°C<br>2,606°F | 567.9°C<br>1,054°F |
| Vapor pressure [ mmHg ] | 1 mmHg @ 394°C<br>760 mmHg @ 765°C | 1 mmHg @ 1,000°C | 10 mmHg @ 656°C |
| References: | Knovel, 2003, Patty, 2001, Perry, 1997 | | |

Table 1. Physical and chemical properties of cadmium and selected cadmium compounds.

**Solid Phase:** Solid cadmium is a soft, blue-white, metal or a gray-white powder. Cadmium metal dust has an auto-ignition temperature of 250°C (482°F). Cadmium is insoluble in water and soluble in ammonium nitrate, nitric acid, and hot sulfuric acid. Cadmium is combustible and burns in air or oxygen, producing cadmium oxide. In moist air, cadmium slowly oxidizes to form cadmium oxide.

**Liquid Phase;** At atmospheric pressure, cadmium metal melts at 321°C (610°F) and forms a liquid. Liquid cadmium has a relatively high vapor pressure, as shown in Table 1. A release of molten cadmium is expected to cool quickly to ambient temperatures and solidify. If the ambient atmosphere is above the melting point, 321°C (610°F), then a fraction of the material is expected to transfer to the vapor phase based on the vapor pressure of cadmium at the ambient temperature.

**Vapor Phase:** Cadmium is a vapor at atmospheric pressure and temperatures above approximately 765°C (1,410°F). In ambient air, cadmium vapor rapidly oxidizes to produce cadmium oxide. Cadmium vapor or fumes may contain elemental cadmium or cadmium oxide as particulate matter in the respirable range. In addition, cadmium vapor reacts with other typical stack gases, as shown in Table 2.

| | | | |
|---|---|---|---|
| Cadmium vapor in the presence of: | ambient air | forms: | cadmium oxide (CdO) |
| | carbon dioxide ($CO_2$) | | cadmium carbonate ($CdCO_3$) |
| | water vapor ($H_2O$) | | cadmium hydroxide ($Cd[OH]_2$) |
| | sulfur dioxide ($SO_2$) | | cadmium sulfite ($CdSO_3$) |
| | sulfur trioxide ($SO_3$) | | cadmium sulfate ($CdSO_4$) |
| | hydrogen chloride (HCl) | | cadmium chloride ($CdCl_2$) |
| These compounds may be formed in stacks and emitted to the environment. | | | |
| Reference: Patty, 2001 | | | |

Table 2. Compounds formed by atmospheric cadmium vapor.

**Reactions:** Combustion reaction: Cadmium metal burns in air to form cadmium oxide (CdO (s)). Combustion processes produce very fine cadmium-containing particles. Cadmium metal does not react with water.

The overall hazards presented by cadmium are listed in the International Chemical Safety Cards (ICSC, 2005). Cadmium accumulates in the liver and kidneys and has a biological half-life, from 17-30 years in man. After uptake from the lung or the gastrointestinal tract, cadmium is transported in blood plasma initially bound to albumin. Cadmium bound to albumin is preferentially taken up by the liver. Cadmium is widely distributed in the body, with the major portion of the body burden located in the liver and kidney. Liver and kidney cadmium concentrations are comparable after short-term exposure, but the kidney concentration exceeds the liver concentration following prolonged exposure.

The concentration of cadmium in the liver of occupationally exposed workers generally increases in proportion to intensity and duration of exposures to values up to 100 µg/g. The concentration of cadmium in the kidney rises more slowly than in the liver after exposure and begins to decline after the onset of renal damage at a critical concentration of 160-285 µg/g.

Most non-occupationally exposed people are exposed to cadmium primarily through the diet. Cadmium can be detected in virtually all tissues in adults from industrialized countries, with greatest concentrations in the liver and kidney. Average cadmium concentrations in the kidney are at birth near zero, and rise roughly linearly with age to a peak (typically around 40-50 µg/g wet weight) between ages 50 and 60, after which kidney concentrations plateau or decline. Liver cadmium concentrations also begin near zero at birth, increase to typical values of 1-2 µg /g wet weight by age 20-25, then increase only slightly thereafter. (Corrosion-doctors, 2011)

Cadmium is classified as an "A2-Suspected Human Carcinogen" by the American Conference of Governmental Industrial Hygienists (ACGIH). In addition, cadmium poses a chemical toxicity hazard to kidneys, lungs, and/or liver if ingested or inhaled. Repeated or prolonged exposure to cadmium can damage target organs. Severe over-exposure can result in death. Skin contact is considered an irritant or sensitizer, and eye contact is considered an irritant.

The main routes of non-occupational exposure to cadmium are inhalation and ingestion of fumes or dust; absorption through the skin is negligible. The major routes of occupational exposure to cadmium are inhalation of dust and fumes and incidental ingestion of dust from contaminated hands, cigarettes, or food. Inhalation includes (1) inhalation of cadmium-containing particles from ambient air; (2) inhalation of cigarette smoke; and (3) occupational inhalation of cadmium fumes and dusts.

Ingestion includes (1) consuming food -- food is the major source of non-occupational cadmium exposure for nonsmokers, with the largest contribution from grain cereal products, potatoes, and other vegetables; (2) drinking water -- drinking water normally has very low levels of cadmium; and (3) incidental ingestion of cadmium-contaminated soil or dust. Table 3 lists the harmful effects of exposure to cadmium.

| | |
|---|---|
| Short Term (Acute) Exposure | Cadmium is much more dangerous by inhalation than by ingestion. High exposures to cadmium that may be immediately dangerous to life or health occur in jobs where workers handle large quantities of cadmium dust or fume; heat cadmium-containing compounds or cadmium-coated surfaces; weld with cadmium solders; or cut cadmium-containing materials such as bolts. Severe exposure may occur before symptoms appear. Early symptoms may include mild irritation of the upper respiratory tract, a sensation of constriction of the throat, a metallic taste, and/or a cough. A period of 1–10 hours may precede the onset of rapidly progressing shortness of breath, chest pain, and flu-like symptoms with weakness, fever, headache, chills, sweating and muscular pain. Acute pulmonary edema usually develops within 24 hours and reaches a maximum by three days. If death from asphyxia does not occur, symptoms may resolve within a week. |
| Long Term (Chronic) Exposure | Repeated or long-term exposure to cadmium, even at relatively low concentrations, may result in kidney damage and an increased risk of cancer of the lung and of the prostate. |
| CFR, 2007, 29 CFR 1910.1027, Subpart Z, (2007) | |

Table 3. Harmful effects of cadmium exposure.

## 4. Source term evaluation

### 4.1 Mass transfer during the spilling phase

The amount of cadmium droplets and vapor which is transferred to the argon gas as a result of the electrorefiner cracking open and spilling the liquid cadmium may be estimated using the five-factor formula from DOE-HDBK-3010-94, (2000). This formula takes into account the material which sprays, evaporates and droplets which splash back up from the floor.

$$ST = MAR \times DR \times ARF \times RF \times LPF \tag{1}$$

where ST = Source term [mass entrained in the gas phase]
MAR = Material-at-risk = mass, spilled cadmium
DR = Damage ratio = fraction of mass involved
ARF = Airborne release fraction = stays in gas
RF = Respirable fraction = fraction assumed inhaled
LPF = Leak path factor = fraction which reaches people

The airborne release fraction, ARF, is evaluated from DOE-HDBK-3010-94, (2000) Volume 1, page 3-33 using the classification of a concentrated heavy metal solution. DOE-HDBK-3010-94 (2000), Section 3.2.3.1 discusses free-fall spill results from heights of 1 and 3 m for aqueous solutions with densities of approximately 1.0 (uranine) and approximately 1.3

(uranium in nitric acid - UNH). Section 3.2.3.1 further states that any solution containing heavy metal salts where the liquid alone has a density in excess of approximately 1.2 g/cm$^3$ is considered a "concentrated heavy metal solution" for assigning ARF and RF values and assigns a bounding ARF/RF of 2x10$^{-5}$/1.0 to concentrated heavy metal solutions.

Liquid cadmium has a density of 7.78 g/cm$^3$, a surface tension of 600 mN/m, and a viscosity of 2.5 cP (Crawley, 1972). For comparison, water at 20°C has a density of 1.0, a surface tension of 73 mN/m, and a viscosity of 1 cP. It is, therefore, reasonable to conclude that molten cadmium will have an ARF and RF more like concentrated heavy metal solutions than like aqueous solutions with densities of approximately 1.0 because it is more viscous, much more dense, and has a much higher surface tension than $H_2O$.

Each of DR, RF, and LPF are set equal to one which is very conservative. The LPF for a specific accident should be estimated as part of physical model development for the movement of the argon gas with the entrained cadmium. For unmitigated release models, the LPF is conservatively assigned as 1.0 (the maximum value).

Ballinger and Hodgson (1986) and Sutter, et. al. (1981) state that viscosity, density, and surface tension are interrelated, and that increases in all three cause a decrease in the mass airborne since more viscous solutions require more energy to break up the liquid into microscopic particles. (It is worth noting, though, that those properties which contribute to a lower ARF also contribute to a higher RF, so that nearly all of the material that becomes airborne is respirable. This issue was addressed by conservatively setting the RF equal to 1.0.) The resulting source term is included in Table 4.

| MAR | DR | ARF | RF | LPF | $ST_{Cd,\ spill}$ = MAR×DR×ARF×RF×LPF |
|---|---|---|---|---|---|
| 587.4 kg | 1 | 2.0E-5 | 1 | 1 | 1.175E-2 kg = 11.75 g |

Table 4. Source Term Parameters for a Release of Cadmium Vapor.

### 4.2 Evaporation mass transfer during the solidification

Experiments have shown that the spill of a liquid takes place quickly, then the liquid spreads out on the floor uniformly to form a thin liquid layer. It's spread is limited by the surface tension of the liquid on the floor or by some obstruction which keeps it from flowing further. The area is estimated from the volume of liquid spilled and the thickness of the liquid layer. In the case of the cadmium, it spills onto a steel lined concrete floor, the high conductivity of the cadmium and the steel causes it to solidify quickly. Calculations, shown later, indicate in less than 3 seconds. The heat transfer can be represented by transient one dimensional heat transfer in the vertical direction, and the mass transfer can be estimated from mass transfer for flow over a flat plate.

The molar vapor generation rate per unit area from an evaporating pool of liquid to an ambient gas above it is a function of the saturation mole fraction of the evaporating liquid at the temperature of the liquid, the mole fraction of the contaminant in the ambient gas, and the mass transfer coefficient between the evaporating liquid and the ambient. For simplification, the properties of air are used as the ambient gas instead of argon. The flux rate by convection is given by (Bird, et. al., 1960) as

The mass transfer coefficient for flow over a smooth flat plate is taken from Reinke and Brosseau, (1997, equations 3A and 4A) and depends upon the Reynolds Number, the Schmidt Number, and the gas velocity over the liquid pool. It is given by

$$k = 0.664 \times Re_L^{(-0.5)} \times u \times Sc^{(-2/3)} \text{ for } Re_L < 15{,}000 \tag{3a}$$

$$k = 0.036 \times Re_L^{(-0.2)} \times u \times Sc^{(-2/3)} \text{ for } 15{,}000 \le Re_L \le 300{,}000 \tag{3b}$$

where $Re_L$ = Reynolds Number [ unitless ]
u = air velocity over liquid pool [ m / s ]
Sc = Schmidt Number [ unitless ]

The velocity in the argon cell is estimated at 0.5 m/s. All the properties in the above expression are evaluated at standard cell conditions of $T_{Cell}$ = 35°C and $P_{Cell}$ = 630 mmHg (from Perry, 1997) which is the standard operating conditions of the argon cell. The value of L is taken to be 2.75 m which is the smallest value of length considered reasonable. The mass transfer coefficient varies inversely to L so the smaller value of L is conservative in producing more mass flux.

$$Re = \frac{uL\rho_{air}}{\mu_{air}} = \frac{0.5\frac{m}{s} * 2.75m * 0.95\frac{kg}{m^3}}{1.74 * 10^{-5}\frac{kg}{m\,s}} = 74{,}900 \tag{4}$$

$$Sc = \frac{\mu_{air}}{\rho_{air} * D_{Cd,air}} = \frac{1.74 * 10^{-5}\frac{kg}{m\,s}}{0.95\frac{kg}{m^3} * 1.70 * 10^{-5}\frac{m^2}{s}} = 1.08 \tag{5}$$

Correlation 3b is chosen due to the value of Reynolds number so substituting these values into Equation 3b, the mass transfer coefficient is

$$k = 0.00181 \text{ m/s}$$

Reinke and Brosseau (1997) give two other models for estimating the mass transfer. The penetration model gives numbers that are a bit larger (10%) than the flat plate. It is much more reasonable to use a well researched correlation like the flow over a flat plate so that is used here. It is realized that the accuracy aim for developing a model for the source term is to be accurate at best by a factor of 2.

The heat balance on the cadmium should contain a loss for the vaporization. This term has been neglected in the evaluation of the heat transfer which will produce a higher temperature and a higher mass flux. So this neglect is conservative. Following is the prediction of the transient temperature of the cadmium. This is followed by an evaluation of the mass transferred to the gas phase during solidification using the temperature profile and cadmium vapor pressure.

### 4.2.1 Cadmium transient temperature

To determine the transient temperature of the cadmium pooled on the floor of FCF, it is necessary to estimate its depth. This analysis assumes that the entire quantity of molten cadmium present in the MK-IV ER, 587.4 kg, spills on the floor. The volume of material released, $V_{pool}$ is calculated from mass and density (mass/volume) as:

$$V_{pool} = \frac{m}{\rho} = \frac{587.4\,kg}{7778\,\frac{\text{kg}}{\text{m}^3}} = 0.0755\,m^3$$

Three different depths of the spill (pool) were considered: 1.27 cm, 1 cm, and 0.39 cm. Dividing the volume by these three depths gives the respective areas of 5.95 m$^2$, 7.55 m$^2$, and 19.36 m$^2$. The deeper the pool, the longer the cadmium takes to solidify. The second depth is recommended by Reinke and Brosseau,1997. The third depth is calculated from a model presented in Solbrig and Clarksean, 1993 which determines the depth from a balance between gravity and surface tension forces to estimate when a spill will stop spreading. The first depth is taken as a conservative estimate of the thickest depth and is used to determine the temperature time and solidification time for the cadmium. The third depth produces the largest area of the spill. These two values are used in conjunction to produce the largest flux (1.27 cm) and the largest area 19.36 m$^2$ to estimate the total cadmium evaporated.

The results of the thermal analysis of the cadmium spilling directly onto the steel liner is shown in Figure 6. On the upper surface of the cadmium, the heat transfer coefficient is 8.72 W/m K convecting to the 35°C cell gas, and the emissivity is 0.8 radiating to a 35°C cell environment. The cadmium starts at 500°C, cools down and solidifies at 321°C. The cadmium thickness, 1.27 cm, is the same as the steel liner. The cadmium is completely solidified by 3 seconds.

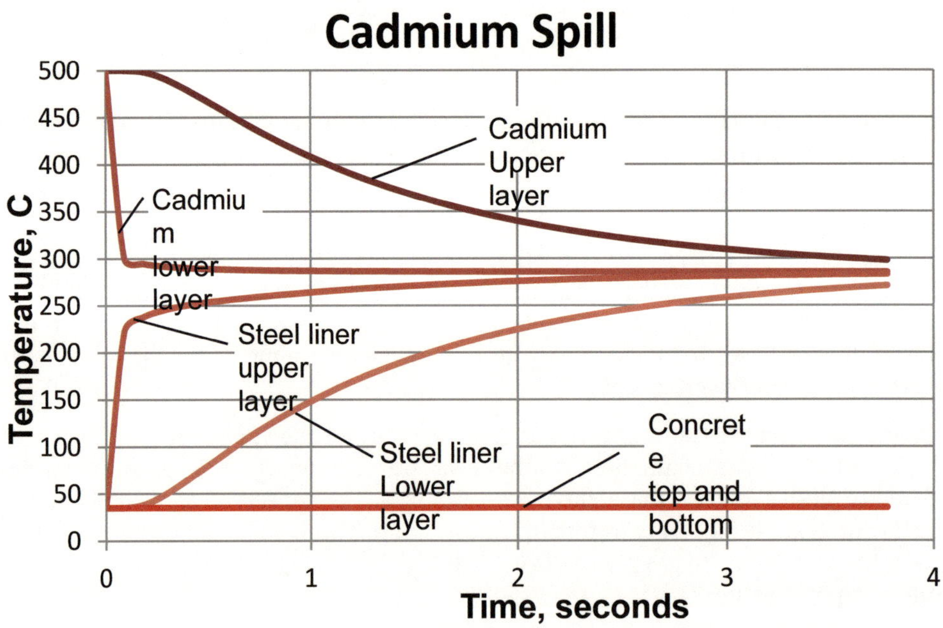


Fig. 6. Temperatures from spilling 500°C cadmium on 35°C steel-lined concrete floor

The upper surface temperature of the cadmium projects a vapor pressure which determines the mass transfer to the ambient.

#### 4.2.2 Vapor generation determination

The vapor pressure curve for cadmium is shown in Figure 7. Due to lack of data below 420°C, the curve is conservatively approximated from 321°C to 420°C.

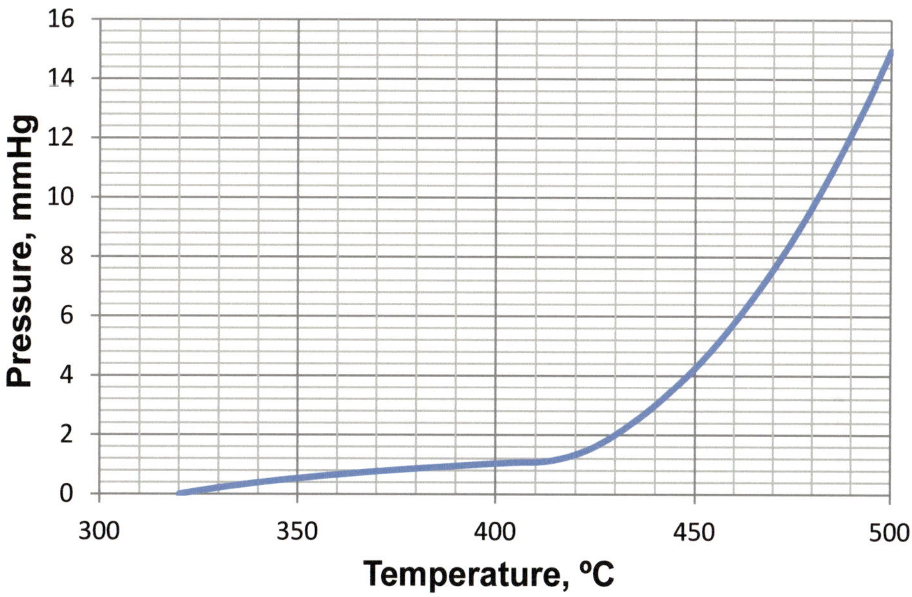


Fig. 7. Vapor Pressure of Cadmium

The temperature data in Figure 6 and the vapor pressure in Figure 7 can be combined to yield the vapor pressure of the cadmium versus time as shown in Figure 8.

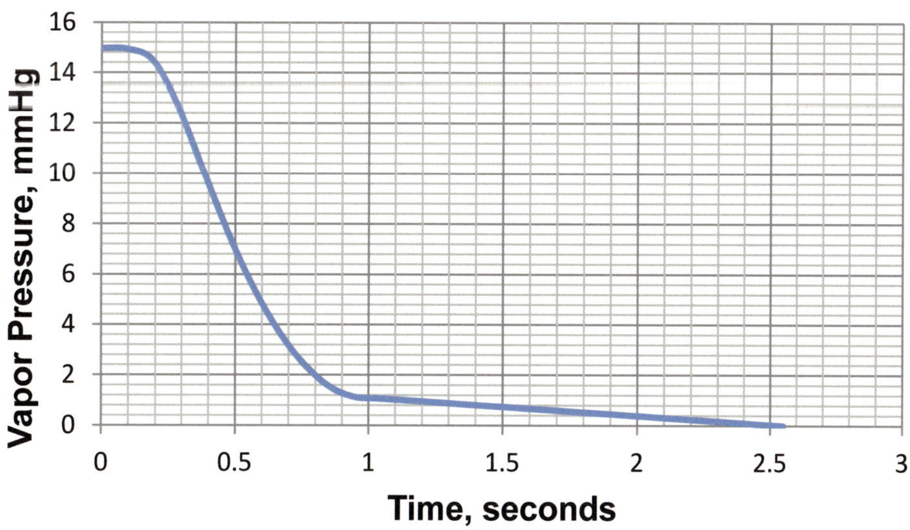


Fig. 8. Vapor Pressure during Solidification

Now, Equation 2 can be used to estimate the molar flux rate leaving the surface by utilizing Dalton's law which states that the saturation mole fraction may be evaluated as the ratio of the saturation pressure to the ambient pressure since it is applicable when contaminant concentrations are low and pressure is near atmospheric.

$$y_{Cd,sat} = \frac{P_{Cd,sat}}{P_{cell}}$$

The mole fraction in the ambient is set equal to zero to maximize the calculated flux. Thus Equation 2 reduces to

$$\frac{dn}{dt} = k\tilde{\rho}\left(\frac{P_{Cd,sat}}{P_{cell}}\right)$$

Thus, the molar flux rate is proportional to the vapor pressure. By substituting in the values for k, into this equation, the molar flux can be determined and is shown in Figure 9.

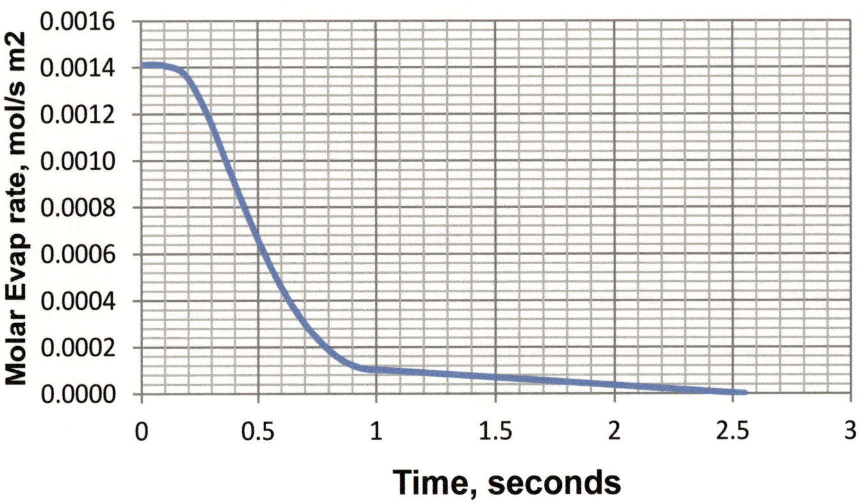


Fig. 9. Molar flux of Cadmium

The total molar flux over 3 seconds is obtained by integrating the instantaneous molar flux in Figure 9 to obtain 7.42x10$^{-4}$ mol/m$^2$. The mass flux is obtained by multiplying this by the molecular weight of cadmium and converting from kg to mg, to obtain 83.5 mg/m$^2$. The total mass over the whole surface is obtained by multiplying this by the 19.36 m$^2$ area to obtain $ST_{Cd,Evap}$ = 1620 mg.

$$ST_{Cd,Evap} = 7.42 * 10^{-4}\frac{gmol}{m^2} * 0.112\frac{kg}{gmol} * 19.36m^2 = 1620\,mg$$

## 5. Evaluation of airborne concentrations

### 5.1 Airborne concentrations in the argon cell operating corridor

Indoor airborne concentrations are evaluated based on the source terms described in Section 4 assuming an instantaneous release with rapid mixing throughout the available argon cell plus operating corridor volume and no subsequent losses, such as those due to leakage, ventilation, plate-out, or deposition.

The release occurs in the argon cell with subsequent release to the argon cell operating corridor through the small area opened up in the cell wall by the earthquake assuming no resistance for interchange between the cell and the corridor, the volume over which the cadmium is dispersed is the combined volume of the argon cell and the argon cell operating corridor. Assume (1) the mass vapor generation rate remains constant over the release duration; (2) rapid mixing occurs throughout the available dispersion volume; and (3) there are no subsequent losses, such as those due to leakage, ventilation, plate-out, or deposition.

$$V_D = 3{,}568 \; [\; m^3 \;]$$

Corridor Airborne Concentration for the spilling phase: The cadmium source term determined in section 4.1 is

$$C_{Cd,indoor,spill} = ST_{Cd,\,spill} \;/\; V_D$$

where
$C_{Cd,indoor,spill}$ = indoor airborne concentration of cadmium [ mg/m$^3$ ]
$ST_{Cd,\,spill}$ = airborne respirable source term of cadmium [ mg ]
$V_D$ = available dispersion volume [ m$^3$ ]

Corridor Airborne Concentration for the Evaporation Phase: The cadmium source term determined in section 4.2 is

$$C_{Cd,indoor,spill} = ST_{Cd,\,spill} \;/\; V_D$$

where
$C_{Cd,indoor,evap}$ = indoor airborne concentration of cadmium [ mg/m$^3$ ]
$ST_{Cd,\,evap}$ = airborne respirable source term of cadmium [ mg ]
$V_D$ = available dispersion volume [ m$^3$ ]

### 5.2 Outside airborne concentrations

Outside airborne concentrations are evaluated based on the source terms described in Section 4 and applicable dispersion coefficients, assuming a release of material to the outside atmosphere with no losses within the building. Thus the source term is actually used twice. Once, assuming all the cadmium stays in the FCF building. The second time, it is assumed that all the material leaves the building and is dispersed in the atmosphere. Three separate locations are tabulated for the possible dose to an individual at that location. These are:

100 m Collocated worker
280 m The distance from the FCF to the bus parking lot is 280 m (0.2 miles).
5,000 m Off-site public

The distance from the FCF to the nearest site boundary is 5,000 m (3 miles). Consequences to the off-site public are evaluated at 5,000 m (3 miles) although no one would actually park and stay at the site boundary.

**Dispersion Coefficient** ($\chi/Q$)An effective dispersion coefficient ($\chi/Q$), based on variable parameters such as release height, distance to receptor, and receptor height, is applied when evaluating outside releases and environmental exposures. The factor ($\chi/Q$)describes Gaussian plume dispersion and may be hand calculated or determined using Radiological Safety Analysis Computer Program (RSAC, 2003) based on meteorological and other input parameters. The dispersion coefficient is used to estimate the quantity of contaminants at a location of interest as the result of an airborne release, based on Pasquill stability class and distance from point of release to receptor. In calculations, it is used as a reduction factor to account for the effects of transporting the contaminant from the release point to the receptor location. For the purposes of this evaluation, the maximally exposed individual (MEI) is assumed to be at the distance corresponding to the maximum dispersion coefficient. This analysis uses very non dispersive meteorology which would persist 95% or greater for outdoor dispersion estimates. The result is an extremely conservative evaluation. For this reason, an actual release scenario is expected to result in consequences less severe than those estimated in this chapter.

These calculations conservatively neglects the particle deposition that would normally occur as the material is transported from the point of release to the receptor location, which would cause a further reduction of the airborne concentration at the receptor location. Note that dispersion methodologies are not precise and provide only estimates of the consequences of an accidental release. The ($\chi/Q$) values used in this evaluation were developed using INL's Radiological Safety Analysis Computer Program (RSAC, 2003) and are listed in Table 5. Higher dispersion coefficients mean that the weather is calmer so that less dispersion takes place. For conservatism, the larger, ($\chi/Q$) values obtained using the Markee method are used for all dose consequence calculations resulting in higher estimates of the cadmium exposure.

| Distance from release point to receptor | 100 m | 280 m | 5,000 m |
|---|---|---|---|
| $\chi/Q$ using Hilsmeier-Gifford method (release duration <15 minutes) | $3.217 \times 10^{-2}$ s/m$^3$ | $5.638 \times 10^{-3}$ s/m$^3$ | $5.697 \times 10^{-5}$ s/m$^3$ |
| $\chi/Q$ using Markee method (15 minutes ≤ release duration ≤60 minutes) | $4.081 \times 10^{-3}$ s/m$^3$ | $1.309 \times 10^{-3}$ s/m$^3$ | $4.120 \times 10^{-5}$ s/m$^3$ |
| Reference: RSAC, 2003 | | | |

Table 5. Dispersion coefficients for the Fuel Conditioning Facility

**Outside Airborne Concentration – Spilling Phase**

Outside airborne concentrations are evaluated based on the source terms described in Section 4, release duration, and applicable dispersion coefficients, assuming a release of material to the outside atmosphere with no losses within the building, and the entire quantity of material is released over the release duration (t). Although the release takes place very quickly, in order to compare it to the limits, the average rate over 60 minutes must be used.

**Outside Airborne Concentration – Spilling Phase**

$C_{Cd,outside,spill} = ST_{Cd,spill} \times (\chi/Q) / t$

where $C_{Cd,outside,spill}$ = outside airborne concentration of cadmium [ mg/m$^3$ ]

$ST_{Cd,spill}$= airborne respirable source term of cadmium [ mg ]

t = release duration [ s ]=15 min

Q = dispersion coefficient [ s/m$^3$ ]

**Outside Airborne Concentration -- Evaporation Phase**

$C_{Cd,outside,evap} = ST_{Cd,evap} \times (\chi/Q) / t$
where $C_{Cd,outside,evap}$ = outside airborne concentration of cadmium [ mg/m$^3$ ]
$ST_{Cd,evap}$= airborne respirable source term of cadmium [ mg ]

## 6. Toxicological exposure parameters

The dose guidelines that are used in this chapter are based upon Temporary Emergency Exposure Limits (TEEL) Protective Action Criteria (PAC) for exposure to the guideline concentration for 60 minutes that are defined at the DOE website (DOE, 2011). The PACs (PACs, 2009) used in this chapter are defined as follows:

**TEEL-0**: This is the threshold concentration below which most people will experience no appreciable risk of health effects. This PAC is always based on TEEL-0

**PAC-1**: The maximum concentration in air below which it is believed nearly all individuals could be exposed without experiencing other than mild transient adverse health effects or perceiving a clearly defined objectionable odor.

**PAC-2**: The maximum concentration in air below which it is believed nearly all individuals could be exposed without experiencing or developing irreversible or other serious health effects or symptoms that could impair their abilities to take protective action.

**PAC-3**: The maximum concentration in air below which it is believed nearly all individuals could be exposed without experiencing or developing life-threatening health effects.

These are:

1. The dose a person could be exposed to for an eight hour day without experiencing adverse health effects 0.005 mg/m$^3$.
2. The maximum concentration in air below which it is believed nearly all individuals could be exposed without experiencing other than mild transient adverse health effects or perceiving a clearly defined objectionable odor. 0.03 mg/m$^3$.
3. The maximum concentration in air below which it is believed nearly all individuals could be exposed without experiencing or developing irreversible or other serious health effects or symptoms that could impair their abilities to take protective action. 1.25 mg/m$^3$.
4. The maximum concentration in air below which it is believed nearly all individuals could be exposed without experiencing or developing life-threatening health effects for 60 minutes. 9.0 mg/m$^3$

A 60 minute exposure or longer to the dose of 9.0 mg/m$^3$ would be expected to produce life-threatening health effects. Based on the above guidelines, Table 6 lists airborne concentrations that bound consequence categories.

| Persons Receiving Dose | Negligible if less than: | Low if less than: | Moderate if less than: | High if greater than |
|---|---|---|---|---|
| All workers | 0.03 [ mg / m$^3$ ] | 1.25 [ mg / m$^3$ ] | 9 [ mg / m$^3$ ] | 9 [ mg / m$^3$ ] |
| Off-site public | 0.002 [ mg / m$^3$ ] | 0.03 [ mg / m$^3$] | 1.25 [mg / m$^3$] | 1.25 [mg / m$^3$] |

Table 6. Chemical Toxicity Consequence Thresholds For Cadmium.

Since 9 mg/m³ is the concentration 60 minute exposure limit for morbidity, then other exposures may be compared to the product of these two so the morbidity concentration time limit is 540 mg minutes/m³. Thus, an exposure of 1.00 mg/m³ for 9 hours or an exposure of 18 mg/m³ for 30 minutes would meet the criteria for life-threatening health effects as well.

## 7. Consequences of a vapor release due to liquid cadmium spill

The consequences are based on the DOE protective action criteria (PAC) described above, and are in the form of maximum allowable airborne concentrations that correspond to consequence levels of negligible, low, moderate and high. The evaporation phase is also very short. Heat transfer calculations show that the cadmium solidifies in less than 3 seconds. This mass is assumed to be uniformly distributed over the FCF argon cell and the operating corridor In addition at the same time, this same mass is assumed to be released to the atmosphere to expose workers on site, workers at the bus area, and members of the people at the site boundary. This release is averaged over 15 minutes. The morbidity PAC limit of 9 mg/m³ is an airborne concentration which is breathed for 60 minutes, it is really the integral of the concentration times the time increment at that concentration over time. Using an averaged value over 15 minutes is appropriate in analyzing the effects of a much higher release concentration which occurs only over 3 seconds. Applying these conservative methods result in the following doses and consequences

The airborne concentration levels are calculated in Table 7 assuming that the release occurs over a 15 minute period.

| Receptor | | Concentration (mg/m³) | Consequences |
|---|---|---|---|
| Facility worker | $11.75 \times 10^3$ mg ÷ 3568 m³ | 3.29 | MODERATE |
| On Site (100m) | $11.75{\times}10^3$ mg /900s×$4.081{\times}10^{-3}$s/m³ | 0.0533 | LOW |
| Bus Area (280 m) | $11.75{\times}10^3$ mg /900s×$1.309{\times}10^{-3}$ s/m³ | 0.0153 | NEGLIGIBLE |
| Public (5,000 m) | $11.75{\times}10^3$ mg /900s×$4.120{\times}10^{-5}$ s/m³ | 0.00054 | NEGLIGIBLE |

Table 7. Consequences of Cadmium Vapor Release Due To MK-IV ER Spill – Spilling

The dose consequence of the evaporation phase is included in Table 8. . The total evaporated mass is conservatively estimated to be 1620 mg in less than 3 seconds. As in the spilling phase, the evaporation rate is average over 15 minutes.

| Receptor | | Concentration (mg/m³) | Consequences |
|---|---|---|---|
| Facility worker | 1620 mg/3568 m³ | 0.454 | LOW |
| Site worker 100 m | 1620 mg/900s × $4.081 \times 10^{-3}$ s/m³ | 0.0073 | NEGLIGIBLE |
| Bus Area (280 m) | 1620 mg/900s × $1.309 \times 10^{-3}$ s/m³ | 0.0023 | NEGLIGIBLE |
| Public (5,000 m) | 1620 mg/900s × $4.120 \times 10^{-5}$ s/m³ | 0.00007 | NEGLIGIBLE |

Table 8. Consequences of Cadmium Vapor Release Due To MK-IV ER Spill – Evaporation

The sum of the two phases is included in Table 9 along with the consequences.

| Receptor | Concentration (mg/m$^3$) | Concentration (mg/m$^3$) | Concentration (mg/m$^3$) | Consequences |
|---|---|---|---|---|
| Facility worker | 3.29 | 0.073 | 3.74 | MODERATE |
| Site worker (100 m) | 0.0533 | 0.358 | 0.0606 | LOW |
| Bus Area (280 m) | 0.0153 | 0.115 | 0.0186 | NEGLIGIBLE |
| Public (5,000 m) | 5.38E-04 | 0.003 | 0.00061 | NEGLIGIBLE |

Table 9. Consequences of Cadmium Vapor Release Due To MK-IV ER Spill - Combined Spilling and Evaporation

So in conclusion, with the many conservative assumptions involved in this analysis, the consequences of this possible accident are extremely small for collocated workers and the public. For a facility worker who remains in the building for 60 minutes the exposure is only moderate, however, staying there for 180 minutes would be detrimental to his health and could cause death. Without the failure of the SES, there would be negligible doses to all workers and the general public since all releases would be through two sets of seismically qualified HEPA filters which would reduce the releases outside the building by a factor of four and eliminate the dose to the facility workers.

## 8. Acknowledgement

Work supported by the U.S. Department of Energy, Office of Nuclear Energy (NE) under DOE Idaho Operations Office Contract DE-AC07-05ID14517.

## 9. References

Ballinger, M.Y., and Hodgson, W.H., (1986), "Aerosols Generated by Spills of Viscous Solutions and Slurries," NUREG/CR-4658, Nuclear Regulatory Commission, Battelle Pacific Northwest Labs, Richland, WA, Dec 1986.

Bird, R..B., Stewart, W.E., and Lightfoot, E.N., Transport Phenomena, P. 655 to 659 ff, John Wiley & Sons, 1960.

Corrosion-doctors, (2011), http://corrosion-doctors.org/Elements-Toxic/Cadmium-distribution.htm

Chemicool, (2011), http://www.chemicool.com/elements/cadmium.html

Crawley, A. F. (1972), "Densities and Viscosities of Some Liquid Alloys of Zinc and Cadmium," *Metallurgical Transactions B*, Volume 3, No. 4, p. 971-975, 1972.

CFR, 2007, 29 CFR 1910.1027, Subpart Z, (2007), "Toxic and Hazardous Substances, Cadmium, Appendix A, Substance Safety Data Sheet, Cadmium," *Code of Federal Regulations*, Office of the Federal Register, July 2007. Reference: 29 CFR, 2007.

DOE, 2011, http://www.hss.energy.gov/HealthSafety/WSHP/chem_safety/teel.html

DOE-HDBK-3010-94, (2000), Change 1, "Airborne Release Fractions/Rates and Respirable Fractions for Nonreactor Nuclear Facilities," U.S. Department of Energy, 2000.

ICSC (2005), "International Chemical Safety Cards," Cadmium, National Institute for Occupational Safety and Health, Validated April 2005,

NUREG/CR-6410, *Nuclear Fuel Cycle Faciility Accident Analysis Handbook*, Division of Fuel Cycle Safety and Safeguards, Office of Nuclear Material Safety and Safeguards, U. S. Nuclear Regulatory Commission, Washington, DC, March 1998.

PACs (2009), "Table 2: Protective Action Criteria (PAC) Rev 24B based on applicable60-minute AEGLs, ERPGs, or TEELs (Chemicals listed in alphabetical order)," June 2009, http://www.atlintl.com/DOE/teels/teel/Revision_24B_Table2.pdf http://www.cdc.gov/niosh/ipcsneng/neng0020.html

Perry, 1997, *Perry's Chemical Engineers' Handbook*, 7th Edition, McGraw-Hill, 1997.

Reinke, P. H. and Brosseau, L. M., (1997), "Development of a Model to Predict Air Contaminant Concentrations Following Indoor Spills of Volatile Liquids," British Occupational Hygiene Society, *Annals of Occupational Hygiene*, Volume 41, No. 4, p. 415-435, 1997.

RSAC 6.2, (2003), Radiological Safety Analysis Computer Program, Version 6.2, Idaho National Laboratory, December 2003.

Solbrig, C. W., and Clarksean, R. L., (1993), "Determination of the Shape of a Plutonium Deposit from a Leaking Crucible," Proceedings of the Second International Conference of Nuclear Engineering, San Francisco, Ca., USA, March 1993, ASME/JSME Nuclear Engineering Conference, Vol. 1, pp. 671-682.

Sutter, S.L., Johnston, J.W., Mishima, J., (1981), "Aerosols Generated by Free Fall Spills of Powders and Solutions in Static Air," NUREG/CR-2139, Nuclear Regulatory Commision, Battelle Pacific Northwest Labs, Richland, WA, Dec, 1981.

Till, C. E., and Chang, Y. I., "*Evolution of the Liquid Metal Reactor: The Integral Fast Reactor (IFR) Concept*," American Power Conference, Chicago, IL, April 24-28, 1989.

Till, C. E., Chang, Y. I., and Hannum, W. H., "*The Inegral Fast Reactor-an Overview*," Progress in Nuclear Energy, Vol 31, P. 3, 1997.

# 14

# Radioactive Waste Management of Fusion Power Plants

Luigi Di Pace[1], Laila El-Guebaly[2],
Boris Kolbasov[3], Vincent Massaut[4] and Massimo Zucchetti[5]
[1]*EURATOM/ENEA Fusion Association, ENEA C.R Frascati*
[2]*University of Wisconsin-Madison, Madison, Wisconsin*
[3]*Kurchatov Institute, Moscow*
[4]*SCK – CEN, Mol*
[5]*EURATOM/ENEA Fusion Association, Politecnico di Torino, Torino*
[1,5]*Italy*
[2]*USA*
[3]*Russia*
[4]*Belgium*

## 1. Introduction

This chapter outlines the attractive environmental features of nuclear fusion, presents an integral scheme to manage fusion activated materials during operation and after decommissioning, compares the volume of fusion and fission waste, covers the recycling, clearance, and disposal concepts and their official radiological limits, and concludes with a section summarizing the newly developed strategy for fusion power plants.

As fusion plays an essential role in the future energy market providing an environmentally attractive source of nuclear energy (Ongena & Van Oost, 2001), it is predictable that there will be tens of fusion power plants commissioned worldwide on an annual basis by the end of the 21$^{st}$ century. The ability of these fusion power plants to handle the radioactive waste stream during operation and after decommissioning suggests re-evaluating the underground disposal option at the outset before considering the environmental impact statement needed for licensing applications. Adopting the 1970s preferred approach of disposing the activated materials in geological repositories after plant decommissioning is becoming difficult to envision because of the limited capacity of existing repositories, difficulty of building new ones, tighter environmental control, and radwaste burden for future generations. Alternatively, fusion scientists are currently promoting a new strategy: avoid underground disposal as much as possible, implement at the maximum extent the recycling of activated materials within the nuclear industry, and/or the clearance and release to commercial markets if materials contain traces of radioactivity. This strategy requires a major rethinking and strong R&D program, hoping all fusion developing countries will be strongly supportive of the proposed recycling and clearance approaches.

Ever since the development of nuclear fusion designs in the early 1970s, most of the related studies and experiments have been devoted to the deuterium (D) and tritium (T) fuel cycle – the easiest way to reach ignition and the preferred cycle (feasible with current technology) for the first generation of fusion power facilities. Nevertheless, the stress on fusion safety has stimulated worldwide research on fuel cycles other than D-T, based on advanced reactions with a much lower neutron level. The focus of this chapter is on fusion power plants fuelled with D-T where the reaction can be expressed as follows:

$$ {}^{2}_{1}\mathrm{D} + {}^{3}_{1}\mathrm{T} \rightarrow {}^{4}_{2}\mathrm{He}\,(3.5\ \mathrm{MeV}) + {}^{1}\mathrm{n}\,(14.1\ \mathrm{MeV}) \qquad (1) $$

Fig. 1(a) shows a schematic of the D-T fusion reaction, while Fig. 1(b) compares the D-T reaction with other potential fusion reactions according to their reaction rate [1].

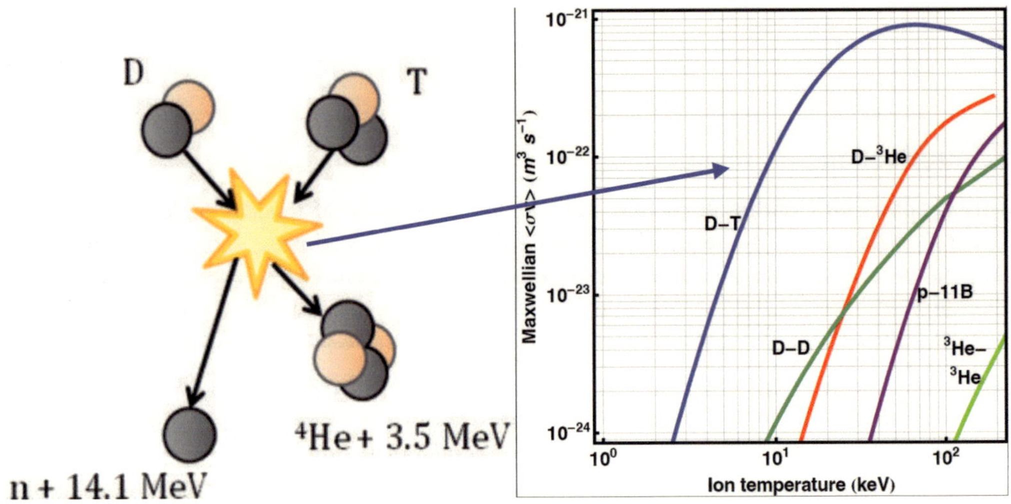


(a) (b)

Fig. 1. (a) D-T fusion reaction
(from http://en.wikipedia.org/wiki/File:Deuterium-tritium_fusion.svg)
(b) Fusion reaction rates
Courtesy of J. Santarius (University of Wisconsin, USA)

[1] The field of plasma physics deals with phenomena of electromagnetic nature that involve very high temperatures. It is customary to express temperature in electronvolts (eV) or kiloelectronvolts (keV), where 1 eV = 11605 K. That derives from the equivalence energy-temperature expressed by the Boltzmann constant $k = 1.3807 \times 10^{-23}$ J·K$^{-1}$, which corresponds to $8.6175 \times 10^{-5}$ eV/K, as 1 eV = $1.6022 \times 10^{-19}$ J. Hence 1 eV = 1/ $8.6175 \times 10^{-5}$ = 11605 K

As noted, the D-T reaction rate peaks at a lower temperature (about 70 keV, or 800 million Kelvin) with a higher value than other reactions commonly considered for future fusion devices with advanced fuel cycles. Deuterium can be easily extracted from seawater while tritium can be produced through neutron interaction with lithium (a readily available light metal in the earth's crust). Although the products of the D-T fusion reaction (helium and neutrons) are not radioactive, neutrons are absorbed/captured by structural materials and fluids surrounding the plasma. The 14.1-MeV energetic neutrons can transmute some elements of the structural materials and produce radioactive isotopes. These materials belong principally to the in-vessel components (e.g. blanket, shield and divertor of a tokamak[2] fusion plant). Furthermore, a small percentage of the D-T fuel is consumed and some tritium (the one not reacting with deuterium and not extracted from the plasma chamber) could escape and contaminate the plasma facing components by various mechanisms (diffusion, implantation and co-deposition). Hence, the issue of fusion radioactive waste handling is not only linked to the safe and environmentally friendly management of activated materials, but also to the detritiation and treatment of contaminated components.

## 2. The attractive environmental features of fusion

Fusion devices, although being nuclear installations, have certain characteristics as to make them environmentally friendly devices. Prior to analyzing the management scheme of fusion activated materials, it is worthwhile to highlight what makes fusion energy safe and environmentally attractive compared to other nuclear energy sources:

- There is no chain nuclear reaction.
- A small amount of fuel circulates (order of grams) in the reaction chamber which maintains the D-T reaction for only few seconds.
- The power density in a fusion reactor is much lower than that of fission reactors and it can be limited by design in such a way to moderate the consequences of most severe accidents.
- The main radioactive inventory is generated by neutron activation of plasma surrounding components. This activation process, indeed, depends strongly on the type of irradiated materials and the careful choice of material constituents.

These and other factors corroborate the hypothesis that fusion power, with a safety-oriented design and a smart choice of its constituting materials, can be intrinsically safe with very low probability of severe accidents (and even in case of accident, without important impact on the surrounding population) and minimal environmental impact (Gulden at al., 2000). These attractive features are defined as the "intrinsically safe" characteristics of fusion.

## 3. Comparison with nuclear fission radioactive waste management

As noticed, differences exist between fission and fusion in terms of fuels, reaction products, activated material type, activity levels, half-life, radiotoxicity, etc. The quantity of activated material originating from the fusion power core is larger than that from the fission core (per

[2] A tokamak is a toroidal device that employs magnetic fields to confine the plasma in the shape of a torus.

unit of electricity produced) (El-Guebaly et al., 2008). The main differences between fission and fusion waste are related to their radiotoxicity (much higher in fission for waste originating from the fuel cycle) and waste form for their final disposal. When recycling is conceived, fission has a large share of highly radioactive and radiotoxic liquid secondary waste from spent fuel reprocessing, which has to be solidified by cementation or vitrification. Fusion waste in terms of volume is mostly solid and does not require those processes in extensive way. However, fusion solid waste too requires treatment (decontamination, detritiation, cutting, compacting) and conditioning (stabilizing e.g. by grout, packaging, etc.) which will generate some secondary waste requiring solidification. It is worthwhile to mention that tritiated water at low tritium concentration will be produced as well from the Fuel Cycle Systems requiring treatment and in some cases conditioning. Most importantly, the fusion generated waste is not intrinsic to the fusion reaction, and therefore is more controllable. Thus, providing prudent and intelligent selection of materials and processes (avoiding noxious impurities), fusion reactors can avoid generating high level and long-lived waste streams. This is probably the most important difference between fusion and fission radioactive waste, and this will have an important impact on their management.

Nuclear weapon proliferation issue of a nuclear device - such as a tokamak-based fusion power plant - needs to be thoroughly addressed. If future fusion power plants can utilize advanced fuel cycles (such as D-$^{3}$He, or $^{3}$He-$^{3}$He), the fuel cycle will practically be tritium-free. However, using the D-T reaction, two main proliferation aspects have to be addressed:

1. Tritium is a weapon proliferation relevant material. However, the emphasis of the NPT (Non-Proliferation Treaty) is on fissionable substances and technologies that are related to U and Pu bombs, as the Treaty excludes fusionable materials.
2. The presence of intense neutron fluxes may bring their use to irradiate uranium in order to breed plutonium. It would also be possible to breed another fissile material, $^{233}$U, through the irradiation of thorium.

Concerning the second point, a possible proliferation-relevant technique could involve an infrequent replacement of a tritium-breeding blanket with modules breeding fissile-fuel. In a fusion power plant it would be much easier to enforce safeguard because one would be looking for fissile or fertile material in an environment where few quantities of it or not at all should be present, in contrast to looking for small discrepancies in the large inventories of a fission power plant. To conclude, the proliferation relevance of a tokamak-based fusion power plant would pose solvable problems from the safeguards viewpoint.

## 4. Previous results of back end studies for fusion power plants

Ever since the late 1990s, some studies have been carried out at international level to analyse waste management issues related to operation of future commercial power plants, focused on the three scenarios for managing fusion active materials: disposal, recycling, and clearance (i.e. declassification to non-radioactive material). They have been applied to selected U.S. and European fusion power plant studies: SEAFP (Raeder (ed.), 1995), ARIES (El-Guebaly, 2007), and PPCS (Maisonnier et al., 2005) (mostly tokamak-based designs, with the sole exception on ARIES-CS – a compact stellarator[3]). In general, these

[3] A stellarator is a device using only external magnetic coils to confine hot plasma and to sustain a controlled nuclear fusion reaction

studies estimated the amount of potential radioactive waste generated by the fusion power plants of different design concepts and their share amongst the different potential routes of management. These studies made use of classification and categorization approaches and management schemes similar to those of fission waste management. As a general conclusion, it was possible to outline the feasibility of recycling and clearance in extensive way.

These approaches became more technically feasible in recent years with the development of radiation-resistant remote handling (RH) tools and the introduction of the clearance category for slightly radioactive materials by the International Atomic Energy Agency (IAEA, 2004) and other national nuclear agencies (US-NRC, 2003; EC-RP, 2000; 2003; NRB, 1996). A great deal of the decommissioning materials (up to 80%) has a very low activity concentration and can be cleared from regulatory control, especially when a duration period (up to 100 y) of interim storage is anticipated (US-NRC, 2003). The remaining 20% of the active materials could be disposed of as low level waste (LLW) or preferably recycled using a combination of advanced and conventional RH equipment. Most fusion active materials contain tritium that could introduce complications to the recycling process. A detritiation treatment prior to recycling is, then, imperative for fusion components with high tritium content (El-Guebaly et al., 2008).

## 5. Revision of clearance and recycling concepts and limits

"Clearance" (unrestricted release from regulatory control) means that the material complying with the requirements defined by the national regulatory authorities can be handled as if it contains no radioactivity significantly higher than naturally occurring. Under this option, solid material can be reused without restriction, recycled into a consumer product, or disposed off in any industrial landfill.

The main requirement for the unconditional clearance is that the Clearance Index CI must be below unity. CI is given by the following relationship:

$$CI = \sum_i \left( A_i / L_i \right) \quad (2)$$

where $A_i$ is the specific activity of a nuclide after storage, $L_i$ is the clearance limit and $i$ represents the different nuclides contained in the material.

The clearance concepts and limits have been under development since the early 1950s. In 1996, the IAEA prepared an interim report [TECDOC-855 (IAEA, 1996)] on recommended clearance limits for 1650 radionuclides of interest to fission and fusion applications. However, these recommendations were never endorsed by all the IAEA member states. Solely Russia included 297 of the limits recommended by the IAEA interim report into its Radiation Safety Regulations of 1996 (NRB, 1996) as "minimally significant specific activities" (MSSA). These regulations were revised twice in 1999 and 2009 (NRB, 1999, 2009), but the MSSA values were not changed, although the number of the radionuclides and natural radioactive elements covered by these regulations was increased up to 300.

A set of documents on the same topic was published by European Commission (EC) (EC-RP, 2000, 2003) and U.S. Nuclear Regulatory Commission (NRC) NUREG-1640 (US-NRC, 2003). IAEA published in 2004 the revised clearance standards (IAEA, 2004) for 277 radionuclides, claiming to take into account the U.S. NUREG-1640 document and European Commission evaluations. The majority of the new clearance limits recommended by the IAEA were notably lower (down to 4 orders of magnitude for some radionuclides) than the values proposed in 1996 by the TECDOC-855 (IAEA, 1996).

All the standards under consideration take the limit for the annual individual effective dose of 10 μSv as the basis for clearance of solids from regulatory control. Nevertheless, a difference by 1-2 orders of magnitude between the last clearance limits recommended by the IAEA and U.S. NRC is observed for many radionuclides. Some clearance limits differ even by 3-4 orders of magnitude. Discrepancy between values offered by the IAEA and EC is less (not greater than an order of magnitude for most radionuclides).

The recent Russian sanitary regulations (OSPORB, 2010) approved in 2010 have introduced into practice, along with MSSA corresponding to the IAEA recommendations of 1996 (IAEA, 1996), clearance limits coincident with the values recommended by the IAEA in 2004. There are no restrictions for utilization of materials and products (except foodstuffs, drinking water and fodder) if clearance index is below unity. The materials with specific activity between the clearance limit and MSSA can have limited use if the annual individual effective radiation dose at their utilization will not exceed 10 μSv. If the utilization of such materials is impossible or inexpedient, they should be disposed of in non-radioactive industrial landfill type facilities. The document (OSPORB, 2010) also contains clearance limits for some long-lived radionuclides in metals (intermediate between general clearance limits and MSSA).

The clearance limits for selected radionuclides encountered in fusion applications, according to the standards and guidelines cited are shown in Table 1.

| Nuclide | IAEA (IAEA, 2004) | United States NUREG-1640 (US-NRC, 2003) (steel / Cu / concrete) | Russia (NRB, 2009; OSPORB, 2010) (general / metals / MSSA) | European Union EC RP 122 (EC-RP, 2000) |
|---|---|---|---|---|
| $^{3}$H | 100 | 526 / 1e5 / 152 | 100 / - / $10^6$ | 100 |
| $^{14}$C | 1 | 313 / 4.17e4 / 83 | 1 / - / $10^4$ | 10 |
| $^{22}$Na | 0.1 | 0.238 / 8.33 / 0.0417 | 0.1 / - / 10 | 0.1 |
| $^{40}$K | 10 | 2.94 / 153.8 / 0.526 | 10 / - / 100 | 1 |
| $^{41}$Ca | --- | 47.6 / 9.1e3 / 13.9 | -- | --- |
| $^{45}$Ca | 100 | 5e3 / 7e4 / 909 | 100 / - / $10^4$ | 100 |
| $^{53}$Mn | 100 | 1.14e4 / 7.1e5 / 6.67e3 | 100 / - / $10^4$ | 1000 |
| $^{54}$Mn | 0.1 | 0.625 / 23.26 / 0.118 | 0.1 / 1 / 10 | 0.1 |
| $^{55}$Fe | 1000 | 2.17e4 / 2.33e5 / 4.76e3 | $10^3$ / - / $10^4$ | 100 |
| $^{59}$Fe | 1 | 0.476 / 22.7 / 0.114 | 1 / - / 10 | 0.1 |
| $^{58}$Co | 1 | 0.588 / 28.57 / 0.133 | 1 / - / 10 | 0.1 |
| $^{60}$Co | 0.1 | 0.192 / 9.1 / 0.035 | 0.1 / 0.3 / 10 | 0.1 |

| Nuclide | IAEA (IAEA, 2004) | United States NUREG-1640 (US-NRC, 2003) (steel / Cu / concrete) | Russia (NRB, 2009; OSPORB, 2010) (general / metals / MSSA) | European Union EC RP 122 (EC-RP, 2000) |
|---|---|---|---|---|
| $^{59}$Ni | 100 | 2.17e4 / 3.57e5 / 4.76e3 | 100 / - / $10^4$ | 100 |
| $^{63}$Ni | 100 | 2.13e4 / 1.85e5 / 4.76e3 | 100 / - / $10^5$ | 100 |
| $^{64}$Cu | 100 | --- | 100 / - / 100 | --- |
| $^{94}$Nb | 0.1 | 0.333 / 11.5 / 0.059 | 0.1 / 0.4 / 10 | 0.1 |
| $^{99}$Mo | 10 | --- | 10 / - / 100 | 1 |
| $^{99}$Tc | 1 | 6.25 / 1.05e3 / 1.64 | 1 / - / $10^4$ | 1 |
| $^{108m}$Ag | --- | 0.345 / 18.18 /0.0588 | -- | 0.1 |
| $^{110m}$Ag | 0.1 | 0.192 / 10.3 / 0.0357 | 0.1 / 0.3 / 10 | 0.1 |
| $^{125}$Sb | 0.1 | 1.41 / 62.5 / 0.23 | 0.1 / 1.6 / 100 | 1 |
| $^{152}$Eu | 0.1 | 0.455 / 16.4 /0.083 | 0.1 / 0.5 / 10 | 0.1 |
| $^{154}$Eu | 0.1 | 0.455 / 16.67 /0.071 | 0.1 / 0.5 / 10 | 0.1 |
| $^{182}$Ta | 0.1 | 0.435 / 16.95 /0.091 | 0.1 / - / 10 | 0.1 |
| $^{192}$Ir | 1 | 0.91 / 52.63 /0.172 | 1 / - / 10 | 0.1 |
| $^{186}$Re | 1000 | --- | 1000 / - / 1000 | 100 |

Table 1. IAEA, U.S., Russian, and EC clearance limits (in Bq/g) for some fusion-relevant nuclides (partly taken from TABLE I of ref. Zucchetti et al., 2009)

The disagreement between clearance limits in different standards and recommendations is due to the choice of different scenarios to model the effective individual dose rates and different approximations adopted to compute the clearance limits from the effective dose rates. For instance, the U.S. studies incorporated realistic modeling of the current U.S. industrial practices and current data on the living habits in the United States in order to minimize unnecessary conservatism in the dose rate estimates.

Consistency of the clearance standards is certainly desirable, particularly for materials that may end up in the international market. However, given the complexity of the scenarios used to develop the clearance standards with so much efforts having gone into these studies over the past 25 years, it seems unlikely that additional, reasonable effort will be able to reduce dramatically in the short run the differences and explain the technical reasons for the major disagreements.

Recycling includes storage in permanently monitored facilities, segregation of various materials, crushing, melting, refabrication and some other processes (Massaut et al., 2007).

In the European Power Plant Conceptual Study (PPCS) (Maisonnier et al., 2005) a simplified categorization of active material recycling criteria was used. A conclusion of the PPCS analysis was that for all five considered plant models[4] (Models A, B C, D and

[4] All five of the plant models PPCS A to D and AB are based on the tokamak concept. PPCS Model A and Model B are based on limited extrapolations in plasma physics performance compared to the design basis of ITER. In PPCS A and PPCS B, the blankets are based, respectively, on the "water-cooled lithium-lead" and the "helium (He) cooled pebble bed" concepts, studied in the European fusion program. Both concepts are based on the use of a low-activation martensitic steel. PPCS Model C and

AB) , if a full use of the potential to recycle radioactive materials is made, there would be no material requiring permanent burial after a decay storage period from a few decades to 100 y, except for a small amount of secondary waste from reprocessing (Forrest, 2005). In other words, the recycling and clearance strategy would appear to have great potential, since its application could strongly reduce the amount of radioactive waste to be disposed of. The vast majority of radioactive materials (87% in PPCS Model AB and 84% in PPCS Model B) can be cleared or recycled with low handling difficulties after a medium term duration decay period. Only small amounts (a few hundred tons) of plasma-facing tungsten and breeders will require specific remote handling mechanisms. Whether or not such recycling operations would be feasible and economically viable, for all the candidate materials had yet to be determined.

In the U.S. ARIES studies (ARIES Project), the technical feasibility of recycling is based on the dose rate to advanced RH equipment capable of handling at 10 kGy/h or more (El-Guebaly et al, 2008). Such dose rates are present at routine operations in the reprocessing of fission reactor fuel and at the outside surfaces of radioactive goods during their weighing, welding, cleaning, contamination monitoring, and transfer to containers. Corresponding equipment for fusion applications is now under development, since it is needed for removing the replaceable components from the vacuum vessel of the International Thermonuclear Experimental Reactor (ITER), being under construction now in the south of France, and moving them to the hot cell.

A comprehensive survey of remote procedures in the nuclear industry was performed in the framework of an international collaborative Study on the Back End of the Fusion Materials Cycle (SBEFMC) carried out under the auspices of the International Energy Agency (IEA) and documented in ref. (Zucchetti et al., 2009). Some participants of this study found that the remote handling criterion used in the PPCS was unduly conservative. They did not ascertain the upper limit of the dose rate for the RH feasibility, stating that the only upper limit for the RH feasibility seems to be the decay heat density (2 $kW/m^3$) and active wet cooling needs.

In this study it was assumed that:

- no active cooling is needed (only natural ventilation) when decay heat density is <10 $W/m^3$;
- dry cooling (e.g. , active ventilation) is required when decay heat density is >10 $W/m^3$ but <2 $kW/m^3$;
- active wet cooling (e.g. actively cooled storage pond) is necessary when decay heat density is >2 $kW/m^3$, coinciding with the definition of high-level radioactive waste.

---

Model D are based on successively more advanced concepts in plasma configuration and in materials technology. Their technology stems, respectively, from a "dual-coolant" blanket concept (He and lithium-lead coolants with steel structures and silicon carbide (SiC) insulators) and a "self-cooled" blanket concept (lithium-lead coolant with a silicon carbide structure). In PPCS C the divertor is the same concept as for Model B. In the most advanced concept, PPCS D, the divertor is cooled with lithium-lead like the blanket. PPCS Model AB is a combination of the concepts for Model A and B, in detail it is based on He-cooled lithium-lead blanket and He-cooled divertor. The blanket is based on the use of EUROFER as structural material, of Pb-17Li (Li at 90% in 6Li) as breeder, neutron multiplier and tritium carrier, and of helium as coolant.

Participants of the SBEFMC expressed doubt that any recycling operations can be performed until the decay heat density decreases to levels not requiring active cooling; hence, interim storage in this case is the only option available. Another conclusion of this study was that for the recycling in foundries, one can for the moment take an activity limit of 1 kBq/g.

The above-considered criteria for feasibility of remote recycling are useful only for the first approach to conceptual design features. They do not show actual recycling expediency, since the processes to be used, type of material and component, economics of fabricating remotely complex forms, and the physical properties of the recycled products also affect the recycling advisability. Furthermore, the acceptability of the recycled materials to nuclear industry has to be considered in parallel with contact dose rate, decay heat and activity content levels.

## 6. Radioactive material generated during the fusion power plant life cycle

As fusion is expected to play an essential role in supplying clean, environmentally-friendly energy in the second half of the 21st century, the ability of power plant designers to keep up with the persistent demands of controlling the radwaste stream becomes extremely important for fusion as well as for fission. Fusion generates only low level waste (LLW) that requires near-surface, shallow-land burial, if clearance or recycling would not be feasible, as all materials are carefully chosen to minimize the long-lived radioactive products.

Fig. 2 displays the reduction in the fusion power core (FPC) volume over the past 2-3 decades by clever designs that apply more advanced technology and physics operating regimes. The volumes include the actual volumes of power core components (from plasma facing components up to magnets), excluding the bioshield.

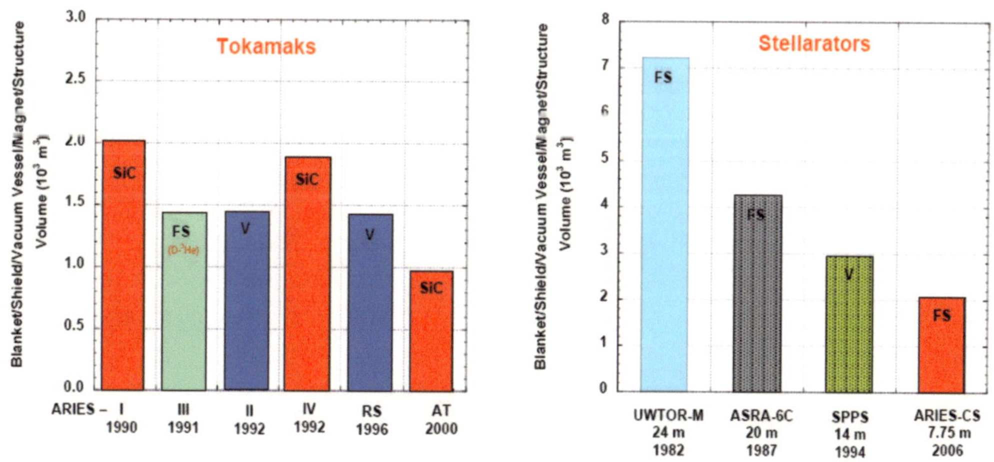


SiC = silicon carbide composite structure; V= vanadium alloy structure; FS = ferritic steel structure

Fig. 2. Evolution of fusion power core (FPC) volumes for U.S. tokamaks and stellarators developed over the past 30 years (actual volumes of power core components, no compactness, no replacements).

As noticed, the tokamak volume was halved over 10 y study period, while the stellarator FPC volume dropped by 3-fold over 25 y study period. Most of these designs were developed in the US by the ARIES team (ARIES Project). Other fusion institutions in Europe and Japan delivered several tokamak designs over the past decade.

To put matters into perspective, Fig. 3 compares the volumes of FPCs of ITER (ITER Project), the advanced tokamak ARIES-AT (Najmabadi et al., 2006), the European PPCS Model C (Maisonnier et al., 2005), the Japanese VECTOR tokamak (Nishio et al., 2004), and the compact stellarator ARIES-CS (Najmabadi et al., 2008) to the fission core and vessel of the Economic Simplified Boiling Water Reactor (ESBWR), a GEN-III$^{+}$ advanced fission reactor.

In recent years, fusion designers have paid more attention to the waste management issues associated with the sizable volume of activated materials discharged from fusion power plants. Specifically, they have striven to minimize the activated materials volume problem, not only by developing advanced designs, but also by reshaping the fusion radwaste management scenario, maximizing the reuse of activated materials through recycling and clearance, avoiding the disposal option.

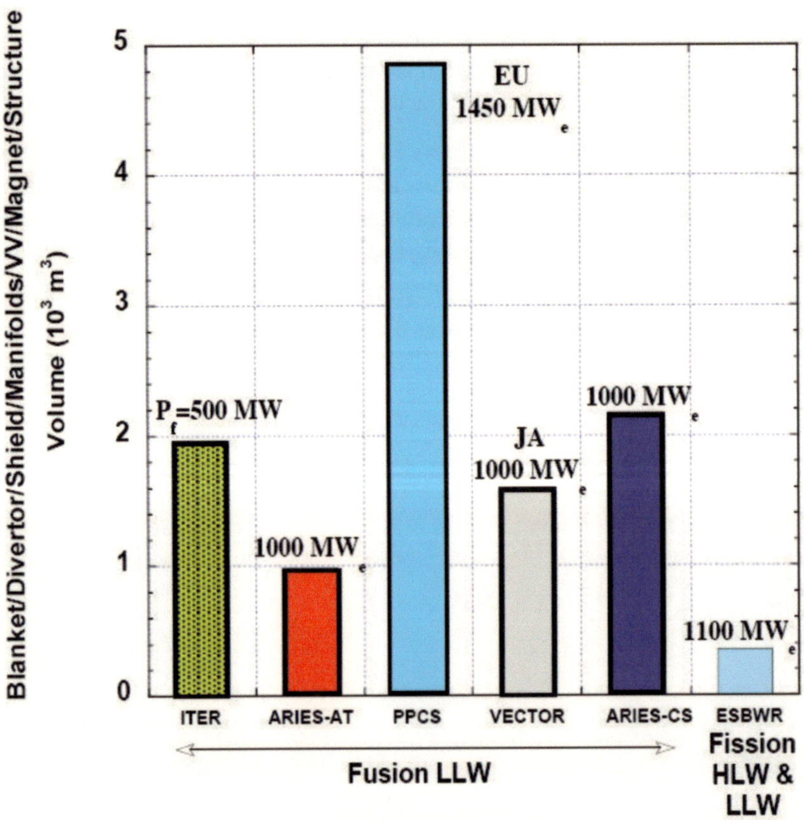


Fig. 3. Volume of fusion power core. Economic Simplified Boiling Water Reactor (ESBWR) fission core and vessel included for comparison (taken from Fig. 1 of ref. El-Guebaly et al., 2008)

Another source of activated materials is the biological shield. As in the fission plant, it surrounds the power core to essentially protect the public and workers against radiation. It is made of 2-3 -m thick steel-reinforced concrete and constructed to withstand natural phenomena, such as earthquakes, tornados, floods, and an airplane crash. Being away from the plasma source, the bioshield is subject to low radiation level in fusion facility and contains very low radioactivity. Since burying such a large volume of slightly activated materials in underground repositories is impractical, the US-NRC and IAEA suggested the clearance concept where such components could temporarily be stored for the radioactivity decay, then released to the commercial market for reuse as shielding blocks for containment buildings of licensed nuclear facilities, concrete rubble base for roads, deep concrete foundations, non-water supply dams for flood control, etc.

Most of the radioactive materials generated during fusion power plant operation are activated solid metallic materials from the main machine components (blanket, divertor, shield, vacuum vessel, and magnets) and concrete from the bioshield. Liquid breeders (such as LiPb, Li, etc.) are normally refurbished for reuse during operation and in future power plants. Even though the dominant radioactive material stream is generated during the decommissioning phase (if one includes the bioshield), a notable amount - as far as radioactive inventory is concerned - is also produced during plant life by routine blanket and divertor replacements. This replacement is necessary due to neutron-induced damage to structural components operating under a 14-MeV neutron flux and due to the need to refurbish the ceramic breeding materials if used in the blanket.

## 7. Clearance and recycling as viable options for managing activated materials from fusion power plants

Clearance and recycling of structural materials are viable for fusion as the half-life of most radioisotopes contained in such a potential waste can be limited to about 5-10 years, meaning that after a period of about 100 years the radioactivity drops down to one millionth of its initial value. A clever choice of constituents and alloying elements can strongly limit the effects of neutron activation, reducing the concentration of long-lived isotopes (Zucchetti et al., 2007). On the other hand, most fusion active materials contain tritium that could introduce serious complications to the recycling process. Thus a detritiation treatment prior to recycling is imperative for fusion components with high tritium content, as written previously, or in alternative to fulfill the waste acceptance criteria for LLW disposal, if recycling might not be feasible. This excludes the need for deep underground repositories.

### 7.1 Clearance issues for activated material from fusion power plant

The issue of clearance of fusion radioactive materials is associated to two main aspects:

a. **The definition of clearance levels** which, including radionuclides specific to fusion, should be accepted, as far as possible, at the national and international levels; currently, there is already a certain movement in this direction, as the IAEA guidelines (IAEA, 2004) have been accepted by the Member States (but are not applied as such in the different countries);

b. **The public acceptance of cleared materials**. Even though the fusion community and nuclear industry favor some form of clearance standards, many industries and environmental groups do not allow slightly radioactive solids to enter the commercial market, no matter how low the clearance levels might be. Some European countries and Russia have nevertheless introduced this concept in their regulations.

From the operative point of view the clearance can be divided in:

- Clearance (unconditional, unrestricted release);
- Conditional clearance.

Unconditional clearance means that the material is handled as if it was no longer containing radioactive species above natural or insignificant level. Under this option, the material can be reused without restriction, recycled into a consumer product, or disposed of in any industrial landfill. The compliance with the limits for the clearance levels defined by the national regulatory authorities must be verified.

Conditional clearance means that the material may be recycled or the component reused in a specified application and subject to continuing regulatory control until specific conditions are met to allow unconditional clearance. For example, slightly radioactive metal released under conditional clearance can be melted in licensed melting facilities to produce metal ingots for using them for making railroad tracks or nuclear casks. Another example is related to fabricate concrete rubble that could be used for road construction. In other words, conditional clearance is a restricted release of slightly radioactive material from regulatory control under certain conditions, in particular for its first re-use. What is mostly considered as "conditional release" is the clearance from regulatory control providing certain paths of reuse are guaranteed (and followed up).

In the U.S. there is very limited support for the unconditional clearance, no matter how restrictive the clearance standards might be. No support for the clearance option exists in the steel and concrete industries. In absence of such a clearance market, the conditional clearance represents a viable option – an alternative to disposal. In this conditional clearance category, the slightly radioactive materials are not recycled into a consumer product, but rather released to dedicated nuclear-related facilities under continuing regulatory control or to specific applications where contact for exposure of the general public is minimal. Such slightly contaminated materials have been released since the early 1980s and continue to be released in the U.S. under existing practices on a case-by-case basis using existing NRC guidance and a specific provision contained in the facility's license. While the conditional clearance process has been ongoing in the U.S. for a few decades, a more formal and uniform process would be highly desirable in particular prior to the decommissioning of operating fission reactors. Three facts support this argument: the limited capacity of existing LLW repositories, the political difficulty of building new ones, and the rising cost of geological disposal with tighter environmental control.

From the European perspective (in Europe several countries already apply unconditional clearance for materials coming from decommissioning of fission reactors and nuclear facilities), the conditional clearance is an interesting option, as it can relax the conditions under which materials can be released. Nevertheless, its application is complicated by the fact that the regulatory control, or at least a control and monitoring of the first re-use of the material has to be performed. That increases the cost of the material management.

Some industries adopted a "zero tolerance" policy, expressing concerns that the presence of radioactive materials in their products could damage their markets, erode public confidence in the safety of their products, and negatively affect their sales. On the other hand, however, some industries would support a restricted use scenario in which cleared materials would be limited to selected purposes (e.g., nuclear facilities or radioactive waste containers) and subject to control by the nuclear regulatory agencies.

There is no uniform or harmonized regulation on clearance even in the European Union. Although the European Commission (EC) has published several guidelines on clearance of materials from regulatory control, see for example Ref. (EC-RP, 2000), each European country can issue its own regulation. Since the 1990s several countries have already issued regulations on clearance and projects have cleared materials in industrial quantities (mostly metals and concrete rubble) for their decommissioning program and related projects, The effective dose limit of 10 μSv/y (1 mrem/y) per practice for cleared solids was adopted in these cases. It is widely accepted by the IAEA, U.S., Russian and EU organizations, it is very small in comparison with the allowable annual dose limit for the public (1 mSv/y). According to the United Nations recommendations, the annual effective radiation dose above background level to members of the public from radiation sources other than medical exposures should not exceed 1 mSv (100 mrem). That means the 10 μSv/y dose limit for cleared solids is 1% of the total allowable excess dose, < 0.5% of the radiation received each year from natural background (2.4-3.6 mSv/y), and significantly less than the amount of radiation that we receive from radioactive $^{40}K$ located in our own body (0.18 mSv/y), from routine medical procedures (0.55 mSv/y), from living in a brick house (70 μSv/y), or from flying across the U.S. (25 μSv).

### 7.2 Recycling issues for activated material from fusion power plant

Pursuing recycling of fusion materials has two main justifications: one is environmental, to limit the amount of waste to be disposed of in repositories, and the other is economical and resources-related, to meet the need for a more efficient and effective use of natural resources including expensive materials (Be, V, W, etc.) envisaged for future fusion power plants.

Recycling levels used in previous EU studies (Raeder (ed.), 1995; Maisonnier et al., 2005) were based on a coarse derivation from a summary of waste categories levels (Rocco & Zucchetti, 1998) extended/extrapolated to a recycling classification, based only on contact dose and decay heat rate levels. Although useful for the first approach of conceptual design features, it did not take into account actual recycling feasibility. Indeed, it appeared that dose rate levels were not a severe constraint for recycling and that the activity content and decay heat removal had to be considered in parallel with the type of material and components to be recycled (Massaut et al., 2007, Pampin et al., 2007). At present, the fission spent nuclear fuel is reprocessed in hot cells with complete remote handling systems and active cooling. These facilities can handle materials with dose rates of up to 1500 Gy/h. Moreover, to remove the components from any tokamak, it is foreseen to use remote handling systems. Advanced radiation-hardened remote handling equipment is available in the fission industry and can be applied to fusion materials recycling. The remote handling needs for recycling are normally less stringent than the ones for removal and handling components in the plant.

Aiming to define the recycling features in the context of a fusion-oriented approach to the back-end of the fusion materials cycle, the following recycling handling categories have been proposed:

- HOH (Hands-On Handling). Contact dose rate (DR) <10 μSv/h.
- S-HOH (Shielded Hands-On Handling). Contact DR < 2 mSv/h.
- RH (Remote Handling). Contact DR >2mGy/h, it can be dealt with by remote handling equipments, without active cooling: decay heat is <2000 W/m$^3$.
- ACM (Active Cooling Material). This requires active cooling and it is unlikely that any recycling operations can be performed until its decay heat decreases to levels not requiring active cooling, hence interim storage with cooling is the only option available.

One of the main tasks of the latest EU study in this field (Ooms, 2007) was to overcome the previous classification and propose realistic routes and management processes for the materials of the PPCS plants, which would assist the design process of fusion plants and provide guidelines for important R&D needs. Distinction is made between "routes" and "radiological requirements" for handling, cooling, transport, etc. "Routes" define actual, applicable management paths and processes to treat the activated and tritium contaminated materials. Radiological requirements reflect limitations posed by the radioactive nature of the materials. The EU study exemplified these by the categories in Table 2.

<table>
<tr><td>Limit</td><td colspan="2">< 10 μSv/h</td><td colspan="2">< 2 mSv/h</td><td>< 2000 W/m<sup>3</sup></td></tr>
<tr><td>Handling</td><td colspan="2">HOH</td><td colspan="2">SHOH</td><td>RH</td></tr>
<tr><td>Routes</td><td>Clearance</td><td colspan="2">Recycle in foundries (1)</td><td colspan="2">Processes to define</td></tr>
<tr><td>Limit</td><td>CI < 1</td><td colspan="2">< 1000 Bq/g</td><td colspan="2">< 2000 W/m<sup>3</sup></td></tr>
</table>

(1) For metals

Table 2. EU management routes for fusion radioactive materials (taken from Table 1 of ref. Zucchetti et al., 2008)

Management routes were generically categorized in clearance (unconditional and conditional), recycling in foundries (this applies only to metals) and more complex recycling for which the processes still have to be defined and/or developed, providing the decay heat remains below 2000 W/m$^3$.

Specific levels can be set for these three main categories, but further descriptions are given in the next sections:

- For the unconditional clearance, the Clearance Index (CI) must be lower than unity
- For the conditional clearance, this would depend upon local regulations
- For the recycling in foundries, one can for the moment take an activity limit of 1000 Bq/g
- For the other recycling possibilities, the only limit seems to be the decay heat and active cooling needs limit.

More recently, it has been proposed (Pampin & O'Brien, 2007) to override these classification criteria with a scoring scheme, rating the difficulty of operations on active material. The radiological scoring overcomes the requirements for the contact dose and includes other aspects (e.g.: cooling at the moment, more if necessary in the future). It is based on actual requirements and procedures such as handling (contact dose rates), cooling (decay heat rates), routes, and the radiological levels derived from EU work reviewing industrial experience (Massaut et al., 2007; Ooms & Massaut, 2005). An important element for a credible management strategy is the capability to assess the technical difficulty of recycling or waste conditioning treatments and operations, despite of the route pursued.

It is desirable the capability to assess and compare the radiological characteristics of the irradiated materials, evaluate generic technical hitches posed by their radioactive nature, and ascertain storage decay times, facilitating the processes envisaged for recycling or disposal. For this purpose, a rudimentary scheme has been developed based on two main aspects: handling equipment/procedures, and cooling requirements.

For handling, three main types are foreseen:

a. Unshielded hands-on handling by qualified radiation workers, HOH, when contact dose levels are below 10 μSv/h
b. Shielded hands-on handling by qualified radiation workers, SHOH, when contact dose rates are below 2 mSv/h; equipment such as shielded glove boxes can be conceived under this category
c. Remote handling when contact dose levels are above 2 mSv/h.

As for cooling requirements, the following levels are envisaged:

a. No active cooling needed (only natural ventilation) when decay heat power is < 10 W/m$^3$
b. Dry cooling (e.g. active ventilation) when decay heat power is > 10 W/m$^3$ but < 2000 W/m$^3$
c. Wet cooling (e.g. actively cooled storage pond) when decay heat power is > 2000 W/m$^3$ – coinciding with the definition of HLW.

Besides the radiation protection aspects given above, the EU recycling study and approach has also addressed the potential routes for the recycled materials. Indeed, even if the material can be handled hands-on or remotely, it makes no sense to go further if no processing routes can be found for this material even without evaluating the economic attractiveness and the potential market. Addressing the routing issue, various scenarios have been analyzed, mostly for metals and materials to be removed from the tokamak core and the immediate surroundings. For material with sufficiently low activity to be unconditionally or conditionally released, usual ways of recycling (often using remelting of the metal components) can be foreseen. Once freely released or conditionally released, the material can follow the existing industrial recycling streams, providing some monitoring of its use for conditional clearance. For material above the release limits, or material for which the measurement of characteristics is difficult, or material for which the treatment would act as a decontamination process (like metal melting for detritiation for instance), recycling within the "nuclear regulated" foundries is currently used at the European and international level. Depending on their license, these foundries can accept plus or minus contaminated or activated materials. So far, the levels of accepted and licensed activity remain very low (on the order of hundreds of Bq/g). At any rate, melting helps to homogenize the activity concentration, overcoming the problem of activity measurements on piece of equipments with complex geometries. Other recycling scenarios with no melting have to be considered as well. For instance, refractory metal (like tungsten) components, made by powder metallurgy process, whether they are still in good condition, might be re-used within the nuclear industry. Other recycling scenarios can also be developed for exotic materials, like the (liquid or solid) breeder materials. The same approach can also be expected for the superconductor material. But all these approaches need to be investigated and developed in more detail.

It is important to develop advanced rad-hard RH equipment that can handle components presenting dose rates up to 10000 Gy/h (10000 Sv/h) or more. This equipment is already needed for removing the replaceable components from the vacuum vessel of a tokamak and moving them to the hot cell. The proposed high dose rates are not far from the present technology; e.g. in ITER design some RH equipment will have to withstand 1500 Gy/h (and even 15000 Gy/h) with a total dose of 5-10 MGy. Such a high dose rate is reached in fusion power plants within a few years after blanket/divertor replacement and arises mostly from radionuclides originating from the main materials and alloying elements, not from impurities.

The question of reprocessing of radioactive (non-clearable) materials in dedicated facilities in order to separate noxious radionuclides is another challenge. The result of this operation is a small quantity of concentrated radioactive waste, plus a processed material that may be either "clearable" or "non-clearable, to be recycled within nuclear industry", if the separation process is viable and effective.

The development of methods to reprocess the activated alloy to extract radiotoxic nuclides is a long and complicated task, but the possibility to eliminate the need for numerous repositories, so minimizing the burden for future generations, apart from the small volumes required to store the secondary waste, is very attractive and worth pursuing.

Examining several fusion designs revealed that the internal components (blanket, divertor, shield, and vacuum vessel) are not clearable even after an extended cooling period of 100 y (El-Guebaly et al., 2008, Zucchetti et al., 2009). Controlling the Nb and Mo impurities in the low-activation steel stucture may help clear the outer vacuum vessel components. Fortunately, the bioshield (that represents the largest single component of the decommissioned radwaste) along with some magnet constituents qualifies for clearance, especially when a long period (up to 100 y) of interim storage is anticipated. This represents a great deal of the decommissioning materials (70-80%). The remaining 20-30% of the active materials could be recycled using a combination of advanced and conventional remote handling equipment (El-Guebaly et al., 2008; Zucchetti et al., 2009).

## 8. Radioactive waste classification

The radioactive waste classification differs appreciably in different countries. Below it is given a brief summary of different waste classifications adopted in some countries, but starting with the IAEA recent classification, as it is an international guideline.

### 8.1 IAEA classification

The IAEA developed and published in 2009 a safety guide containing general scheme for classifying radioactive waste that identifies the conceptual boundaries between different classes of waste and provides guidance on their definition on the basis of long term safety considerations (IAEA, 2009). Six classes of radioactive waste are considered as the basis for the classification scheme in this safety guide:

1. Exempt waste (EW) that meets the criteria for clearance, exemption or exclusion from regulatory control for radiation protection purposes as described in Ref. (IAEA, 2004). In reality, however, once such waste has been cleared from regulatory control, it is not considered as radioactive waste any more.
2. Very short-lived waste (VSLW) that can be stored for decay over period of up to a few years and consequently cleared from regulatory control. In general, VSLW contains radionuclides with half-lives of the order of 100 days or less.
3. Very low level waste (VLLW) that does not need a high level of containment and isolation and is suitable for disposal in near surface landfill type facilities with limited regulatory control (e.g. soil and rubble with low levels of activity concentrations). In terms of radioactive waste safety, a radionuclide with a half-life of less than about 30 years is considered to be short-lived. Concentrations of longer lived radionuclides in VLLW are generally very limited.
4. Low level waste (LLW) that is above clearance levels, but with limited amounts of long-lived radionuclides. Such waste requires robust isolation and containment for periods of up to a few hundred years and is suitable for disposal in engineered near surface facilities.
5. Intermediate level waste (ILW) that may contain long-lived radionuclides, in particular, alpha emitting radionuclides that will not decay to a level of activity concentration acceptable for near surface disposal during the time for which institutional control can be relied upon (in some countries up to around 300 years). Therefore, ILW requires disposal at depths of the order of tens meters to a few hundred meters. However, ILW needs no provision, or only limited provision, for heat dissipation during its storage and disposal.
6. High level waste (HLW) with levels of activity concentrations high enough to generate significant quantities of radioactive decay heat or with large amounts of long-lived radionuclides that need to be considered in the design of a disposal facility for such waste. Disposal in deep, stable geological formations usually several hundred meters or more below surface is generally recognized option for disposal of HLW.

Contact radiation dose rate is not used to distinguish waste classes in the new IAEA classification scheme. The guide assumes that detailed quantitative boundaries taking into account broad range of parameters may be developed in accordance with national programs and requirements.

### 8.2 U.S. classification

The Nuclear Regulatory Commission has established classifications for waste generated by nuclear power industries, university research laboratories, manufacturing and food irradiation facilities, and hospitals. Low and high level wastes are classified according to the activity concentration and types of radioisotopes. For each level, there is a specific disposal requirement according to the US-NRC 10CFR61 document (US-NRC, 1982) so that the waste is disposed properly and safely. At present, there is no low and intermediate level waste (LILW) category in the U.S.

For nuclear power plants, the NRC classification is based largely on radionuclides that are important to fission facilities. In a nuclear fusion system, the isotopes are different because of the different materials being considered and the different decay products that are generated. In the early 1990s, analyses (Fetter et al, 1990) were performed to determine the Class C specific activity limits for all long-lived radionuclides of interest to fusion using a methodology similar to that used in 10 CFR 61. Although Fetter's calculations carry no regulatory acceptance, they are useful because they include fusion-specific isotopes. The ARIES approach requires all components to meet both NRC and Fetter's limits for LLW until the NRC develops official guidelines for fusion waste.

### 8.3 Russian classification

According to the Russian Basic Sanitary Regulations Ensuring Radiation Safety (OSPORB, 2010), radioactive waste includes matters, materials, mixtures and products that are not subject to further utilization and in which $\Sigma_i(A_i/MSSA_i) > 1$, where $A_i$ is specific activity of a technogenic radionuclide *i* and MSSA is minimally significant specific activities, as defined in previous paragraph 5. The waste with unknown radionuclide composition is considered as radioactive if the total specific activity of technogenic radionuclides in them exceeds 100 Bq/g for beta emitters, 10 Bq/g for alpha emitters, and 1.0 Bq/g for transuranium radioactive nuclides. There are three categories of radioactive waste: low level waste (LLW), intermediate level waste (ILW) and high level waste (HLW) depending on their specific activity as given in Table 3. In the case when different radioactive nuclides relate to different categories, the waste relates to the highest category.

| Category | Tritium | β Emitters | α Emitters except Transuranium Nuclides | Transuranium Nuclides |
|---|---|---|---|---|
| LLW | $10^6$-$10^7$ | $<10^3$ | <100 | <10 |
| ILW | $10^7$-$10^{11}$ | $10^3$-$10^7$ | $10^2$-$10^6$ | 10-$10^5$ |
| HLW | $>10^{11}$ | $>10^7$ | $>10^6$ | $>10^5$ |

Table 3. Specific activity (Bq/g) of different categories of liquid and solid waste (OSPORB, 2010)

### 8.4 Italian classification

Italian regulations deal with national laws on radioactive materials and with Technical Guides from the Italian nuclear regulatory committee ("Guida Tecnica n. 26" - Technical Guide No. 26 - and others). The waste is classified into three categories (the first category = low level waste, the second category = intermediate level waste, the third category = high level waste) on the basis of the radioisotope characteristics (half-life and radiotoxicity) and concentration limits, and considering the possible options for final disposal. Without going into details, the boundary between the second and third category, for activated metallic materials, is a concentration of 370 Bq/g for long-lived nuclides ($t_{1/2} > 100$ y), 37000 Bq/g for medium-lived nuclides (5 y < $t_{1/2}$ < 100 y) and 37x10[6] Bq/g for short-lived nuclides. This limit deals with waste that has been conditioned and treated for disposal.

A regulation concerning the *"allontanamento"* (Italian word for "clearance") of solid radioactive spent materials has been issued in Italy recently. This regulation is necessary for the ongoing decommissioning activities of four Italian fission reactors. Concentration limits are issued for each relevant nuclide, however, they may be partially summarized - for our purposes - as follows: a non-alpha-emitter metallic material may be cleared, if its specific activity is less than 1 Bq/g (including tritium). For materials other than metallic and concrete, the limit is 0.1 Bq/g. For concrete, the limit is almost halfway, depending on the type of nuclides. These limits are applicable if only one nuclide is present in the waste, otherwise the criterion $\Sigma_i(A_i/L_i) < 1$, where $A_i$ is the mass (Bq/g) or superficial (Bq/cm$^2$) activity concentration of the nuclide $i$, and $L_i$ is the related activity concentration limit, must be respected. Both mass and superficial specific activity limits must be met. Recycling in Italy is permitted for cleared materials only.

### 8.5 French classifcation

The waste classification in France (LOI 739, 2006; Décret 357, 2008) is managed by the French Agency for the Management of Radioactive Waste (ANDRA). There are four different types of waste:

- TFA (Très Faible Activité) or very low level waste;
- FMAVC (Faible et Moyenne Activité, Vie Courte) corresponding to low and short-lived (< 31 years) intermediate level waste;
- MAVL (Moyenne Activité à Vie Longue) corresponding to long-lived (> 31 years) intermediate level waste;
- HAVL (Haute Activité à Vie Longue) corresponding to long-lived (> 31 years) high level waste, with thermal effect.

**Very low level activity (TFA) waste criteria**

It should be mentioned that the French regulation does not recognize the clearance concept. Therefore, it was decided to create a category for VLLW and an evacuation route for this category of wastes. The Centre de Stockage TFA (CSTFA) at Morvilliers is the final disposal for Very Low Level Waste (TFA) since summer 2004. There are also specific tritium thresholds to be respected such as: tritium specific activity lower than 1000 Bq/g and tritium degassing rate lower than 200 Bq/m$^3$/day and 10 Bq/m$^3$/day, for HTO and HT, respectively. The acceptance of a batch of waste depends on an index considering the nuclide specific activity and the nuclides class (depending on the nuclide radiotoxicity). This radiological acceptance index in storage ("Indice Radiologique d'Acceptabilité de Stockage" (IRAS)) is defined as:

$$\sum \frac{A_i}{10^{C_i}} \qquad (3)$$

Where $A_i$ is the specific activity of the nuclide (in Bq/g) and $C_i$ is the nuclide class (0, 1, 2, 3), depending on the nuclide radiotoxicity.

A waste batch can be accepted if it complies simultaneously with the 2 following conditions: IRAS index < 1 and IRAS index of the different packages within the batch lower than 10.

**Low Level Activity (FMA) Waste Criteria**

FMA waste is disposed of in surface repositories. The Centre de Stockage de l'Aube (CSA) is the current final disposal for this type of waste. For acceptance in the CSA the activity of selected nuclides has to be evaluated and declared if their specific activities are higher than the declaration threshold. Waste containing nuclides above the embedding threshold needs to be fixed with a matrix having containment properties. Otherwise, a blocking matrix can be used to allow waste immobilization.

The maximum tritium degassing rate for packages stored in CSA is 2 Bq/g/day due to occupational radiation exposure.

**Intermediate Level Activity (MAVL) Waste Criteria**

The waste, which cannot be stored as FMA or TFA, has been considered as MAVL waste since no acceptance criteria has yet been defined, except a decay heat per package limited to 13 W. Studies are currently performed, in France, to define the best strategy for MAVL management. Geological disposal is one of the possibilities studied. A recent law has been voted on radioactive waste: "Programme relatif à la gestion durable des matières et des déchets radioactifs" (program related to durable management of materials and radioactive waste) and describes the objective of such storage site.

**High Level Activity (HAVL) Waste Criteria**

High level waste is stored temporarily in tanks before being calcined in the form of a powder and then incorporated into a molten glass.

As prescribed by article 3 of the June 28th 2006 Planning Act (LOI-739, 2006), ANDRA is developing, as in the case of MAVL, a 500-metre deep disposal concept for HAVL. The outcome of this study is the commissioning of a repository by 2025 in Meuse/Haute-Marne, subject to government approval and after a public debate. Pending the commissioning of the deep repository, HLW is stored at production sites, La Hague (AREVA), Marcoule (CEA) and Cadarache (CEA).

## 9. Integrated active fusion material management strategy

Given all the above considerations, in order to overcome previous classifications and propose realistic routes and management processes for the materials, a distinction has been made between the Regulatory Route (unconditional clearance, conditional clearance, no-clearance) and the Management Route (recycling/reuse, disposal) as summarized in Table 4. Recycling/reuse "routes" define actual, applicable management paths and processes to treat the activated (and tritium contaminated) materials. Radiological requirements reflect limitations posed by the radioactive nature of the materials. The rationale of the proposal is matching handling categories with feasible recycling routes for fusion radioactive materials.

| Regulatory Route | Management Route | |
|---|---|---|
| | Recycling/Reuse | Disposal |
| Clearance (unconditional) | **Outside the nuclear industry**. All final destinations are feasible [*this can be after a certain decay storage time) this can happen within a licensed facility until specific conditions are met to allow clearance (i.e. in melting facilities to produce metal ingots)*] | In a landfill (*for urban, special or toxic waste, depending on chemical toxicity of the waste*) |
| Conditional Clearance | **Within the nuclear industry** or in general industry for specific applications. Continuous regulatory control. [*Examples include: building concrete rubbles for base road construction or as an additive for manufacturing new concrete buildings; or metal used for making shielding blocks and containers*] | In special industrial (and/or toxic) landfill |
| No-clearance (No-release) | **Within the nuclear industry** (it can be direct reuse, **or after processing**) | In a licensed repository for radioactive waste (after an interim storage if applicable) |

Table 4. An integrated approach to fusion radioactive materials management (taken from Table IX of ref. Zucchetti et al., 2009)

The integration of the recycling and clearance processes in fusion power plants is at an early stage of development.

The principal elements of the recycling/clearance process are depicted in Fig. 4. At any rate, by examining the various management step of fusion material at the back end, one might predict the following steps:

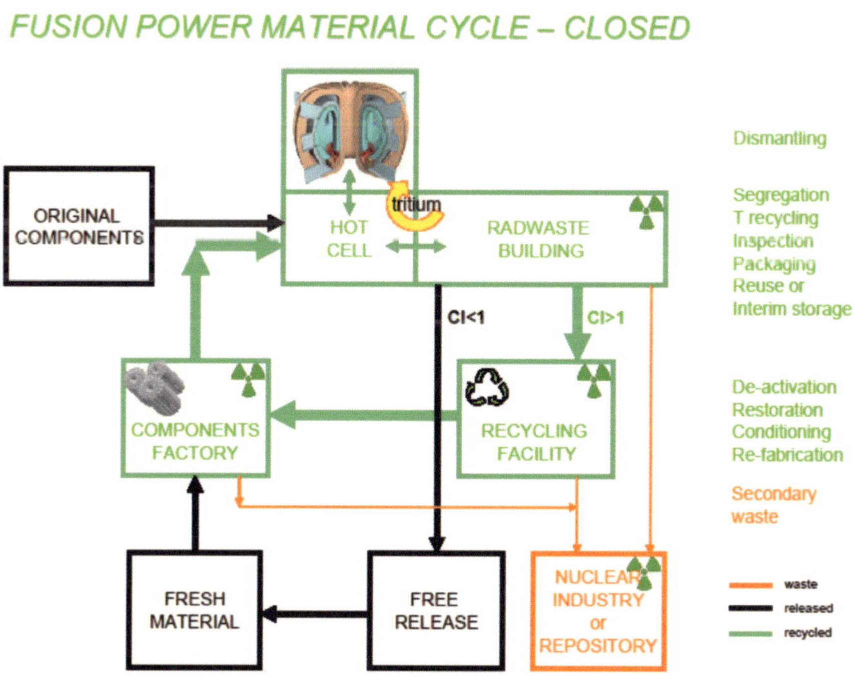


Fig. 4. Diagram of recycling and clearance processes (taken from Fig. 2 of ref. Massaut et al., 2007)

1. After extraction from the power fusion plant core, components are taken to the hot cell to disassemble and remove any parts that will be reused, separate into like materials, detritiate, and consolidate into a condensed form. This is probably one of the most challenging steps.
2. Ship materials to a temporary storage onsite (or to a centralized facility) to store for several years.
3. If the Clearance Index (CI) does not go down to unity in less than e.g. 100 y, transfer the materials to a recycling center to refabricate remotely into useful forms. Fresh supply of materials could be added as needed.
4. If the CI can go down to unity in less than e.g. 100 y, store the materials for 1-100 y then release to the public sector to reuse without any restriction.

Due to the lack of experience, it is almost impossible to state how long it will take and how it will cost to refabricate the replaceable components (blanket and divertor) out of radioactive materials. This is probably the key element for defining a complete waste management strategy. In addition, many efforts should be put on developing these technologies. The minimum time that one would expect is one year temporary storage and two years for fabrication, assembly, inspection, and testing. All processes must be done remotely with no personnel access to fabrication premises.

## 10. Conclusions

In summary, the need for a study on fusion radioactive waste arose to examine the "back-end of the fusion materials cycle" as an important stage in maximizing the environmental benefits of fusion as an energy provider.

An integrated approach to the management procedures for active materials following the change-out of replaceable components and decommissioning of fusion facilities has been proposed. The attractive environmental features of fusion have been put into evidence, and the question of proliferation relevance of fusion power plants has been briefly analyzed.

The reference is towards previous European and U.S. assessments of the back-end for fusion power plant studies, stressing this important result: most materials can be cleared or recycled, and/or disposed of as low level waste. More significantly, a new radioactive materials management strategy has been proposed for the clearance, recycling, and disposal approaches.

Concerning the clearance:

- Two alternatives are feasible: unconditional clearance, conditional clearance. Conditional clearance seems to be a viable option in the absence of a market receiving unconditional clearance materials.
- The problem of public acceptance of clearance and thus recycled material has been analyzed: how to improve the confidence of actual market towards cleared and then recycled materials ?
- Experience gained from the clearance of radioactive waste in Germany, Sweden, Spain, and Belgium might be considered useful.
- A brief review of the IAEA, U.S., Russian, and EU clearance guidelines, highlighting the similarities and differences has been performed.

Concerning the recycling approach:

- A brief review of previous approaches to recycling of fusion active materials has been presented.
- Lessons learnt from the fission experience related to hot cell performances and operations with highly radioactive materials must be used.
- For certain materials, re-use is a solution to reduce active materials inventory and the cost of producing new materials.
- The melting process tends to decontaminate the melt, segregating the slag, dust, and fumes. After slag removal and composition adjustments, the metal alloys could have properties very similar to, or equal to, those of fresh alloys.
- Economic viability of recycling has to be considered in deciding its put in practice.
- Contact dose rate, decay heat rate, radioactivity concentration are important radioactive quantities to be reduced.
- Hands-on, simple shielded and remote handling approaches for handling activated materials, have been discussed.
- Recycling outside the nuclear industry, recycling within nuclear-specific foundries, other recycling scenarios without melting as viable approaches, to answer the routing question, have been considered as viable options.
- The question of reprocessing of radioactive (non-clearable) materials in special facilities in order to separate noxious radionuclides has been mentioned. It is probably the key aspect to be developed in the near future.

Given all the above considerations, an integrated activated materials management strategy has been proposed: it divides the fusion activated materials according to the Regulatory Route (unconditional clearance, conditional clearance, no-clearance) and the Management Route (recycling/reuse, disposal) with a matrix linking the two routes. Furthermore, an approach to the technical difficulty of recycling or waste conditioning, adopting a scoring system, depending on the handling and cooling requirements of the components and materials, completed the approach.

In conclusion, the parameters that govern the back-end of the fusion materials cycle were clearly defined. A new fusion-specific approach for the entire back-end cycle of fusion materials is required. The proposal is for a comprehensive one: it takes into account the evacuation routes for the waste and materials, the handling difficulties, as well as the critical issues and challenges facing all three options: recycling, clearance, and disposal. This approach includes all the procedures necessary to manage radioactive materials from fusion facilities, from the removal of the components from the device to their reuse through recycling/clearance, or to the disposal of the waste in shallow underground repositories. Such an approach requires further refinement, approval of the national authorities, and more important a dedicated R&D program to address the identified critical issues. Nevertheless, it allows a complete attention to most of the parameters involved in such a complex management system. Also, it allows investigating and comparing different plant designs and material compositions, in view of their environmental impact.

As a matter of fact, it is important to clearly define the parameters governing the management procedures for radioactive materials following the change-out of replaceable components and decommissioning of fusion facilities. In that respect, recycling and clearance (i.e. declassification to non-radioactive material) still play the role as the two recommended options for reducing the amount of fusion waste, while disposal as LLW could be an alternative route.

## 11. References

ARIES Project. (n.d.), http://aries.ucsd.edu/ARIES/

Décret 357. (2008). Décret n° 2008-357 du 16 avril 2008 pris pour l'application de l'article L. 542-1-2 du code de l'environnement et fixant les prescriptions relatives au Plan national de gestion des matières et des déchets radioactifs (JO n° 92 du 18 avril 2008). (in French)

EC-RP. (2000). *Radiation Protection 122 - Practical Use of the Concepts of Clearance and Exemption, Guidance on General Clearance Levels for Practices.* European Commission, Directorate General for the Environment. Available at: http://ec.europa.eu/energy/nuclear/radioprotection/publication/doc/122_part1_en.pdf

EC-RP. (2003). *Radiation Protection 134 - Evaluation of the Application of the Concepts of Exemption and Clearance for Practices according to Title III of Council Directive 96/29/Euratom of 13 May 1996 in EU Member States.* European Commission, Directorate General for Energy and Transport. Available at: http://ec.europa.eu/energy/nuclear/radioprotection/publication/doc/134_en.pdf

El-Guebaly L. (2007). Evaluation of disposal, recycling and clearance scenarios for managing ARIES radwaste after plant decommissioning. *Nuclear Fusion*, Vol. 47, No. 7, July 2007, pp (485-488)

El-Guebaly L., Massaut V., Tobita K. & Cadwallader L. (2008). Goals, challenges, and successes of managing fusion activated materials, *Fusion Engineering and Design*, Vol. 83, No. 7-9, December 2008, pp. (928–935)

ESBWR Design. (n.d.) http://en.wikipedia.org/wiki/Economic_Simplified_Boiling_Water_Reactor

Fetter S., Cheng E.T . & Mann F.M. (1990). Long term radioactive waste from fusion reactors: Part II, *Fusion Engineering and Design* Vol. 13, No. 2, November 1990, pp. (239-246)

Forrest R.A., Taylor N.P., & Pampin R. (2005). Categorisation of Active Material from PPCS Model Power Plants, *Proceedings of the 1st IAEA Technical Meeting on First Generation of Fusion Power Plants: Design and Technology*, Vienna, 5–7 July 2005, TM-27424, http://www-pub.iaea.org/MTCD/publications/PDF/P1250-cd/datasets/index.htm

Gulden W., Cook I., Marbach G., Raeder J., Petti D., Seki Y. (2000). An update of safety and environmental issues for fusion, *Fusion Engineering and Design*, Vol. 51-52, November 2000, pp. (419-427)

IAEA. (1996). *Clearance Levels for Radionuclides in Solid Materials: Application of the Exemption Principles, Interim Report for Comment*, TECDOC-855, 1996. International Atomic Energy Agency, Vienna. http://www-pub.iaea.org/MTCD/publications/PDF/te_855_web.pdf

IAEA. (2004). *Application of the Concepts of Exclusion, Exemption and Clearance*, Safety Standards Series, No. RS-G-1.7, International Atomic Energy Agency, Vienna. http://www-pub.iaea.org/MTCD/publications/PDF/Pub1202_web.pdf

IAEA. (2009). *Classification of Radioactive Waste.* General Safety Guide, Safety Standards Series No. GSG-1, (2009). http://www-pub.iaea.org/MTCD/publications/PDF/Pub1419_web.pdf

ITER Project. (n.d.), http://www.iter.org/a/index_nav_1.htm

Maisonnier D., Cook I., Sardain P., Andreani R., Di Pace L. & Forrest R. (2005). *A Conceptual Study of Commercial Fusion Power Plants. Final Report of the European Fusion Power Plant Conceptual Study (PPCS).* Report EFDA-RP-RE-5.0

LOI-739. (2006). French law 2006-739, 28th June 2006

Massaut V., Bestwick R., Brodén K., Di Pace L., Ooms L. & Pampin R. (2007). State of the art of fusion material recycling and remaining issues. *Fusion Engineering and Design,* Vol. 82, No. 15-24, October 2007, pp. (2844–2849)

Najmabadi F., Abdou A., Bromberg L., Brown T., Chan V.C., Chu M.C. et al. (2006), The ARIES-AT advanced tokamak, advanced technology fusion power plant. *Fusion Engineering and Design,* Vol. 80, No. 1-4, January 2006, pp. (3-23)

Najmabadi F., Raffray A.R., Abdel-Khalik S., Bromberg L.. Crosatti L., El-Guebaly L. et al. (2008). The ARIES-CS compact stellarator fusion power plant. *Fusion Science and Technology,* Vol. 54,No. 3, October 2008, pp. (655-672)

Nishio S., Tobita K., Tokimatsu K., Shiniya K. & Senda I. (2004). Technological and environmental prospects of low aspect ratio tokamak reactor "VECTOR", *Proceedings of the 20th IAEA Fusion Energy Conference,* Villamoura, Portugal, 1-6 November 2004, FT/P7-35. http://wwwnaweb.iaea.org/napc/physics/fec/fec2004/datasets/FT_P7-35.html

NRB. (1996). *Radiation Safety Regulations (NRB-96): Hygienic Regulations,* approved by the Head state sanitary inspector of the Russian Federation, cancelled with adoption of NRB-99. Informational and publishing centre of the State committee for sanitary-and-epidemiologic supervision of Russian Federation, Moscow. Available on the Internet at: http://www.complexdoc.ru/ntd/485696 (in Russian)

NRB. (1999). *Radiation Safety Regulations (NRB-99).* Sanitary Regulations (SP 2.6.1.758.99). Informational and publishing centre of the Ministry of Health of Russia, Moscow (in Russian)

NRB. (2009) *Radiation Safety Regulations (NRB-99/2009).* Hygienic Regulations and Standards (SanPiN 2.6.1.2523-09/2009), approved by the Head state sanitary inspector of the Russian Federation, Ministry of Health of Russia, Moscow; available on the Internet at: http://www.complexdoc.ru/ntd/534510 (2009) (in Russian)

Ongena J. & Van Oost G. (2001), Energy for Future Centuries - Will Fusion be an Inexhaustible, Safe and Clean Energy Source?, EFDA-JET-RE(00)01, January 2001

Ooms L. & Massaut V. (2005). Feasibility of Fusion Waste Recycling, TW4-TSW-001-D1, Report SCK•CEN R-4056, February 2005, Mol, Belgium

Ooms L. & Boden S. (2007). Recycling Paths for Fusion Power Plant Materials and Components, TW5-TSW-001 D7, Final Report, SCK•CEN-R-4402 TW5, February 2007, Mol, Belgium

OSPORB. (2010) *Basic Sanitary Regulations Ensuring Radiation Safety (OSPORB 99/2010.* SP 2.6.1.2612-10. Federal Service for Supervision in the Field of Protection of Consumers and Well-Being of People, Moscow; available on the Internet at: http://rghost.ru/2971264 (2010) (in Russian)

Pampin R., Massaut, V., Taylor, N.P. (2007). Revision of the inventory and recycling scenario of active material in near-term PPCS models, *Nuclear Fusion,* Vol. 47 , No. 7, 1 July 2007, Article number S10, pp. (S469- S476)

Pampin R. & O'Brien M.H. (2007). Irradiated Material Management In Near-Term Fusion Power Plants, *IEA. Co-Operative Programme on the Economic, Safety and Environmental Aspects of Fusion Power, Collaborative Study on the Back-End of Fusion Materials,* Final Report UKAEA FUS 548, November 2007

Raeder J. (Ed). (1995), *Report of the Safety and Environmental Assessment of Fusion Power (SEAFP) Project,* European Commission, DGXII, EURFUBRU XII-217(95)

Rocco P. & Zucchetti M. (1998). Advanced management concepts for fusion waste, *Journal of Nuclear Materials,* Vol. 258-263, PART 2 B, October 1998, pp. (1773-1777)

US-NRC. (1982). *10CFR61, Licensing requirements for land disposal of radioactive waste,* U.S. Nuclear Regulatory Commission Federal Register 47 FR 57463, Dec. 27, 1982 and subsequent amendments
http://www.nrc.gov/reading-rm/doc-collections/cfr/part061/

US-NRC. (2003). *Radiological Assessments for Clearance of Materials from Nuclear Facilities,* U.S. Nuclear Regulatory Commission Main Report NUREG-1640, Washington, D.C. http://www.nrc.gov/reading-rm/doc-collections/nuregs/staff/sr1640/

Zucchetti M., El-Guebaly L., Forrest R.A., Marshall T.D., Taylor N.P., & Tobita K. (2007). The feasibility of recycling and clearance of active materials from a fusion power plant, *Journal of Nuclear Materials,* Vol. 367-370 B, SPEC. ISS., 1 August 2007, pp. (1355-1360)

Zucchetti M., Di Pace L., El-Guebaly L., Kolbasov B., Massaut V., Pampin R., Wilson P. (2008). An integrated approach to the back-end of the fusion materials cycle, *Fusion Engineering and Design,* Vol. 83, No. 10-12, December 2008, pp. (1706–1709)

Zucchetti M., Di Pace L., El-Guebaly L., Kolbasov B.N., Massaut V., Pampin R. & Wilson P.(2009). The back end of the fusion materials cycle, *Fusion Science and Technology,* Vol. 55, No. 2, February 2009, pp. (109-139)

# Section 3

## Disposal Activities

15

# Diffusion of Radionuclides in Concrete and Soil

Shas V. Mattigod[1], Dawn M. Wellman[1], Chase C. Bovaird[1], Kent E. Parker[1], Kurtis P. Recknagle[1], Libby Clayton[2] and Marc I. Wood[3]
[1]*Pacific Northwest National Laboratory, Richland, WA*
[2]*BP Exploration, Anchorage, AK*
[3]*CH2M Hill Plateau Remediation Company, Richland, WA*
*USA*

## 1. Introduction

One of the methods being considered for safely disposing of low-level radioactive wastes (LLW) is to encase the waste in concrete. Such concrete encasement would contain and isolate the waste packages from the hydrologic environment and would act as an intrusion barrier. The current plan for waste isolation consists of stacking low-level waste packages on a trench floor, surrounding the stacks with reinforced steel, and encasing these packages in concrete. These concrete-encased waste stacks are expected to vary in size with maximum dimensions of 6.4 m long, 2.7 m wide, and 4 m high. The waste stacks are expected to have a surrounding minimum thickness of 15 cm of concrete encasement. These concrete-encased waste packages are expected to withstand environmental exposure (solar radiation, temperature variations, and precipitation) until an interim soil cover or permanent closure cover is installed, and to remain largely intact thereafter.

Any failure of concrete encasement may result in water intrusion and consequent mobilization of radionuclides from the waste packages. The mobilized radionuclides may escape from the encased concrete by mass flow and/or diffusion and move into the surrounding subsurface environment. Therefore, it is necessary to assess the performance of the concrete encasement structure and the ability of the surrounding soil to retard radionuclide migration. The retardation factors for radionuclides contained in the waste packages can be determined from measurements of diffusion coefficients for these contaminants through concrete and fill material. Some of the mobilization scenarios include 1) potential leaching of waste form before permanent closure cover is installed; 2) after the cover installation, long-term diffusion of radionuclides from concrete waste form into surrounding fill material; 3) diffusion of radionuclides from contaminated soils into adjoining concrete encasement and clean fill material. Additionally, the rate of diffusion of radionuclides may be affected by the formation of structural cracks in concrete, the carbonation of the buried waste form, and any potential effect of metallic iron (in the form of rebar) on the mobility of radionuclides.

The radionuclides iodine-129 ($^{129}I$), technetium-99 ($^{99}Tc$), and uranium-238 ($^{238}U$) are identified as long-term dose contributors in LLW (Mann et al. 2001; Wood et al. 1995). Because of their anionic nature in aqueous solutions, $^{129}I$, $^{99}Tc$, and carbonate-complexed

$^{238}U$ may readily leach into the subsurface environment (Serne et al. 1989, 1992a, b, 1993, and 1995). The leachability and/or diffusion of radionuclide species must be measured to assess the long-term performance of waste grouts when contacted with vadose-zone pore water or groundwater.

Although significant research has been conducted on the design and performance of cementitious waste forms, the current protocol conducted to assess radionuclide stability within these waste forms has been limited to the Toxicity Characteristic Leaching Procedure, Method 1311 Federal Registry (Environmental Protection Agency [EPA] 1992) and ANSI/ANS-16.1 leach test (American National Standards Institute [ANSI] 1986). These tests evaluate the performance under water-saturated conditions and do not evaluate the performance of cementitious waste forms within the context of waste repositories which are located within water-deficient vadose zones. Moreover, these tests assess only the diffusion of radionuclides from concrete waste forms and neglect evaluating the mechanisms of retention, stability of the waste form, and formation of secondary phases during weathering, which may serve as long-term secondary hosts for immobilization of radionuclides.

The results of recent investigations conducted under arid and semi-arid conditions (Al-Khayat et al. 2002; Garrabrants et al. 2002, 2003, and 2004; Gervais et al. 2004; Sanchez et al. 2002, 2003) provide valuable information suggesting structural and chemical changes to concrete waste forms which may affect contaminant containment and waste form performance. However, continued research is necessitated by the need to understand the mechanism of contaminant release; the significance of contaminant release pathways; how waste form performance is affected by the full range of environmental conditions within the disposal facility; the process of waste form aging under conditions that are representative of processes occurring in response to changing environmental conditions within the disposal facility; the effect of waste form aging on chemical, physical, and radiological properties, and the associated impact on contaminant release. Recent reviews conducted by the National Academies of Science recognized the efficacy of cementitious materials for waste isolation, but further noted the significant shortcomings in our current understanding and testing protocol for evaluating the performance of various formulations. The objective of our study was to measure the diffusivity of Re, Tc and I in concrete containment and the surrounding vadose zone soil. Effects of carbonation, presence of metallic iron, and fracturing of concrete and the varying moisture contents in soil on the diffusivities of Tc and I were evaluated.

## 2. Concrete composition and fabrication of test specimens

The concrete composition for the burial encasement was indicated in Specification for Concrete Encasement for Contact-Handled LLW. This specification was used as the basis to prepare concrete for fabrication of test specimens. Details of concrete composition and test specimen fabrication has been described in a technical report (Wellman et al. 2006).

### 2.1 Specified concrete composition for encasement

The specified composition includes sulfate-resistant Portland Type I or Type II cement, a pozzolanic material (Class F fly ash), fine and coarse aggregates, and steel fiber. Additional specifications include a water-to-cement ratio of 0.4 and an air content $6.0 \pm 1.5\%$. The nominal proportions and material specifications based on this initial design are listed in the material specifications for field mix column in Table 1.

A laboratory-scale concrete mixture (Table 1) was prepared based on specifications for Portland Type I and Type II cement. Because of the required small dimensions of laboratory test specimens, the size of the coarse aggregate and the dimensions of the steel fiber were proportionately reduced. This was accomplished by decreasing the 2-cm (~0.75 in.) coarse aggregate size in the original specification to a particle size ranging from 2.83 mm to 2 mm in the laboratory mix. The aggregate mixture passing a 7-mesh sieve and retained on a 10-mesh sieve met this particle size specification. The scaled-down steel fibers used in the laboratory mix consisted of Bekaert Dramix brand deformed steel wire fibers that were cut to a nominal length of 8 mm (0.31 in). The deformed end portions were retained for use in the concrete mixture and the straight middle section of the fiber was discarded. Based on these modifications, a concrete mix was prepared that consisted of Portland Cement (Type I & II, ASTM C-150 compliant), Class F fly ash, scaled-down coarse aggregate, fine aggregate, scaled-down deformed steel fiber, and a water-entraining agent (PolyHeed® 997). The water-entraining agent was included in the mix to facilitate the workability of the concrete. The volumes of the PolyHeed® 997 and the air-entraining agent, MB-AE™ 90, were not included in the normalization calculations because of their negligible contribution to the overall mix volume. The material specification and composition for the laboratory-scale concrete mixture is given in Table 1.

| Material | Material Specifications for Field Mix | Normalized Laboratory Design | Material Specifications Used in Revised Laboratory Mix Comparison |
|---|---|---|---|
| Cement | Portland Type I or Type I/II sulfate-resistant cement | 0.27 | Portland Type I & II |
| Fly Ash | Class F fly ash; nominal 15% of cement by volume | 0.01 | Class F fly ash; nominal 20% of cement by volume |
| Coarse Aggregate | No. 676 or equivalent (3/4″ nominal size) | 0.04 | Sieve size +7 to -10 (2.83 - 2 mm size) |
| Fine Aggregate | Sand | 0.51 | Sand -10 sieve size (< 2 mm) |
| Water | Nominal water-to-cement ratio: 0.4 | 0.10 | Water-to-cement ratio: 0.5 |
| Steel Fiber | Deformed Type I, nominal length 2.5 - 3.8 cm (1 - 1.5″) | 0.04 | Deformed, nominal length 8 mm (0.32 in) |
| PolyHeed® 997 | N/A | 0.00375 | Water-entraining agent |
| Air Content | 6.0±1.5% | 6.0±1.5% | N/A |

Table 1. Laboratory-Scale Material Specification and Composition

### 2.1.1 Concrete mold design

The concrete molds for casting specimens were fabricated from Schedule 40 polyvinyl chloride (PVC) piping material. Gaskets were glued to the bottom of the molds and leak tested before use. The PVC forms were pre-treated with form release, a liquid that allows the concrete specimen to release easily from the mold. The first treatment was applied 3 days prior, and the second treatment was applied a few hours before wet concrete was added to the molds.

### 2.1.2 Concrete mix and specimen preparation

Concrete monoliths were prepared with mix components added in this order: water, steel (if applicable), coarse aggregate, fine aggregate, fly ash, cement, PolyHeed® 997, and MB-AE™ 90. The concrete was mixed on medium speed using a Hobart 3-speed bench top mixer in a 4 L steel bowl. The PVC molds were filled in the vertical positions. After filling, the molds were lightly tapped on the laboratory bench until a significant decrease in the release of air bubbles was observed. The forms were stored in plastic bags with damp paper towels to provide moisture while the concrete set.

### 2.2 Concrete-soil diffusion tests

The objective of half-cell tests was to examine the rate of diffusion of key, long-lived, mobile contaminants in unsaturated Hanford Site vadose zone sediment in contact with concrete monolith. These experiments were conducted using half cells of Trench 8 soil (~4-cm diameter and ~10-cm long) in contact with concrete monolith (~4-cm diameter and ~4-cm long). The soil was oven dried, then brought up to 4, 7, or 15% soil moisture content with 18 MΩ cm distilled de-ionized water. Moisture content analysis was performed and the nominal moisture content for 4% cores was 1%, 7% cores were found to be 4%, and 15% were found to be 11%.

For spiked concrete tests, iodine, technetium or rhenium (as Tc surrogate) were first added to water and then mixed into the dry ingredients. For soil spiked tests, the contaminants were first added to water and then added oven-dried soil. Sampling of soil half-cells was conducted at appropriate time intervals by extruding and taking transverse sections of the sediment at about 0.5-cm intervals for the first 4 to 6 cm from the interface and at 1-cm intervals for the remaining part of the half cell. The soil samples were weighed and extracted with either deionized distilled water, and the extracts were analyzed for stable iodide, rhenium and technetium-99 using ICP-MS. The diffusion profile in concrete half-cells were determined by dry-slicing the cells with a circular diamond saw. The resulting thin slices were then crushed and extracted with deionized distilled water and the solutions were analyzed for concentrations of the contaminant of interest.

## 3. Diffusion coefficient calculations

The radionuclide diffusivity is defined by Fick's equation as

$$J = - D_w \, dC_w/dx \tag{1}$$

where, J = flux of radionuclide at a given point, $D_w$ = the diffusivity of water-based radionuclide concentration, and $C_w$ = the radionuclide concentration in the pore water.

Ignoring the possibility that radionuclides might be present in the air within the unsaturated sediment, and recognizing that in the case of a two-phase system (water and soil), Equation 1 can be used in performing a mass balance over a small volume leading to the equation (Crank 1975)

$$dC_w/dt = D_w/\theta *(d^2C_w/dx^2) \quad (2)$$

where θ = the volume fraction water in the soil's pore space. However, the slope on the probit plot provides the diffusivity that solves the equation for diffusion in a homogeneous single phase medium,

$$dC/dt = D*(d^2C/dx^2) \quad (3)$$

The solution for this diffusion equation is provided by

$$C/C_0 = \frac{1}{2} \text{erfc} (x/(\sqrt{4Dt})) \quad (4)$$

Where, $C/C_0$ is the normalized concentration, and erfc is the error function (Crank 1975). The diffusivities in the soil were calculated using a probit analysis approach. The details of the probit analysis are provided in Finney (1971). This technique allows the transformation of a sigmoid curve of concentrations, normalized with respect to the initial concentration ($C_0$) as a function of diffusion distance (x) produced in a half-cell diffusion experiment to a linear plot. The transformed equation is,

$$\text{Probit (p)} = \Phi^{-1} (p) = a - x/\sqrt{(2Dt)} \quad (5)$$

When probit transform of $C/C_0$ is plotted against x, the resulting slope $b = 1/\sqrt{(2Dt)}$ can used to calculate the diffusivity (D) as $D= 1/(2b^2t)$, where t is the sampling time. This approach has been used previously to determine diffusivity in half-cell diffusion experiments (Brown et al. 1964. Lamar 1989).The diffusion coefficient $D_W$ that was calculated from the D obtained from the probit plot based on concentrations in the pore water must then be multiplied by θ. Detailed statistical analysis (Abdel Rahman and Zaki 2011) was not conducted to assess the uncertainities associated with slopes from the linear regression and the calculated diffusion coefficient values.

## 4. Effects of carbonation and metallic iron on the diffusion of technetium and iodine in soils and concrete

Diffusion experiments were conducted using carbonated (by immersion in sodium carbonate solution) concrete-soil half cells. Soil half-cell specimens were spiked with I and Tc to achieve a measurable diffusion profile in the concrete part of the half-cell. The goal of this test was to examine the effect of carbonation and metallic iron addition on diffusion of I and Tc into concrete.

Typical diffusion profiles for Tc and I in concrete and the corresponding Probit plots are shown in figures 1 and 2.

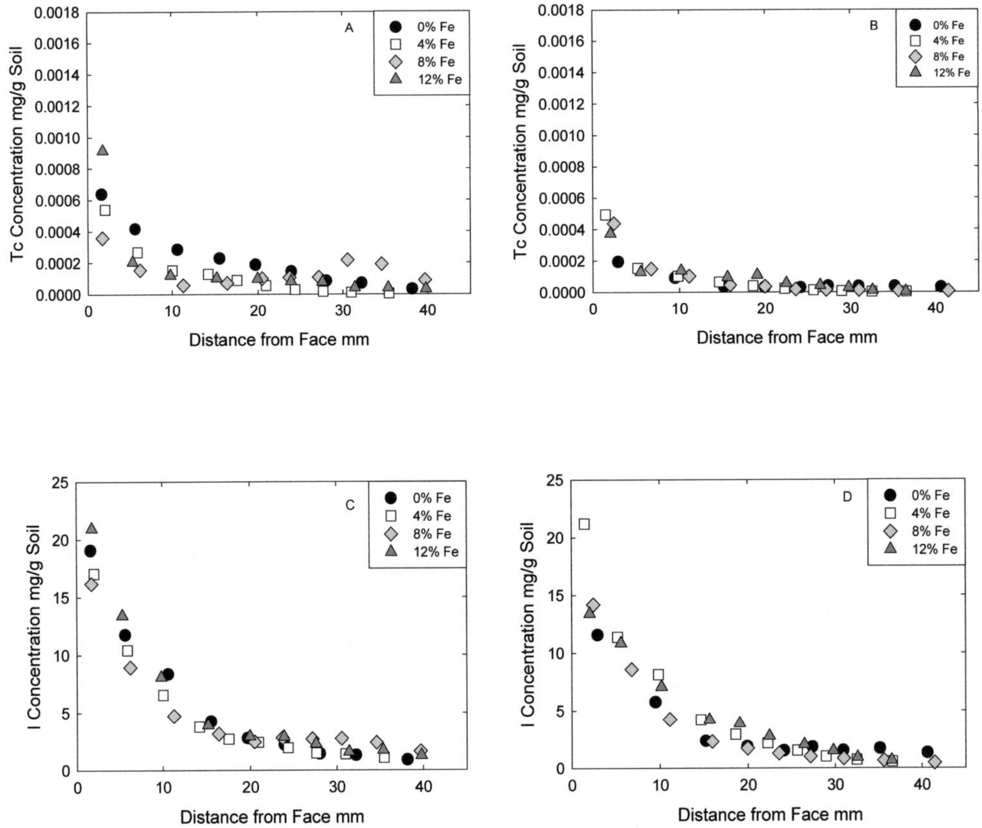


Fig. 1. Concrete half-cell concentration profiles as a function of iron content:
Top left: Tc concentration for non-carbonated concrete at 4% soil moisture;
Top right: Tc concentration for carbonated concrete at 4% soil moisture;
Bottom left: I concentration for non-carbonated concrete at 4% soil moisture;
Bottom right: I concentration for carbonated concrete at 4% soil moisture.

The results indicated diffusivities ranging from $1 \times 10^{-10}$ to $1.9 \times 10^{-9}$ $cm^2/s$ for Tc, and $1.3 \times 10^{-10}$ to $2.3 \times 10^{-9}$ $cm^2/s$ for I (Table 2). The highest Tc and I diffusivities were observed in all uncarbonated, Fe-free concrete cores contacting spiked soils at all three moisture contents (4%, 7%, and 15%). However, the diffusivities of both Tc and I (except in one case) were significantly attenuated in all carbonated concrete cores. The reduction of Tc diffusivities ranged from 55% to 72%. Meanwhile, I diffusivities were reduced by 61% at soil moisture content of 4% and by 58% at soil moisture content of 15%. However, I diffusivity showed an anomalous increase of ~38% in an uncarbonated, Fe-free concrete core in contact with a soil core with 7% moisture content.

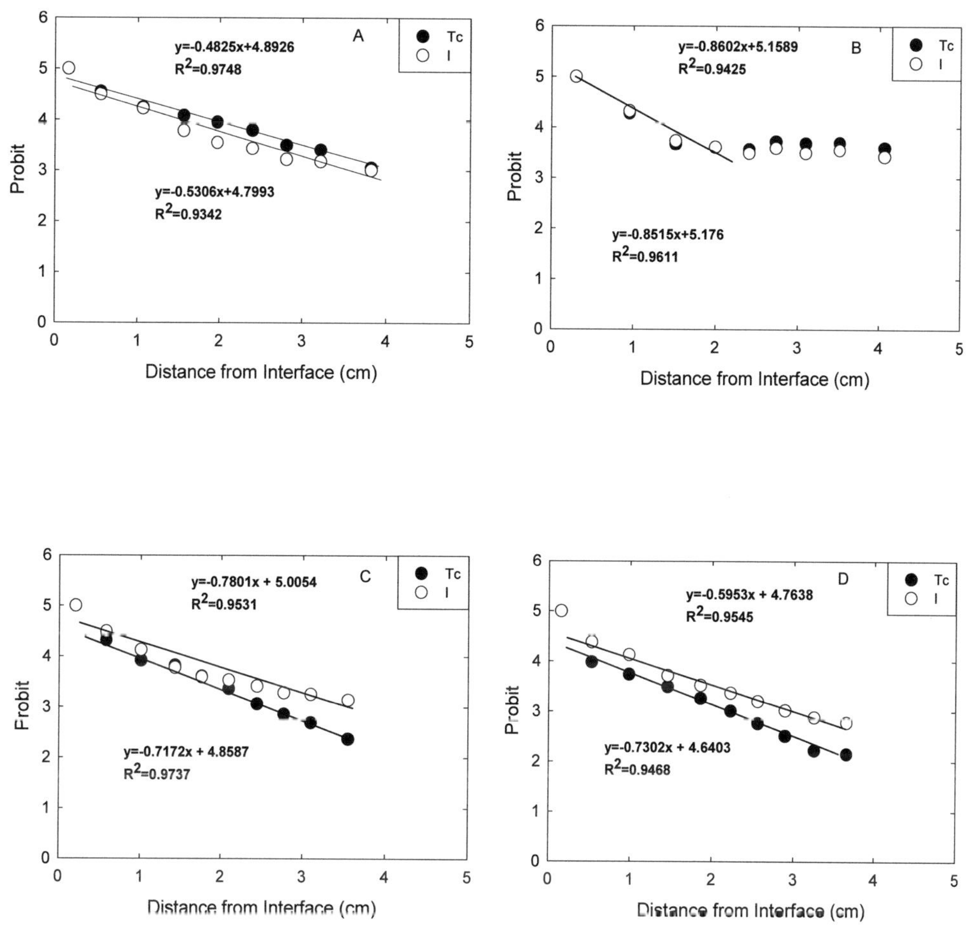


Fig. 2. Probit plots for concrete cores. Top left: 4% moisture content, no Fe, no carbonation; Top right: 4% moisture content, no Fe, carbonated; Bottom left: 4% moisture content, 4% Fe, no carbonation; Bottom right: 4% moisture content, 4% Fe, carbonated.

Similar attenuation in Tc and I diffusivities were also observed in carbonated concrete half cells containing various quantities of Fe. The diffusivities of Tc showed reduction typically ranging from 25 to 81%, except enhanced diffusivities (~40 - 30%) were observed in two concrete cores containing 4% Fe and in contact with soil with 4 and 15% moisture contents. In these same cores, similar increases in I diffusivities were also observed (10% - 6%). However all other Fe-containing carbonated concrete cores exhibited reduction in I diffusivities that ranged from 9% to 76%.

| Core ID | Initial Moisture Content (Wt %) | Carbonation | Fe (wt %) | Tc Diffusivity ($cm^2/s$) | I Diffusivity ($cm^2/s$) |
|---|---|---|---|---|---|
| Tc-C-08-3-0-325 | 4 | N | 0 | 1.7E-09 | 1.4E-09 |
| Tc-C-08-3-0-329 | 7 | N | 0 | 1.9E-09 | 4.0E-10 |
| Tc-C-08-3-0-330 | 15 | N | 0 | 1.3E-09 | 1.2E-09 |
| Tc-C-08-3-0-332 | 4 | Y | 0 | 5.3E-10 | 5.4E-10 |
| Tc-C-08-3-0-333 | 7 | Y | 0 | 5.4E-10 | 5.5E-10 |
| Tc-C-08-3-0-334 | 15 | Y | 0 | 5.9E-10 | 5.1E-10 |
| Tc-C-08-3-4-350 | 4 | N | 4 | 7.7E-10 | 6.5E-10 |
| Tc-C-08-3-4-351 | 7 | N | 4 | 6.9E-10 | 2.3E-09 |
| Tc-C-08-3-4-353 | 15 | N | 4 | 3.6E-10 | 7.9E-10 |
| Tc-C-08-3-4-357 | 4 | Y | 4 | 7.4E-10 | 1.1E-09 |
| Tc-C-08-3-4-359 | 7 | Y | 4 | 2.6E-10 | 4.2E-10 |
| Tc-C-08-3-4-360 | 15 | Y | 4 | 7.8E-10 | 5.4E-10 |
| Tc-C-08-3-8-401 | 4 | N | 8 | 1.9E-10 | 7.1E-10 |
| Tc-C-08-3-8-402 | 7 | N | 8 | (a) | 5.6E-10 |
| Tc-C-08-3-8-403 | 15 | N | 8 | (a) | 6.0E-10 |
| Tc-C-08-3-8-407 | 4 | Y | 8 | 5.1E-10 | 4.9E-10 |
| Tc-C-08-3-8-409 | 7 | Y | 8 | 2.2E-10 | 3.4E-10 |
| Tc-C-08-3-8-410 | 15 | Y | 8 | 3.6E-10 | 3.4E-10 |
| Tc-C-08-3-12-425 | 4 | N | 12 | (a) | 4.3E-10 |
| Tc-C-08-3-12-426 | 7 | N | 12 | 2.3E-10 | 4.0E-10 |
| Tc-C-08-3-12-427 | 15 | N | 12 | 4.7E-10 | 7.7E-10 |
| Tc-C-08-3-12-432 | 4 | Y | 12 | 1.1E-10 | 1.3E-10 |
| Tc-C-08-3-12-433 | 7 | Y | 12 | 1.0E-10 | 3.3E-10 |
| Tc-C-08-3-12-435 | 15 | Y | 12 | 4.4E-10 | 3.1E-10 |

(a) Probit analysis could not be performed due to poorly defined diffusion profile

Table 2. Diffusivity of Tc and I in concrete half cells

## 5. Diffusion of iodine, rhenium, and technetium from soil into fractured concrete

A set of diffusion experiments were conducted to examine the effect of fracturing of spiked encasement concrete on the diffusion of radionuclides into soil. Spiked and cured concrete monoliths were encased in shrink wrap and struck with a hammer to prevent the formation of rubble. Each fractured core possessed a single fracture extending the length of the core, perpendicular to the concrete-sediment interface.

In one set of experiments, soil half-cell specimens were spiked with I and Re to achieve a measurable diffusion in the fractured concrete part of the half-cell. The diffusion tests were conducted under unsaturated conditions at 4%, 7%, and 15% (moisture content by weight) with or without Fe addition (4 wt %) and carbonation. In a second set of experiments, Tc-spiked soil half cells at a fixed moisture content were contacted with fractured concrete half cells.

Typical probit plots are shown in Figure 3 and the data is listed in Table 3. Overall, the calculated diffusivities for Re ranged from 3.8 x $10^{-12}$ to 2 x $10^{-9}$ $cm^2/s$ and for I ranged from 1.3 x $10^{-10}$ to 2.3 x $10^{-9}$ $cm^2/s$ in fractured concrete. The highest Re and I diffusivities were observed in fractured concrete cores that were in contact with spiked soils with 15% moisture content. Effects of carbonation in enhancing diffusion are noticeable at higher moisture contents (7 and 15%) of the contacting soil. Additions of 4% by weight Fe did not noticeably affect the diffusivities of Re and I in fractured concrete.

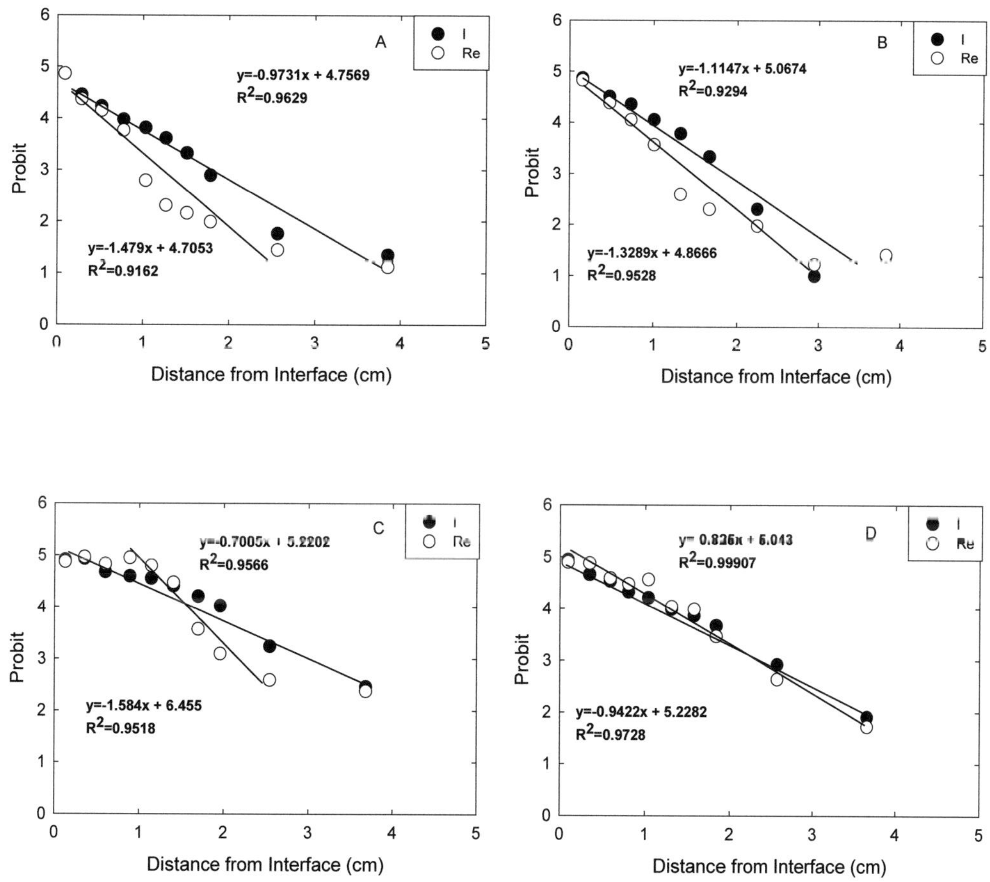


Fig. 3. Probit plots for concrete cores. Top left: 4% moisture content, no Fe, no carbonation; Top right: 4% moisture content, 4% Fe, no carbonation; Bottom left: 4% moisture content, no Fe, carbonated; Bottom right: 4% moisture content, 4% Fe, carbonated half cells.

| Core ID | Moisture Content wt% | Carbonation | Fe (wt %) | Re Diffusivity ($cm^2/s$) | I Diffusivity ($cm^2/s$) |
|---|---|---|---|---|---|
| C-5-0-2 | 4 | N | 0 | 1.36E-10 | 3.13E-10 |
| C-5-4-26 | 4 | N | 4 | 1.68E-10 | 2.39E-10 |
| C-5-0-1 | 4 | Y | 0 | 1.18E-10 | 5.90E-10 |
| C-5-4-21 | 4 | Y | 4 | 3.34E-10 | 4.35E-10 |
| C-5-0-7 | 7 | N | 0 | 3.75E-12 | 1.25E-10 |
| C-5-4-27 | 7 | N | 4 | 8.88E-11 | 1.93E-10 |
| C-5-0-5 | 7 | Y | 0 | 2.60E-10 | 2.77E-10 |
| C-5-4-23 | 7 | Y | 4 | 1.69E-10 | 4.00E-10 |
| C-5-0-10 | 15 | N | 0 | 2.03E-09 | 2.33E-09 |
| C-5-4-30 | 15 | N | 4 | 4.02E-10 | 1.76E-09 |
| C-5-0-6 | 15 | Y | 0 | 6.90E-10 | 5.10E-10 |
| C-5-4-24 | 15 | Y | 4 | 6.72E-10 | 1.21E-09 |

Table 3. Diffusivity data for Re-I in fractured concrete half cells

In a second set of experiments, Tc-spiked soil half cells at a fixed moisture content were contacted with fractured concrete half cells. The resulting probit analysis for Tc diffusivity in concrete cores is listed in Table 4. These values ranged from $3.1 \times 10^{-11}$ $cm^2/s$ to $3.6 \times 10^{-10}$ $cm^2/s$. The data indicated that carbonation in all cases enhanced Tc diffusivity in concrete cores. Significant increases in Tc diffusivities were noted when cores with higher concentrations of Fe (8% and 12%) were carbonated. These data indicate that carbonation of Fe containing concrete cores may enhance micro-cracking of concrete resulting in an increase in Tc diffusivity. Similar phenomena have been noted in previous studies in which Fe containing carbonated concrete cores was in contact with Tc-spiked soil cores with 4% moisture content. Also, the beneficial effect of Fe on reducing Tc diffusivity in non-carbonated specimens is not observable until the Fe content is at least 8% by mass. These data indicated that diffusivity in concrete is controlled by carbonation and Fe content and the presence of a single fracture had no measureable effect.

| Core ID | Carbonation | Fe (wt %) | Tc Diffusivity ($cm^2/s$) |
|---|---|---|---|
| Tc-C-10-5-0-101 | N | 0 | 1.1E-10 |
| Tc-C-10-5-0-102 | Y | 0 | 1.4E-10$^{0}$ |
| Tc-C-10-5-4-105 | N | 4 | 2.1E-10 |
| Tc-C-10-5-4-106 | Y | 4 | 3.6E-10 |
| Tc-C-10-5-8-107 | N | 8 | 3.1E-11 |
| Tc-C-10-5-8-108 | Y | 8 | 1.9E-10 |
| Tc-C-10-5-12-110 | N | 12 | 8.4E-11 |
| Tc-C-10-5-12-111 | Y | 12 | 2.1E-10 |

Table 4. Diffusivity of Tc in fractured concrete cores

## 6. Diffusion of iodine and rhenium from fractured concrete into soil

Two sets of diffusion experiments were initiated using fractured concrete-soil half cells. The objective was to examine the diffusion from spiked and fractured concrete cores into soil half cells. In one set, Re and I spiked concrete specimens were prepared with and without carbonation and additions of 0, 4, 8, and 12% Fe. After curing, these concrete specimens were fractured and half cells were prepared by placing these specimens in contact with soil at 4, 7, and 15% initial moisture content. Typical probit plots are shown in Figure 4 and the results of probit analysis for these sets of diffusion data are shown in Table 5.

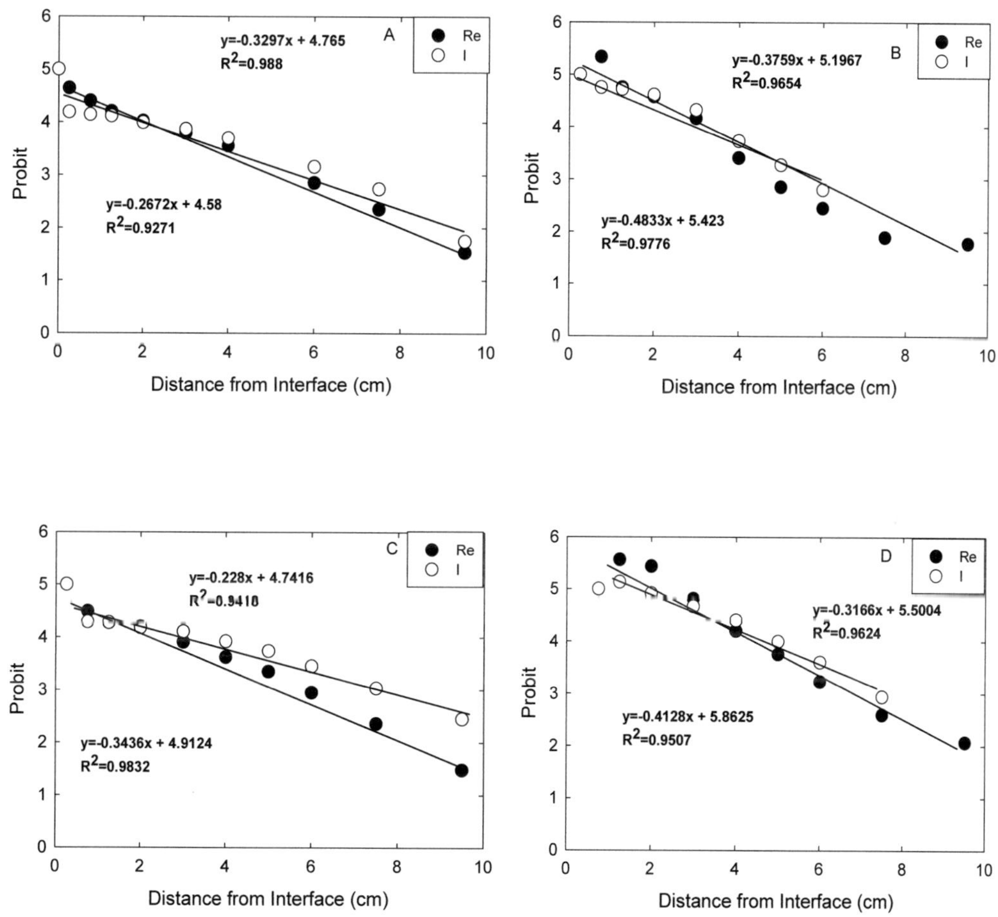


Fig. 4. Probit analysis for Re and I in soil cores. Top left: 4% moisture content, no Fe, no carbonation; Top right: 4% moisture content, no Fe, carbonated; Bottom left: 4% moisture content, 4% Fe, no carbonation; Bottom right: 4% moisture content, 4% Fe, carbonated.

The calculations showed that at 4 % moisture content, with no iron present, carbonation reduced the diffusivities of both Re and I in soils, roughly by half. The addition of 4% by mass Fe into un-carbonated concrete did not significantly affect Re and I diffusivities in soil. In carbonated specimens, adding 4% Fe seemed to increase only I diffusivity without significantly affecting Re diffusivity. When the concentrations of Fe in concrete cores were increased to 8%, significantly enhanced diffusivities in soils were observed. For instance, in soils in contact with un-carbonated concrete cores, the presence of 8% Fe in the concrete increased Re and I diffusivities by ~17 and 36%, respectively. With carbonation, however, the diffusivities of both Re and I more than doubled as compared to carbonated specimens with no Fe. Similarly, increasing the Fe content to 12% by mass in both un-carbonated and carbonated concrete increased Re diffusivities soils by about a third as compared to soils contacting cores with no added Fe. In soils contacting similarly treated concrete cores, the I diffusivities did not change significantly (carbonation) or decreased by ~20% (no carbonation).

In soil cores with higher moisture content (7%) and in contact with carbonated concrete cores containing no iron, the diffusivity of Re was ~20% lower than in soils in contact with un-carbonated concrete cores. The addition of 4% Fe by mass to carbonated concrete cores increased Re and I soil diffusivities by ~1.5 and ~3 times as compared to diffusivities in soil cores contacting concrete half cells with no Fe content. Rhenium diffusivity in soil cores in contact with noncarbonated concrete half cells with 12% Fe was about an order of magnitude lower than in soil cores in contact with similar concrete half cells containing no Fe.

| Core ID | MC (Wt %) | Carbonation | Fe (wt %) | Re Diffusivity ($cm^2/s$) | I Diffusivity ($cm^2/s$) |
|---|---|---|---|---|---|
| C-08-5-0-501 | 4 | N | 0 | 2.2E-09 | 3.3E-09 |
| C-08-5-0-502 | 7 | N | 0 | 2.1E-07 | NA[a] |
| C-08-5-0-504 | 4 | Y | 0 | 1.0E-09 | 1.7E-09 |
| C-08-5-0-505 | 7 | Y | 0 | 1.7E-07 | 1.2E-07 |
| C-08-5-4-526 | 4 | N | 4 | 2.0E-09 | 4.5E-09 |
| C-08-5-4-530 | 4 | Y | 4 | 1.4E-09 | 2.3E-09 |
| C-08-5-4-531 | 7 | Y | 4 | 2.6E-07 | 3.6E-07 |
| C-08-5-8-552 | 4 | N | 8 | 2.5E-08 | 4.4E-08 |
| C-08-5-8-555 | 4 | Y | 8 | 2.4E-09 | 3.8E-09 |
| C-08-5-12-576 | 4 | N | 12 | 1.5E-09 | 2.6E-08 |
| C-08-5-12-580 | 4 | Y | 12 | 1.4E-09 | 1.7E-09 |

[a] Probit analysis could not be performed due to poorly defined diffusion profile

Table 5. Diffusivity analysis Re and I in soil half cells

In soil cores (7% moisture content) contacting Fe-free concrete with and without carbonation, the Re and I diffusivities were more than an order of magnitude higher than in soils containing 4% moisture content.

At the end of the experiment lasting over a year, Re and I diffusion in soil cores with 15% moisture content had proceeded to the degree that no distinct concentration gradients were present; therefore, diffusivity values could not be ascertained. These data suggest that diffusivities in soil are affected mainly by its moisture content. State of carbonation, Fe content and the presence of fracture in source- concrete do not significantly affect soil diffusivities except mass transfer of contaminants.

## 7. Modeling the diffusion of iodine and technetium from a reservoir through a concrete enclosure and into surrounding soil

To model the diffusion of iodine and technetium through 6 inches of concrete into the surrounding soil, pairs of high- and low-diffusivity values were selected from the experimental data. The values selected are listed in Table 6.

| Diffusing Component | Material | Low Diffusivity, $cm^2/s$ | High Diffusivity, $cm^2/s$ |
|---|---|---|---|
| Iodine | Concrete | 1.30E-10 | 2.30E-09 |
| | Soil | 1.70E 09 | 2.50E-08 |
| Tc | Concrete | 3.10E-11 | 3.60E-10 |
| | Soil | 1.30E-08 | 5.80E-08 |

Table 6. Diffusivity values for concrete and soil used in the simulations

### 7.1 Modeling approach

The computational fluid dynamics (CFD) code, STAR-CD[1] was used for the calculations. STAR-CD is a commercial CFD code that solves the finite volume formulations for conservation of mass, momentum, and energy for general-purpose thermal-fluids simulations. STAR-CD was used to simulate the iodine and technetium species diffusion through the concrete encasement box and into the soil using the analogy of thermal conduction,

$$\frac{dT}{dt} = \frac{k}{\rho C_p} \frac{d^2T}{dx^2} \tag{6}$$

and mass diffusion:

$$\frac{dC}{dt} = D\frac{d^2C}{dx^2} \tag{7}$$

[1] STAR-CD, *Version 4.14 Methodology*2010: Computational Dynamics Ltd.

In the heat conduction equation (6) $T$ is the absolute temperature in Kelvin, $\rho$ is the density, and $C_p$ is the specific heat of the material. In the mass diffusion equation (7) $C$ is the specie concentration and $D$ is the specie diffusivity. In STAR-CD we can simulate the specie diffusion through the porous concrete and soil by setting the coefficients of (6) and (7) as equal and solving the thermal conduction problem through the two materials.

The following assumptions were made for performing the simulations:

- Constant specie concentration reservoir exists adjacent to inside wall of concrete encasement box.
- 1-dimensional diffusion through encasement box wall and into the surrounding soil.
- No additional resistance between concrete and soil to decrease the specie diffusion.

A one-dimensional (1-D) finite volume model was constructed to represent the species transport through the concrete encasement wall and into the soil. A 1-D rectangular mesh was used for transport through the encasement wall. As the species exit the concrete and enter the surrounding soil, the environment is more cylindrical in nature as the species are free to diffuse radially into the soil. Thus the mesh for the concrete wall was mated to a cylindrical mesh extending out into the surrounding soil as shown schematically in Figure 5. The model domain is shown outlined with a dashed line in the figure. In the model, the concrete wall is assumed to be in contact with a constant specie reservoir (at location 1 in the figure). The contact between concrete and soil (at location 2 in the figure) is assumed to be perfect with no additional resistance there for specie diffusion. The outer radius of the model was established far away from the reservoir such that specie concentration remained zero for the times examined (1000, 3000, 10000, and 30000 years). Simulations were performed for diffusion of Iodine and Tc under low-diffusivity and high-diffusivity conditions (Table 6).

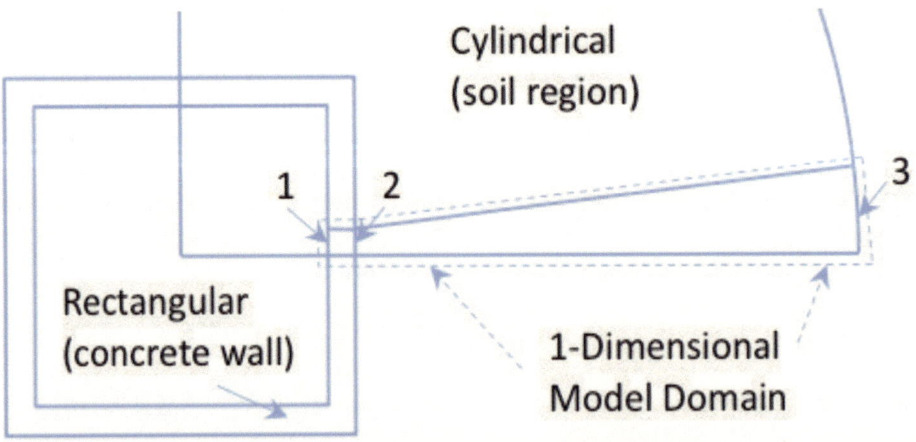


Fig. 5. Plan-view schematic of the model domain (dashed outline) where the rectangular concrete encasement wall is in contact with the surrounding soil region into which the species diffuse radially. 1. Concrete wall in contact with specie reservoir. 2. Concrete in contact with soil. 3. Outer radius is established far enough away from the reservoir that species concentration is zero at long times (i.e., 30,000 years).

### 7.2 Results

The results of the simulations are listed in 7 and 8 and the diffusion profiles are shown in Figures 6 through 9.

The normalized concentrations ($C/C_0$) at the concrete (~15 cm thick) soil interface as a function of time are listed in Table 7. The data indicates that under low diffusivity conditions, the $C/C_0$ for both I and Tc at all simulated times are very low (0 - 0.07). Under high diffusivity conditions, the $C/C_0$ values for Tc ranged from 0 – 0.03, whereas, iodine had $C/C_0$ values ranging from 0.05 – 0.22 indicating deeper penetration into concrete.

| | Time (Y) | | | |
|---|---|---|---|---|
| Diffusivity | 1,000 | 3,000 | 10,000 | 30,000 |
| | $C/C_0$ | | | |
| I Low | 0.00 | 0.00 | 0.02 | 0.07 |
| I high | 0.05 | 0.10 | 0.17 | 0.22 |
| Tc Low | 0.00 | 0.00 | 0.00 | 0.00 |
| Tc High | 0.00 | 0.01 | 0.02 | 0.03 |

Table 7. Normalized concentration values for I and Tc at the concrete-soil interface

The depths of penetration, defined as $C/C_0$ of ~0.005 for various simulated time periods, are listed in Table 8. Under low diffusivity conditions, iodine at the end of 30,000 years is predicted to penetrate about 69 cm into soil (Table 8 and Figure 6), whereas Tc appears to be confined within the encasing of ~15-cm thick concrete (Table 8 and Figure 8). Under higher diffusivity conditions, the simulations indicated higher depths of penetration of I and Tc into soil. For instance, iodine is predicted to penetrate from ~ 39 – 398 cm into the soil for the time period ranging from 1,000 – 30,000 years (Table 8, Figure 7). Technetium under high diffusivity conditions is predicted to penetrate ~112 –286 cm into the soil at the end of 10,000 and 30,000 years respectively (Table 8, Figure 9).

| | Time (Y) | | | |
|---|---|---|---|---|
| | 1,000 | 3,000 | 10,000 | 30,000 |
| Diffusivity | Depth of Soil Penetration (cm) at $C/C_0$ = ~0.005 | | | |
| I Low | 0 | 0 | 18 | 69 |
| I high | 39 | 99 | 215 | 398 |
| Tc Low | 0 | 0 | 0 | 0 |
| Tc High | 0 | 0 | 112 | 286 |

Table 8. Depth of penetration of I and Tc into soil as a function of time

The experimental data indicates that low diffusivity for I is engendered when carbonated encasing concrete is surrounded by soil with very low moisture contents. For Tc, low diffusivity is observed when carbonated and Fe-containing concrete is in contact with very low moisture content sediments. Comparatively, higher diffusivities of I and Tc are typically observed when higher moisture content soil is in contact with uncarbonated, Fe-free encasing concrete.

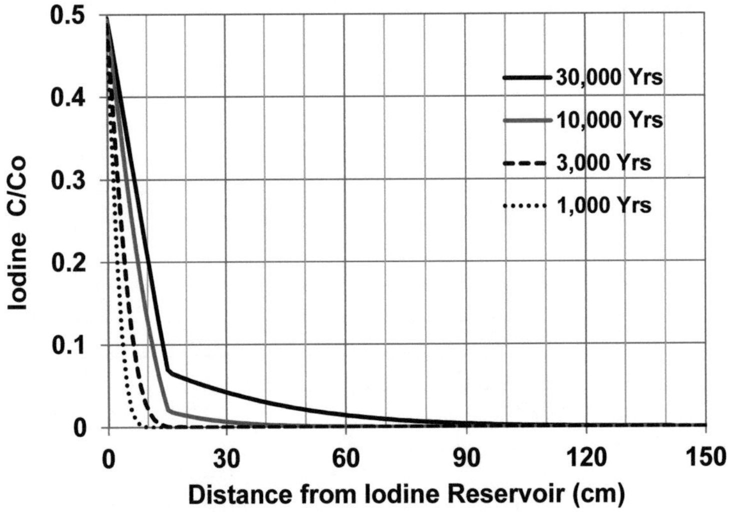


Fig. 6. Iodine concentration profiles for the low diffusivity case. Zero to 15 cm represents concentrations within the concrete encasement wall. Fifteen cm beyond are concentrations in the surrounding soil. $C/C_0 = 0.005$ was about 69 cm into the soil at 30,000 years.

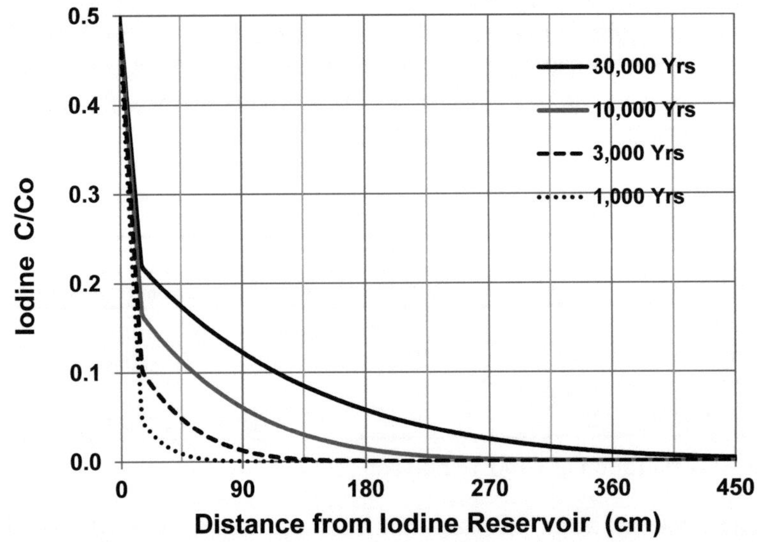


Fig. 7. Iodine concentration profiles for the high diffusivity case. Zero to 15 cm represents concentrations within the concrete encasement wall. Fifteen cm and beyond are concentrations in the surrounding soil. $C/C_0 = 0.005$ was about 398 cm into the soil depth at 30,000 years.

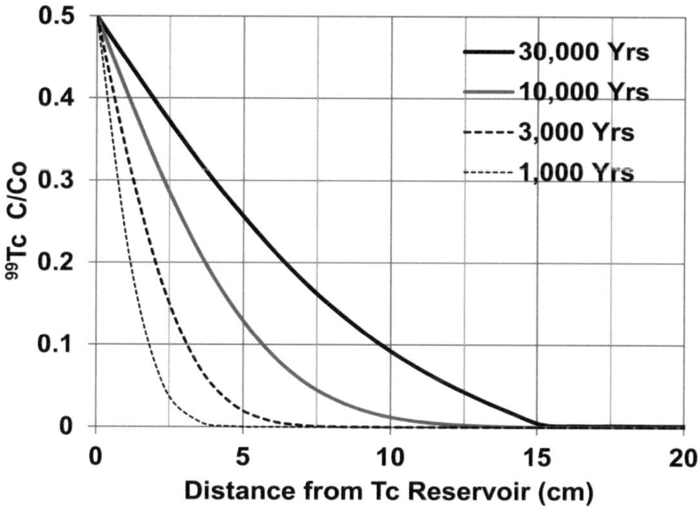


Fig. 8. Technetium concentration profiles for the low diffusivity case. Zero to 15 cm represents concentrations within the concrete encasement wall. Fifteen cm and beyond are concentrations in the surrounding soil. $C/C_0 = 0.005$ was about to the outer edge of the concrete encasement (15 cm total penetration depth) at 30,000 years.

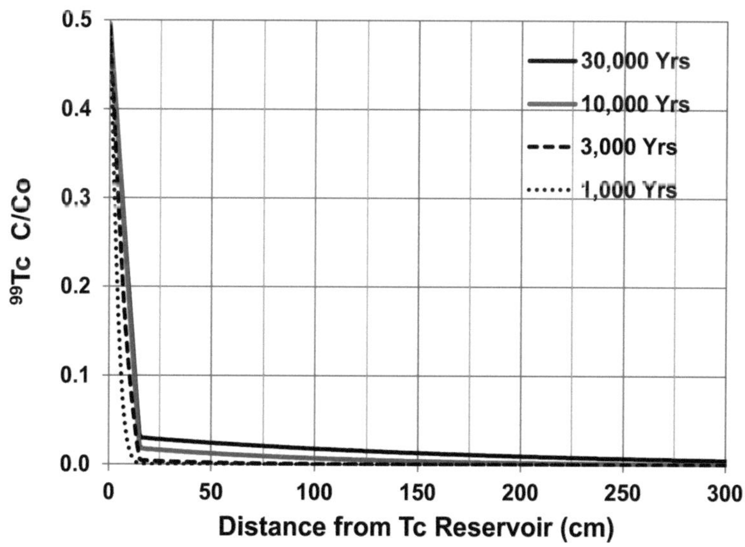


Fig. 9. Technetium concentration profiles for the high diffusivity case. Zero to 15 cm represents concentrations within the concrete encasement wall. Fifteen cm and beyond are concentrations in the surrounding soil. $C/C_0 = 0.005$ was about 286 cm into the soil at 30,000 years.

## 8. Conclusions

Tests were conducted to measure the diffusion of Tc, Re (as a surrogate for Tc), and I in unsaturated sediments and in concrete that encases low-level radioactive wastes. In these tests, effects of varying sediment moisture content, carbonation, metallic iron addition, and fracturing of concrete on diffusivity of these nuclides were also evaluated. The conclusions from these tests are as follows.

1. Regardless of the variables introduced in the analysis (e.g., water content of the sediment, level of iron additives and carbonation of concrete), diffusion coefficients for both Tc and I varied over a range of about 2 orders of magnitude with Tc being slightly more retarded than I.
2. Degradation of concrete through fracturing and carbonation will not strongly influence diffusion rates primarily because of expected low moisture contents in the waste environment (vadose zone sediments).
3. Within the range of measurements, increasing moisture content of the sediment routinely increased diffusion rates.
4. The addition of metallic iron appeared to have competing effects on diffusion. In some cases, limited reduction of Tc may have occurred. On the other hand, metallic iron appears to have facilitated microcracking during carbonation.
5. In the context of a 1,000 year compliance period, concrete encasement of waste provides a significant delay in radionuclide release into the subsurface.

## 9. Acknowledgment

This work was conducted at Pacific Northwest National Laboratory, operated by Battelle Memorial Institute for the U.S. Department of Energy under contract DE-AC05-76RL01830. Funding for this project was provided by CH2M Hill Plateau Remediation Company and the Department of Energy Environmental Management Office of Technology Innovation and Development. We would like to thank Steven Baum, Kenton Rod, Sara Rither, John Nelson, and Danielle Saunders for assisting in the sectioning of the concrete soil half-cells. We also thank the anonymous reviewer for helpful comments.

## 10. References

Abdel Rahman RO, and AA Zaki. 2011. "Comparative study of leaching conceptual models: Cs leaching from different ILW cement based matrices". *Chemical Engineering Journal. doi:10.1016/j.cej.2011.08.038.*

Al-Khayat H, MN Haque, and NI Fattuhi. 2002. "Concrete Carbonation in Arid Climate." *Materials and Structures* 35(7):421-426.

ANSI. 1986. *Measurement of the Leachability of Solidified Low-Level Radioactive Wastes by a Short-Term Test Procedure,* ANSI/ANS-16.1, American Nuclear Society, Chicago.

Brown DA, BE Fulton and RE Phillips. 1964. "Ion Exchange Diffusion: I. A Quick-Freeze Method for the Measurement of Ion Diffusion in Soil and Clay Systems." *Soil Sci. Soc. Am. Proc.* 28:628-632.

Crank J. 1975. *The Mathematics of Diffusion.* Second Edition. Oxford University Press, New York.

EPA—U.S. Environmental Protection Agency. 1999. *Toxicity Characteristics Leaching Procedure. Test Methods for Evaluating Solid Wastes – Physical and Chemical Methods SW-846.* Method 1311.

Finney DJ. 1971. *Probit Analysis; A Statistical Treatment of the Sigmoid Response Curve*. Third edition, Cambridge University Press, New York.

Garrabrants AC, and DS Kosson. 2003. "Modeling Moisture Transport from a Portland Cement-Based Material During Storage in Reactive and Inert Atmospheres." *Drying Technology* 21(5):775-805.

Garrabrants AC, F Sanchez, C Gervais, P Moszkowicz, and DS Kosson. 2002. "The Effect of Storage in an Inert Atmosphere on the Release of Inorganic Constituents during Intermittent Wetting of a Cement-Based Material." *Journal of Hazardous Materials* 91(1-3):159-185.

Garrabrants AC, F Sanchez, and DS Kosson. 2004. "Changes in Constituent Equilibrium Leaching and Pore Water Characteristics of a Portland Cement Mortar as a Result of Carbonation." *Waste Management* 24(1):19-36.

Gervais C, AC Garrabrants, F Sanchez, R Barna, P Moszkowicz, and DS Kosson. 2004. "The Effects of Carbonation and Drying During Intermittent Leaching on the Release of Inorganic Constituents from a Cement-Based Matrix." *Cement and Concrete Research* 34(1):119-131.

Lamar DA. 1989. *Measurement of Nitrate Diffusivity in Hanford Sediments via the Half-Cell Method.* Letter Report to Westinghouse Hanford Company, Pacific Northwest National Laboratory, Richland, Washington.

Mann FM, RJ Puigh II, SH Finfrock, EJ Freeman, R Khaleel, DH Bacon, MP Bergeron, PB McGrail, and SK Wurstner. 2001. *Hanford Immobilized Low-Activity Waste Performance Assessment: 2001 Version*, DOE/ORP-2000-24, Rev. B. Pacific Northwest National Laboratory, Richland, WA.

Sanchez F, C Gervais, AC Garrabrants, R Barna, and DS Kosson. 2002. "Leaching of Inorganic Contaminants from Cement-Based Waste Materials as a Result of Carbonation during Intermittent Wetting." *Waste Management* 22(2):249-260.

Sanchez F, AC Garrabrants and DS Kosson. 2003. "Effects of Intermittent Wetting on Concentration Profiles and Release from a Cement-Based Waste Matrix." *Environmental Engineering Science* 20(2):135-153.

Serne RJ, WJ Martin, VL LeGore, CW Lindenmeier, SB McLaurine, PFC Martin, and RO Lokken. 1989. *Leach Tests on Grouts Made with Actual and Trace Metal-Spiked Synthetic Phosphate/Sulfate Waste.* PNL-7121, Pacific Northwest Laboratory, Richland, WA

Serne RJ, RO Lokken, and LJ Criscenti. 1992a. "Characterization of Grouted LLW to Support Performance Assessment." *Waste Management* 12(2-3):271-287.

Serne, RJ, LL Ames, PF Martin, VL LeGore, CW Lindenmeier, and SJ Phillips. 1992b. *Leach Testing of in Situ Stabilization Grouts Containing Additives to Sequester Contaminants.* PNL-8492, Pacific Northwest Laboratory, Richland, Washington.

Serne RJ, JL Conca, VL LeGore, KJ Cantrell, CW Lindenmeier, JA Campbell, JE Amonette, and MI Wood. 1993. *Solid-Waste Leach Characterization and Contaminant-Sediment Interactions. Volume 1, Patch Leach and Adsorption Tests and Sediment Characterization.* PNL-8889, Pacific Northwest Laboratory, Richland, WA.

Serne RJ, WJ Martin, and VL LeGore. 1995. *Leach Test of Cladding Removal Waste Grout Using Hanford Groundwater*. PNL-10745, Pacific Northwest Laboratory, Richland, WA.

Wellman DM, SV Mattigod, GA Whyatt, L Powers, KE Parker, and MI Wood. 2006. *Diffusion of Iodine and Rhenium in Category 3 Waste Encasement Concrete and Soil Fill Material.* PNNL-16268, Pacific Northwest National Laboratory, Richland, WA.

Wood, MI, R Khaleel, PD Rittman, AH Lu, SH Finfrock, TH DeLorenzo, RJ Serne, and KJ Cantrell. 1995. *Performance Assessment for the Disposal of Low-Level Waste in the 200 West Area Burial Ground.* WHC-EP-0645, Westinghouse Hanford Company, Richland, Washington

# 16

# Hydrogeologic Characterization of Fractured Rock Masses Intended for Disposal of Radioactive Waste

Donald M. Reeves, Rishi Parashar and Yong Zhang
*Desert Research Institute*
*USA*

## 1. Introduction

There are currently 441 nuclear power reactors in operation or under construction distributed over 30 countries (International Atomic Energy Agency, 2011). The global radioactive waste inventory reported as storage in 2008 was approximately 17.6 million cubic meters: 21% short-lived, low- and intermediate-level waste, 77% long-lived, low- and intermediate-level waste and 2% high-level waste (International Atomic Energy Agency, 2011). There is a consensus among most of the scientific community that geologic repositories offer the best solution for the long-term disposal of radioactive waste. In the United States, for example, geologic disposal is considered the only technically feasible, long-term strategy for isolating radioactive waste from the biosphere without active management (Long & Ewing, 2004; National Research Council, 2001; Nuclear Energy Agency, 1999).

The process of selecting geologic repositories is an issue for many countries with radioactive waste, and several countries including Belgium, Canada, Finland, France, Japan, Republic of Korea, Sweden, Switzerland and the United States have underground research laboratories to conduct in-situ tests related to radioactive waste disposal (International Atomic Energy Agency, 2001). The Swedish Nuclear Fuel and Waste Management Company (SKB) recently selected Östhammar, Sweden as the final spent fuel geologic repository site after a nearly 20 year selection process. Once operational, this facility will be the first repository in the world designated for long-term disposal of high-level radioactive waste.

Repositories intended for long-term storage of high-level radioactive waste are comprised of both engineered and geologic barriers to either isolate from or impede the release of radioactive elements (radionuclides) to the biosphere. The engineered barrier serves as the primary barrier to radionuclide transport and describes systems that consist of (a) form of the waste, such as radionuclides contained within a solid, vitric matrix rather than in an aqueous phase, (b) waste canister, (c) backfill and buffer, and (d) tunnel grouting. In the event of radionuclides circumventing the engineered barrier, the geologic barrier serves as a secondary impediment to the release of radionuclides to the biosphere. The geologic barrier relies on the intrinsic ability of the host geologic medium to limit the transport of radionuclides.

The intrinsic ability of the host rock to limit radionuclide migration is a complex interplay between advective-dispersive motion of radionuclides in dissolved and colloidal form in groundwater, and retention processes such as adsorption and molecular diffusion into low velocity zones. These processes are present for both unsaturated and saturated hydrologic conditions. Thus, hydrogeologic characterization of a potential repository must provide estimates of advective transport velocity based on permeability and porosity of the host rock; delineate the surrounding groundwater flow system including regional flow directions, hydraulic gradients, and zones of recharge and discharge; and determine the potential significance of radionuclide retention mechanisms.

In this Chapter, we focus our attention on the hydrogeologic characterization of fractured rock masses intended for the disposal of radioactive waste. The emphasis on fractured rock masses is two-fold. First, many repository sites proposed for high-level radioactive waste disposal are comprised of low-permeability hard rock (e.g., volcanic, igneous and sedimentary), although softer geologic materials such as salt domes and thick clay sequences are also under consideration. Low-permeability rock masses have little or negligible matrix porosity and permeability, with connected networks of discontinuous fractures imparting secondary porosity and permeability. Second, predicting radionuclide transport in fractured media is a formidable challenge as the spatially-discontinuous nature of fracture networks, along with high degrees of heterogeneity within fracture properties, lead to highly anisotropic flow systems with complex patterns of fluid flow and subsequent radionuclide migration (de Dreuzy et al., 2001; Klimczak et al., 2010; Neuman, 2005; Reeves et al., 2008a;b;c; Schwartz et al., 1983; Smith & Schwartz, 1984). Detailed radionuclide transport predictions are typically reliant on numerical simulations that incorporate site-specific fracture data (e.g., Arnold et al., 2003; Bodvarsson et al., 2003; Cvetkovic et al., 2004; Pohll et al., 1999; Pohlmann et al., 2002; 2004; Reeves et al., 2010; Robinson et al., 2003; Smith et al., 2001), though some analytical techniques for first-cut approximations have been developed (Reeves et al., 2008b; Zhang et al., 2010).

This Chapter is not designed to provide a full treatise on the characterization of rock masses for development of complex numerical models used to predict radionuclide transport. Rather, we present an alternative approach where site-specific fracture network properties can be used to infer flow and transport characteristics of fracture networks by expanding the framework proposed by Reeves et al. (2008c). This framework can be used to qualitatively evaluate the suitability of candidate rock masses intended for the disposal of high-level radioactive waste based on fracture statistics. In stark contrast to current evaluation approaches that rely on costly field investigations to supply data to numerical models for radionuclide transport predictions, this type of evaluation promotes both time and economic savings by screening candidate fractured rock masses according to relatively simple criteria obtained from fracture characterization efforts.

## 2. Fractured rock characterization

Fractures are spatially discontinuous features that exhibit strong variability in geometrical and hydraulic properties. This variability is a result of the complex interplay between current and past stress fields, rock mechanical properties (i.e, Young's modulus, Poisson's ratio), mechanical fracture interaction and distributions of flaws or weakness in a rock mass. Fractured rock masses are typically characterized during field campaigns that measure fracture attributes from a number of sources including boreholes, rock outcrops, road cuts and tunnel complexes. Seismic techniques can also be used to image fault structures in the subsurface. Hydraulic properties of fractured media can be either inferred from fracture aperture or hydraulic tests performed on boreholes.

Full characterization of fractured rock masses is not possible since known fracture locations and their attributes consist of an extremely small sample of the overall fracture network, i.e., any fracture characterization effort grossly undersamples a field site due to limited accessibility to the fractures themselves. Fracture data, however, can be used to generate representative, site-specific fracture networks through the derivation of probabilistic descriptions of fracture location, orientation, spacing, length, aperture, hydraulic conductivity/transmissivity and values of network density (Figure 1). With the exception of adsorption, these are the statistical properties that form the basis of the geologic repository screening framework. Statistical analysis of fracture attributes will be extensively covered in this section.

### 2.1 Orientation

Fracture networks will typically have two or more fracture sets characterized by fracture orientation (e.g., Barton, 1995; Bonnet et al., 2001; Bour & Davy, 1999; Pohlmann et al., 2004; Reeves et al., 2010). The presence of at least two intersecting sets of fractures reflects the physics of rock fracture propagation where two sets of fractures can arise from a single stress field (Jaeger et al., 2007; Twiss & Moores, 2007). Unless fractures are very long such as regional-scale faults, it is important from a flow perspective to have at least two intersecting fracture sets to promote connectivity across a rock mass.

The orientation of fracture planes is denoted by strike and dip convention. Analysis of fracture orientation begins with projecting the poles to fracture planes onto a stereonet and using contours of pole density to identify fracture sets (Figure 2). Upon identification of fracture sets, mean orientation and the variability of fracture poles for each fracture set can be determined.

The distribution of fracture orientation is usually modeled using a Fisher distribution (Fisher, 1953):

$$f(x) = \frac{\kappa \cdot \sin x \cdot e^{\kappa \cdot \cos x}}{e^{\kappa} - e^{-\kappa}} \tag{1}$$

where the divergence, $x$ (degrees), from a mean orientation vector is symmetrically distributed $(-\frac{\pi}{2} \leq x \leq \frac{\pi}{2})$ according to a constant dispersion parameter, $\kappa$. The Fisher distribution is a special case of the Von Mises distribution, and is similar to a normal distribution for spherical data (Mardia & Jupp, 2000). The extent to which individual fractures cluster around the mean orientation is described by $\kappa$ where higher values of $\kappa$ describe higher degrees of clustering. It is our experience that values of $\kappa$ are commonly in the range of $10 \leq \kappa \leq 50$ for natural fracture networks. Stochastic simulation of Fisher random deviates in the discrete fracture networks following this section is based on the method proposed by Wood (1994).

The Bingham distribution provides an alternative to the Fisher distribution for cases in which fracture strike and dip are asymmetrically clustered around mean fracture orientations (Bingham, 1964):

$$f(x) = \frac{exp\left[\kappa_1(M_1x)^2 + \kappa_2(M_2x)^2 + \kappa_3(M_3x)^2\right]}{4\pi \cdot F(1/2;3/2;\mathbf{E})} \tag{2}$$

where $\kappa_1, \kappa_2$ and $\kappa_3$ are dispersion coefficients that satisfy the condition: $\kappa_1 \leq \kappa_2 \leq \kappa_3 = 0$, $M_1$, $M_2$ and $M_3$ are the column vectors of matrix $\mathbf{M}$, $\mathbf{E}$ is the eigenvector matrix, and $F(1/2;3/2;\mathbf{E})$ is a hypergeometric function of the matrix argument. The probability distribution function described by (2) can especially occur for faults that exhibit a greater range in deviations in strike than dip. A shortcoming of the Bingham distribution, however, is that it is not

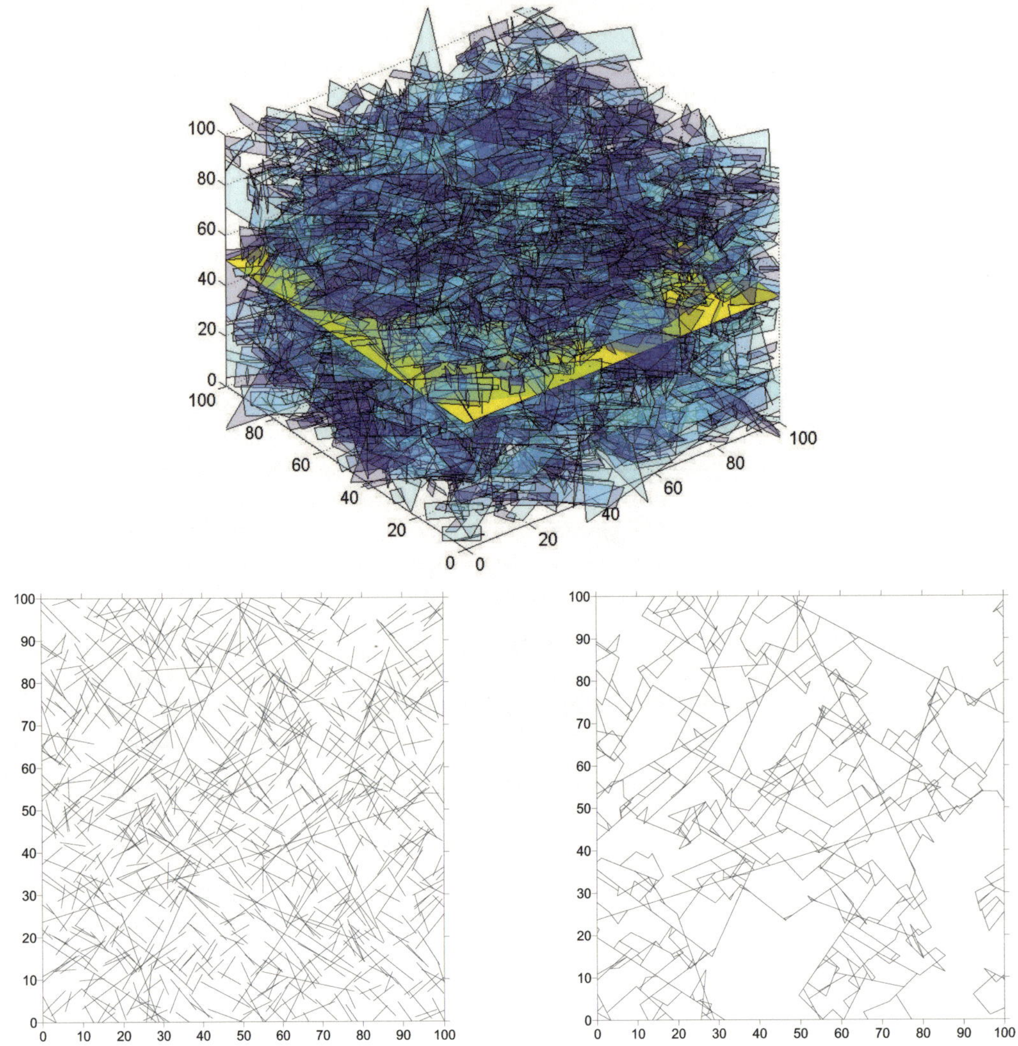


Fig. 1. Illustration showing the correspondence between two- and three-dimensional fracture networks. The three-dimensional network (top) is generated according to two fracture sets with significant variability about mean fracture orientations, a power-law length distribution exponent of $a = 2.0$ and a relatively sparse density. The two-dimensional network at the bottom left is computed by projecting all fractures onto the yellow horizontal slice located in the center of the three-dimensional DFN. The two-dimensional network on the bottom right is the result of identifying the hydraulic backbone by eliminating all dead-end fracture segments and non-connected clusters. Once the hydraulic backbone is identified, flow and particle transport can then be computed for the network.

mathematically possible to use (2) for the stochastic generation of asymmetric deviates. The authors are currently developing a method to simulate asymmetric deviations from mean fracture set orientations.

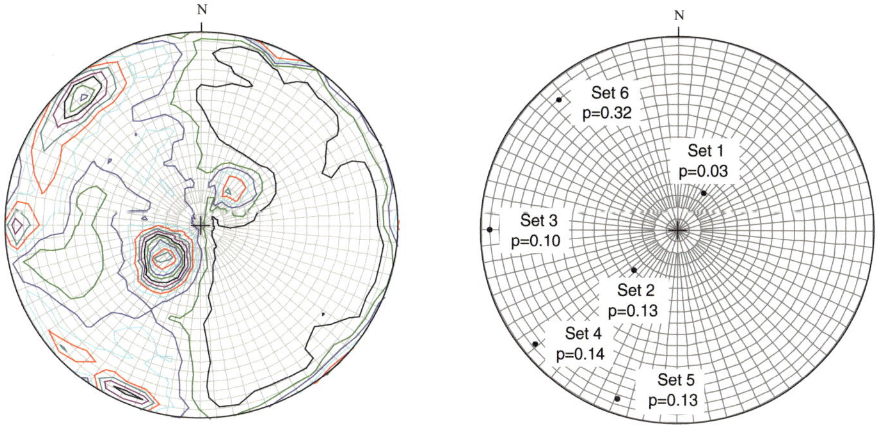


Fig. 2. Stereonet plots of poles to fracture planes with contour plots of all poles (left) and identified fracture sets along with prior probability (right). From Reeves et al. (2010).

## 2.2 Spacing

Fracture spacing refers to the linear distance between fractures. This distance also provides a length scale for unfractured matrix blocks. Fracture spacings from a data set require a correction (Terzaghi, 1965):

$$D = D' sin(\alpha) \tag{3}$$

to convert the apparent spacing $D'$ measured along a transect to true fracture spacing $D$. Values of $\alpha$ denote the angle of the transect relative to the mean fracture orientation or a pre-determined reference direction (Figure 3). Apparent spacing is equal to true spacing if the transect is perpendicular to the mean fracture orientation or reference direction. If $\alpha = 90°$ the Terzaghi correction factor $f = sin(\alpha)$ reduces to 1.

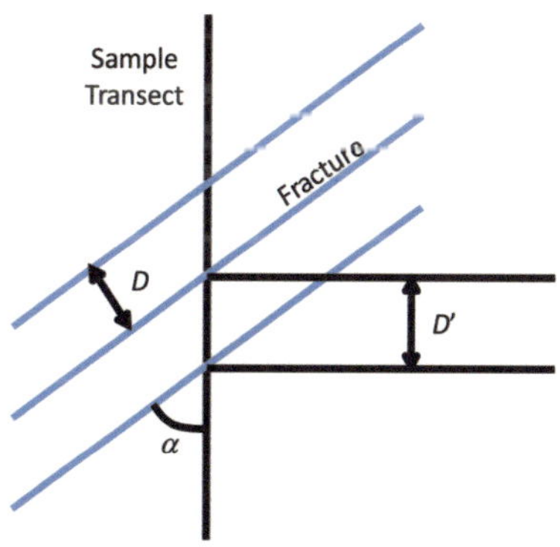


Fig. 3. Illustration showing how the Tergazhi correction accounts for the bias between apparent spacing $D'$ and true spacing $D$ based on the orientation of the sample transect in relation to mean fracture orientation. After http://www.rocscience.com/downloads/dips /WebHelp/dips/Terzaghi_Weighting.htm.

Once the spacing between fractures is corrected, values of fracture frequency and average fracture spacing can be computed. Fracture frequency [units of inverse length, $L^{-1}$] is defined as the total number of fractures along the distance of a transect. Average fracture spacing [units of length, L] is simply the inverse of fracture frequency, and defines the size of the unfractured matrix block. This metric and its influence on radionuclide retention will be discussed in a later section.

Spacing in natural fracture networks is most commonly an exponentially distributed random variable. This can be tested by plotting the inverse empirical cumulative distribution function of fracture spacing, also known as a survival function (Ross, 1985). If spacing is exponential, the probability decay of the tail of the empirical fracture spacing distribution will exhibit a straight line on a semi-log plot. A Poisson point process defined by independent and identically distributed uniform deviates provides easy generation of exponential spacing (Ross, 1985).

Other possible distributions of fracture spacing include uniform (Rives et al., 1992) and fractal clustering (Barton, 1995; Darcel et al., 2003), both of which are considered extreme end members. Uniform spacing may occur in thin geologic layers which restrict fracture growth in the vertical direction and promote long horizontal fracture growth with nearly constant spacing (Rives et al., 1992). Exact causes of fractal clustering are less known and may be related to the role of mechanical fracture interaction during propagation that likely controls fracture length and spacing (Ackermann & Schlische, 1997; Darcel et al., 2003; Olson, 1993; Segall & Pollard, 1983). Networks with fractal clustering can be generated via a multiplicative cascade process (Mandelbrot, 1974; Schertzer & Lovejoy, 1987).

### 2.3 Length

Fracture length denotes the horizontal trace length of a fracture. There is a consensus in recent literature that fracture lengths above a lower length cutoff, $l_{min}$, are power-law:

$$P(L > l) = Cl^{-a} \tag{4}$$

with a power law exponent, $a$, that ranges between 1 and 3 in natural fracture networks (Bonnet et al., 2001; Bour & Davy, 1997; 1999; Renshaw, 1999). $C$ is a constant based on $l_{min}$ and $a$. Though lognormal distributions of fracture length have been reported in the literature, they are a result of improper sampling of the largest fractures within a sampling window. Lognormal distributions easily arise in data sets with power-law tails if the largest values are censored.

Determination of the distribution of fracture length is similar to that of fracture spacing and involves the analysis of an inverse empirical cumulative distribution function. Fracture lengths that are power-law will exhibit linear trends on a log-log plot for the tail of the distribution. In this example, the tail of the distribution refers to the greatest 5-10% of length values. The slope of the power-law trend of the data is equal to $-a$.

Truncations can frequently occur in fracture length data due to constraints imposed by the finite scale of the sampling window. For example in Reeves et al. (2010), the longest fracture measured in a tunnel drift was parallel to the drift and was approximately two-thirds of the total drift length. Instead of choosing between a traditional power-law or lognormal distribution, an upper truncated Pareto (power law) model (Aban et al., 2006):

$$P(L > l) = \frac{\gamma^a (l^{-a} - \nu^{-a})}{1 - (\frac{\gamma}{\nu})^a} \tag{5}$$

was used to compute the power-law trend in the data, where $L_{(1)}, L_{(2)}, \ldots, L_{(n)}$ are fracture lengths in descending order and $L_{(1)}$ and $L_{(n)}$ represent the largest and smallest fracture lengths, $\gamma$ and $\nu$ are lower and upper fracture length cutoff values, and $a$ describes the tail of the distribution. Truncated power law models like (5) can also be useful for imposing an upper length scale to the generation of stochastic networks at the regional scale. Lacking evidence of domain-spanning faults (with the exception of bounding faults of the stock itself) for a 5 km wide granitic stock, Reeves et al. (2010) assigned an upper limit of 1 km in the stochastic generation of fault networks.

### 2.4 Hydraulic conductivity

Boreholes are commonly used to characterize fractured media. Borehole geophysics can provide useful information about fractures within the rock including fracture frequency, orientation, aperture and mineral infilling. Fracture aperture, defined as the width of the void space normal to fracture walls, can be used to infer hydraulic properties of fractures. Fracture apertures at land surface have low confining stresses that are not representative of subsurface confining stresses within a rock mass. We therefore recommend that fracture aperture values used to compute flow are measured in boreholes where in-situ stress is preserved.

The cubic law, a solution to the Navier-Stokes equation for laminar, incompressible flow between two parallel plates, describes a general relationship between fluid flow and fracture aperture (Snow, 1965):

$$Q = -\frac{\rho g}{12\mu} b^3 \nabla h \tag{6}$$

where fluid discharge per unit width, $Q$ [$L^2/t$], is proportional to the cube of the hydraulic aperture, $b$. Similar to Darcy's Law, the cubic law (6) assigns discharge through a fracture as a linear function of the hydraulic gradient, $\nabla h$. The relationship between hydraulic aperture and transmissivity ($T$), hydraulic conductivity ($K$) and permeability ($k$) is described by: $T = \frac{\rho g}{12\mu} b^3$, $K = \frac{\rho g}{12\mu} b^2$ and $k = \frac{b^2}{12}$, respectively. Fluid-specific properties density, $\rho$, and viscosity, $\mu$, allow for conversions between permeability and hydraulic conductivity/transmissivity. As a note of caution, the relationship between mechanical aperture, the physical distance between fracture walls, and hydraulic aperture, the equivalent aperture for a given flow rate, is unclear. As a general rule, hydraulic aperture is typically smaller than mechanical aperture (Chen et al., 2000; Cook et al., 1990; Renshaw, 1995; Zimmermann & Bodvarsson, 1996). Discrepancies between mechanical aperture and hydraulic aperture are attributed to surface roughness, flow path tortuosity, and stress normal to the fracture. Though empirical correction factors have been used to correlate mechanical and fracture apertures (Bandis et al., 1985; Cook et al., 1990; Renshaw, 1995), no method is reliable for a wide range of aperture values.

Hydraulic testing of boreholes yields reliable estimates of fracture $T$ and $K$. While there are many different hydraulic testing techniques, the isolation of specific intervals during testing with the use of dual-packer systems provides the best data to characterize the distribution of transmissivity/hydraulic conductivity. These tests yield flow rate information for applied fluid pressures, which also allows for the inverse computation of hydraulic aperture using (6). These aperture values, in addition to $T$ and $K$ estimates, are useful for parameterizing flow and transport models.

Studies in highly characterized rock masses have shown that fracture $K$ is extremely heterogeneous and may encompass 5 to 8 orders of magnitude (Andersson et al., 2002a;b;

Guimerá & Carrera, 2000; Paillet, 1998). Often the distribution of $K$ (and $T$) is thought to be lognormal:

$$p(x) = \frac{1}{x\sqrt{2\pi\sigma^2}} exp\left[\frac{-(\log x - \mu)^2}{2\sigma^2}\right] \tag{7}$$

where $x$ is the mean and $\sigma$ is the standard deviation. Values of $log(\sigma_K)$ are typically around 1 for fractured media (Andersson et al., 2002a;b; Stigsson et al., 2001). However, other studies suggest power law distributions (Gustafson & Fransson, 2005; Kozubowski et al., 2008) and that these lognormal distributions, similar to length, could be caused by censoring flow data possibly due to instrument limitations. Additionally, flow through rough-walled fractures can be non-Darcian (Cardenas et al., 2007; Qian et al., 2011; Quinn et al., 2011). This may further complicate the estimation of hydraulic conductivity in field hydraulic tests as flow is no longer linearly proportional to a pressure gradient as described by (6).

### 2.5 Density

Fracture networks consist of two-dimensional planes embedded within a rock matrix (Figure 1). The lack of access to the total rock volume makes it impossible to directly measure the three-dimensional fracture density of a rock mass. Instead, three-dimensional density for discrete fracture networks is estimated from density measurements of lower dimensions, i.e., one-dimensional fracture frequency from boreholes and/or tunnel drifts or two-dimensional fracture density from outcrops and fracture trace maps. Definitions of fracture density according to dimension are: one-dimensional density (also known as fracture frequency), $\rho_{1D}$ [$L^{-1}$], is expressed as the ratio of total number of fractures, $f_i$, to transect length, $L$: $\rho_{1D} = L^{-1}\sum_{i=1}^{n} f_i$; two-dimensional fracture density, $\rho_{2D}$ [$L/L^2$], is expressed as the ratio of the sum of fracture lengths, $l$, to area, $A$: $\rho_{2D} = A^{-1}\sum_{i=1}^{n} l_i$; and three-dimensional fracture density, $\rho_{3D}$ [$L^2/L^3$], is expressed as the ratio of the sum of fracture plane area, $A_i$, to rock volume, $V$: $\rho_{3D} = V^{-1}\sum_{i=1}^{n} A_i$.

Numerical techniques can be used to upscale one-dimensional fracture frequency [$L^{-1}$] estimates to a three-dimensional spatial density [$L^2/L^3$] (Holmén & Outters, 2002; Munier, 2004). For example, one-dimensional transects can be used to upscale two-dimensional networks by adding fractures until the one-dimensional transect density is satisfied along several transects placed along the two-dimensional network. Three-dimensional networks can be generated in a similar fashion by either generating fractures until the frequency along one-dimensional boreholes is satisfied or by projecting fractures onto sampling planes (e.g., Figure 1) until a two-dimensional density criterion is satisfied.

Fracture density is highly dependent on the distribution of fracture lengths in a model domain where the density at the percolation threshold increases with increasing values of $a$ (Darcel et al., 2003; Reeves et al., 2008b; Renshaw, 1999). This will become apparent in the fracture network examples in the next section.

## 3. Flow and advective transport properties of fracture networks

A central theme of this Chapter is that flow and radionuclide transport characteristics of fractured media can be inferred from fracture statistics. The previous section discussed in detail how fractured rock masses can be analyzed according to statistics of fracture orientation, spacing, length, hydraulic conductivity and values of fracture density. This section contains detailed explanations on how these attributes can provide *a priori* insight into flow and transport properties of a fractured rock mass. We begin by defining network structure, and then discuss how the structure relates to transport characteristics such as shape of breakthrough curves, potential for early arrivals, and variability of individual breakthroughs about the ensemble. Simulations of flow and transport in two-dimensional discrete fracture networks (DFN) with physically realistic parameters are used to illustrate specific concepts.

### 3.1 Network structure and transport

The structure of natural fracture networks is the end result of the complex interplay between stress fields and their anisotropy, mechanical properties of the rock, mechanical fracture interaction and distribution of initial flows in a rock mass. The reliance on probabilistic descriptions of fracture attributes reflects our lack of ability to accurately construct fracture networks based on mapping studies alone. The limited accessability to the network leaves an incomplete understanding of the patterns of fracturing within a rock mass that can often be improved through visual inspection of representative networks generated according to site-specific statistics.

A total of three different network types are generated from two fracture sets with power law distribution of lengths with exponent values in the range $1.0 \leq a \leq 3.0$, fracture density, $1.0 \leq \rho_{2D} \leq 2.0$ m/m$^2$, orientations of $\pm 45°$ with variability described by $\kappa$=20 and a lognormal hydraulic conductivity distribution with $\log(\sigma_K) = 1$ (Figures 4–5). Once a network is generated, the hydraulic backbone is identified by eliminating dead-end segments and isolated clusters. This is accomplished in our model through both geometric and flow techniques. The hydraulic backbone represents the interconnected subset of a fracture network that is responsible for conducting all flow and transport across a domain. Hence, analysis of backbone characteristics can provide insight into these processes.

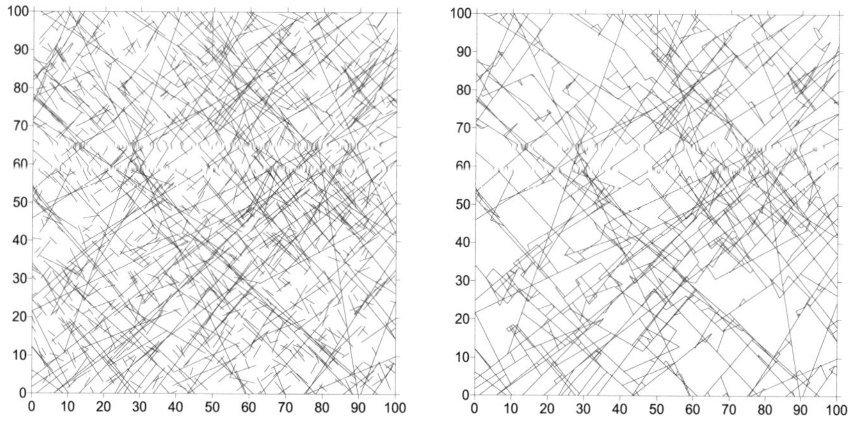


Fig. 4. Discrete fracture network realization (left) and hydraulic backbone (right) generated from two fracture sets with power distribution of lengths with exponent $a = 1.0$, $\rho_{2D} = 1.0$ m/m$^2$, $l_{min} = 2.0$ m and orientations at $\pm 45°$ with variability described by $\kappa = 20$. Note that the hydraulic backbone is dominated by long fractures.

The generated networks in this study do not explore the full parameter space for fractured media. However, the wide range of fracture length exponents provides sufficient variability and produces three distinct types of hydraulic backbones. Networks generated with $a = 1.0$

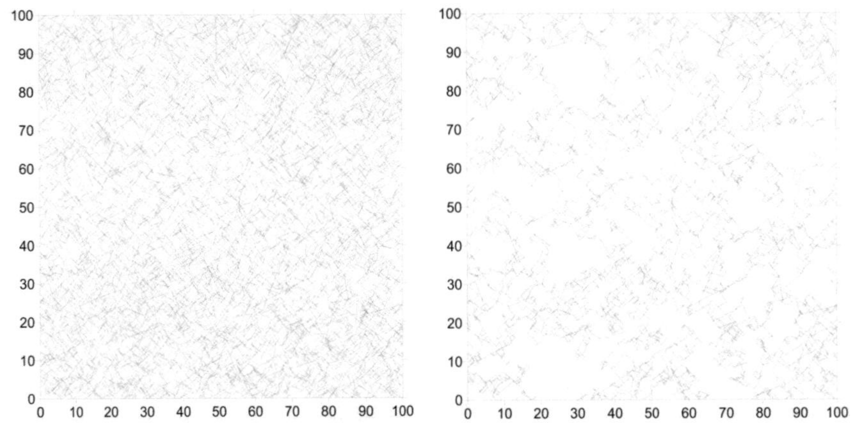

Fig. 5. Discrete fracture network realization (left) and hydraulic backbone (right) generated from two fracture sets with power distribution of lengths with exponent $a = 3.0$, $\rho_{2D} = 2.0$ m/m$^2$, $l_{min} = 2.0$ m and orientations at $\pm 45^\circ$ with variability described by $\kappa = 20$. Note that the hydraulic backbone is dominated by short fractures.

produce backbone structures dominated by long fractures (Figure 4), and networks with $a = 3.0$ produce backbone structures dominated by short fractures (Figure 5). Backbones with a mixture of short and long fractures are produced for networks generated with $a = 2.0$ (Figure 6). Another feature of these networks is that density of the network increases from $\rho_{2D}$=1.0 m/m$^2$ to $\rho_{2D}$=2.0 m/m$^2$ as the value of $a$ increases from 1.0 to 3.0. This increase in density is necessary to maintain a percolating backbone. For example, the density values assigned to $a = 1.0$ and $a = 2.0$ ($\rho_{2D}$ =1.0 and 1.5 m/m$^2$, respectively) result in a non-percolating networks if used with $a = 3.0$. Conversely, networks generated with $a = 1.0$ and $\rho_{2D} = 2.0$ m/m$^2$ (assigned to $a = 3.0$) produces unrealistically dense networks.

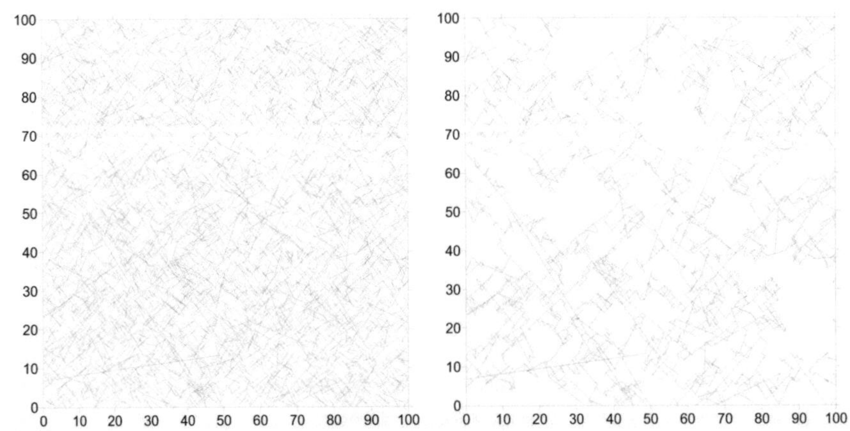

Fig. 6. Discrete fracture network realization (left) and hydraulic backbone (right) generated from two fracture sets with power distribution of lengths with exponent $a = 2.0$, $\rho_{2D} = 1.5$ m/m$^2$, $l_{min} = 2.0$ m and orientations at $\pm 45^\circ$ with variability described by $\kappa = 20$. Note that the hydraulic backbone is a mix of short and long fractures.

Computation of flow in two-dimensional discrete fracture networks involves solving for hydraulic head at all internal nodes (intersection point of two or more fractures) inside the domain according to Darcy's law (de Dreuzy & Ehrel, 2003; Klimczak et al., 2010; Priest, 1993). In the simulations presented, boundary conditions are applied to all nodes on the domain to induce flow from top to bottom according to a linear hydraulic gradient of 0.01. Fluid flow through the backbone is then solved iteratively at each node via a biconjugate gradient method (Figure 7). A total of 500,000 conservative (non-sorbing) radionuclide particles are then introduced at the top of the domain, traced through the network, and recorded as they exit the model domain. The velocity along each fracture segment is constant, and radionuclide particles are moved by advection from node to node. Radionuclide particles change direction at nodes proportional to the flux of the down gradient fracture segments. This process is repeated to produce 50 statistically equivalent networks and resultant breakthrough curves for each of the three network types.

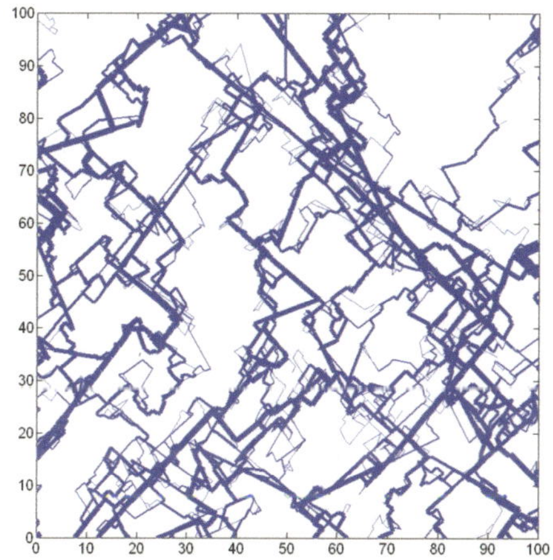


Fig. 7. Discrete fracture network generated according to parameters specified in the caption to Figure 6 and a lognormal distribution of hydraulic conductivity with $\log(\sigma_K) = 1.0$. Line thickness is proportional to flow through each segment. Note the large contrasts in flow, and hence transport velocities, between different segments of the hydraulic backbone. Particles sample segments of different velocity at intersections where particles are routed proportional to segment flux.

The radionuclide breakthroughs reflect the characteristics of the fracture networks (Figure 8). First, fracture length defines the correlation structure of the network. Lower values of $a$ place more probability weight on extreme values of fracture length, and consequently these networks consist of very long fractures that often span the model domain. Radionuclides traveling through these networks only change direction at fracture intersections, and the combination of long fractures and sparse density promotes fast transport. This can be observed in Figure (8) where mean ensemble breakthrough occurs at 1.5 years for a transport distance of 100 meters. Networks with $a = 2.0$ assign less probability weight on extreme values and produce backbones with a mixture of short and moderately long fracture lengths. This backbone structure requires radionuclides to move through a mixture of short and long fractures and sample a wider variety of velocities in the process of migrating across the model domain. Consequently, mean ensemble breakthrough for networks with $a = 2.0$ occurs at 11 years, which is approximately 7 times slower than for $a = 1.0$ (Figure 8). The slowest transport occurs for networks with $a = 3.0$ which consist of backbones dominated by short fracture segments. The short segments and higher density causes radionuclide particles to sample a very large number of fracture segments and velocities in the process of migrating across the domain. The mean breakthrough for $a = 3.0$ networks occurs at 26 years, which is approximately 17 and 2.4 times slower than the $a = 1.0$ and $a = 2.0$ networks (Figure 8).

Second, the breakthrough curves for individual realizations all show a high degree of variability (Figure 8). In fact, this variability is the reason why breakthroughs are presented in a cumulative form, as the shape of the breakthroughs can range from sharp peaks (more prevalent for networks with $a = 1.0$) to more broad "Gaussian-like" plumes (more prevalent for networks with $a = 3.0$). Networks with $a = 2.0$ represent an intermediate case where breakthroughs for individual realizations can encompass both scenarios. All of the networks exhibit heavy late-time tails which are briefly discussed in the following section. The variability in fracture networks is quantified by comparing deviations of individual realizations from the ensemble. Specifically, the standard deviation of breakthroughs is computed for cumulate probabilities ranging from 0.05 to 0.40 in intervals of 0.05 (Figure 9). This metric is then normalized by the ensemble mean for comparison to the other network types, and termed realization variability. Trends of realization variability are highest for $a = 1.0$ and lowest for $a = 3.0$. The largest realization variability values occur for early breakthroughs (0.05 to 0.10 of total mass) for $a = 1.0$ and $a = 2.0$. This can be attributed to the fact that these networks, especially $a = 1$, produce early arrivals that vary greatly in time. Reeves et al. (2008c) studied deviations in spatial distribution of individual solute plumes from the ensemble, and their results support our findings that networks with $a = 1.0$ are inherently unpredictable, and predictability in transport increases as mean fracture length decreases.

## 4. Radionuclide retention mechanisms

The previous section illustrates how statistics of fractured massses can be used to infer advective characteristics of radionuclide transport. In this section, we address the retentive properties of fractured rock masses which include adsorption of radionuclides onto fracture walls and within matrix blocks, and diffusional mass exchange of radionuclides between fractures and low-velocity zones. Adsorption is a chemical process that is related to the mineralogy of the host rock, and cannot be inferred from fracture network statistics. One must also consider that several radionuclides, such as $^{3}H$, $^{14}C$, $^{36}Cl$, $^{39}Ar$ and $^{85}Kr$, are conservative and not subject to sorption processes. The dominant retention mechanism for conservative radionuclides is diffusional mass exchange. The retention characteristics of diffusional mass exchange of radionuclides between fractures and matrix blocks can be inferred from fracture network properties and will receive the majority of attention in this section. A short discussion on sorption is included for completeness in the context of the screening framework. A network-scale retention mechanism arising from velocity contrasts among conductive fracture segments and is also introduced for additional consideration.

### 4.1 Diffusional mass exchange

Molecular diffusion promotes the exchange of radionuclides between active fracture flow paths and low-velocity zones such as stagnant zones within fractures, fault gauge and unfractured matrix blocks. Among these low-velocity regions, the influence of diffusional mass exchange of radionuclides in unfractured matrix blocks is much greater than diffusional

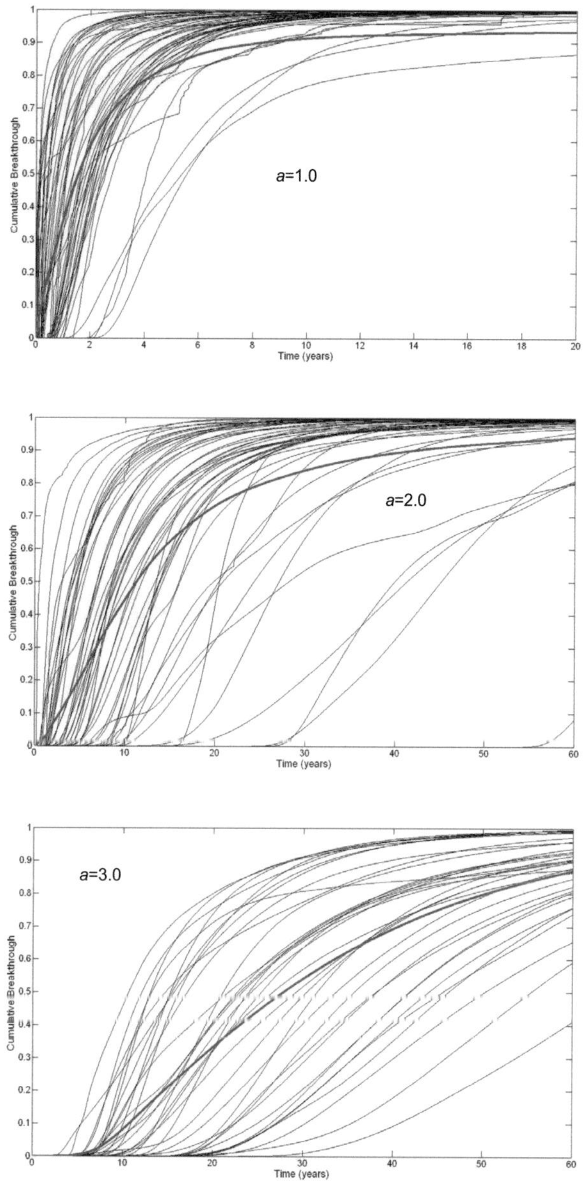


Fig. 8. Normalized conservative particle breakthroughs for 100 m scale transport in two-dimensional networks with power-law exponents $a = 1.0$ (top), $a = 2.0$ (middle) and $a = 3.0$ (bottom) for 50 transport realizations. Individual realizations are represented as thin black lines with the blue curve representing the ensemble. Note that mean arrival times increase with increases in power-law exponent, where $a = 1.0$, $a = 2.0$ and $a = 3.0$ lead to mean arrival times of 1.5, 11 and 26 years, respectively. Deviations in breakthroughs also increase with increasing values of $a$. The smooth breakthroughs in Figure 8 reflects the high number of particles.

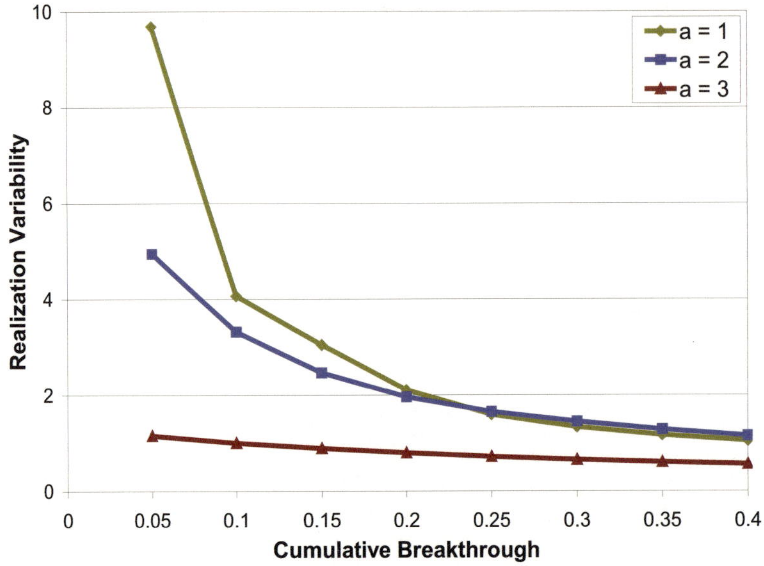


Fig. 9. Plot of realization variability in particle breakthroughs about ensemble breakthough for networks generated according to power law length distribution exponents of $a = 1.0$, $a = 2.0$, and $a = 3.0$. Note that realization variability decreases as the value of the power law length exponent decreases.

exchange within fracture stagnant zones and fault gauge (Andersson et al., 2002a;b; Grisak & Pickens, 1980; Neretnieks, 1980; Sudicky & Frind, 1982; Winberg et al., 2000). The reason for this is simple: the time a radionuclide molecule spends trapped in an immobile zone is proportional to the size of that zone, and the scale of matrix blocks is typically on the order of centimeters to meters while within fracture stagnant zones and fault gauge are on the order of micrometers to millimeters. Diffusion into matrix blocks was found to be a major retention mechanism at the time scales of the Tracer Retention Understanding Experiments conducted at the Äspö underground laboratory at both single fracture (Winberg et al., 2000) and block scales (Andersson et al., 2002a;b). For these in-situ experiments, the prevalence of molecular diffusion as a retention mechanism is exhibited by late-time breakthroughs with slopes equal to -3/2 for both conservative and non-conservative radionuclides.

Fracture spacing is a network parameter that denotes unfractured matrix block size and provides a length scale for molecular diffusion. The influence of fracture spacing on contaminant breakthroughs is illustrated in Figure 10 for $^3H$. The figure is generated from an exact analytical solution to conservative transport through a series of 40 m parallel-plate fractures with 110 $\mu$m apertures subject to dual-domain mass transfer through fractures and matrix blocks (Sudicky & Frind, 1982). The mean age of $^3H$ breakthroughs versus fracture spacing ($2B$) is shown in Figure 11. Figures 10 and 11 show that increases in mean breakthrough time for conservative radionuclides are highly correlated to fracture spacing with a power-law (nearly linear) trend. These results indicate that the retention potential for a candidate repository rock mass can be inferred from fracture spacing, and that larger average spacing leads to greater retention due to diffusional mass transfer between fractures and matrix.

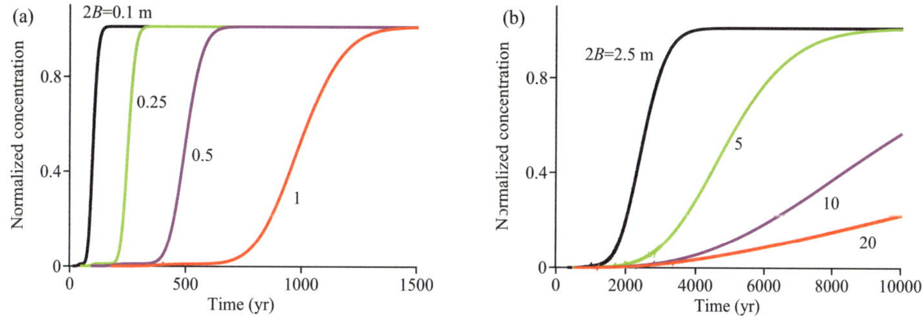


Fig. 10. Normalized $^{3}$H breakthroughs for a series of 40 m parallel-plate fractures with 110 $\mu$m apertures ($K = 1.0 \times 10^{-9}$m/s) for a variety of matrix block sizes ($2B$). A diffusion coefficient for tritiated water at 25°C ($D = 2.4 \times 10^{-08}$ m$^2$/s) with a matrix tortuosity and porosity of 0.25 and 0.1, respectively, were used to generate the breakthroughs from analytical solutions (Sudicky & Frind, 1982). Note the dramatic affect of fracture spacing on radionuclide breakthrough due to diffusional mass exchange between fractures and matrix blocks.

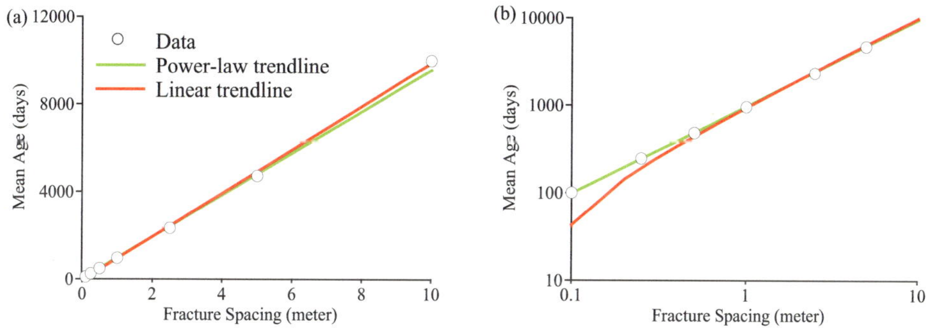


Fig. 11. Mean age ($A$) of $^{3}$H exiting the fracture versus fracture spacing ($2B$) from Figure 10 with best-fit linear (red) and power law (green) trendlines. The slightly non-linear trend in mean age versus spacing is more apparent in log-log axes (b). The best-fit power-law trendline is: $A = 971(2B)^{0.9934}$.

Although Figures 10 and 11 are helpful in illustrating and partially quantifying the influence of fracture spacing on radionuclide retention by diffusional mass transfer, it is important to acknowledge that fractured media is inherently much more complex and many simplifying assumptions were made in this analysis. First, cystalline rocks typically have very low porosities and this void space may become disconnected further into the matrix block, giving rise to the concept of maximum penetration depth which describes a cutoff to the distance radionuclides can diffuse into the matrix block (e.g., Neretnieks, 1980). Thus, for crystalline rock masses, average fracture spacing would have to reflect penetration depth which is usually unknown. Second, fracture networks are discontinuous in nature and exhibit power-law fracture lengths which promote a broad distribution of matrix block sizes. For these networks, a "characteristic" or average fracture spacing does not exist and the retention of radionuclides in matrix blocks of different sizes can result in a continuum of exchange rates that are both faster and slower than the what would be predicted using the arithmetically computed average block size. Third, weathering processes at the interface of a fracture and the host rock can create higher porosity rim zones that enhance diffusion (Cvetkovic, 2010). Despite these complicating factors, it can be generally stated that rock masses with greater average fracture spacing have greater potential for radionuclide retention.

### 4.2 Radionuclide adsorption

There may be cases where potential repositories have similar fracture characteristics, the most important being the distribution of fracture length, network density and average fracture spacing, and the sorption capacity of the host rock may serve as a secondary screening criterion. Adsorption of radionuclides is a reversible chemical process that can include both cation exchange and surface complexation. It can often be difficult to distinguish in-situ between retention processes as molecular diffusion plays a role in introducing sorbing radionuclides to adsorptive sites within fractured rock environments. The use of multiple tracers, including conservative and non-conservative with varying sorbing rates, can be used to ascertain the relative contributions of retention from diffusional mass transfer and adsorption (e.g., Andersson et al., 2002b).

Adsorption in fractured rock can occur along fracture walls and within fault gauge and matrix blocks. Sorption along fracture walls is a surface process rather than a volumetric process, like sorption within matrix blocks. Similar to the discussion on molecular diffusion, the TRUE-1 and TRUE Block Scale tests conducted at Äspö Hard Rock Laboratory indicate that sorption in the unfractured rock matrix is a dominant process over adsorption onto fracture walls and fault gauge (Andersson et al., 2002a;b). This is because matrix blocks have a greater surface area and length scale, and hence, a greater number of available sorption sites than fracture walls and fault gauge.

Sorption is typically characterized using equilibrium isotherm models parameterized via laboratory batch experiments containing soil, sediment or crushed rock from a site of interest (e.g., Langmuir, 1997; Zheng & Bennet, 2002). This approach to modeling sorption assumes that rates of sorption/desorption between radionuclides and the sorptive medium are orders of magnitude more rapid than the advective groundwater velocity. However, recent studies have shown that desorption rates of U and $^{237}$Np may yield a 3-4 order of magnitude range of desorption rate constants despite uniform flow fields approximating equilibirum conditions (Arnold et al., 2011; Dean, 2010). We recommend from a cost savings perspective that candidate rock masses be screened on the basis of sorption coefficients determined from laboratory batch experiments. However, the findings of large desorption rate constants questions the usefulness of characterization techniques relying on equilibrium assumptions for use in radionuclide transport models.

### 4.3 Network-scale retention

Numerical investigations have shown that velocity contrasts within fracture segments on the hydraulic backbone can lead to solute retention within the backbone itself (Berkowitz and Scher, 1997; Painter & Cvetkovic, 2002; Reeves et al., 2008c). Painter & Cvetkovic (2002) and Reeves et al. (2008c) found that distributions of inverse velocity, $1/v$, a surrogate for retention, exhibited power law decay trends at late times. Values of $\gamma$ measured from the tail of the distribution of $1/v$ versus time were in the range $1.1 \leq \gamma \leq 1.8$ for Painter & Cvetkovic (2002) and approximately $\gamma$=1 for Reeves et al. (2008c). Analysis of the survival function of ensemble breakthrough versus time for the $a = 1.0$ networks in this study yielded a value of $\gamma = 1.5$ (Figure 12). Values of $\gamma$ for $a = 2.0$ and $a = 3.0$ networks were not computed due to incomplete breakthroughs that leave the tail of the distribution undefined. None of these studies found a correlation between distributions of fracture length and $\gamma$; however, all of these studies used transmissivity distributions encompassing several orders of magnitude. It is entirely possible that this type of retention results from the interaction of power law distributions of trace length and distributions of transmissivity that vary by several orders of magnitude, such as lognormal or power law. More work in this area is needed to provide further insight into this retention process in the context of fracture statistics.

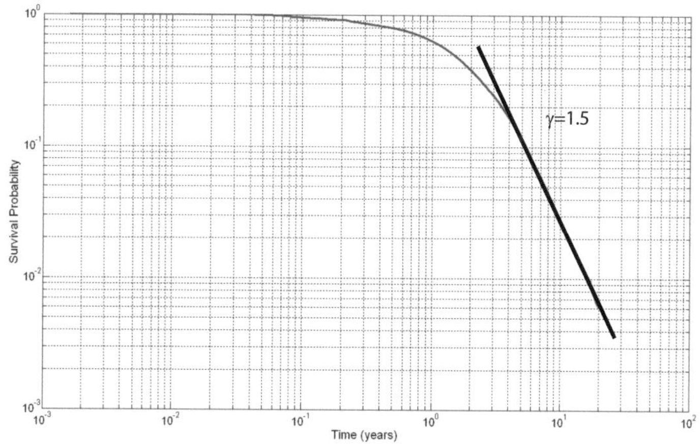


Fig. 12. Survival function (inverse empirical cumulative distribution function) of particle breakthrough versus time for the $a = 1.0$ networks generated in this study (blue) along with best-fit tail trendline (black). The trendline scales according to $-1 - \gamma$ where a slope of -2.5 corresponds to $\gamma$=1.5.

## 5. Geologic repository screening framework

The geologic barrier relies on the intrinsic ability of the host rock to limit the release of radionuclides to the biosphere. The discussion up to this point has included both advective and retentive characteristics of fractured media. Desirable qualities for a potential repository rock mass from a radionuclide transport perspective include: low potential for early arrivals, longer bulk breakthroughs, and low variability in predicted results. Simulated particle breakthroughs show a clear trend that rock masses with networks dominated by short fractures exhibit advective transport characteristics that are most desirable for geologic repositories. These networks have the lowest potential for early arrivals, breakthroughs occur at much longer timescales than networks consisting of longer fractures, and the variability of individual realizations about the ensemble are lower than for networks generated using $a = 2.0$ and $a = 3.0$. These conclusions are supported by the study of Reeves et al. (2008c).

Retention of radionuclides via molecular diffusion favors networks with large fracture spacings, as retention time for conservative radionuclides in a low-velocity matrix block is proportional to block size. This suggests that the sparsest fracture networks are the most desirable. Given the link between fracture density and the distribution of trace length, networks dominated by long fractures (e.g., $a < 2.0$) would be the optimum choice.

However, these networks produce breakthroughs that exhibit the earliest arrivals and bulk transport times, and have the greatest variability. It is our opinion that advective transport characteristics should be weighted higher than retention since slower advective transport rates through a network decreases the propensity for fast radionuclide transport and early arrivals. With these network types, retention processes such as molecular diffusion and adsorption will further retard the relatively slow migration rates of radionuclides. Note that molecular diffusion has the ability to significantly retard contaminant migration even for relatively small matrix block sizes (Figures 10 and 11). Networks with identical length distributions can be distinguished from another on the basis of average fracture spacing, a secondary criterion. The rock mass with larger values of average spacing (assuming similar host rock porosity) will have a greater retention potential via molecular diffusion. Sorption characteristics of a rock mass may serve as a tertiary criterion, and this consideration must take into the account the waste itself and the prevalence of non-conservative radionuclides.

This screening framework is intended to be cost effective as flow and transport characteristics of fractured rock masses are inferred solely on the basis of fracture statistics. The selection of an ideal geologic repository for high-level radioactive waste disposal is a complex endeavor that must include considerations of geologic stability (e.g., lack of recent seismic events, volcanic activity, high geothermal gradients), hydrology, rock mechanical strength, geochemistry and proximity to large population centers and natural resources (e.g., National Research Council, 1978). It is assumed that these factors will be considered at some point in the selection process.

## 6. References

Aban, I.B.; Meerschaert, M.M. & Panorska, A.K. (2006). Parameter estimation methods for the truncated Pareto distribution, *J. Amer. Stat. Assoc.*, Vol. 101, 270–277.

Ackermann, R.V. & Schlische, R.W. (1997). Anticlustering of small normal faults around larger faults, *Geol.*, Vol. 25, No. 12, 1127–1130.

Andersson, P.; Byegärd, J.; Dershowitz, B.; Doe, T.; Hermanson, J.; Meier, P.; Tullborg, E.-L.; & Winberg, A. (2002a). *Final Report on the TRUE Block Scale Project: 1. Characterization and Model Development*, Swedish Nuclear Fuel and Waste Management Company (SKB), Technical Report TR-02-13, Stockholm, Sweden.

Andersson, P.; Byegärd, J.; & Winberg, A. (2002b). *Final Report on the TRUE Block Scale Project: 2. Tracer Tests in the Block Scale*, Technical Report TR-02-14, Swedish Nuclear Fuel and Waste Management Co. (SKB), Stockholm, Sweden.

Arnold, B.W.; Kuzio, S.P.; & Robinson, B.A. (2003). Radionuclide transport simulation and uncertainty analysis with the saturated-zone site-scale model at Yucca Moutain, Nevada, *J. Contam. Hydrol.*, Vol 62–63, doi:10.1016/S0169-7722(02)00158-4.

Arnold, B.; Reimus P.; James, S. (2011). *Flow and Transport in Saturated Media: FY2011 Status Report*, prepared fro the U.S> Department of Energy Used Fuel Disposition Campaign, FCRD-USED-2011-0003111, Sandia National Laboratories and Los Alamos National Laboratory.

Bandis, S.C.; Makurat A. & Vik, G. (1985). Predicted and measured hydraulic conductivity of rock joints, *Proceedings of the International Symposium on Fundamentals of Rock Joints*, Björdkliden, Norway, Septemeber 15-20.

Barton, C.C. (1995). Fractal analysis of scaling and spatial clustering of fractures, *Fractals in the Earth Science*, Barton, C.C. & P.R. LaPointe, eds., pp. 141–178, Plenum, New York.

Berkowitz, B. & Scher, H. (1997). Anomalous transport in random fracture networks, *Phys. Rev. Lett.*, Vol. 79, No. 20, 4038–4041.

Bingham, C. (1964). *Distributions on the sphere and the projective plane*, Ph.D. Dissertation, Yale University.

Bodvarsson, G.S.; Wu Y.-S.; & Zhang, K. (2003) Development of discrete flow paths in unsaturated fractures at Yucca Mountain, *J. Contam. Hydrol.*, Vol 62–63, doi:10.1016/S0169-7722(02)00177-8.

Bonnet, E.O.; Bour, O.; Odling, N.; Davy, P.; Main, I.; Cowie, P. & Berkowitz, B. (2001). Scaling of fracture systems in geologic media, *Rev. Geophys.*, Vol. 39, No. 3, 347–383.

Bour, O. & Davy, P. (1997). Connectivity of random fault networks following a power law fault length distribution, *Water Resour. Res.*, Vol. 33, 1567–1583.

Bour, O. & Davy, P. (1999). Clustering and size distributions of fault patterns: Theory and measurements, *Geophys. Res. Lett.*, Vol. 26, No. 13, 2001–2004.

Cardenas, M.B.; Slottke, D.T.; Ketcham, R.A. & Sharp, J.M. (2007). Navier-Stokes flow and transport simulations using real fractures shows heavy tailing due to eddies, *Geophys. Res. Lett.*, Vol. 34, No. L14404, doi:1029/2007/GL030554.

Chen, Z.; Neuman, S.P.; Yang, Z. & Rahman, S.S. (2000). An experimental investigation of hydraulic behavior of fractures and joints in granitic rock, *Int. J. Rock. Mech. Min. Sci.*, Vol. 37, 267–273.

Cook, A.M.; Myer, L.R.; Cook, N.G.W. & Doyle, F.M. (1990). The effect of tortuosity on flow through a natural fracture, *In: Rock Mechanics Contributions and Challenges, Proceedings of the 31st U.S. Symposium on Rock Mechanics*, W.A. Hustrulid and G.A. Johnson, eds., A.A. Balkema, Rotterdam.

Cvetkovic, V.; Painter, S.; Outters, N.; & Selroos, J.-O. (2004) Stochastic simulation of radionuclide migration in discretely fractured rock near the Äspö Hard Rock Laboratory, *Water Resour. Res.*, Vol. 40, doi:10.1029/2003WR002655.

Cvetkovic, V. (2010). Significance of fracture rim zone heterogeneity for tracer transport in crystalline rock, *Water Resour. Res.*. Vol. 46, W03504, doi:10.1029/2009WR007755.

Darcel, C.; Davy, P.; Bour, O & de Dreuzy, J.R. (2003). Connectivity properties of two-dimensional fracture networks with stochastic fractal correlation, *Water Resour. Res.*, Vol. 39, No. 10, doi:10.1029/2002WR001628.

de Dreuzy, J.R. & Erhel, J. (2003). Efficient algorithms for the determination of the connected fracture network and the solution to the steady-state flow equation in fracture networks, *Comput. Geosci.*, Vol. 29, No. 1, 107–111.

de Dreuzy, J.R.; Davy, P. & Bour, O. (2001). Hydraulic properties of two-dimensional random fracture networks following a power-law length distribution: 1. Effective connectivity, *Water Resour. Res.*, Vol. 37, No. 8, 2065–2078.

Dean, C.A. (2010). *Long-term desorption of uranium and neptunium in heterogeneous volcanic tuff materials*, Ph.D. Dissertation, University of New Mexico, Albuquerque, NM.

Fisher, R. (1953). Dispersion on a sphere, *Proc. R. Soc. Lond. Ser. A.*, 217, 295–305.

Guimerá, J. & Carrera, J. (2000). A comparison of hydraulic and transport parameters measured in low-permeability fracrtured media, *J. Contam. Hydrol.*, Vol. 41, 261-281.

Grisak, G.E. & Pickens, J.F. (1980). Solute transport through fractured media: 1. The effect of matrix diffusion, *Water Resour. Res.*, Vol. 16, No. 4, 719–730.

Gustafson, G. & Fransson, A. (2005). The use of the Pareto distribution for fracture transmissivity assessment, *Hydrogeol. J.*, Vol. 14, 15-20, doi:10.1007/s10040-005-0440-y.

Holmén, J.G. & Outters, N. (2002). *Theoretical study of rock mass investigation efficiency*, TR-02-21, Swedish Nuclear Fuel and Waste Management Company (SKB), Stockholm, Sweden.

International Atomic Energy Agency, (2011). *Nuclear Technology Review 2011*, IAEA/NTR/20011, Vienna, Austria, http://iaea.org/Publications/Reports/ntr2011.pdf.

International Atomic Energy Agency, (2001). *The Use of Scientific and Technical Results from Underground Research Laboratory Investigations for the Geological Disposal of Radioactive Waste*, IAEA-TECDOC-1243, Vienna, Austria.

Jaeger, J.C.; Cook, N.G.W. & Zimmerman, R.W. (2007). *Fundamentals of Rock Mechanics*, 4th Ed., Blackwell, Malden, MA.

Klimczak, C.; Schultz R.A.; Parashar, R. & Reeves, D.M. (2010), Cubic law with correlated aperture to length and implications for network scale fluid flow, *Hydrol. J.*, doi:10.1017/s10040-009-0572-0.

Kozubowski, T.J.; Meerschaert, M.M. & Gustafson, G. (2008), A new stochastic model for fracture transmissivity assessment, *Water Resour. Res.*, Vol. 44, No. W02435, doi:10.1029/2007WR006053.

Langmuir, D. (1997). *Aqueous Environmental Chemistry*, Prentice Hall, Upper Saddle River, New Jersey.

Long, J.C.S. & Ewing, R.C. (2004) Yucca Mountain: Earth-science issues at a geologic repository for high-level nuclear waste, *Annu. Rev. Earth Planet Sci.*, Vol. 32, 363–401, doi:10.1146/annurev.earth.32.092203.122444.

Mandelbrot, B. (1974). Intermittent turbulence in self-similar cascades: Divergence of high moments and dimension of the carrier, *J. Flud Mech.*, Vol. 62, 331–350.

Mardia, K.V. & Jupp, P.E. (2000). *Directional Statistics*, Wiley, New York.

Munier, R. (2004). *Statistical analysis of fracture data adopted for modeling discrete fracture networks – Version 2.*, Rep. R. 04-66, Swedish Nuclear Fuel and Waste Management Company (SKB), Stockholm, Sweden.

National Research Council (1978). *Geological Criteria for Repositories for High-Level Radioactive Wastes*, National Academy Press, Washington, D.C.

National Research Council (2001). *Disposition of High-Level Waste and Spent Nuclear Fuel, The Continuing Societal and Technical Challenges*, National Academy Press, Washington, D.C., ISBN 0-309-07317-0.

Neretnieks, I. (1980). Diffusion in the rock matrix: An important factor in radionuclide retention?, *J. Geophys. Res.*, Vol 85, No. B8, 4379–4397.

Neuman, S.P. (2005). Trends, prospects and challenges in quantifying flow and transport through fractured rocks, *Hydrogeol. J.*, Vol. 13, 124–147, doi:10.1007/s10040-004-0397-2.

Nuclear Energy Agency (1999). *Confidence in the Long-Term Safety of Deep Geological Repositories. Its Development and Communication*, Organisation for Economic Cooperation and Development, NEA 01809, http://www.oecd- nea.org/rwm /reports/1999/confidence.pdf.

Olson, J.E. (1993). Joint pattern development: Effects of subcritical fracture growth and mechanical crack interaction, *J. Geophys. Res.*, Vol. 98, No. B9, 12,225–12,265.

Paillet, F.L. (1998). Flow modeling and permeability estimation using borehole flow logs in heterogeneous fractured formation, *Water Resour. Res.*, Vol. 34, No. 5, 997–1010.

Painter, S.; V. Cvetkovic; & J.-O. Selroos (2002). Power-law velocity distributions in fracture networks: Numerical evidence and implications for tracer transport, *Geophys. REs. Lett.*, Vol. 29, No. 14, doi:10.1029/2002GL014960.

Pohll, G.; Hassan, A.E.; Chapman, J.B.; Papelis, C. & Andricevic, R. (1999). Modeling ground water flow and radionuclide transport in a fractured aquifer, *Ground Water*, Vol. 37, No. 5,770–784.

Pohlmann, K.; Hassan, A.E. & Chapman, J.B. (2002). Modeling density-driven flow and radionuclide transport at an underground nuclear test: Uncertainty analysis and effect of parameter correlation, *Water Resour. Res.*, Vol. 38, doi:10.1029/2001WR001047.

Pohlmann, K.; Pohll, G.; Chapman, J.; Hassan, A.E.; Carroll, R; & Shirley, C. (2004). Modeling to support groundwater contaminant boundaries for the Shoal undergound nuclear test, *Desert Research Institute Report No. 45184.*

Pohlmann, K; Ye, M.; Reves, D.M.; Zavarin, M.; Decker, D. & Chapman J. (2007) Modeling of groundwater flow and radionuclide transport at the Climax Mine sub-CAU, Nevada Test Site, *DOE/NB/26383-05, Desert Research Institute Report No. 45226.*

Qian, J.; Chen, Z.; Zhan, H. & Guan, H. (2011). Experimental study of the effect of roughness and Reynolds number on fluid flow in rough-walled single fractures: a check of the local cubic law, *Hydrol. Process.*, Vol. 25, No. 4, 614–622, doi:10.1002/hyp.7849.

Quinn, P.M.; Cheery, J.A.; & Parker, B.L. (2011). Quantification of non-Darcian flow observed during packer testing in fractured sedimentary rock, *Water Resour. Res.*, Vol 47, No. W09533, doi:10.1029/2010WR009681.

Priest, S.D. (1993). *Discontinuity Analysis of Rock Engineering*, Chapman and Hall, London.

Reeves, D.M.; Benson, D.A. & Meerschaert, M.M. (2008a). Transport of conservative solutes in simulated fracture networks: 1. Synthetic data generation, *Water Resour. Res.*, Vol. 44, No. W05401, doi:10.1029/2007WR006069.

Reeves, D.M.; Benson, D.A.; Meerschaert, M.M. & Scheffler, H.-P. (2008b). Transport of conservative solutions in simulated fracture networks: 2. Ensemble solute transport and the correspondence to operator-stable limit distributions, *Water Resour. Res.*, Vol. 44, No. W05410, doi:10.1029/2008WR006858.

Reeves, D.M.; Benson, D.A. & Meerschaert M.M. (2008c). Influence of fracture statistics on advective transport and implications for geologic repositories, *Water Resour. Res.*, Vol. 44, W08405, doi:10.1029/2007WR006179.

Reeves, D.M.; Pohlmann, K.; Pohll, G.; Ye, M. & Chapman, J. (2010). Incorporation of conceptual and parametric uncertainty into radionuclide flux estimates from a fractured granite rock mass, *Stoch. Environ. Res. Risk Assess.*, doi:10.1007/s00477-010-0385-0.

Renshaw, C.E. (1995). On the relationship between mechanical and hydraulic apertures in rough walled fractures, *J. Geophys. Res.*, Vol. 100, No. B12, 24,629–24,363.

Renshaw, C.E. (1999). Connectivity of joint networks with power law length distributions, *Water Resour. Res.*, Vol. 35, No. 9, 2661–2670.

Rives, T.M.; Razack, M.; Petit, J.-P. & Rawnsley, K.D. (1992). Joint spacing: Analogue and numerical simulations, *J. Struct. Geol.*, Vol. 14, 925–937.

Robinson, B.A.; Li, C. & Ho, C.K. (2003). Performance assessment model development and analysis of radionuclide transport in the unsaturated zone, Yucca Mountain, Nevada, *J. Contam. Hydrol.*, Vol. 62-63, doi:10.1016/S0169-7722(02)00166-3.

Ross, S.M. (1985). *Introduction to Probility Models*, 3rd Ed., Academic Press, Orlando, FL.

Schertzer, D. & Lovejoy, S. (1987). Physical modeling and analysis of rain and clouds by anisotropic scaling multiplicative processes, *J. Geophys. Res.*, Vol. 85, No. D8, 9693–9714.

Schwartz, F.W.; Smith, L.; & Crowe, A.S. (1983). A stochastic analysis of macroscopic dispersion in fractured media, *Water Resour. Res.*, Vol. 19, No.5, 1253–1265.

Segall, P. & Pollard, D.D. (1983). Joint formation in granitic rock in the Sierra Nevada, *Geol. Soc. Am. Bull.*, Vol. 94, 563–575.

Smith, P.A.; Alexander, W.R.; Kickmaier, W.; Ota, K.; Frieg, K. & McKinley, I.G. (2001). Development and testing of radionuclide transport models for fractured rock: examples for the Nagra/JNC Radionuclide Migration Programme in the Grimsel Test Site, Switzerland, *J. Contam. Hydrol.*, Vol. 47, No. 2-4, 335–348.

Smith, L. & Schwartz, F.W. (1984). An analysis of the influence of fracture geometry on mass transport in fractured media, *Water Resour. Res.*, Vol. 20, No. 9, 1241–1252.

Snow, D.T. (1965). *A parallel plate model of fractured permeability media*, Ph.D. Dissertation, University of California, Berkeley.

Stigsson, M.; Outters, N. & Hermanson, J. (2001). *Äspö Hard Rock Laboratory, Prototype Repository Hydraulic DFN Model no. 2*, IPR-01-39, Swedish Nuclear Fuel and Waste Management Company (SKB), Stockholm, Sweden.

Sudicky, E.A. & Frind, O.E. (1982). Contaminant transport in fractured porous media: Analytical solutions for a system of parallel fractures, *Water Resour. Res.*, Vol. 18, No. 6, 1634–1642.

Terzaghi, R. (1965). Sources of error in joint surveys, *Geotechnqiue*, Vol. 15, No. 3, 287–304.

Twiss, R.J. & Moores, E.M. (2007). *Structural Geology*, 2nd Ed., W.H. Freeman, New York.

Winberg, A; Andersson P.; Hermanson J.; Byegard J.; Cvetkovic, V.; & Birgersson, L. (2000). *Aspo Hard Rock Laboratory, Final Report of the First Stage of the Tracer Retention Understanding Experiments*, Technical Report TR-00-07, Swedish Nuclear Fuel and Waste Management Co. (SKB), Stockholm, Sweden.

Wood, A.T.A. (1994). Simulation of the Von Mises distribution, *Commun. Stat.-Sim.*, Vol. 21, No. 1, 157–164.

Zhang Y.; Baeumer, B. & Reeves, D.M. (2010). A tempered multiscaling stable model to simulate transport regional-scale fractured media, *Geophys. Res. Lett.*, Vol. 37, No. L11405, doi:10.1029/2010GL043609.

Zheng, C. & Bennett, G.D. (2002). *Applied Contaminant Transport Modeling*, John Wiley and Sons, Inc., New York.

Zimmermann, R.W. & Bodvarsson, G.S. (1996). Hydraulic conductivity of rock fractures, *Trans. Porous Media*, Vol. 23, 1–30.

# 17

# Statistical Analyses of Pore Pressure Signals in Claystone During Excavation Works at the Mont Terri Underground Research Laboratory

Rachid Ababou[1], Hassane Fatmi[1],
Jean-Michel Matray[2], Christophe Nussbaum[3] and David Bailly[1,2]
[1]*IMFT, Institut de Mécanique des Fluides de Toulouse*
[2]*IRSN, Institut de Radioprotection et de Sûreté Nucléaire*
[3]*SWISSTOPO, Mont Terri Consortium*
[1,2]*France*
[3]*Switzerland*

## 1. Introduction

In many countries (such as Belgium, Germany, France, Japan, Switzerland, and United Kingdom), deep argillaceous formations are considered as potential host rocks for geological disposal of high-level and intermediate-level long-lived radioactive wastes. Some of these countries are investigating the suitability of high compacted clay-rich rocks at depths down to around 500 m below the ground surface. The general disposal concept comprises a network of drifts and tunnels linked to the surface by shafts and ramps, all artificially ventilated. Research is ongoing in Underground Research Laboratories, like the Mont Terri site in the Swiss Jura, to assess and ensure the safety of the repositories for the full decay life of the radioactive waste, i.e. the capacity of the hypothetical repository to prevent the migration of radionuclides towards the biosphere.

One of the important safety considerations regarding nuclear waste disposal repositories are damages and disturbances induced by excavation works to the rock-mass in the vicinity of the openings. These excavations can alter the initial and favorable isolation properties of the host rock. The most obvious problem is that of changes in the stress field around the tunnel during excavation that induces a fracture network, namely the Excavation Damaged Zone or EDZ, consisting mainly in unloading joints and shrinkage cracks (Bossart et al., 2002) and therefore characterized by irreversible changes. This phenomenon creates an increase in permeability of potentially several orders of magnitude higher than that of the undamaged host rock (Blümling et al., 2007). The EDZ and its raised permeability could possibly affect the performance and safety of the repository as it could act as a preferential pathway, allowing radionuclides to migrate to the surface, bypassing bentonite seals along the drifts.

However, excavation works may also contribute to reversibly disturb the geological medium in the so-called Excavation disturbed Zone (EdZ). This term (EdZ) designates a zone that is thought to be disturbed by elastic hydro-mechanical variations of pressure and stress, but without irreversible changes in flow and transport properties. Within the EdZ, there are no negative effects on the long-term safety, but short-term effects may exist as the response to hydro-mechanical coupling. For instance, mine-by test experiments performed during excavation works often show an increase in pore pressures in the vicinity of the newly excavated gallery. This increase may exceed 1 MPa (10 bars), for instance during excavation of Ga03 in the Tournemire URL (Massmann, 2009). This pore pressure increase appears either instantaneous or delayed depending on the position of the measurement sections with respect to the excavation front.

In short, the objective of this text is to study the isolation properties of a geologic repository against the migration of radionuclides through the host rock and eventually towards the biosphere. In this context, the hydro-geologic properties of the host rock must be characterized at different scales (e.g., near field vs. far field).

The permeability is a very important property in terms of isolation, but in this work, we focus on novel methods for identifying two other related properties: (i) specific storativity (as the elastic response to earth tides), and (ii) the effective dynamic porosity (using the barometric effect). Indeed, the analysis of pore pressure and atmospheric pressure time series makes it possible in some cases to determine hydraulic parameters such as specific storage coefficient and effective porosity using simplified groundwater flow and compressibility models (e.g., Marsaud et al. 1993). These models have already been applied to obtain a preliminary characterization of the compacted clay rock of the Tournemire URL in Aveyron, France (Fatmi et al. 2005).

More precisely, our main goal in this text is to characterize the above mentioned hydraulic and mechanical properties of the clay rock from analyses of measured atmospheric and pore pressure signals, and also, to study the response of the clay rock's "Excavation disturbed Zone" ("EdZ") during the excavation of galleries.

The data analysis techniques rely strongly on statistical concepts (theory of random processes) and on the decomposition of signals on various bases:

- Harmonic basis (sine, cosine): Fourier spectrum S(f) vs. frequency (f).
- Wavelet multiresolution bases (Daubechies): wavelet components Cj(t) at different time scales "j".

These techniques are used for interpreting the proposed hydromechanical models, in order to identify the hydrogeologic properties of the claystone from measured signals (possibly disturbed by excavation works):

- Wavelet decomposition of pore pressure signals (selection of semi-diurnal earth tide component), and statistical envelope analysis (effect of moving excavation front): earth tide effect model → specific storativity;
- Spectral cross-analysis (spectral gain) of air pressure vs. pore pressure signal: barometric effect model → porosity.

The study presented in this text is entirely based on data and time series collected at the excavated geologic claystone site of the Mont Terri URL in the Swiss Jura (Mont Terri Project - international consortium), and we focus on claystone properties at "near field" scales, within meters or tens of meters from excavated galleries at the Mont Terri site. Similar techniques are currently being developed for other datasets collected at the Tournemire URL (Aveyron, France) operated by the "Institut de Radioprotection et de Sûreté Nucléaire" (France).

In this work, in order to achieve the above mentioned objectives, we use pressure signals obtained over a period of ten years at Mont Terri (1996/2005), from which records as long as *one year or more* will be analysed in this text (see ***Section 4***). But, to analyse the effects of a moving excavation front, we also focus on a period of 5 months of "*syn-excavation*" period for gallery "Ga98", and on a much shorter period of fast pressure changes lasting only a couple of weeks (see ***Section 5***). In total, we study both long and short time scale records, from years to weeks.

Some results of pressure signal pre-processing and long time scale analyses were presented in *Fatmi et al. (2007, 2008)* at an early stage of this study. The present text provides a significant update of the results of long time scale analyses, and presents also new types of results on the analysis of excavation effects on pore pressures at very short time scales (weeks).

Accordingly, the outline of this chapter is as follows.

***Section 2*** describes the hydro-geologic site and the pore pressure measurements at the Mont Terri site (in the framework of experiment LP14).

***Section 3*** is devoted to the mathematical and statistical methods of signal pre-processing (e.g. reconstruction of missing data) and of signal analysis (time-lag correlation functions, frequency spectra and cross-spectra, wavelets, envelopes). For hydro-geologic interpretation of pressure signals, the methods involve, not only statistical treatments, but also simplified elastic hydro-mechanical models: the concept of a specific storativity coefficient "Ss" (in relation to earth tides); and the barometric efficiency (involving both "Ss" and porosity "Φ") in relation to air pressure fluctuations. These models are briefly described at the end of the section (***Box N°2***) and their application is developed in the remaining sections of the text (see below).

***Section 4*** develops the implementation of "long time scale" statistical analyses and hydro-geologic interpretations of signals, in the case where the objective is to obtain a global estimate of the two hydro-geologic coefficients (Ss,Φ) in two steps, first "Ss" from semi-diurnal earth tide effects using multi-resolution wavelet analysis, and secondly "Φ" using the barometric spectral gain at diurnal frequency.

***Section 5*** develops, in contrast, a somewhat more sophisticated implementation over "short time scales" with a strong non stationary trend in the pore pressure signal. This occurs in the case of a moving excavation front passing nearby the pressure sensor. The objective is to identify the effect of the evolving "EdZ" at the location of the pressure sensor, and in particular, its effect on the evolution of Ss(t). The time scale of analysis is typically a couple of weeks, and a new tool has been developed (based on statistical envelopes) in order to identify the modulation in time of the amplitude of earth tide fluctuations.

***Section 6*** provides a summary of the methods and results, and suggests possible modifications and extensions of this work concerning statistical methods (multi-cross analyses) and hydro-geologic interpretations (enhanced hydro-mechanical models and inverse problems): see the "outlook" in ***Sub-Section 6.2***.

The text is completed by an appendix (for abbreviations and symbols), and by a list of references.

## 2. Hydrogeologic setting and pressure signal measurements

### 2.1 Hydrogeologic setting at Mont Terri (Opalinus clay rock)

The Mont Terri Rock Laboratory was excavated within the "Opalinus Clay" of the Mont-Terri anticline, which has a very low permeability. The geological cross section shown in

**Figure 1** shows the asymmetrical, dome-shaped fold of the Mont Terri anticline, where sediments of Triassic and Jurassic age are exposed.

The Opalinus clay is about 180 million years old (Mesozoic age, Jurassic period, Lower Dogger, Aalenian). It is a shallow marine mud deposit, extending over wide areas in Central Europe. It is underlain by the Jurensis Marls (Toarcian) and overlain by the "Lower Dogger" limestones (Bajocian). Three facies have been identified within this clay rock: sandy, carbonate-rich sandy, and shaly. The latter facies is studied as an analogue for hypothetical disposal of radioactive waste in a naturally isolating geologic formation (*the Mont Terri site is a research site only, not destined to host nuclear wastes or any radioactive products whatsoever*). For more details on geology and hydrogeology of the site: see *Fatmi et al. (2008), Fatmi (2009)* and *Schaeren and Norbert (1989)*. See also, more recently, *Nussbaum et al. (2011)*.

We emphasize again that the Opalinus clay rock at Mont Terri has very low permeability or hydraulic conductivity: $K \approx 10^{-13}$ to $5\ 10^{-13}$ m/s (1 E-13 to 5 E-13 m/s), ± 1 order of magnitude, in the intact zone outside the *Excavated Damaged Zone*. This clay rock also displays a significant self-sealing capacity due to a high proportion of swelling minerals (about 10%). In contact with water, it swells and tends to seal the cracks and fractures that occurred after excavations (galleries, niches). This self-sealing reduces permeability and participates in restoring initial rock properties.

These factors, among others, explain the choice of this type of clay rock formation by the Swiss authorities - and by the international Mont Terri consortium - as a potential host rock for high level radioactive wastes. The Mont Terri URL (URL stands for Underground Research Laboratory ) is managed by SWISSTOPO. The Mont Terri consortium includes Swiss, French, British, Canadian, Spanish, Belgian, U.S., Japanese, and other nations' public authorities and their national research institutions.

Other types of geological formations were also studied by the Swiss authorities, including a crystalline rock (granite) at the Grimsel site in the Swiss Alps (e.g., FEBEX drift *in situ* experiment). Thermo-Hydro-Mechanical upscaling and modelling of the FEBEX drift *in situ* experiment was studied in *Cañamón*'s thesis (*2006*), later published as *Cañamón (2009)*. These references also include statistical signal analyses of the FEBEX thermo-migration laboratory experiment, or "mock-up" experiment (see also *Cañamón et al. (2004)*. The in situ and mock-up FEBEX experiments were performed under funding by the Spanish nuclear authorities *ENRESA*.

In the remainder of this text, we focus exclusively on the analysis of pressure signals collected *in situ* in Mont Terri's Opalinus Clay rock site.

### 2.2 Pressure signal measurements and pre-processing (Mont Terri BPP-1 borehole)

The pressure signals used in this work were obtained over a period of 10 years with half-hourly time steps (on average), at the Mont Terri URL, in the framework of the LP14 project (*Long term Pressure monitoring project - phase 14*). Concerning Mont Terri's LP project, see *Marschall et al. (2009)*. The BPP-1 borehole (101 mm diameter, and 20 m length) was dug in the year 1996, from niche PP located in Mont Terri's "exploration gallery". The measurement sections or sensors (PP-1 and PP-2) are located at the end of the BPP-1 borehole, outside the excavation damaged zone (EDZ) of the exploration gallery, in the silty-shaly facies of the Opalinus clay.

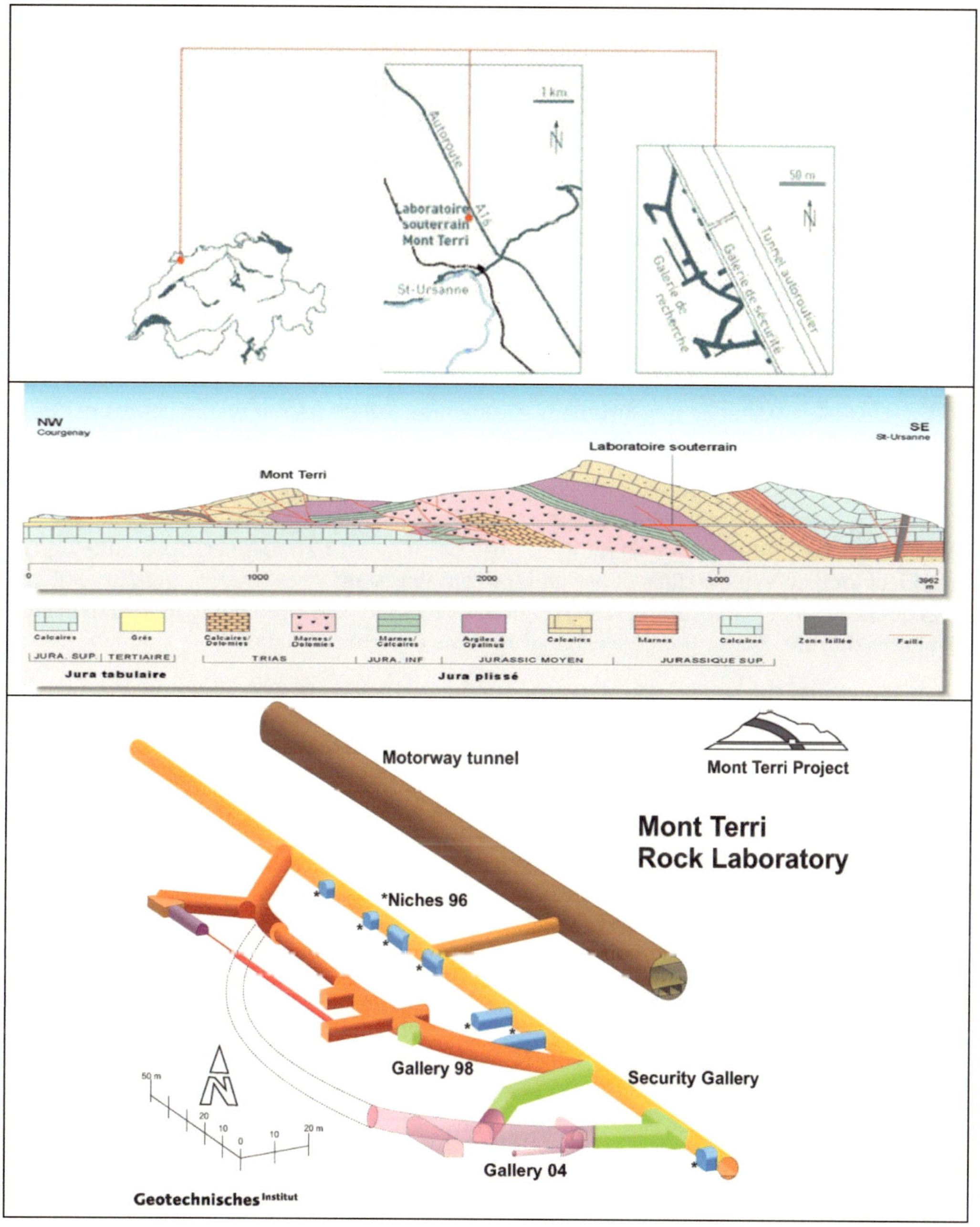


Fig. 1. **(a) Above.** Location of the Mont Terri Underground Rock Laboratory in Switzerland, with its main experimental galleries. **(b) Middle**: Geologic cross-section of the folded Jura showing the Mont Terri anticline and the 4 km long motorway tunnel giving access to the Opalinus Clay rock (Freivogel M. 2001, Bâle University, after Schaeren G. & Norbert J. 1989: Tunnels du Mont Terri et du Mont Russelin. La traversée des "roches à risques": marnes et marnes à anhydrite. *Soc. Suisse Ing. Arch.*, Doc. SIA D 037, 19–24). **(c) Below:** Schematic 3D perspective view of some of the Mont Terri galleries and the motorway tunnel.

Absolute pore pressures P(t) were measured from boreholes (as explained further below), and atmospheric pressure $P_{AIR}(t)$ or $P_{ATM}(t)$ was measured underground in the exploration gallery. Absolute pore pressures were then transformed into relative pressures, defined as: $P_{REL}(t) = P(t)-P_{ATM}(t)$. For more technical details on the BPP-1 borehole and its pressure sensors, see ***Figure 3*** further below, and see *Thury and Bossart (1999)*. Finally, see also the previous ***Figure 1*** for 3D perspective views of the tunnels, galleries, and boreholes.

### 2.2.1 Signal pre-processing, record lengths, missing data and other issues

The BPP-1 borehole was initially selected for this study because it provided the longest pore pressure times series over a period of about 10 years (from 17/12/1996 to 30/06/2005). However, after pre-processing for outliers, data gaps, and variable steps, the 10 year long sequences from the PP1 and PP2 pore pressure sensors (and from the air pressure sensor) had to be fractioned into shorter "reconstituted" contiguous sequences. For relative pressure $P_{REL}(t)$, the resulting sequences were roughly on the order of a year (*no longer than 1.5 year at best*).

It should be emphasized that, in order to obtain relative pressure sequences ($P_{REL}(t) = P(t)-P_{ATM}(t)$), the pore pressure P(t) and air pressure $P_{ATM}(t)$ had to be processed "jointly" (*Fatmi et al. 2008*). Joint pre-processing of pore pressure and atmospheric pressure signals get rid of outlier values, data gaps, and/or variable time steps in *both* signals, but the consequence is that the resulting "clean signal" of relative pressure is only available in sub-sequences that are shorter than the original 10 years of recorded data.

It was found that the longest un-processed contiguous sequence of relative pressure $P_{REL}(t)$, directly available from the raw data without need for preprocessing, was only 1 month long (subsequence 02/08/2002 - 04/09/2002). Pre-processing was necessary in order to increase the length of the $P_{REL}(t)$ records beyond one month. After pre-processing, the longest $P_{REL}(t)$ record for sensor PP1 has a duration of about 14 to 15 months (29/01/2004-12/04/2005), or 20736 half-hourly time steps. This longer, partially reconstructed signal, will be analysed in ***Section 4.2***.

Statistical tests, involving recursions, were used to detect possible outliers (transformed into missing data). In the presence of missing data, for partial reconstitutionS, we used:

- decompositions into trends and residuals using moving average filters (and other filters),
- linear interpolation techniques (usually only for relatively short data gaps),
- a bidirectional first order autoregressive model (going forward and backward in time)
- statistical tests and validations of reconstituted series based on moments and covariances

Time step homogenization is also a kind of partial reconstitution. We used for this a linear interpolation/extrapolation method, with a constant time step "$\Delta$to" selected by the user. Here, all processed signals have, after "homogenization", a half-hourly time step ($\Delta$to = 30 mn). This choice is in agreement with the nominal acquisition step (also the most frequently encountered time step in the collected data).

***Figure 2*** shows a simplified flow chart presenting these Pre-Processing tasks and methods. For more details on pre-processing methods, see *Fatmi (2009, Chap.3)* and *Fatmi et al. (2007, 2008)*.

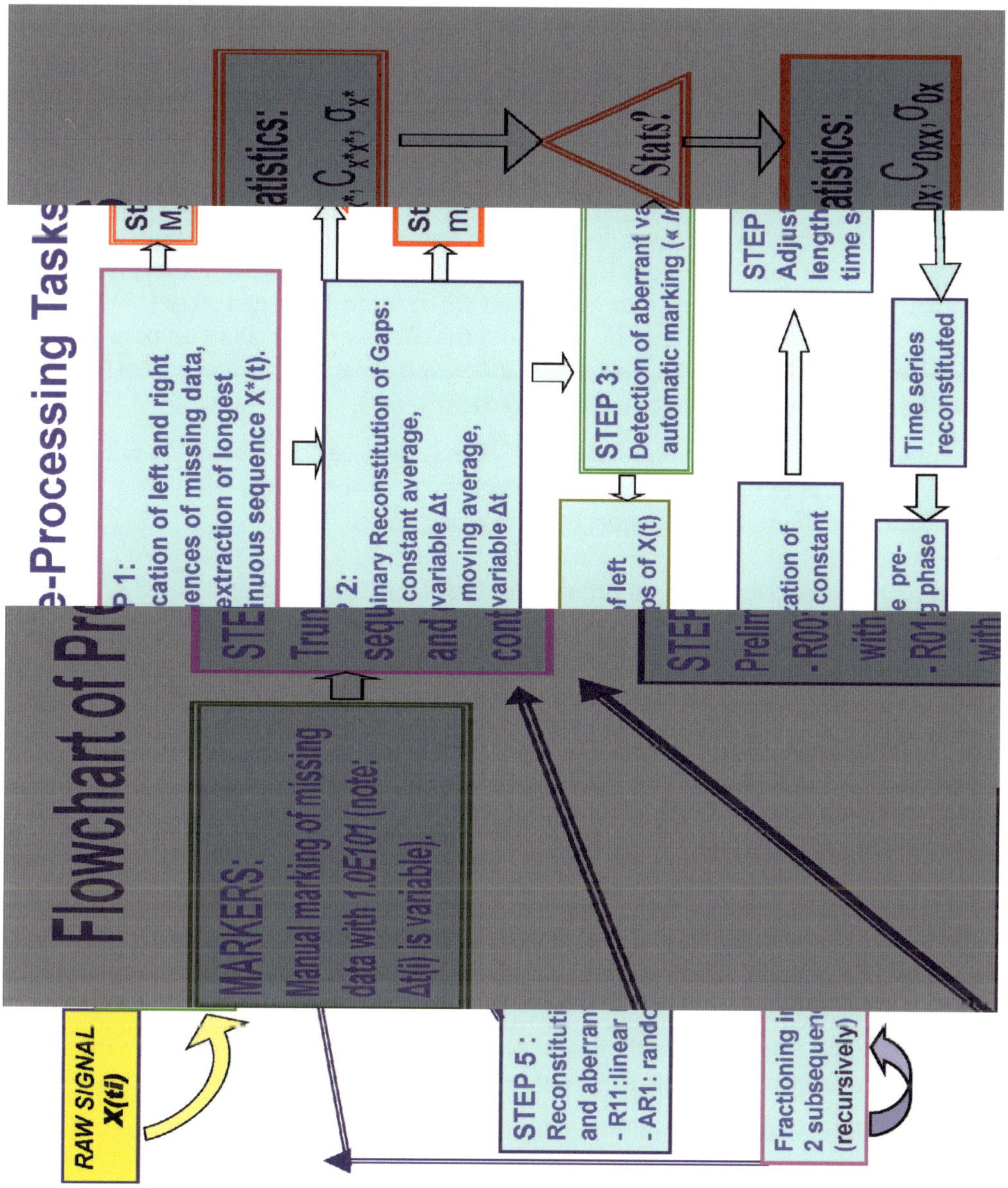


Fig. 2. Flowchart of Pre-Processing tasks and methods (programmed and implemented as a MATLAB TOOLBOX). The pre-processing tasks contain some iterations and recursions, particularly when computing statistics to detect outliers or to fill data gaps. Only some of the iterative procedures are explicitly shown here. The symbols (Mx*, Cxx*, σx*) and (mox, Coxx, σox) represent estimates of the mean, the covariance function, and the standard deviation of the signal at various stage of pre-processing. The symbol Δto is the constant time step chosen by the user for time step homogenisation. At the end of the pre-processing stage, the processed (transformed) signal is ready for statistical analyses. The processed signal may be shorter than the initial "raw" signal.

### 2.2.2 Pressure sensors location with respect to the excavated Gallery 1998

Since we are especially interested here in the effects of excavation on the hydrogeologic properties of the clay rock (EDZ), let us focus here on a brief description of the excavation of gallery Ga98.

The Ga98 gallery was dug at Mont Terri during the year 1998, over a period of 5 months (more precisely, from 17/11/1997 to 23/04/1998).

Recall that the main goal of this work is to study not only the properties of the claystone, but also, their evolution in response to excavation (Excavation Damaged Zone). This will be done, for the case of gallery Ga98, by analysing the effects of excavation on pore pressures collected at sensors PP1 and PP2, knowing that they are distant only a few meters (4.4 m and 4.5 m) from the excavated gallery Ga98 ***(Figure** 3)*.

Thus, due to their close location to the 1998 gallery, the sensors of borehole BPP-1 can be used to monitor pore pressures during the excavation of the 1998 gallery: we will exploit this particular configuration in this paper, as in a *"mine-by test"*.

### 2.2.3 Note on related analyses of other signals at Mont Terri and at other clay sites

While we focus in this work on air and pore pressures at Mont Terri, for the sake of completeness, we present in this section a brief note on other signal analyses: e.g., air temperature, relative humidity, and crack aperture at Mont Terri, and also, signals recorded at other URL sites in clay rock (Bure and Tournemire, in addition to Mont Terri). Thus:

The Mont Terri "BPP-1" pore pressure signals were compared and cross-analysed with air pressure, air temperature, and relative air humidity measured in the exploration gallery of the Mont Terri URL.

In addition, similar signals were analysed and cross-analysed at two other clay rock URL's (Underground Research Laboratories) in addition to the Mont Terri URL:

- the clay rock URL of Tournemire (Aveyron, France), operated by the French "Institut de Radioprotection et de Sûreté Nucléaire" (IRSN), and
- the clay rock URL of Bure (Meuse / Haute Marne, France), operated by the French national agency on the management of nuclear waste (ANDRA).

Concerning these sites (Tournemire and Bure), the reader is referred to the Ph.D. thesis of *Fatmi (2009)*, available on line. Also, it should be noted that despite their similarities, the three clay rock sites have different tectonic, structural and hydro-mechanical characteristics (e.g., see *Nussbaum et al. 2007*, comparing Bure and Mont Terri).

- Furthermore, crack aperture time series measurements were also developed at the Mont Terri site (Cyclic Deformation phase 14 project "CD14"). The crack displacement signal was statistically analysed. The air temperature, pressure, and relative humidity were also measured and cross-analysed together with crack displacements. Some of the statistical methods used in the present study (cf. ***Section** 3*) were also applied to the "cross-analysis" of crack aperture response to hydro-meteorological fluctuations at various time scales. The results of these crack aperture investigations are reported in *Möri et al. (2012, in press at the time of this writing)*.

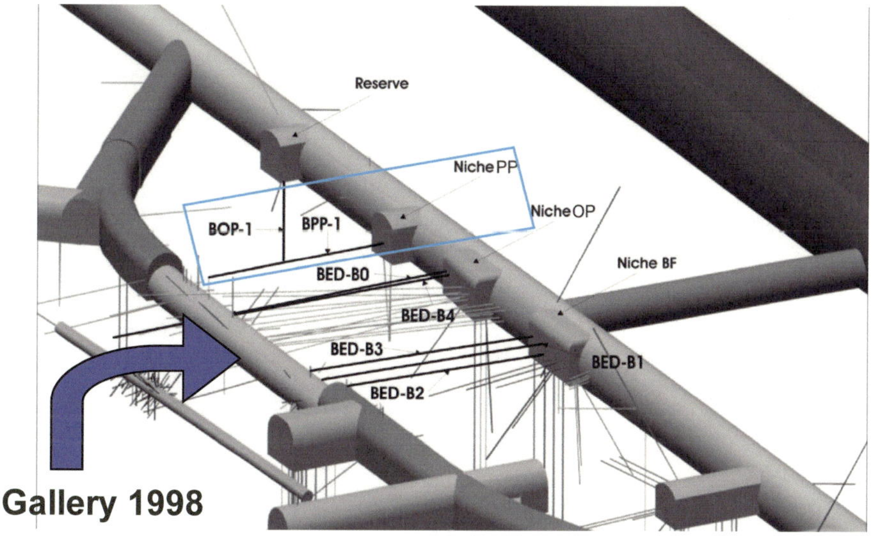


Fig. 3. Tunnels, galleries, and boreholes at the Mont Terri URL, showing Gallery 1998 ("Ga98"), and borehole BPP-1 dug from the PP niche. This niche is located in the reconnaissance gallery ("exploration gallery") parallel to the main road tunnel (dark grey). The pore pressure sensors used in this study are located near the end of borehole BPP-1 (lower left corner of blue rectangular frame), only 4 m away from Ga98. Their response during excavation of Ga98 is studied in this work.

### 2.3 Previous results and scope of statistical analyses this work (long and short time scales analyses of pore pressures)

The general objective of this work, in the context of radioactive waste disposal in geologic clay sites, was to exploit pressure measurements in order to quantify the phenomena affecting pore pressure (earth tides, barometric pressures,...), and *in fine*, to estimate the hydro-geologic or hydro-mechanical properties of the clay rock : elastic compressibility and/or specific storativity (Ss), effective elastic porosity($\Phi$), ....

Some pore pressure analyses methods and results (*Fatmi et al. 2007, 2008)*, were previously developed for characterizing clay rock properties in the *undisturbed state*. These analyses used long time scale records (as long as possible), in the absence of excavation works, or prior to excavation works. The resulting estimates of Ss and $\Phi$ from pressure signals in the selected borehole at the Mont Terri URL compared favorably with those estimated by other methods, such as pulse tests on shorter time scales. Similar analyses were also carried out for other clay rock URL's: Tournemire and Bure (*Fatmi 2009*). In the present work (***Sections 4 and*** 5), we provide a synthetic, updated, and upgraded presentation of the above-mentioned results, that is:

- statistical analyses and interpretations of *long pressure records*, leading to *global* estimates of average clay rock properties (Ss and $\Phi$), but also...
- statistical analyses and interpretations of *short non stationary pressure records*, focusing on the effects of excavation works on pore pressures, and leading to estimates of *evolutionary* and *local* clay rock properties, Ss(t) and $\Phi$(t).

All these analyses, as mentioned above, exploit the effects of natural perturbations of pore pressure, caused by hydro-meteorological as well as geophysical oscillations:

- *hydrometeorology*: atmospheric pressure oscillations;
- *planetary geophysics*: earth tides.

Thus, in the short time scale analyses (excavation works) we consider also the effects of changes in hydro-geological properties due to human activities in the URL site. The excavation of an underground gallery serves as a "*mine by*" test. The objective is to monitor the response of pore pressure (measured at a fixed location) to the evolution of the excavation front. The effect of the excavation (if any) should be superimposed on the natural perturbations of pore pressure (atmospheric and earth tide effects).

We will see that it is possible to detect, statistically, the effects of the evolving excavation front on the component of pore pressure fluctuations due to earth tides. The time scale of analysis here is relatively short, because we focus on a particular "non stationary" event, the approach of the excavation front close enough to the pressure sensor(s). This is to be distinguished from the "long time sale analyses", where we try to use records that are as long as possible, and not too much influenced by local trends.

Along these lines, for short time scale "excavation effects", our main objective is as follows:

1. to analyse the effects of drift excavation on pore pressures during excavation (syn-excavation) [1], and if possible, to analyse this effect continuously as a function of time (statistically, this boils down to analysing a signal with strongly non stationary fluctuations).

In the case at hand, we focus on the passage of an excavation front of gallery GA98 at Mont Terri (gallery 1998), and we examine closely its possible influence on the pore pressures measured in pressure sensors (PP1, PP2) of the BPP-1 borehole (*here, we will narrow this down to the single pressure sensor PP1, closest to the excavation front*). The specific objectives are, essentially:

- to quantify the evolution of elastic specific storativity Ss(t) as a function of time, during the different phases (pre-, syn-, post-excavation phases), and partiuclarly, the maximum change of Ss(t) during the syn-excavation phase;
- to interpret the evolution of Ss(t) and to compare it with the distance D(t) of the moving excavation front (and the moving "EdZ") relative to the location of the fixed pressure sensor.

## 3. Methods for statistical analyses and interpretations of pressure signals

Pre-processing procedures were described in the previous introductory sections (see flowchart in ***Figure 2***). We now focus on methods of statistical analyses, and also, on the simplified hydro-mechanical "models" to be used for hydro-geologic interpretation of pore & air pressure analyses.

---

[1] More generally, for comparison purposes, we have also analysed pre- post-excavation phases; however the results will only be briefly mentioned here for lack of space (see *Fatmi 2009* for details).

### 3.1 Overview of methods of statistical analysis (and interpretation)

After the preliminary stage of signal pre-processing, we are left with several records of contiguous time series with constant time steps (some of these signals have been pre-processed to some extent, and possibly, they have been partially reconstituted). These pre-processed pressure time series are then statistically analysed in order to characterize the hydraulic and hydro-mechanical behaviour of the porous clay rock formation (Ss, Φ). To achieve this, the following signal analysis methods were used and applied to Mont Terri signals [2]:

i. Auto-correlation and cross-correlation analyses of pressure signals versus time lag, and also (closely related), deconvolution or identification of input/output transfer function, e.g., between atmospheric pressure and relative pore pressure.
ii. Fourier spectral analyses in the frequency domain, based on Fourier transforms of the correlation functions. This leads to identification of dominant frequencies, and also, cross-spectrum analysis and identification of the spectral gain between two signals (such as atmospheric pressure and relative pressure).
iii. Orthogonal multi-resolution wavelet analysis was used to decompose the signals in the time/scale domain (two-parameter domain). Thus, for any selected dyadic time scale (e.g., for the semi-diurnal time scale of earth tides), we isolated and extracted the corresponding "wavelet component" of the original signal. Also, we used wavelets in some cases as a filter, decomposing the original signal into a filtered quantity (the wavelet "approximation") and its residual signal (the wavelet residual) at any given dyadic time scale (the scale of the filter).
iv. Estimation of the statistical envelope of a modulated random signal based on Cramer-Leadbetter statistical envelope theory, implemented with the Hilbert transform. This technique is used to characterize the modulation, in time, of the signal extracted from semi-diurnal wavelet component analysis of pore pressure during excavation.

Each of the above-mentioned techniques was used at one stage or another in our study of Mont Terri pressure signals on various time scales.

The correlation and cross-correlation function analyses versus time lag (I), as well as the corresponding Fourier spectral analyses in frequency space (II), are both based on the theory of random processes: the reader is referred to *Bras et al. (1985); Box et al. (1976); Papoulis et al. (2002), Yaglom (1987), Yevjevich (1972)*. Concerning spectral estimation and analysis (II), see also: Blackman & Tukey (1958), Max (1980), Max et al. (1996), Priestley (1981), Vanmarcke (1983). Time domain deconvolution and input/output analyses, causal or non-causal, linear or non linear, are presented in *Labat et al. (1999c; 2000a)* for the case of rainfall-runoff signals.

Multi-resolution wavelet analysis (III) is an orthogonal representation of signals in "scale-time" space (a two-parameter space). The mathematical theory is presented in *Mallat (1989)*. See *Labat, Ababou et al. (1999a,b; 2000b; 2002)* concerning applications of wavelets to rainfall/runoff signals in karst hydrology. In multi-resolution wavelet analyses, "time" is

---

[2] The reader interested mostly in the applications, and not in the theoretical signal analysis methods, may jump at first reading to ***Sections 4-5***, where the methods are implemented for Mont Terri pressure data.

discrete in the usual way ($t(n) = n\times\Delta t$), but wavelet "scales" are dyadic (powers of 2 : $2^m\times\Delta t$). Wavelet "scales" are *analogous* to inverse frequencies, but still, they should not be confused with Fourier frequencies.

The theoretical tools III (wavelets) and IV (envelopes) will be combined and used together in ***Section** 5 (5.2+5.3).* The goal is to assess, on a short time scale of a few weeks, the influence of excavation on the evolution of pore pressure and, finally, of clay rock properties in the "EdZ".

The mathematical and statistical methods cited above have well known theoretical bases, however, they can be implemented in different ways in practice. For this study, we have devised an integrated set of programs, written in MATLAB and organized in "TOOLBOXES" as follows:

- Signal Pre-Processing TOOLBOX, and
- Signal Analyses TOOLBOX.

The two TOOLBOXES are described in *Fatmi 2009's* PhD thesis, available on line.[3]

It should be noted that some of the methods and techniques were specifically designed or adapted by us for this study, e.g.: estimations of cross-spectra and spectral gains; identification of time domain transfer functions; filtering techniques (comparisons between moving averages and wavelet approximations); and also, an original implementation of the *Cramer-Leadbetter* envelope, performed with the *Hilbert Transform*, and applied to a selected wavelet component.

Other auxiliary studies not reported here include the design of a number of synthetic tests to validate statistical methods (cross-correlations, cross-spectra, envelopes, comparison of filtering techniques, etc). For details on some of these tests, the reader is referred to *Fatmi (2009).*

***Figure 4*** shows a simplified flowchart presenting some of the signal analysis methods used in this work. These methods were programmed in a MATLAB TOOLBOX for SIGNAL ANALYSES (distinct from the PRE-PROCESSING TOOLBOX described in ***Figure*** 2).

### 3.2 Univariate (auto) correlation functions, bivariate (cross) correlation functions, and temporal transfer functions

Correlation function analysis is a way to study the temporal structure of the signals, in the time lag domain. The univariate analysis of a single signal leads to a study of its auto-correlation function. The bivariate analysis of two signals jointly leads to both auto- and cross-correlation functions.

[3] These MATLAB TOOLBOXES were developed mainly by H. Fatmi, supervised by R. Ababou, with the aid of the other co-authors of this chapter and of other researchers and collaborators, including: A. Mangin, senior researcher at the Moulis CNRS laboratory (who initiated several of the signal analysis procedures); doctoral students Y. Wang and K. Alastal (who tested spectral and wavelet applications at the Institute of Fluid Mechanics); and intern students A. Mallet, A. Moulia and Ch. Joly who participated in various validation tests using synthetic signals.

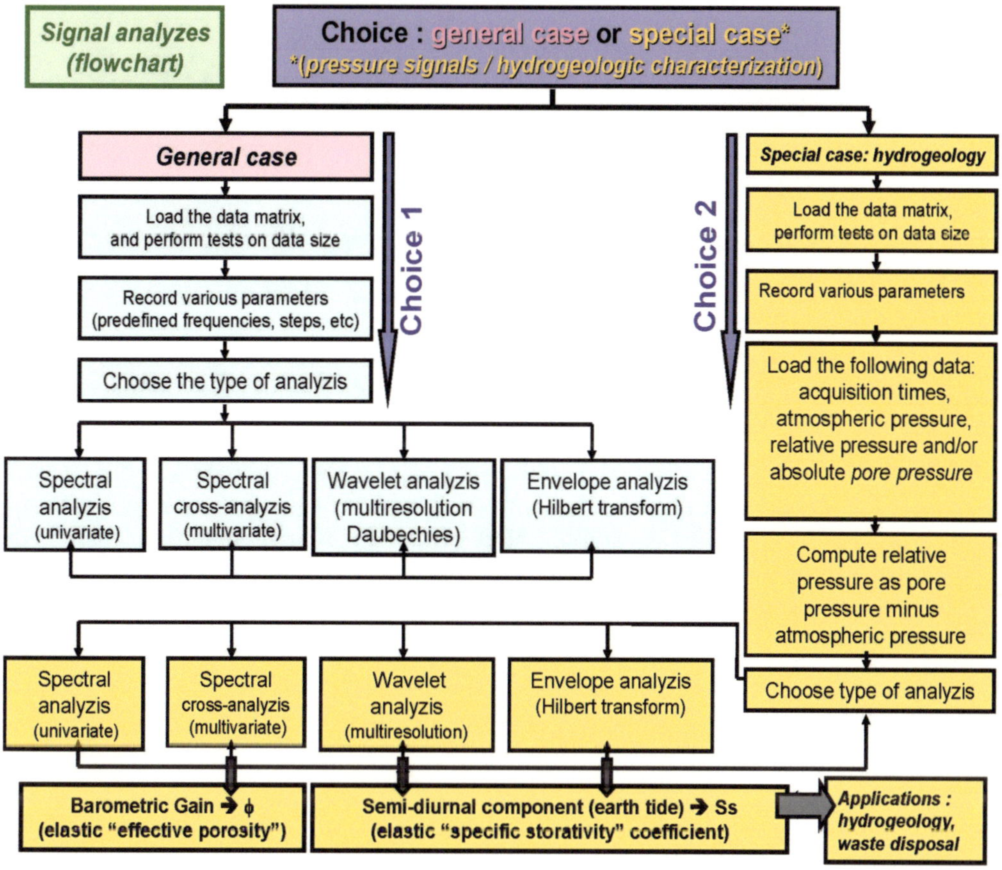


Fig. 4. Simplified flowchart presenting some of the signal analysis methods used in this work. These methods were programmed in a MATLAB TOOLBOX for SIGNAL ANALYSES.

More precisely, the temporal structure of a signal X(t), or of two signals X(t) and Y(t) jointly, "is captured" by a temporal structure function that depends on two points in time (t, t+τ). The stationarity hypothesis simplifies this analysis by assuming that the structure function depends only on the time lag (τ). In addition, the ergodicity hypothesis (which is notoriously hard to validate in practice) further assumes that the time average of the signal is equivalent to the ensemble mean (mathematical expectation) as the record length becomes infinitely long.

Thus, for two statistically stationary and ergodic processes, X(t) and Y(t), the cross-covariance function is supposed to depend only on the time lag (τ) between X(t) and Y(t+τ) (and it does not depend on "t"). The cross-covariance Cxy (and the cross-correlation Rxy) can be estimated as follows, from a single data record of the two processes (X(t),Y(t))[4]:

---

[4] The auto-covariance functions Cxx(τ) and Cyy(τ) are defined in a similar way; for instance, Cxx(τ) can be obtained simply by replacing XY by XX in the cross-covariance function Cxy(τ). The auto-correlation function Rxx(τ) is simply Cxx(τ)/Cxx(0) where Cxx(0) = Var(X(t)) = $\sigma_X^2$.

$$C_{XY}(j) = \frac{1}{N} \sum_{i=1}^{i=N-j} \left( X(t_i) - \bar{X} \right) \left( Y(t_{i+j}) - \bar{Y} \right) \text{ and } R_{XY}(j) = \frac{C_{XY}(j)}{\sigma_X . \sigma_Y} \qquad (1)$$

In this equation:

- "i" is the discrete time (t(i))
- "j" is the discrete time lag (τ(j))
- $R_{XY}(j)$ is the cross-correlation (function of discrete time lag "j"),
- $C_{XY}(j)$ is the cross-covariance (function of discrete time lag "j"),
- N: total number of data in the observed time series.

One also defines "M", the maximum time lag to be retained for correlation function analyses. The discrete time lag index "j" runs through $j = \pm\{0,1,2,..., M\}$. It is necessary to keep $M \leq N-1$ (in practice $M \leq N/3$ or much less).

A sub-sampling step $k_0$ can be defined, with $k_0 = 1$ or 2 or 3 or ... M (take $k_0 = 1$ if "no sub-sampling"). The time lag in physical time units is: $\tau_j = j.\Delta t$ (if $k_0 = 1$) or more generally $\tau_j = j.k_0.\Delta t$ ($k_0 \geq 1$) where ko is the sub-sampling step. In our case, we choose the obvious, i.e., the smallest sub-sampling step ($k_0$=1) in order to sweep all the samples of the time series. For very long series or unduly small time steps, sub-sampling ($k_0 \geq 2$) may be advisable, but this is not our case here.

The above ***Eq. 1*** gives a *biased* estimate of the cross-covariance function Cxy(τ). To obtain an *unbiased* estimate, the denominator (N) should be replaced by (N-j-1) in the above estimator, although some authors also use (N-j).

Finally, it is noted that Cxy(τ) is directly related to (but not equal to) the temporal transfer function Fxy(τ) between processes X(t) and Y(t). See *Labat et al. (1999c; 2000a)* for a detailed analysis and discussion of temporal transfer functions in hydrogeology & hydrology (*Instantaneous Unit Hydrographs*). Also, it is interesting to note that the transfer function Fxy(τ) can be interpreted like the "slope" of a generalized linear regression between the two signals X(t) and Y(t). The identification of this transfer function in time is an inverse problem; it is also a particular type of optimal estimation problem, also known as the "deconvolution problem".

### 3.3 Spectral and cross-spectral analyses, frequency gain

Fourier spectral analysis decomposes the signal (or its structure function) into periodic functions, expressed in the frequency domain using the Fourier Transform: e.g. the Fast Fourier Transform (FFT) implemented in MATLAB and many other software's.

More precisely, univariate spectral analysis consists in studying the frequency spectrum of only one signal (its auto-spectrum $S_{XX}(f)$), while bivariate or cross-spectral analysis, treats simultaneously two signals, leading to the frequency cross-spectrum $S_{XY}(f)$ of the two signals.

Other related quantities are the co-spectrum and the phase spectrum (*not used here*), and the frequency gain (*used below*). The frequency gain can be interpreted as a spectral transfer function, particularly in the case where one of the two signals (X(t)) is the "cause" of the other signal (Y(t)). This frequency gain is (in a sense) the spectral version of the temporal transfer function between the "input" X(t) and the "output" Y(t).

The spectrum Sxx(f) of a signal X(t) can be defined and estimated in various ways (continuous/discrete time, infinite/finite time window, with/without filtering). Below, the spectrum of X(t) is expressed as a function of frequency, for the case of a discrete time process, within a limited time window, and with filtering…

### 3.3.1 Estimation of the Fourier spectrum (Wiener-Kinchine method, filtered)

The Fourier auto-spectrum of signal X(t) is estimated here as the Fourier transform of the correlation function Rxx(τ), based on the *Wiener-Khinchine theorem*, valid for stationary random processes. The final expression proposed here includes, also, a time-lag filter "D(τ)".

The dimensional estimated spectrum S(f) is then of the form:

$$S_{XX}(f_i) = 2\Delta t\left[C_{XX}(0) + 2\sum_{j=1}^{M} D_j.C_{XX}(j).\cos\left(2\pi j\Delta t\, f_i\right)\right] \quad (2a)$$

and the reduced dimensionless spectrum s(f) is of the form:

$$s_{XX}(\hat{f}_i) = 2\left[1 + 2\sum_{j=1}^{M} D_j.R_{XX}(j).\cos\left(2\pi j\, \hat{f}_i\right)\right] \quad (2b)$$

where:

$$R_{XX}(j) = C_{XX}(j)/\sigma_{XX}^2\ , (R_{XX}(0) = 1)\,. \quad (3a)$$

$C_{XX}(\tau)$ is the auto-covariance function and $R_{XX}(\tau)$ is the auto-correlation function, or normalized covariance function. Similarly, $S_{XX}(f)$ is the auto-spectrum, and "$s_{XX}$" is the reduced dimensionless spectrum (the reduced spectrum is normalized by the variance of the signal and expressed in terms of dimensionless frequency $\hat{f}_i$). Furthermore, in order to obtain a frequency dependent "measure" of fluctuations analogous to the notion of standard deviation, we define the "Root Mean Square spectrum" as follows (it has the same units as the signal X(t) and $\sigma_X$):

$$S_{XX}^{RMS}(f_i) = 2\sqrt{S_{XX}(f_i)\Delta f}\ . \quad (3b)$$

The frequencies and dimensionless frequencies are defined as follows:

$f_i$ : frequency (Herz) (1 Hz=1 s-1); $T_i = 1/f_i$ : period (seconds)

Let M be either the number of time steps in the data, or the maximum number of lags allowed in the Tuckey filter D(j) (*see next section* 3.3.2). If the Tuckey filter is turned off, one can let M = N-1 where N is the total number of points in the observed signal. Then, we have:

$$f_i = \frac{1}{2M\Delta t_i} : \text{period (seconds)}$$

According to Shannon's sampling theorem (see ***Figure*** 5), here are the maximum and minimum frequencies that can be "identified" given the time step and the duration of the signal[5]:

$$f_{max} = \frac{1}{2\Delta t} = \frac{1}{T_{min}} \text{ [Hz]} \tag{4a}$$

$$f_{min} = \frac{1}{T_{max}} = \frac{1}{2t_{max}} = \frac{1}{2M\Delta t} \text{ [Hz]}, \tag{4b}$$

where $t_{max} = M\Delta t$ (size of the time window). In these relations, Tmin and Tmax are, respectively, the smallest and largest periods that can be represented with this time series ($T_{max} \geq 2T_{min}$). The dimensionless frequency $\hat{f}_i$ is defined by:

$$\hat{f}_i = f_i \times \Delta t \;\Rightarrow\; \hat{f}_i = \frac{i}{2 \times M \times k_0}; \tag{4c}$$

where ko is the sub-sampling step in time (we always use full sampling with ko = 1 in this work).

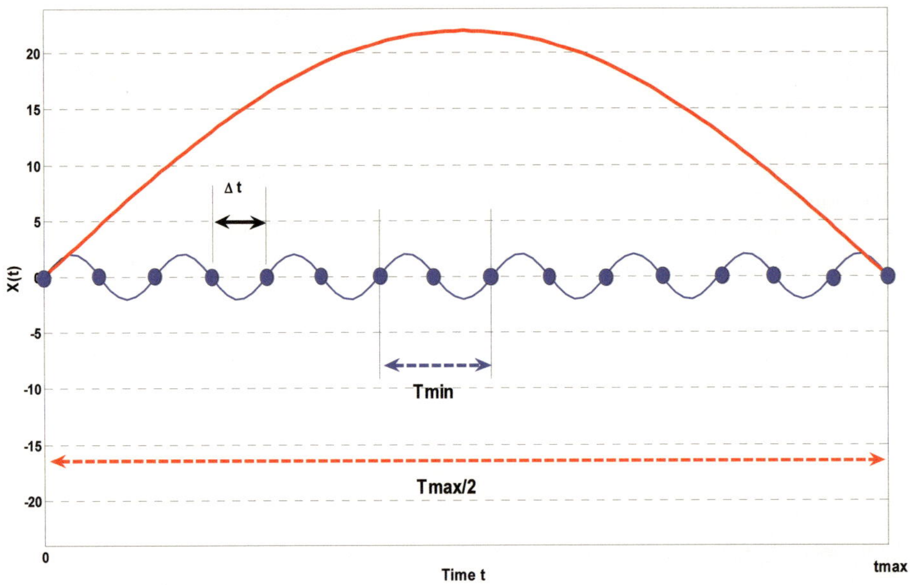


Fig. 5. Schematic illustration of some aspects of signal sampling in discrete time within a limited observation window (which may be due to truncation of a longer time series). $\Delta t$ : time step (seconds); $t_{max} = M\Delta t$ : size of observation window. $T_{min}$ : smallest period that can be identified (Shannon). $T_{max}$ : largest period that can be identified. $f_{max} = 1/T_{min}$ is Shannon's maximum frequency (sampling theorem).

[5] Note: Shannon's sampling theorem concerns in fact only the maximum frequency; but nevertheless, there is also a minimum frequency theorem, as shown below.

The dimensionless frequency $\hat{f}_i$ belongs to the interval [0, 1/2], that is: $0 \le \hat{f}_i \le 0.5$. Thus, for $k_0 = 1$, we have:

$$\hat{f}_0 = 0 \text{ and } \hat{f}_M = 0.5 \tag{4d}$$

The index "j" in the Fourier spectrum/covariance ***Eq. 2*** is the label of the *time lag*, defined as follows:

$$j = 0, 1, 2, \ldots, M \le N\text{-}1$$

where the zero time lag (j = 0) is taken into account separately outside the sum in ***Eq. 2***. On the other hand, the index "i" is the label of discrete frequency, or of dimensionless frequency ($\hat{f}_i$):

$$i = 0, 1, \ldots, M \le N\text{-}1 \text{ or (see the note just below): } i = 0, 1, \ldots, M, \ldots, N\text{-}1.$$

The number "M" represents the discrete size of the filter truncation window (number of Δt's in the maximum lag considered), while "N" is the total number of data of the observed signal (full time window), and Δt is the constant time step of measurements (possibly after homogenization of time steps). In general, for a filtered spectrum, M ≪ N (at most, M=N-1 for the unfiltered spectrum).

*Note.* Concerning the discrete number of frequencies represented in the filtered spectrum (index "i"), it is possible to impose either a "coarsened" representation of the spectrum with frequencies running from i=1 to i=M (where usually M≪N), or alternatively, to maintain a fine representation of the spectrum with frequencies running from i = 0, 1, …N-1, regardless of the maximum time lag "M" used in the Tuckey filter "D". In the latter case, one should replace "M" by "N-1" in all of ***Eqs. 4*** (index "i").

### 3.3.2 The Tuckey filter ("hanning" filter) for spectral estimation

Finally, note that "s(f)" in ***Eq. 2*** is an estimator of the spectrum, with a filter function "D". More precisely, "Dj" represents a time-lag filter function; it is applied to the correlation function within the Fourier transform. The weight function D(j) included inside the Fourier transform represents a "smoothing function" or "filter" of Rxx(j) in time lag space (j), where "j" is the discrete index of the dimensional lag time τ(j). Thus, we can write equivalently D(j) or D(τ(j)).

Our particular choice here the "Tukey-2" filter (also called a "*hanning*" filter):

$$D_j = \frac{\left(1 + \cos\left(\frac{\pi j}{M}\right)\right)}{2} \tag{5}$$

where the index "j" runs generally over (j = -M,…,-1,0,+1,…,+M), or here over (j = +1,…,+M), in the particular discrete Fourier Transform sum written in ***Eq. 2***.

The filter D(j) is necessary in most cases. Indeed, it is necessary to apply smoothing to Rxx(τ) or Rxx(j) because of the sampling errors on the estimated Rxx(j), which is obtained on a finite time window. Thus, it can be shown precisely that the estimation error on the covariance Cxx(j) increases with time lag $\tau(j)$: see *Yevjevich (1972, Chap.3)* and *Bartlett (1946)*. For example, in the case of an autoregressive process with exponential covariance function $C_{XX}(\tau) = \sigma_X^2 \exp(-|\tau|/\tau_{COR})$, it has been shown (*Ababou et al. 1988: Appendix 2*) that the variance of sampling error, defined by $\varepsilon_{RMS}^2(\tau) = Var\left(\hat{C}_{XX}(\tau)\right)/C_{XX}^2(\tau)$, is given by:

$$\varepsilon_{RMS}^2(\tau) = 2\frac{\tau_{COR}}{t_{MAX}}\left(1 - \frac{1}{2}\frac{\tau}{t_{MAX}} - \frac{1}{4}\frac{\tau_{COR}}{t_{MAX}} + \frac{\tau}{\tau_{COR}}\left(1 - \frac{1}{2}\frac{\tau}{t_{MAX}}\right)\right). \tag{6}$$

in terms of continuous lag time ($\tau$). The two parameters involved in this analytical example are the signal's correlation time ($\tau_{COR}$) and the size of the observation window ($t_{MAX}$). As can be seen, the relative RMS error $\varepsilon_{RMS}$ is proportional to $\sqrt{\tau_{COR}/t_{MAX}}$ at zero lag ($\tau$ = 0), and $\varepsilon_{RMS}$ goes to 100% as $\tau \to t_{MAX}$. This illustrates why it is necessary to apply a filter like $D(\tau(j))$ in order to smooth out the estimated $Cxx(\tau(j))$ or $Rxx(\tau(j))$ at large time lags, and also to smooth out the resulting frequency spectra obtained by Fourier transform of $Cxx(\tau(j))$ or $Rxx(\tau(j))$.

The filter we have chosen here ("Tukey-2", or "*hanning*" filter) adequately smoothes out large time scales (or noise at the scale of low frequency increments in the frequency domain). Experience with hydrological signals indicates that this filter has less variance bias than others (according to *Mangin 1984*, it over-estimates the total variance by no more than 8%). However, other filters may be considered in future: in particular, it should be noted that a "minimum bias" spectral filter has been proposed by *Papoulis & Pillai (2002)*.

Finally, note that the Fourier transform of our time-lag filter $D(\tau(j))$ is itself a "filter" in frequency space (D(f)). Its spectral band should contain a large part of the spectral band of the signal, in order not to waste too much information in this filtering procedure when estimating the spectrum. In our applications, the maximum lag "M" in the $D(\tau(j))$ filter (***Eq.*** 5) was adjusted depending on each specific application and objective (time scales and frequencies of interest).

### 3.3.3 Cross-spectrum, frequency gain, and coherence function

The (dimensional) cross-spectrum of two signals (X(t), Y(t)) is a complex function of frequency which can be described as follows:

$$S_{XY}(f_i) = |S_{XY}(f_i)| e^{\varphi_{XY}(f_i)} \text{ or } S_{XY}(f_i) = S_{\mathrm{Re}}(f_i) + S_{\mathrm{Im}}(f_i) \tag{7a}$$

where

$$|S_{XY}(f_i)| = \sqrt{S_{\mathrm{Re}}^2(f_i) + S_{\mathrm{Im}}^2(f_i)} \tag{7b}$$

and

$$\varphi_{XY}(f_i) = \tan^{-1}\left( S_{\mathrm{Im}}(f_i) \Big/ S_{\mathrm{Re}}(f_i) \right), \tag{7c}$$

are respectively the amplitude (cross)-spectrum and the phase (cross)-spectrum. The latter is defined only in an interval of size π, such as [-π/2,+π/2]. More precisely, however, the phase spectrum can be identified unambiguously within the full interval [-π, +π] with the equation:

$$\varphi_{XY}(f_i) = sign\left(S_{\mathrm{Im}}(f_i)\right)\cos^{-1}\left( S_R(f_i) \Big/ \left|S_{XY}(f_i)\right| \right), \tag{7d}$$

as will be demonstrated shortly with a simple test.

Given the complex cross-spectrum Sxy(f), one can also define its real part (or "co-spectrum"), and its imaginary part (or "quadrature spectrum"). These are defined and estimated using the Wiener-Khinchine representation (similar to that used in ***Sec. 3.3.1*** for the auto-spectrum). The result is expressed below in terms of sine and cosine transforms (where a Tuckey filter "Dj" is inserted):

$$S_{\mathrm{Re}}(f_i) = 2\Delta t\,\sigma_X\sigma_Y\left[ R_{XY}(0) + \sum_{j=1}^{M}\left(R_{XY}(j) + R_{YX}(j)\right)D_j\cos(2\pi j\Delta t\, f_i\,)\right] \tag{8a}$$

$$S_{\mathrm{Im}}(f_i) = 2\Delta t\,\sigma_X\sigma_Y\left[ \sum_{j=1}^{M}\left(R_{XY}(j) - R_{YX}(j)\right)D_j\sin(2\pi j\Delta t\, f_i)\right] \tag{8b}$$

It is useful to define also the dimensional Root Mean Square cross amplitude spectrum, as follows (it is important to note that this RMS cross-spectrum has the units of $\sigma_X \times \sigma_Y$):

$$\left|S_{XY}^{RMS}(f_i)\right| = 2\sqrt{\left|S_{XY}(f_i)\right|\Delta f} \tag{9}$$

Finally, we define a spectral coherence function:

$$Coh_{XY}^{RMS}(f_i) = \frac{\left|S_{XY}^{RMS}(f_i)\right|}{\sqrt{S_{XX}^{RMS}(f_i)\,S_{YY}^{RMS}(f_i)}} \tag{10}$$

and a spectral gain, or frequency gain function:

$$G_{XY}^{RMS}(f_i) = \frac{\left|S_{XY}^{RMS}(f_i)\right|}{S_{XX}^{RMS}(f_i)} \tag{11}$$

In the latter equation defining the "gain", it should be noted that the two signals are treated differently: X(t) is considered the input signal, and Y(t) the output signal. Thus, Gxy(f) is the input→output spectral gain. It quantifies, for each frequency "f", the effect G of the signal X(t) on the output signal Y(t). This effect is called the gain.

### 3.3.4 Synthetic example of cross-spectral analysis on two harmonic signals

Two "synthetic" harmonic signals are used in order to test the auto- and cross-spectral estimates described above (spectrum of each signal, then cross amplitude spectrum, phase spectrum, frequency gain).[6] The two signals, input $X(t)=X_1(t)$ and output $Y(t)=X_2(t)$, are defined as follow:

$$X_1(t) = A_0 + \sum_{k=1}^{4} A_k \cos(\omega_k t + \phi_k)\,;\; X_2(t) = B_0 + \sum_{k=1}^{4} B_k \cos(\omega_k t + \gamma_k) \qquad (12a)$$

Both signals have four harmonic components, with the same four frequencies $\omega_i = 2\pi/T_i$ ($i = 1,2,3,4$) with periods $T_i$ = 6 months, 8 days, 24 hours and 12 hours, respectively. In addition, each signal has its own constant mean ($A_0$ or $B_0$). The mean may be viewed as the zero frequency harmonic ($i = 0$). The four amplitudes and phases are different for the two signals. Here is a summary or the values used in this test:

- **Signal $X_1(t)$:** Mean $A_0 = 50$ [*units of $X_1$*]; amplitudes $(A_i) = (12, 9, 4, 1)$ [*units of $X_1$*]; phases $(\varphi_i) = (\pi/4, 0, \pi/8, \pi/6)$ [*rad*].
- **Signal $X_2(t)$:** Mean $B_0 = 50$ [*units of $X_2$*]; amplitudes $(B_i) = (1, 4, 9, 12)$ [*units of $X_2$*]; phases $(\gamma) = (0, 0,0, 0)$ [*rad*].

In the test presented below, the two synthetic signals are generated during a time span of 1 year (in comparison, the longest period in the signal is 6 months). The time step is $\Delta t = 900$ s = 15 mn (in comparison, the shortest period in the signal is 12 hours). Spectral estimations are carried out "unfiltered" (the Tukey filter "Dj" is not applied: we take M = N-1 as the maximum time lag).

The spectral estimation results, shown in ***Figure 5 & Figure 6***, are exactly as expected: they coïncide very closely with the exact theoretical values that we have inferred directly from cross-spectral analysis of these harmonic signals (assuming infinite record $t_{MAX}\to\infty$, and continuous time $\Delta t\to 0$). For reference, here are some theoretical results for each discrete frequency $f(i)=\omega_i/2\pi$:

$$\begin{aligned} &Coh_{XY}^{RMS}(f_k) = 1;\; \left|S_{XY}^{RMS}\right|(f_k) = \sqrt{A_k}\sqrt{B_k} \\ &G_{XY}^{RMS}(f_k) = \sqrt{B_k / A_k}\,;\; \varphi_{XY}(f_k) = \phi_k - \gamma_l \end{aligned} \qquad (12b)$$

For instance, focusing on the frequency of the 24 *hour* harmonic, we have the following results (we use *unbiased* covariances, except for phase & gain spectra which are estimated using *biased* covariances):

- the RMS auto-spectrum of $X_1(t)$ should be exactly 4.00 [*units of $X_1$*], and we find 3.996 from the estimated spectrum (***Fig.7.a***);
- the RMS auto-spectrum of $X_2(t)$ should be exactly 9.00 [*units of $X_2$*], and we find 8.999 for the estimated spectrum (***Fig.7.b***);

[6] Note: this example was developed recently, based on similar tests developed earlier by the authors and collaborators (see acknowledgments).

- the RMS coherence spectrum [*dimensionless*] should be exactly 1 for each of the four harmonics,
  and we do find that result for the estimated coherence spectrum (***Fig.7.c***);
- the RMS amplitude cross-spectrum should be exactly 6.00 [*units of* $\sqrt{X_1}\sqrt{X_2}$], and we find 5.997 for the estimated cross spectrum amplitude (***Fig.7.d***);
- the phase spectrum should be exactly $\pi/8$ [*rad*], and we find 0.391 [*rad*] for the estimated phase (***Fig.6.a***);
- the spectral gain between $X_1(t)$ and $X_2(t)$ should be exactly $\sqrt{9/4} = 3/2$ [*units of* $\sqrt{X_2}/\sqrt{X_1}$], and we find 1.50, the exact result with two significant digits (***Fig.6.b***).

As a final note on spectral estimation, it should be emphasized that the Wiener-Kinchine method used in this work relies on the Fourier transform of auto- and cross-correlation functions, which are themselves estimated in the time domain assuming that the processes (X(t), Y(t)) are jointly statistically stationary. Alternatively, however, the power spectrum Sxx(ω) of a signal X(t) could be estimated from the "*periodogram*" Pxx(ω), which is the direct Fourier transform of the signal $X(t_i)$ itself (over a finite time window, in discrete time). Filtering or smoothing will be again necessary, as in the Wiener-Khinchine method. The difference is that the required smoothing can now be performed in frequency space (ω). For instance, the periodogram can be smoothed out using a moving average filter in frequency space, to finally obtain a filtered estimate of Sxx(ω). For more details on this, the reader may consult *Blackman and Tuckey (1958), Yevjevich (1972, Chap.3), Papoulis and Pillai (2002, Chap.12)*, and *Priestley (1981).*

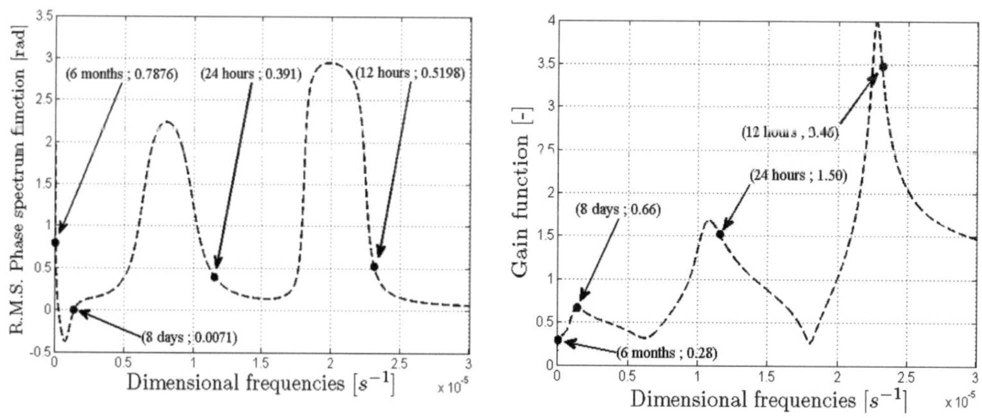


(a) (b)

Fig. 6. Synthetic test with 4 harmonics: **(a)** the RMS phase spectrum function [phase difference between $X_1(t)$ and $X_2(t)$, in *radians*]; **(b)** the RMS spectral gain function [gain between $X_1(t)$ and $X_2(t)$, in *units of* $\sqrt{X_2}/\sqrt{X_1}$].

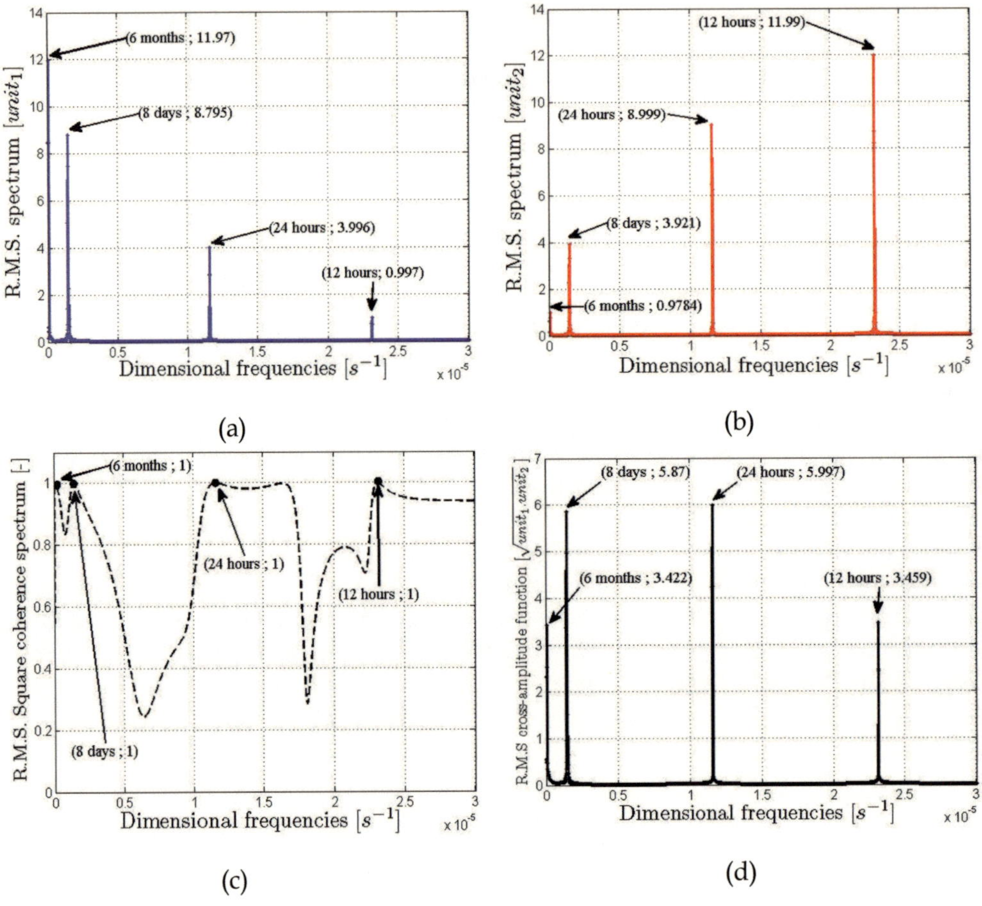


Fig. 7. **(a)** Estimated RMS spectrum of $X_1$; **(b)** Estimated RMS spectrum of $X_2$; **(c)** Estimated RMS square coherence spectrum; **(d)** Estimated RMS cross-amplitude spectrum.

### 3.4 Multiresolution wavelet analysis

#### 3.4.1 Introduction on wavelets

The analysis of a signal in terms of multi-resolution wavelets uses an orthogonal basis $\left\{\psi_{j,k}(t)\right\}$ where "k" represents the temporal translation (position), and "j" represents the dilation or the contraction of time scale, based on a dyadic time discretization of the signal X(t).

The basis being orthogonal, the components of different "scales" are independent from each other. Moreover, each component at a given scale can be traced according to the "time" parameter or "temporal position". This is to be compared with the traditional Fourier analysis, where each Fourier coefficient C(f) represents the overall contents of the signal at frequency « f ». In other words, Fourier coefficients C(f) encapsulate the "f" frequency content of the signal X(t) for all times "t", without representing the possible temporal localization of the frequency content.

On the other hand, the orthogonal wavelet approach makes it possible to clearly highlight non-stationary behaviour and time-localization phenomena (*Labat et al. 2000b*). It also makes it possible to follow the relationship between the various scale levels as a function of time, and to isolate each independent component to study its own evolution. In this sense, these analyses can lead to a better identification of the process responsible for the observed variations. For an extension to wavelet multivariate analyses, or cross-wavelet analyses, see *Labat et al. (2002)*.

We choose here to conduct multi-resolution analysis with discrete orthogonal wavelets, using the wavelet bases of Daubechies. The software employed in this work combines a set of wavelet functions programmed in MATLAB and available on line at the University of Stanford (WAVELAB), and other functions contained in MATLAB's own WAVELET TOOLBOX (some of which are used by Stanford's WAVELAB package).

### 3.4.2 Wavelet decomposition of a signal on an orthogonal multiresolution basis

In this subsection, we present the decomposition of a signal on a wavelet basis, first for continuous wavelets (non orthogonal bases), and secondly for dyadic multi-resolution wavelets. In the remainder of this text, we will focus essentially on the latter wavelets, for which a compactly supported orthogonal wavelet basis can be constructed (Daubechies basis). The term "dyadic" refers to the fact the scales of multiresolution wavelets vary discretely as powers of "2". For example, if the length of the recorded signal is $t_{MAX} = 64\Delta t$, then there are exactly 7 wavelet scales $2^j\Delta t$ (j=0,1,…,6); the shortest scale is $2^0\Delta t = \Delta t$, the longest scale is $2^6\Delta t = 64\Delta t = t_{MAX}$. The signal can be decomposed in terms of these dyadic time scales. We will define later (*in sub-section* **3.4.3**) the concepts of wavelet "detail", "approximation" and "residual", at any given dyadic scale.

### 3.4.3 Continuous wavelets (a brief introduction)

A signal X(t) is represented formally via wavelet coefficients $C^X_{s,\tau}$ in continuous time as follows:

$$X(t) = \frac{1}{K_\psi} \int_{\sigma=0}^{\sigma=+\infty} \frac{1}{\sigma^2} \left( \int_{\tau=-\infty}^{\tau=+\infty} C^X_{\sigma,\tau}\, \psi_{\sigma,\tau}(t) d\tau \right) d\sigma \qquad (13a)$$

$$\psi_{\sigma,\tau}(t) = \frac{1}{\sqrt{\sigma}} \psi\left(\frac{t-\tau}{\sigma}\right) \; ; t\in \mathbb{R};\ \tau\in \mathbb{R};\ \sigma\in \mathbb{R}\text{-}\{0 \qquad (13b)$$

$$C^X_{\sigma,\tau} = \int_{-\infty}^{+\infty} X(t)\psi^*_{\sigma,\tau}(t)dt \equiv \left\langle X(t), \psi^*_{\sigma,\tau}(t) \right\rangle \qquad (13c)$$

where "t" is time, "τ" is a time-shift parameter, and "σ" is a scale parameter (a time scale parameter). It is assumed, for simplicity, that the signal is known for $t\in[-\infty,+\infty]$. ***Eq. 13a*** gives the decomposition of signal X(t). ***Eq. 13b*** expresses the (dilated) wavelet function at a given scale σ. ***Eq. 13c*** gives the wavelet coefficients of the signal: they are the scalar products of X(t) with the (conjugate) dilated and translated wavelet functions $\psi_{\sigma,\tau}{}^*(t)$.

A discrete time analogue of these equations can be constructed as follows: ***(a)*** replace "t" with t(i) = i×Δt (i = …,-2,-1,0,+1,+2,…) where "i" is the time index; ***(b)*** replace "τ" with τ(k)= k×Δt (k =…,-2,-1,0,+1,+2,…) where "k" is the time-shift index; and ***(c)*** replace "σ" with σ(s) = s×Δt (s = 0,1,2,3…) where "s" is the scale index (*discrete but not dyadic -- see further below*). Similarly, a finite time window analogue can also be constructed. Thus, one can deal with the practical case of discrete time step Δt and finite window [$-T_{MAX}/2$, $+T_{MAX}/2$] or [0, $T_{MAX}$].

Different wavelet bases can be obtained depending on the choice of the "mother wavelet" ψ(t). However, in the framework of "continuous" wavelets, it is not possible to find a set of wavelet functions $\psi_{s,k}(t)$ that form an orthogonal basis. For example, the *Morlet* wavelet basis is not orthogonal, implying that there are redundancies in the *Morlet* wavelet decomposition.

### 3.4.4 Multi-resolution wavelets and dyadic time scales

In contrast, with dyadic multi-resolution wavelets, an orthogonal wavelet basis with compact support can be constructed (*Daubechies 1988, 1991*), leading to multi-resolution wavelet analysis (*below*)[7]. This is made possible by the use of dyadic scales of the form $\sigma(j) = 2^j \times \Delta t$ *(powers of 2)*. Thus, in what follows, let "i" be the time index; "j" the dyadic scale exponent (power); and "k" the time-shift index. The decomposition of X(t) on the multi-resolution wavelet basis takes the form:

$$X(i) = \sum_{j=0}^{j=+\infty} \sum_{k=-\infty}^{k=+\infty} C_{j,k}^{X} \psi_{j,k}(i) \quad \text{(where "i" can be replaced by "}t_i\text{")} \tag{14a}$$

$$\psi_{j,k}(t_i) = 2^{j/2} \psi(2^j \times t_i - \tau_k) \tag{14b}$$

$$C_{j,k}^{X} = \sum_{i=-\infty}^{i=+\infty} X(t_i) \psi_{j,k}^{*}(t_i) \equiv \left\langle X(t_i), \psi_{j,k}^{*}(t_i) \right\rangle \tag{14c}$$

where:

$j \geq 0$ : exponent of dyadic scale dilatation "$2^j$×•" (or compression if $j \leq 0$);
k: discrete time shift ↔ τ(k)= k×Δt; i: discrete time ↔ t(i)= i×Δt;
$i \in \mathbb{Z}$; $k \in \mathbb{Z}$; $j \in \mathbb{N}$ (the latter is the discrete dyadic scale)

The $\psi_{j,k}(t)$ basis functions form an orthonormal basis, with their images obtained by time translations and scale dilations/compressions of the mother wavelet ψ(t) [8]. Thus, we have:

---

[7] Two functions f(t) and g(t) are "orthogonal" if their scalar product <f(t),g(t)> is null, where we define <f(t),g(t)> as the integral of f(t)×g(t) over the time domain. A function f(t) has compact support if it is null outside a finite interval [a,b].

[8] In the literature, the decomposition of X(t) is formulated in such a way that only dilatations (or "dilations") are retained, starting from the smallest scale Δt at j = 0, and increasing "j"; but it is possible to formulate this differently in terms of both dilatations and compressions.

$$\int_{-\infty}^{+\infty} \psi_{m,n}(t)\psi_{m',n'}(t)\,dt = \delta_{m,m'}\,\delta_{n,n'} \tag{15}$$

where δ is the classical Kronecker symbol defined by $\delta_{ij}=1$ if i=j and $\delta_{ij}=0$ if i≠j. ***Eq. 14a*** shows that the discrete time signal X(i) or X(t(i)), of finite energy, can be broken up into a linear combination of translations and dyadic dilatations of the basic functions with adequate coefficients C(j,k) given by ***Eq. 14c***. Note that ***Eq. 14a*** can be viewed as the synthesis of a signal based on wavelet coefficients C(j,k). It can also be interpreted as a succession of approximations of the signal X(i) (in the least-squares sense) by a sequence {Xn(i): n=1,.....N} defined by:

$$X_n(i) = \sum_{j=0}^{j=n-1} \sum_{k=-\infty}^{k=+\infty} C_{j,k}^{X} \psi_{j,k}(i) \;\; \left(X_n(i) \to X(i) \text{ as } n \to \infty\right). \tag{16}$$

The above equations constitute the conceptual basis of multi-resolution wavelet analysis. For more details on the theory, see *Mallat (1989)*. Below, we focus on some useful consequences of the orthonormal wavelet decomposition above, introducing the wavelet approximation + residual, and interpreting the wavelet "approximation" as a filter of given dyadic scale.

### 3.4.5 Multiresolution wavelet decomposition: approximation A(t), residual R(t), and detail D(t) or component C(t)

In this subsection, we define the important concept of wavelet "detail" or "component", i.e., and we show how to extract from the signal X(t) a wavelet "component" with any given dyadic scale. It is important to note that only dyadic scales can be extracted; however, this is the price to pay for the other "good" properties (compact support and orthogonality of the decomposition). We also present the decomposition of signal X(t) into an "approximation" A(t) plus a "residual" R(t), both defined at any given cut-off scale (again, the cut-off scale can only be dyadic). Remarkably, the wavelet "approximation" can be viewed as a "filtered" version of the original signal: it contains only longer time scale components (longer than the given dyadic cut-off scale).

### 3.4.6 Orthogonal decomposition of X(t) - wavelet components, approximation, residual

Let us assume for simplicity that the (finite) discrete signal $X(t_i)$ has a dyadic length ($N=2^M$), the number of time steps is a power of 2 (dyadic). The multi-resolution wavelet decomposition above can be expressed as a succession of nested "*approximations*" $A_m$, corresponding to increasing *scales* "*m*" (thus, "m" can designate a chosen dyadic time scale cut-off).

The difference between the signal $X(t_i)$ and its approximation of order "m" is the "*residual*". The difference between the approximations of order (m) and (m+1) is called the "*detail*" of order (m+1).

Intuitively, the *approximation* corresponds to the smoothed image of the signal, while the "*detail*" is a particular *wavelet component* that highlights the irregularities of the signal at a certain scale.

Starting from ***Eq. 14a***, we obtain at each selected dyadic scale "m" the "wavelet detail" $D_m(t_i)$:

$$D_m^X(t_i) = \sum_{k=-\infty}^{k=+\infty} C_{m,k}^X \psi_{m,k}(t_i) \Leftrightarrow D_X^m(i) = \sum_{k=-\infty}^{+\infty} \langle X, \psi_{m,k} \rangle . \psi_{m,k}(i) \qquad (17a)$$

where "t(i)" can also be replaced by "i". The approximation $A_m$ of the signal $X(t_i)$ at dyadic scale resolution "m" (cut-off scale "m"), and the residual $R_m$ at the same scale "m", are then given by:

$$R_m^X(t_i) = \sum_{j \le m-1} D_j^X(t_i);\ A_m^X(t_i) = X(t_i) - R_m^X(t_i) \Rightarrow A_m^X(t_i) = \sum_{j \ge m} D_j^X(t_i) \qquad (17b)$$

Thus, at any dyadic scale "m", the signal is the sum of its approximation and residual:

$$X(t_i) = A_m^X(t_i) + R_m^X(t_i) \qquad (17c)$$

Using the latter equations to calculate the approximation of order (m-1) leads to:

$$A_{m-1}^X(t_i) = A_m^X(t_i) + D_{m-1}^X(t_i) \Leftrightarrow A_{m+1}^X(t_i) = A_m^X(t_i) - D_m^X(t_i) \qquad (17d)$$

As "m" increases, the $A_m$'s become coarser and coarser approximations. The above equation shows that the "coarser" approximation $A_{m+1}$ is obtained from $A_m$ by subtracting the detail $D_m$. Note that the zero order residual $R_0$ is null by construction. The zero order approximation $A_0$ is identical with the original signal. Thus, we have, with our index conventions:

$$R_0^X(t_i) = 0;\ R_1^X(t_i) = D_0^X(t_i);\ ...;\ R_M^X(t_i) = X(t_i) - D_M^X(t_i);\ R_{M+1}^X(t_i) = X(t_i) \qquad (17e)$$

$$A_0^X(t_i) = X(t_i);\ A_1^X(t_i) = X(t_i) - D_0^X(t_i);...;\ A_M^X(t_i) = D_M^X(t_i) \qquad (17f)$$

where M is the exponent of the largest dyadic scale. For example, M = 6 if the signal length N is in the range $64 \le N \le 127$. The dyadic wavelet decomposition is performed without loss of information if N is a power of "2", such as N = 64; but if N = 127 the loss is maximal.

### 3.4.7 Theoretical interpretation of multiresolution wavelets (nested spaces and filters)

For completeness, note that a scale function $\varphi(t)$ is also introduced. It can be shown that the "approximation" can be directly decomposed orthogonally over the set of dilated/translated scale functions $\varphi_{j,k}(t)$ (*not shown here*). The scale function $\varphi(t)$ is also called the *father wavelet,* to be distinguished from the *mother wavelet* $\psi(t)$ used above in the decomposition of ***Eq. 17a***.

Theoretically, approximations $A_m$ and details $D_m$ are each decomposed into bases of $\varphi$'s and $\psi$'s, respectively, which form nested spaces called $V_m$ and $W_m$, respectively. Considering a given scale "m", and recalling that $m \ge 1$ ("m" increases for increasing dyadic scale $2^m \times \Delta t$), it can be seen that the "detail" $D_m$ represents the wavelet component of scale "(m)", the residual $R_m$ incorporates the sum of all scales strictly smaller than "m", and the approximation $A_m$ incorporates all scales greater or equal to "m" (*note: the approximation "m" includes the scale "m" itself*).

The nested spaces V and W are related by $V_{m+1} = V_m - W_m$, as indicated by ***Eq. 17d***. In fact, ***Eq. 17d*** describes a sequence of multi-resolution filters (the $A_m$'s) and their residuals (the $R_m$'s).

### 3.4.8 Remarks on multi-resolution approximation as a filter

The process of selecting a given wavelet scale "m" is similar to passing a "band pass" filter on the signal. The wavelet "approximation" $A_m$ can be viewed as a "low pass filter" of given dyadic scale. This filter is somewhat similar to a moving average filtering of the original signal. It was shown empirically in some cases that the two filters are quite close. See for instance *Wang et al. (2010)* concerning water level signals in coastal & beach hydrodynamics. The advantage of the wavelet approximation as a "filter" is that it has the property of being orthogonal to its residual.

### 3.4.9 Final remarks on wavelets notations and terminology (pitfalls)

The presentation above is certainly too brief to be complete. Therefore, let us emphasize a few more points in order to avoid any possible confusion with multiresolution wavelets.

i. The wavelet "detail" $D_m(t_i)$ defined above can also be seen as a wavelet "component", to be denoted $C_m(t_i)$: this notation designates the wavelet component (C) at the $m^{th}$-dyadic scale (m). The sum of all components (over all scales "m") yields the signal itself. ***Note***: from now on, we will prefer to use this notation, component $C_m(t_i)$, rather than $D_m(t_i)$.
ii. For each given scale "m", the wavelet "component" $C_m(t_i)$ is a discrete time signal. This signal (component C(t)) *should not be confused* with the "wavelet coefficients $C_{jk}$", defined earlier, where "k" designates a time shift parameter. The "content" of signal $X(t_i)$, at dyadic scale "j", is the sum over all time shifts "k" of the product $C_{jk}\times\psi_{jk}(t_i)$...The coefficient $C_{jk}$ alone is not a time signal.
iii. Multi-resolution wavelets can be presented differently in other texts, with other index conventions and notations. For example, the dyadic time scale labeled "m" can either increase or decrease with index "m" (*here we have chosen that the scale increases with "m"*). Also, some authors choose (*1/scale*) instead of (*scale*) in their visualisations of the wavelet decomposition in scale-time space.

### 3.4.10 Example application: Multiresolution wavelet analysis of a synthetic signal

Given the limitations of Fourier analysis, multiresolution analysis is particularly useful as a tool for analysis, synthesis, and/or compression, of complex *non stationary* signals. For this reason, we present briefly a synthetic example involving a signal that has a simple harmonic component added to a strongly non stationary mean or trend, as follows:

$$X(t_i) = \left| \begin{array}{ll} 1+0.5\sin(2\pi t_i/8) & \text{if } t_i < 15 \\ 1-(t-16)/8+0.5\sin(2\pi t_i/8) & \text{if } t_i \geq 15 \end{array} \right. ; \; t_i = i\Delta t \, ; \; i = (0,1,2,\ldots,31) \, ; \; \Delta t = 1 \quad (18)$$

This signal has an exactly dyadic length N = 32 = $2^5$ (the dyadic exponent is M = 5).

***Figure 8 (a,b,c,d)*** shows the signal X(t) and three different decompositions using the wavelet bases of Haar (Daubechies 2), Daubechies 4, and Daubechies 20. In each case, the six wavelet components $C_i(t)$ are shown (i = 0, 1, ..., 5). It can be seen that component $C_2(t)$ tends to capture the sinusoïd of period T=8. The corresponding dyadic scale is $S_j = 2^j\Delta t = 4$ with j=2. Therefore, we see that the wavelet scale corresponds in this case to the half-period of the signal. On the other hand, the non stationary mean level is captured by component $C_4(t)$, with scale $S_j = 2^j\Delta t = 16$, which corresponds to the half-duration of the non stationary signal. Note also that the components become smoother as the degree of the wavelet basis increases. The Haar wavelet is well suited to capture abrupt changes of mean level, but not for capturing smoother type of non stationarity.

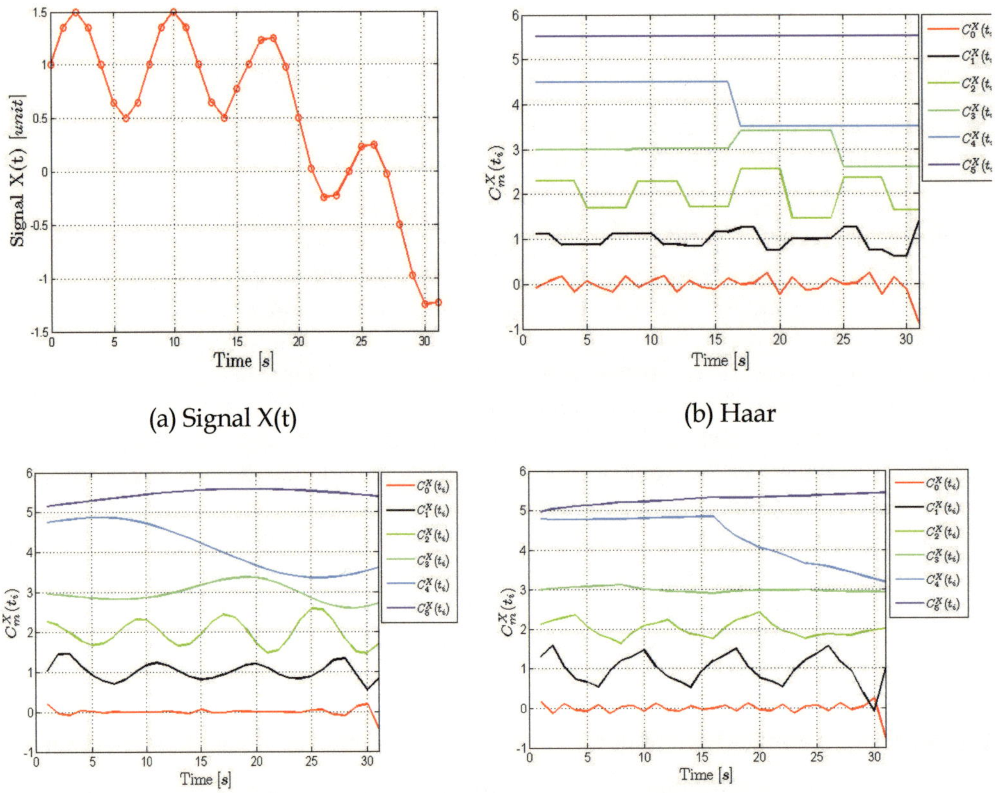


(a) Signal X(t) (b) Haar

(d) Daubechies 20 (c) Daubechies 4

**Note**: in graphs (**b, c, d**), each component Ci(t) is shown on a shifted ordinate axis positioned at y = i (i=0,1,2...); this allows for a better visualization of the different components on a single graph; apart from this shift, the ordinate scales are not distorted (same for all components Ci(t)).

Fig. 8. The signal X(t) and its dyadic components Ci(t) extracted from three different multiresolution bases (Haar, Daubechies 4 and Daubechies 20). **(a):** The synthetic non stationary signal X(t) to be analysed. **(b):** The decomposition of X(t) into components Ci(t) using Haar (Daubechies 2) mother wavelet. **(c)**: The decomposition of X(t) into components Ci(t) using Daubechies 4 mother wavelet. **(d):** The decomposition of X(t) into components Ci(t) using Daubechies 20 mother wavelet.

***Figure 9 (a,b)*** shows a construction of the successive approximations and/or residuals of signal X(t) for the lowest degree wavelet (the Haar wavelet).

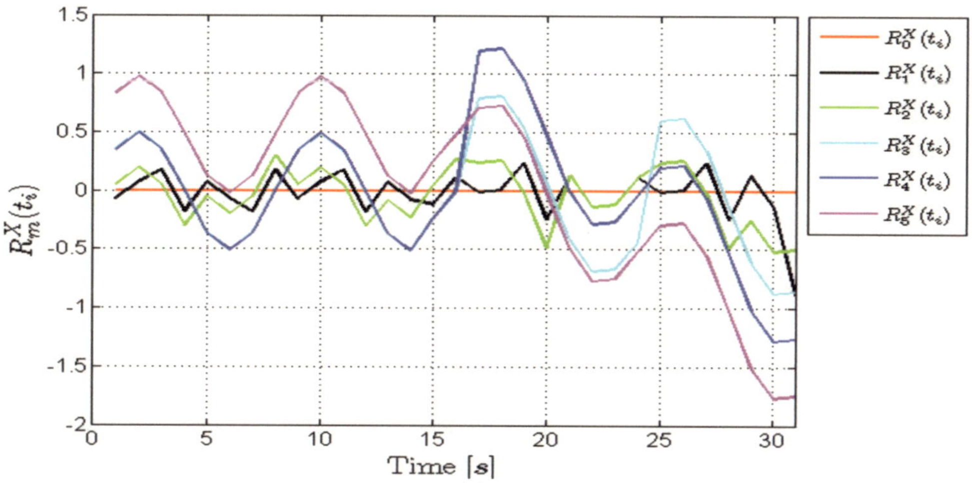


(a)

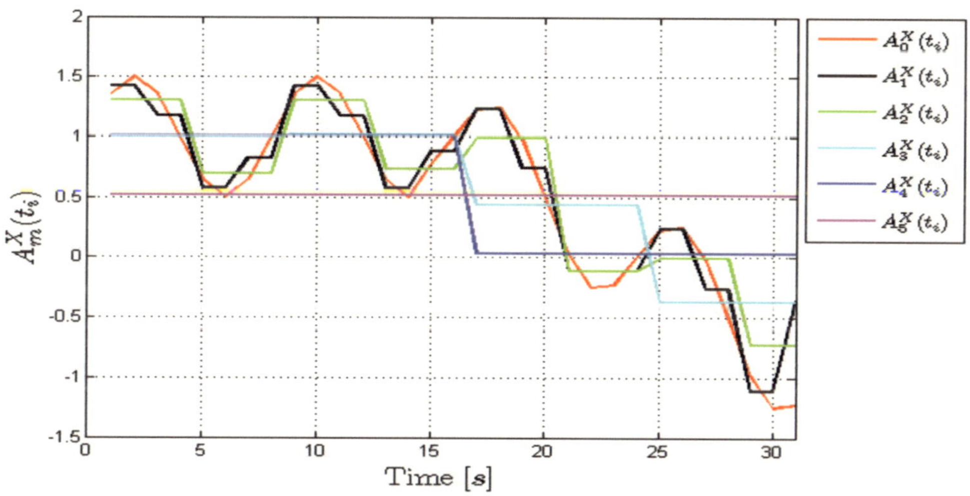


(b)

Fig. 9. Plots of successive residuals and approximations using the Haar mother wavelet (lowest degree wavelet). **(a):** Plot of successive residuals $R_0(t) = 0$, $R_1(t) = C_1(t)$, ..., and $R_5(t) = X(t)-m_X = X(t)-A_5(t)$ (note that R6(t) = X(t), not shown). **(b):** Plot of successive approximations $A_0(t) = X(t)$, $A_1(t) = X(t)-C_1(t)$, ..., and $A_5(t) = X(t)$..

#### 3.4.11 Methods and techniques of wavelet analyses and cross-analyses

The above multi-resolution wavelet theory can be used to study different aspects of the signals, depending on the objectives:

- One can study the "*detail*" or (perhaps more interestingly) the "*residual*" of a signal X(t) at any selected dyadic scale "m". The analysis can be performed for several signals X(t), Y(t). One can compare for instance the details of air and pore pressure at scale 24h.
- One can select dyadic components from one or several signals, and these can be studied and compared as a function of dyadic scale ($2^m * \Delta t$) across the whole spectrum of available scales, from finest scale $\Delta t$ to largest dyadic scale (which is never smaller than half the record length).

Wavelets can also be used in different ways for cross-analyses of two signals like atmospheric and pore pressure. For instance:

- It is possible to develop a somewhat sophisticated approach for orthogonal cross-analyses of signals based on the theory of wavelet spectra and cross-spectra (see for instance *Labat 2005* and references therein), but this theory will not be used here.
- Simpler analyses can be developed by studying the cross-correlation between atmospheric pressure (Patm) and pore pressure (P). The wavelet component of interest is treated as a signal, and cross-component correlation analysis can be conducted at the chosen time scale: see *Labat et al. (1999 a,b)*; *Wang et al. (2010); Möri et al. (2011);* for applications of cross-wavelet analyses in karst hydrology, beach hydrodynamics, and clay rock hydrogeology, respectively.
- Yet another practical approach is to compare directly the amplitudes of the wavelet components of the two signals X(t) and Y(t) at the same dyadic scale "m".

The latter alternative may be viewed as a "*simplified cross-wavelet approach*"**.** It is a way to combine two "simple wavelet analyses" to finally obtain a simplified but practical cross-analysis of the two signals at the selected scale. In this work, we will use this type of approach *to identify semi-diurnal earth tide effects* in pore pressure data, leading to an estimate of the elastic "Ss" as the ratio of tidal strain / pore pressure amplitudes in terms of semi-diurnal wavelet components.[9]

In ***Sections 4 & 5***, this simplified cross-wavelet approach will be applied to obtain a constant global "Ss" and also to quantify the effect of excavation on a more local estimate of Ss(t). To achieve this, we will need to characterize the time varying amplitude of a semi-diurnal wavelet component (time scale of moon induced earth tides). In his way, we will be able to "see" the effect of the moving excavation front on the semi-diurnal wavelet "signal".

For this task, we need a new tool that can evaluate statistically the "time-modulated amplitude" of a signal. This tool is available already (although we apply it in an unusual way, in combination with wavelets): it is the theory of "*statistical envelopes*", presented in the next sub-section below.

### 3.5 Statistical envelope analysis of a signal ("modulated amplitude")

The *Cramer-Leadbetter* theory[10] and the related *Hilbert Transform*[11], are the theoretical tools that enable us to define, mathematically and statistically, the "modulated amplitude" (or the

[9] In fact, our current implementation of this general idea to the Mont Terri site is further simplified, using constant tidal amplitude because of the lack of direct earth dilatation record at the site.

[10] Cramer-Leadbetter theory in a nutshell: see ***Box N°1***.

"envelope") of a fluctuating and evolutionary pressure process. The theory of statistical envelopes is well established for zero mean stationary random processes (*Vanmarcke 1983; Veneziano 1979*). This theory, together with the associated Hilbert Transform, is particularly useful for characterizing relatively narrow-band signals, e.g. ocean & coastal waves *(Huang et al. 1999).*

But the present study (see ***Section 5***) suggests that the "statistical envelope" can also be useful in a variety of other cases in geophysics, provided it is combined with carefully selected "components" of the signals (e.g., here, the semi-diurnal wavelet component of pore pressure). In fact, the mathematical theory of statistical envelopes is established only for stationary processes, but as will be seen in ***Section 5***, using wavelets, we will get rid of the strong non stationary mean trend due to excavation. This is naturally accomplished by using multi-resolution wavelets at specific dyadic scales. The dominant scale of the mean trend is ~11days, while the wavelet scale selected for analysis is semi-diurnal (scale ≈ 12hours). The effects of the 11 day trend due to excavation are still "visible" as a time varying "modulated" amplitude in the semi-diurnal wavelet component $C_{12HOURS}(t)$, which we capture using envelope theory. Thus, our use of the theory of envelopes and Hilbert Transform in the present context, with semi-diurnal pressure fluctuations superimposed on a strong nonlinear trend due to excavation, seems rather new in the literature.

**Cramer-Leadbetter theory in a nutshell.**

For a zero mean (detrended) stationary random process X(t), one defines first a dephased "sister" process S(t). Based on X(t) and S(t), one obtains a symmetric double envelope ±R(t) such that:

$R(t)=(X^2(t)+S^2(t))^{1/2}$.

The envelope is usually much smoother than the original process X(t).

In the special case where X(t)=sin(ωt), we obtain S(t)=cos(ωt), and then $R(t) = \sin^2(\omega t)+\cos^2(\omega t) = 1$, as one may expect. Indeed, it should be obvious that the envelope of the sinusoïdal signal X(t)=sin(ωt) is just a pair of horizontal lines of ordinates -1 and +1.

**Hilbert transform in a nutshell.**

The Hilbert transform is an integral transform of the form:

$$\hat{X}(t) = H(X(t)) = (h \otimes X)(t) = \int_{-\infty}^{+\infty} X(t)h(t-\tau)d\tau = \frac{1}{\pi}\int_{-\infty}^{+\infty} \frac{X(t)}{t-\tau}d\tau .$$

For a real-valued signal X(t), H(X(t) is essentially the convolution of the signal X(t) by the kernel, or transfer function, h(t) = 1/(πt). Note: X(t) = sin(t) ⇒ H(X(t)) = -cos(t).

The Hilbert transform can be implemented in a complex framework as follows (as in the Matlab software): let X(t) the real valued process; let Y(t) = H(X(t) its real Hilbert transform. Then the complex Hilbert transform of signal X(t) is defined as Z(t) = X(t)+**i**.Y(t) where **i** is the pure imaginary number (**i** = √-1). The above defined envelope of X(t) is then obtained as:

$$R(t) = \pm\sqrt{X(t)^2 + H\left(X(t)\right)^2} = \pm\sqrt{X(t)^2 + Y(t)^2} = \pm\left|Z(t)\right| .$$

The Hilbert transform is implemented numerically for discrete time signals in MATLAB.

Box 1. Envelope theory in a nutshell

[11] Hilbert transform in a nutshell see ***Box N°1***.

Rather than present here the full theory of statistical envelopes, we refer the reader to the *above-cited references* for the theory, and to the *summary* presented here in Box ***N°1.*** The reader who is not directly interested in the theory can move on to ***Section 5*** for application of statistical envelope + wavelet analysis, to the non stationary evolution of semi-diurnal fluctuations of (relative) pore pressure during the excavation of Mont Terri gallery Ga98 ("*mine by test*").

As mentioned earlier, the final objective and result of this analysis is to estimate the evolution of specific storativity "Ss(t)" in the EdZ of the clay rock. This is done by using the tidal dilatation model of Bredehoeft relating tidal strain / relative pore pressure. The originality of this "envelope" approach should be emphasized. The amplitude of the semi-diurnal wavelet component is estimated as a modulated envelope, which varies more smoothly than the signal itself. The resulting storativity coefficient is therefore a time-varying coefficient, Ss(t), which reflects the mechanical disturbance due to the evolving excavation front, over short time scales (days, weeks).

### 3.6 Physically-based hydro-mechanical models to characterize the clay rock from pressure signals (earth tides and barometric effect)

The statistical methods of signal analysis are complemented or combined with physically based, but *"simplified" hydro-mechanical models.* These are used for hydrogeologic interpretation of the pressure signals (pore pressure and atmospheric pressure), and they are also related to inferred earth tide strain fluctuations, with the aim of characterizing quantitatively some of the hydrogeologic properties of the porous clay rock (Ss, $\Phi$).

Specifically, the simplified hydro-mechanical models used here relay on assuming both water and the prous matrix (bulk) to be elastically compressibility, and on accepting Terzaghi's 1936 effective stress concept. The corresponding models can be traced back to *Jacob (1940)* for the barometric efficiency model; and to *Bredehoeft (1967)* for the earth tides model (also *Jacob 1940* and *Cooper 1966)*. These physically-based models both assume that the total stress can be decomposed additively into effective stress between grains, and pressure (fluid stress), according to the Terzaghi concept. This implies that the grains of the claystone are "incompressible". The more general Biot poroelasticity theory can account for the case where individual grains are compressible (see for instance *Hsieh et al. 1988* concerning specific storativity).

In this work we use Terzaghi based hydro-mechanical models. They are outlined below in ***Box N°2***, where the final results are expressed in terms of:

- The specific (elastic) storativity Ss [$m^{-1}$];
- The effective (elastic or dynamic) porosity $\Phi$ [$m^3/m^3$].

## 4. Long time scales: statistical structure of pressure signals and characterization of time averaged properties (specific storativity Ss; elastic porosity $\phi$)

This section is devoted to the analysis of long records (months to years) of observed air pressure and relative pore pressure signals, for the purpose of characterizing global and average hydro-geologic properties of the clay stone (global in space, and also, average in time)…Statistical analyses based on correlations, spectra, and wavelets, will enable us to detect and characterize in particular the diurnal relation between relative pore pressure and air pressure (leading to a "barometric efficiency"), and the half-diurnal effect of earth tides (leading to a characterization of compressibility and/or specific storativity).

The purpose of this "box" is to summarize the simplified, but physically-based, hydro-mechanical models. These models are used in combination with statistical analyses in order to interpret pressure signals in terms of rock properties (elastic storativity Ss, effective elastic porosity Φ).

**Tidal strain - astronomic and global earth (the semi-diurnal component)**

Astronomic movements of Earth, Moon, Sun, induce fluctuations of the gravimetric field g(t), which has an impact on deformations of the earth crust (displacements, strains). Focusing on the M2 harmonic (Moon semi-diurnal), the volumetric strain fluctuations take the form:

$\varepsilon(t) = \Delta\varepsilon.\sin(\omega_{M2}t + \phi)$ where $\Delta\varepsilon$ is the tidal strain amplitude [$m^3/m^3$].

Concerning M2, it is the principal lunar earth tide, and its period is more precisely 12h 25 mn. *Takeuchi (1950)* gives: $\Delta\varepsilon = 0.5 \times W_2/(ag)$, where W2 is the tidal potential of M2, "a" is the radius of the earth, and "g" is gravity. One may take $W2/g \approx 10$ for M2. The constant 0.5 was computed by Takeuchi from earth crust mechanics data (Lamé stiffness moduli, Love number). At the scale of a geologic unit (as opposed to the global scale of the earth), *Melchior (1978)* estimates $\Delta\varepsilon \approx 2\times10^{-8}$ $[m^3/m^3]$ (value adopted *in this work* in the absence of direct measurements).

**Tidal strain and elastic specific storativity Ss of a geologic unit - the *Bredehoeft* model**

Specific storativity Ss : Ss is in units of [$m^{-1}$] ↔ Ss/(ρg) is in units of [(m3/m3)/Pa].

$S_S\,[m^{-1}] = (\delta V_{MP}/V_{MP})/\delta H$ where $V_{MP}$ is the volume of deformable bulk Porous Matrix,

and δH is total head variation, with head H defined by: $H = \dfrac{p - p_{ATM}}{\rho_w g} + z$. Equivalently:

$S_S/(\rho g)\,[Pa^{-1}] = (\delta V_{MP}/V_{MP})/\delta p_{REL}(t)$, with $p_{REL}(t) = p(t) - p_{ATM}(t)$ (relative pressure).

Using Terzaghi's 1936 effective stress concept, assuming constant total stress, and assuming elastic compressibility of water (β=1/Kw) and of the bulk porous matrix (α=1/Kb...) leads to:

$S_S = \rho_w g(\alpha + \Phi\beta) = \rho_w g\left(1/K_B + \Phi/K_W\right)$ (where Kw □ 2 GigaPa for water).

The *Bredehoeft (1967)* model relates tidal strain to relative pressure fluctuations (or □H) via Ss:

$\delta\varepsilon_{TIDAL} \approx S_S\delta H$ , or equivalently: $\delta\varepsilon_{TIDAL} \approx (S_S/(\rho g))\,\rho g\delta H = (S_S/(\rho g))\,\delta p_{REL}$.

**Barometric efficiency B - classical model (*compressibility & effective stress*)**

Without going into details (see *Freeze and Cherry 1979*), we present the result obtained with the effective stress hypothesis (*Terzaghi 1936*), in the case where the total stress is held constant, and assuming here also that our hydro-geologic unit behaves like a (*partially*) confined aquifer:

$\delta p_{REL}(t) = -B \times \delta p_{ATM}(t)$ , with the barometric efficiency B given by: $B = \beta\Phi/(\alpha + \beta\Phi)$.

Using the previous expression for Ss in terms of (α,β) compressibilities $\Rightarrow B = \beta\Phi/(S_s/(\rho g))$.

In this work, barometric efficiency is re-interpreted statistically as the spectral gain at frequency 1/24h. The above equation shows that barometric efficiency B, specific elastic storativity Ss, and effective elastic porosity Φ are interrelated (the compressibility β=1/Kw of water is known).

Box 2. Summary of physically based hydro-mechanical models

Long records with many time steps (tens of thousands or much more) usually have non stationary structures at various scales. In this section, the undesired non stationary structures are partially eliminated by the selection of specific peak frequencies (in Fourier spectra), and/or by the selection of pertinent dyadic "scales" via multi-resolution wavelet analysis. We seek here only global characteristics of the signals (global amplitudes, etc.).

As explained earlier in the methodology ***Section 3***, in order to achieve these goals (hydrogeologic characterization), we use simplified hydro-mechanical "models" to account for the porosity "Φ", the compressibility of water, and the compressibility of the bulk porous matrix. This leads also to the related concept of elastic specific storativity "Ss", based on Terzaghi's decomposition of total stress into effective stress+pressure. The "Ss" coefficient represents an equivalent elastic coefficient for the water filled poro-elastic rock. See the "outlook" comments in ***Section 6*** for a discussion on the limitations and possible extensions of these hydro-mechanical models.

### 4.1 A first example: statistical analyses and interpretations of "clean" records of raw (un-processed) pressure signals over~1 month

The above-mentioned statistical techniques are first illustrated here using only relatively short sections of the pore pressure signal PP1(t) at Mont Terri, with constant time step and no data gaps (no reconstitution required): see ***Figures 10, 11, 12, 13***.

The unprocessed pore pressure signal is analysed in terms of:

1. its frequency spectrum (shown in ***Figure 11***), and
2. its wavelet component at nearly semi-diurnal time scale (shown in ***Figure 13***).

#### 4.1.1 Specific storativity from earth tides (using semi-diurnal wavelets)

We estimate in this subsection the month-scale specific storativity "Ss" from semi-diurnal wavelet component of relative pore pressure (strain/pressure -- earth tide effect).

The specific storage coefficient quantifies the volume of water stored per unit volume of porous domain and per unit variation of hydraulic head. It can be estimated from earth tide amplitude using the relation of Bredehoeft (1967), based on a simplified model that takes into account bulk deformation effects due to earth tides in the poro-elastic water filled porous medium:

$$S_s = \frac{|\Delta\varepsilon|}{|\Delta h|} \tag{19}$$

where **Ss** is specific storativity ($m^{-1}$), $|\Delta\varepsilon|$ is the amplitude of volumetric strain fluctuations related to the M2 semi-diurnal earth tide, and $|\Delta h|$ is the amplitude of relative pressure head fluctuations (estimated from the semi-diurnal wavelet component of relative pressure). The strain amplitude $|\Delta\varepsilon|$ was estimated by *Takeuchi (1950)* based on astronomic earth tides and earth elasticity, then later on (in the 1960's) by Melchior and others for hydro-geologic formations: see *Melchior (1978)* (originally in *1960), Bredhoeft (1967),* and others: $|\Delta\varepsilon| = 2 \times 10^{-8}\ m^3/m^3$.

Note that the site specific signal of volumetric strain $\Delta\varepsilon(t)$ is not available, so its amplitude is only estimated indirectly (from data surveys in the literature). However, our method can be extended to the case where $\Delta\varepsilon(t)$ is available, by applying a semi-diurnal wavelet analysis on the $\Delta\varepsilon(t)$ signal.

The amplitude of relative pressure head $\Delta h(t)$ is obtained from multi-resolution wavelet analysis of $P(t)-P_{ATM}(t)$. The wavelet component with dyadic time scale close to semi-diurnal is isolated, and its amplitude is evaluated using either of the following "norms":

- the RMS (Root-Mean-Square) standard deviation: "*norm 2*" or $\| \|_2$ ; or else
- the absolute mean deviation ("*norm 1*" or $\| \|_1$).

Note:

$$\sigma^{(p)} = \left(\frac{1}{N}\sum\left|X(i)-\overline{X}\right|^p\right)^{1/p} \Rightarrow \sigma^{(1)} = \frac{1}{N}\sum\left|X(i)-\overline{X}\right| \text{ and } \sigma^{(2)} = \sqrt{\frac{1}{N}\sum\left(X(i)-\overline{X}\right)^2} \quad (20)$$

***Figure 12*** shows the evolution of the components according to time and scale. The goal is to isolate the component corresponding to earth tides (the semi-diurnal component): as indicated on this figure, we have chosen the component of dyadic scale T = 8h (close to 12h). Note: in other calculations, we have attempted to improve the accuracy of semi-diurnal scale selection by computing the mean of the dyadic 8h and 16h wavelet components: $(C_{8H}(t)+C_{16H}(t))/2$.

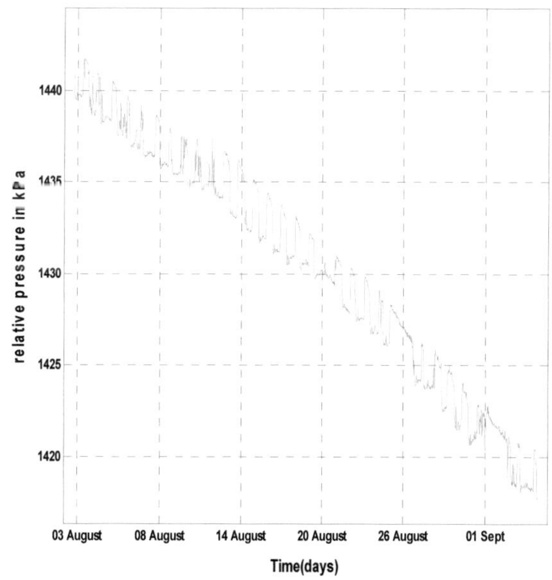


Fig. 10. Signal of the raw relative pore pressure PP1(t) (kPa) at Mont Terri (from 02/08/2002 to 04/09/2002). Time span: 15 month. Time step of raw pressure signal: $\Delta t$ = 30 min (and k=1).

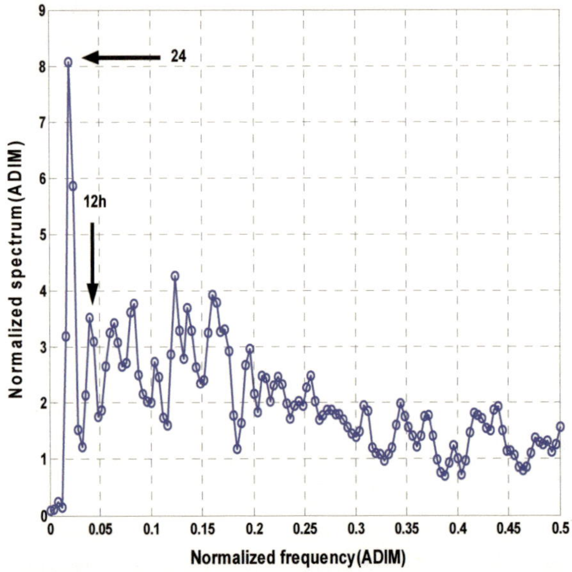


Fig. 11. Reduced spectrum of ΔPP1(t), the first difference of relative pore pressure (kPa), at Mont Terri. Time span: ~1 month (from 02/08/2002 to 04/09/2002). Time step of raw pressure signal: Δt = 30 min (and k=1).

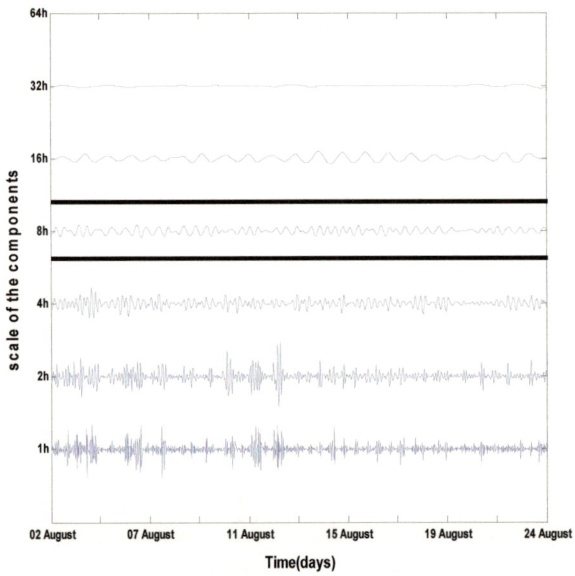


Fig. 12. Multiresolution wavelet analysis : several dyadic time scale components are shown for the raw relative pore pressure PP1(t) at Mont Terri. Duration: 02/08/2002 to 22/08/2002. Time step: Δt = 30 min (and sampling step: k=1).

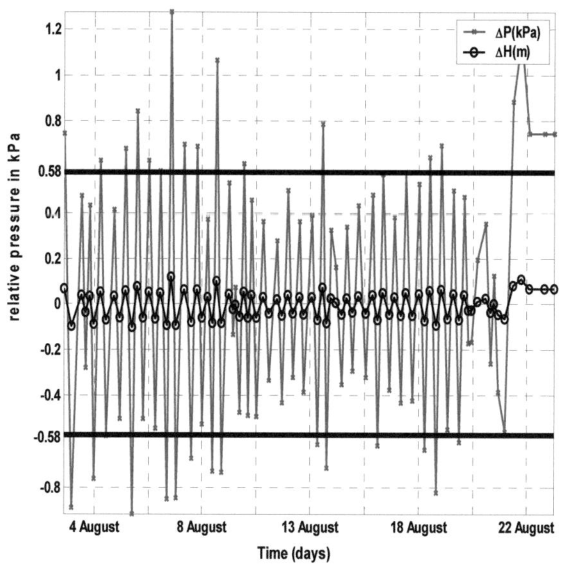


Fig. 13. Multiresolution wavelet analysis : dyadic time scale component 8h (near 12h) for the raw signal of relative pore pressure PP1(t) at Mont Terri. Duration: 02/08/2002 to 22/08/2002. Time step: $\Delta t = 30$ min (k=1). Amplitude of fluctuations: $\Delta h = 0.058$ m.

***Figure 13*** shows the evolution of the selected wavelet component (in this case, $C_{8H}(t)$). We have determined numerically the amplitude of the fluctuations of this wavelet component, e.g., its absolute mean deviation. The result can be expressed in terms of relative pressure head: for instance it was found that the absolute mean amplitude ("norm1") of semi-diurnal pressure head is $\Delta h = 0.058$ m. We can then calculate specific storativity Ss by applying the previous relation (***Eq. 19***). The results are the following, using either "norm1" or "norm2" for $\Delta h$:

$\Delta h$(norm 1)=**0.058 m**; $S_s$(norm 1)= **$3.4\times10^{-7}$ m$^{-1}$**
$\Delta h$(norm 2)=**0.064 m** ; $S_s$(norm 2)=$3.1\times10^{-7}$ **m$^{-1}$**

### 4.1.2 Effective elastic porosity Φ from barometric efficiency (using spectral gain)

We now estimate month-scale effective porosity from the diurnal frequency gain between relative pore pressure and air pressure (barometric efficiency).

The diurnal (and other) components would be due in part to barometric pressure effects. The effect of barometric pressure fluctuations on observation wells in closed, confined aquifers, is well known. The piezometric head in the aquifer is related linearly to atmospheric pressure (assuming linear poro-elastic behavior) via a coefficient named "barometric efficiency" (B or $B_E$).

Although the geometry, boundary conditions, and hydraulic configuration of the Mont Terri claystone appear quite different from those of a confined aquifer, we nevertheless propose to retain the same linear relationship. This makes it possible to calculate effective porosity, according to the relations described in ***Section 3.6*** and Box N°2 (the following is *Jacob's 1940* relation*):*

$$\Phi = \frac{E_w \times B}{\rho \times g} \times S_S \tag{21}$$

where $\Phi$ or $\Phi_{EFF}$ is the effective elastic porosity ($m^3/m^3$), B the barometric efficiency which expresses the elastic response of the system, ρ the density of water, g the acceleration of gravity, $E_W$ (or $K_W$) is the stiffness modulus of water ($E_W = K_W \approx 2.05$ GigaPascals).

In order to calculate barometric efficiency B, a statistical cross-spectral analysis of the relation between atmospheric pressure and relative pressure was developed, with 30 minute time steps, to determine the reality of the effect of barometric pressure. We carried out the cross-spectral analyses between barometric pressure and relative pressure in borehole BPP-1 (from the PP niche, which contains two rooms, PP1 and PP2).

The estimated spectral gain function (in frequency space) was finally used to determine barometric efficiency (B) by looking at the value of the spectral gain at diurnal frequency, as shown in ***Figure 14***. From this spectral gain, barometric efficiency is determined, and porosity is calculated with ***Eq. 21*** just above, using the previous estimate of specific storativity (Ss) from semi-diurnal wavelet analysis. We obtain in this fashion the following results:

$$\phi_{EFFECTIVE}\,(\text{norm } 1) \approx 6\%;\ \phi_{EFFECTIVE}\,(\text{norm } 2) \approx 5.4\%$$

Recall that this procedure to obtain "Φ" (by spectral analysis of barometric gain at diurnal frequency) is combined with the previous procedure to estimate "Ss" (by wavelet analysis of semi-diurnal earth tide effects). In fact, the two procedures do not need to be implemented sequentially. They can be viewed as a system of two equations (***Eq. 19*** and ***Eq. 21***) with two unknowns (Ss, Φ). The two unknowns could have been determined both at once (alternative approach).

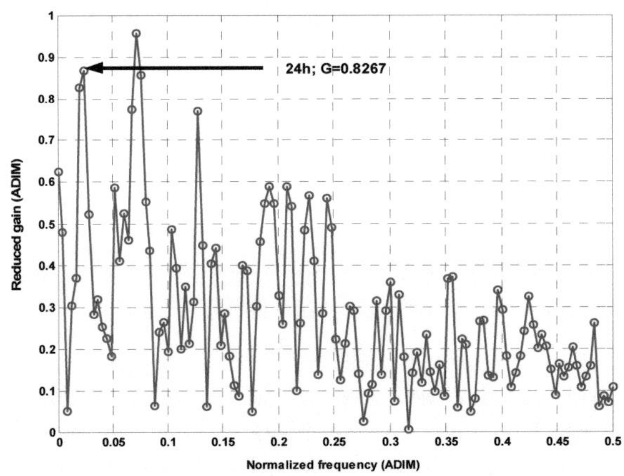


Fig. 14. Reduced spectral gain *vs.* normalized frequency f×Δt (the maximum value of f×Δt is 0.5, in agreement with Shannon's sampling theorem). This spectral gain function was obtained by cross-spectral analysis of the raw signals of atmospheric pressure Patm(t) [kPa] and relative pore pressure PP1(t)-Patm(t) [kPa]. Signal duration: about 1 month (from 02/08/2002 to 04/09/2002). Time step: 30 mn. The gain at diurnal frequency (1/24h) is indicated by the arrow; it is used to estimate barometric efficiency.

### 4.2 Long term analysis of pressure signals over more than 1 year (after pre-processing)

The same techniques illustrated in the previous ***Section 4.1*** are now being applied to a much longer - but *pre-processed* - joint record of absolute pore pressure and atmospheric pressure ($p(t)$, $p_{ATM}(t)$). The time span of the processed pair of signals is about 15 months *(from 29 January 2004 to 12 April 2005).*

Note that, initially, only shorter pieces of "clean" signals were available during that time window. Note also that the pre-processing procedures (reconstructing, homogenizing, etc.) must be implemented jointly on $p(t)$ and on $p_{ATM}(t)$ over the same time window, since the relative pressure signal must be obtained from $p_{REL}(t) = p(t)-p_{ATM}(t)$, and cross-analysis of ($p_{REL}(t)$, $p_{ATM}(t)$) must be implemented to identify the barometric effect. A flow-chart depicting pre-processing tasks and methods was shown earlier in ***Sub-Section 2.2.1*** (***Figure 2***).

The results of statistical analyses of the 15 month long pre-processed pressure signals ($p_{REL}(t)$, $p_{ATM}(t)$) are shown below in ***Figures 15, 16, 17, 18*** for the PP1 pore pressure sensor.

***Figure 15*** shows $p_{REL}(t)$ over the entire 15 month time span. Similarly to the previous ***Section 4.1***, we analysed the cross-spectral gain of ($p_{REL}(t)$, $p_{ATM}(t)$) at frequency 1/24h (*not shown here*), and the wavelet component at nearly semi-diurnal time scale (***Figure 17***). In addition, ***Figure 16*** shows the spectrum of $\Delta p_{REL}(t)$, the first order difference of $p_{REL}(t)$. Differencing enhances the peak frequencies, and allows one to detect a dominant 1/24h frequency, and a somewhat less dominant 1/12h frequency (subdominant but still visible). We see that these two characteristic frequencies can be detected in spite of the obvious non stationarity (large scale non linear trend) of the 15 month pore pressure signal (***Figure 15***).

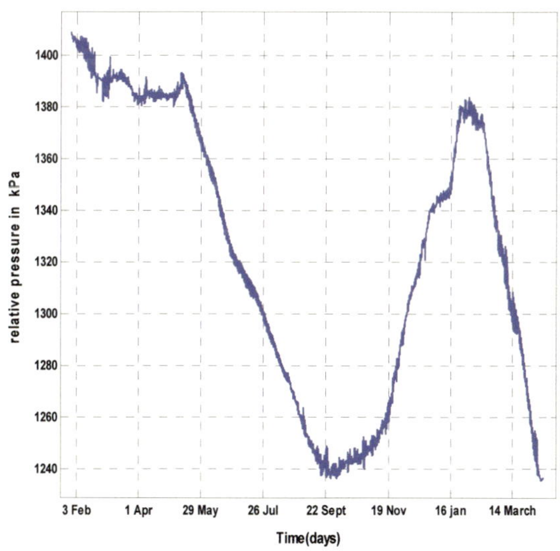


Fig. 15. Signal of the relative pore pressure PP1(t) (kPa) at Mont Terri Time span: 15 month (29/01/2004 to 12/04/2005 ). Time step of pre-processed signal: $\Delta t$ = 30 min, and sampling step k=1 (i.e. no under-sampling).

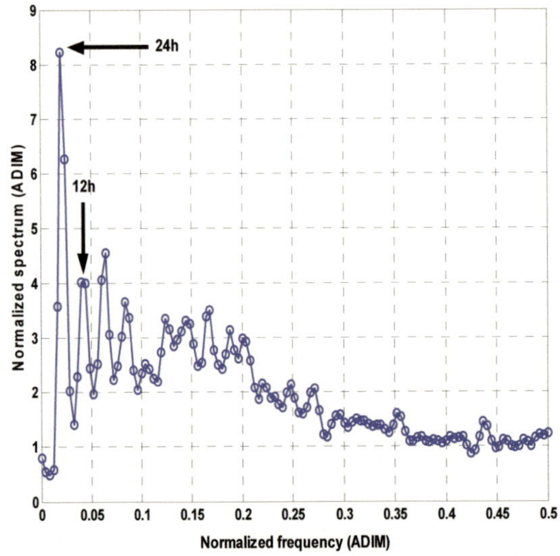


Fig. 16. Reduced spectrum of $\Delta PP1(t)$ ($\Delta p_{REL}(t)$), first difference of relative pore pressure PP1(t) [kPa] at Mont Terri. Time span ~15 months (29/01/2004 to12/04/2005). Time step of pre-processed signal: $\Delta t$=30 min (no sub-sampling).

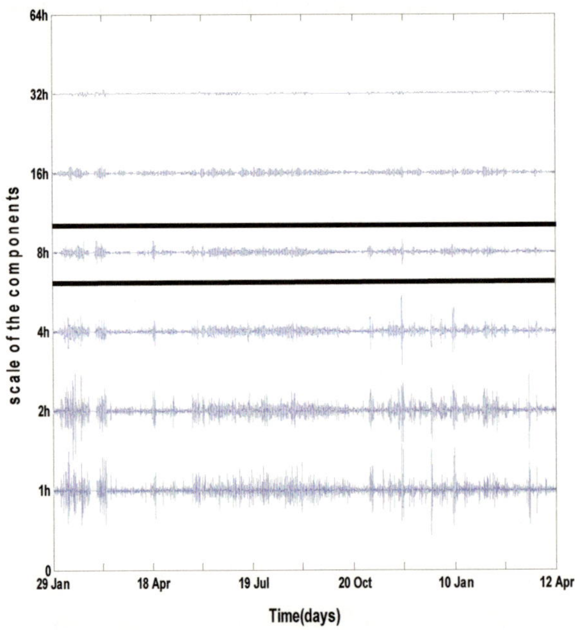


Fig. 17. Multi-resolution wavelet scale-time plot. Several dyadic components are shown here for the signal of relative pore pressure PP1(t) or $p_{REL}(t)$ (pre-processed) at the Mont Terri site. Signal duration: from 29/01/2004 to 12/04/2005; time step $\Delta t$ = 30 min; no sub-sampling (k=1).

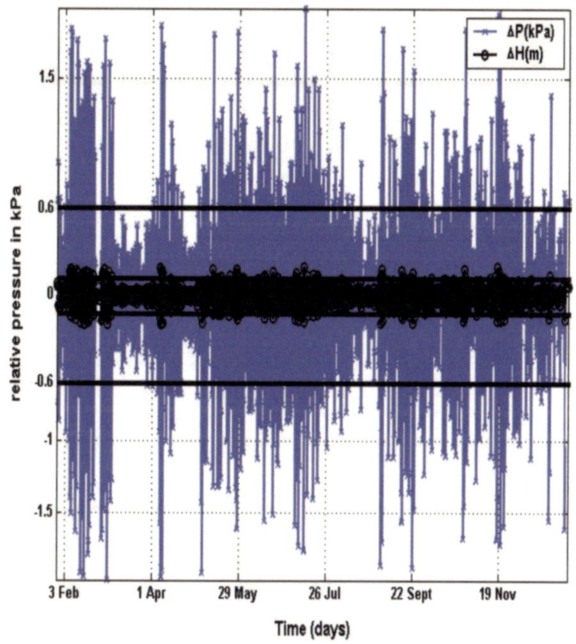


Fig. 18. Multi-resolution wavelet analysis: graph of the 8h dyadic scale component (near 12h) of the pre-processed relative pore pressure, PP1(t) or $p_{REL}(t)$. Record length ~15 months (from 29/01/2004 to 12/04/2005), $\Delta t$ = 30 min, k=1 (no sub-sampling). The amplitude of semi-diurnal pressure fluctuations is $\Delta h \approx 0.06$ m equivalent water height.

The methods used here are again the same as those applied in the first example of ***Section 4.1*** where we analysed shorter, unprocessed pressure signals. We obtain here successively the specific storativity "Ss" and the effective elastic porosity "Φ". The final results are shown below for two different methods of quantifying the amplitude of semi-diurnal wavelet pressure fluctuations:

$|\Delta h|$ (*norm 1*) ≈ **0.06m**; **Ss** (*norm 1*) ≈ $\mathbf{3.3\times10^{-7}m^{-1}}$ **Φ** (*norm 1*) ≈ **1.0%**
$|\Delta h|$ (norm 2) ≈ **0.08m** ; **Ss** (norm 2) ≈ $\mathbf{2.6\times10^{-7}m^{-1}}$ **Φ** (*norm 2*) ≈ **0.9%**

where "*norm 1*" stands for the absolute mean, and "*norm 2*" stands for the Root Mean Square (or Standard Deviation).

## 5. Short time scale effects of gallery excavation on the evolution of pore pressure and of elastic properties (specific storativity Ss; elastic porosity)

### 5.1 Excavation of a gallery: observed effects on the evolution of pore pressure (syn-excavation phase, Mont Terri gallery Ga98)

We focus here on a total duration of ~5 months, corresponding to the syn-excavation period of gallery Ga98 (17th of November 1997 to 23rd of April 1998). The pre- and post-excavation periods were also analysed for comparison (*Fatmi 2009*), but it is sufficient to treat here the 5 month "syn-excavation" period. Furthermore, we will also focus, within this 5 month period, on a shorter phase of about 2 weeks of "fast" pressure changes (*see below*).

Upon inspection of ***Figure 19***, it is clear that the pore pressure signal in sensor PP1 exhibits a strong response to the passage of the excavation front. This indicates the existence of a moving EdZ ("Excavation disturbed Zone") as the excavation front passes near the pressure sensors. Pressure rises gradually at first, then very rapidly ("fast" 11 day period). After this rise, pressure seems to reach a stable value but in fact, it starts a long and slow relaxation process (confirmed from analyses of longer records). The sharp rise of pore pressure can be explained as a hydro-mechanical disturbance of the geologic porous formation due to excavation works.

More precisely, ***Figure 19*** shows the pore pressure data from the PP1 sensor, in comparison with the evolving distance D(t) of the excavation front. The pre- and post-excavation phases are not shown. The 5 month "syn-excavation" phase was determined to run from the 17th of November 1997 to the 23rd of April 1998, and it is shown entirely in this figure (*see caption for other details and comments*). Clearly, within the 5 month syn-excavation phase, there is a remarkable sub-phase of "fast" pressure rise, lasting about two weeks only (~11 days). Similar effects were observed for the other sensor PP2 in the same borehole (*not shown here*).

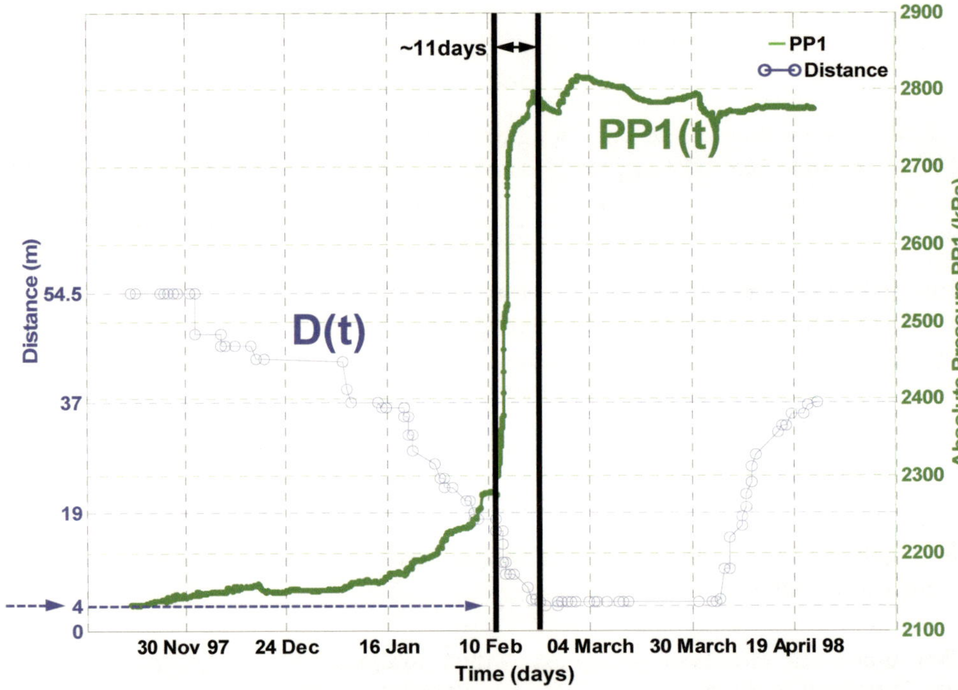


Fig. 19. Comparison of pore pressure and excavation front versus time over a 5 month period (syn-excavation). The green curve "PP1(t)" represents the time evolution of absolute pore pressure p(t) in sensor PP1. The blue curve with circles represents the distance D(t) between measurement chamber PP1 and the excavation front of gallery Ga98. The arrow at left indicates the minimal distance achieved (4 m), after which excavation was stopped for about a month (February/March 1998), before restarting around the 2nd week of April 1998. The sharp rise of pressure PP1(t) from 2250 to 2800 kPa ($\Delta p \approx +5.5$ bars) indicates the "fast" part of the syn-excavation phase, which lasted only 11 days (or roughly two weeks).

A study of the signal on longer time scales (*not shown in this text*) confirms in fact that the pore pressure disturbance was indeed reversible (hence the name "Excavation *disturbed* Zone", as explained earlier). Indeed, a slow relaxation of pore pressure was observed during a long period, somewhat less than two years, subsequent to the end of excavations works (there are almost two years counting from the start of excavation till the end of pressure relaxation: 30/11/1997-18/02/2000). The pore pressure at the end of this period reached approximately the value it had at the beginning of the period (pre-excavation) and it continued to decrease to reach finally the pre-pumping test value (*see Fatmi 2009, Fig.28, p.93, for more details*).

These observations justify our initial assumption and remark on the distinction between "EDZ" (*irreversibly damaged* "Excavation Damaged Zone"), and "EdZ" (*Excavation disturbed Zone*): the latter is thought to be a *reversibly disturbed* zone due to quasi-elastic hydro-mechanical variations of pressure and stress. These reversible variations occurred in the zone of our pressure sensor, and they caused the observed variations of pressure during the passage of the excavation front (which came within 4 m of the pressure sensor at the closest point).

Now, looking at the short and fast rise period of 11 days in ***Figure 19***, it may seem "impossible" at first sight to detect a pattern of semi-diurnal pressure fluctuations due to earth tides, and even less, to follow possible changes in these fluctuations during the phase of rapid pressure change. Yet, although earth tides are partially masked, they can still be detected indirectly using proper filtering techniques. For example, applying differential filters can enhance subdominant frequencies, moving average filters can be used to look at residuals, and multi-resolution *wavelet* analysis can be used to extract precisely the semi-diurnal earth tide component (*as will be seen*).

Below, we will take advantage of the precise way in which multi-resolution wavelets can capture the semi-diurnal component of the original signal, and we will then analyse the new "12h" signal using envelope techniques, to finally obtain time-varying hydraulic parameters of the claystone during the passage of the excavation front nearby the pore pressure sensor.

### 5.2 Multi-resolution wavelet analysis of relative pressure signals during excavation: extraction of earth tide semi-diurnal component

First, we obtain the relative pore pressure $p_{REL}(t)$ from the absolute pore pressure signal p(t) [*shown in **Fig. 19** above as "PP1(t)"*] by subtracting the atmospheric pressure signal $p_{ATM}(t)$ (*not shown*), whence: $p_{REL}(t) = p(t)-p_{ATM}(t)$.

Obviously, one should not expect $p_{ATM}(t)$ to be influenced by excavation works (rather, $p_{ATM}(t)$ contains barometric fluctuations, with peak frequency 1/24h, part of these having an impact on pore pressure).

The procedure to identify the elastic specific storativity Ss is then similar to that illustrated in ***Section 4.1***, except that here, after extracting the semi-diurnal wavelet component "$C_{12H}(t)$" from the relative pore pressure signal $p_{REL}(t)$, we perform an additional analysis of the modulated amplitude, or "envelope" of $C_{12H}(t)$, in a statistical framework.

Note, from now on, the semi-diurnal wavelet component "$C_{12H}(t)$" is treated as a signal. The wavelet signal $C_{12H}(t)$ and its statistical envelope are now presented and analysed in the next section below (***Section*** 5.3), in order to derive, from that information, the time-varying specific storativity "Ss(t)" during the syn-excavation phase.

### 5.3 Statistical envelope of semi-diurnal wavelet of relative pressure during excavation (Hilbert transform & method of maxima)

In this section, we present the wavelet component signal "$C_{12H}(t)$" of relative pore pressure, and we analyse its modulated amplitude by computing its Cramer-Leadbetter statistical envelope using MATLAB's Hilbert Transform (Section 3.5: Statistical envelope analysis...).

In addition, in some cases, we compare this theoretical envelope with an empirical one computed by our own "method of maxima". Briefly, the latter method generates an empirical envelope that has irregular time steps. The result of the empirical method agrees generally with the theoretical envelope except for isolated discrepancies (this is a minor point, but the interested reader can consult Fatmi 2009 for more details).

Figure 20 shows (in blue) the semi-diurnal signal "$C_{12H}(t)$", i.e. the 12h wavelet component of relative pressure in sensor PP1, and it shows also (superimposed in red) the statistical Cramer-Leadbetter envelope of "$C_{12H}(t)$" (envelope computed with the Hilbert Transform). Recall our gal is to analyse the effects of excavation of gallery Ga1998. Here, the entire "syn-excavation" phase of 5 months is shown (approximately from mid-november 1997 to mid-april 1998).

Figure 21 shows a temporal zoom of Figure 20 over a much shorter period. The time window is approximately one month, and it is centered around the "fast" sub-phase of syn-excavation, which lasts only 11 days.

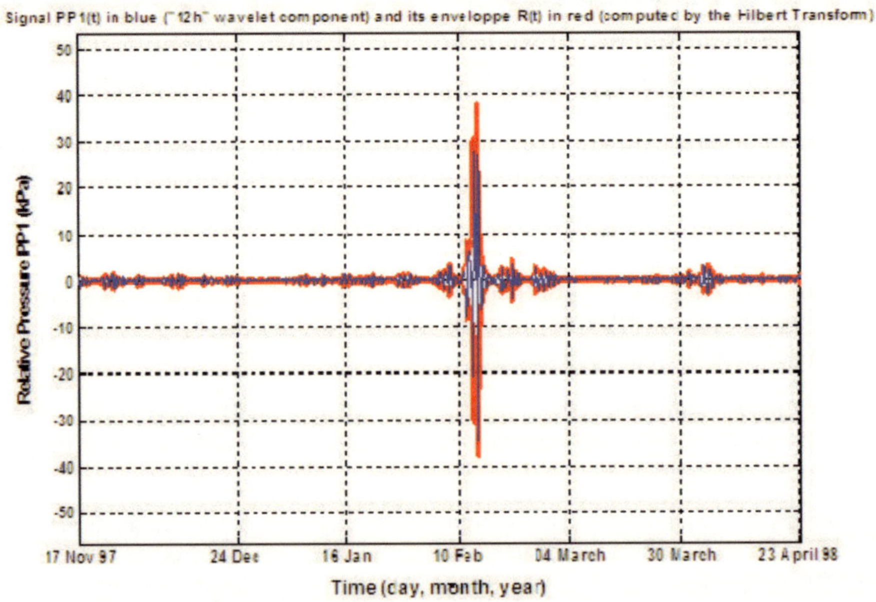


Fig. 20. Semi-diurnal wavelet component $C_{12H}(t)$ of relative pore pressure in PP1 (blue curve) and its envelope (red curve) during the "syn-excavation" phase of gallery Ga98 at Mont Terri (5 months, from 17 Nov 1997 to 23 April 1998, with $\Delta t$=30 mn time steps).

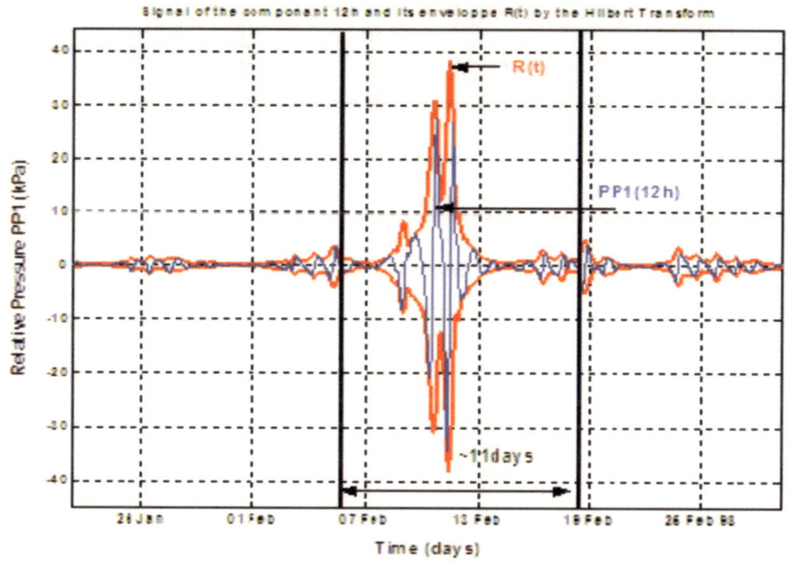


Fig. 21. Temporal zoom of the previous figure, shown over a time window of approximately 1 month (instead of 5 months in the previous figure). The (blue) curve designated as "PP1 (12h)" is the wavelet component signal $C_{12H}(t)$ extracted from PP1 relative pressure. The envelope is the (red) curve designated as "R(t)". The ~ 11 day period of fast pressure change occurs between the dates of 7th and 19th of February 1998.

Comparing ***Figure 21*** to ***Figure 19***, it is now clear (and remarkable) that the 11 day phase of rapidly increasing pore pressure (***Figure 19)*** corresponds to a significant increase of the *amplitude* of the semi-diurnal component of pore pressure (***Figure 21)***. Recall, from ***Fig. 19***, that this 11 day phase corresponds to the approach of the excavation front nearest to the PP1 pressure sensor (distance D(t) ~ 4 m at the closest point).

### 5.4 Calculation of specific storage Ss(t) as a function of time based on Brodohooft's simplified hydro-mechanical model (strain / pressure)

By analogy with the procedure described in ***Section 4.1-4.1.1***, we estimate again the elastic specific storativity Ss from the semi-diurnal wavelet component $C_{12H}(t)$ of relative pore pressure (using now the average of the 2 components at dyadic scales 8h and 16h, respectively). And again, we invoke the Bredehoeft earth tide strain/pressure model (see ***Eq. 19*** earlier).

Only, this time, instead of taking a constant amplitude for $C_{12H}(t)$, we use instead the time-modulated amplitude of $C_{12H}(t)$ represented by its statistical envelope "R(t)". On the other hand, since we still do not have direct *in situ* measurements of the volumetric strain $\Delta\varepsilon(t)$, we must replace $\Delta\varepsilon(t)$ by an indirect estimate of its amplitude based on data surveys: accordingly, we use again $|\Delta\varepsilon| = 2\times10^{-8}\ m^3/m^3$ (as in ***Sections 4.1 and 4.2***). In this way, we obtain, instead of ***Eq. 19***:

$$S_S(t) = \frac{|\Delta\varepsilon|}{R(t)} \tag{19b}$$

***Figure* 22** shows the time evolution of $\log_{10}$ Ss(t), where Ss(t) was estimated using the empirical envelope R(t) computed by the method of maxima, rather than the theoretical Cramer-Leadbetter envelope computed by the Hilbert Transform. The evolution of the excavation front distance D(t) is also shown on the same graph.

***Figure* 23** shows the time evolution of $\log_{10}$ Ss(t), using the theoretical Cramer-Leadbetter envelope (computed by the Hilbert Transform) to estimate Ss(t). The evolution of the excavation front distance D(t) is also shown on the same graph.

Comparing ***Figure* 22** and ***Figure* 23**, it is somewhat re-assuring that the two methods of estimating the wavelet envelope yield approximately the same results. Thus, we focus mostly on results obtained with the Hilbert Transform method for the remaining plots and analyses.

The time scale of ***Figure* 23** is about 1.5 month, including the interesting 11 day window of fast changes. It can be seen that the specific storativity *Ss(t)* decreases by one to two orders of magnitude during just *a few days*, as the excavation front approaches within roughly *6 m* to the PP1 pressure sensor (before reaching an even closer distance of *4 m*).

***Figure* 24** shows a zoomed view of Ss(t) with a cartesian ordinate rather than semi-log. The zoomed time window is only about 2 weeks or so, to see more details in the 11 day period of fast change.

Finally, ***Figure* 25** gives a broader picture, depicting the time evolution of $\log_{10}$ Ss(t) over the entire "syn-excavation" phase of 5 months (including, again, the 11 day period of fast changes).

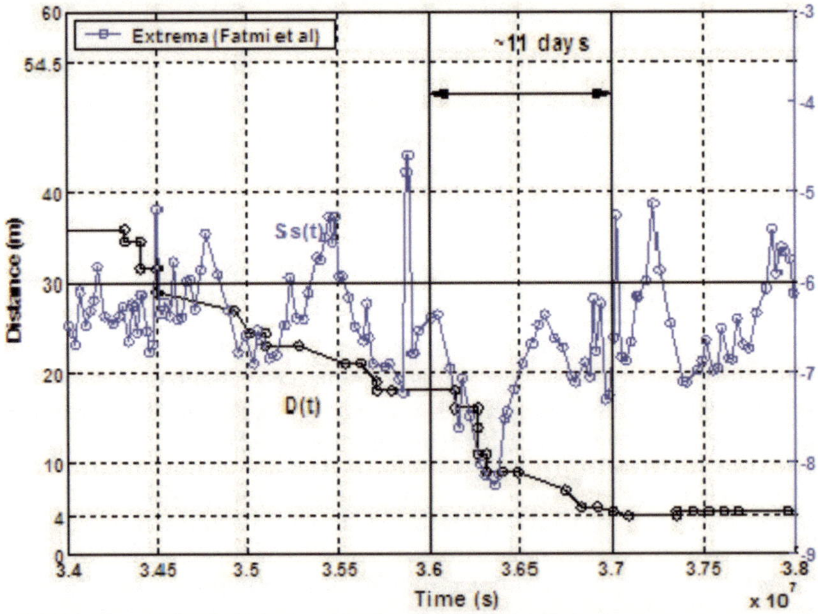


Fig. 22. Evolution of the decimal logarithm of Ss(t). The time window here is about 1.5 month or so, and it includes the 11 day period of fast change (*double arrow*). Here, Ss(t) is calculated from an empirical envelope (estimated with the lethod of maxima rather than the Hibert transform) of the semi-diurnal dyadic component of relative pore pressure PP1 during excavation (gallery Ga98, excavation from 17/11/1997 to 23/04/1998, time step $\Delta t$=30min).

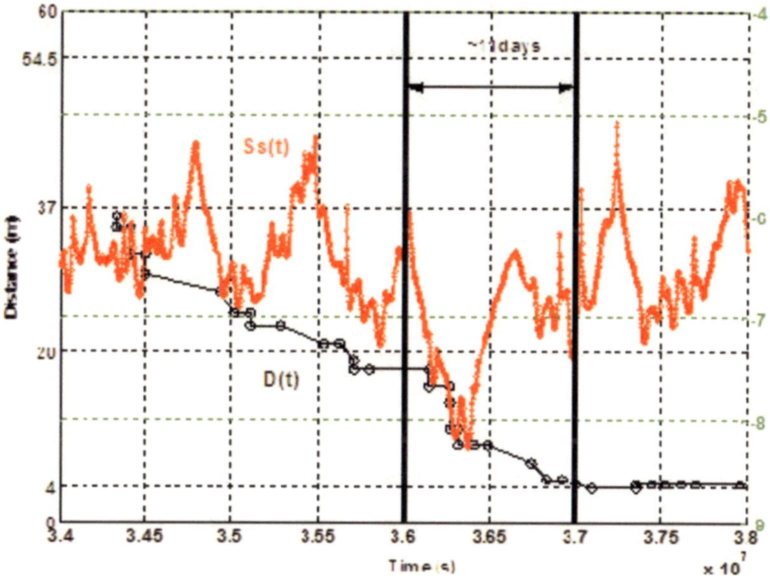


Fig. 23. Evolution of the decimal logarithm of Ss(t), on the same time window as the previous figure (1.5 month). Here, the red curve shows Ss(t) calculated from the theoretical envelope (Hilbert Transform, Cramer-Leadbetter envelope), applied to the semi-diurnal dyadic component of relative pore pressure PP1 during excavation of gallery Ga98. The distance D(t) of the excavation front from the PP1 pressure sensor is shown as the black curve with circles "o".

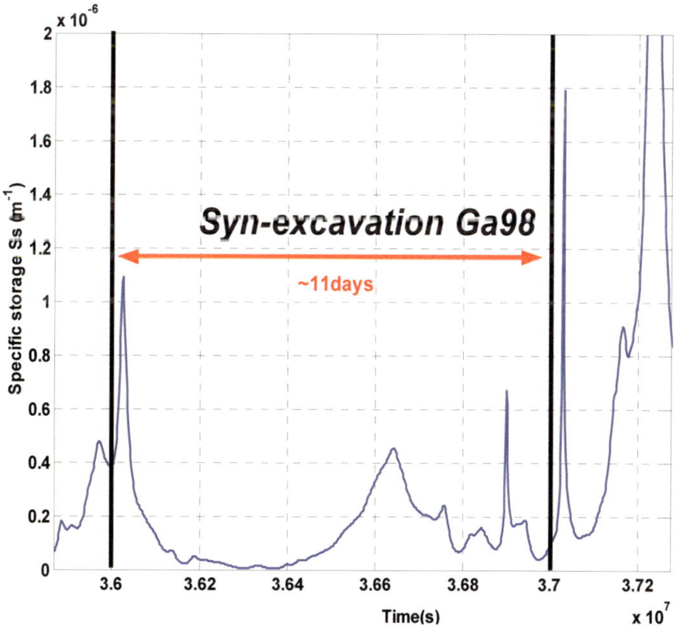


Fig. 24. Zoomed view of the evolution of Ss(t) on a cartesian ordinate (rather than logarithmic). The plot is centered on the 11 day period of fast pressure change, corresponding to the approach of the excavation front of gallery Ga98 to the pressure sensor.

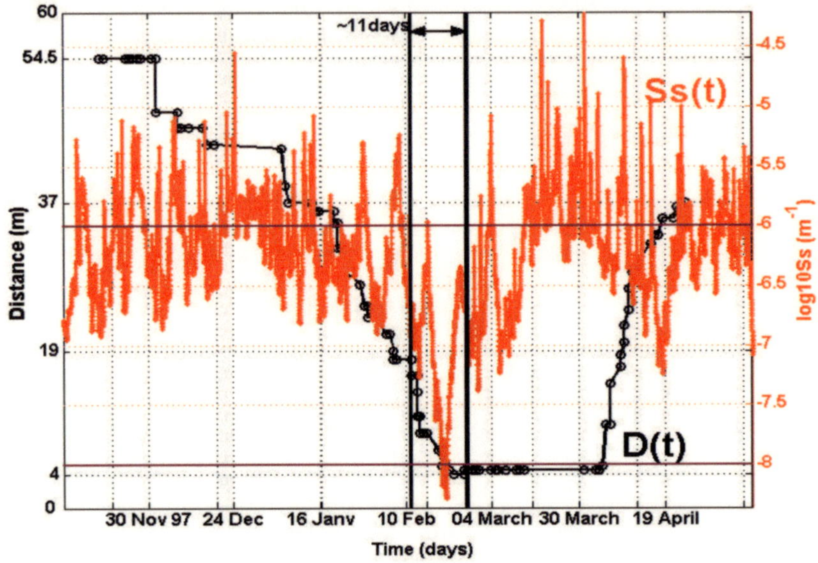


Fig. 25. A broader view of the evolution of log Ss(t) (red curve, computed with the Hilbert Transform envelope). Here the time window is the entire "syn-excavation" phase of 5 months (including, again, the 11 day period of very fast changes). The black curve with circles "o" represents the distance D(t) of the excavation front from the PP1 pressure sensor.

### 5.5 Calculation of elastic porosity Φ(t) from barometric effect (diurnal gain $p_{REL}/p_{ATM}$) and from the previously estimated Ss(t)

Finally, from the previously estimated evolution of Ss(t), and using the spectral barometric gain at 1/24h frequency (*as in **Section 4.1.2***), it is possible to apply again the simplified hydro-mechanical model of ***Eq. 21,*** in order to estimate the time-varying elastic porosity Φ(t):

$$\Phi(t) = \frac{E_w \times B}{\rho \times g} \times S_S(t) \tag{21b}$$

where *Ss(t)* is the time varying specific storativity obtained earlier from the strain/pressure envelope (**Eq.11b,** ***Section*** **5.4**).

It should be observed that, in this analysis, the elastic porosity Φ(t) varies in time proportionally to Ss(t). It can be seen that a constant barometric efficiency "B" is assumed; the diurnal gain "g" between atmospheric and relative pore pressure is assumed constant during excavation. But, because pore pressure is influenced by excavation (even if air pressure is not), the assumption of constant barometric gain ("B" or "g") during excavation is debatable and should be verified.

In conclusion, a revised version of the procedure should be to analyse the diurnal barometric gain, like the semi-diurnal earth tide effects, in a wavelet framework rather than in a Fourier spectral framework. This would provide a tool to either capture or eliminate, depending on purpose, any non stationary / evolutionary phenomena.

## 6. Summary, conclusions, and outlook

### 6.1 Summary and conclusions

We have presented in this chapter a set of statistical methods for pre-processing and analysing multivariate hydro-geo-meteorological time series, and we have developed hydro-mechanical interpretations, in order to infer hydro-geological properties such as Ss and Φ. The aim of this study is to quantify some of the properties of the clay rock in terms of elastic response to barometric and earth tides fluctuations, both *globally* and *locally* in space-time. We focused on atmospheric and pore water pressure signals measured in Opalinus clay stone, at the Mont Terri underground rock laboratory (URL), in the framework of Mont Terri's LP 14 experiment.

Pre-processing methods were developed in order to detect and correct some defects common to most time series data, like aberrations or outliers, variable time steps, and missing data. These pre-processing steps are necessary, in most cases, to obtain signals that are both "clean" and long enough for useful analyses. Durations of a couple of weeks to a few months are needed for short scale analyses related to excavation works. Ideally, several years are required for global characterizations, and even more for climate trend analyses.

Once the pressure signals were pre-processed (reconstructed), statistical analyses were performed with cross-correlation functions and cross-spectra, and with multi-resolution wavelet analyses. These techniques were then used to determine "globally" some hydro-mechanical parameters of the Opalinus clay: elastic specific storativity (Ss), and effective elastic porosity (Φ). The values compare favourably with data from the literature, obtained independently with completely different methods. In particular, the simplified hydro-mechanical model of *Bredehoeft* relating $p_{REL}(t)$ to earth tides was applied to the "12h" dyadic component of $p_{REL}(t)$, leading to a global estimated Ss value of $10^{-6}$ $m^{-1}$ (1E-6 $m^{-1}$) in excellent agreement with that deduced from hydraulic tests at the Mont Terri site.

After this "global" characterization of the clay from long term pore pressure measurements, we turned our attention to the (significant) impact of excavation works on pore pressure signals. We devised an original method to infer the time evolution of clay rock properties (Ss(t)) due to the excavation disturbance. The pressure disturbance coïncides (with some lead time or negative delay) with the arrival near the pressure sensor of the excavation front D(t) (or of the "EdZ" around and ahead of the front).

Indeed, pressure signals recorded in the PP1 chamber of the BPP-1 borehole during excavation of the Ga98 gallery, have shown a net increase of absolute pressure (+5.5 bars) over the 11 day period of fast change (or as much as +7 bars over the total 5 month "*syn-excavation*" phase). This is evidence of a direct hydro-mechanical disturbance of the rock during the passage of the excavation front.

A study of the signal on longer time scales (from pre- to post-excavation stages, *not shown in this text for lack of space*) confirmed that the disturbance due to excavation was indeed reversible. A slow relaxation of pore pressure was observed during a period of about two years (or somewhat less) subsequent to the end of excavation, reaching approximately the value it had before excavation. This observation appears to justify our initial remark on the distinction between "EDZ" (*irreversibly damaged* "Excavation Damaged Zone"), and "EdZ" (*Excavation disturbed Zone*). The latter "EdZ" is a *reversibly disturbed* zone due to quasi-elastic hydro-mechanical variations of pressure and stress.

In summary, the signal analysis and processing methods applied at various (long/short) time scales, exploit the influence of natural "forcings" like earth tides and barometric fluctuations on pore pressure signals. Combining multi-resolution wavelet and he Cramer/Hilbert envelope analysis reveals a time modulation (disturbance) of the "12h" dyadic wavelet component, suggesting a modification in time of the specific storativity Ss(t) during the excavation of gallery passing nearby the pressure sensor. During the passage of excavation, there is a net decrease of Ss(t) by almost two orders of magnitude, from $10^{-6}$ (1E-6) $m^{-1}$, down to $10^{-8}$ (1E-8) $m^{-1}$. This minimum occurs during a short period of just a few days, corresponding to front crossing (actually, in advance of front crossing by a couple of meters).

This result demonstrates how statistical analyses on pore pressures can be a valuable tool in the assessment of potential clay repositories for the purpose of high level radioactive waste disposal in deep, isolating geologic formations.

### 6.2 Outlook

A useful extension of this work would involve the development of a multivariate or "*multi-cross* analysis" of time series data {Xi(t), i = 1 , 2, ..., K}, where the Xi's are either reconstructed or raw signals, involving possibly a number of pore pressures measured in different boreholes at different measurement sections, as well as other types of signals such as air pressure, air temperature, air humidity (relative or absolute), wall crack displacements, and moisture contents in the rock.

Accordingly, in multi-cross analysis, the results would be expressed in matrix form $\boldsymbol{M}_{XiXj}(s)$ where "s" is the relevant parameter (time lag, frequency, etc.): thus $\boldsymbol{M}_{XiXj}(s)$ may represent a K×K matrix of cross-correlation functions, or of Fourier cross-spectra, and/or, of multi-resolution cross-wavelets (or their cross-correlations at given scales). These analyses will be useful for assessing the added value of *all* signals collected at the site, in terms of the "information" gained concerning the hydro-geologic properties of clay rock at different time scales (hourly, daily, yearly).

The purpose of such multi-cross analyses might be to integrate the complete set of all available signals in order to solve an *inverse problem*, namely, the identification of hydro-geologic properties from observations of state variables. Thus, in the work presented here, the state variables were the pore pressure at different sensors, atmospheric pressure, and (indirectly) tidal strain. The properties to be identified were (Ss, Φ) - but this set of unknown coefficients could be extended in future, in a generalized treatment of the "hydro-geologic identification problem".

For instance, techniques to estimate the hydraulic conductivity "K" from natural signals sampled at different positions might be developed as a complement to classical well pumping tests (*e.g. pulse tests*). Indeed "K" is an important property for a geologic radioactive waste repository, as "K" must be low to efficiently isolate radioactive waste from the environment. Furthermore, it is well known that "K" may vary greatly in geologic media, depending on the spatial scale of analysis (e.g. *Ababou 2008, Sahimi 1993, Gelhar 1993, Ababou et al. 1990,*). With the "multi-cross" concept, the spatial scales of analysis can be associated with the distances between sensors.

The multi-cross analysis of these signals will also serve as a complement to physically based models, with some of the signals considered as inputs, and others considered as outputs of the models. It will be very useful to link the statistical approaches with physically-based models that can account for the hydro-mechanical coupling between pore pressure and *exogeneous* fluctuations (piezometric variations in adjacent aquifers, barometric fluctuations, and earth tides).

In addition, some aspects of the procedures developed in this work could be modified in future:

- Thus, we developed a sequential procedure to estimate first "Ss" then "Φ". First, we estimated "Ss" from wavelet analysis of semi-diurnal earth tide effects, and secondly, we estimated "Φ" from spectral analysis of barometric gain at diurnal frequency (given the previously estimated "Ss"). In reality, these 2 steps need not be implemented sequentially: they can be viewed as a system of 2 equations with 2 unknowns (Ss, Φ) to be determined both at once.
- To estimate the global "Ss" and (also) the evolutionary "Ss(t)" during excavation, we used a fixed value, $|\Delta\varepsilon| = 2\times10^{-8}\ m^3/m^3$, for the amplitude of the volumetric strain related to the M2 semi-diurnal earth tide (*based on Melchior 1978, Bredhoeft 1967, and others*). This is because the site specific signal of volumetric strain $\Delta\varepsilon(t)$ was not available, so we had to estimate its amplitude indirectly from data surveys. Nevertheless, our method can be extended to the case where the strain signal $\Delta\varepsilon(t)$ is available. The new procedure would be to apply semi-diurnal wavelet analysis, then envelope analysis, to the $\Delta\varepsilon(t)$ signal, in a way similar to that already implemented for relative pore pressure $p_{REL}(t)$.

Another extension worth considering concerns the simplified hydromechanical "models" which we have used so far for interpreting the pressure signals. These models need to be enhanced:

- Thus, it was noted that "Ss" represents an equivalent elastic coefficient for the water filled poro-elastic rock. However, this concept rests on a simplified view of poro-elastic deformations, usually applied to confined aquifers. Its interpretation for the hydrogeologic configuration and boundary conditions of the claystone at Mont Terri is not straightforward and requires further analyses.
- Also, a more "physical" hydromechanical approach based on Biot's poro-elastic theory (*Biot 1956*) could be developed, as in the coupled T-H-M models developed in *Ababou et al. (2005)* and *Canamon (2009)*. Such models should be upgraded with explicit coupling of pore pressure to external "forcing" functions (air pressure, earth tides, etc.), in order to better interpret the effects of these signals.
- Finally, the hydro-mechanical model could be constructed to predict space-time propagations explicitly. Thus, *Mallet et al. (2007, 2008)* developed a simple model to illustrate pore pressure fluctuations and propagation under boundary forcing in a compressible hydrogeologic medium.

In summary, the simplified hydro-mechanical models used in this study can be replaced, in future, with models based on PDE systems. The statistical "interpretations" of pressure signals presented in this text will then appear as the solution of PDE inverse problems (in a statistical setting). This point of view sheds a new light on the coefficients (Ss, Φ) obtained in the present study. They can be viewed as the solution of an inverse problem based on spectral gain ($p_{REL}/p_{ATM}$) and on wavelet amplitude ratio ($|\Delta\varepsilon| / |\Delta p_{REL}|$). For this reason, it is thought that the statistical methods used here can be adapted for solving an inverse problem involving a hydro-mechanical PDE model forced by pressure signals and other measurements (humidity, temperature, etc.).

## 7. Acknowledgments

This chapter presents data analyses developed within Mont Terri's LP 14 experiment (Long term Pressure monitoring project, phase 14), as part of the international research programs of the MONT TERRI CONSORTIUM on the properties of claystone as a potential geologic repository for radioactive waste. The authors acknowledge the support of institutions participating in the LP 14 project at Mont Terri (ANDRA, NAGRA, IRSN, NWMO, SWISSTOPO). In addition, the authors gratefully acknowledge the help and advice provided by the following persons: Paul Bossart (SWISSTOPO); Martin Cruchaudet (ANDRA); David Jaeggi (SWISSTOPO); Herwig Mueller (NAGRA); Eric Sykes (NWMO). The authors wish also to thank Alain Mangin (CNRS-UPS Toulouse, Laboratoire Souterrain de Moulis) who initiated some of the statistical tools and methods used in this project, and others who provided scientific advice on various issues (cracks, geophysics, signals, hydrogeology, porous media): Andrea Möri (SWISSTOPO); Jacques Delay (ANDRA, LSM/HM Bure); Israel Cañamón Valera (Universidad Politecnica de Madrid); Moumtaz Razak (Université de Poitiers, HYDRASA); and Michel Quintard (IMFT, Toulouse).

## 8. Appendix: Symbols and abbreviations

| ABBREVIATIONS, ACRONYMS | |
|---|---|
| ANDRA | Agence Nationale pour la gestion des Déchets RAdioactifs (*France*) |
| BPP | Borehole instrumented for Pore Pressure measurements (*Mont Terri site*) |
| EdZ | Excavation disturbed Zone (reversibly disturbed zone around an excavation in rock, elastic hydro-mechanical disturbance) |
| EDZ | Excavation Damaged Zone (irreversibly damaged, fissured and fractured zone around an excavation in rock) |
| IRSN | Institut de Radioprotection et de Sûreté Nucléaire (France) |
| LP14 | Long term Pressure monitoring project - phase 14 (Mont Terri Project) |
| NAGRA | National Cooperative for the Disposal of Radioactive Waste (*Switzerland*) |
| NWMO | Nuclear Waste Management Organization (*Canada*) |
| PP | Pore Pressure, i.e., absolute pore water pressure (*Mont Terri site*) |
| SWISSTOPO | Swiss Federal Office of Topography (*and Mont Terri Project*) |
| T-H-M or THM | Thermo-Hydro-Mechanical (coupled models) |
| URL | Underground Rock Laboratory (*or:* Underground Research Laboratory) |

## MATHEMATICAL SYMBOLS: LATIN

| | |
|---|---|
| $A_m(t)$ | Wavelet approximation of order "m" (multi-resolution dyadic scale "m") |
| $B$, $BE$, $B_E$ | Barometric efficiency, [dimensionless] or [Pa/Pa] (relative pressure / atmospheric pressure effects) |
| $C_{12H}(t)$ | Wavelet component of scale 12h (semi-diurnal) |
| $C_m(t)$ | Wavelet component of dyadic scale "m" (time scale = $2^m$□□t) |
| $C_{XY}(\tau)$ | Cross-covariance function of X(t) and Y(t), versus time-lag [units of Cxy] = [units of X(t)] □ [units of Y(t)] |
| D(j) or D(τ(j)) | Tuckey-2 filter function of time-lag (used for estimation of Fourier spectra) |
| $D_m(t)$ | Wavelet "detail" at dyadic scale "m" (*see also: wavelet component $C_m(t)$*) |
| $f$ | Frequency "f" in [Hz] or [Number of cycles / second]; "f" is related to angular frequency "ω" by: f = ω / 2π. |
| $g_{XY}(f)$ | Reduced spectral gain *vs.* frequency, obtained from cross-spectrum. The complex-valued reduced gain is defined as: $g_{XY}(f) = s_{XY}(f)/s_{XX}(f)$. |
| $H$ | Total hydraulic head [m], related to relative pressure by: $\rho g \delta H = \delta p_{REL}$ |
| $k$ | Permeability (intrinsic Darcy permeability) in [m$^2$] |
| $K$ | Hydraulic conductivity [m/s] (sometimes unduly called "permeability") |
| $M2$, $M_2$ | Moon semi-diurnal component of astronomic tides (here, earth tides) |
| $p$, $p_{ABS}$ | Pore pressure (or "absolute" pore pressure), [Pa]. (*) |
| $p_{ATM}$ | Atmospheric pressure (or air pressure), [Pa]. (*) |
| $p_{REL}$ | Relative pore pressure [Pa], defined as : $p_{REL} = p - p_{ATM}$ (*) |
| $R(t)$ | Statistical envelope of a signal (the envelope is in fact ±R(t)). |
| $R_{XY}(\tau)$ | Cross-correlation function of X(t) and Y(t), versus time-lag; Rxy is the normalized version of Cxy: [units of Rxy] = [dimensionless] |
| $s_{XY}(f)$ | Reduced cross-spectrum of signals X(t) and Y(t), *vs.* frequency "f". It is equal to the cross-spectrum $S_{XY}(f)$ divided by $(\sigma_X \times \sigma_Y)$. The cross-spectrum is complex-valued (even if the signals are real-valued). |

| | |
|---|---|
| $s_{XX}(f)$ | Reduced auto-spectrum of X(t) *vs.* frequency "f"; it is equal to the spectrum $S_{XX}(f)$ divided by the variance $\sigma_X^2$. The (auto)-spectrum is always real-valued for a real-valued process X(t). |
| $S_S$ | Specific elastic storativity, [$m^{-1}$] or [*$m^3/m^3$ per meter of head $\delta H$*] |

(*) Pressures "p" are sometimes denoted with a capital "P" within the text.

| MATHEMATICAL SYMBOLS: GREEK | |
|---|---|
| $\Delta\varepsilon$, $\Delta\varepsilon_{TIDAL}$ | Tidal dilatation, volumetric strain [$m^3/m^3$] (here, this notation is used for the *amplitude* of the tidal strain). |
| $\Delta t$ | Time step |
| $\tau$ | Time-lag or "delay", used in correlation function analyses: if t' and t" are two time instants, the time lag is $\tau$ = t"-t'. |
| $\omega$ | Angular frequency, or "pulsation", in [radians/s]; "$\omega$" is related to frequency "f" [Hz] by : $\omega = 2\pi f$. |
| $\Phi$ | Effective elastic porosity [$m^3/m^3$] |

## 9. References


Ababou R. (2008): *Quantitative Stochastic Hydrogeology: the Heterogeneous Environment.* Chap. 8 in Part III of "Overexploitation & Contamination of Shared Groundwater Resources: Management, (Bio)Technological, and Political Approaches to Avoid Conflicts." NATO-ASI: Advanced Studies Institute Series, C.J.G. Darnault (ed.), Springer Science & Business Media BV, January 2008 (pp. 119-182).

Ababou R., I. Cañamón, F. J. Elorza (2005): Thermo-Hydro-Mechanical simulation of a 3D fractured porous rock: preliminary study of coupled matrix-fracture hydraulics. Proceedings, *FEMLAB 2005 Conference: "Une conférence multiphysique"*. Paris, France, 15 November 2005, pp.193-198.

Ababou R., L. W. Gelhar (1990): *Self-Similar Randomness and Spectral Conditioning: Analysis of Scale Effects in Subsurface Hydrology.* Chap. XIV in "Dynamics of Fluids in Hierarchical Porous Media", J. Cushman (editor), Academic Press, New York, 1990 (pp. 393-428).

Ababou R., L.W. Gelhar, D. McLaughlin (1988): *Three-Dimensional Flow in Random Porous Media.* Tech. Report No. 318, Ralph Parsons Laboratory for Water Resources & Hydrodynamics, Department of Civil Engineering, Massachusetts Institute of Technology (MIT), Cambridge, Massachusetts, USA. (2 vols., 833 pp., March 1988). [See also *Ababou R., 1988, PhD thesis, MIT :* http://libraries.mit.edu/ http://dspace.mit.edu/handle/1721.1/14675].

Biot M.A. (1956): General Solutions of the Equations of Elasticity and Consolidation for a Porous Material. *Journal of Applied Mechanics* 23: pp. 91-96. 1956.

Blackman R.B., Tukey J.W. (1958): *The Measurement of Power Spectra*. Dover Publications. NY, 1958.

Bossart P., Meier P.M., Moeri A., Trick T., and Mayor J.-C. (2002). Geological and hydraulic characterisation of the excavation disturbed zone in the Opalinus Clay of the Mont Terri Rock Laboratory. *Engineering Geology* 66:19-38.

Blümling P., Frederic B., Patrick L., Martin C.-D. (2007). The excavation damaged zone in clay formations time-dependent behaviour and influence on performance assessment. *J. Phys. Chem. Earth* 32:588–599.

Box, G.E.P., G.M. Jenkins (1976): *Time Series Analysis, Forecasting, and Control*. Revised Edition. San Francisco, CA: Holden-Day Publishers.

Bras R., I. Rodriguez-Iturbe (1985): Random Functions in Hydrology. Dover, NY, 1985 (re-edited: 1993).

Bredehoeft J.D. (1967): *Response of Well-aquifer Systems to Earth Tides*. U.S Geological Survey, Washington, D.C., 20242.

Cañamón I., F.J. Elorza, A.Mangin, P.L. Martin, R. Rodriguez (2004): Wavelets and statistical techniques for data analysis in a mock-up high-level waste storage experiment. *International Journal of Wavelets, Multiresolution and Information Processing (IJWMIP)*: Volume: 2, Issue: 4 (2004), pp. 351-370. DOI: 10.1142/S0219691304000585.

Cañamón Valera I., 2009: *Coupled phenomena in 3D fractured media (analysis and modeling of thermo-hydro-mechanical processes)*. VDM Verlag Dr. Müller, 2009, 212 pp. (ISBN: 3639189973, 978-3-639-18997-1).

Cañamón Valera I., 2006: *Analysis and Modeling of Coupled Thermo-Hydro-Mechanical Phenomena in 3D Fractured Media (Analyse et Modélisation des Phénomènes Couplés Thermo-Hydro-Mécaniques en Milieux Fracturés 3D)*. Thèse de doctorat (en anglais). IMFT - Institut National Polytech. de Toulouse & Universidad Politécnica de Madrid (Escuela Técnica Superior de Ingenieros de Minas). Thesis supervisors: R. Ababou (INPT-IMFT), J. Elorza (UPM-DMAMI).

Daubechies I. (1988): Orthonormal bases of compactly supported wavelets. *Comm. Pure & Appl. Math.*, XLI, pp. 909-996, 1988.

Daubechies I. (1991): "Ten lectures on wavelets". *CBMS-NSF Series Appl. Math.*, SIAM, 1991.

Fatmi H., R. Ababou, J.-M. Matray (2008):"Statistical pre-processing and analyses of hydrogeo meteorological time series in a geologic clay site (methodology and first results for Mont Terri's PP experiment)". *Journal of Physics & Chemistry of the Earth (JPCE)*, Special Issue «Clays in Natural & Engineered Barriers for Radioactive Waste Confinement» (CLAY'2007): 33(2008) S14-S23.

Fatmi, H., Ababou R., Matray J.M., Joly C. (2010) "Hydrogeologic characterization and evolution of the 'Excavation Damaged Zone' by statistical analyses of pressure signals: Application to galleries excavated at the claystone sites of Mont Terri (Ga98) and Tournemire (Ga03)". *4th Internat. Meet. "Clays in Natural & Eng. Barriers for Rad. Waste Confin."*, Nantes 2010. Poster P/EDZ/CH/03: Book of abstracts: pp. 809-810.

Fatmi H. (2009). *Méthodologie d'analyse des signaux et caractérisation hydrogéologique: Application aux chroniques de données obtenues aux laboratoires souterrains du Mont Terri, Tournemire et Meuse/Haute-Marne*. Ph.D. thesis, Institut National Polytechnique de Toulouse, Toulouse & Fontenay-aux-Roses (France), June 2009, 248 pp., in french. [*Available on line at* INPT].

Fatmi H., R. Ababou, J.-M. Matray, Ch. Nussbaum (2011): "Statistical analyses of pressure signals, hydrogeologic characterization and evolution of *Excavation Damaged Zone* (claystone sites of Mont Terri and Tournemire)." *Proceedings MAMERN11*: 4th Internat. Conf. Approx. Methods and Numer. Modelling in Envir. & Natural Resources, Saidia (Morocco), May 23-26, 2011. B.Amaziane, D.Barrera, H.Mraoui, M.L.Rodriguez & D.Sbibih (eds.), Univ. de Granada, ISBN:078-84-338-5230-4 (2011), pp.325-329.

Fatmi H., R Ababou, J.-M. Matray (2007) : "Méthodologie de prétraitement et d'analyse du signal. Application aux chroniques de données multivariées obtenues au Mont Terri". Rapport IRSN DEI/SARG/n°2007-035 (rapport d'avancement No.1 pour 2006/07), Fontenay-aux-Roses, 27 pp., 2007.

Freeze R.A. & J.A. Cherry (1979): *Groundwater*. Prentice Hall, Englewood Cliffs NJ, 604 pp.

Gelhar L.W. (1993): *Stochastic Subsurface Hydrology*. Prentice Hall, Englewood Cliffs, New Jersey, 390 pp., 1993.

Hsieh P.A., Brededhoeft J.D., Rojstaczer A. (1988). Response of well aquifer to earth tides: problem revisited. *Water Resources Research*, 24, 3, pp 468-472.

Huang N. E., Zheng Shen, Long S. R. (1999): A new view of nonlinear water waves: The Hilbert spectrum. *Annual Review of Fluid Mechanics*, Vol. 31, pp. 417-457 (2p.3/4), 1999. ISSN:0066-4189.

Jacob C.E. (1940): On the flow of water in an artesian aquifer. Transactions of the American Geophysical Union, 2, pp. 574-786, 1940.

Labat D. (2005): Recent advances in wavelet analyses: Part 1. A review of concepts. *Journal of Hydrology*: 314 (2005) 275-288.

Labat D., R. Ababou, A. Mangin, 1999a: Wavelet analysis in karstic hydrology. 1rst part: Univariate analysis of rainfall rates and karstic spring runoffs. *C.R.Acad.Sci. de Paris*, Sci. de la Terre et des Planètes (Earth & Planetary Sciences), 1999, 329, pp.873-879.

Labat D., R. Ababou, A. Mangin, 1999b: Wavelet analysis in karstic hydrology. 2nd part: Rainfall-runoff cross-wavelet analysis. *C.R. Acad.Sci. de Paris*, Sciences Terre et Planètes (Earth & Planetary Sciences), 1999, 329, pp.881-887.

Labat D., R. Ababou, A. Mangin (1999c): Linear and Nonlinear Models Accuracy in Karstic Springflow Prediction at Different Time Scales. *SERRA - Stochastic Environmental Research & Risk Assessment*, 13(1999):337-364, Springer-Verlag.

Labat, R. Ababou, A. Mangin (2000a): Rainfall-runoff relations for karstic springs – Part I : Convolution and spectral analyses. *Journal of Hydrology*, Vol. (238), Issues 3-4 (5 Dec.2000): pp. 123-148.

Labat D., R. Ababou, A. Mangin (2000b): Rainfall-runoff relations for karstic springs - Part II : Continuous wavelet and discrete orthogonal multiresolution analyses. *Journal of Hydrology*, Vol. (238), Issues 3-4 (5 Dec.2000): pp. 149-178.

Labat D., R. Ababou, A. Mangin (2002): Analyse multirésolution croisée de pluies et débits de sources karstiques. *C. R. Geosciences* (Elsevier), 334 (2002) 551-556.

Mallat, S. (1989): A Theory For Multiresolution Signal Decomposition: The Wavelet Representation. *IEEE Trans. on Pattern Anal. & Mach. Int.*, 11(7), pp. 674-693, 1989.

Mallet, A., R. Ababou, J.-M. Matray (2007): Multidimensional modeling of a poro-elastic medium: identification of the behavior of a clay formation from a stochastic representation of piezometric fluctuations. Poster Abstract, Book of Abstracts, *Internat. Conf. CLAY 2007*, Lille, France, 2007.

Mallet A., J.-M. Matray, R. Ababou (2008) : *Etude des fluctuations piézométriques à l'intérieur d'un milieu poreux saturé, et détermination des paramètres caractéristiques du milieu poreux.* Note Technique IRSN/DEI/SARG/2008-024 (Version 1) - Technical Note of the Institut de Radioprotection et Sûreté Nucléaire (IRSN), Fontenay-aux-Roses, France, 2008.

Mangin A. (1984). Pour une meilleure connaissance des systèmes hydrologiques à partir des analyses corrélatoires et spectrales. *J. Hydrol.*, 67, 25-43, 1984.

Marsaud B., A. Mangin, F. Belc (1993): Estimation des caractéristiques physiques d'aquifères profonds à partir de l'incidence barométrique et des marées terrestres. *Journal of Hydrology*, 144 (1993) 85-100.

Marschall P., Ababou R., Bossart P., Cruchaudet M., Fatmi H., Matray J.-M., T. Tanaka & P. Vogel (2009). Chap. 8: Hydrogeology experiments. In: "*Report of the Swiss Geological Survey N°3: Mont Terri Rock Laboratory-Project, Program 1996 to 2007 & Results*". Bossart P. & Thury M. (eds.), Federal Office of Topography - SwissTopo (2008/09), ISBN 978-3-302-40016-7, Wabern, Swizerland, pp. 95-106.

Massmann J. (2009). *Modelling of Excavation Induced Coupled Hydraulic-Mechanical Processes in Claystone.* Dissertation, Institut ffir Strômungsmchanik and Elektron. Rechnen im Bauwesen der Leibniz Univ. Hanover, Hanover, Germany.

Max J., (1980) : *Méthodes et techniques de traitement du signal et application aux mesures physiques*, Tome I (354 pp.) et Tome II (454 pp.), Chap.I-XXVIII, Masson, Paris.

Max J., J.L. Lacoume, (1996) : *Méthodes et techniques de traitement du signal.* Dunod, 5ème ed., 355 pp.

Melchior P. (1978): *The tides of the planet earth.* Pergamon Press, Paris, 609 pp.

Möri A., Bossart P., Matray J.-M., Müller H., Frank H, & Ababou R. & Fatmi H. (2012): "Mont Terri project, cyclic deformations in the Opalinus clay". Special Issue "Clays in Natural & Engineered Barriers for Radioactive Waste Confinement" (4th Internat. Meeting Clays 2010, Nantes, 29 March - 1rst April 2010). *Journal of Physics & Chemistry of the Earth* (ISSN 1474-7065, DOI: ---); *in press (2012).*

Nussbaum C., Bossart P., Amann F. and Aubourg C. (2011). Analysis of tectonic structures and excavation induced fractures in the Opalinus Clay, Mont Terri underground rock laboratory (Switzerland). *Swiss Journal of Geosciences*: Volume 104, Issue 2 (2011), Pages 187-210. DOI: 10.1007/s00015-011-0070-4.

Nussbaum, C., Wileveau, Y., Bossart, P., Möri, A., Armand, G., (2007). "Why are the geometries of the EDZ fracture networks different in the Mont Terri and Meuse/Haute-Marne Rock Laboratories? Structural Approach." *CLAY'2007: Clays In Natural & Engineered Barriers for Radioactive Waste Confinement*, Lille, France, 17-18 Sept. 2007. (2 pp., extended abstract).

Papoulis A., S.U. Pillai (2002): *Probability, Random Variables, and Stochastic Processes.* Mc Graw-Hill Book Company, New York. 2002 (4th ed, 16 Chapters, 852 pp.).

Priestley M.B. (1981). *Spectral analysis and time series.* Academic Press, 890 pp.

Sahimi M. (1993): Flow phenomena in rocks: from continuum models to fractals, percolation, cellular automata, and simulated annealing. Reviews of Modern Physics, Vol. 65, N°4, pp. 1393-1533 (1993).

Schaeren, G.and Norbert, J. (1989): Tunnels du Mont Terri et du Mont Russelin. La traversée des "roches à risques": marnes et marnes à anhydrite. *Soc. Suisse Ing. Arch.*, Doc. SIA D 037, 19–24.

Takeuchi, H (1950): On the earth tide of the compressible earth of variable density and elasticity. *Transactions American Geophysical Union*: 31, 651-689, 1950.

Terzaghi, V.K. (1936): "The Shearing Resistance of Saturated Soils and the Angle between the Planes of Shear". First International Conference of Soil Mechanics, Vol. 1: pp. 54-56. Harvard University. 1936.

Thury M. & Bossart P. (1999): "Mont Terri rock laboratory. Results of hydrogeological, Geochemical and geothechnical Experiments Performed in 1996 and 1997". *Swiss National Hydrological and Geological Survey*. Geologic Report N°23. Bern, Switzerland, 1999.

Vanmarke E. (1983): *Random Fields Analysis and Synthesis*. The MIT Press, Cambridge MA, USA, 382 p.

Veneziano D. (1979): Envelopes of Vector Random processes and their Crossing Rates. *The Annals of Probability*: 1979, Vol.7, No.1, 62-74.

Wang Y., R. Ababou, M. Marcoux (2010): "Signal Processing of Water Level Fluctuations in a Sloping Sandy Beach Modeled in a Laboratory Wave Canal", in: *Experimental & Applied Modeling of Unsaturated Soils GSP202* (Proc.2010 GeoShanghai International Conference, Part II: Applied Modeling and Analyses. L. R. Hoyos, X. Zhang, A. J. Puppala, eds.). American Society of Civil Engineers: *ASCE Geotech. Special Publi. (GSP):* Vol.376, No.41103, pp.204-210. http://dx.doi.org/10.1061/41103(376)26

Yaglom, A. M. (1987). *Correlation Theory of Stationary and Related Random Functions: Basic Results*. Springer-Verlag, New York.

Yevjevich V., (1972): *Stochastic Processes in Hydrology*. Water Resources Publications, Fort Collins, Colorado, USA, 8 Chapters, 276 pp.

# 18

# Particulate Phases Possibly Conveyed from Nuclear Waste Repositories by Groundwater

Constantin Marin

*"Emil Racoviţă" Institute of Speleology of the Romanian Academy, Bucharest Romania*

## 1. Introduction

Among the most severe sources of potential concern in terms of natural environment contamination with antropogenic radionuclides, a huge potential is associated to the radioactive waste resulting from the processing of the nuclear fuel used for energy generation purposes and from military and medical actvities. Although it represents about 3% of the total radioactive waste existing nowadays in the world, its total radioactivity amounts to more than 95% of that of the low & intermediate level and high-level radioactive waste taken together (Hu et al., 2010). By the present time, deep geologic disposal is considered to be the safest method for isolating highly radioactive and long-lived waste over millennia (Acero et al., 2010; Kim et al., 2011; Kurosawa & Ueta, 2001), while for the low and intermediate radioactive waste management, near surface storage is, for economic reasons, the preferred option (Dogaru et al., 2010; IAEA, 2004; Niculae et al., 2009).

It is a well-known fact that groundwater represents the most effective agent by which radionuclides stored in repositories could be transferred to the adjacent environment (Altmann, 2008; Baik et al., 2009; Geckeis & Rabung, 2008; Wersin et al., 2011). This is the reason why, each time when the issue of the radioactive waste repositories is addressed, irrespective of the repository type or of its inventory of stored radionuclides, the primary concern is to assess the local groundwater contamination risk.

Radionuclides may be conveyed by groundwater in a dissolved state, but especially in association with particles of various origins that are carried along with water. Those particles may have various structures and chemical compositions, their possible natures being: inorganic, organic or even micro-organisms. The inorganic particles genesis is closely related to the geochemistry of the environment with which groundwater is in contact, while particles of organic and biological nature are, as a general rule, allochtonous.

Particles dimensions may be very different, ranging from those of the colloids, up to microscopic particles, generically designated as "suspended particulate matter". Their reactivity and their ability of binding and conveying contaminants is to a certain extent controlled by their dimension, namely their reactivity is progressively enhanced as the particulate phase is more finely dispersed (Kersting & Zavarin, 2011; Wigginton et al., 2007).

This is also the reason why an increased interest is dedicated to the radionuclides behaviour in the presence of the colloidal matter.

The role played by the colloidal phase is intensely investigated not only for groundwater, but also for terrestrial surface water, namely for streams and lakes (Aleksandrova et al., 2010; Bondareva, 2011; Matsunaga et al., 2004; Monte, 2010; Monte et al., 2009; Ollivier et al., 2011; Semizhon et al., 2010), for estuarine water (Barros et al., 2004; Eyrolle & Charmasson, 2004; Porcelli et al., 1997), for sea water (Bowie et al., 2010; Otosaka et al., 2006; Scholten et al., 2005) or for soil water (Goryachenkova et al., 2009; Maity et al., 2011; Matisoff et al., 2011; Rachkova et al., 2010; Seliman et al., 2010; Xu et al., 2011). The colloid-radionuclide interaction is important not only from the perspective of their migration. For instance, the colloidal systems build-up is utilized for the radionuclides separation (Mansur & Mushtaq, 2011) and there has been noticed that they have interfering effects in the quantitative determination of the radio-isotopes. (Constantinou & Pashalidis, 2010; 2011; Kiliari & Pashalidis, 2010; Kyriakou & Pashalidis, 2011).

The radionuclides migration associated with the particulate phase in general, and with colloids in particular, that flow along with groundwater, is a topic of utmost importance for assessing the safety of the radioactive waste repositories. For assessing that overall process, there are conducted both field observations and experimental simulations in laboratory. The *in situ* investigation and the modeling of the role played by the colloidal matter and by the suspended particulate matter are mainly concerned with the saturated region of the water-bearing structures (e.g., Baik et al., 2010; Grambow, 2008; Kelkar et al., 2010; Laverov et al., 2010; Mal'kovskii, 2011; Malkovsky, 2011; Mazurek et al., 2011; Pourret et al., 2010; Severino et al., 2007; Utsunomiya et al., 2009). The modeling of the radionuclides migration in the unsaturated zone is addressed, as a general rule, by laboratory experiments (e.g., Ku et al., 2009; Massoudieh & Ginn, 2007). The latter range into two broad categories, batch tests and flow-through column experiments. In the first case, a solution spiked with the investigated radionuclide is mixed for a certain time-interval with the solid of interest, then the solution and/or the solid are analyzed (e.g., Anderson et al., 2009; Bradbury et al., 2005; Hu, Cheng, Zhang & Yang, 2010; Lujanienė et al., 2010; Rabung et al., 2005; Singer, Maher & Brown Jr, 2009). The column experiments investigate the radiocolloids migration characteristics, by simulating the groundwater flow conditions (e.g., Bryan et al., 2005; Li et al., 2011; Mibus et al., 2007; Solovitch-Vella et al., 2006).

Radionuclide migration with groundwater colloids through porous media (e.g., Bradford & Bettahar, 2006; Delos et al., 2008; Grolimund et al., 2001; Grolimund & Borkovec, 2001; 2006; Ilina et al., 2008; Kretzschmar et al., 1997; Li et al., 2010; Panfilov et al., 2008; Santos & Barros, 2010), or across fractured rock systems (e.g., Hu & Mori, 2008; Jeong et al., 2011; Kosakowski, 2004; Kurosawa & Ueta, 2001; Malkovsky & Pek, 2009b; Schindler et al., 2010; Tang & Weisbrod, 2009; 2010; Yamaguchi et al., 2008) are intensely investigated topics. At the same time, extensive research is conducted for assessing the humic and fulvic colloids effect on the radionuclides migration across the underground environment (e.g., Bouby et al., 2011; Geraedts & Maes, 2008; Joseph et al., 2011; Lippold et al., 2005; Lippold et al., 2005; Pshinko, 2009; Pshinko et al., 2009; Schmeide & Bernhard, 2010; Singh et al., 2009; Singhal et al., 2009; Yoshida & Suzuki, 2006).

The present work intends to succinctly review and to critically analyze the most recent contributions of the scientific literature in which there are addressed certain issues concerning the part that particulate phases play in the migration of the radionuclides characteristic to the radioactive waste stored in geological repositories, across more or less deep aquifer structures. An attempt is performed to provide an adequate definition of the particulate phases, and the main issues related to the radionuclides speciation are discussed, with particular emphasis on the radiocolloids development and distribution. A topic that is widely addressed in the present work concerns the radioparticles fractionation and their chemical characterization.

## 2. Radioparticles in groundwater

### 2.1 Particulate phases

It is a well-known fact that in a natural subsurface water-rock system, elements are distributed among the three constitutive phases, namely: (1) the solid phase, consisting of the mineral substratum and of the sediments; (2) the aqueous phase, within which elements are considered to occur as dissolved species, and (3) the particulate phase. It is by now fully demonstrated and unanimously accepted that much higher concentrations of contaminants are being carried by the particulate phases, as compared to concentrations of the corresponding dissolved species which are carried in the aqueous solution (Kalmykov & Denecke, 2011; Morel, 1983; Stumm & Morgan, 1996).

The particulate phase conveyed by groundwater is highly dynamic. Particles are continuously produced as a result of physical erosion and of chemical alteration of the mineral substratum, or ensuing to precipitation from super-saturated solutions. They undergo permanent composition changes, and are continuously removed from the water by dissolution, coagulation, and finally by deposition or binding to the solid phase.

Particles existing in groundwater may be both of inorganic and of organic nature (Wolthoorn et al., 2004). As a general rule, the inorganic particles composition mirrors to a large extent the nature of the mineral substratum in contact with water. Specifically, those particles may include solid fragments dislocated from the substratum, clay minerals, Fe(III), Mn(III,IV) and Al(III) oxihydroxide microparticles, silicates, carbonates, complexes and polymers of certain elements, etc. Organic particles are, at least in principle, of allochtonous origin, and they include fragments of degraded organic matter, macromolecules of organic substances which are specific to the soil (ex. humic or fulvic acids), but also organisms which are either alive or decomposing, microorganisms, viruses, as well as their exudates.

Aquatic particulate phases exhibit a continuous particle size distribution. In spite of that, practical reasons require that distinction is made between constituents that are dissolved, and those existing as particles on the one hand, and the various types of particles on the other. From a thermodynamic perspective, "dissolved" refers to a constituent for which a chemical potential can be defined (Stumm & Morgan, 1996).

In terms of dimension, shape and characteristics of transfer across the environment, taken as a whole, the particles range in two broad categories, namely colloids and suspended particles. A distinction between those two categories made by taking into account only the dimension criterion fails to be entirely satisfactory, although it is widely accepted and operationally useful. Opinions fail to be unanimous even as far as the dimensional boundary separating the two categories is concerned, so that according to various authors, the corresponding dimension may be 0.20 μm, 0.45 μm or 1 μm (Malkovsky & Pek, 2009a). In fact, the indicated values represent the pore dimensions of the filtering membranes which are used for separating the particles. The lower boundary of the colloidal particles dimension is accepted to be, as a general rule, $\approx 1$ nm, a circumstance which authorizes, from a certain perspective, the assimilation of colloids to nanoparticles (Geckeis et al., 2011; Wigginton et al., 2007). An important characteristic of the aquatic colloids, useful in distinguishing them from the suspended particles, consists in the fact that their vertical movement is not significantly affected by gravitational settling (Gustafsson & Gschwend, 1997; Stumm & Morgan, 1996).

### 2.2 Speciation of radionuclides in the environment

The migration of an element in a natural environment is basically conditioned by the way it speciates in that environment. By definition, the chemical species of a certain element are its specific appearances, defined as electronic or oxidation states, as complex or molecular structures (Templeton et al., 2000). At a given moment, the same element may occur in the hydrosphere under various physical-chemical appearances, including forms associated to the particles in suspension or dissolved forms, such as simple inorganic species, organic complexes, metallic ions adsorbed to the colloidal matter, etc.

Accounting for the radionuclides speciation is a fundamental step toward describing the geo- and bio-chemical processes in which they are involved, and above all for understanding how they migrate through the environment. (Hu et al., 2010; Salbu, 2009; Salbu et al., 2004). The definitions of the terms addressing the elements speciation, recommended by International Union for Pure and Applied Chemistry (IUPAC) (Templeton et al., 2000), have been adapted for the radionuclides by Salbu & Skipperud (2009) as follows:

*Radionuclide species* are defined according to their physicochemical properties such as nominal molecular mass, charge properties and valence, oxidation state, structure and morphology, density, degree of complexation.

The *speciation of radionuclides* is the distribution of a radionuclide amongst defined chemical radionuclide species in a system.

Colloids and suspended particles cannot be considered, under the above definition, species, they are, alternatively, chemical fractions. The *chemical fraction* is a group of chemical entities which have common physical (e.g. size), or chemical (e.g. reactivity) properties, that group being operationally outlined by means of an analytical process. The concerned analytical process is termed *fractionation*. Chemical fractions are not mutually exclusive, i.e. rather than identifying the involved chemical species, they indicate a specific behaviour. In terms of involved experimental techniques, fractionation is much more accessible that the detailed determination of chemical species. For instance, the results of the *in situ* fractionation of water samples derived from The Fen Central Complex (southern Norway), one of the world's largest natural reservoirs of thorium ($^{232}$Th), have shown that radionuclides occur mainly as colloids and chemical species of low molecular mass (Popic et al., 2011).

There is a certain dependence of the chemical reactivity of particles carried by water on their dimension (Wigginton et al., 2007), namely reactivity is, as a general rule, enhanced, as the particle dimension decreases. This behaviour is mainly a result of the fact that with decreasing particle dimension, the ratio between the constitutive atoms and the surface area of the particle increases. This is one of the reasons why a large interest is being recently dedicated to the investigation of how radionuclides are transferred by means of colloidal particles. At the same time, the colloids mobilization and transfer can occur both in a saturated, and in an unsaturated flow regime, this quite important issue requiring a careful consideration for an appropriate management of the radioactive waste repositories safety. The part that suspended particles, and macro-particles in general, play in the radionuclides migration is not a negligible one, yet it is worth considering it only in a saturated flow regime.

## 3. Formation of the radiocolloids

Entities formed by the coupling of colloidal particles with radionuclides and conveyed as such by groundwater, and which commonly are also termed *radiocolloids*, are classified in two large groups, according to their origin: (1) *intrinsic-*, *eigen-colloids* or "true" colloids; and (2) *carrier-* or *pseudo-colloids* (Geckeis et al., 2011; Malkovsky et al., 2009; Malkovsky & Pek, 2009a).

### 3.1 Intrinsic radiocolloids

Intrinsic-colloids are formed spontaneously, as a result of the polymerization of complexes derived from the hydrolysis of metal ions. Under certain environmentally-controlled circumstances, an increased tendency to form such colloids is displayed by Pu, Am, Np, for which the dimensions of the resulting aggregates and their number per unit volume are, as a general rule, proportional to the total concentrations of the concerned actinides in the solution (Murakami et al., 2005).

It is a well-known fact that tetravalent plutonium has a strong predilection to develop polymeric complexes and colloids. There has been however noticed that small polymers such as dimers, trimers and tetramers include mixed oxidation states of Pu (Walther et al., 2009). In natural waters with pH ranging between 6 and 8, Pu(IV) prevalently occurs under the form of true colloids. The latter exhibit a predisposition for getting bound to the rocks surface, a behaviour which with increasing particles dimension, becomes more and more obvious. It was found, for instance, that when the particles dimensions exceed 220 nm, Pu(IV) is virtually quantitatively sorbed on the rock surface (Perevalov et al., 2009). At the same time, equilibrium distribution of Pu(IV) polymers depending on the total Pu(IV) concentration in the solution was analyzed theoretically by Kulyako et al. (2008).

### 3.2 Pseudo-colloids

Pseudo-colloids are formed through the binding of radionuclides to the pre-existing colloidal particles of the groundwater. Any mineral fragment, either crystalline or amorphous (hydrated Al, Fe and Mn oxides), organic compounds (humic and fulvic acids), but also biota consisting prevalently of viruses and bacteria may act as carrier particles for the radio-nuclides. As compared to the first group of radiocolloids, pseudo-colloids are much more abundant in groundwater, and therefore they exert a much more extensive control on the radionuclides transfer.

Experiments have been conducted which were aimed at establishing the conditions under which Fe(III), Cr(III) and Zr(IV) build up pseudo-colloids together with colloidal silica. There was thus noticed that in a Fe(III) solution with a concentration of $\approx 1 \times 10^{-7}$ M, under pH conditions normally met in the hydrosphere, $Fe^{3+}$ cations and mononuclear $Fe(OH)_n^{3-n}$ hydroxo complexes mostly occur, and pseudo-colloids may form by the binding of Fe(III) species to colloidal silica (Davydov et al., 2003). In Cr(III) $\approx 1 \times 10^{-6}$ M synthetic solutions, when pH >4, in solution there are prevailing the $Cr(OH)^{2+}$ and $Cr(OH)_2^+$ species, with chromium displaying an obvious predilection toward forming pseudo-colloids with silica (Davydov et al., 2006). At the same time, in a solution of $\approx 1 \times 10^{-13}$ M concentration, Zr(IV) occurs under hydrated form as $Zr(OH)^{3+}$ and $Zr(OH)_2^{2+}$ . At pH 2–12 Zr(IV) participates in formation of stable pseudocolloid particles (Davydov et al., 2006).

By using the surface complexation model, several investigators have modeled the radionuclides adsorption on the surface of the colloidal particles to form pseudo-colloids. For instance Batuk et al. (2011) interpreted in this way the sorption behavior and speciation of U on silica colloids, Degueldre & Bolek (2009) modeled plutonium adsorption on hydrous metal oxide solids, Del Nero et al. (2004) interpreted Np(V) sorption on amorphous Al and Fe silicates, and uranyl ions on Al-hydroxide (Froideval et al., 2006). Sorption of Np(V), Pu(V), and Pu(IV) on colloids of Fe(III) oxides and hydrous oxides and $MnO_2$ was studied over wide ranges of solution pH and ionic strength by Khasanova et al. (2007). The surface complexation model assumes that the adsorbing ion forms a surface complex with the adsorbing site, similar to the formation of a dissolved complex.

A box model has been proposed in order to interpret the kinetics of the radionuclides uptake on suspended particulate matter (Barros & Abril, 2005; 2008). At the same time, there has been investigated the kinetics of the Cs(I) sorption on hydrous silica (Pathak & Choppin, 2006), and $Am^{3+}$ on suspended silica as a function of pH and ionic strength in the presence of complexing anions, humic acid and metal ions (Pathak & Choppin, 2007).

Living organisms with dimensions similar to colloids, like for instance pathogenic bacteria or viruses, are present in groundwater naturally, to form a distinct group named biocolloids (Bekhit et al., 2009). Since they are living organisms they migrate in the subsurface porous medium, being subject to a complex of biological, physical and chemical processes. Those micro-organisms act as pseudo-colloids, since their surfaces are often negatively charged, thus having the ability to bind and carry radionuclides through the subsurface environment (Johnsson et al., 2008; Luk'yanova et al., 2008; Seiler et al., 2011; Singer, Farges & Brown Jr, 2009; Wilkins et al., 2006; 2010).

### 3.3 Colloids generated by engineered barriers

Besides the two already mentioned colloid groups, Malkovsky et al. (2009) and Malkovsky & Pek (2009a) distinguish an additional third group, which they designate as "primary colloids". The latter are colloidal particles derived as a result of groundwater leaching the isolating materials utilized in the storage of low and intermediate level radioactive wastes, as well as of high-level radioactive wastes in geological disposal.

Action taken in order to prevent, as much as possible, the contamination of the geological environment with radionuclides stored in a radioactive waste repository, irrespective whether the latter is located at the ground surface or in the underground, makes use of the so-called *"barriers"*. The broadest meaning for barrier is "a physical obstruction that prevents or delays the movement of radionuclides or other material between components in a system, for example a waste repository" (IAEA, 2003). Such barriers may be either natural (i.e. geological), or constructed, in that latter case being called *"engineered barriers"*. In most instances there are used *"multiple barriers"*, namely "two or more natural or engineered barriers used to isolate radioactive waste in, and prevent migration of radionuclides from, a repository" (IAEA, 2007). Such barriers include the glass or ceramic matrixes to which the liquid waste is usually converted (Anderson et al., 2009; Curti et al., 2009), the steel canisters used for isolating the waste, the buffering backfill materials placed between the containers and the walls of the repository cells. The most adequate filling material for this purpose is bentonitic clay (Akgün et al., 2006; Ferrage et al., 2005; Galamboš et al., 2009; 2011; Gaucher et al., 2004; Pérez del Villar et al., 2005), since it has a low permeability and its coefficient of radionuclides diffusion is quite small (Arcos et al., 2008; Bradbury & Baeyens, 2011; Hu, Xie, He, Sheng, Chen, Li, Chen & Wang, 2010; Missana & García-Gutiérrez, 2007; Wang et al., 2005). When in contact with groundwater, all those barriers can release colloidal particles that in terms of both their chemical, and their mineralogical composition, are not characteristic to the concerned geological environment (Cadini et al., 2010; De Windt et al., 2004; Filby et al., 2008; Kunze et al., 2008; Wieland et al., 2004).

The effect of engineered barriers in terms of radiocolloids production is an outstandingly important research topic. It is worth mentioning that especially bentonite barriers in contact with weakly mineralized groundwater generate a highly concentrated colloidal phase which is liable to carry radionuclides (Albarran et al., 2008; 2011; Kalmykov et al., 2011; Kurosawa & Ueta, 2001; Missana et al., 2008; Sabodina et al., 2006; Tertre et al., 2005; Vilks et al., 2008).

### 3.4 The colloids stability

The colloids stability in groundwater is primarily controlled by the processes through which they agglomerate; at their turn, those processes are ruled by the colloid surface charge and by the solution composition, namely by its pH and ionic strength. (Geckeis et al., 2011; Schelero & von Klitzing, 2011). Accordingly, the groundwater chemistry plays a fundamental part in controlling the stability of the colloidal particles. Those particles ability of remaining in suspension in an aqueous environment depends on the interactions that are established between them when they reach close to one another. Colloids become stabilized through the formation of an electric double layer strong enough for preventing agglomeration. Yet this layer may be destroyed and the colloidal particles consequently coagulate and leave the system, along with the increase in ionic strength. There has been noticed that a reverse correlation exists between the colloids concentration in the solution and the ionic strength of the latter (Deepthi Rani & Sasidhar, 2011; Loux, 2011).

## 4. Fractionation and radioparticles charcterization

Analytical tools and detection methods used to characterize radioparticles in groundwater may be categorized as a function of the parameter to be determined as follows: (1) size fractionation; (2) size distribution; (3) surface area characterisation; (4) chemical and radiochemical analysis (May et al., 2008). Common size fractionation methods include ultrafiltration, tangential/cross-flow ultrafiltation (TFF/CFF), centrifugal-ultrafiltration (e.g., Gimbert et al., 2005; Liu et al., 2006; Pourret et al., 2007) and field-flow fractionation (FFF). Normally, those techniques are followed by the chemical analysis of the separated fractions. They will be discussed in detail in the following sections, since they are, taken together, the most frequently used experimental investigation approaches.

The size distribution of radioparticles can be investigated by a wide variety of techniques, such as laser light scattering (LLS) (e.g., Dreissig et al., 2011), diffuse light scattering (DLS) (e.g., Lahtinen et al., 2010), laser-induced breakdown detection (LIBD) (e.g., Baik et al., 2007), or photoelectron spectroscopy (Laverov et al., 2010). For the same purpose, atomic force microscopy (AFM) and transmission electron microscopy (TEM) (e.g., Doucet et al., 2005; 2004) are used as well.

The specific surface area (SSA) of the particles is the parameter describing the interdependence between the particles dimensions and their chemical or mineralogical composition. The SSA determination is frequently conducted by means of the Brunauer-Emmett-Teller (BET) gravimetric method. In an indirect way, information about the particles surface area may be obtained by means of AFM or TEM measurements.

In order to characterize the radionuclides speciation, a series of investigators have resorted to sequential chemical extraction experiments, conducted in accordance with pre-established work protocols (Bondareva, 2011; Bondareva & Bolsunovskii, 2008). As a general rule, ensuing to a scheme of sequential extraction of radionuclides from colloidal matter of groundwater, the following products will result: (1) water-soluble; (2) exchangeable, by using as reactant a 0.5 M $Ca(NO_3)_2$ solution at pH 5.5; (3) associated with carbonates, in the presence of 0.1 M $NH_4Ac$ solution, pH 4.8; (4) associated with organic matter, with 0.1 M NaOH solution, pH 10; (5) amorphous oxides, by using a mixture of 0.18 M $(NH_4)_2C_2O_4$ and 0.1 M $H_2C_2O_4$ solutions, at pH 3.5; (6) the residue digestion by means of HF (Novikov et al., 2009; Novikov et al., 2009).

### 4.1 Suspended particulate matter (SPM) fractionation

The SPM analysis techniques have been developed and are frequently conducted for surface, estuarine or sea water samples, but they are equally utilized as well for the analysis of radionuclide-contaminated groundwater (Katasonova & Fedotov, 2009; Stepanets et al., 2009).

Two different approaches are utilized in order to determine the concentrations of heavy metals and radionuclides bound on the SPM, one which is direct, and the other indirect. The direct determination method consists in separating the suspensions on filtering membranes of various porosities, followed by subsequent digestion and quantitative assessment of the contaminants from the separated material, by means of an adequate spectrometric technique (Blo et al., 2000; Nordstrom et al., 1999; Ödman et al., 1999; 2006; Ollivier et al., 2011; Yeager et al., 2005). Through the indirect method, both the filtered and the unfiltered water samples are analyzed in parallel, and the resulting concentration difference is considered to represent the concentration of the element bound on the SPM (Cidu & Frau, 2009; Cortecci et al., 2009; Gammons et al., 2005; Pokrovsky & Schott, 2002).

In a comparative study, Butler et al. (2008) have demonstrated that congruent results were obtained when the two methods were applied in parallel. Potential artifacts induced by filtration, such as contamination and/or adsorption of metals within the membrane have been investigated for different membrane materials, metals, etc. (Hedberg et al., 2011).

### 4.2 Ultrafiltrafiltration

Ultrafiltration became the usual technique for separating colloidal particles from any type of natural water. The separation is performed by using a filtering membrane of a nominal size, often reported in a molecular size cut-off, in dalton (Da) units. Dalton is non-SI unit accepted for being utilized in the International System of Units whose values in SI units must be obtained experimentally. One dalton unit is equivalent to one atomic mass unit (amu) and is used in ultrafiltration in order to determine the approximate size of particles for which a rigorous molecular mass cannot be indicated.

Single filtration is frequently used for radioparticles separation (Caron & Smith, 2011; Novikov et al., 2009). Many investigators resort however to sequential (cascade) filtration, which involves the utilization of a series of cells with mixing, where each cell contains a membrane and a drainage system connected to measuring devices (Bauer & Blodau, 2009; Dreissig et al., 2011; Eyrolle & Charmasson, 2004; Graham et al., 2011; Pourret et al., 2010; Stepanets et al., 2009). In most cases, before starting the ultrafiltration operation, the SPM is separated by filtering the water samples on filtering membranes of 0.45 μm or 0.22 μm (e.g., Graham et al., 2008; Novikow et al., 2009; Singhal et al., 2009).

Tangential-flow ultrafiltration (TFF/CFF) is a common method for size fractionation in natural waters which has also been applied for colloids separation (e.g., Andersson et al., 2001; Buesseler et al., 2009; Goveia et al., 2010; Hassellöv et al., 2007; Ohtsuka et al., 2006). The main advantage of TFF is its use as a preparative fractionation method that allows for processing of large volumes of sample - even water samples reaching, each one, up to several hundreds of liters. It is also well known that the size distribution of colloids in natural waters can easily change due to aging, changes in pH, ionic strength or redox conditions (Hedberg et al., 2011). The processes that can potentially alter the size distributions of the colloids include coagulation, adsorption to surfaces, hydrolysis and precipitation. In addition to these processes, associated trace constituents are also affected by sorption processes, solution complexation and redox precipitations (Katasonova & Fedotov, 2009; Salbu, 2009).

### 4.3 Field-Flow Fractionation

Several recently published reviews emphasize the efficiency of the Field-Flow Fractionation (FFF) techniques for separating and estimating physical parameters of different materials: biopolymers, biological cells, microorganisms, and colloidal and solid particles (Bouby & Geckeis, 2011; Dubascoux et al., 2010; Kowalkowski et al., 2006; Qureshi & Kok, 2011; Stolpe et al., 2005; Williams et al., 2011).

Particles separation by means of the FFF techniques is achieved by a combined action of the non-uniform flow velocity profile of a carrier liquid and a transverse physical field applied perpendicularly to this carrier. Carrier liquid flowing along the channel forms a nearly parabolic flow velocity profile across the channel. The sample to be investigated is dissolved or suspended in a carrier fluid and is pumped through a thin, not filled, channel. At the present time, FFF comprises a family of separation devices with a great number of sub-techniques used mainly for the separation and characterization of particulate species in the size range from $10^{-3}$ µm to $10^{2}$ µm. Highly popular among those sub-techniques is the flow field-flow fractionation (Fl-FFF). The version of the manufactured separation system for which the channel conveying the carrier liquid had an asymmetric shape (As-Fl-FFF) proved to be the most efficient, as it enabled nanoparticles ranging form 1 nm to 100 µm to be separated. As a general rule, an Inductively Coupled Plasma Mass Spectrometry (ICP-MS) is used as on-line detector, which allows reaching low detection limits, high sensitivity, large dynamic range and ability to simultaneously measure a large number of elements (Table 1).

| Method | Analyt | Detection method | Details | References |
|---|---|---|---|---|
| As-Fl-FFF | Cs, Eu, Th, U | ICP-MS | Interaction of bentonite colloids with metals in presence of HA | Bouby et al. (2011) |
| As-Fl-FFF | minor elements | UV detector, ICP-MS | Analysis of colloids released from bentonite and crushed rock | Lahtinen et al. (2010) |
| As-Fl-FFF | U | UV detector, ICP-MS | U complexation by groundwater dissolved organic C | Ranville et al. (2007) |
| As-Fl-FFF | Cs, La, Ce, Eu, Th, U | ICP-MS | Quantitative characterization of natural colloids | Bouby et al. (2008) |
| Fl-FFF | U(VI) | ICP-MS | U(VI) sorption to nanoparticulate hematite | Lesher et al. (2009) |
| Fl-FFF | $^{57}Fe$, $^{65}Cu$, $^{127}I$, $^{184}W$, $^{88}Sr$, $^{238}U$ | ICP-MS | Chemical and colloidal analyses of natural seep water | Cizdziel et al. (2008) |
| As-Fl-FFF | | LIBD, ICP-MS | Characterization of aquatic groundwater colloids | Baik et al. (2007) |
| As-Fl-FFF | | | Cm(III) complexation behavior of isolated HA and FA derived from Opalinus clay | Claret et al. (2005) |
| As-Fl-FFF | As, Cd, Sb, Se, Sn, Pb | ICP-MS | Soil leachate | Dubascoux et al. (2008) |

Table 1. Some field-flow fractionation procedures and detection method used to charactherize colloid particles.

A large number of recent works contributing to the development of the FFF techniques significantly widened their applicability range in particle size analysis (Ahn et al., 2010; Baalousha et al., 2006; 2005; Dubascoux et al., 2008; Gascoyne, 2009; Isaacson & Bouchard, 2010; Otte et al., 2009; Pifer et al., 2011).

### 4.4 Passive sampling techniques

For a given element, the overwhelming majority of its species are unstable chemical forms that occur under precarious equilibria. As a general rule, these equilibria are disturbed during the routine operations of collection, transport and storage of the samples, this fact resulting in most cases in erroneous information about the considered system. This is the reason why recently, in the elements speciation analysis, passive sampling techniques have received increasingly large acceptance Vrana et al. (2005).

In the most general meaning, passive sampling is that particular sampling technique which relies on the free transfer of the analyte from the sampled environment to a receiving phase in a sampling device, by the effect of the difference between the chemical potentials that the analyte has in the two environments. The transfer of the analyte from one environment toward the other continues until equilibrium is reached within the system, or until the sampling is stopped by an operator. In the first case, it is said that the passive sampling device operates in an equilibrium regime, while in the second one, it is said that it operates in a kinetic regime. In both situations, the sampling occurs without the involvement of any source of energy other than the indicated difference of chemical potential.

Among the passive sampling techniques, the diffusive gradients in thin films (DGT) technique, introduced by Davison & Zhang (1994), is highly ranked as a consequence of its ability to determine labile species in natural waters, sediments and soils. The DGT technique theoretical background relies on the Fick's first law of diffusion. For aqueous systems determinations a passive sampling device is used, which consists of a plastic piece in the shape of a piston, on which two gel discs and a filtering membrane are installed. The first gel, impregnated with binder material, is used for retaining the analytes. The second gel disc has a pre-determined porosity and its role consists in maintaining a constant concentration during the analyte diffusion between the solution and the binder material. The typical binder material is the Chelex-100 resin, while the material used for the diffusion control is the acrylamide/agarose hydrogel. In the end, the binder gel is eluted and the resulting solution is analyzed by means of ICP-MS French et al. (2005); Garmo et al. (2008; 2006), multi-collector ICP-MS (Malinovsky et al., 2005), thermal ionization mass-spectrometry (TIMS) (Dahlqvist et al., 2005), or directly through the gel analysis by laser-ablation ICP-MS (Pearson et al., 2006, Warnken et al., 2004). A comparative study between the DGT techniques and ultrafiltration has been conducted by Forsberg et al. (2006).

By simultaneously immersing several devices which have diffusion gels of various thicknesses or porosities, there is possible to obtain information about the nature of the complexes which are present in various categories of natural waters, and also about the kinetics of the geochemical reactions in which those complexes are involved (Zhang & Davison, 2000; 2001), including in porewater (Leermakers et al., 2005; Wu et al., 2011). DGT has been successfully utilized in order to monitor the radionuclides migration (Chang et al., 1998; Duquène et al., 2010; Gao et al., 2010; Gregusova & Docekal, 2011; Li et al., 2007; 2009; 2006; Salbu, 2007).

The same class of techniques also encompasses the diffusive equilibration in thin films (DET). In this latter case, the sampling device includes only a single layer of gel. This gel layer is maintained in contact with the environment to be analyzed, until equilibrium is reached between the analyte concentration in the environment, and the corresponding concentration in the gel. The technique is mainly utilized in sedimentary environments (Dočekalová et al., 2002). The total content of the analyte in the gel mirrors the ability of the concerned species to penetrate the gel, being controlled by its dimension. In the case of elements associations with colloidal matter, by combining information provided by DET on the equilibration, with information provided by DGT on the species dynamics, a much more appropriate description is obtained in terms of elements speciation (Fones et al., 2001; Gao et al., 2006; van der Veeken et al., 2008; Vandenhove et al., 2007).

In spite of being quite simple to handle, the DGT/DET techniques require a very cautious approach in terms of interpretation. One must take into account that the elements behaviour is not identical with respect to the binder gels, and that equilibration is largely controlled by the pH and the ionic strength of the solution. In addition, potential artifacts may be introduced during the devices preparation, elution and determination processes.

## 5. Conclusion

All the energy generation, industrial, medical, or military activities which utilize radioactive substances are producing low and intermediate level, and high-level radioactive waste, which needs to be isolated from the biosphere in order to protect the future generations from the hazards potentially induced by the associated radioactivity. As a function of the radioactivity level and of the half life length of the radioisotopes existing in their inventory, either near surface disposal facilities or deep geological repositories are assigned to the storage of that waste.

It is necessary that a radioactive waste repository location is selected very carefully in terms of its hydrogeological environment, since it is a well-known fact that groundwater is the most important vector involved in the transfer of the contaminants. The radionuclides migration across the geosphere takes place as water-dissolved constituents, but especially bound to the particulate phases carried by groundwater. Generally speaking, elements in a "dissolved" state speciate, i.e. they are distributed among forms defined by electronic states, oxidation states, isotopic compositions, as well as complex or molecular specific structures, while when they occur as particles, it is said that they belong to chemical fractions. A chemical fraction is a group of chemical entities which have common physical (e.g. size), or chemical (e.g. reactivity) properties, that group being operationally outlined by means of an analytical process.

Many authors rightly believe that the part played in the radionuclides transfer by particulate phases in general, but especially by the colloidal matter, is so important, that modeling approaches which do not take it into account are unrealistic. In an aqueous environment, radiocolloids occur as intrinsic-colloids and pseudo-colloids. The first category is specific to several transuranic elements which under certain conditions (e.g., pH, ionic strength), possess the capacity of forming structures with colloidal properties. The second type of colloids forms through the attachment of the radionuclide to particles pre-existing in groundwater. Those particles are spontaneously formed in the aquifer structures, but they might also derive from engineered barriers that are built in order to stop radionuclides from migrating out of the repositories. In this respect, bentonitic clays used as a buffer are the most important source of colloidal particles.

In order to identify and to characterize the particulate phases-radionuclides associations, a set of specific separation and determination methods is resorted to. Among the separation techniques, the filtration/ultrafiltration is the most frequently used, while for the colloidal matter characterization, much more efficient are the techniques belonging to the field-flow fractionation category. The latter have the advantage that they may be hyphenated with quantitative determination techniques, among which the inductively coupled plasma mass spectrometry is the most widely used. Recently, into the radiocolloids analysis domain there additionally included the "diffusive gradients in thin films (DGT)" and "diffusive equilibration in thin films (DET)" techniques, which appear to be outstandingly promising for the study of the radionuclides migration across the environment.

## 6. Acknowledgements

The present study will be conducted in the framework of the research project MIGRELEMENT (Project No. 32112/2008), financially supported by The Executive Agency for Higher Education, Research, Development and Innovation Funding (UEFISCDI).

## 7. References

Acero, P., Auqué, L. F., Gimeno, M. J. & Gómez, J. B. (2010). Evaluation of mineral precipitation potential in a spent nuclear fuel repository, *Environ. Earth Sci.* 59: 1613–1628.

Ahn, J. Y., Kim, K. H., Lee, J. Y., Williams, P. S. & Moon, M. H. (2010). Effect of asymmetrical flow field-flow fractionation channel geometry on separation efficiency, *J. Chromatogr. A* 1217(24): 3876–3880.

Akgün, H., Koçkar, M. K. & Aktürk, O. (2006). Evaluation of a compacted bentonite/sand seal for underground waste repository isolation, *Environ. Geol.* 50: 331–337.

Albarran, N., Alonso, Ú. Missana, T., García-Gutiérrez, M. & Mingarrom, M. (2008). Evaluation of geochemical conditions favourable for the colloid-mediated uranium migration in a granite fracture, *e-Terra* 5: 1–8.

Albarran, N., Missana, T., García-Gutiérrez, M., Alonso, U. & Mingarro, M. (2011). Strontium migration in a crystalline medium: effects of the presence of bentonite colloids, *J. Contam. Hydrol.* 122(1-4): 76– 85.

Aleksandrova, O. N., Schulz, M. & Matthies, M. (2010). Natural remediation of surface water systems contaminated with nuclear waste via humic substances in South Ural, *Water, Air, Soil Pollut.* 206: 203–214.

Altmann, S. (2008). 'Geo'chemical research: A key building block for nuclear waste disposal safety cases, *J. Contam. Hydrol.* 102(3-4): 174–179.

Anderson, E. B., Rogozin, Y. M., Smirnova, E. A., Bryzgalova, R. V., Malimonova, S. I., Andreeva, N. R., Shabalev, S. I. & Savonenkov, V. G. (2009). Effect of colloidal component in solutions on adsorption of actinides [Am(III), Pu(IV)] from simulated groundwater by glass and granodiorite, *Radiochemistry* 51: 542–550.

Andersson, P. S., Porcelli, D., Gustafsson, O., Ingri, J. & Wasserburg, G. J. (2001). The importance of colloids for the behavior of uranium isotopes in the low-salinity zone of a stable estuary, *Geochim. Cosmochim. Acta* 65(1): 13–25.

Arcos, D., Grandia, F., Domenech, C., Fernandez, A. M., Villar, M. V., Muurinen, A., Carlsson, T., Sellin, P. & Hernan, P. (2008). Long-term geochemical evolution of the near field repository: Insights from reactive transport modelling and experimental evidences, *J. Contam. Hydrol.* 102(3-4): 196–209.

Baalousha, M., Kammer, F. V. D., Motelica-Heino, M. & Hilal, H. S.and Le Coustumer, P. (2006). Size fractionation and characterization of natural colloids by flow-field flow fractionation coupled to multi-angle laser light scattering, *J. Chromatogr. A* 1104: 272–281.

Baalousha, M., Kammer, F. V. D., Motelica-Heino, M. & Le Coustumer, P. (2005). Natural sample fractionation by FlFFF-MALLS-TEM: Sample stabilization, preparation, pre-concentration and fractionation, *J. Chromatogr. A* 1093: 156–166.

Baik, M. H., Kim, S. S., Lee, J. K., Lee, S. Y., Kim, G. Y. & Yun, S. T. (2010). Sorption of $^{14}C$, $^{99}Tc$, $^{137}Cs$, $^{90}Sr$, $^{63}Ni$, and $^{241}Am$ onto a rock and a fracture-filling material from the Wolsong low- and intermediate-level radioactive waste repository, Gyeongju, Korea, *J. Radioanal. Nucl. Chem.* 283: 337–345.

Baik, M. H., Lee, S. Y. & Shon, W. J. (2009). Retention of uranium(VI) by laumontite, a fracture-filling material of granite, *J. Radioanal. Nucl. Chem.* 280: 69–77.

Baik, M.-H., Yun, J.-I., Bouby, M., Hahn, P.-S. & Kim, J.-I. (2007). Characterization of aquatic groundwater colloids by a laser-induced breakdown detection and ICP-MS combined with an asymmetric flow field-flow fractionation, *Korean J. Chem. Eng.* 24: 723–729.

Barros, H. & Abril, J. M. (2005). Constraints in the construction and/or selection of kinetic box models for the uptake of radionuclides and heavy metals by suspended particulate matter, *Ecol. Model.* 185(2-4): 371–385.

Barros, H. & Abril, J. M. (2008). Kinetic box models for the uptake of radionuclides and heavy metals by suspended particulate matter: equivalence between models and its implications, *J. Environ. Radioactivity* 99: 146–158.

Barros, H., Laissaoui, A. & Abril, J. M. (2004). Trends of radionuclide sorption by estuarine sediments. Experimental studies using $^{133}Ba$ as a tracer, *Sci. Total Environ.* 319(1-3): 253–267.

Batuk, D. N., Shiryaev, A. A., Kalmykov, S. N., Batuk, O. N., Romanchuk, A., Shirshin, E. A. & Zubavichus, Y. V. (2011). Sorption and speciation of uranium on silica colloids, *in* S. N. Kalmykov & M. A. Denecke (eds), *Actinide Nanoparticle Research*, Springer Berlin Heidelberg, pp. 315–332.

Bauer, M. & Blodau, C. (2009). Arsenic distribution in the dissolved, colloidal and particulate size fraction of experimental solutions rich in dissolved organic matter and ferric iron, *Geochim. Cosmochim. Acta* 73(3): 529–542.

Bekhit, H. M., El-Kordy, M. A. & Hassan, A. E. (2009). Contaminant transport in groundwater in the presence of colloids and bacteria: Model development and verification, *J. Contam. Hydrol.* 108(3-4): 152–167.

Blo, G., Contado, C., Fagioli, F. & Dondi, F. (2000). Size-elemental characterization of suspended particle matter by split-flow thin cell fractionation and slurry analysis-electrothermal atomic absorption spectrometry, *Analyst* 125(7): 1335–1339.

Bondareva, L. (2011). The relationship of mineral and geochemical composition to artificial radionuclide partitioning in Yenisei river sediments downstream from Krasnoyarsk, *Environ. Monit. Assess.* Online First: 1–17.

Bondareva, L. G. & Bolsunovskii, A. Y. (2008). Speciation of artificial radionuclides $^{60}Co$, $^{137}Cs$, $^{152}Eu$, and $^{241}Am$ in bottom sediments of the Yenisei river, *Radiochemistry* 50: 547–552.

Bouby, M. & Geckeis, H. (2011). Characterization of colloid-borne actinides by flow field-flow fractionation (FlFFF) multidetector analysis (MDA), *in* S. N. Kalmykov & M. A. Denecke (eds), *Actinide Nanoparticle Research*, Springer Berlin Heidelberg, pp. 105–135.

Bouby, M., Geckeis, H. & Geyer, F. W. (2008). Application of asymmetric flow field-flow fractionation (AsFlFFF) coupled to inductively coupled plasma mass spectrometry (ICPMS) to the quantitative characterization of natural colloids and synthetic nanoparticles, *Anal. Bioanal. Chem.* 392: 1447–1457.

Bouby, M., Geckeis, H., Lutzenkirchen, J., Mihai, S. & Schafer, T. (2011). Interaction of bentonite colloids with Cs, Eu, Th and U in presence of humic acid: A flow field-flow fractionation study, *Geochim. Cosmochim. Acta* 75(13): 3866–3880.

Bowie, A. R., Townsend, A. T., Lannuzel, D., Remenyi, T. A. & van der Merwe, P. (2010). Modern sampling and analytical methods for the determination of trace elements in marine particulate material using magnetic sector inductively coupled plasma-mass spectrometry, *Anal. Chim. Acta* 676(1-2): 15–27.

Bradbury, M. H. & Baeyens, B. (2011). Predictive sorption modelling of Ni(II), Co(II), Eu(III), Th(IV) and U(VI) on MX-80 bentonite and Opalinus Clay: A bottom-up approach, *Appl. Clay Sci.* 52(1-2): 27–33.

Bradbury, M. H., Baeyens, B., Geckeis, H. & Rabung, T. (2005). Sorption of Eu(III)/Cm(III) on Ca-montmorillonite and Na-illite. Part 2: Surface complexation modelling, *Geochim. Cosmochim. Acta* 69(23): 5403–5412.

Bradford, S. A. & Bettahar, M. (2006). Concentration dependent transport of colloids in saturated porous media, *J. Contam. Hydrol.* 82(1-2): 99–117.

Bryan, N. D., Barlow, J., Warwick, P., Stephens, S., Higgo, J. J. W. & Griffin, D. (2005). The simultaneous modelling of metal ion and humic substance transport in column experiments, *J. Environ. Monit.* 7: 196–202.

Buesseler, K. O., Kaplan, D. I., Dai, M. & Pike, S. (2009). Source-dependent and source-independent controls on plutonium oxidation state and colloid associations in groundwater, *Environ. Sci. Technol.* 43(5): 1322–1328.

Butler, B. A., Ranville, J. F. & Ross, P. E. (2008). Direct versus indirect determination of suspended sediment associated metals in a mining-influenced watershed, *Appl Geochem.* 23: 1218–1231.

Cadini, F., De Sanctis, J., Girotti, T., Zio, E., Luce, E. & Taglioni, A. (2010). Monte Carlo estimation of radionuclide release at a repository scale, *Ann. Nucl. Energy* 37(6): 861–866.

Caron, F. & Smith, D. (2011). Fluorescence analysis of natural organic matter fractionated by ultrafiltration: Contrasting between urban-impacted water, and radio-contaminated water from a near-pristine site, *Water, Air, Soil Pollut.* 214: 471–490.

Chang, L.-Y., Davison, W., Zhang, H. & Kelly, M. (1998). Performance characteristics for the measurement of Cs and Sr by diffusive gradients in thin films (DGT), *Anal. Chim. Acta* 368(3): 243 – 253.

Cidu, R. & Frau, F. (2009). Distribution of trace elements in filtered and non filtered aqueous fractions: Insights from rivers and streams of Sardinia (Italy), *Appl. Geochem.* 24(4): 611–623.

Cizdziel, J. V., Guo, C., Steinberg, S. M., Yu, Z. & Johannesson, K. H. (2008). Chemical and colloidal analyses of natural seep water collected from the exploratory studies facility inside Yucca Mountain, Nevada, USA, *Environ. Geochem. Health* 30(1): 31–44.

Claret, F., Schäfer, T., Rabung, T., Wolf, M., Bauer, A. & Buckau, G. (2005). Differences in properties and Cm(III) complexation behavior of isolated humic and fulvic acid derived from Opalinus clay and Callovo-Oxfordian argillite, *Appl. Geochem.* 20(6): 1158–1168.

Constantinou, E. & Pashalidis, I. (2010). Uranium determination in water samples by liquid scintillation counting after cloud point extraction, *J. Radioanal. Nucl. Chem.* 286: 461–465.

Constantinou, E. & Pashalidis, I. (2011). Thorium determination in water samples by liquid scintillation counting after its separation by cloud point extraction, *J. Radioanal. Nucl. Chem.* 287: 261–265.

Cortecci, G., Boschetti, T., Dinelli, E., Cidu, R., Podda, F. & Doveri, M. (2009). Geochemistry of trace elements in surface waters of the Arno River Basin, northern Tuscany, Italy, *Appl. Geochem.* 24(5): 1005–1022.

Curti, E., Dähn, R., Farges, F. & Vespa, M. (2009). Na, Mg, Ni and Cs distribution and speciation after long-term alteration of a simulated nuclear waste glass: A micro-XAS/XRF/XRD and wet chemical study, *Geochim. Cosmochim. Acta* 73(8): 2283–2298.

Dahlqvist, R., Andersson, P. S. & Ingri, J. (2005). The concentration and isotopic composition of diffusible Nd in fresh and marine waters, *Earth Planet. Sci. Lett.* 233(1-2): 9–16.

Davison, W. & Zhang, H. (1994). *In-situ* speciation measurements of trace components in natural waters using thin-film gels, *Nature* 367(6463): 546–548.

Davydov , Y. P., Voronik, N. I., Davydov, D. Y. & Titov, A. S. (2006). Speciation of Cr(III) radionuclides in solutions, *Radiochemstry* 48(4): 365–368.

Davydov, Y. P., Davydov, D. Y. & Zemskova, L. M. (2006). Speciation of Zr(IV) radionuclides in solutions, *Radiochemstry* 48(4): 358–364.

Davydov, Y. P., Grachok, M. A. & Davydov, D. Y. (2003). Speciation of Fe(III) radionuclides in aqueous solutions, *Radiochemistry* 45: 40–46.

De Windt, L., Pellegrini, D. & van der Lee, J. (2004). Coupled modeling of cement/claystone interactions and radionuclide migration, *J. Contam. Hydrol.* 68(3-4): 165–182.

Deepthi Rani, R. & Sasidhar, P. (2011). Stability assessment and characterization of colloids in coastal groundwater aquifer system at Kalpakkam, *Environ. Earth Sci.* 62: 233–243.

Degueldre, C. & Bolek, M. (2009). Modelling colloid association with plutonium: The effect of pH and redox potential, *Appl. Geochem.* 24(2): 310–318.

Del Nero, M., A., A., Madé, B., Barillon, R. & Duplâtre, G. (2004). Surface charges and Np(V) sorption on amorphous al and fe silicates, *Chem. Geol.* 211(1-2): 15–45.

Delos, A., Walther, C., Schäfer, T. & Büchner, S. (2008). Size dispersion and colloid mediated radionuclide transport in a synthetic porous media, *J. Colloid Interface Sci.* 324(1-2): 212–215.

Dogaru, D., Niculae, O., Terente, M., Jinescu, G. & Duliu, O. G. (2010). Complementary safety indicators of Saligny radioactive waste repository, *Romanian Reports Phys.* 62(4): 811–820.

Doucet, F. D., Maguire, L. & Lead, J. R. (2005). Assessment of cross-flow filtration for the size fractionation of freshwater colloids and particles, *Talanta* 67: 144–154.

Doucet, F. J., Maguire, L. & Lead, J. R. (2004). Size fractionation of aquatic colloids and particles by cross-flow filtration: analysis by scanning electron and atomic force microscopy, *Anal. Chim. Acta* 522: 59–71.

Dočekalová, H., Clarisse, O., Salomon, S. & Wartel, M. (2002). Use of constrained DET probe for a high-resolution determination of metals and anions distribution in the sediment pore water, *Talanta* 57(1): 145–155.

Dreissig, I., Weiss, S., Hennig, c., Bernhard, G. & Zänker, H. (2011). Formation of uranium(IV)-silica colloids at near-neutral pH, *Geochim. Cosmochim. Acta* 75(2): 352–367.

Dubascoux , S., Von Der Kammer, F., Le Hécho, I., Gautier, M. P. & Lespes, G. (2008). Optimisation of asymmetrical flow field flow fractionation for environmental nanoparticles separation, *J. Chromatogr. A* 1206: 160–165.

Dubascoux, S., Le Hécho, I., Hassellöv, M., Von Der Kammer, F., Gautier, M. P. & Lespes, G. (2010). Field-flow fractionation and inductively coupled plasma mass spectrometer coupling: History, development and applications, *J. Anal. At. Spectrom.* 25(5): 613–623.

Dubascoux, S., Le Hécho, I., Gautier, M. P. & Lespes, G. (2008). On-line and off-line quantification of trace elements associated to colloids by As-Fl-FFF and ICP-MS, *Talanta* 77: 60–65.

Duquène, L., Vandenhove, H., Tack, F., Van Hees, M. & Wannijn, J. (2010). Diffusive gradient in thin FILMS (DGT) compared with soil solution and labile uranium fraction for predicting uranium bioavailability to ryegrass, *J. Environ. Radioactivity* 101(2): 140–147.

Eyrolle, F. & Charmasson, S. (2004). Importance of colloids in the transport within the dissolved phase (< 450 nm) of artificial radionuclides from the Rhône river towards the Gulf of Lions (Mediterranean Sea), *J. Environ. Radioact.* 72(3): 273–286.

Ferrage, E., Tournassat, C., Rinnert, E. & Lanson, B. (2005). Influence of pH on the interlayer cationic composition and hydration state of Ca-montmorillonite: Analytical chemistry, chemical modelling and XRD profile modelling study, *Geochim. Cosmochim. Acta* 69(11): 2797–2812.

Filby, A., Plaschke, M., Geckeis, H. & Fanghánel, T. (2008). Interaction of latex colloids with mineral surfaces and Grimsel granodiorite, *Journal of Contaminant Hydrology* 102(3-4): 273–284.

Fones, G. R., Davison, W., Holby, O., Jorgensen, B. B. & Thamdrup, B. (2001). High-resolution metal gradients measured by in situ DGT/DET deployment in Black Sea sediments using an autonomous benthic lander, *Limnol. Oceanogr.* 46(4): 982–988.

Forsberg, J., Dahlqvist, R., Gelting-Nyström, J. & Ingri, J. (2006). Trace metal speciation in brackish water using diffusive gradients in thin films and ultrafiltration: Comparison of techniques, *Environ. Sci. Technol.* 40(12): 3901–3905.

French, M. A., Zhang, H., Pates, J. M., Bryan, S. E. & Wilson, R. C. (2005). Development and performance of the diffusive gradients in thin-films technique for the measurement of technetium-99 in seawater, *Anal. Chem.* 77(1): 135–139.

Froideval, A., Del Nero, M., Gaillard, C., Barillon, R., Rossini, I. & Hazemann, J. (2006). Uranyl sorption species at low coverage on Al-hydroxide: TRLFS and XAFS studies, *Geochim. Cosmochim. Acta* 70(21): 5270–5284.

Galamboš, M., Kufčáková, J. & Rajec, P. (2009). Adsorption of cesium on domestic bentonites, *J. Radioanal. Nucl. Chem.* 281: 485–492.

Galamboš, M., Rosskopfová, O., Kufčáková, J. & Rajec, P. (2011). Utilization of Slovak bentonites in deposition of high-level radioactive waste and spent nuclear fuel, *J. Radioanal. Nucl. Chem.* 288: 765–777.

Gammons, C. H., Nimick, D. A., Parker, S. R., Cleasby, T. E. & McCleskey, R. B. (2005). Diel behavior of iron and other heavy metals in a mountain stream with acidic to neutral pH: Fisher Creek, Montana, USA, *Geochim. Cosmochim. Acta* 69: 2505–2516.

Gao, Y., Baeyens, W., Galan, S. D., Poffijn, A. & Leermakers, M. (2010). Mobility of radium and trace metals in sediments of the Winterbeek: Application of sequential extraction and DGT techniques, *Environ. Poll.* 158(7): 2439–2445.

Gao, Y., Leermakers, M., Gabelle, C., Divis, P., Billon, G., Ouddane, B., Fischer, J.-C., Wartel, M. & Baeyens, W. (2006). High-resolution profiles of trace metals in the pore waters of riverine sediment assessed by DET and DGT, *Sci. Total Environ.* 362(1-3): 266–277.

Garmo, Ø. A., Davison, W. & Zhang, H. (2008). Effects of binding of metals to the hydrogel and filter membrane on the accuracy of the diffusive gradients in thin films technique, *Anal. Chem.* 80(23): 9220–9225.

Garmo, Ø. A., Lehto, N. J., Zhang, H., Davison, W., Røyset, O. & Steinnes, E. (2006). Dynamic aspects of dgt as demonstrated by experiments with lanthanide complexes of a multidentate ligand, *Environ. Sci. Technol.* 40(15): 4754–4760.

Gascoyne, P. R. C. (2009). Dielectrophoretic-field flow fractionation analysis of dielectric, density, and deformability characteristics of cells and particles, *Anal. Chem.* 81(21): 8878–8885.

Gaucher, E. C., Blanc, P., Matray, J.-M. & Michau, N. (2004). Modeling diffusion of an alkaline plume in a clay barrier, *Appl. Geochem.* 19(10): 1505–1515.

Geckeis, H. & Rabung, T. (2008). Actinide geochemistry: From the molecular level to the real system, *J. Contam. Hydrol.* 102(3-4): 187–195.

Geckeis, H., Rabung, T. & Schäfer, T. (2011). Actinide-nanoparticle interaction: Generation, stability and mobility, *in* S. N. Kalmykov & M. A. Denecke (eds), *Actinide Nanoparticle Research*, Springer Berlin Heidelberg, pp. 1–30.

Geraedts, K. & Maes, A. (2008). Determination of the conditional interaction constant between colloidal technetium(IV) and Gorleben humic substances, *Appl. Geochem.* 23: 1127–1139.

Gimbert, L., Haygarth, P., Beckett, R. & Worsfold, P. (2005). Comparison of centrifugation and filtration techniques for the size fractionation of colloidal material in soil suspensions using sedimentation field-flow fractionation, *Environ. Sci. Technol.* 39(6): 1731–1735.

Goryachenkova, T. A., Kazinskaya, I. E., Kuzovkina, E. V.and Novikov, A. P. & Myasoedov, B. F. (2009). Association of radionuclides with colloids in soil solutions, *Radiochemistry* 51: 201–210.

Goveia, D., Lobo, F. A., Burba, P., Fraceto, L., Dias Filho, N. L. & Rosa, A. H. (2010). Approach combining on-line metal exchange and tangential-flow ultrafiltration for in-situ characterization of metal species in humic hydrocolloids, *Anal. Bioanal. Chem.* 397(2): 851–860.

Graham, M. C., Oliver, I. W., MacKenzie, A. B., Ellam, R. M. & Farmer, J. G. (2008). An integrated colloid fractionation approach applied to the characterisation of porewater uranium-humic interactions at a depleted uranium contaminated site, *Sci. Total Environ.* 404(1): 207–217.

Graham, M. C., Oliver, I. W., MacKenzie, A. B., Ellam, R. M. & Farmer, J. G. (2011). Mechanisms controlling lateral and vertical porewater migration of depleted uranium (DU) at two UK weapons testing sites, *Sci. Total Environ.* 409(10): 1854–1866.

Grambow, B. (2008). Mobile fission and activation products in nuclear waste disposal, *J. Contam. Hydrol.* 102(3-4): 180–186.

Gregusova, M. & Docekal, B. (2011). New resin gel for uranium determination by diffusive gradient in thin films technique, *Anal. Chim. Acta* 684(1-2): 142–146.

Grolimund, D., Barmettler, K. & Borkovec, M. (2001). Release and transport of colloidal particles in natural porous media 2. Experimental results and effects of ligands, *Water Resour. Res.* 37(3): 571–582.

Grolimund, D. & Borkovec, M. (2001). Release and transport of colloidal particles in natural porous media 1. Modeling, *Water Resour. Res.* 37(3): 559–570.

Grolimund, D. & Borkovec, M. (2006). Release of colloidal particles in natural porous media by monovalent and divalent cations, *J. Contam. Hydrol.* 87(3-4): 155–175.

Gustafsson, O. & Gschwend, P. M. (1997). Aquatic colloids: Concepts, definitions, and current challenges, *Limnol. Oceanogr.* 42(3): 519–528.

Hassellöv, M., Buesseler, K. O., Pike, S. M. & Dai, M. (2007). Application of cross-flow ultrafiltration for the determination of colloidal abundances in suboxic ferrous-rich ground waters, *Sci. Total Environ.* 372(2-3): 636–644.

Hedberg, Y., Herting, G. & Wallinder, I. O. (2011). Risks of using membrane filtration for trace metal analysis and assessing the dissolved metal fraction of aqueous media - A study on zinc, copper and nickel, *Environ. Poll.* 159(5): 1144–1150.

Hu , Q.-H., Weng, J.-Q. & Wang, J.-S. (2010). Sources of anthropogenic radionuclides in the environment: a review, *J. Environ. Radioact.* 101(6): 426–437.

Hu, B., Cheng, W., Zhang, H. & Yang, S. (2010). Solution chemistry effects on sorption behavior of radionuclide 63ni(ii) in illite-water suspensions, *J. Nucl. Mater.* 406(2): 263–270.

Hu, J., Xie, Z., He, B., Sheng, G., Chen, C., Li, J., Chen, Y. & Wang, X. (2010). Sorption of Eu(III) on GMZ bentonite in the absence/presence of humic acid studied by batch and XAFS techniques, *Sci. China, Chemistry* 53: 1420–1428.

Hu, Q. & Mori, A. (2008). Radionuclide transport in fractured granite interface zones, *Phys. Chem. Earth, Pts. A/B/C* 33(14-16): 1042–1049.

IAEA (2003). Radioactive waste management glossary, *Technical Report IAEAL 03-00320*, International Atomic Energy Agency, Vienna.

IAEA (2004). Surveillance and monitoring of near surface disposal facilities for radioactive waste, *Technical Report Safety reports series, No. 35*, International Atomic Energy Agency, Vienna.

IAEA (2007). IAEA safety glossary. Terminology used in nuclear safety and radiation protection, *Technical Report IAEAL 07-00481*, International Atomic Energy Agency, Vienna.

Ilina, T., Panfilov, M., Buès, M. & Panfilova, I. (2008). A pseudo two-phase model of colloid transport in porous media, *Transp. Porous Med.* 71: 311–329.

Isaacson, C. W. & Bouchard, D. (2010). Asymmetric flow field flow fractionation of aqueous $C_60$ nanoparticles with size determination by dynamic light scattering and quantification by liquid chromatography atmospheric pressure photo-ionization mass spectrometry, *J. Chromatogr. A* 1217(9): 1506–1512.

Jeong, M.-S., Hwang, Y. & Kang, C.-H. (2011). Modeling of the pseudo-colloids migration for multi-member decay chains with a arbitrary flux boundary condition in a fractured porous medium, *J. Radioanal. Nucl. Chem.* 289: 287–293.

Johnsson, A., Ödegaard Jensen, A., Jakobsson, A.-M., Ekberg, C. & Pedersen, K. (2008). Bioligand-mediated partitioning of radionuclides to the aqueous phase, *J. Radioanal. Nucl. Chem.* 277: 637–644.

Joseph, C., Schmeide, K., Sachs, S., Brendler, V., Geipel, G. & Bernhard, G. (2011). Sorption of uranium(VI) onto Opalinus Clay in the absence and presence of humic acid in Opalinus Clay pore water, *Chem. Geol.* 284(3-4): 240–250.

Kalmykov, S. N. & Denecke, M. A. (eds) (2011). *Actinide Nanoparticle Research*, Springer-Verlag, Berlin.

Kalmykov, S. N., Zakharova, E. V., Novikov, A. P., Myasoedov, B. F. & Utsunomiya, S. (2011). Effect of redox conditions on actinide speciation and partitioning with colloidal matter, *in* S. N. Kalmykov & M. A. Denecke (eds), *Actinide Nanoparticle Research*, Springer Berlin Heidelberg, pp. 361–375.

Katasonova, O. & Fedotov, P. (2009). Methods for continuous flow fractionation of microparticles: Outlooks and fields of application, *J. Anal. Chem.* 64(3): 212–225.

Kelkar, S., Ding, M., Chu, S., Robinson, B. A., Arnold, B., Meijer, A. & Eddebbarh, A.-A. (2010). Modeling solute transport through saturated zone ground water at 10 km scale: Example from the Yucca Mountain license application, *J. Contam. Hydrol.* 117(1-4): 7–25.

Kersting, A. B. & Zavarin, M. (2011). Colloid-facilitated transport of plutonium at the Nevada Test Site, NV, USA, *in* S. N. Kalmykov & M. A. Denecke (eds), *Actinide Nanoparticle Research*, Springer Berlin Heidelberg, pp. 399–412.

Khasanova, A. B., Shcherbina, N. S., Kalmykov, S. N., Teterin, Y. A. & Novikov, A. P. (2007). Sorption of Np(V), Pu(V), and Pu(IV) on colloids of Fe(III) oxides and hydrous oxides and $MnO_2$, *Radiochemistry* 49: 419–425.

Kiliari, T. & Pashalidis, I. (2010). The effect of physicochemical parameters on the separation recovery of plutonium and uranium from aqueous solutions by cation exchange, *J. Radioanal. Nucl. Chem.* 286: 467–470.

Kim, J.-S., Kwon, S.-K., Sanchez, M. & Cho, G.-C. (2011). Geological storage of high level nuclear waste, *KSCE J. Civil Eng.* 15: 721–737.

Kosakowski, G. (2004). Anomalous transport of colloids and solutes in a shear zone, *J. Contam. Hydrol.* 72(1-4): 23–46.

Kowalkowski, T., Buszewski, B., Cantado, C. & Dondi, F. (2006). Field-flow fractionation: Theory, techniques, applications and the challenges, *Crit. Rev. Anal. Chem.* 36(2): 129–135.

Kretzschmar, R., Barmettler, K., Grolimund, D., Yan, Y., Borkovec, M. & Sticher, H. (1997). Experimental determination of colloid deposition rates and collision efficiencies in natural porous media, *Water Resour. Res.* 33(5): 1129–1137.

Ku, T. L., Luo, S., Goldstein, S. J., Murrell, M. T., Chu, W. L. & Dobson, P. F. (2009). Modeling non-steady state radioisotope transport in the vadose zone - A case study using uranium isotopes at Peña Blanca, Mexico, *Geochim. Cosmochim. Acta* 73(20): 6052–6064.

Kulyako, Y. M., Mal'kovskii, V. I., Trofimov, T. I., Myasoedov, B. F., Fujiwara, A. & Tochiyama, O. (2008). Formation of polymeric Pu(IV) hydroxide structures in aqueous solutions, *Radiochemistry* 50: 594–600.

Kunze, P., Seher, H., Hauser, W., Panak, P. J., Geckeis, H., Fanghänel, T. & Schäfer, T. (2008). The influence of colloid formation in a granite groundwater bentonite porewater mixing zone on radionuclide speciation, *J. Contam. Hydrol.* 102(3-4): 263–272.

Kurosawa, S. & Ueta, S. (2001). Effect of colloids on radionuclide migration for performance assessment of HLW disposal in Japan, *Pure Appl. Chem.* 73(12): 2027–2037.

Kyriakou, M. & Pashalidis, I. (2011). Application of different types of resins in the radiometric determination of uranium in waters, *J. Radioanal. Nucl. Chem.* 287: 773–778.

Lahtinen, M., Hölttä, P., Riekkola, M. L. & Yohannes, G. (2010). Analysis of colloids released from bentonite and crushed rock, *Phys. Chem. Earth, Pts. A/B/C* 35: 265–270.

Laverov, N. P., Velichkin, V. I., Malkovsky, V. I., Tarasov, N. N. & Dikov, Y. P. (2010). A comprehensive study of the spread of radioactive contamination in the geological medium near Lake Karachai, Chelyabinsk oblast, *Geol. Ore Deposits* 52: 5–13.

Leermakers, M., Gao, Y., Gabelle, C., Lojen, S., Ouddane, B., Wartel, M. & Baeyens, W. (2005). Determination of high resolution pore water profiles of trace metals in sediments of the Rupel River (Belgium) using DET (Diffusive Equilibrium in Thin Films) and DGT (Diffusive Gradients in Thin Films) techniques, *Water Air Soil Pollut.* 166(1–4): 265–286.

Lesher, E. K., Ranville, J. F. & Honeyman, B. D. (2009). Analysis of pH dependent uranium(VI) sorption to nanoparticulate hematite by flow field-flow fractionation - inductively coupled plasma mass spectrometry, *Environ. Sci. Technol.* 43(14): 5403–5409.

Li, W., Li, C., Zhao, J. & Cornett, R. J. (2007). Diffusive gradients in thin films technique for uranium measurements in river water, *Anal. Chim. Acta* 592(1): 106 – 113.

Li, W., Wang, F., Zhang, W. & Evans, D. (2009). Measurement of stable and radioactive cesium in natural waters by the diffusive gradients in thin films technique with new selective binding phases, *Anal. Chem.* 81(14): 5889–5895.

Li, W., Zhao, J., Li, C., Kiser, S. & Cornett, R. J. (2006). Speciation measurements of uranium in alkaline waters using diffusive gradients in thin films technique, *Anal. Chim. Acta* 575(2): 274–280.

Li, Y., Tian, S. & Qian, T. (2011). Transport and retention of strontium in surface-modified quartz sand with different wettability, *J. Radioanal. Nucl. Chem.* 289: 337–343.

Li, Z., Zhang, D. & Li, X. (2010). Tracking colloid transport in porous media using discrete flow fields and sensitivity of simulated colloid deposition to space discretization, *Environ. Sci. Technol.* 44(4): 1274–1280.

Lippold , H., Mansel, A. & Kupsch, H. (2005). Influence of trivalent electrolytes on the humic colloid-borne transport of contaminant metals: competition and flocculation effects, *J. Contam. Hydrol.* 76(3-4): 337–352.

Lippold, H., Müller, N. & Kupsch, H. (2005). Effect of humic acid on the pH-dependent adsorption of terbium (III) onto geological materials, *Appl. Geochem.* 20(6): 1209–1217.

Liu, J., Andya, J. D. & Shire, S. J. (2006). A critical review of analytical ultracentrifugation and field flow fractionation methods for measuring protein aggregation, *AAPS J.* 8: E580–E589.

Loux, N. T. (2011). Simulating the stability of colloidal amorphous iron oxide in natural water, *Water, Air, Soil Pollut.* 217: 157–172.

Lujanienė, G., Beneš, P., Štamberg, K., Šapolaité J., Vopalka, D., Radžiūtė, E. & Ščiglo, T. (2010). Effect of natural clay components on sorption of Cs, Pu and Am by the clay, *J. Radioanal. Nucl. Chem.* 286: 353–359.

Luk'yanova, E. A., Zakharova, E. V., Konstantinova, L. I. & Nazina, T. N. (2008). Sorption of radionuclides by microorganisms from a deep repository of liquid low-level waste, *Radiochemistry* 50: 85–90.

Maity, S., Mishra, S., Bhalke, S., Pandit, G., Puranik, V. & Kushwaha, H. (2011). Estimation of distribution coefficient of polonium in geological matrices around uranium mining site, *J. Radioanal. Nucl. Chem.* 290(1): 75–79.

Malinovsky, D., Dahlqvist, R., Baxter, D. C., Ingri, J. & Rodushkin, I. (2005). Performance of diffusive gradients in thin films for measurement of the isotopic composition of soluble Zn, *Anal. Chim. Acta* 537(1-2): 401 – 405.

Mal'kovskii, V. I. (2011). Deposition of heterogeneous radiocolloid from groundwater on enclosing rocks, *Dokl. Earth Sci.* 436: 39–41.

Malkovsky, V. (2011). Theoretical analysis of colloid-facilitated transport of radionuclides by groundwater, *in* S. N. Kalmykov & M. A. Denecke (eds), *Actinide Nanoparticle Research*, Springer Berlin Heidelberg, pp. 195–243.

Malkovsky, V. I., Dikov, Y. D., Kalmykov, S. N. & Buleev, M. I. (2009). Structure of colloid particles in groundwaters on the territory of the Mayak production association and its impact on the colloid transport of radionuclides in subsoil environments, *Geochem. Int.* 47: 1100–1106.

Malkovsky, V. I. & Pek, A. A. (2009a). Effect of colloids on transfer of radionuclides by subsurface water, *Geol. Ore Deposits* 51: 79–92.

Malkovsky, V. I. & Pek, A. A. (2009b). Effect of elevated velocity of particles in groundwater flow and its role in colloid-facilitated transport of radionuclides in underground medium, *Transp. Porous Med.* 78: 277–294.

Mansur, M. S. & Mushtaq, A. (2011). Separation of yttrium-90 from strontium-90 via colloid formation, *J. Radioanal. Nucl. Chem.* 288: 337–340.

Massoudieh, A. & Ginn, T. R. (2007). Modeling colloid-facilitated transport of multi-species contaminants in unsaturated porous media, *J. Contam. Hydrol.* 92(3-4): 162–183.

Matisoff, G., Ketterer, M. E., Rosen, K., Mietelski, J. W., Vitko, L. F., Persson, H. & Lokas, E. (2011). Downward migration of Chernobyl-derived radionuclides in soils in Poland and Sweden, *Appl. Geochem.* 26(1): 105–115.

Matsunaga, T., Nagao, S., Ueno, T., Takeda, S., Amano, H. & Tkachenko, Y. (2004). Association of dissolved radionuclides released by the Chernobyl accident with colloidal materials in surface water, *Appl. Geochem.* 19(10): 1581–1599.

May, C. C., Worsfold, P. J. & Keith-Roach, M. J. (2008). Analytical techniques for speciation analysis of aqueous long-lived radionuclides in environmental matrices, *Trends Anal. Chemi.* 27(2): 160–168.

Mazurek, M., Alt-Epping, P., Bath, A., Gimmi, T., Waber, H. N., Buschaert, S., De Cannière, P., De Craen, M., Gautschi, A., Savoye, S., Vinsot, A., Wemaere, I. & Wouters, L. (2011). Natural tracer profiles across argillaceous formations, *Appl. Geochem.* 26(7): 1035–1064.

Mibus, J., Sachs, S., Pfingsten, W., Nebelung, C. & Bernhard, G. (2007). Migration of uranium(IV)/(VI) in the presence of humic acids in quartz sand: A laboratory column study, *J. Contam. Hydrol.* 89(3-4): 199–217.

Missana, T., Alonso, U., García-Gutiérrez, M. & Mingarro, M. (2008). Role of bentonite colloids on europium and plutonium migration in a granite fracture, *Appl. Geochem.* 23(6): 1484–1497.

Missana, T. & García-Gutiérrez, M. (2007). Adsorption of bivalent ions (Ca(II), Sr(II) and Co(II)) onto FEBEX bentonite, *Phys. Chem. Earth* 32: 559–567.

Monte, L. (2010). Modelling multiple dispersion of radionuclides through the environment, *J. Environ. Radioact.* 101(2): 134–139.

Monte, L., Periañez, R., Boyer, P., Smith, J. T. & Brittain, J. E. (2009). The role of physical processes controlling the behaviour of radionuclide contaminants in the aquatic environment: a review of state-of-the-art modelling approaches, *J. Environ. Radioact.* 100(9): 779–784.

Morel, F. M. M. (1983). *Principles of Aquatic Chemistry*, John Wiley and Sons, New York.

Murakami, T., Sato, T., Ohnuki, T. & Isobe, H. (2005). Field evidence for uranium nanocrystallization and its implications for uranium transport, *Chem. Geol.* 221(1-2): 117–126.

Niculae, O., Andrei, V., Ionita, G. & Duliu, O. G. (2009). Preliminary safety concept for disposal of the very low level radioactive waste in Romania, *Appl. Radiation Isotopes* 67: 935–938.

Nordstrom, D. K., Alpers, C. N., Coston, J. A., Taylor, H. E., McCleskey, R. B., Ball, J. W., Ogle, S., Cotsifas, J. S. & Davis, J. A. (1999). Geochemistry, toxicity, and sorption properties of contaminated sediments and pore waters from two reservoirs receiving acid mine drainage, *U.S. Geological Survey Toxic Substances Hydrology Program. Proceedings Of The Technical Meeting, Charleston, South Carolina, March 8-12, 1999. U.S. Geological Survey Water-Resources Investigations Report 99-4018a. Volume 1. Contamination From Hard-Rock Mining*, Vol. 1, pp. 289–296.

Novikov , A. P., Kalmykov, S. N., Kuzovkina, E. V., Myasoedov, B. F., Fujiwara, K. & Fujiwara, A. (2009). Evolution of actinide partitioning with colloidal matter collected at "Mayak" site as studied by sequential extraction, *J. Radioanal. Nucl. Chem.* 280: 629–634.

Novikov, A. P., Kalmykov, S. N., Goryachenkova, T. A., Kazinskaya, I. E., Barsukova, K. V., Lavrinovich, E. A., Kuzovkina, E. V. & Myasoedov, B. F. (2009). Association of radionuclides with the colloidal matter of underground waters taken from observation wells in the zone of impact of Lake Karachai, *Radiochemistry* 51: 644–648.

Novikow, A., Goryachenkova, T. A., Kalmykov, S. N., Vlasova, I. E., Kazinskaya, I. E., Barsukova, K. V., Lavrinovich, E. A., Kuzovkina, E. V., Tkachev, V. V. & Myasoedov, B. F. (2009). Speciation of radionuclides in colloidal matter of underground waters taken from observation wells in the zone of impact of Lake Karachai, *Radiochemistry* 51: 649–653.

Ödman, F., Ruth, T. & Pontér, C. (1999). Validation of a field filtration technique for characterization of suspended particulate matter from freshwater. Part I. Major elements, *Appl. Geochem.* 14: 301–317.

Ödman, F., Ruth, T., Rodushkin, I. & Pontér, C. (2006). Validation of a field filtration technique for characterization of suspended particulate matter from freshwater. Part II. Minor, trace and ultra trace elements, *Appl. Geochem.* 21(12): 2112–2134.

Ohtsuka, Y., Yamamoto, M., Takaku, Y., Hisamatsu, S. & Inaba, J. (2006). Cascade ultrafiltering of $^{210}Pb$ and $^{210}Po$ in freshwater using a tangential flow filtering system, *J. Radioanal. Nucl. Chem.* 268: 397–403.

Ollivier, P., Radakovitch, O. & Hamelin, B. (2011). Major and trace element partition and fluxes in the Rhône River, *Chem. Geol.* 285(1-4): 15–31.

Otosaka, S., Amano, H., Ito, T., Kawamura, H., Kobayashi, T., Suzuki, T., Togawa, O., Chaykovskaya, E. L., Lishavskaya, T. S., Novichkov, V. P., Karasev, E. V., Tkalin, A. V. & Volkov, Y. N. (2006). Anthropogenic radionuclides in sediment in the Japan Sea: distribution and transport processes of particulate radionuclides, *J. Environ. Radioact.* 91(3): 128–145.

Otte, T., Brüll, R., Macko, T., Pasch, H. & Klein, T. (2009). Optimisation of ambient and high temperature asymmetric flow field-flow fractionation with dual/multi-angle light scattering and infrared/refractive index detection, *J. Chromatogr. A* 1217(5): 722–730.

Panfilov, M., Panfilova, I. & Stepanyants, Y. (2008). Mechanisms of particle transport acceleration in porous media, *Transp. Porous Med.* 74: 49–71.

Pathak, P. N. & Choppin, G. R. (2006). Kinetic and thermodynamic studies of cesium(I) sorption on hydrous silica, *J. Radioanal. Nucl. Chem.* 270: 299–305.

Pathak, P. N. & Choppin, G. R. (2007). Sorption of $Am^{3+}$ cations on suspended silicate: Effects of pH, ionic strength, complexing anions, humic acid and metal ions, *J. Radioanal. Nucl. Chem.* 274: 517–523.

Pearson, D., Nowell, G., Widerlund, A. & Davison, W. (2006). Quantifying micro-niche behaviour in sediments: Precise and accurate measurements of DGT gels by laser ablation high resolution laser ablation ICPMS, *Geochim. Cosmochim. Acta* 70(18, Supplement 1): A478–A478.

Perevalov, S. A., Kulyako, Y. M., Vinokurov, S. E., Myasoedov, B. F., Fujiwara, A. & Tochiyama, O. (2009). Sorption of Pu(IV) in the polymeric colloidal form on rock typical of Mayak Production Association area, *Radiochemistry* 51: 373–377.

Pérez del Villar, L., Delgado, A., Reyes, E., Pelayo, M., Fernández-Soler, J. M., Cózar, J. S., Tsige, M. & Quejido, A. J. (2005). Thermochemically induced transformations in Al-smectites: A Spanish natural analogue of the bentonite barrier behaviour in a radwaste disposal, *Appl. Geochem.* 20(12): 2252–2282.

Pifer, A. D., Miskin, d. R., Cousins, S. L. & Fairey, J. L. (2011). Coupling asymmetric flow-field flow fractionation and fluorescence parallel factor analysis reveals stratification of dissolved organic matter in a drinking water reservoir, *J. Chromatogr. A* 1218(27): 4167–4178.

Pokrovsky, O. S. & Schott, J. (2002). Iron colloids/organic matter associated transport of major and trace elements in small boreal rivers and their estuaries (NW Russia), *Chem. Geol.* 190: 141–179.

Popic, J. M., Salbu, B., Strand, T. & Skipperud, L. (2011). Assessment of radionuclide and metal contamination in a thorium rich area in Norway, *J. Environ. Monit.* 13: 1730–1738.

Porcelli, D., Andersson, P. S., Wasserburg, G. J., Ingri, J. & Baskaran, M. (1997). The importance of colloids and mires for the transport of uranium isotopes through the Kalix River watershed and Baltic Sea, *Geochim. Cosmochim. Acta* 61(19): 4095–4113.

Pourret, O., Dia, A., Davranche, M., Gruau, G., Hénin, O. & Angée, M. (2007). Organo-colloidal control on major- and trace-element partitioning in shallow groundwaters: Confronting ultrafiltration and modelling, *Appl. Geochem.* 22(8): 1568–1582.

Pourret, O., Gruau, G., Dia, A., Davranche, M. & Molénat, J. (2010). Colloidal control on the distribution of rare earth elements in shallow groundwaters, *Aquatic Geochem.* 16(1): 31–59.

Pshinko, G. N. (2009). Impact of humic matter on sorption of radionuclides by montmorrilonite, *J. Water Chem. Technol.* 31: 163–171.

Pshinko, G. N., Timoshenko, T. G. & Bogolepov, A. A. (2009). Effect of fulvic acids on Th(IV) sorption on montmorillonite, *Radiochemistry* 51: 91–95.

Qureshi, R. N. & Kok, W. T. (2011). Application of flow field-flow fractionation for the characterization of macromolecules of biological interest: a review, *Anal. Bioanal. Chem.* 399(4): 1401–1411.

Rabung, T., Pierret, M. C., Bauer, A., Geckeis, H., Bradbury, M. H. & Baeyens, B. (2005). Sorption of Eu(III)/Cm(III) on Ca-montmorillonite and Na-illite. Part 1: Batch sorption and time-resolved laser fluorescence spectroscopy experiments, *Geochim. Cosmochim. Acta* 69(23): 5393–5402.

Rachkova, N. G., Shuktomova, I. I. & Taskaev, A. I. (2010). The state of natural radionuclides of uranium, radium, and thorium in soils, *Eurasian Soil Sci.* 43: 651–658.

Ranville, J. F., Hendry, M. J., Reszat, T. N., Xie, Q. & Honeyman, B. D. (2007). Quantifying uranium complexation by groundwater dissolved organic carbon using asymmetrical flow field-flow fractionation, *J. Contam. Hydrol.* 91(3-4): 233–246.

Sabodina, M. N., Kalmykov, S. N., Artem'eva, K. A., Zakharova, E. V. & Sapozhnikov, Y. A. (2006). Behavior of Cs, Np(V), Pu(IV), and U(VI) in pore water of bentonite, *Radiochemistry* 48: 488–492.

Salbu, B. (2007). Speciation of radionuclides – analytical challenges within environmental impact and risk assessments, *J. Environ. Radioact.* 96(1-3): 47–53.

Salbu, B. (2009). Fractionation of radionuclide species in the environment, *J. Environ. Radioact.* 100(4): 283–289.

Salbu, B., Lind, O. C. & Skipperud, L. (2004). Radionuclide speciation and its relevance in environmental impact assessments, *J. Environ. Radioact.* 74(1-3): 233–242.

Salbu, B. & Skipperud, L. (2009). Speciation of radionuclides in the environment, *J. Environ. Radioact.* 100(4): 281–282.

Santos, A. & Barros, P. H. L. (2010). Multiple particle retention mechanisms during filtration in porous media, *Environ. Sci. Technol.* 44(7): 2515–2521.

Schelero, N. & von Klitzing, R. (2011). Correlation between specific ion adsorption at the air/water interface and long-range interactions in colloidal systems, *Soft Matter* 7: 2936–2942.

Schindler, M., Fayek, M. & Hawthorne, F. C. (2010). Uranium-rich opal from the Nopal I uranium deposit, Peña Blanca, Mexico: Evidence for the uptake and retardation of radionuclides, *Geochim. Cosmochim. Acta* 74(1): 187–202.

Schmeide, K. & Bernhard, G. (2010). Sorption of Np(V) and Np(IV) onto kaolinite: Effects of pH, ionic strength, carbonate and humic acid, *Appl. Geochem.* 25(8): 1238–1247.

Scholten, J. C., Fietzke, J., Mangini, A., Stoffers, P., Rixen, T., Gaye-Haake, B., Blanz, T., Ramaswamy, V., Sirocko, F., Schulz, H. & Ittekkot, V. (2005). Radionuclide fluxes in the Arabian Sea: the role of particle composition, *Earth Planet. Sci. Lett.* 230(3–4): 319–337.

Seiler, R. L., Stillings, L. L., Cutler, N., Salonen, L. & Outola, I. (2011). Biogeochemical factors affecting the presence of $^{210}Po$ in groundwater, *Appl. Geochem.* 26(4): 526–539.

Seliman, A. F., Borai, E. H., Lasheen, Y. F., Abo-Aly, M. M., DeVol, T. A. & Powell, B. A. (2010). Mobility of radionuclides in soil/groundwater system: Comparing the influence of EDTA and four of its degradation products, *Environ. Pollut.* 158(10): 3077–3084.

Semizhon, T., Röllin, S., Spasova, Y. & Klemt, E. (2010). Transport and distribution of artificial gamma-emitting radionuclides in the River Yenisei and its sediment, *J. Environ. Radioact.* 101(5): 385–402.

Severino, G., Cvetkovic, V. & Coppola, A. (2007). Spatial moments for colloid-enhanced radionuclide transport in heterogeneous aquifers, *Adv. Water Resour.* 30(1): 101–112.

Singer, D. M., Farges, F. & Brown Jr, G. E. (2009). Biogenic nanoparticulate $UO_2$: Synthesis, characterization, and factors affecting surface reactivity, *Geochim. Cosmochim. Acta* 73(12): 3593–3611.

Singer, D. M., Maher, K. & Brown Jr, G. (2009). Uranyl-chlorite sorption/desorption: Evaluation of different U(VI) sequestration processes, *Geochim. Cosmochim. Acta* 73(20): 5989–6007.

Singh, B. K., Jain, A., Kumar, S., Tomar, B. S., Tomar, R., Manchanda, V. K. & Ramanathan, S. (2009). Role of magnetite and humic acid in radionuclide migration in the environment, *J. Contam. Hydrol.* 106(3-4): 144–149.

Singhal, R. K., Karpe, R., Muthe, K. P. & Reddy, A. V. R. (2009). Plutonium-239+240 selectivity for pseudo-colloids of iron in subsurface aquatic environment having elevated level of dissolved organic carbon, *J. Radioanal. Nucl. Chem.* 280: 141–148.

Solovitch-Vella, N., Garnier, J.-M. & Ciffroy, P. (2006). Influence of the colloid type on the transfer of $^{60}Co$ and $^{85}Sr$ in silica sand column under varying physicochemical conditions, *Chemosphere* 65(2): 324–331.

Stepanets, O. V., Ligaev, A. N., Borisov, A. P., Travkina, A. V., Shkinev, V. M., Danilova, T. V., Miroshnikov, A. Y. & Migunov, V. I. (2009). Geoecological investigations of the Ob-Irtysh River Basin in the Khanty-Mansi Autonomous region: Yugra in 2006–2007, *Geochem. Int.* 47: 657–671.

Stolpe, B., Hassellöv, M., Andersson, K. & Turner, D. R. (2005). High resolution ICPMS as an on-line detector for flow field-flow fractionation; multi-element determination of colloidal size distributions in a natural water sample, *Anal. Chim. Acta* 535(1-2): 109–121.

Stumm, W. & Morgan, J. J. (1996). *Aquatic Chemistry: Chemical Equilibria and Rates in Natural Waters*, 3rd edn, John Wiley & Sons, Ltd.

Tang, X.-Y. & Weisbrod, N. (2009). Colloid-facilitated transport of lead in natural discrete fractures, *Environ. Pollut.* 157(8-9): 2266–2274.

Tang, X.-Y. & Weisbrod, N. (2010). Dissolved and colloidal transport of cesium in natural discrete fractures, *J. Environ. Qual.* 39(3): 1066–1076.

Templeton, D. M., Ariese, F., Cornelis, R., Danielsson, L.-G., Muntau, H., Leeuwen, H. P. & Lobiński, R. (2000). Guidelines for terms related to chemical speciation and fractionation of elements. Definitions, structural aspects, and methodological approaches (IUPAC Recommendations 2000), *Pure Appl. Chem.* 72(8): 1453–1470.

Tertre, E., Berger, G., Castet, S., Loubet, M. & Giffaut, E. (2005). Experimental sorption of $Ni^{2+}$, $Cs^{+}$ and $Ln^{3+}$ onto a montmorillonite up to 150°C, *Geochim. Cosmochim. Acta* 69(21): 4937–4948.

Utsunomiya, S., Kersting, A. B. & Ewing, R. C. (2009). Groundwater nanoparticles in the far-field at the Nevada test site: Mechanism for radionuclide transport, *Environ. Sci. Technol.* 43(5): 1293–1298.

van der Veeken, P. L. R., Pinheiro, J. P. & van Leeuwen, H. P. (2008). Metal speciation by DGT/DET in colloidal complex systems, *Environ. Sci. Technol.* 42(23): 8835–8840.

Vandenhove, H., Antunes, K., Wannijn, J., Duquène, L. & Van Hees, M. (2007). Method of diffusive gradients in thin films (DGT) compared with other soil testing methods to predict uranium phytoavailability, *Sci. Total Environ.* 373(2-3): 542–555.

Vilks, P., Miller, N. & Vorauer, A. (2008). Laboratory bentonite colloid migration experiments to support the Äspö colloid project, *Phys. Chem. Earth, Pts. A/B/C* 33(14-16): 1035–1041.

Vrana, B., Allan, I. J., Greenwood, R., Mills, G. A., Dominiak, E., Svensson, K., Knutsson, J. & Morrison, G. (2005). Passive sampling techniques for monitoring pollutants in water, *Trends Anal. Chem.* 24(10): 845–868.

Walther, C., Rothe, J., Brendebach, b., Fuss, M., Altmaier, M., Marquardt, C. M., Büchner, S., Cho, H.-R., Yun, J.-I. & Seibert, A. (2009). New insights in the formation processes of Pu(IV) colloids, *Radiochim. Acta* 97: 199–207.

Wang, X., Tan, X., Ning, Q. & Chen, Q. (2005). Simulation of radionuclides $^{99}$Tc and $^{243}$Am migration in compacted bentonite, *Appl. Radiat. Isot.* 62(5): 759–764.

Warnken, K. W., Zhang, H. & Davison, W. (2004). Analysis of polyacrylamide gels for trace metals using diffusive gradients in thin films and laser ablation inductively coupled plasma mass spectrometry, *Anal. Chem.* 76(20): 6077–6084.

Wersin, P., Stroes-Gascoyne, S., Pearson, F. J., Tournassat, C., Leupin, O. X. & Schwyn, B. (2011). Biogeochemical processes in a clay formation *in situ* experiment: Part G - Key interpretations and conclusions. implications for repository safety, *Appl. Geochem.* 26(6): 1023 – 1034.

Wieland, E., Tits, J. & Bradbury, M. H. (2004). The potential effect of cementitious colloids on radionuclide mobilisation in a repository for radioactive waste, *Appl. Geochem.* 19: 119–135.

Wigginton, N. S., Haus, K. L. & Hochella Jr, M. F. (2007). Aquatic environmental nanoparticles, *J. Environ. Monit.* 9(12): 1306–1316.

Wilkins, M. J., Livens, F. R., Vaughan, D. J. & Lloyd, J. R. (2006). The impact of Fe(III)-reducing bacteria on uranium mobility, *Biogeochemistry* 78: 125–150.

Wilkins, M. J., Livens, F. R., Vaughan, D. J., Lloyd, J. R., Beadle, I. & Small, J. S. (2010). Fe(III) reduction in the subsurface at a low-level radioactive waste disposal site, *Geomicrobiol. J.* 27(3): 231–239.

Williams, S. K. R., Runyon, J. R. & Ashames, A. A. (2011). Field-flow fractionation: Addressing the nano challenge, *Anal. Chem.* 83(3): 634–642.

Wolthoorn, A., Temminghoff, E. J. M., Weng, L. & van Riemsdijk, W. (2004). Colloid formation in groundwater: effect of phosphate, manganese, silicate and dissolved organic matter on the dynamic heterogeneous oxidation of ferrous iron, *Appl. Geochem.* 19(4): 611–622.

Wu, Z., He, M. & Lin, C. (2011). In situ measurements of concentrations of Cd, Co, Fe and Mn in estuarine porewater using DGT, *Environ. Poll.* 159(5): 1123–1128.

Xu, C., Zhang, S., Ho, Y.-F., Miller, E. J., Roberts, K. A., Li, H.-P., Schwehr, K. A., Otosaka, S., Kaplan, D. I., Brinkmeyer, R., Yeager, C. M. & Santschi, P. H. (2011). Is soil natural organic matter a sink or source for mobile radioiodine ($^{129}I$) at the Savannah River Site?, *Geochim. Cosmochim. Acta* 75(19): 5716–5735.

Yamaguchi, T., Nakayama, S., Vandergraaf, T. T., Drew, D. J. & Vilk (2008). Radionuclide and colloid migration experiments in quarried block of granite unde in-situ coditions at a depth of 240 m, *J. Power Energy Syst.* 2: 186–197.

Yeager, K. M., Santschi, P. H., Phillips, J. D. & Herbert, B. E. (2005). Suspended sediment sources and tributary effects in the lower reaches of a coastal plain stream as indicated by radionuclides, Loco Bayou, Texas, *Environ. Geol.* 47: 382–395.

Yoshida, T. & Suzuki, M. (2006). Migration of strontium and europium in quartz sand column in the presence of humic acid: Effect of ionic strength, *J. Radioanal. Nucl. Chem.* 270: 363–368.

Zhang, H. & Davison, W. (2000). Direct in situ measurements of labile inorganic and organically bound metal species in synthetic solutions and natural waters using diffusive gradients in thin films, *Anal. Chem.* 72(18): 4447–4457.

Zhang, H. & Davison, W. (2001). *In situ* speciation measurements. Using diffusive gradients in thin films (DGT) to determine inorganically and organically complexed metals, *Pure Appl. Chem.* 73(1): 9–15.

# 19

# Modelling Groundwater Contamination Above High-Level Nuclear-Waste Repositories in Salt, Granitoid and Clay

Michal O. Schwartz
*MathGeol*
*Germany*

## 1. Introduction

Modelling the groundwater contamination in the aquifer above a repository for high-level nuclear waste is a complex task. The usual procedure is based on modelling the transport of radionuclides in one or two dimensions only (NAGRA, 2002; Kosakowski, 2004; Keesmann et al., 2006; SKB, 2006; Nykyri et al., 2008). However, modern computing techniques allow to perform simulations in three dimensions, which are a more realistic representation of the geosphere or far field. This is shown for candidate repositories in salt (Gorleben, Germany) and granitoid (Olkiluoto, Finland). To date, is not possible to conduct similar three-dimensional simulations for a repository in clay because the necessary input data are not available. Nevertheless, it can be shown that the results obtained from the simulations with a granitoid-hosted repository are useful for evaluating the performance of a clay-hosted repository. The objective of this comparative study is to provide tools that facilitate the identification of disposal sites and enhance the transparency of decision-making processes.

## 2. Computer codes for flow-transport models

The generation of a gas phase is an important issue for a repository in salt. Therefore, both one-phase (liquid only) and two-phase (liquid and gas) simulations are performed for the salt-hosted Gorleben repository. The two-phase simulations are performed with the TOUGHREACT code (Xu et al., 2005). This is a numerical simulation program that calculates chemically reactive non-isothermal flows of multi-phase fluids and geochemical transport in porous and fractured media. The program was developed by introducing reactive chemistry into the multi-phase flow code TOUGH2 (Pruess et al., 1999). The code performs a fully coupled simultaneous solution of mass and energy balance for multicomponent fluids, combined with a multiphase extension of Darcy's Law and Fickian diffusion. The program consists of a number of functional units with flexible and transparent interfaces. The governing equations are discretised using integral finite difference for space and fully implicit first-order finite difference in time. The resulting non-linear algebraic equations are solved by Newton-Raphson iteration. The simulations are performed with the ECO2 module, which treats the aqueous phase as a mixture of water and salt. The dependence of density, viscosity, enthalpy and vapour pressure on salt concentration is taken into account, as well as the effects of salinity on gas solubility in the liquid phase and related heat of solution.

The one-phase scenario for Gorleben is calculated with TOUGH2-MP (Zhang et al., 2008), the parallel version of the TOUGH2 code. The EOS7R module is used, which calculates the transport of a parent radionuclide and a daughter radionuclide. The aqueous phase is treated as a mixture of water and brine.

For a granitoid-hosted repository, the potential gas release is negligibly small. Therefore, only a one-phase scenario is calculated for the Olkiluoto repository. The TOUGHREACT code with the ECO2 module is used for these simulations.

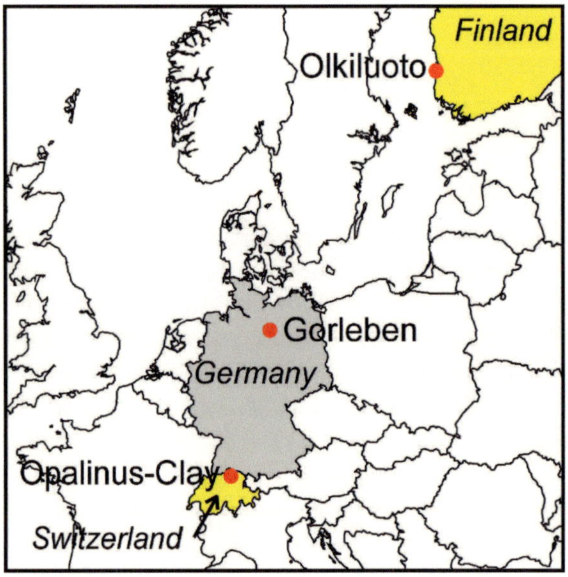


Fig. 1. Location of the candidate repositories Gorleben (Germany), Olkiluoto (Finland) and Opalinus Clay (Switzerland)

## 3. The salt-hosted repository at Gorleben

### 3.1 Repository layout

The repository (Figs. 1 and 2) will accommodate 5,440 waste containers with a total of 8,550 t U of burnt fuel (Keesmann et al., 2005). The waste containers, which are made of Mn-Ni steel, have a length of 5.5 m and a diameter of 1.6 m (Javeri, 2006). They will be emplaced in horizontal disposal drifts, which will be connected to two or more shafts by a system of access drifts. The envisaged disposal depth is about 800 m. There are 260-280 m Quaternary-Tertiary siliciclastic sediments above the present mine workings within the Zechstein salt dome.

### 3.2 Near field

The near field is the hydrochemical regime of the backfilled underground workings in the salt dome and the far field is hydrochemical regime of the caprock and cover sediments. The far-field simulations use the near-field releases as input. To date, only two near-field scenarios have been published. The first considers a single-phase case with liquid-phase transport only. The second considers a two-phase case with both liquid-phase and gas-phase transport.

The near-field release of the single-phase scenario (Fig. 3; Keesmann et al., 2005) is the latest update of the Early-Intrusion-Case of Storck et al. (1988). About 2,000 $m^3$ of brine is assumed to intrude into the repository and dissolve the waste up to the solubility limits shown in Table 1. The brine is squeezed out of the repository due to the convergence of the elastically deforming rock salt. The exit point of the contaminated brine is the contact between Quaternary sediments and the Main Anhydrite (Hauptanhydrit) of the Zechstein salt dome at the borehole location GoHy3020 (Fig. 2) on the -250-m level.

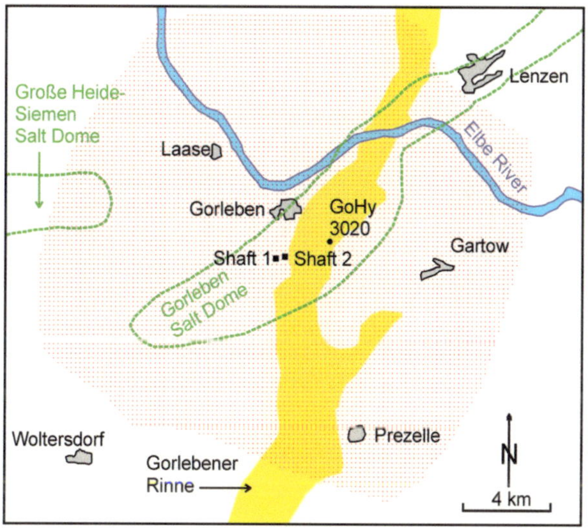


Fig. 2. Map of the Gorleben area with model mesh

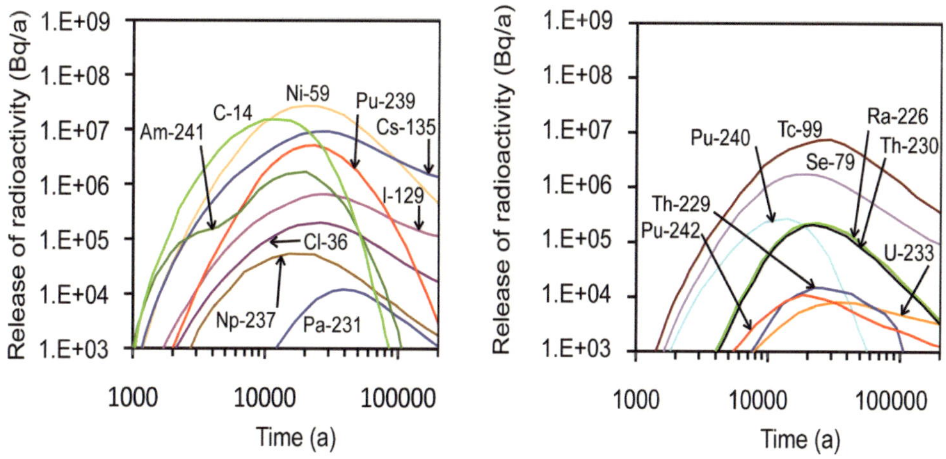


Fig. 3. Gorleben model. Near-field release of radionuclides (Bq/a) for the one-phase scenario

The two-phase scenario for the release of C-14 in the presence of a gas phase is based on a case described by Javeri (2006). Brine squeezed out from a 1,000 m³ brine pocket intrudes a backfilled disposal drift during 2,000 years following the closure of the repository. The disposal drift hosts 33 waste containers. At low oxygen fugacities, the container wall is subjected to corrosion dominated by the reaction

$$3Fe + 4H_2O = Fe_3O_4 + 4H_2. \quad (1)$$

The corrosion process releases hydrogen gas at a rate that increases from zero to 3.84 kg/a in the period from year zero to year 100 and decreases from 3.84 kg/a to zero in the period from year 20,000 to year 40,000. The hydrogen gas leaves the repository via the backfilled shaft according to the release rates shown in Figure 4.

| Solubility (mol/L) | | | | | |
|---|---|---|---|---|---|
| | **One-phase scenario** | **Two-phase scenario** | | **One-phase scenario** | **Two-phase scenario** |
| Am | $1.0x10^{-5}$ | - | Pa | $1.0x10^{-6}$ | - |
| C | $1.0x10^{-2}$ | unlimited | Pu | $1.0x10^{-6}$ | - |
| Cl | unlimited | - | Ra | $1.0x10^{-6}$ | - |
| Cs | unlimited | - | Se | $1.0x10^{-4}$ | - |
| I | unlimited | - | Tc | $1.0x10^{-4}$ | - |
| Ni | $1.0x10^{-4}$ | - | Th | $1.0x10^{-6}$ | - |
| Np | $1.0x10^{-5}$ | - | U | $1.0x10^{-4}$ | - |

Table 1. Gorleben model. Liquid-phase solubility in the near field

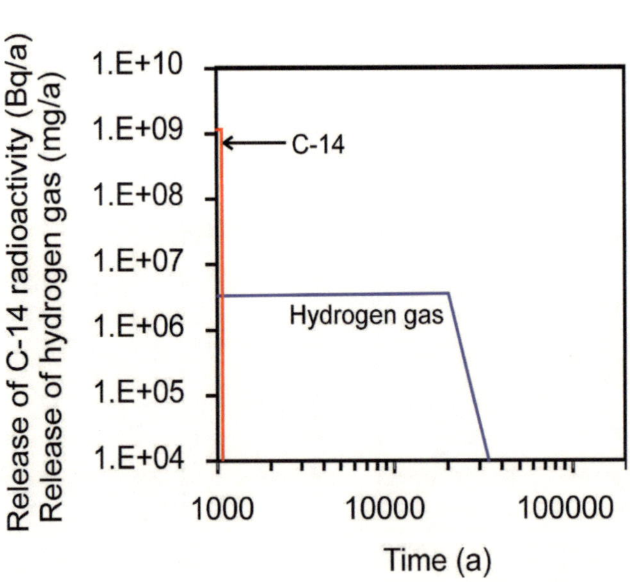


Fig. 4. Gorleben model. Near-field release of hydrogen gas (mg/a) and C-14 (Bq/a) for the two-phase scenario

Fig. 5. Gorleben model. Block diagram. Vertical exaggeration ratio 12:1

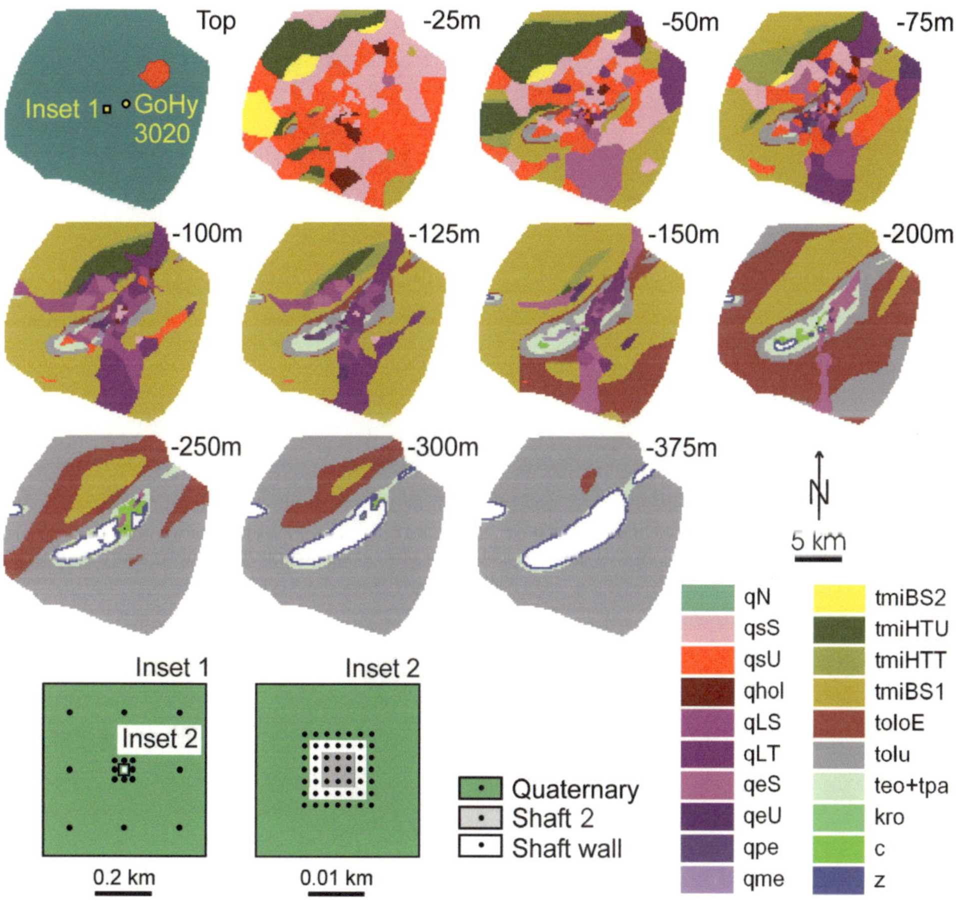


Fig. 6. Gorleben model. Hydrostratigraphic units (Table 3) of the top, -25 m, -50 m, -75 m, -100 m, -125 m, -150 m, -200 m, -250 m, -300 m and -375 m layer. The location of the borehole GoHy3020 and the structure of the mesh in the area surrounding Shaft 2 (Insets 1 and 2) are also shown

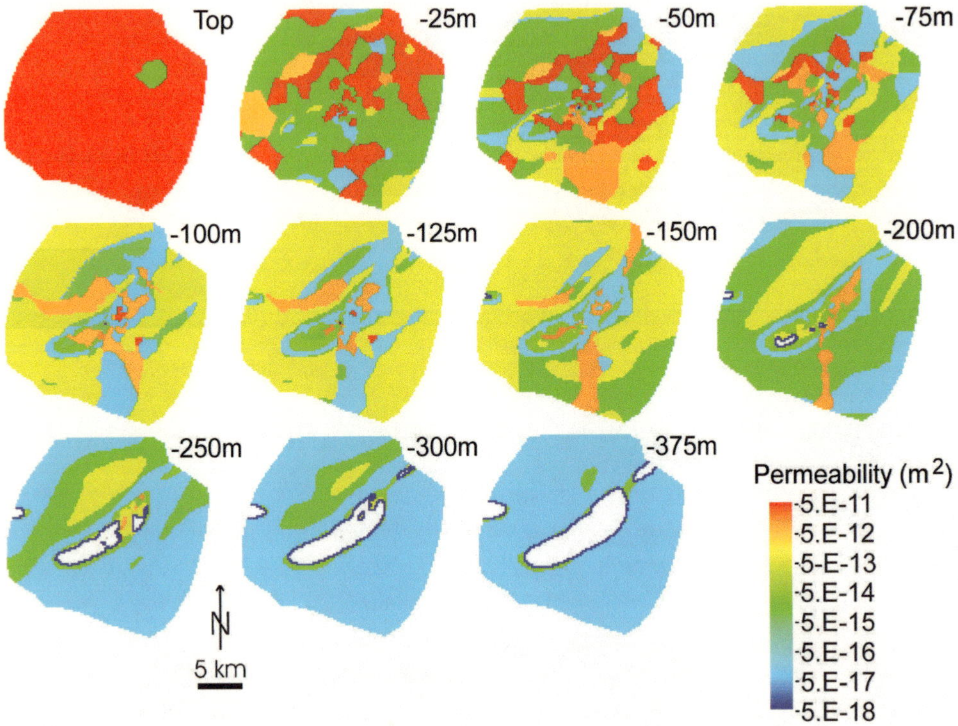


Fig. 7. Gorleben model. Permeability ($m^2$) of the top, -25 m, -50 m, -75 m, -100 m, -125 m, -150 m, -200 m, -250 m, -300 m and -375 m layer

Together with hydrogen, C-14 species are released. The average waste canister contains C-14 with an activity of 3 x $10^{10}$ Bq (Keesmann et al., 2005). A fraction of 15 % is available for rapid release upon failure of the canister. This is assumed to result in a release of 13 % of the initial C-14 inventory out of the repository during a 100-year period starting in the year 1,000 after the closure of the repository. The corresponding release from the total of the 33 waste containers is 1.3 x $10^9$ Bq/a. C-14 is the only radionuclide considered in the two-phase scenario because it is the only long-lived radionuclide that strongly partitions into the gas phase.

TOUGHREACT allows for injection of a single non-condensable gas, which is hydrogen in the case of the present study. Other gaseous species are introduced as secondary species via the corresponding primary liquid species according to a dissociation reaction such as

$$HCO_3^- + H^+ = CO_{2(gas)} + H_2O. \quad (2)$$

The primary species are $HCO_3^-$, $H^+$ and $O_{2(aq)}$. The secondary species are $CO_{2(gas)}$, $CH_{4(gas)}$, $CH_{4(aq)}$, $CO_{2(aq)}$ and $CO_3^{-2}$. The initial and boundary solutions have a $H^+$ activity of $10^{-7}$ and an $O_{2(aq)}$ activity of $10^{-67}$. The source for the two-phase model is the near-field release shown in Figure 4. Hydrogen gas is injected into the primary central cell of shaft 2 on the -237.5-m level whereas a boundary solution with $HCO_3^-$ is injected from a secondary boundary cell that is connected to this primary cell.

| General properties | | | | |
|---|---|---|---|---|
| Model length/width/height (m) | 22,400/21,000/400 | | | |
| Model temperature (°C) | 25 | | | |
| Simulation period (a) | 200,000 | | | |
| Rock density (kg/dm³) | 2.5 | | | |
| | | | | |
| **Saturated hydraulic properties (Table 3)** | | | | |
| Permeability (m²) | $5x10^{-11}$-$1x10^{-18}$ | | | |
| Porosity (-) | 0.03-0.3 | | | |
| | | | | |
| **Unsaturated hydraulic properties** | | | | |
| Residual liquid saturation | 0. | | | |
| Residual gas saturation | 0. | | | |
| Van Genuchten parameter α (Pa) | $4.46x10^4$ | | | |
| Van Genuchten parameter *m* (-) | 0.21 | | | |

| Distribution coefficients (dm³/kg) | | | | | |
|---|---|---|---|---|---|
| | **One-phase scenario** | | | | **Two-phase scenario** |
| | **Low-salinity water (<10 g/L salt)** | | **High-salinity water (≥10 g/L salt)** | | **Low/high-salinity water** |
| | Sand | Silt/clay | Sand | Silt/clay | Sand/silt/clay |
| Am | 300 | 20,000 | 3,000 | 20,000 | - |
| C | 0.2 | 2 | 0.2 | 2 | 0. |
| Cl | 0.1 | 0.1 | 0.1 | 0.1 | - |
| Cs | 70 | 400 | 2 | 70 | - |
| I | 2 | 2 | 0.1 | 2 | - |
| Ni | 20 | 300 | 6 | 90 | - |
| Np | 10 | 300 | 10 | 300 | - |
| Pa | 600 | 6,000 | 600 | 6,000 | - |
| Pu | 100 | 3,000 | 100 | 3,000 | - |
| Ra | 40 | 300 | 2 | 40 | - |
| Se | 1 | 1 | 1 | 1 | - |
| Tc | 1 | 6 | 1 | 1 | - |
| Th | 200 | 2,000 | 200 | 200 | - |
| U | 2 | 80 | 0.6 | 20 | - |

| Aqueous phase diffusivity (m²/s) | |
|---|---|
| Salt | $1.0x10^{-7}$ |
| Hydrogen | $1.0x10^{-9}$ |
| All radionuclides | $1.0x10^{-9}$ |

Table 2. Set-up of the Gorleben far-field model

| Quaternary | | | |
|---|---|---|---|
| **Symbol** | **Rock** | **Permea-bility ($m^2$)** | **Poros-ity (-)** |
| qN | Holocene, Weichsel glacial (sand) | $5x10^{-11}$ | 0.3 |
| qsS | Saale glacial (sand) | $4x10^{-11}$ | 0.3 |
| qsU | Saale glacial (silt, clay, drift marl) | $1x10^{-14}$ | 0.15 |
| qhol | Holstein interglacial (silt, clay) | $1x10^{-16}$ | 0.03 |
| qLS | Elster glacial/Lauenburger-Ton-Komplex (sand) | $1x10^{-11}$ | 0.3 |
| qLT | Elster glacial/Lauenburger-Ton-Komplex (clay, silt) | $1x10^{-16}$ | 0.03 |
| qeS | Elster glacial [excluding Lauenburger-Ton-Komplex] (sand) | $1x10^{-11}$ | 0.3 |
| qeU | Elster glacial [excluding Lauenburger-Ton-Komplex] (silt, drift marl) | $1x10^{-14}$ | 0.15 |
| qpe | Bavel-Cromer-Komplex (silt) | $1x10^{-14}$ | 0.15 |
| qme | Menap glacial (sand) | $1.5x10^{-11}$ | 0.3 |
| **Tertiary/Cretaceous/Zechstein** | | | |
| **Symbol** | **Rock** | **Permea-bility ($m^2$)** | **Poros-ity (-)** |
| tmiBS2 | Lower Miocene/Obere Braunkohlensande (sand) | $5x10^{-12}$ | 0.3 |
| tmiHTU | Lower Miocene/Hamburger-Ton-Komplex (silt, clay, sand) | $1x10^{-14}$ | 0.15 |
| tmiHTT | Lower Miocene/Hamburger-Ton-Komplex (clay, silt) | $1x10^{-16}$ | 0.03 |
| tmiBS1 | Lower Miocene/Untere Braunkohlensande [includ. Neochat] (sand, silt) | $5x10^{-13}$ | 0.3 |
| toloE | Upper Oligocene/Eochat (clay, silt) | $1x10^{-14}$ | 0.15 |
| tolu | Lower Oligocene/Rupelton (clay, silt) | $1x10^{-16}$ | 0.03 |
| teo+tpa | Eocene, Paleocene (clay, silt, sand) | $3x10^{-14}$ | 0.15 |
| kro | Cretaceous (limestone, marlstone, sand) | $1x10^{-14}$ | 0.15 |
| c | Caprock (former Zechstein salt) | $5x10^{-13}$ | 0.3 |
| z | Zechstein salt | $1x10^{-18}$ | 0.1 |

| Structural material | | |
|---|---|---|
| | **Permeability ($m^2$)** | **Porosity (-)** |
| Backfill of Shaft 2 | $3x10^{-13}$ - $3x10^{-12}$ | 0.3 |
| Wall of Shaft 2 | $1x10^{-16}$ | 0.1 |

Table 3. Saturated hydraulic properties of rocks and structural material of the Gorleben far-field model

### 3.3 Geometry of the far-field model and initialisation

The irregular mesh has 290,435 elements and measures 22,400 m x 21,000 m x 400 m (Figs. 5-7; Table 2). Three sets of nodal distances in the x and y direction are used. The backfilled shaft 2, which measures 6 x 6 m horizontally, and the area immediately surrounding shaft 2 (shaft wall and bedrock) have horizontal nodal distances of 2 m. An intermediate zone (only bedrock) has horizontal nodal distances of 20 m. The remaining part of the primary mesh (only bedrock) has nodal distances of 200 m.

The top layer has nodal z values representing the head (between 12.7 m and 22.1 m altitude; Fig. 8). The second layer has nodes located 0.1 m vertically below those of the top layer. The vertical nodal distance between the following 32 layers is 12.5 m. The top layer, which serves to maintain constant pressure and salinity, exclusively consists of infinite-volume boundary elements ($10^{45}$ $m^3$). Other boundary cells are the rock-salt cells and the bottom-layer cells of the -375-m level. They have a volume of $10^{14}$ $m^3$. This volume is large enough for maintaining nearly constant salinity throughout the simulation period but is simultaneously flexible enough to account for pressure adjustments. The volume of the remaining cells is calculated according to their nodal positions. A secondary boundary cell is connected to the primary cell in the centre of shaft 2 at the top of the salt dome (-237.5-m level). The boundary cell is used for injecting $HCO_3^-$ whereas the primary cell is used for injecting hydrogen gas.

The initialisation starts with a pressure of $10^5$ Pa in all models cells. The infinite-volume elements of the top layer have zero salinity. The large-volume elements ($10^{14}$ $m^3$) for the rock-salt unit and the bottom layer have an initial salinity of $X_{brine} = 1$ ($X_{salt} = 0.25$). Based on these boundary conditions, the diffusivity of brine is estimated in a trial-and-error procedure. Suitable vertical salinity gradients at the end of the 100,000-year initialisation period are obtained with a diffusivity $10^{-7}$ $m^2/s$ (Fig. 9). These salinity gradients, which are close to the present-day values (Klinge et al., 2007), are maintained with insignificant variations throughout the following 7,000-year and 200,000-year simulation periods.

### 3.4 Hydrostratigraphy and hydraulic properties

The hydrostratigraphic information of 197 drill hole logs is transferred to the nearest model node at the corresponding z-position (Figs. 6 and 8; Ludwig et al., 1989; Beushausen & Ludwig, 1990; Ludwig et al., 1993; Klinge et al., 2001). Based on this information, the hydrostratigraphic units for the remaining nodes are assigned by interpolation and extrapolation.

The saturated hydraulic properties of the hydrostratigraphic units of the caprock and cover sediments (Table 3) are assigned according to Vogel (2005). This scheme is based on the geometric mean of empirical maximum and minimum values. No decision has been taken yet with respect to the saturated hydraulic properties of the backfill of shaft 2. Therefore, various permeability cases are calculated (base-case permeability of $1 \times 10^{-12}$ ; sensitivity cases of $3 \times 10^{-13}$, $5 \times 10^{-13}$, $2 \times 10^{-12}$ and $3 \times 10^{-12}$ $m^2$). The porosity is 0.3 for all cases.

The unsaturated permeabilities are calculated with the relative-permeability function of Corey (1954; quoted by Brooks & Corey, 1964). The parameters for the capillary-pressure function of Van Genuchten (1980) are those of the Apache Leap tuff reported by Rasmussen (2001).

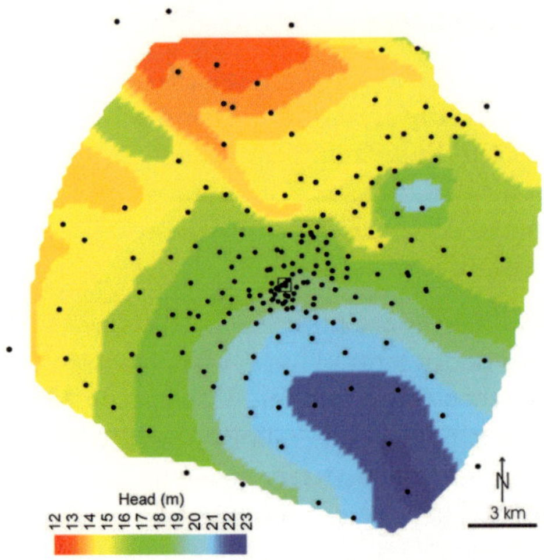


Fig. 8. Gorleben model. Head (m), location of Shaft 2 (open black square) and boreholes (solid black circles)

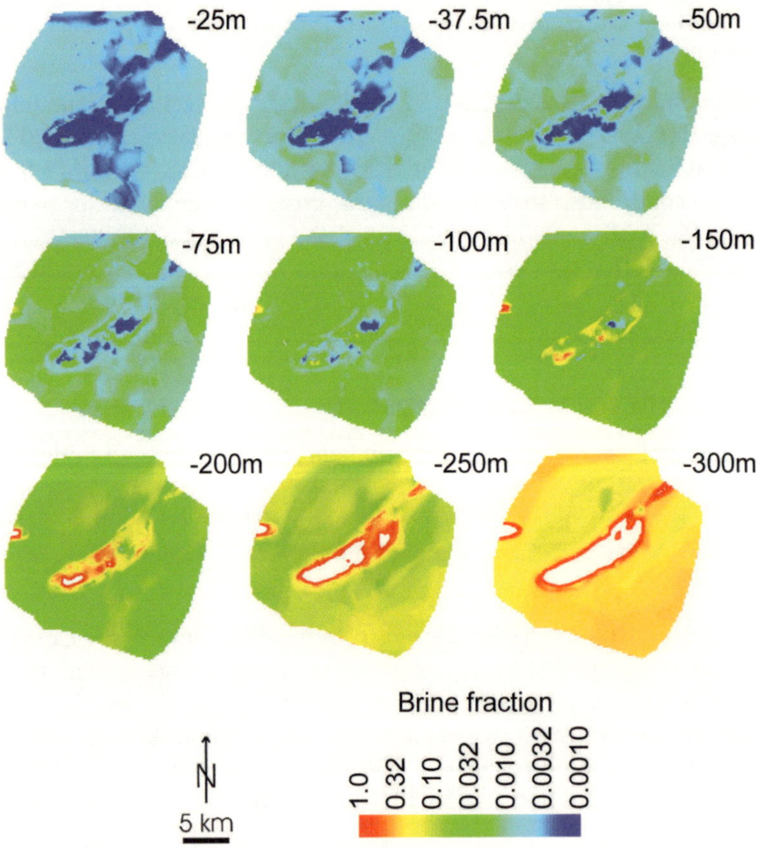


Fig. 9. Gorleben model. Salinity (brine fraction) of the -25 m, -37.5 m, -50 m, -75 m, -100 m, -150 m, -200 m, -250 m, and -300 m layer

| | Inventory (Bq/t U) | Half-life (a) | Dose conversion factor (Sv/a:Bq/L) |
|---|---|---|---|
| Am-241 (parent of Np-237) | $2.05\times10^{14}$ | 432 | $3.4\times10^{-4}$ |
| C-14 | $1.89\times10^{10}$ | 5,700 | $5.9\times10^{-5}$ |
| Cl-36 | $3.43\times10^{8}$ | $3.0\times10^{5}$ | $9.5\times10^{-6}$ |
| Cs-135 | $2.18\times10^{10}$ | $2.0\times10^{6}$ | $3.7\times10^{-6}$ |
| I-129 | $2.02\times10^{9}$ | $1.6\times10^{7}$ | $2.1\times10^{-4}$ |
| Ni-59 | $6.02\times10^{10}$ | $7.5\times10^{4}$ | $5.4\times10^{-8}$ |
| Np-237 | $1.34\times10^{10}$ | $2.1\times10^{6}$ | $7.5\times10^{-4}$ |
| Pa-231 | $1.27\times10^{6}$ | $3.3\times10^{4}$ | $2.0\times10^{-3}$ |
| Pu-239 (parent of U-235) | $2.16\times10^{13}$ | $2.4\times10^{4}$ | $3.5\times10^{-4}$ |
| Pu-240 (parent of U-236) | $5.11\times10^{13}$ | $6.5\times10^{3}$ | $3.5\times10^{-4}$ |
| Pu-242 (parent of U-238) | $3.78\times10^{11}$ | $3.7\times10^{5}$ | $3.2\times 0^{-4}$ |
| Ra-226 | $1.06\times10^{4}$ | $1.6\times10^{3}$ | $1.8\times10^{-4}$ |
| Se-79 | $1.75\times10^{10}$ | $1.1\times10^{6}$ | $6.1\times10^{-7}$ |
| Tc-99 | $2.54\times10^{11}$ | $2.1\times10^{5}$ | $6.3\times10^{-7}$ |
| Th-229 | $6.45\times10^{3}$ | $7.9\times10^{3}$ | $6.7\times10^{-4}$ |
| Th-230 (parent of Ra-226) | $3.92\times10^{6}$ | $7.5\times10^{4}$ | $1.1\times10^{-4}$ |
| U-233 | $3.62\times10^{6}$ | $1.6\times10^{5}$ | $4.7\times10^{-5}$ |
| U-235 | $4.97\times10^{8}$ | $7.0\times10^{8}$ | $4.5\times10^{-5}$ |
| U-236 | $8.43\times10^{9}$ | $2.3\times10^{7}$ | $4.4\times10^{-5}$ |
| U-238 | $1.22\times10^{10}$ | $4.5\times10^{9}$ | $4.2\times10^{-5}$ |

Table 4. Inventory of radionuclides in the candidate repository at Gorleben, half-life of radionuclides and dose conversion factor (DCF)

### 3.5 Sorption and diffusion

The distribution coefficients for the radionuclides in the one-phase scenario are taken from Suter et al. (1998). C-14 in the two-phase scenario is assumed to be non-sorbing under low oxygen fugacities compatible with the release of hydrogen gas. The aqueous phase diffusivity is $10^{-9}$ $m^2/s$ for all species (Keesmann et al., 2005).

### 3.6 Simulated far-field transport

The simulation period is from year 1,000 to year 200,000 for the single-phase scenario and from year 1,000 to year 8,000 for the two-phase scenario. The input for the one-phase simulations is the near-field release of 17 radionuclides, of which five are the parents for daughter radionuclides (Am-241, Pu-239, Pu-240, Pu-242 and Th-230; Fig. 3; Table 4). The biosphere is represented by the -37.5-m level, which approximately is the limit between fresh water and saline water (1 g/L salt or 0.0034 brine fraction). The radioactive plume reaches its highest vertical extent in the year 200,000 (end of the simulation period) and in the year 7,000 for the one-phase and two-phase scenario, respectively (Figs. 10 and 11).

The calculation of the radioactive dose is based on Dose Conversion Factors (DCFs; Table 4; Hirsekorn et al., 1991). The annual dose $D_D$ (Sv/a) is calculated according to the equation

$$D_D=\sum X_iC_i \text{ [Sv/a]}, \quad (3)$$

where $X_i$ is the concentration (Bq/L) of the radionuclide $i$ and $C_i$ is the DCF (Sv/a:Bq/L) of the radionuclide $i$.

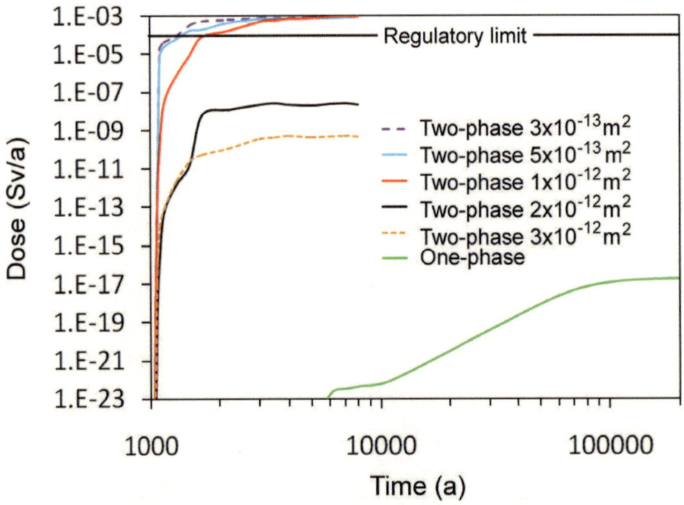


Fig. 10. Gorleben model. Peak values of the radioactive dose (Sv/a) on the -37.5 m level for the one-phase scenario and two-phase scenario (base case with a permeability of $1 \times 10^{-12}$ m$^2$ for the backfill of Shaft 2 and sensitivity cases with a permeability of $3 \times 10^{-13}$ m$^2$, $5 \times 10^{-13}$ m$^2$, $2 \times 10^{-12}$ m$^2$ and $3 \times 10^{-12}$ m$^2$)

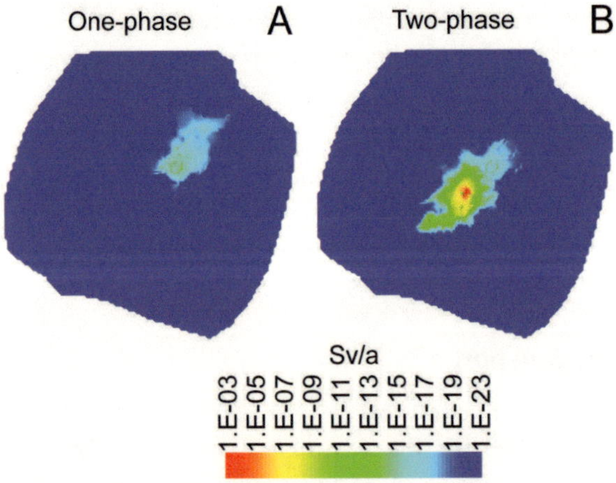


Fig. 11. Gorleben model. (A and B) Maps of the -37.5-m layer with radioactive dose (Sv/a) in the simulation year 200,000 for the one-phase scenario and year 7,000 for the base case of the two-phase scenario

For the single-phase scenario, the peak doses on the -37.5-m level is $1 \times 10^{-17}$ Sv/a. For the two-phase scenario, the peak dose varies strongly depending on the assumptions for the permeability of the backfill of shaft 2. For the base-case permeability ($1 \times 10^{-12}$ m$^2$) and lower permeabilities, the peak dose is above the German regulatory limit of $1 \times 10^{-4}$ Sv/a.

## 4. The granitoid-hosted repository at Olkiluoto

### 4.1 Repository layout

The planned repository is located in granitoid at 420 m depth at the island of Olkiluoto off the south-west coast of Finland (POSIVA, 2009). It can accommodate 3,000 containers with a total of 5,800 t U of burnt fuel (Löfman, 2005). According to the KBS-3V concept, the containers will be emplaced in vertical deposition holes driven from the floor of deposition tunnels, which are connected to a nearly horizontal system of access tunnels and transport tunnels. The containers that host the waste have a 5 cm thick corrosion-resistant copper shell around an inner deformation-resistant cast-iron canister; they are embedded in a 35 cm thick bentonite buffer. The horizontal dimensions of the repository are 1,600 m x 1,400 m.

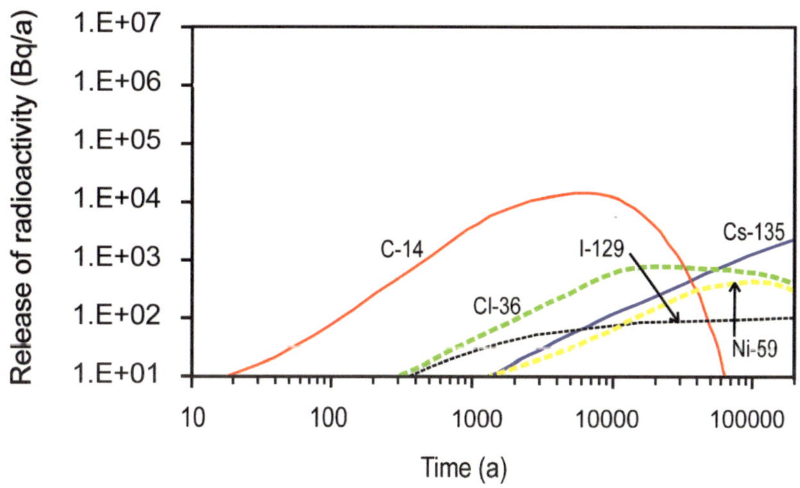


Fig. 12. Olkiluoto model. Near-field release of radionuclides (Bq/a)

### 4.2 Near-field

The near field is the hydrochemical regime of the underground workings and engineered barrier around the radioactive waste, consisting of the container wall and bentonite buffer. The reference failure mode for a defective canister and release of radioactivity from the near field to the undisturbed rock of the far field (geosphere) is the near-field base case (Nykyri et al., 2008). The canister is assumed to have a production defect with a diameter of 1 mm through which water enters into the canister causing dissolution of the fuel up to the solubility limits shown in Table 5. The near-field release of radionuclides (Fig. 12) is used as source for the contaminant injection cells in the far-field model.

As opposed to the easily corrodible steel canisters in the salt repository (see above), the copper shell of the canisters in the granitoid repository (KBS-3V concept) has an extraordinary chemical stability (Schwartz, 2008). Among the various candidate materials for waste packages, copper has unique oxidation characteristics. The conversion from native metal (Cu°) to metal oxide ($Cu_2O$) occurs in a mildly oxidising to mildly reducing environment (positive Eh for pH <7.7 at 25°C; Fig. 13) whereas the conversion of steel, titanium or nickel-chromium alloy already is possible under strongly reducing conditions, i.e., below the stability field of liquid water. Native copper is present as natural occurrence in the Earth's crust. The largest deposits are located on the Keweenaw Peninsula of Michigan, where they have been mined to a depth of 2.2 km (Schwartz, 1996). These deposits formed one billion years ago at temperatures of 100-200°C. The mineralising fluids were slightly less reducing but had much higher copper concentrations than common present-day groundwater.

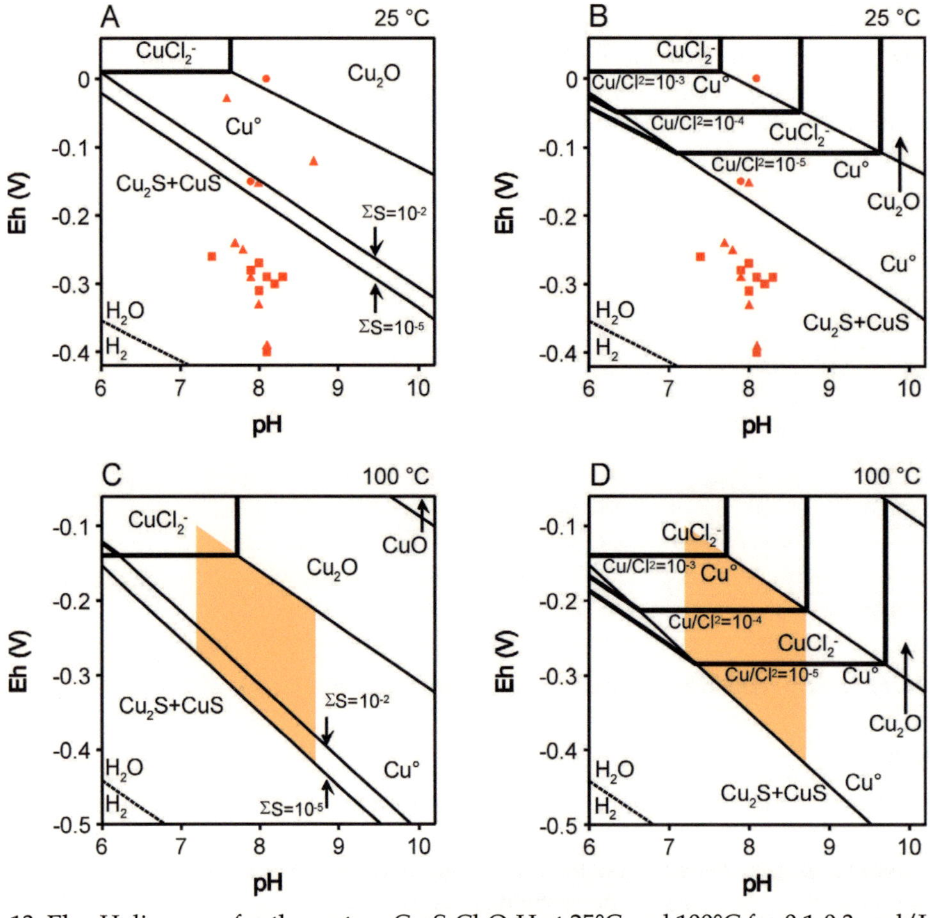


Fig. 13. Eh-pH diagrams for the system Cu-S-Cl-O-H at 25°C and 100°C for 0.1-0.3 mol/L Cl (Schwartz, 2008). Shaded area = Eh-pH range in which the Keweenaw native-copper deposits have formed (Schwartz, 1996). Point symbols = groundwater in crystalline rock of envisaged disposal sites in Sweden, Finland and Switzerland. (A and C) Diagrams for $\Sigma S$ of $10^{-2}$ and $10^{-5}$, and $\Sigma Cu/Cl^2 = 10^{-3}$. (B and D) Diagrams for $\Sigma S = 10^{-5}$ and $\Sigma Cu/Cl^2$ of $10^{-3}$, $10^{-4}$ and $10^{-5}$

Under the envisaged repository conditions, $CuCl_2^-$ is the dominant aqueous copper species. Accordingly, the thermodynamically relevant parameter is the $\Sigma Cu/Cl^2$ ratio, which refers to the sum of activities of all dissolved copper species divided by the $Cl^-$ activity raised to the second power. Mine waters in the Keweenaw district have $\Sigma Cu/Cl^2$ ratios in the range from $10^{-4}$ to $10^{-5}$. Similar conditions are achievable in the bentonite buffer around the copper container if the bentonite contains artificial admixtures of fine-grained native copper. This is a technically feasible variant of the KBS-3V concept, which simulates the environment of a natural native-copper deposit. Thus thermodynamic equilibrium between container and fluid on the high Eh side of the native-copper stability field is implemented in the repository-container system as analogue to the natural counterparts.

The low-Eh and low-pH side of the native-copper stability field is the copper-sulphide phase boundary ($CuS$ and $CuS_2$). The position of this boundary depends on the total activity of aqueous sulphur species ($\Sigma S$). The groundwater in the envisaged repositories in Finland, Sweden and Canada has $\Sigma S$ in the range from $<10^{-5}$ to $10^{-2}$. This implies conditions partly below the stability field of native copper, pH being in the range of 7-9 and Eh in the range from -0.03 to -0.4 V. Provided the bentonite contains no sulphur-bearing minerals, the corrosion of the copper container is constrained by the diffusion of aqueous sulphide from the granitoid aquifer through the bentonite to the container surface. This situation is quite different from corrosion of other package materials such as steel, titanium or nickel-chromium alloy, which all corrode by reaction with omnipresent water. It is not possible to control corrosion of these materials by a diffusion barrier around the container.

The reference buffer of the KBS-3V concept is the pyrite-bearing MX-80 bentonite (0.3 % sulphide sulphur). Another option is the Avonlea bentonite, which has no sulphides but a high content of gypsum (2 %). Neither the presence of pyrite ($FeS_2$) nor gypsum ($CaSO_4 \cdot 2H_2O$) is desirable because $\Sigma S$ of the system should be as low as possible. The Czech EKOBENT bentonite (Prikryl & Woller, 2002), which has a total sulphur concentration of less than 0.05 % and does not contain sulphides, would be the better choice although it has not been tested as thoroughly as MX-80 or Avonlea. The original idea behind the use of a sulphide-bearing bentonite in the KBS-3V repository was to accelerate the transition from oxidising conditions (during the construction phase) towards a final anoxic regime. This scheme, however, has been called in question (SKI, 2006). There are alternatives such as the implementation of saturated conditions in the shortest possible time. This can be achieved by incremental flooding of those parts of the slightly inclined emplacement tunnels where waste packages are already in place.

In the case of a sulphur-free buffer, general corrosion of the copper container can be determined by a simple calculation. The experimentally derived diffusion coefficient of aqueous sulphide in bentonite is approximately $7 \times 10^{-8}$ $cm^2/s$ (King & Stroes-Gascoyne, 1995; King et al., 2002). The groundwater of the envisaged repositories in Finland, Sweden and Canada has sulphide concentrations in the range from $<3 \times 10^{-7}$ to $3 \times 10^{-5}$ mol/L. Assuming sulphide concentrations at the upper limit, $6 \times 10^{-17}$ mol/(s $cm^2$) are transported through the 35 cm thick bentonite to the container surface. This is equivalent to a general corrosion rate of $3 \times 10^{-8}$ cm/a, when $Cu^{\circ}$ is transformed to $Cu_2S$. Allowing for localised corrosion by multiplying the general rate with a correction factor of 5, the maximum corrosion depth is less than 0.2 cm within one million years. A service time for the 5-cm thick copper shell beyond one million years is a realistic possibility (Schwartz, 2008). The mass-transport assumption in this case sets an upper corrosion limit because some of the diffusing sulphide reacts with disseminated native-copper in the buffer and, therefore, is not available for corrosion of the copper container. Thus the near-field release for a canister that has no production defect and has not suffered mechanical damage is assumed to be zero for the 200,000-year simulation period.

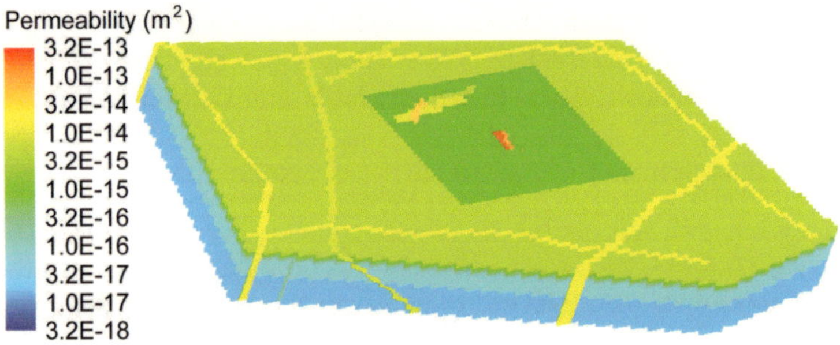


Fig. 14. Olkiluoto model. Block diagram

| | |
|---|---|
| C | unlimited |
| Cl | unlimited |
| Cs | unlimited |
| I | unlimited |
| Ni | $4.3 \times 10^{-3}$ |

Table 5. Olkiluoto model. Liquid-phase solubility (mol/L) in the near-field

### 4.3 Geometry of the far-field model and initialisation

The model has a total of 323,657 elements. The irregular rectangular primary mesh is the fracture continuum, which measures 7,700 m x 5,500 m x 750 m (Figs. 14-16) and consists of 27 layers with 168,048 cubic or cuboid cells. The nodal distance in the x and y direction is 66.6 m. The top layer has nodal z values representing the head (-8 m to 10 m altitude). The second layer has nodes located 0.1 m vertically below those of the top layer. The vertical nodal distance between the following 25 layers is 30 m. The top layer, which serves to maintain constant temperature, pressure and salinity, exclusively consists of infinite-volume boundary elements ($10^{45}$ m$^3$). The bottom layer cells have a volume of $10^{14}$ m$^3$. This volume is large enough for maintaining nearly constant temperature and salinity throughout the simulation period but is simultaneously flexible enough to account for pressure adjustments. The volume of the remaining cells is calculated according to their nodal positions. The secondary matrix continuum comprises the layers 2 to 26. The detailed geometric relationships between fracture and matrix continuum are determined on the basis of a parameter estimation procedure (see "Parameter estimation").

The infinite-volume elements of the top layer have a pressure of $10^5$ Pa, a temperature of 6°C and zero salinity. The bottom-layer large-volume elements ($10^{14}$ m$^3$) have an initial temperature of 14°C and an initial salinity of 0.025 mass fraction salt (corresponding to 0.44 mol/L Cl). The remaining layers have initial values along the gradients defined by the top and bottom layers. These boundary conditions and initial conditions, together with a high diffusivity of salt ($10^{-3}$ m$^2$/s), allow to maintain suitable vertical gradients throughout the simulation, unless disturbed by radioactive heat sources. These gradients include temperature and salinity gradients that are close to the present-day values (Pitkänen et al., 2004; Löfman, 2005).

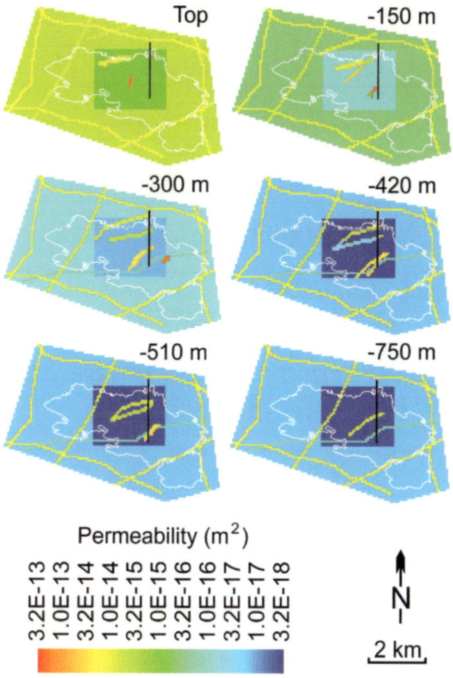


Fig. 15. Olkiluoto model. Permeability ($m^2$) of the top, -150 m, -300 m, -420 m, -510 m and 750 m layer. The position of the parameter-estimation model (black solid line orientated N-S) is also shown

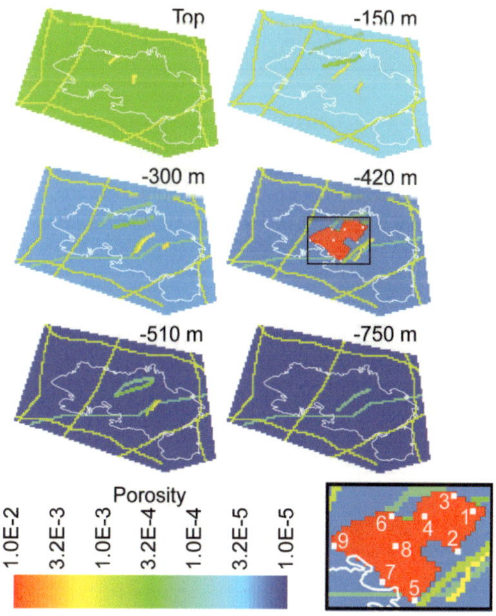


Fig. 16. Olkiluoto model. Flow porosity of the top, -150 m -300 m, -420 m, -510 m and -750 m layer. The repository area at the -420 m level is the red-coloured zone (indicating 0.01 porosity). Inset: The location of the source elements (1-9) in the repository area at the -420 m level

| **General properties** | | | | |
|---|---|---|---|---|
| Model length/width/height (m) | 7,700/5,500/750 | | | |
| Simulation period (a) | 200,000 | | | |
| Rock grain density ($kg/dm^3$) | 2.7 | | | |
| Rock grain specific heat (J/kg°C) | 800 | | | |
| Formation heat conductivity (W/m°C) | 3.65 | | | |
| Matrix porosity for anions (-) | 0.001 | | | |
| Matrix porosity for neutral species and cations (-) | 0.005 | | | |
| Flow porosity [average/range] (-) | $6.6x10^{-5}/10^{-5}-10^{-3}$ | | | |
| Fracture spacing (m) | 5 | | | |
| | | | | |
| Inventory | | | | |
| C-14 (Bq/t U) | $2.78x10^{10}$ | | | |
| Cl-36 (Bq/t U) | $1.04x10^{9}$ | | | |
| Cs-135 (Bq/t U) | $2.15x10^{10}$ | | | |
| I-129 (Bq/t U) | $1.14x10^{9}$ | | | |
| Ni-59 (Bq/t U) | $1.32x10^{11}$ | | | |
| | | | | |
| | **Distance factor (-)** | | **Retardation factor for the fracture continuum (-)** | |
| | **15-continua models** | **Dual-continuum models** | **15-continua models** | **Dual-continuum models** |
| C-14 | 1 | 0.04 | 1 | 1 |
| Cl-36 | 1 | 0.45 | 1 | 1 |
| Cs-135 | 1 | 0.022 | 1 | 50 |
| I-129 | 1 | 0.46 | 1 | 1 |
| Ni-59 | 1 | 0.019 | 1 | 70 |
| | | | | |
| | **Distribution coefficient ($dm^3/kg$)** | | **Effective diffusivity ($m^2/s$)** | |
| C-14 | 0 | | $1.0 \times 10^{-13}$ | |
| Cl-36 | 0 | | $1.0 \times 10^{-14}$ | |
| Cs-135 | 50 | | $1.0 \times 10^{-13}$ | |
| I-129 | 0 | | $1.0 \times 10^{-14}$ | |
| Ni-59 | 100 | | $1.0 \times 10^{-13}$ | |

Table 6. Set-up of the Olkiluoto model

The simulation period is 200,000 years. During the first 10,000 years, the terrain is subjected to glacio-isostatic uplift (POSIVA, 2006; Löfman & Poteri, 2008) and the transient head conditions are translated into transient mesh geometries. The head reference point, which is the vertical projection of the north-western corner of the repository, is fixed at a constant value (2 m). The z values of the top layer change from simulation year zero to year 1,000 and 10,000 according to the z values in Figure 17. The rising terrain implies a retreating shoreline of the Baltic Sea, indicated by the <0-m head contours moving away from the present island of Olkiluoto.

There are nine contaminant-injection cells evenly distributed over the repository area on the -420 m level (Fig. 16). These source cells have only one connection to a fracture element. The near-field releases of the radionuclides shown in Figure 12 are used as input (see "Near field"). Transport paths and, hence, travel times for the radionuclides strongly depend on the location within the repository from which they are released. An ideal simulation would include all repository cells but this is too expensive computationally. Scoping calculations indicate that the relevant results such as median, mean and maximum doses change little when more than nine source cells are used.

### 4.4 Hydraulic properties

The geometry of the Hydrogeological Zones (HZs) of high permeability and surrounding Sparsely Fractured Rock (SFR) of low permeability corresponds to the hydrogeological structure model of Vaittinen et al. (2009). Permeability and flow porosity (Figs. 15 and 16) are based on POSIVA (2009). The only exception is the repository area on the -420 m level, where the permeability is based on Vaittinen et al. (2009) and POSIVA (2009) but the flow porosity is 0.01; this flow porosity value is derived from a backfilled repository volume of 1,310,000 $m^3$ (Kirkkomäki, 2007) and an average backfill porosity of 0.35.

### 4.5 Heat sources

The heat production of the radioactive waste in a single canister is shown for the first 2,000 years after deposition in Figure 18. These values are multiplied by a factor of 8.8, which is the number of canisters in a model cell, and are used as heat-generation data for the repository area on the -420 m level.

### 4.6 Parameter estimation

The retention of radionuclides in a fractured medium is caused by diffusion from the fracture into the stagnant fluid of the matrix adjacent to the fracture. An ideal simulation of a fracture media problem would require the discretisation of the matrix continuum (multiple-interacting continua or MINC model; Pruess, 1983; Pruess & Narasimhan, 1985; Pruess, 1992; Xu & Pruess, 2001; Liou, 2007).

Ten or more matrix continua are necessary for accurately simulating diffusive transport between fracture and matrix as the comparisons between numerical and analytical solutions have shown (Tang et al., 1981; Sudicky & Frind, 1982; Wu & Pruess, 2000; Diersch, 2009). However, the resulting problem size would be too large in the case of a three-dimensional simulation. Therefore, the problem size has to be reduced by replacing the model with a finely discretised matrix by an equivalent dual-continuum model (one matrix continuum). This is achieved by parameterising two variables among the three variables that control the retention of radionuclides; these are the diffusive transport distance between fracture and matrix, fluid volume of the fracture and effective diffusivity. The equivalent model is designed as conventional double-porosity model (Hantush et al., 2000, 2002; Lichtner, 2000).

The critical step in designing the model with a single matrix continuum (equivalent dual-continuum model) is the estimation of the parameters that justify to replace the corresponding finely discretised model (with a multi-continua matrix). The MINC pre-processor program of the TOUGHREACT code (Pruess et al., 1999) is suitable for models with up to 34 matrix continua. Scoping calculations have shown that models with 14 matrix continua have nearly the same retention properties as those of a model with the maximum number of continua. Thus, a two-dimensional 15-continua model has been designed that contains the fracture continuum of a 2,000-m long vertical subsection of the three-dimensional model enclosing the radioactive source element 1 (Figs. 15 and 16). A row of 14 matrix elements is connected to each fracture element with the exception of the fracture elements of the top and bottom layers.

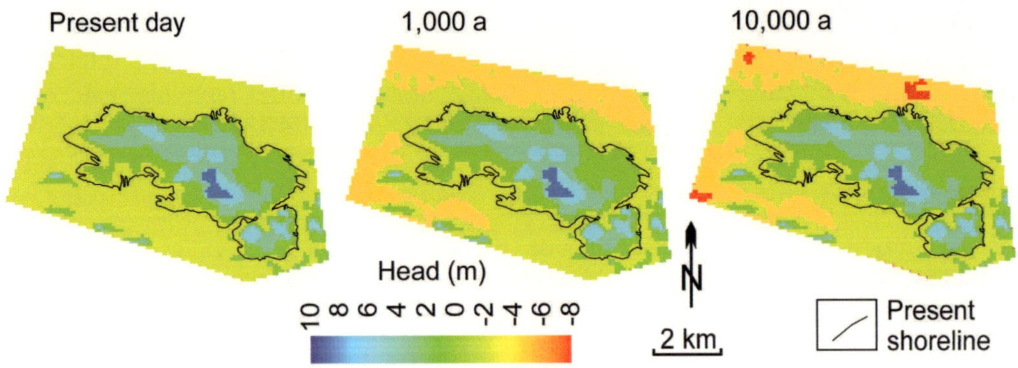


Fig. 17. Olkiluoto model. Vertical nodal positions (z coordinates; m) of the top layer in the simulation year zero, 1,000 and 10,000

The detailed geometry of the MINC mesh is calculated with the ONE-D proximity function of the MINC pre-processor program of the TOUGHREACT code. The fracture spacing is set at 5.0 m, which is the reciprocal of the weighted mean of the fracture intensity (Nykyri et al., 2008). The element volume fraction of the matrix continuum is derived from a matrix porosity of 0.001 for anions (Cl-36 and I-129) and 0.005 for neutral species and cations (C-14, Cs-135 and Ni-59) (Nykyri et al., 2008). According to the anion exclusion concept, anions and cations do not occupy the same pore system in low-porosity rocks, due to electrical forces caused by the usually negative charge of the mineral surfaces. Anions experience repulsion close to the mineral surfaces whereas cations are adsorbed on the surfaces (Rasilainen, 1997).

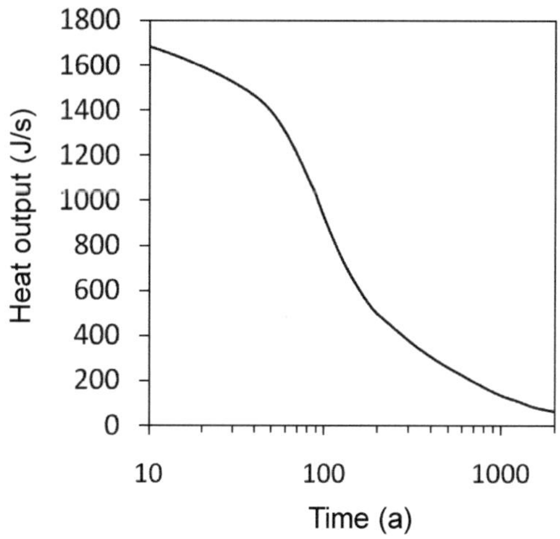


Fig. 18. Olkiluoto model. Heat output (J/s) from a canister with 1.93 t U burnt fuel from year 10 to year 2,000 after disposal (Löfman, 2005)

The near-field releases of radionuclides (Fig. 12) are used as input. The decay, diffusion and sorption data for the radionuclides (Tables 4 and 6; Nykyri et al., 2008) are the same as those for the three-dimensional far-field model.

Sorption of radionuclides takes place in the matrix continuum of the 15-continua model but not in the fracture continuum. The retardation factor $R_{m,i}$ of the radionuclide $i$ (dimensionless) for the matrix continuum is calculated according to the equation

$$R_{m,i}=1+((1-n_m)K_{d,i}d)/n_m, \qquad (4)$$

where $n_m$ is the matrix porosity (dimensionless), $d$ is the rock density (kg/dm$^3$) and $K_{d,i}$ is the distribution coefficient (dm$^3$/kg) of the radionuclide $i$ for the rock matrix (Table 6).

TOUGHREACT calculates diffusion according to Fick's first law. The retention of radionuclides is controlled by the diffusivity, the fluid volume and transport distance between fracture and matrix. A dual-continuum model can almost perfectly substitute the 15-continua model if only two variables are parameterised; e.g., the fluid volume of the fracture together with the transport distance between fracture and matrix (as applied here) or the fluid volume of the fracture together with the diffusivity (producing similar results).

The fluid volume of the fracture continuum is adjusted by parameterising the retardation factor of the fracture continuum (which equals unity in the 15-continua model). The diffusive transport distance is parameterised by multiplying the fracture spacing of the 15-continua model (5 m) with a distance factor. The parameters are determined by the log-least-squares method. This method minimises the squared deviations between the annual far-field releases of the dual-continuum model and those of the 15-continua model (Fig. 19; Table 6). Note that the annual far-field release of a radionuclide (Bq/a) is the annual transport of the radionuclide from the second topmost layer to the topmost layer of the two-dimensional model.

The parameters obtained from this optimised two-dimensional model are used for the three-dimensional model that has one fracture continuum and one matrix continuum. This equivalent dual-continuum model is used for the subsequent transport simulations.

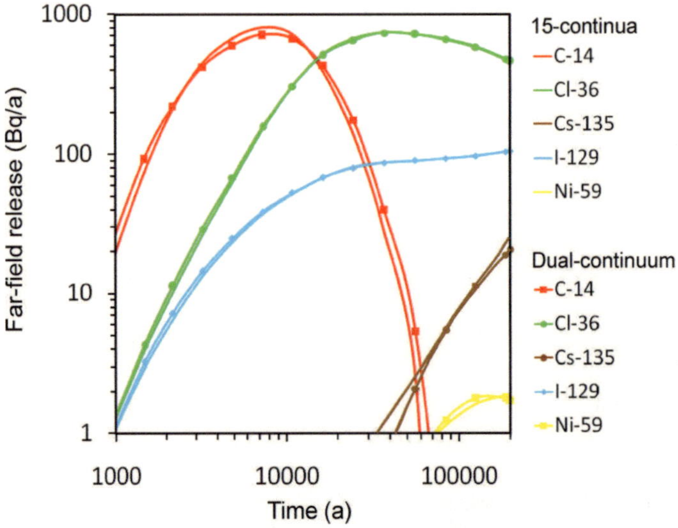


Fig. 19. Parameter estimation for the Olkiluoto model. The far-field release of radionuclides is shown for a 15-continua model and the equivalent dual-continuum model (optimised model)

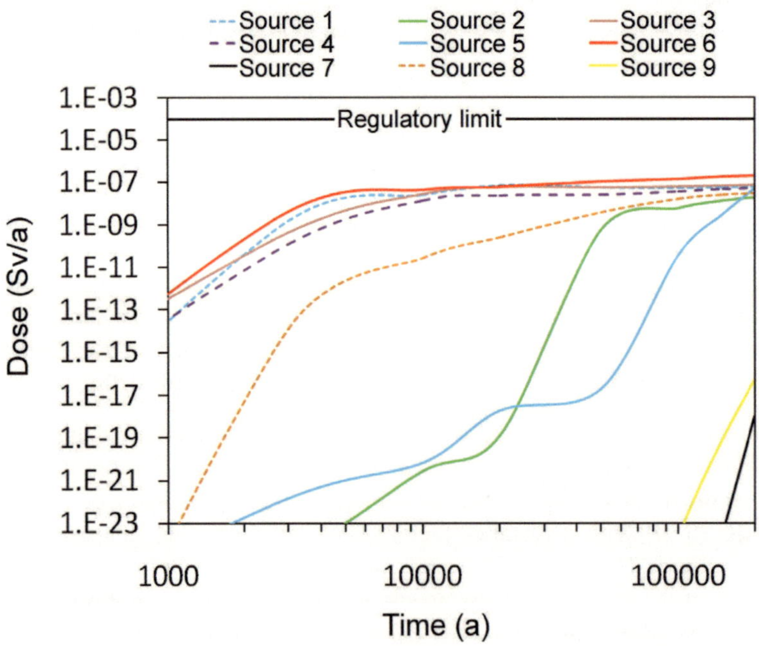


Fig. 20. Olkiluoto model. Peak values of the radioactive dose (Sv/a) for the second topmost layer for the sources 1-9

### 4.7 Simulated far-field transport

The radionuclide releases of the near-field base case are used as input for the far-field simulations. The biosphere is represented by the second topmost layer of the far-field model, which has fresh-water quality for most of the simulation period. The near-field base case assumes that one canister randomly chosen in the repository layout has a defect. There is a strong variability of radioactivity released to the biosphere depending on the location of the defective canister (Fig. 20). The canister locations 1-4, 6 and 8 in the north-eastern and central part of the repository imply a relatively large amount of radioactivity released to the biosphere whereas canisters located at the positions 5, 7 and 9 in the south-western part imply low or insignificant amounts of radioactivity. Figures 21A and 21B show the contaminant plumes for the source 3 at the NE corner and source 5 at the SE corner of the repository, respectively. The plumes emanating from the remainder of the canister locations have shapes that are intermediate between those shown. The maximum dose, which is produced by the canister position 6 ($1.2 \times 10^{-7}$ Sv/a), is far below the Finish regulatory limit of $1 \times 10^{-4}$ Sv/a.

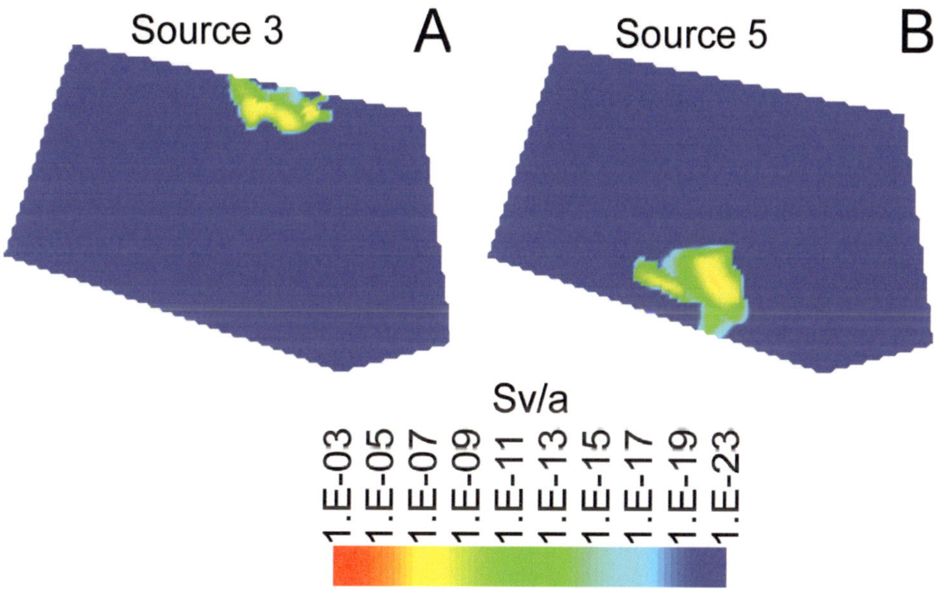


Fig. 21. Olkiluoto model. (A and B) Maps of the second topmost layer with radioactive dose (Sv/a) in the simulation year 200,000 for the sources 3 and 5

## 5. The Opalinus-Clay repository

The current scenarios assume that the repository will be located in the Jurassic Opalinus Clay in the Zürcher Weinland region, Switzerland (Fig. 1; NAGRA 2002). The claystone formation is gently dipping and has a thickness of 105-125 m. It is overlain by Middle Jurassic through Pleistocene sediments of approximately 600 m thickness. About 2,800 waste canisters containing 3,200 t U burnt fuel will be deposited in near-horizontal emplacement tunnels, which are connected to an access ramp via operational tunnels.

Two types of waste containers are considered for the repository. The first option is a steel canister, which is similar to that envisaged for the Gorleben repository. The second option is a container with a corrosion-resistant copper shell, such as the canisters for the Olkiluoto repository. So far, only the near-field release for the first option has been published (NAGRA, 2002; Kosakowski, 2004). For the second option, which considers improved safety standards, there are no near-field data available. The hydrochemical regime of a clay repository is similar to that of a granitoid repository but totally different from the highly corrosive regime of a salt repository. The performance of a copper-cast iron canister in a clay repository is as good as the performance of such a canister in a granitoid repository. Except for the case of a production defect or mechanical damage, the canister can assumed to be isolating the waste for a period of more than one million years. The permeability of a clay host rock is less than the permeability of a granitoid host rock. Thus it can be assumed that the potential groundwater contamination caused by a defective canister in the Opalinus-Clay repository is not higher than that for the Olkiluoto granitoid repository if the same type of waste canister is used.

## 6. Conclusions

The quality of models for groundwater contamination above high-level repositories in granitoid, salt and clay is not uniform. The most problematic case is a salt repository for two main reasons. First, the plastic behaviour of the host rock (salt) is a variable that introduces a high degree of uncertainty in any model attempt. Second, there is the inevitable generation of hydrogen gas caused by the corrosion of the steel waste canisters in high-salinity environments at low oxygen fugacities. The conditions of gas release constitute a great source of uncertainty. There is the possibility of unsteady or explosive release of hydrogen gas transporting short-lived radionuclides (Schulze, 2002). Furthermore, the permeability of the backfill of the shaft, through which the gas is assumed to flow, has a strong influence on the performance of the repository.

For a steady (non-explosive) gas release, a backfill permeability of $\leq 10^{-12}$ m$^2$ results in a radioactive dose of about 1 mSv/a. A higher permeability would decrease the radioactive dose to values below the German regulatory limit (0.1 mSv/a). A relatively high permeability favours the flow of liquid with respect to the flow of gas whereas a relatively low permeability has the opposite effect. A low permeability is the appropriate safety concept for a one-phase scenario (only liquid); however, a high permeability is the appropriate concept for a two-phase scenario (liquid and gas). As both scenarios are equally realistic, there is a typical conflict of targets. Such problems are not an important issue for the standard concept of a granitoid-hosted repository and the non-standard concept of a clay-hosted repository (both with copper-cast iron waste containers).

## 7. References

Beushausen, M. & Ludwig, R. (1990). Hydrogeologische Gliederung der oberoligozänen und miozänen Schichten. Bundesanstalt für Geowissenschaften und Rohstoffe, Hannover, Germany, Report Archive No. 106036, various paginations.

Brooks, R.H. & Corey, A.T. (1964). Hydraulic properties of porous media. *Hydrology Papers Colorado State University Fort Collins*, Vol. 3, pp. 1-27.

Corey, A.T. (November 1954). The interrelation between gas and oil relative permeabilities. *Producer's Monthly*, Vol. 19, No. 1 (quoted by Brooks and Corey 1964).

Diersch, H.J. (2009). *FEFLOW reference manual*. DHI-WASY GmbH, Berlin, Germany. URL: www.feflow.info/manuals.html

Hantush, M.M., Mariño, M.A. & Islam, M.R. (2000). Models for leaching of pesticides in soils and groundwater. *Journal of Hydrology*, Vol. 227, pp. 66-83.

Hantush, M.M., Govindaraju, R.S., Mariño, M.A. & Zhang, Z. (2002). Screening model for volatile pollutants in dual porosity soils. *Journal of Hydrology*, Vol. 260, pp. 58-74.

Hirsekorn, R.P., Nies, A., Rausch, H. & Storck, R. (1991). Performance assessment of confinements for medium-level and a-contaminated waste (PACOMA Project). GSF report 12/9 (GSF-Forschungszentrum für Umwelt und Gesundheit, Braunschweig, Germany), various paginations.

Javeri, V. (2006). Three-dimensional analyses of coupled gas, heat and nuclide transport in a repository including rock salt convergence. *Proceedings of the TOUGH Symposium 2006, 15-17 May 2006*, pp. 1-9, Lawrence Berkeley National Laboratory, Berkeley, California, U.S.A.. URL: http://esd.lbl

Keesmann, S., Nosek, U., Buhrmann, D., Fein, E. & Schneider, A. (2005). Modellrechnungen zur Langzeitsicherheit von Endlagern in Salz- und Granit-Formationen. GRS - Gesellschaft für Anlagen- und Reaktorsicherheit, Braunschweig, Germany, Report GRS-206, various paginations.

King, F. & Stroes-Gascoyne, S. (1995). Microbially influenced corrosion of nuclear fuel waste disposal containers. *Proceedings. Int. Conf. on Microbially Influenced Corrosion*, NACE International and American Welding Society, pp. 35/1-35/14.

King, F., Ahonen, L., Taxen, C., Vuorinen, U. & Werme, L. (2002). Copper corrosion under expected conditions in a deep geologic repository. *POSIVA Report* 2002-01. ULR: www.posiva.fi

Kirkkomäki, T. (2007). Design and stepwise implementation of the final repository (in Finnish, English abstract). *POSIVA Työraportti* 2006-92. URL: www.posiva.fi

Klinge, H., Boehme, J., Grissemann, C., Houben, G., Ludwig, R., Rübel, A., Schelkes, K., Schildknecht, F. & Suckow, A. (2007). Standortbeschreibung Gorleben Teil 1: Die Hydrogeologie des Deckgebirges des Salzstocks Gorleben. *Geologisches Jahrbuch*, Vol. C71, pp. 5-147.

Klinge, H., Margane, A., Mrugalla, S., Schelkes, K. & Söfner, B. (2001). Hydrogeologie des Untersuchungsgebietes Dömitz-Lenzen. Bundesanstalt für Geowissenschaften und Rohstoffe, Hannover, Germany, Report Archive No. 0121207, various paginations.

Kosakowski, G. (2004). Time-dependent flow and transport calculations for project Opalinus Clay (Entsorgungsnachweis). *NAGRA Technical Report* 03-10. URL: www.nagra.ch

Lichtner, P.C. (2000). Critique of dual continuum formulations of multicomponent reactive transport in fractured porous media. *Los Alamos National Laboratory Report* LA-UR-00-1097. URL: www.osti.gov

Liou, T.S. (2007). Numerical analysis of a short-term tracer experiment in fractured sandstone. *Terrestrial Atmospheric Oceanic Sciences*, Vol. 18, pp. 1029-1050.

Löfman, J. (2005). Simulation of hydraulic disturbances caused by the decay heat of the repository in Olkiluoto. *POSIVA Report* 2005-07. URL: www.posiva.fi

Löfman, J. & Poteri, A. (2008). Groundwater flow and transport simulations in support of RNT-2008 analysis. *POSIVA Working Report* 2008-52. URL: www.posiva.fi

Ludwig, R., Mandl, J. & Uhlig, A. (1989). Zwischenbericht über die Diskretisierung hydrogeologischer Strukturen im Bereich Gorleben. Bundesanstalt für Geowissenschaften und Rohstoffe, Hannover, Germany, Report Archive No. 0105061, various paginations.

Ludwig, R., Schneider, M. & Sewing, D. (1993). Projekt Gorleben: Hydrogeologische Gliederung der quartärzeitlichen Schichtenfolge. Bundesanstalt für Geowissenschaften und Rohstoffe, Hannover, Germany, Report Archive No. 110256, various paginations.

NAGRA (2002). Project Opalinus Clay - safety report. *NAGRA Technical Report* 02-05. URL: www.nagra.ch

Nykyri, M., Nordman, H., Marcos, N., Löfman, J., Poteri, A. & Hautojärvi, A. (2008). Radionuclide release and transport - RNT-2008. *POSIVA Report* 2008-06. URL: www.posiva.fi

Pitkänen, P., Partamies, S. & Luukkonen, A. (2004). Hydrochemical interpretation of baseline groundwater conditions at the Olkiluoto site. *POSIVA Report* 2003-07. URL: www.posiva.fi

POSIVA (2006). Expected evolution of a spent nuclear fuel repository at Olkiluoto. *POSIVA Report* 2006-05. URL: www.posiva.fi

POSIVA (2009). Olkiluoto site description 2008. *POSIVA Report* 2009-1. URL: www.posiva.fi

Prikryl, R. & Woller, F. (April 2002). Going underground - a new market for Czech bentonite in nuclear waste disposal. *Industrial Minerals*, pp. 72-77.

Pruess, K. (1983). GMINC - a mesh generator for flow simulations in fractured reservoirs. *Lawrence Berkeley National Laboratory Report* LBL-15227. URL: http://esd.lbl

Pruess, K. & Narasimhan, T.N. (February 1985). A practical method for modeling fluid and heat flow in fractured porous media. *Society Petroleum Engineers Journal*, pp. 14-26.

Pruess, K. (1992). Brief guide to the MINC-method for modeling flow and transport in fractured media. *Lawrence Berkeley National Laboratory Report* LBL-32195. URL: http://esd.lbl

Pruess, K., Oldenburg, C. & Moridis, G. (1999). TOUGH2 user's guide, version 2.0. *Earth Sciences Division, Lawrence Berkeley National Laboratory, University of California*, Berkeley, U.S.A.. URL: http://esd.lbl

Rasilainen, K. (1997). Matrix diffusion model. *VTT Publication* 331 (ESPOO, Helsinki, Finland). www.vtt.fi/inf/pdf/publications/1997/P331.pdf

Rasmussen, T.C. (2001). Pressure wave vs. tracer velocities through unsaturated fractured rock. *American Geophysical Union Geophysical Monograph*, Vol. 42, pp. 45-52.

Schulze, O. (2002). Auswirkung der Gasentwicklung auf die Integrität geringdurchlässiger Barrieregesteine. Bundesanstalt für Geowissenschaften und Rohstoffe, Hannover, Germany, Report Archive No. 0122442, 142 pp.

Schwartz, M.O. (1996). Native copper deposits and the disposal of high-level waste. *International Geology Review*, Vol. 38, pp. 33-41.

Schwartz, M.O. (2008). High-level waste disposal, ethics and thermodynamics. *Environmental Geology*, Vol. 54, pp. 1485-1488.

SKB (2006). Long-term safety for KBS-3 repositories at Forsmark and Laxemar - a first evaluation. *SKB Technical Report* TR-06-09. URL: www.skb.se

SKI (2006). Engineered barrier system - assessment of the corrosion properties of copper canisters. *SKI Report* 2006:11. URL: www/ski.se

Storck, R., Aschenbach, J., Hirsekorn, R.P., Nies, A. & Stelte, N. (1988). PAGIS performance assessment of geological isolation systems for radioactive waste: disposal in salt formations. *Office for Official Publications of the European Communities*, Luxembourg.

Sudicky, E.A. & Frind, E.O. (1982). Contaminant transport in fractured porous media: Analytical solutions for a system of parallel fractures. *Water Resources Research*, Vol. 18, pp. 1634-1642.

Suter, D., Biehler, D., Blasser, P. & Hollmann, A. (1998). Derivation of a sorption data set for the Gorleben overburden. *Proceedings DisTec'98, September 9-11, 1998*, pp. 581-584, Hamburg, Germany.

Tang, D.H., Frind, E.O. & Sudicky, E.A. (1981). Contaminant transport in fractured porous media: Analytical solution for a single fracture. *Water Resources Research*, Vol. 17, pp. 555-564.

Vaittinen, T., Ahokas, H. & Nummela, J. (2009) Hydrogeological structure model of the Olkiluoto site - update in 2008. *POSIVA Working Report* 2009-15. URL: www.posiva.fi

Van Genuchten, M.T. (1980). A closed-form equation for predicting the hydraulic conductivity of unsaturated soils. *Soil Science Society America Journal*, Vol. 44, pp. 892-898.

Vogel, P. (2005). Orientierende 3-D-Grundwassermodellrechnungen auf den Strukturen eines hydrogeologischen Modells Gorleben. Bundesanstalt für Geowissenschaften und Rohstoffe, Hannover, Germany, Report Archive No. 0125865, 70 pp.

Wu, Y.S. & Pruess, K. (2000). Numerical simulation of non-isothermal multiphase tracer transport in heterogeneous fractured porous media. *Advances in Water Resources*, Vol. 23, pp. 699-723.

Xiao, Z., Gammons, C.H. & Williams-Jones, A.E. (1998). Experimental study of copper(I) chloride complexing in hydrothermal solutions at 40 to 300°C and saturated vapor pressure. *Geochimica et Cosmochimica Acta*, Vol. 62, pp. 2949-2964.

Xu, T. & Pruess, K. (2001). Modeling multiphase non-isothermal fluid flow and reactive geochemical transport in variably saturated fractured rocks: 1. Methodology. *American Journal Science*, Vol. 301, pp. 16-33.

Xu, T., Sonnenthal, E., Spycher, N. & Pruess, K. (2005). TOUGHREACT user's guide: a simulation program for non-isothermal multiphase reactive geochemical transport in variably saturated geologic media. *Earth Sciences Division, Lawrence Berkeley National Laboratory, University of California*, Berkeley, U.S.A.. URL: http://esd.lbl

Zhang, K., Wu, Y. & Pruess, K. (2008). User's guide for TOUGH2-MP - a massively parallel version of the TOUGH2 code. *Earth Sciences Division, Lawrence Berkeley National Laboratory, University of California*, Berkeley, U.S.A.. URL: http://esd.lbl

# 20

# An Assessment of the Impact of Advanced Nuclear Fuel Cycles on Geological Disposal

Jan Marivoet and Eef Weetjens
*Belgian Nuclear Research Centre (SCK•CEN)*
*Belgium*

## 1. Introduction

After a strong growth in the 1970s and 1980s, the deployment of nuclear energy has stagnated. Only a few nuclear power plants became operational during the last 20 years. More recently, rising oil prices, increasing energy demands and the need for carbon-free energy to limit global warming have resulted in a revival of interests in nuclear energy. New nuclear power plants are now under construction in, e.g., Finland, France, Japan, Korea, the Russian Federation, China and India. Especially in these last 2 countries a considerable increase of the contribution of nuclear energy to the electricity generation is expected (Nuclear Energy Agency [NEA], 2008). The nuclear renaissance made that advanced nuclear fuel cycles are being studied worldwide. They aim at making more efficient use of the available resources, reducing the risk of proliferation of nuclear weapons, and facilitating the management of the radioactive waste. However, the Fukushima accident in 2011 strongly tempered the revival of nuclear energy in several countries.

During the last 20 years the introduction of partitioning and transmutation (P&T) techniques in future nuclear fuel cycles has often been presented as a promising option that might strongly facilitate the long-term management of high-level radioactive waste, because transmutation of the actinides, which are present in spent nuclear fuel, in fast neutron reactors can strongly reduce the long-term radiotoxicity of the radioactive waste. On the other hand, the relevance of this reduction of radiotoxicity for radioactive waste management is questioned by the geological disposal community.

The impact of advanced fuel cycles on radioactive waste management and geological disposal has been studied in some national and international projects between 2003 and 2010. Highly relevant projects were the Expert Group on the Impact of P&T on Radioactive Waste Management of the Nuclear Energy Agency (NEA, 2006) and the Red-Impact project of the European Commission (von Lenza et al., 2008; Marivoet et al., 2009). In the following sections the impact of advanced fuel cycles on various aspects of radioactive waste management and, more specifically, of geological disposal in a clay formation will be investigated. The presented results were mainly obtained from SCK•CEN's participation in the Red-Impact project (Marivoet et al., 2007; Weetjens & Marivoet, 2008).

## 2. Considered fuel cycles and repository system

Making an assessment of the impact of advanced fuel cycles on radioactive waste management and geological disposal requires that a number of representative fuel cycle scenarios are identified. As a basis for comparison, various indicators are presented in the following sections by taking disposal in Boom Clay as example to develop the case-study. Results obtained in international projects (NEA, 2006; Marivoet et al., 2009) show that most conclusions that can be drawn from the evaluations of disposal in Boom Clay are largely valid for disposal in other clay formations and to a large extent for disposal in hard rock formations.

### 2.1 Considered fuel cycles

On the basis of an NEA report on advanced fuel cycles (NEA, 2002), 5 fuel cycle scenarios (3 industrial scenarios, which are based on presently available technology, and 2 innovative scenarios) were selected for detailed analyses in the Red-Impact project:

- fuel cycle A1: the reference fuel cycle, which is the "once through" cycle based on pressurised water reactor (PWR) plants with uranium oxide (UOX) fuel;
- fuel cycle A2: fuel cycle based on PWR plants with uranium oxide fuel, the generated Pu is recycled once as mixed oxide (MOX) fuel;
- fuel cycle A3: fuel cycle based on a sodium cooled fast neutron reactors with MOX fuels, in which Pu is multi-recycled;
- fuel cycle B1: fuel cycle based on a sodium cooled fast neutron reactors with MOX fuels, in which all the actinides are recycled;
- fuel cycle B2: fuel cycle based on PWR plants with uranium oxide fuel, the generated Pu is recycled once as MOX fuel, all the minor actinides and the Pu in the spent MOX fuel are recycled in a fast neutron accelerator driven system (ADS).

Results obtained for a variant of fuel cycle B1 (called B1V), in which it is assumed that the heat generating Cs and Sr are separated from the high-level waste (HLW), are also presented.

For each selected fuel cycle mass flow schemes have been prepared and the corresponding neutronic calculations have been made (von Lensa et al., 2008). The estimated waste inventories strongly depend on the assumed separation efficiency of the considered partitioning processes and the applied waste conditioning methods. It was assumed that in future reprocessing plants the reprocessing losses will be limited to 0.1% for all the actinides. For the volatile fission and activation products it was assumed that 1% of the $^{129}I$ and $^{36}Cl$ and 10% of the $^{14}C$ remain in the high-level waste. Specific and optimised waste matrices for conditioning the high-level waste that will arise from advanced fuel cycles have not yet been developed. Therefore, it was assumed that the conditioning methods for the waste from advanced fuel cycles will be very similar to the ones applied today in existing reprocessing facilities.

The volumes of high-level (HLW) and intermediate-level waste (ILW) that are generated in the considered fuel cycles have been estimated in the Red-Impact project (Cunado, 2007) as follows. For the 2 fuel cycles that are applied today (A1 and A2) realistic estimates of the generated waste volumes could be made on the basis of data provided by the participating waste agencies. For the future fuel cycles (A3, B1 and B2) the volumes of the HLW were estimated on the basis of the mass of the generated fission products and assumed reprocessing losses, and by assuming that the waste will be conditioned as vitrified HLW with the same waste loading as applied today for cycle A2. The volumes of ILW were estimated on the basis of the information available on the cladding and structural components of the fuel assemblies, whereas for the technological waste from reprocessing it was assumed that future reprocessing plants will generate similar waste volumes as the present ones.

The amount of natural uranium that is needed in each of the considered fuel cycles to produce 1 TWh(e) electricity is given in Table 1 (Marivoet & Weetjens, 2008). This indicator illustrates how the introduction of advanced fuel cycles increases the efficiency of the use of uranium. The recycling of Pu in fuel cycle A2 results in a small increase, of about 10%, of the efficiency of the use of uranium for electricity production. Recycling Pu and introducing an ADS in a nuclear reactor park consisting of PWRs (fuel cycle B2) results in an increase of the generated electricity with about 25%. However, replacing PWRs by fast neutron reactors results in a drastic increase of the amount of electricity produced per kg uranium, because natural or depleted uranium can be used instead of enriched uranium for the production of the reactor fuel. In principle, fully recycling all the actinides in a fast reactor (fuel cycle B1) can result in an increase of the efficiency of the use of natural uranium with a factor 100 or more.

| **Fuel cycle** | | **A1** | **A2** | **A3** | **B1** | **B2** |
|---|---|---|---|---|---|---|
| Consumption of natural uranium | kg/TWh(e) | 20723 | 18448 | 986 | 106 | 15766 |

Table 1. Amount of natural uranium needed to produce 1 TWh(e) electricity (Marivoet & Weetjens, 2008)

## 2.2 Considered repository system

The repository considered for the impact evaluations is the SAFIR 2 repository concept that has been developed by the Belgian radioactive waste management organisation ONDRAF/NIRAS (2002). This reference repository is assumed to be excavated in the Boom Clay formation at the Mol site in NE Belgium. At the considered site the host formation is about 100 m thick, of which an 80 m thick central zone consists of very homogeneous clay (in terms of its flow and transport parameters).

The considered repository will have a central access facility consisting of at least two vertical transport shafts and two horizontal transport galleries. The disposal galleries will be excavated perpendicular upon the transport galleries. The HLW canisters will be placed one after the other or with some spacing between two canisters to respect temperature limitations, along the axis of the disposal galleries. A clayey buffer will be placed between the HLW canisters and the gallery walls. Because the Boom Clay is a plastic clay, a concrete lining is required to avoid convergence of the gallery walls during the operational phase of the repository. Figure 1 shows a scheme of a disposal gallery configuration for uranium oxide spent fuel.

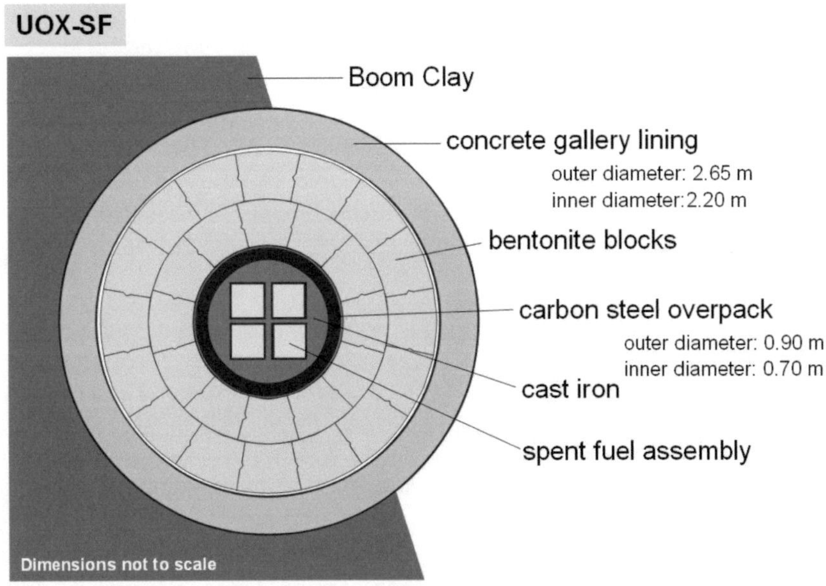


Fig. 1. Gallery configuration for disposal of uranium oxide spent fuel (4 assemblies per container)

## 3. Assessment of the impact of advanced fuel cycles

### 3.1 Impact on waste volumes

The total volumes of the waste canisters (these are the primary waste containers) containing the generated HLW and ILW are given in Table 2 (Cunado, 2007). These results show that the volume of the HLW generated in fuel cycle B1 is 5 times smaller than the volume of the spent fuel from fuel cycle A1. However, when the volume of the ILW is also considered, the volume reduction is limited to a factor 2.

| **Fuel cycle** | | **A1** | **A2** | **A3** | **B1** | **B2** |
|---|---|---|---|---|---|---|
| HLW volume | $m^3/TWh(e)$ | 2.216 | 0.806 | 0.420 | 0.397 | 0.464 |
| ILW volume | $m^3/TWh(e)$ | - | 0.387 | 0.825 | 0.825 | 0.518 |
| Total volume | $m^3/TWh(e)$ | 2.216 | 1.193 | 1.245 | 1.223 | 0.982 |
| Relative volume | - | 1.000 | 0.539 | 0.562 | 0.552 | 0.443 |

Table 2. Volumes of the generated HLW and ILW (packed in canisters) estimated for the considered fuel cycles

The actual volume of HLW is not considered a meaningful indicator for geological disposal. If waste volumes were important, it would be relatively simple to reduce the volume of spent fuel by removing the void between the fuel pins by dismantling the assemblies. However, the dimensions of a geological repository for HLW disposal are mainly imposed by the thermal output of the waste and not by its volume. Therefore, the thermal output and the dimensions of the repository are discussed in the following section.

### 3.2 Impact on repository dimensions

Figure 2 gives the evolution of the thermal output of the HLW generated in the considered fuel cycles. After a 50 years cooling time the thermal output of the vitrified HLW from cycle A2 is only 57% of the one of spent fuel from cycle A1; however, if the MOX spent fuel from cycle A2 has to be disposed of as well, the reduction of the thermal output is only 3%. In the case of advanced fuel cycle B1 the thermal output of the HLW is reduced with a factor 3 in comparison with the reference cycle A1. It is also interesting to note that the thermal output of the HLW from fuel cycle B1 is 3 times lower after a 100 years cooling time than after a 50 years cooling time; over the same time span the thermal output of the spent fuel from cycle A1 is only reduced with a factor 2.

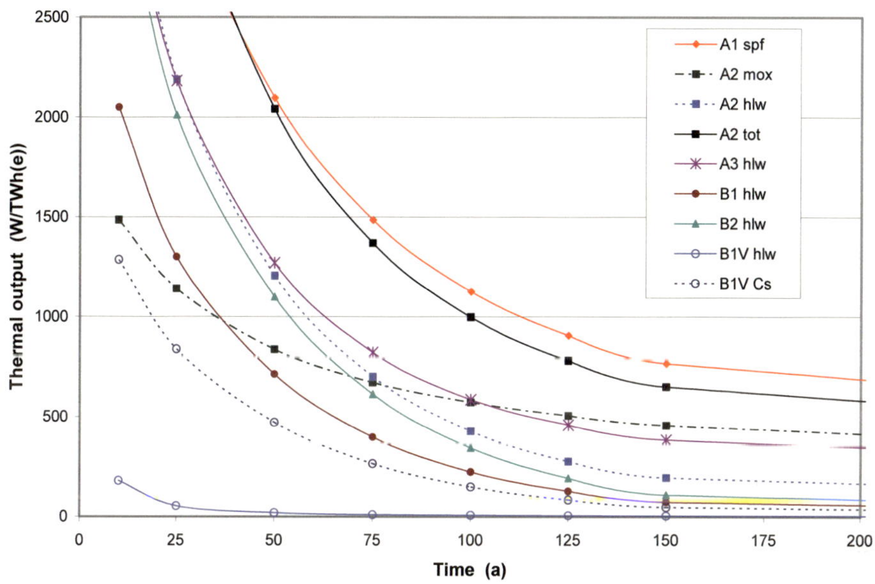


Fig. 2. Evolution of the thermal output of the high-level waste resulting from the considered fuel cycles (cycle B1V refers to cycle B1 with removal of Cs and Sr from the HLW) (Marivoet & Weetjens, 2009)

Removal of the heat generating Cs and Sr from the HLW results in a strong reduction of the thermal output of that waste; e.g., in the case of fuel cycle B1 the thermal output of the HLW is reduced with a factor 40. However, in this case the strongly heat generating Cs and Sr waste has to be stored during at least a century in a cooling facility to have a significant reduction of the thermal output, and eventually the Cs-waste, which contains the very long-lived $^{135}Cs$ ($T_{½}$ = 2.3 Ma), has to be disposed of in a geological repository.

The minimum lengths of the galleries for disposal of HLW are derived from heat dissipation calculations by ensuring that the temperature limitations are respected; for the considered disposal concept the main temperature limitation is that the temperature at the interface between the gallery liner and the host clay formation should remain below 100 °C. An overview of the estimated lengths of the HLW disposal galleries for a 50 years cooling time is given in Table 3.

| Fuel cycle | | A1 | A2 | A3 | B1 | B1V | B2 |
|---|---|---|---|---|---|---|---|
| Allowable thermal loading (50 a) | W/m | 353 | 332-376 | 365 | 379 | - | 379 |
| HLW gallery length | m/TWh(e) | 5.92 | 5.74 | 3.48 | 1.88 | 0.44 | 2.89 |
| Relative HLW gallery length | - | 1.00 | 0.97 | 0.59 | 0.32 | 0.074 | 0.49 |

Table 3. Allowable thermal loading per metre gallery and estimated minimum lengths of the HLW disposal galleries for a 50 years cooling time (cycle B1V refers to a variant of cycle B1 with removal of Cs and Sr from the HLW; a 100 years cooling time is assumed for the Cs-waste) (Marivoet & Weetjens, 2009)

For HLW containing only traces of actinides, which is the case for fuel cycles B1 and B2, the maximum allowable linear thermal loading at disposal time of the disposal gallery is 379 W/m; in the case of the MOX spent fuel from cycle A2, which contains a considerable fraction of actinides, this maximum thermal loading is reduced to 332 W/m. Because of the relatively small difference (about 12%) between the 2 extreme values of the allowable thermal loading, the minimum lengths of the disposal galleries are about proportional to the thermal output of the waste at disposal time. Multi-recycling of the Pu in fast reactors (cycle A3) results in a reduction of the needed length of the disposal galleries of about 40%, a PWR reactor park combined with an ADS for burning the minor actinides and the multi-recycled Pu (cycle B2) gives a reduction of about 50%, and full recycling of all actinides in fast reactors gives a reduction of about 68%. By assuming that the Cs-waste is cooled during 100 years, the removal of Cs and Sr from the HLW gives a reduction of the needed gallery length of more than 90% in comparison with the once through cycle A1.

### 3.3 Evolution of radiotoxicity in the waste

Radiotoxicity is a measure of how nocuous a radionuclide is to health. The type and energy of rays, absorption in the organism, residence time in the body, etc. influence the degree of radiotoxicity of a radionuclide. In the context of radioactive waste management radiotoxicity is defined as the sum over all radionuclides present in the waste of the products of the activity of a radionuclide with its dose coefficient for ingestion; values of the latter are published by the International Commission on Radiological Protection (ICRP, 1996).

The introduction of P&T techniques in a fuel cycle, i.e. recycling of all actinides in a fast neutron spectrum facility (fast reactor or ADS), results in a drastic reduction of the radiotoxicity in the high-level waste (see Figure 3).

After about 30 years cooling time the minor actinides, mainly Pu, Am and Cm isotopes, are the main contributors to the radiotoxicity. In the case of the reference fuel cycle A1 it takes more than 200 000 years before the radiotoxicity in the spent fuel has dropped below the level present in the natural uranium that was needed to produce that fuel. For fuel cycle A2 only a small reduction of the radiotoxicity can be observed because all the minor actinides and the Pu of the MOX spent fuel are present in the disposed waste. Recycling of all the generated Pu (cycle A3) makes that the radiotoxicity of the natural uranium is reached after about 20 000 years, and when all the actinides are recycled (cycles B1 and B2) it is reached already after about 400 years.

The strong reduction of the time after which radiotoxicity reaches the level of natural uranium is one of the main arguments used to justify the introduction of P&T in future fuel cycles. However, as we will see in the following section, the actinides are very efficiently confined in the geological disposal system and their releases into the biosphere are about negligible, because most actinides are poorly soluble in the chemical conditions prevailing in a geological repository in clay and, furthermore, all actinides are strongly sorbed on the clay minerals (or immobile organic matter) of the host formation. As a consequence, there is no relationship between radiotoxicity and the radiological consequences of a geological repository.

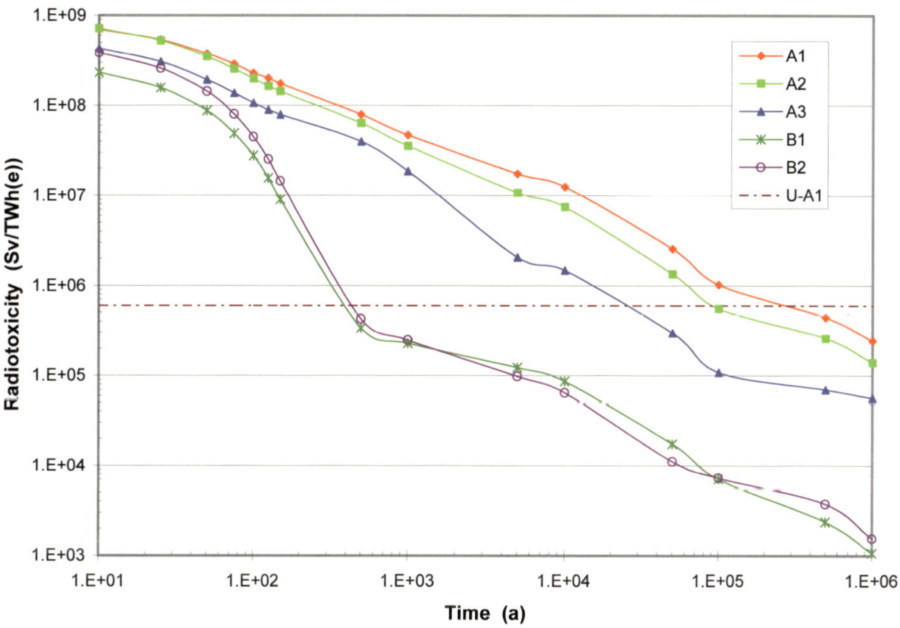


Fig. 3. Evolution of the radiotoxicity present in the high-level waste arising from the 5 considered fuel cycles (the reference line U-A1 gives the radiotoxicity present in the natural uranium needed to generate 1 TWh(e) in cycle A1) (Marivoet & Weetjens, 2009)

### 3.4 Impact on long-term dose in the case of the reference scenario

The main indicator to evaluate the safety of a geological repository for the disposal of radioactive waste is the effective annual dose to a representative person of the potentially most exposed group in the case of the reference (i.e. expected evolution of the repository system) scenario. The International Commission on Radiological Protection (ICRP, 2000) recommends to use a dose constraint of 0.3 mSv/a, or lower, as reference value for the effective dose. Most national radiological protection authorities are now imposing a dose constraint in the range 0.1 - 0.3 mSv/a.

The reference scenario assumes that the barriers of the disposal system will function as expected. For the considered repository system it was assumed that the overpack will have a lifetime of 2000 years, and that the waste matrices of the vitrified HLW and spent fuels will dissolve in a 100 000 and 200 000 years period respectively. The solubility limits in the near field are taken into account. The main barrier of the disposal system is the natural barrier provided by the host clay formation; because water flow in the clay is negligible, the radionuclide transport through the clay layer is essentially diffusive; furthermore, many radionuclides are strongly sorbed on the clay minerals. After migration through the clay layer, only a small fraction of the disposed radionuclides reach the aquifers overlying the host formation and eventually the biosphere via a small river draining the contaminated part of the aquifer.

The total dose (normalised per produced electricity) and its main contributors calculated for the disposal of spent fuel from fuel cycle A1 in case of the reference scenario are shown in Figure 4. In case of direct disposal of spent fuel the main contributor to the total dose is $^{129}I$. Other important contributors to the dose are the long-lived fission products $^{79}Se$, $^{99}Tc$ and $^{126}Sn$ and the activation product $^{36}Cl$. Actinides give contributions to the total dose after 2 million years; the maximum actinide dose is more than 2 orders of magnitude lower than the maximum $^{129}I$ dose.

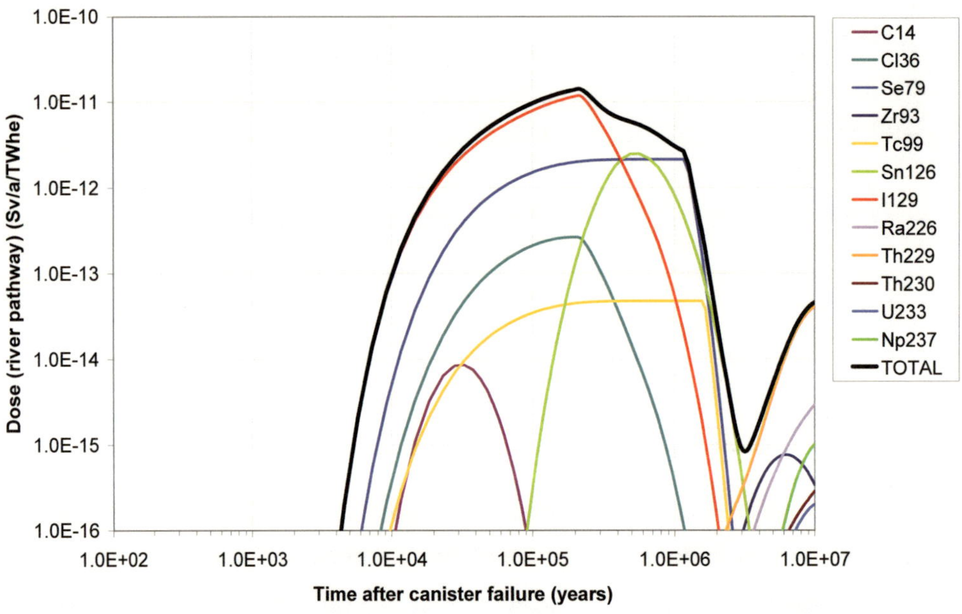


Fig. 4. Evolution of the total effective dose via the river pathway due to the disposal of spent fuel arising from fuel cycle A1 in the case of the reference scenario

Figure 5 gives the doses calculated for the disposal of the high-level and intermediate-level waste (ILW) in case of the reference scenario for the 5 considered fuel cycles. The calculated total doses (estimated by multiplying the normalised doses with the electricity generated within the existing national nuclear programme; the present Belgian nuclear programme is expected to generate about 1700 TWh(e)) are a few orders of magnitude below the dose constraint.

The impact of the transmutation of the actinides in the advanced fuel cycles on the maximum doses is negligible, because the maximum dose is essentially due to mobile long-lived fission and activation products. The amount of generated fission products is about proportional to the heat produced by nuclear fission in the reactors. The difference in the calculated maximum doses is mainly explained by the amount of $^{129}I$ that is present in the disposed waste. In the case of spent fuel disposal (cycle A1, and partially in the case of cycle A2 in which the MOX spent fuel is assumed to be directly disposed of), all the iodine is going into the repository, whereas in the case of reprocessing only a small fraction of the iodine, which is estimated to be about 2% (1% in the HLW and 1% in the ILW), remains in the waste; the bulk of the iodine escapes as gas from the HLW stream in the dissolver at the reprocessing plant. On the very long-term, i.e. after a few million years for disposal in clay, calculated doses are lower in the case of advanced fuel cycles, because smaller amounts of actinides are present in the HLW arising from these fuel cycles.

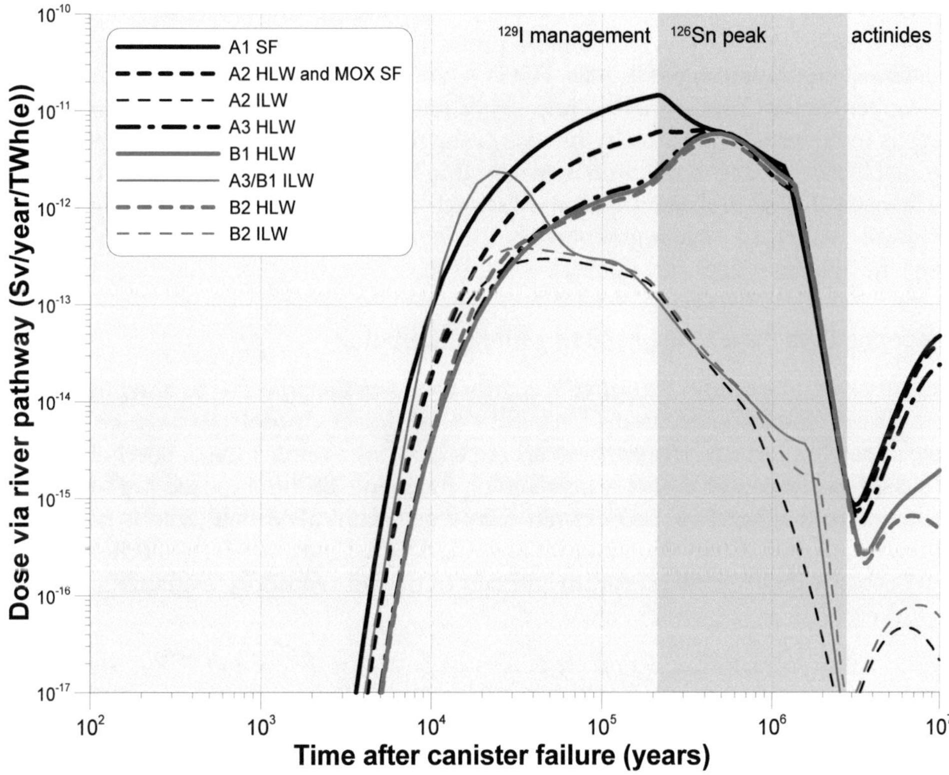


Fig. 5. Evolution of the total effective dose due to the disposal of the high-level and intermediate-level waste arising from the 5 considered fuel cycles (Marivoet & Weetjens, 2008)

Figure 5 also gives the doses due to the disposal of the ILW arising from the considered fuel cycles. For fuel cycle A2 the maximum ILW dose is more than one order of magnitude lower than the maximum HLW dose. However, the ILW dose due to cycles A3 and B1, which are based on fast reactors, shows a relatively high initial dose. This dose is due to the considerable amounts of $^{14}C$ which are calculated to be generated in fast reactors.

### 3.5 Impact on long-term dose in other cases and altered evolution scenarios

In the safety assessments made for the SAFIR 2 report and for the upcoming Belgian safety case a number of altered evolution scenarios have been analysed as well. Also for those scenarios, the total dose is mainly due to mobile fission and activation products. However, actinides can play a more important role in safety assessments of disposal in other host formations, such as hard rock (granite) and volcanic rocks (tuff or basalt).

In the case of Yucca Mountain in the United States (Department of Energy [DoE], 2008), the spent fuel is assumed to be disposed above the groundwater table in a mountain consisting of welded tuff located in an arid area. A very low amount of water is flowing through the repository. The oxidising chemical conditions in the disposal galleries result in a relatively high solubility of some actinides. As a consequence the total dose calculated for Yucca Mountain is due to mobile fission and activation products and to actinides.

Disposal in hard rock formations is studied in, e.g., Sweden and Finland and safety cases have been submitted to the radiological protection authorities by the respective waste management organisations (SKB, 2006; POSIVA, 2010). In these cases the use of a very long-lived copper canister makes that no releases of radionuclides are expected during the first hundreds of thousands of years. In the case of the reference scenario in which the failure of some canisters is assumed, the dose is mainly due to $^{129}I$ and to a lesser extent to $^{135}Cs$ and $^{14}C$. However, for several variant scenarios, such as penetration of oxidising glacial melt water in the near field or a degradation of the bentonite buffer, $^{226}Ra$, $^{231}Pa$ and $^{239}Pu$ can become the dominant dose contributors.

### 3.6 Impact on the dose in the case of a human intrusion

A few very low probability disruptive scenarios can have the potential to bring man directly in contact with the disposed waste. One of the most drastic disruptive scenarios that might be considered is a human intrusion scenario in which it is assumed that a borehole is drilled across the repository and that cores containing fragments of the disposed high-level waste are taken from that borehole and examined by a geotechnical worker, who is not aware of the presence of radioactive material (Smith et al., 1987). The dose to a geotechnical worker and its contributors calculated for an intrusion into the repository containing spent fuel from fuel cycle A1 are shown in Figure 6.

Up to 100 000 years minor actinides, successively $^{241}Am$, $^{240}Pu$ and $^{239}Pu$, are the main contributors to the dose to a geotechnical worker; at the long term, i.e. after 100 000 years, the main contributors to the dose are $^{222}Rn$ and $^{229}Th$.

The doses to a geotechnical worker calculated for an intrusion into the repository containing the HLW arising from the considered fuel cycles are shown in Figure 7. This figure also gives 3 reference lines, which correspond to the ICRP intervention levels of 10 and 100 mSv and to the dose calculated for a geotechnical worker who is handling and analysing cores that are assumed to be taken from the uranium mine at Cigar Lake (Canada).

Figure 7 shows that only for 3 HLW types, arising from the advanced fuel cycles B1 and B2, the dose to a geotechnical worker drops within 1 million years under the dose associated with an intrusion into a Cigar Lake uranium ore body. For the other high-level waste and spent fuel types the doses to a geotechnical worker remain (much) higher than the dose associated with an intrusion into the Cigar Lake ore body. Consideration of the ICRP intervention levels of 10 and 100 mSv (ICRP, 2000) can be used to contextualise the dose values calculated for the various spent fuel and vitrified HLW types. Applying a simple criterion of ranking when the calculated dose for each HLW type is bounded by the ICRP intervention levels allows a "ranking" of the high-level waste types.

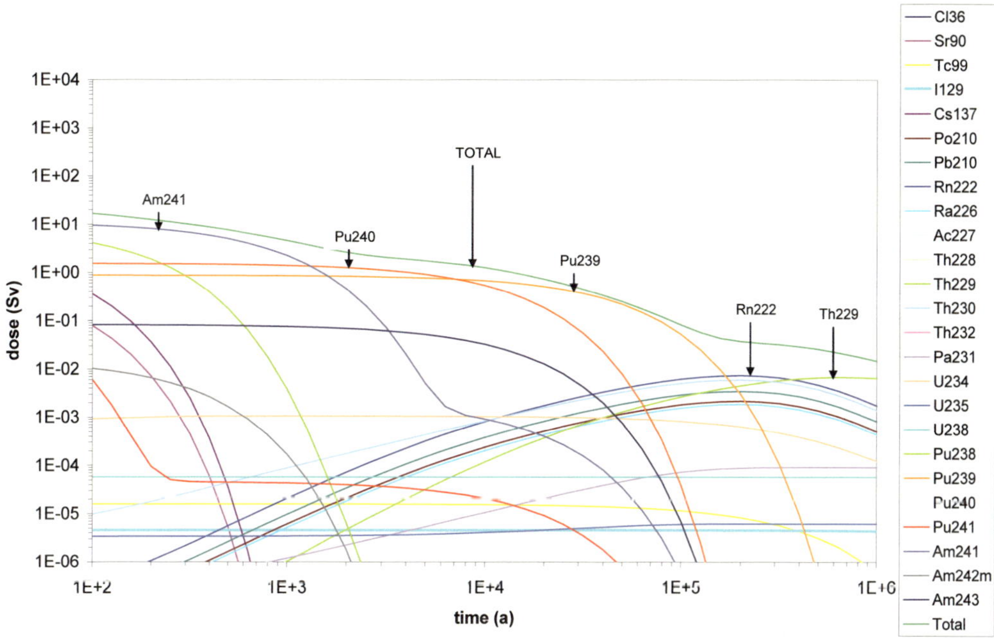


Fig. 6. Evolution of the dose to a geotechnical worker and its contributors calculated for an intrusion into a repository containing spent fuel from fuel cycle A1

Table 4 gives the estimated "required isolation times" for the main spent fuel or vitrified HLW types arising from each of the 5 considered fuel cycle scenarios. In order to verify whether the radiotoxicity present in the disposed waste can be used as an alternative indicator, Table 4 also gives the time after which the radiotoxicity in the disposed waste drops under the radiotoxicity in fresh uranium required for the fuel production in fuel cycle A1 (see Fig. 7). Comparison of the times obtained in Table 4 from the 100 mSv intervention level with the times derived from the radiotoxicity shows that these times are often of the same order of magnitude.

In the case of fuel cycles generating different HLW types, one or more waste types can occur in which a lot of radioactivity is concentrated, but of which only a very small number of waste packages are generated; this is the case for the MOX spent fuel in fuel cycle A2 and the vitrified HLW arising from the pyro-reprocessing of ADS spent fuel in fuel cycle B2. The results given in Table 4 show that the "required isolation times" obtained for the 100 mSv intervention level for the waste types of which the largest quantities are generated correspond much better with the times estimated on the basis of the radiotoxicity arising from the considered fuel cycle.

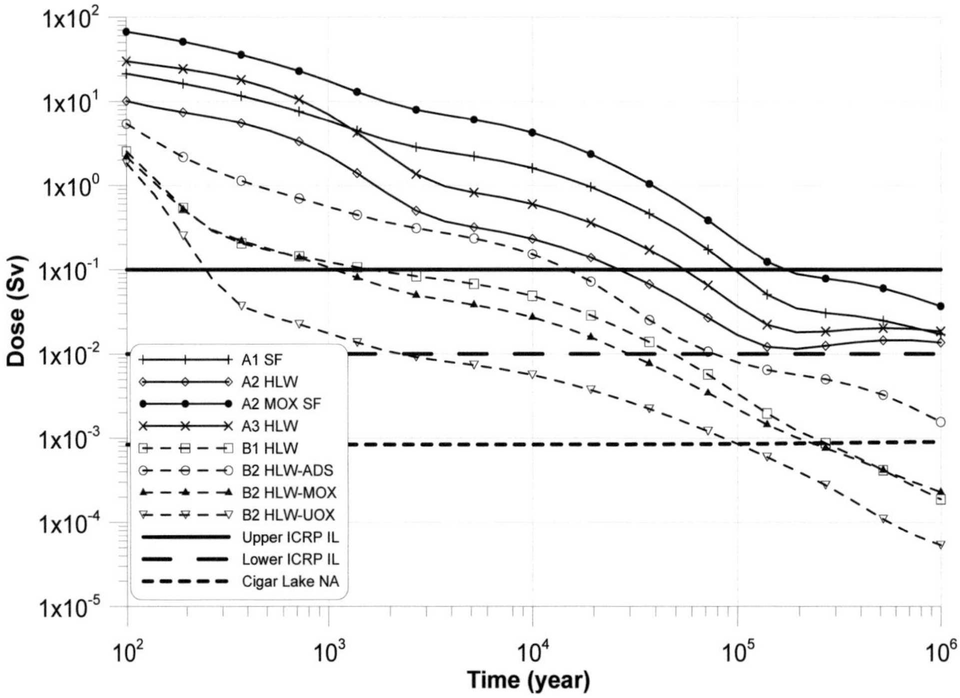


(IL: intervention level; NA: natural analogue)

Fig. 7. Evolution of the dose to a geotechnical worker for the 5 considered fuel cycles (Marivoet & Weetjens, 2009)

| Fuel cycle | Comparator / Waste type | ICRP 10 mSv intervention level | ICRP 100 mSv intervention level | Radiotoxicity |
|---|---|---|---|---|
| Fuel cycle B1 | HLW | ~ 40 000 a | ~ 1000 a | ~ 300 a |
| Fuel cycle B2 | HLW-UOX | ~ 2000 a | ~ 250 a | ~ 300 a |
| | HLW-ADS | ~ 70 000 a | ~ 13 000 a | |
| Fuel cycle A3 | HLW | > 1 Ma | ~ 70 000 a | ~ 24 000 a |
| Fuel cycle A2 | HLW | ~ 100 000 a | ~ 20 000 a | ~ 90 000 a |
| | MOX SF | > 1 Ma | ~ 200 000 a | |
| Fuel cycle A1 | UOX SF | > 1 Ma | ~ 100 000 a | ~ 200 000 a |

Table 4. Ranking of high-level waste and spent fuel types, on basis of calculated dose from geotechnical worker scenario in comparison with ICRP intervention levels; for comparison the time after which the radiotoxicity in the disposed waste drops under the radiotoxicity present in the fresh uranium required for the fuel production of fuel cycle A1 is also given (Marivoet et al., 2009)

From a geological disposal point-of-view, it should be noted that the relevance of a scenario such as inadvertent human intrusion in an assessment of the post-closure consequences of a geological repository is contentious (National Academy of Sciences [NAS], 1995). For example, the considered geotechnical worker scenario assumes that, in spite of the fact that the drill-bit has to perforate a thick-wall metallic container, the geotechnical workers is not aware of the presence of very exotic materials in the underground and will not take precautions when examining the extracted cores. Recently, representatives of various radiological protection authorities discussed the treatment of human intrusion at an international workshop (NEA, 2010). They concluded that "*the results of human intrusion scenarios could also be used in a safety case to demonstrate the robustness of the repository system and the safety concept. The consequences to the intruder who comes into direct contact with the waste should not be required to meet regulatory protection goals. These consequences may need to be calculated but they should neither be evaluated against quantitative limit values nor be used as a crucial criterion for the repository optimization process.*"

### 3.7 Other waste management aspects

#### 3.7.1 Presentation of fast spectrum reactors as minor actinide burners

Fast reactors or ADSs will not burn all the minor actinides in one single cycle. Only a fraction will be "burned" and multi recycling of the actinides will be necessary to reach a significant reduction. Each cycle consists of cooling of the irradiated fuel, reprocessing, fuel fabrication and irradiation in a fast neutron facility. Dufek et al. (2006) estimated that it will take about 100 years to reach a substantial decrease of the actinides.

Present PWRs are capable of achieving burn-ups up to 60 GWd/tHM. Thereafter, build-up of fission products poisons the chain reaction and the reactor must be refuelled. Much higher burn-ups, in the range of 130 to 200 GWd/tHM, are considered for fast reactors and ADSs. The spent fuels that will arise from those high burn-ups will be much "hotter" (thermal output, gamma and neutron radiation, etc.) than the spent fuels presently considered for geological disposal. As a consequence, those fuels cannot be disposed of in the repository concepts that have been developed for disposal of spent fuel. In the advanced fuel cycles it is (tacitly) assumed that the spent fuels will be infinitely recycled. It has not yet been studied how such fuel cycles can be efficiently ended in the case of a nuclear phase out.

The long duration of the P&T process spread over several generations makes it a difficult issue for decision makers (Commission Nationale d'Evaluation [CNE], 2006; LaTourette et al., 2010). On such long timescales it cannot be excluded that alternatives for nuclear fission will become available such as nuclear fusion or alternative energy resources.

#### 3.7.2 Management of volatile fission products

In the reprocessing plants at La Hague (France) and Sellafield (UK) the iodine released from the spent fuel is discharged, together with other volatile fission products, into the sea. On the other hand, the reprocessing plant at Rokkasho (Japan) captures the iodine on Ag-filters. However, a possible final waste matrix, which remains stable during several hundreds of thousands of years in disposal conditions, has not yet been developed. Continuing the discharge of iodine into the sea cannot be considered as a sustainable solution.

### 3.7.3 Generation of activation products

The inventory of the activation products was calculated within the Red-Impact project (Cunado, 2007), whereas most other P&T projects focus on actinide and fission product inventories. The generation of huge amounts of $^{14}C$ were calculated for the advanced fuel cycles. The large amounts of stainless steel used in the considered fast reactor design (i.e. European Fast Reactor) led to the generation of a considerable $^{14}C$ inventory. Nitride fuel was considered as the reference fuel in the ADS fuel cycle (B2); however, because of the large amounts of $^{14}C$ that were calculated to be generated, it was decided to also consider an oxide fuel as alternative fuel.

## 4. Conclusion

The introduction of advanced fuel cycles, based on fast neutron reactors, for electricity production can increase the efficiency of the use of natural uranium with more than a factor 100, because these reactors can burn all the uranium isotopes and not only the $^{235}U$ as it is the case in a PWR.

The volume of the high-level waste arising from advanced fuel cycles is about 5 times smaller than the volume of the spent fuel from the reference once-through fuel cycle. However, the dimensions of a geological repository are mainly imposed by the thermal output of the waste. Therefore, it is more important that the high-level waste arising from advanced fuel cycles generates about 3 times less heat than the spent fuel from the once-through fuel cycle. As a consequence, the needed length of the disposal galleries as well as the footprint of the geological repository is significantly smaller in the case of advanced fuel cycles.

The total doses due to geological disposal of spent fuel, vitrified high-level and intermediate-level waste in clay (or granite) formations are dominated by contributions of mobile fission and activation products. Geological repositories, in whose near field reducing chemical conditions prevail, very efficiently confine the actinides, because of their high sorption on clay minerals and low solubility. As a consequence the transmutation of actinides in fast reactors or ADS would have little impact on the doses due to the disposal of the radioactive waste in the case of the expected evolution scenario. On the other hand, in the case of altered evolution scenarios, e.g., in which the chemical conditions in the near field of the repository might become oxidising, some actinides can be dominant dose contributors.

The removal of $^{129}I$ upon reprocessing has a strong impact on the maximum dose. The development of very stable matrices for the immobilisation of iodine is highly desirable.

The radiotoxicity present in the high-level waste from advanced fuel cycles, after a few centuries, is drastically reduced by the transmutation of the actinides. This would lead to a considerable reduction of the potential hazard in the case of some disruptive scenarios such as an inadvertent human intrusion. However, the use of this low probability scenario as a decision making tool for long-term radioactive waste management is highly debatable.

Considerable amounts of mobile activation products, such as $^{14}C$, are calculated to be generated in fast neutron facilities and, consequently, to give a significant contribution to the total dose; this observation confirms the importance of developing low activation fuel matrices and construction materials for future fast neutron facilities.

## 5. Acknowledgment

The authors would like to thank the European Commission for partially funding SCK•CEN's participation in the Red-Impact project as part of the 6th EURATOM Framework Programme for nuclear research and training activities (2002-2006) under contract FI6W-CT-2004-002408. We also thank our colleagues Lionel Boucher from CEA (Cadarache, France) for the neutronic calculations and Miguel Cunado from Enresa (Madrid, Spain) for the preparation of the waste inventories, which form the basic data for the presented analyses.

## 6. References

CNE (2006) *Rapport Global d'Evaluation.* Commission Nationale d'Evaluation, Paris

Cunado, M. (2007) *Final Report on Waste Composition and Waste Package Description for Each Selected Equilibrium Scenario.* Red-Impact Project, Deliverable 3.7

DoE (2008) Yucca Mountain Repository License Application: Safety Analysis Report. Chapter 2 *Repository Safety after Permanent Closure.* US Department of Energy, Washington

Dufek, J., Arzhanov, V. & Gudowski, W. (2006) *Nuclear Spent Fuel Management Scenarios: Status and Assessment Report.* SKB, Stockholm, report R-06-61

ICRP (1996) *Age-Dependent Doses to Members of the Public from Intake of Radionuclides: Part 5 Compilation of Ingestion and Inhalation Dose Coefficients.* ICRP Publication 72, Annals of the ICRP, 26(1)

ICRP (2000) *Radiation Protection Recommendations as Applied to the Disposal of Long-lived Solid Radioactive Waste.* ICRP Publication 81, *Annals of the ICRP*, 28(4)

LaTourette, T., Light, T., Knopman, D. & Bartis, J.T. (2010) *Managing Spent Nuclear Fuel: Strategy Alternatives and Policy Implications.* Rand Corporation, Santa Monica

Marivoet, J., Peiffer, F., Cunado, M. & Vokal, A. (2007) *Final Report on Waste Management and Disposal.* Red-Impact Project, Deliverable 4.3

Marivoet, J. & Weetjens, E. (2008) Impact of Advanced Fuel Cycles on Geological Disposal in a Clay Formation. *Nuclear Technology*, Vol. 163, No. 1, pp. 74-84

Marivoet, J., Cunado, M., Norris, S. & Weetjens, E. (2009) Impact of Advanced Fuel Cycle Scenarios on Geological Disposal. *Proceedings of Euradwaste '08: 7th EC Conference on the Management and Disposal of Radioactive Waste*, 20-22 October 2008, Luxembourg. EC, Luxembourg, report EUR 24040, pp. 141-151

Marivoet, J. & Weetjens, E. (2009) Impact of Advanced Fuel Cycles on Geological Disposal. *Proceedings of Scientific Basis for Nuclear Waste Management Symposium XXXIII*, Saint-Petersburg, 25-29 May 2009. Materials Research Society, Warrendale, Volume 1193, pp. 117-126

NAS (1995) *Technical Bases for Yucca Mountain Standards.* National Academy of Sciences, National Academy Press, Washington

NEA (2002) *Accelerator-driven Systems (ADS) and Fast Reactors (FR) in Advanced Nuclear Fuel Cycles: A Comparative Study.* Nuclear Energy Agency, Paris

NEA (2006) *Impact of Advanced Nuclear Fuel Cycle Options on Waste Management Policies.* Nuclear Energy Agency, Paris, report No. 5990

NEA (2008) *Nuclear Energy Outlook 2008.* Nuclear Energy Agency, Paris, report No. 6348

NEA (2010) *Towards Transparent, Proportionate and Deliverable Regulation for Geological Disposal: Main Findings from the RWMC Regulators' Forum Workshop*, Tokyo, 20-22 January 2009. Nuclear Energy Agency, Paris, report No. 6825

ONDRAF/NIRAS (2001) *SAFIR 2: Safety Assessment and Feasibility Interim Report.* ONDRAF/NIRAS, Brussels, Belgium, report NIROND 2001-06 E

POSIVA (2010) *Interim Summary Report of the Safety Case 2009*. POSIVA Oy, Olkiluoto, report 2010-02

SKB (2006) *Long-term Safety for KBS-3 Repositories at Forsmark and Laxemar – A First Evaluation. Main Report of the SR-Can Project*. SKB, Stockholm, report TR-06-09

Smith, G.M., Fearn, H.S., Delow, C.E., Lawson, G. & Davis, J.P. (1987) *Calculations of the Radiological Impact of Disposal of Unit Activity of Selected Radionuclides.* National Radiological Protection Board, Chilton, UK, report R205

von Lenza, W., Nabbi, R. & and Rossbach, M. (Eds.) (2008) *Red-Impact: Impact of Partitioning, Transmutation and Waste Reduction Technologies on the Final Nuclear Waste Disposal.* FZ Jülich, report vol. 15

Weetjens, E. & Marivoet, J. (2008) *Impact of advanced fuel cycles on radioactive waste disposal in a clay formation*. SCK•CEN, Mol, report ER-63